A History of Scientific Journals

A History of Scientific Journals
Publishing at the Royal Society, 1665–2015

Aileen Fyfe, Noah Moxham, Julie McDougall-Waters
and Camilla Mørk Røstvik

First published in 2022 by
UCL Press
University College London
Gower Street
London WC1E 6BT

Available to download free: www.uclpress.co.uk

Text © Aileen Fyfe, Noah Moxham, Julie McDougall-Waters and Camilla Mørk Røstvik 2022
Images © Authors and copyright holders named in captions, 2022

The authors have asserted their rights under the Copyright, Designs and Patents Act 1988 to be identified as the authors of this work.

A CIP catalogue record for this book is available from The British Library.

This book contains third-party copyright material that is not covered by the book's Creative Commons licence. Details of the copyright ownership and permitted use of third-party material is given in the image (or extract) credit lines. If you would like to reuse any third-party material not covered by the book's Creative Commons licence, you will need to obtain permission directly from the copyright owner.

This book is published under a Creative Commons Attribution-Non-Commercial 4.0 International licence (CC BY-NC 4.0), https://creativecommons.org/licenses/by-nc/4.0/. This licence allows you to share and adapt the work for non-commercial use providing attribution is made to the author and publisher (but not in any way that suggests that they endorse you or your use of the work) and any changes are indicated.

Fyfe, A. et al. 2022. *A History of Scientific Journals: Publishing at the Royal Society, 1665–2015*. London: UCL Press. https://doi.org/10.14324/111.9781800082328

Further details about Creative Commons licences are available at http://creativecommons.org/licenses/

ISBN: 978-1-80008-234-2 (Hbk.)
ISBN: 978-1-80008-233-5 (Pbk.)
ISBN: 978-1-80008-232-8 (PDF)
ISBN: 978-1-80008-235-9 (epub)
DOI: https://doi.org/10.14324/111.9781800082328

Contents

List of figures	vii
List of tables	xi
List of abbreviations	xii
Contributor roles	xiv
Acknowledgements	xv
Introduction: Origin myths	1

Part I: Invention, 1665–1750

1 The first *Philosophical Transactions*, 1665–1677 *Noah Moxham and Aileen Fyfe*	19
2 Repeated reinventions, 1677–1696 *Noah Moxham and Aileen Fyfe*	51
3 Stabilising the *Transactions*, 1696–1752 *Noah Moxham and Aileen Fyfe*	82
4 The *Transactions* and the wider world, c. 1700–1750 *Noah Moxham and Aileen Fyfe*	115

Part II: Maturity and institutionalisation, 1750–1820

5 For the use and benefit of the Society, 1750–1770 *Noah Moxham and Aileen Fyfe*	149

6 Sociability and gatekeeping, 1770–1800   183
Noah Moxham and Aileen Fyfe

7 Circulating knowledge, c. 1780–1820   217
Noah Moxham and Aileen Fyfe

Part III: The professionalisation of science, 1820–1890

8 Reforms, referees and the *Proceedings*, 1820–1850   257
Noah Moxham and Aileen Fyfe

9 Editing the journals, 1850s–1870s   296
Julie McDougall-Waters and Aileen Fyfe

10 Scientific publishing as patronage, c. 1860–1890   331
Julie McDougall-Waters and Aileen Fyfe

Part IV: The growth of science, 1890–1950

11 The rise of the *Proceedings*, 1890–1920s   363
Julie McDougall-Waters and Aileen Fyfe

12 Keeping the publications afloat, 1895–1930   403
Julie McDougall-Waters and Aileen Fyfe

13 Why do we publish? 1932–1950   436
Camilla Mørk Røstvik and Aileen Fyfe

Part V: The business of publishing, 1950–2015

14 Selling the journals in the 1950s and 1960s   475
Camilla Mørk Røstvik and Aileen Fyfe

15 Survival in a shrinking, competitive market, c. 1970–1990   514
Camilla Mørk Røstvik and Aileen Fyfe

16 Money and mission in the digital age, 1990–2015   549
Camilla Mørk Røstvik and Aileen Fyfe

Reflections: Learning from 350 years   592

Bibliography   607
Index   627

List of figures

1.1	Title page of the first volume of the *Transactions*, 1665–6	23
1.2	'Introduction' to the first issue of the *Transactions*, 1665	25
1.3	An item about laudanum, 1674	32
2.1	Title page of Robert Hooke's *Lectures and Collections*, 1678	58
2.2	Tyson's dissection of a rattlesnake, 1683	64
3.1	Meteorological observations from Petersburgh, sent to James Jurin by Thomas Consett, 1724	98
3.2	Editorial interventions on Cromwell Mortimer's 1734 paper on viper bites	107
3.3	Proof corrections to William Derham's 1733 'Abstract of meteorological diaries …'	108
4.1	William Nettleton's letter about inoculation, to James Jurin, 11 November 1723	131
4.2	John Lowthorp's *Abridgement* (1705–8)	136
5.1	William Arderon's copy of John Hill's *Review*, 1751	156
5.2	The role of the 1752 Committee of Papers in the editorial process	163

5.3	The 'advertisement' prefaced to volume 47 of the *Transactions*	168
5.4	Locations of institutions that were sent gifts of the *Transactions* in 1765	176
6.1	The formal editorial process for the *Transactions* in the late eighteenth century, and the informal influence of the Society's officers	186
6.2	Decisions of the Committee of Papers, 15 June 1780	187
6.3	Charles Blagden's annotations on an 1800 paper on luminescence in fish	204
6.4	Manuscript of paper by Olof Swartz, 1787	207
7.1	Signatures of those who had claimed copies of Part 2 of the *Transactions* for 1801	230
7.2	Institutions to which presents of the *Transactions* were sent in 1816	233
8.1	The first issue of the *Proceedings*, printed in February 1831, but reporting the meeting of 18 November 1830	266
8.2	Printed flyer, 13 March 1832, inviting advance subscriptions to the retrospective *Abstracts*, 1800–30	268
8.3	Estimates of printing costs received by the Finance Committee, December 1846	273
8.4	Michael Faraday conveyed his opinion on two unnamed papers, 10 May 1830	275
8.5	The Royal Society editorial system after 1838	280
8.6	The volumes of surviving referees' reports from the 1830s and 1840s	281
9.1	Royal Society election certificate for George Gabriel Stokes, 1851	298

9.2	The Royal Society editorial process in the 1860s	304
9.3	First page of the 'Register of Papers', 1853	305
9.4	The 'Register of Papers' records the submission of a paper by George Hoggan and Frances Elizabeth Hoggan, 1877	309
9.5	Standard printed letter requesting a referee's report, 1883	316
10.1	Global distribution of free copies of the *Transactions* and/or the *Proceedings* in 1878	337
10.2	Lithographic plate of the skull of an Australian fish, 1871	349
10.3	The growth of the *Transactions*, 1790–1890	351
11.1	The editorial workflow after 1896	367
11.2	The number of papers submitted to the Royal Society, 1900–89	376
11.3	Instructions to referees (1899), incorporating questions drafted by treasurer John Evans in 1894	383
11.4	Printed referee report form (1899), with spaces for answers	384
11.5	Pages printed annually in Royal Society journals, 1890–1960	392
12.1	Global distribution of free copies of the *Transactions* and/or the *Proceedings* in 1908	407
12.2	Cardiff Central Library complained about the cuts to the free list, 1921	421
12.3	Global distribution of free copies of the *Transactions* and/or the *Proceedings* after the 1932 cuts	423
12.4	The publication finances in 1930	427
13.1	Compositors at the Cambridge University Press Printing House, early twentieth century	451
13.2	Personal letter from A.V. Hill to Sydney Roberts at Cambridge University Press, 30 June 1936	453

13.3	Printing machines at the Cambridge University Press Printing House, mid-twentieth century	459
13.4	Analysis of editorial progress, prepared for the Royal Society's Council, in August 1947	465
14.1	Subscriptions to Royal Society journals, 1963–72	485
14.2	Draft code for new journals, 1963	492
14.3	Key performance indicators for the Royal Society journals, *Year Book*, 1969	496
14.4	A rare example of a paper with a woman author being evaluated by a woman, 1951	501
14.5	The Royal Society editorial process after 1968	505
15.1	The refereeing flow chart proposed for *Transactions A* in 1989	536
15.2	'There's change in the air ...': promotional material for the 1990 relaunch	537
15.3	The Royal Society editorial process after 1990	538
15.4	'Before and after' cover designs, July 1990	540
16.1	'The structure of the Scientific Information System in the UK', 1993	550
16.2	'New typesetting equipment enables papers for the journals to be processed entirely in-house using the TeX mathematical typesetting system', 1994	557
16.3	Publishing surplus, from Publishing Director's Report to the officers for year-end 2013	571
16.4	Publication times, from Publishing Director's Report to the officers for year-end 2013	576

List of tables

1.1	List of items in the first issue of the *Transactions*	27
7.1	Costs of producing the *Transactions* in 1806	228
8.1	Average 'ordinary' income and expenditure of the Royal Society 1828–33, as reported by the treasurer in 1833	270
8.2	Annual circulation of the *Transactions*, on average for the years 1835–44	271
8.3	Most frequent referees in the period 1830–48	282
9.1	Most frequent referees in the period 1854–72	307
10.1	Circulation of the *Proceedings* in 1863	334
13.1	Print run, sales and distribution of Royal Society periodicals, c. 1935	449
13.2	Final estimates of printing charges, presented to Council, 1936	454
14.1	Publication finances in 1951	480
14.2	Retail prices per volume	483
15.1	Circulation of the Royal Society research journals in 1986	520
15.2	Production costs, 1936 and 1986	522
16.1	Submissions to Royal Society journals, 1988–2013 (scientific research journals only)	552

List of abbreviations

BL	British Library
CUL	Cambridge University Library
CUP	Cambridge University Press [in footnotes: Cambridge University Press archive, held at Cambridge University Library]
JdeS	*Journal des sçavans*
N&R	*Notes and Records of the Royal Society*
ODNB	*Oxford Dictionary of National Biography*, https://www.oxforddnb.com/
Proc	*Proceedings of the Royal Society* (*ProcA*: series A; *ProcB*: series B)
PT	*Philosophical Transactions* (*PTA*: series A; *PTB*: series B)
RI	Royal Institution of Great Britain
RS	Royal Society [in footnotes: Royal Society archive]
AB	Account Books
AE	Alfred Egerton Papers
C	Papers for Council meetings
CAP	Council Agenda and Papers
CMB	Committee Minute Books
CMO	Council Minutes (Originals)
CMP	Council Minutes (Printed)
EC	Election Certificate

EL	Early Letters
GLB	George Lindor Brown Papers
JBO	Journal Book Original
LBO	Letter Book Original
MC	Miscellaneous Correspondence
MM	Miscellaneous Manuscripts
MS	Manuscripts (General)
NLB	New Letter Book
OM	Officers' Meetings
PEC	Publications Executive Committee
PMC	Publications Management Committee
RMA	Records Management Audit (offsite)
RR	Referee Reports
RSC	Royal Society of Chemistry archives
T&F archive	Taylor & Francis archive, held at St Bride Printing Library, London

Supplementary material

Interested readers can find the following supplementary material at our project website https://arts.st-andrews.ac.uk/philosophicaltransactions/ :

- List of printers to the Royal Society
- List of editors of the *Philosophical Transactions*
- List of publishing staff
- Details of price, print run and editorial regime for a year of your choice, 1665–2015

Contributor roles

We have used the Contributor Roles Taxonomy (CRediT, see https://credit.niso.org/) to describe the co-authors' various contributions to this project, and to acknowledge other people who were essential to research.

Aileen Fyfe: 1 (conceptualisation), 4 (funding acquisition), 5 (investigation), 7 (project administration), 10 (supervision), 12 (visualisation), 14 (writing: review and editing).

Noah Moxham: 5 (investigation), 13 (writing: original draft), 14 (writing: review and editing).

Julie McDougall-Waters: 5 (investigation), 13 (writing: original draft).

Camilla Mørk Røstvik: 5 (investigation), 13 (writing: original draft), 14 (writing: review and editing).

Keith Moore, Librarian, the Royal Society: 8 (resources).

Stuart Taylor, Director of Publishing, the Royal Society: 8 (resources).

Matt Harrison: 12 (visualisation).

University of St Andrews research computing team (principally Swithun Crowe and Patrick McCann): 2 (data curation, especially the Key Facts Generator at https://arts.st-andrews.ac.uk/philosophicaltransactions/data/).

Acknowledgements

A book of this length depends on the support, encouragement and practical help of many more people than the authors on its cover. Sometimes, it was a matter of providing access to resources; sometimes, it was the conversations that provided useful insights, or a different perspective. Over the last nine years, there have been lots of conversations: we are grateful to the many historians who have shared their expertise on particular people or events; and we have enjoyed learning from professional publishers and scholarly communications experts, whose practical experience of contemporary journal publishing not only helped us understand the most recent chapters of this book, but generated intriguingly different questions to ask about the past. We are so grateful to all those who have helped, whether they knew it or not.

For financial support, we thank the Arts & Humanities Research Council, whose grant (AH/K001841) funded four years of intensive research, by three postdoctoral researchers, at the archives of the Royal Society, among other places. It also supported a range of activities allowing us to share our findings with other historians, with scientists, with the publishing industry and with activists and policy makers in the field of academic publishing. We also thank the University of St Andrews open access fund, for support with the final publication of this book. We are also grateful to the University of St Andrews and the Royal Society for providing supportive and congenial places to work, think and write.

The idea for this project began with a suggestion from Keith Moore, Head of Library Services at the Royal Society. The process of turning it into a funded reality was made easier by the support of Keith and his former colleague Felicity Henderson; and with advice from St Andrews colleagues Andrew Pettegree, John Hudson and Frances Andrews.

Once the project was under way, we were helped by vast numbers of people at, or associated with, the Royal Society. In the Library, we would particularly like to acknowledge Keith Moore, Rupert Baker, Louisiane Ferlier, Laura Outterside and Fiona Keates. The full cataloguing of the Referee Reports collection, during the later stages of our project, was invaluable. We also got to know the Society's publishing team, who gave us access to modern (often electronic) records that have not yet made it to the archive; who were always willing to chat about how publishing works now and how they perceive it as having changed; and whose real enthusiasm for our project encouraged us to believe that our research was not 'just' for historians, but had wider contemporary relevance in the Society and beyond. We particularly wish to thank Stuart Taylor, Phil Hurst and Charles Lusty.

The Former Staff Association of the Royal Society was enormously helpful in assisting us to find out about and, in some cases, contact former members of the publications staff. We thank Peter Cooper and Peter Collins for facilitating this.

In addition to our archival work and our informal conversations, we undertook a series of interviews with current and former staff of the Royal Society; and with some of the fellows involved in publishing. We are deeply grateful to all those who agreed to be interviewed and quoted for this book.

We thank our project advisory board, whose broad-ranging expertise helped us manage a project that frequently threatened to grow impossibly big. They are: Mordechai Feingold, Alexis Weedon, Tilli Tansey, Nuala Zahedieh and Jonathan Ashmore FRS.

The Royal Society's archives were the focus of our research, for obvious reasons, but we are also grateful to the staff at the various other libraries, archives and organisations which hold material relevant to our project. These include: the British Library (for the personal papers of various people, including Hans Sloane); the St Bride Printing Library (for the Taylor & Francis archive); Cambridge University Library (for the Stokes papers and the Cambridge University Press archive); the Nuffield Foundation (for the 'learned journals' project archives; thanks to Josh Hillman for arranging access, and Michelle Clarke for scanning); the Royal Society of Chemistry (thanks to Richard Kidd and Sharon Ashbrook FRSC); the City of Westminster Archives (for Harrison & Sons).

We are enormously grateful to those who have taken on the heroic tasks of editing the correspondence of key figures in our story. The existence of collected correspondences for Henry Oldenburg, James Jurin, Joseph Banks, Michael Faraday, Charles Darwin, John Tyndall and George

Stokes, among others, gave us a practical way to go beyond the formal institutional archives of the Royal Society. We are especially grateful to those editorial colleagues who proved that scholarly generosity lives on, and were willing to share their (searchable!) electronic files with us: Andrea Rusnock (for the Jurin correspondence); Frank James (for the Faraday correspondence); Bernard Lightman and James Elwick (who granted us access to the electronic files of the Tyndall correspondence ahead of publication). Future scholars will be able to benefit from the inclusion of these collections in the *Early Modern Letters Online* project (http://emlo.bodleian.ox.ac.uk/) and the growing *Epsilon* project for those wishing to explore nineteenth-century scientific correspondence (https://epsilon.ac.uk/).

For stimulating and informative conversations about the Royal Society, scientific journals and academic publishing practices in general, we thank Lisa Jardine, Michael Hunter, Joad Raymond, Anna Marie Roos, Jim Bennett, Sachiko Kusukawa, Sietske Fransen, Katie Reinhart, Simon Werrett, Sally Shuttleworth, Berris Charnley, Sally Frampton, Jon Topham, Frank James, Graeme Gooday, Anna Gielas, Jim Secord, Alison Walker, Brian Balmer, Michael Gordin, Roland Jackson, Michael Barany, Alex Csiszar, Melinda Baldwin, Iain Watts, Paul Ranford, Jenny Beckman, Charlotte Sleigh, Rebekah Higgitt and the *Wunderkammer* reading group. Also Stephen Curry, Cameron Neylon, Caitriona Maccallum, Stuart Lawson, Toby Greene, Robert Campbell, Jan Velterop, Anthony Watkinson, Wendy Hall, Lenna Cumberbatch, Stefan Janusz and Chris Purdon.

Perhaps ironically for a book about publishing, getting this book published was not as straightforward as one might have wished. It is undoubtedly a long book; but it is also the case that the options for academic monograph publishing have changed during the years we were writing this book. We are grateful to Abby Collier for supporting our book in its early phases; and to Lara Speicher and Pat Gordon-Smith at UCL Press for finding a way for us to make this book widely available as an open access publication. And to Grace Patmore, Julie Willis and Rachel Carter for their efficient work and clear communications in the production process; and to Matt Harrison for the lovely maps.

Noah would like to thank Sara Lyons, Ada, and Colette and Richard Moxham.

Camilla would like to thank Mr Albus and Dr Tegg Westbrook for love and support. And to Aileen for managing this large and complex project, and for showing what good leadership in academia can look like.

Aileen is only too conscious that her family have had to live through all the 'behind the scenes' processes of academic publishing that we write about in this book. She could not have done this without Paul's love and support, both practical and intellectual. Our daughters, Lucy and Emily, cannot remember life before 'the book' was in process. Their expectations for a reasonable timeframe for book-production have been influenced over the years by Erin Hunter, J.K. Rowling, Rick Riordan and Suzanne Collins, so it is perhaps not surprising that the excitement they once displayed on discovering that 'Mummy is writing a book!' has been transformed into resignation that 'Mummy is *still* writing her book'. Thank you, girls, for your support, your 'understanding' and your love. (Sorry about the shocking lack of cats, demi-gods or dystopian strife.)

Introduction: Origin myths

In 2015, the Royal Society of London celebrated the 350th anniversary of the *Philosophical Transactions*, the world's first and longest-running scientific journal. The celebrations looked to both past and future. There was an exhibition on the history of the *Transactions*, a series of video stories and a set of specially commissioned commentaries reflecting on the significance of key historic research papers. There was a four-day meeting on the 'Future of Scientific Scholarly Communication', where representatives from publishers, funders and other learned organisations discussed current problems with research evaluation, reproducibility, and the business of publishing, while wondering what the future might look like.[1] Underlying these activities and the associated media coverage was the assumption that the event being celebrated was the invention of the scientific journal.

The story that the scientific journal was invented in 1665 is commonplace. It routinely surfaces in contemporary debates about the future of academic publishing when proposed changes, whether to editorial peer review or to the way that the circulation of knowledge is financed, are dramatically proclaimed to be the biggest change since the seventeenth century. As historians, we recognise the powerful attraction of being able to assert that the publication of research in journals has been intrinsic to science since the birth of the experimental method: we used it ourselves in the 2015 exhibition, as we had earlier used it to make the case for funding a major research project on the *Philosophical Transactions*. But the more we have studied the history of the *Transactions* and its siblings, the more we have come to realise how limiting this origin story is.

In this book, we use the Royal Society archives to uncover the story of how the *Transactions* developed from the speculative commercial side-line of an entrepreneurial scholar in the late seventeenth century, to the official publication and chief business of one of the world's oldest and most influential learned societies in the nineteenth century. We then examine how this historic periodical was adapted to meet the changing needs of scholars during the professionalisation of science and the growth of the scientific enterprise, as its role was challenged by the emergence of an increasing number of other journals offering places for researchers to publish their work (including the Royal Society's own *Proceedings*). Throughout, we examine how editorial decision-making practices changed, and we investigate how the circulation of knowledge was financed, including the changing relationships between learned society publishing and commercial interests.

The message we hope readers will take away from this 350-year story is that the nature, organisation and purpose of scientific journals have a history: the twenty-first-century scholarly journal was not created in 1665, but has been developed from a seed sown then. Just as the practice of science has changed since the seventeenth century, so too have the practices of scientific journals. Their roots may be early modern, but the practices of contemporary journals also bear the legacy of the intervening periods of history. Recognising this historical development is important not just for historians, but for all those involved in thinking about academic journals: practices that may seem now to be written in stone have been different in the past; and there is every reason to believe that they will be different in the future.

The power of myths

In 2015, the president of the Royal Society, Paul Nurse, described 1665 as the moment when 'the first Secretary of the Society, Henry Oldenburg, proposed a way of disseminating and verifying new discoveries in science. As a result, *Philosophical Transactions* came into existence: the world's first science journal.' He explained the significance of this event:

> This was a truly seminal development whereby scientists from all over the world were able to communicate their ideas, establish priority and, most importantly, expose their work to their peers for assessment. *Philosophical Transactions* established the four fundamental principles (registration, verification, dissemination and archiving) still in use by the almost 30,000 science journals

today. But science publishing has remained almost unchanged until a few decades ago and the introduction of the internet.[2]

This telling of the story – in which Henry Oldenburg's *Transactions* established 'four fundamental principles' and that scientific publishing from then on 'remained almost unchanged' – is not simply self-promotion by the Royal Society. For instance, the version told by the Publishers' Association to a UK parliamentary inquiry in 2004 – which has since been widely quoted – also featured Oldenburg and the Royal Society, the same 'four functions' that 'are so fundamental', and the claim that 'all subsequent journals ... have conformed to Oldenburg's model' and thus, little has changed since 1665.[3] One can see why the scientific publishing industry might wish to imply that its activities are at once immaculately descended from the time of the scientific revolution and essential to the practice and stability of contemporary science. And one can see why researchers themselves might find it reassuring to believe that their publishing practices, minimally adjusted for changes in technology, are essentially the same as those of Isaac Newton. But it is not difficult to demonstrate that the myth is wrong in almost all significant respects.

Some basic facts are not in dispute: there was a man called Henry Oldenburg, and he did indeed begin publishing a periodical in London in March 1665 that was mostly dedicated to natural science. And it was called *Philosophical Transactions*. The 'philosophical' part of the title refers to the old term 'natural philosophy', or the study of how the natural world works, while 'transactions' can be understood as 'goings-on', along the lines of 'the business being transacted'. His subtitle revealed that he aimed to report 'the Present Undertakings, Studies and Labours of the Ingenious in Many Considerable Parts of the World'.

There is, however, a basic historical improbability to the idea that the communication needs of scientists in the twenty-first century were anticipated with uncanny precision by an impoverished seventeenth-century émigré, working as an unpaid administrator for the fledgling Royal Society. It also implies that the needs of natural philosophers in the seventeenth century were broadly the same as those of modern scientists, despite the many other ways in which society and science have changed over three and a half centuries.

Unsurprisingly, the periodical that Oldenburg launched in 1665 was very different from modern scientific journals. It is not merely that its typography used the archaic long S, or that its editor used different vocabulary and rhetorical tropes. It differed in the type of content it carried, and in the purposes it served for its editor, authors and readers. Indeed, other than appearing at more or less regular intervals, and disseminating information relating to natural knowledge, it lacked most of the functions that are now routinely associated with scientific journals. The registering of discoveries took place elsewhere; pre-publication peer review was entirely absent; and the integration of the *Transactions* into wider systems for archival record-keeping in the sciences was a lengthy and hesitant process.

Nor did Oldenburg's *Transactions* set the model for all other scientific journals, if by that we understand the peer-reviewed research journal. For one thing, other early periodicals, including the two plausible rival claimants to the title of 'first scientific periodical', conducted themselves quite differently. The national rivalry about whether the *Journal des Sçavans* (Paris, 1665), the *Philosophical Transactions* (London, 1665) or the *Miscellanea Curiosa* (Schweinfurt, 1670) has the better claim to inventing the scientific journal disguises the fact that none of them was actually very much like a scientific research journal in the modern sense. For instance, Oldenburg's *Transactions* contained many different genres of writing, as well as original accounts of new research. And, second, there have been, and are, many different sorts of scientific journal. Oldenburg's *Transactions* is as much the ancestor of the abstract or review journal, or of the scientific news and book reviews now found in the front half of *Nature*, as it is of the peer-reviewed research journal.

Over-emphasising the role of the *Transactions* as the first scientific journal inevitably distorts our understanding of what it actually was, and what it actually did, by framing it with categories defined by things that mattered to scientists and publishers in the late twentieth and early twenty-first centuries. For Oldenburg and his peers, the early *Transactions* had a variety of different functions, not the least of which was the long-cherished hope of earning its compiler a decent living. This was equally true for his successors in the eighteenth and nineteenth centuries, for whom the *Transactions* would become a means of institutional reputation-building and a token in gift-exchange. The present-centred myth impoverishes our understanding of history.

And by linking 1665 so closely to the present, the myth disguises fundamental changes in structure, function, organisation, ownership and principles of access over the intervening 350 years. Here, the continuity of the *Transactions* title (despite several changes to the subtitle) gives a misleading appearance of continuity of enterprise. Publishers and men of science in the nineteenth century were rather more aware of the intervening twists and turns in the history of scientific journals, not least because they were living through the period in which the research journal gained its significance to knowledge and to career-building.[4] A new generation of commercial scientific journals had begun to appear in Britain in the decades around 1800, and they distinguished themselves from the older periodicals, including the *Transactions*. One of their editors lamented the absence of a 'periodical philosophical journal' in Britain, since, 'from about the middle of the eighteenth century', the *Transactions* had begun to consist 'entirely of original papers'.[5] What disqualified the late eighteenth-century *Transactions* from being a scientific journal in the eyes of the early nineteenth century was its now-exclusive emphasis on original research: although it had come much closer to the modern research journal of the myths, some contemporaries regretted this. If we are to understand the historical formation of the social, cultural and economic attitudes that underpin contemporary approaches to publishing science, we need to look at the story of what happened between Oldenburg and now: a 'forgotten middle' that has been almost perfectly hidden from view by the simplistic origin myth.

Going beyond the myths

This book is undoubtedly about the *Philosophical Transactions*, but it is not the history of a single entity, nor even a single title. From the beginning, the *Transactions* was fashioned and constrained by the wider ambit of learned publishing, and by the ambitions of the Royal Society, with which it was closely but ambiguously associated. The *Transactions* took 87 years to become an official publication, and from the 1830s onwards it would be editorially and financially linked with other periodicals published by the Royal Society, particularly the *Proceedings*.

The Society did not originally own the *Transactions*, but the fact that its early editors were all officers of the Society has meant that records of their management of the *Transactions* are frequently mixed in with other Society business, in the correspondence files and in the minute books. From the mid-eighteenth century, there have been dedicated sections of the archive dealing with the *Transactions*. As the decades went by, these came to include editorial and publication committee minutes, printers' bills, editorial correspondence, referee reports, negotiations with publishers and financial ledgers. The archival record is far from complete – the financial records, for instance, are most detailed for the mid-nineteenth century and for the early twenty-first century. Combining local density and chronological breadth, these records represent by far the richest archival resource for the history of periodical publishing in the sciences.

The existence of this archive allows us to study aspects of the history of the *Transactions* and its siblings that are simply impossible for other journals that have no surviving archive at all, let alone one of such chronological depth. For instance, although copies of the scientific weekly magazine-journal *Nature* (a relative newcomer to academic publishing, founded two centuries after the *Transactions* in 1869) can be found in countless university libraries across the world, it has no surviving archive: its history must be told from its printed pages and whatever correspondence or papers survive from key figures associated with it.[6]

For many periodicals, their own surviving printed volumes are the only way we can learn about them. Historians of science have long used printed journals to investigate the activities of specific people, or the development of particular theories, or to trace scholarly controversies. And literary and linguistic scholars have found the long run of the *Transactions* useful for studying changes in the style and structure of scientific writing, and of the English language itself. Their work has explored how scientific articles developed from relatively short, chatty contributions often presented as letters (in the seventeenth century), into more formally structured papers with standardised sections written in impersonal language (by around 1900); and how those articles became increasingly full of technical jargon and acronyms in the twentieth century.[7] Other scholars have studied the wider landscape of scientific publishing, revealing the *Transactions* as one periodical among many, and exploring the various ways in which authors and readers used scientific journals.[8]

The problem with relying on the printed pages is that we see only what the editor or publisher wanted us to see. We see the contributions that were printed, but have no idea what was rejected. We may not even be able to tell which items were commissioned and paid for, and which were submitted unsolicited. We may see an editor's preface, setting out his vision for the journal, but we do not know whether this is the unvarnished truth, a clever marketing ploy or somewhere in between.[9] We may see a cover price or details of how to subscribe, but we are unlikely to find information about who actually did subscribe, or see evidence of how the editor and publisher sought to bring their periodical to its intended readers.

It was scholars trained in library science who began to study scientific journals as objects in themselves, rather than as containers of source material.[10] Their work inspired Bill Brock and Jack Meadows to undertake a pioneering study of the archives of Taylor & Francis, the scientific journal publisher whose origins lie in 1798.[11] In contrast to their story, focused upon the business of printing, the Royal Society's archives grant us much more extensive insight into the editorial processes. Our history is a story of the editors, reviewers and committee members, as well as the treasurers, administrators and managers, whose work – whether voluntary or paid – made the *Transactions* what it was.

Our story has fewer printers, paper-makers, engravers or lithographers than we had expected. Judging from the records they left behind, the Royal Society's officers showed a distinct lack of interest in most improvements in paper-making, printing and typesetting techniques, or even in the relative merits of different providers of these goods and services. The exceptions come from the handful of occasions when the Society considered changing its printer, and we have the most detail for the mid-nineteenth century, when that printer was Taylor & Francis. Other than the new techniques for reproducing illustrations, the industrialisation of printing largely passed the Society by. Its circulation and its production values were, after all, rather different from those of the mass-produced newspapers of Victorian Britain. It is not until the second half of the twentieth century that the Society's officers and staff became actively engaged in new technologies of typesetting, printing and, ultimately, the electronic distribution of texts.

Being able to follow the *Transactions* and its siblings over three and a half centuries has encouraged us to take a broad perspective on big questions. We can see changes that would have been invisible on a canvas of just a few decades, and have a better chance of being able to identify trends and shifts that were truly significant in the long term. Painting the big picture comes at some cost of detail: we have sufficient material to write about most of the episodes in this book at twice or thrice the length we do here; and we are in no doubt that there is plenty more yet to be written about the *Transactions* and Royal Society publishing. But we have tried to keep the aim of a big picture constantly in mind in this book. Those readers who are interested in the underpinning quantitative data (about production costs, prices, print runs and so forth) will find much of it freely and openly available via our project website hosted by the University of St Andrews.[12] We have also written about particular themes (such as peer review, copyright, gender or finances) elsewhere, both in academic journals and in formats accessible to contemporary policy-makers and commentators in the scholarly communications sphere.[13]

Publishing with the legacy of history

We could consider the history of scientific journals in evolutionary terms, pointing to the gradual changes over time, and linking divergences from the ancestral form with adaptation to changed circumstances. We might even suggest that certain innovations or modifications have been more or less successful at surviving in particular intellectual or economic niches. But this naturalistic imagery would miss the human element. One of our key points concerns the amount of effort and hard work – intellectual, organisational and managerial – that goes into creating and running a journal. This emphasis on journals as human-built artefacts, not natural kinds, means that we have found it useful to think about the *Transactions*, and journals more generally, as technologies themselves, not just the products of printing technologies.[14] This offers us three key benefits.

First, it encourages us to focus less on trying to define what a scientific journal *is* (something that is difficult to do, if the definition is to be meaningful over 350 years), and more on what it *does*, and for whom. Our non-exhaustive list of the things journals might do is rather longer than the four functions cited earlier. It includes: circulating scientific news; recording claims about new discoveries or observations; making money for its owner; synthesising the recent work in a given field; enabling authors to build their professional careers; creating a repository or archive of knowledge; accrediting claims to knowledge; enhancing the reputation of the editor or sponsoring institution; keeping readers up to date with the latest publications in their area of interest; providing a form of public accountability for the research community; and creating a community of readers with shared interests. These are all things that the *Transactions* has been used for at some point in its history, but not continuously and never all at once. The changing patterns of use help us to understand the ways in which the scientific enterprise, and its relation to print, have changed.

Second, this framing encourages us to think about the human and technological *systems* that underpin scientific journals, and how those systems adapt to changes in circumstance or technology.[15] Printing presses and the copper plates of engravers are obviously part of this system and, as time goes by, we see these being replaced by lithographs and wood engravings, by offset litho printing machines, and ultimately by XML-first workflows. But there are many more techniques and technologies involved in producing a scientific journal. Content will need to be acquired, which may involve personal and social networks, as well as paper, ink and the postal service. There may need to be processes for selecting appropriate content, and perhaps for revising, improving or standardising it before publication. The costs of paper, printing, distribution and staffing will need to be financed somehow, which may require negotiations with sponsors or patrons, or marketing strategies targeted at potential paying customers.

Over 350 years, both the technological options available and the identity, location and interests of the relevant human players have changed repeatedly. This means that the technical, editorial and organisational systems underpinning the *Transactions* have changed too. New opportunities or elements have been introduced into the journal system. But we have found that the new elements are usually incorporated alongside continuing older practices and techniques. It is rare indeed for a practice to be identified as obsolete, and removed. Thus, journals are simultaneously modern, and carry the vestigial traces of past practices or conditions. We will show that the *Transactions* has been reinvented several times, but it never re-started completely from scratch. The name is just the most visible element to be carried forward.

Third, historical and sociological studies of technology reveal the importance of looking beyond the moment of invention and investigating what happens later. Subsequent innovations may in fact be more influential or successful than the initial invention, and innovation and adaptation are essential for survival in new circumstances.[16] Equally, the way in which technologies are later used may not be what their inventors had imagined; in particular, being used by a different group of people can change the meaning and function of a technology.[17]

When we think about the scientific journal in this light, it becomes clear that the long history of the *Transactions* cannot be taken for granted as an inevitable consequence of Oldenburg's invention in 1665. That long history is something that needs to be interrogated and investigated. We need to think about how the practices of running scientific journals were maintained, adapted, transformed or – in some cases – became obsolete. The fact of the endurance of the *Transactions* becomes, paradoxically, as much an attestation of change as of continuity.

Both the *Transactions*, in particular, and the general concept that it might be useful to print scientific knowledge in periodical format have now survived for over 350 years. Few institutions or technologies last that long without being adapted or reinvented to fit changing social, political and scientific circumstances. And so it was with the *Transactions*, which has been repeatedly adapted and reinvented for new contexts. New opportunities, strategies and technologies were adopted under new demands and pressures. But at the same time, older ways of doing things were seldom utterly discarded. Some were retained and repurposed alongside the newer elements in the journal system. It turns out that the *Transactions* itself, and modern scientific journals more generally, are a fascinating mixture of new and old.

The retention of older elements in modern journal systems, at the *Transactions* and elsewhere, means that the social and historical circumstances of those innovations have a lasting influence. For instance, the tradition that authors of scientific papers are unpaid (unlike authors of books, or contributors to literary magazines) originates in the gentlemanly culture of the eighteenth century, in which natural philosophers presented their observations to the Royal Society as a gift. Similarly, the prevalence of the single-blind form of peer review in the natural sciences (as against the double-blind form that is used in many humanities fields) can be traced to the organisational structures underpinning the meetings and publications of the Royal Society in the mid-nineteenth century.

The nature of the Royal Society as a club or community is central to this story.[18] For instance, the convention by which scientific journals are sustained by the unpaid labour of their contributors, editors, referees and editorial board members owes a great deal to the voluntarist norms that emerged from the relationship of the early scientific periodical to a learned society. On the other hand, our later chapters discuss the challenges facing a publishing system that was originally based upon membership in an exclusive club, as a greater diversity of people (most visibly, in terms of gender, social status and geographic origin) came to be involved in scientific research and authorship. It is sobering to realise how many of our habitual practices as academics derive from the workings of a white, male, upper-class British culture that would not have been particularly welcoming to many of us.

One of the strengths of the scholarly society as publisher turns out to be a surprisingly long-standing philanthropic commitment to the circulation of knowledge within the scientific community. We have discovered that the mid-twentieth-century emergence of a lucrative financial strategy based on institutional subscriptions has obscured a previous (and much longer) tradition of scientific publishing that was largely subsidised by the members and sponsors of the scientific community. The prevalence of the myth of Oldenburg's invention serves as a telling illustration of how easily these kinds of present debts to the past can be lost to view, and the potential significance of longer perspectives for their recovery.

350 years in five phases

In this book, we will trace the successive transformations of the *Transactions* from Oldenburg's printed news-sheet, through its long existence as the most prestigious and authoritative English-language periodical for the publication of research memoirs, to its modern, mostly digital existence, as a pair of thematic reviews complementing the nine research journals published by the Royal Society in 2015. We have identified five broadly chronological phases that inform the structure of our book.

Unsurprisingly, we start with invention. Part I covers the years 1665 to 1750, and it begins with a richer and more complex story of how and why Oldenburg launched the *Transactions*. It will become clear why we are convinced that Oldenburg's *Transactions* had very little in common with modern scientific journals. We also pay particular attention to the role of earlier periodicals, the commercial dimension and print trade conditions in shaping Oldenburg's periodical. We look particularly carefully at what happened to the *Transactions* after Oldenburg's death in 1677: the editorial visions of his successors reveal significant variations in their understanding of the function and form of the *Transactions*. It took decades for a settled concept to emerge of what a scientific periodical was (or was for). We also examine the ambiguous question of the periodical's relationship to the Royal Society. We will evaluate the evidence for Royal Society ownership, editorial control and support in the years prior to the emergence of a formal relationship, and ask to what extent the routine association of the *Transactions* with the Royal Society is justified for the early decades.

Part II covers the years 1750 to 1820. By this time, the *Transactions* had entered what we might call its mature phase and become thoroughly institutionalised. Its identity and purpose were largely settled. Rather than the frequent, scrappy miscellany of the early Oldenburg years, it had become a relatively ponderous vehicle for long, detailed memoirs describing their authors' observations and discoveries. The vast majority of these memoirs had been formally presented to meetings of the Royal Society, consolidating the long-standing informal association between the Society and the *Transactions*. In 1752, the Society took on the ownership, editorial management and financial burden of the *Transactions* (whose costs were heroic throughout this period). In this incarnation, the *Transactions* was imitated by numerous other learned societies and academies in the late eighteenth and early nineteenth centuries. Indeed, 'the transactions of a learned society' became such a recognisable genre that it was the foil against which entrepreneurial publishers in the 1790s launched a new generation of scientific periodicals that aimed to be briefer, more flexible and more frequently published. In some respects, the *Transactions* was at its most influential during this period: it had the prestige of the Royal Society behind it (and helped to create it), and it had relatively few competitors.

Yet, what worked for the gentlemanly culture of eighteenth-century natural philosophy was unlikely to work equally well for the emerging profession of academic (let alone industrial or government) scientists. Thus, since the *Transactions* reached its maturity, it has faced a series of adaptations to the challenges posed by changing circumstances. On the printed page, the changes have not been dramatic, but there have been significant transformations to the editorial, managerial and financial structures, and to the functions played by the *Transactions* in the very different scientific and publishing contexts that have emerged. The challenges faced by the Royal Society and its publications over the last 200 years have not always come along neatly and distinctly separated, but we have grouped them into three broad, and chronological, groups.

Part III covers the years 1820 to 1890, when the main challenge facing the organisers of the *Transactions* was its place amid the professionalisation of science. This involved questions about expertise, the development of career structures, and professional recognition for men (and latterly women) of science. The *Transactions* had spent its first 150 years closely associated with a voluntary association of gentlemanly elites for whom the pursuit of natural knowledge might be a serious matter but was not a career. During the nineteenth century, its editorial practices were substantially transformed and complicated; refereeing began to be used in addition to editorial committees to enable the making of more expert judgements; and a new journal, the *Proceedings*, was established for faster communication of brief results. For scientific authors, publication in the pages of the *Transactions*, and other journals like it, came to carry far more social capital than it had formerly done.

The challenges associated with the professionalisation of science were not completely resolved by the end of the nineteenth century, but a different set of issues became more prominent in the management of the *Transactions* and the *Proceedings*. Part IV covers the years 1890 to 1950, and focuses upon the challenges that can broadly be linked to the substantial growth of the scientific enterprise, driven by the expansion of universities and the development of research training programmes, all of which vastly increased the number of scientific researchers in Britain.

By the late nineteenth century, there was sufficient research to fill the pages of a huge variety of journals, some founded by new discipline-based societies, others by entrepreneurial editors or publishers. In this crowded landscape of increasingly specialised journals, it was no longer quite clear whether the *Transactions* or the *Proceedings* had a distinctive or useful role. At the same time, the expanding page count of both journals was making them an increasing burden on the Royal Society, both financially and in terms of the human resources involved in the editorial review process. From around 1890 to 1950, our story centres on the Royal Society struggling to find a way of successfully adapting its creaking system of journal publishing to the needs of twentieth-century science. The emergence of the *Proceedings* as a full-blown research journal in this period could be seen as a success story, but it meant that the *Transactions* was, for a time at least, relegated to something of an historical artefact.

The rate of growth of science would increase still further after the mid-twentieth century, but the Royal Society would be strangely insulated from this growth until the very end of the century. The dominant theme of Part V of this book, covering the years 1950 to 2015, is the role of learned society publishers in an increasingly commercialised and competitive world of scientific journal publishing. After the Second World War, certain commercial publishers were highly successful at launching journals to cater to new research specialisms, and in making them profitable. Since societies like the Royal Society had been underwriting the costs of their journals for decades (or more), this raised questions about whether private enterprise would be a better solution, and, indeed, whether there was anything unique about learned society journals. It also suggested that learned societies could potentially learn a different business model. Over the second half of the twentieth century, the *Proceedings* and *Transactions* were transformed into sources of income for the Royal Society. They also became more international than ever, seeking authors and purchasers all over the globe. The Society finally came up with a new role for the *Transactions* and, in the early twenty-first century, purposefully utilised its broad cross-disciplinary base to launch an interdisciplinary journal.

The advent of digital technologies of producing, distributing and reading journals may seem like it ought to mark another phase in our history. But, as of 2015, the ways that they had been used in scholarly publishing were far less transformative than one might imagine. Digital journals, including the 11 then published by the Royal Society, are excellent examples of how the new is combined with the old. Despite the possibilities of the technologies, most journals still published articles that visibly resembled their printed predecessors, issued them in numbered parts or volumes, and sold them, in various ways, to university libraries, as they had done since the 1950s. They also used a system of editorial evaluation that had been developed in the 1830s. This is what we mean when we say that the organisational systems underpinning journals, as well as the meanings and functions of those journals, are a mix of the novel and the traditional.

———

This book does not aspire to be the definitive history of the scientific journal. No single periodical could bear that weight: none can claim by itself to have inaugurated every significant development in the history and use of scientific journals. Since the history of *the* scientific journal is the story of its rise to a near-monopoly of the formal communication of new claims to scientific knowledge, it cannot be told by focusing on a uniquely visible or prestigious journal. This is as true of the *Philosophical Transactions* as it is of *Nature*, or the *Philosophical Magazine*, or the *Annales de Chimie*, or the *Göttingische Gelehrte Anzeigen*, or the *Journal des Sçavans*, or any one of a dozen other journals with famous names, long histories, rosters of well-known contributors, and serious claims to eminence or innovation in the sphere of scholarly scientific communication. There are times in its history when the *Transactions* falls far behind the wave of innovation in scientific communication; but even at those times, its historical continuity, and the archive that lies behind it, make it a useful indicator of prevailing tendencies.

At the same time, we believe that no history of the scientific journal in the wide sense is possible without the history we have written. We have sought to investigate what scientific periodicals tell us about the practices and priorities of science in their own time, but we have also sought to paint the big picture. A 350-year history enables us to identify broad shifts and recurring themes. It points to issues that will merit further investigation of other journals and other publishers. And it will provide the framework to situate more detailed studies, and against which the development of scholarly journals in other countries and regions can be compared.

The story we tell here is not the biography of a single entity, but a history of repeated transformation and reinvention, reaction and reform, crisis and complacency. By following the *Transactions* through its many lives, we uncover the complexity of the processes by which an innovation became an institution – and the historical burdens of institutional status. One of the results of those processes is that periodicals retain vestiges of their former identities, and can even preserve and transmit elements of their own histories as contemporary practice; and one of the key effects is that the material and epistemic conditions of scientific publishing in (say) the eighteenth century can be tacitly, even unconsciously, absorbed by and integrated into periodicals founded long afterwards. The modern scientific journal is, in significant ways, predicated on historical conditions that no longer obtain.

We suggest that it is empowering for the academic community of today – and instructive for those who write policies for it or provide services to it – to understand the historical roots of certain practices. It may then become possible to see which features of our current system of academic publishing are indeed essential, and which are historical accidents; or to assess which elements are more or less tightly intertwined.

Despite all our caveats, there is no doubt that something came into being in March 1665 which had not existed before, and which has survived and, in a wider sense, thrived ever since. Periodical publication became, and remained, fundamental to the activity of scientific researchers. But it did not happen immediately, and it was far from being an obvious intellectual or commercial triumph. There is a story to be told of the risks, near-death experiences and recoveries – not to mention prolonged periods of inaction – that link Oldenburg's invention with the scientific journals of today. All of this matters at a time when the structures of academic publishing are being consciously renegotiated.

Notes

1 For the exhibition brochure, see Aileen Fyfe, Julie McDougall-Waters, and Noah Moxham, '*Philosophical Transactions*: 350 years of publishing at the Royal Society (1665–2015)' (London: The Royal Society and the University of St Andrews, 2014). The special issues of *Philosophical Transactions A* and *Transactions B* appeared in April 2015. For the 'Science stories' videos, see Royal Society, 'Science stories', https://royalsociety.org/journals/publishing-activities/publishing350/science-stories; and for the discussion meeting, see Royal Society, 'The future of scholarly scientific communication: conference 2015' (London: Royal Society, 2015).
2 Royal Society, 'The future of scholarly scientific communication', 4.
3 'The origin of the scientific journal and the process of peer review', Annex 1 to 'Memorandum from the Publishers Association', appendix 20 in vol. II (written evidence, HC 399-II) of House of Commons Select Committee on Science and Technology, 'Scientific Publications: free for all? (Tenth Report, 2003–04)' (London: HMSO, 2004).
4 Alex Csiszar, *The Scientific Journal: Authorship and the politics of knowledge in the nineteenth century* (Chicago, IL: University of Chicago Press, 2018); Melinda Baldwin, *Making 'Nature': The history of a scientific journal* (Chicago, IL: University of Chicago Press, 2015).
5 Thomas Thomson, quoted in Jonathan R. Topham, 'Anthologizing the book of nature': The circulation of knowledge and the origins of the scientific journal in late Georgian Britain', in *The Circulation of Knowledge between Britain, India, and China*, ed. Bernard Lightman and Gordon McOuat (Boston, MA: Brill, 2013). See also Topham, 'The scientific, the literary and the popular: Commerce and the reimagining of the scientific journal in Britain, 1813–1825', *N&R* 70 (2016).
6 Baldwin, *Making 'Nature'*.
7 Dwight Atkinson, *Scientific Discourse in Sociohistorical Context: The Philosophical Transactions of the Royal Society of London, 1675–1975* (London: Routledge, 1998); Alan G. Gross, Joseph E. Harmon, and Michael S. Reidy, *Communicating Science: The*

scientific article from the seventeenth century to the present (New York: Oxford University Press, 2002); Hannah Kermes et al., 'The Royal Society corpus: From uncharted data to corpus', in *Proceedings of the LREC 2016* (Portoroz, Slovenia, 2016); David Banks, *The Birth of the Academic Article: Le Journal des Sçavans and the Philosophical Transactions, 1665–1700* (Sheffield: Equinox, 2017).

8 Jonathan R. Topham, 'Scientific publishing and the reading of science in early nineteenth-century Britain: An historiographical survey and guide to sources', *Studies in History and Philosophy of Science* 31A (2000); Csiszar, *The Scientific Journal*; Gowan Dawson et al., eds, *Science Periodicals in Nineteenth-Century Britain: Constructing scientific communities* (Chicago, IL: University of Chicago Press, 2020).

9 For an illustration of the way editors' correspondence changes our view of editorship, see Anna M. Gielas, 'Early sole editorship in the Holy Roman Empire and Britain, 1770s–1830s' (PhD, University of St Andrews, 2019).

10 David A. Kronick, *A History of Scientific and Technical Periodicals: The origins and development of the scientific and technical press, 1665–1790* (Metuchen, NJ: Scarecrow Press, 1976); *Scientific and Technical Periodicals of the Seventeenth and Eighteenth Centuries: A guide* (Metuchen, NJ: Scarecrow Press, 1991); A.J. Meadows, *The Scientific Journal* (London: Aslib, 1979); A.J. Meadows (ed.), *The Development of Science Publishing in Europe* (Amsterdam: Elsevier, 1980).

11 William H. Brock and A.J. Meadows, *The Lamp of Learning: Taylor & Francis and the development of science publishing* (London: Taylor & Francis, 1998).

12 A. Fyfe et al., 'The history of the scientific journal: The economic, social and cultural history of the world's oldest scientific journal, 1665–2015' (2013–), https://arts.st-andrews.ac.uk/philosophicaltransactions.

13 For instance, Aileen Fyfe et al., *Untangling Academic Publishing: A history of the relationship between commercial interests, academic prestige and the circulation of research* (St Andrews: University of St Andrews, 2017), http://doi.org/10.5281/zenodo.546100. On peer review, see Noah Moxham and Aileen Fyfe, 'The Royal Society and the prehistory of peer review, 1665–1965', *Historical Journal* 61, no. 4 (2018); and Aileen Fyfe et al., 'Managing the growth of peer review at the Royal Society journals, 1865–1965', *Science, Technology and Human Values* 45, no. 3 (2020). On copyright, see Aileen Fyfe, Julie McDougall-Waters, and Noah Moxham, 'Credit, copyright, and the circulation of scientific knowledge: The Royal Society in the long nineteenth century', *Victorian Periodicals Review* 51, no. 4 (2018). On gender, see Camilla M. Røstvik and Aileen Fyfe, 'Ladies, gentlemen, and scientific publication at the Royal Society, 1945–1990', *Open Library of Humanities* 4 (2018). On finances, see Aileen Fyfe, 'Journals, learned societies and money: *Philosophical Transactions, ca. 1750–1900*', *N&R* 69, no. 3 (2015); and 'From philanthropy to business: The economics of Royal Society journal publishing in the twentieth century', *N&R* https://doi.org/10.1098/rsnr.2022.0021.

14 Thomas P. Hughes, *Human-Built World: How to think about technology and culture* (Chicago, IL: University of Chicago Press, 2004).

15 Aileen Fyfe, *Steam-Powered Knowledge: William Chambers and the business of publishing, 1820–1860* (Chicago, IL: University of Chicago Press, 2012).

16 Thomas P. Hughes, *Networks of Power: Electrification in Western society* (Baltimore, MD: Johns Hopkins, 1983); Christine MacLeod, 'Strategies for innovation: The diffusion of new technology in nineteenth-century British industry', *Economic History Review* 45 (1992).

17 Wiebe E. Bijker, Thomas P. Hughes, and Trevor J. Pinch, eds, *Social Construction of Technological Systems: New directions in the sociology and history of technology* (Cambridge, MA: MIT Press, 1987); Nelly Oudshoorn and Trevor Pinch, eds, *How Users Matter: The co-construction of users and technology* (Cambridge, MA: MIT Press, 2003); Graeme Gooday, *Domesticating Electricity: Expertise, uncertainty and gender, 1880–1914* (London: Pickering & Chatto, 2008); Fyfe, *Steam-Powered Knowledge*.

18 On journals as clubs, see Jason Potts et al., 'A journal is a club: A new economic model for scholarly publishing', *SSRN* (2016).

Part I
Invention, 1665–1750

1
The first *Philosophical Transactions*, 1665–1677

Noah Moxham and Aileen Fyfe

In November 1664, an already overworked Henry Oldenburg received a letter from Paris, asking him to contribute to a new periodical as its English correspondent. Despite his German birth, and the fact that he had only settled in England in 1660, Oldenburg was a natural choice for the role.[1] He was brilliantly connected, in general and in particular. He had a very wide personal acquaintance among the learned of Europe, gleaned during two decades of Continental travel, and an extraordinary facility for languages, including German, Dutch, French, Latin, Italian, faultless English and (at a pinch) Danish.[2] In London, he had become closely associated with a group of gentlemen who met weekly to discuss natural philosophy and undertake experiments; through charters granted in 1662 and 1663, this group became the Royal Society of London for the Improvement of Natural Knowledge.[3] As their secretary, with responsibility for record-keeping and correspondence, no one was better informed than Oldenburg about the activity and organisation of natural philosophy in England. He was also personally close to the group's brightest ornament, the Anglo-Irish chemist Robert Boyle, for whom he acted as translator, literary agent and London correspondent.[4] It is from one of his weekly letters to Boyle that we learn of the invitation from Paris.

The explicit purpose of the new Parisian publication, the *Journal des Sçavans*, was 'to make known what goes on in the Republic of letters'.[5] It was the initiative of Louis XIV's finance minister, Jean-Baptiste Colbert, who appointed Denis de Sallo as its first compiler and granted him a royal monopoly over printed literary news (where 'literary' is best understood as broadly 'intellectual').[6] It was to consist, first, of 'an exact catalogue of the most important books printed in Europe', explaining 'what they deal with, and what they may be useful for'; second, to supply obituaries of people 'famous for their doctrines or for their works', and lists of their publications; third, 'to make known experiments in Physics and Chemistry, which may help to explain the workings of nature'; and fourth, to disseminate important legal decisions, edicts and warrants of the Church and state. Oldenburg was invited to supply accounts of new books and philosophical discoveries from England.[7] Oldenburg was, he confided to Boyle, 'very unwilling to decline this taske but yet how to undertake it, being so very single, and having so much already charged upon me, I doe not yet know. But I must remember my Motto, Providebit Dominus [the Lord will provide].'[8]

Oldenburg's ambivalence was not simply due to the difficulty of finding time. Earlier in 1664 he had been contemplating launching a newsletter of his own. He had imagined a personalised weekly letter, in manuscript, for wealthy clients, consisting 'both of State and literary news'. Oldenburg had asked Boyle to suggest likely takers, but evidently struggled to find subscribers at his proposed rate of six to ten pounds a year.[9] The project was put on the back burner or abandoned altogether; there is no subsequent reference to it in Oldenburg's surviving correspondence. Yet it had been proposed out of necessity. Oldenburg's labours on behalf of the Royal Society were unremunerated, and his work for Boyle was his only reliable source of income.[10] The invitation to contribute to the *Journal des Sçavans* clearly reminded him of his newsletter idea, but made clear that he would face competition from cheaper, printed products.

The first issue of the *Journal des Sçavans* appeared in Paris in early January 1665, and on 11 January OS, Oldenburg brought a copy before a meeting of the Royal Society.[11] By early February, he was announcing his intention to publish his own philosophical periodical, and Sir Robert Moray, the Royal Society's vice-president, told the Dutch mathematician and horologist, Christiaan Huygens: 'He [Oldenburg] will not concern himself with theology or law; but, in addition to philosophical matters which reach us from overseas, he will publish any experiments, or at least the most important ones, that are carried out here.'[12]

Moray explicitly likened the new periodical to the *Journal des Sçavans*, explaining that it would be similar, 'but much more philosophical in nature'. The intention was that it would appear monthly in English with a quarterly Latin digest.

Oldenburg brought his first issue of *Philosophical Transactions* before the Royal Society's Council for its approval on 1 March 1665.[13]

It was pieced together out of news from his correspondents, short notes drawn from the manuscript records of the Royal Society, and translations from the *Journal des Sçavans*. It was the first of 136 issues Oldenburg would edit until the summer of 1677, when he died.

The *Journal des Sçavans* and the *Philosophical Transactions* have come to be thought of as the world's first learned periodicals. They appeared within a few weeks of each other, and it is clear that their respective founders were intensely aware of one another. The editors of the *Journal* recruited Oldenburg as a regional correspondent before it launched, while Oldenburg's *Transactions* was, in turn, conceived as a more specialised instance of the new type of periodical publishing inaugurated by the *Journal*. The *Journal* surveyed the Republic of Letters, while the *Transactions* focused upon 'the Grand design of improving Natural knowledge, and perfecting all Philosophical Arts, and Sciences'.[14] The two periodicals provided valuable context and content for one another, operating in different but overlapping spheres and each borrowing freely from its counterpart. (The notion of international copyright was almost two centuries in the future.) Both periodicals would prove durable, and the *Philosophical Transactions* now enjoys the distinction of being the oldest English periodical (of any sort) still in print.

Its ultimate longevity can make it difficult to appreciate just how uncertain, contingent and experimental Oldenburg's *Transactions* was (and, indeed, would remain under his immediate successors).[15] Equally, the modern significance of academic and scientific journals can make it difficult to appreciate the early *Transactions* in its own context.[16] This is why we have emphasised its strong international orientation. Several telling points emerge from this emphasis. First, the immediate antecedents and early models for the *Transactions* were news-gathering and review projects. Second, Oldenburg's editorial ambitions were commercial in nature: the project that eventually took shape as the *Transactions* was intended to help him make his living. Third, both Oldenburg's original intention of starting a manuscript newsletter service and his recruitment by the prospective *Journal des Sçavans* testify to his standing and his personal networks within the European learned community. Recognising these points helps us appreciate the purpose and emphasis of Oldenburg's editorial practices, including his care in distinguishing his own efforts from forthcoming ventures; the way in which he highlighted his personal role within the periodical in order to promote it; and his careful positioning of the *Transactions* to appeal both to English and Continental audiences.

As we will see, Oldenburg would face a range of obstacles in establishing and maintaining the *Transactions*. On the editorial side, he had to contend with the need to secure an adequate supply of publishable material; with the sensibilities of prominent, occasionally aggrieved authors; and with the Society's apparent reluctance to trust his periodical as a suitable venue for publishing the knowledge-claims emerging from its own experimental practices. He also had to reckon with the disruptions caused by the second and third Anglo-Dutch wars (1665–67 and 1672–74), which affected communication with the Continent. The plague outbreak of summer 1665 drove most of the senior fellows, though not Oldenburg himself, out of the capital, while the Great Fire of September 1666 forced the Society out of its home in Gresham College and into temporary quarters. Personally, he also had to contend with bereavement, imprisonment in the Tower of London in the summer of 1667 (on suspicion of espionage during the war), and (until his second marriage in 1668) relative poverty.[17]

This chapter explores Oldenburg's vision for the *Transactions*, and investigates how he put it into practice. He referred to himself as its 'author' or 'publisher' (as we see in Figure 1.1), though the role he performed would later come to be known as 'editor'. We start by examining his editorial practice, including both the sourcing and shaping of copy, and what little is known about his arrangements with the print trades. Oldenburg's close association with the Royal Society has meant that historians have sometimes been as confused as contemporaries about the actual relationship between the Society and the *Transactions* in these early years. Oldenburg always insisted that the *Transactions* was his personal, independent enterprise; but we tease out the Society's role in its production, the ways Oldenburg benefited from his association with the Society, and the uncertain position of the periodical in the Royal Society's broader enterprise of 'promoting natural knowledge'. Later chapters will explore how the fellows of the Society came to value the *Transactions*, and found ways to ensure the continuity of a publication they did not clearly own.

Oldenburg's editorial practice

When Oldenburg decided to launch the *Transactions*, what did he think he was doing, and how did he set about doing it? The evidence of how the earliest issues were actually assembled, and contemporary accounts of them, are equally fragmentary, largely because of gaps in Oldenburg's surviving correspondence.[18] Thus, as other scholars have done before

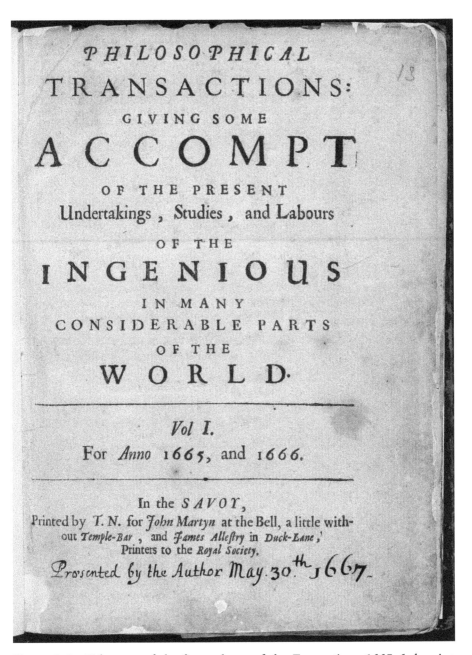

Figure 1.1 Title page of the first volume of the *Transactions*, 1665–6, bearing Oldenburg's inscription describing himself as 'the author' © The Royal Society.

us, we turn to Oldenburg's 'introduction' to his first issue (Figure 1.2). It offers us a public statement of intent, making no mention of Oldenburg's private interests:

> Whereas there is nothing more necessary for promoting the improvement of Philosophical Matters, than the communicating to such, as apply their Studies and Endevours that way, such things as are discovered or put in practise by others; it is therefore thought fit to employ the Press, as the most proper way to gratifie those, whose engagement in such Studies, and delight in the advancement of Learning and profitable Discoveries, doth entitle them to the knowledge of what this Kingdom, or other parts of the World, do, from time to time, afford, as well of the progress of the Studies, Labours, and attempts of the Curious and learned in things of this kind, as of their compleat Discoveries and performances: To the end, that such Productions being clearly and truly communicated, desires after solid and usefull knowledge may be further entertained, ingenious Endeavours and Undertakings cherished, and those, addicted to and conversant in such matters, may be invited and encouraged to search, try, and find out new things, impart their knowledge to one another, and contribute what they can to the Grand design of improving Natural knowledge, and perfecting all Philosophical Arts, and Sciences, All for the Glory of God, the Honour and Advantage of these Kingdoms, and the Universal Good of Mankind.[19]

Oldenburg insisted that the function of his periodical would consist as much in reporting the incomplete, the ongoing, and the provisional, as in accounts of 'complete discoveries and performances'. He emphasised the promotion of natural-philosophical dialogue, as well as the significance of the replication, emulation or extension of others' research, and cast the *Transactions* as a facilitator for all of them. The total effect was to represent the *Transactions* as a scientific news-sheet, an up-to-the-minute account of goings-on in the world of natural philosophy, and something rather different from a series of discrete and fully articulated claims to new knowledge in the manner of a modern academic journal.[20] This vision was manifested in Oldenburg's actual editorial practice, and his frequent inclusion of single-paragraph requests for confirmation of rumoured discoveries, or announcements of experiments intended to be performed.

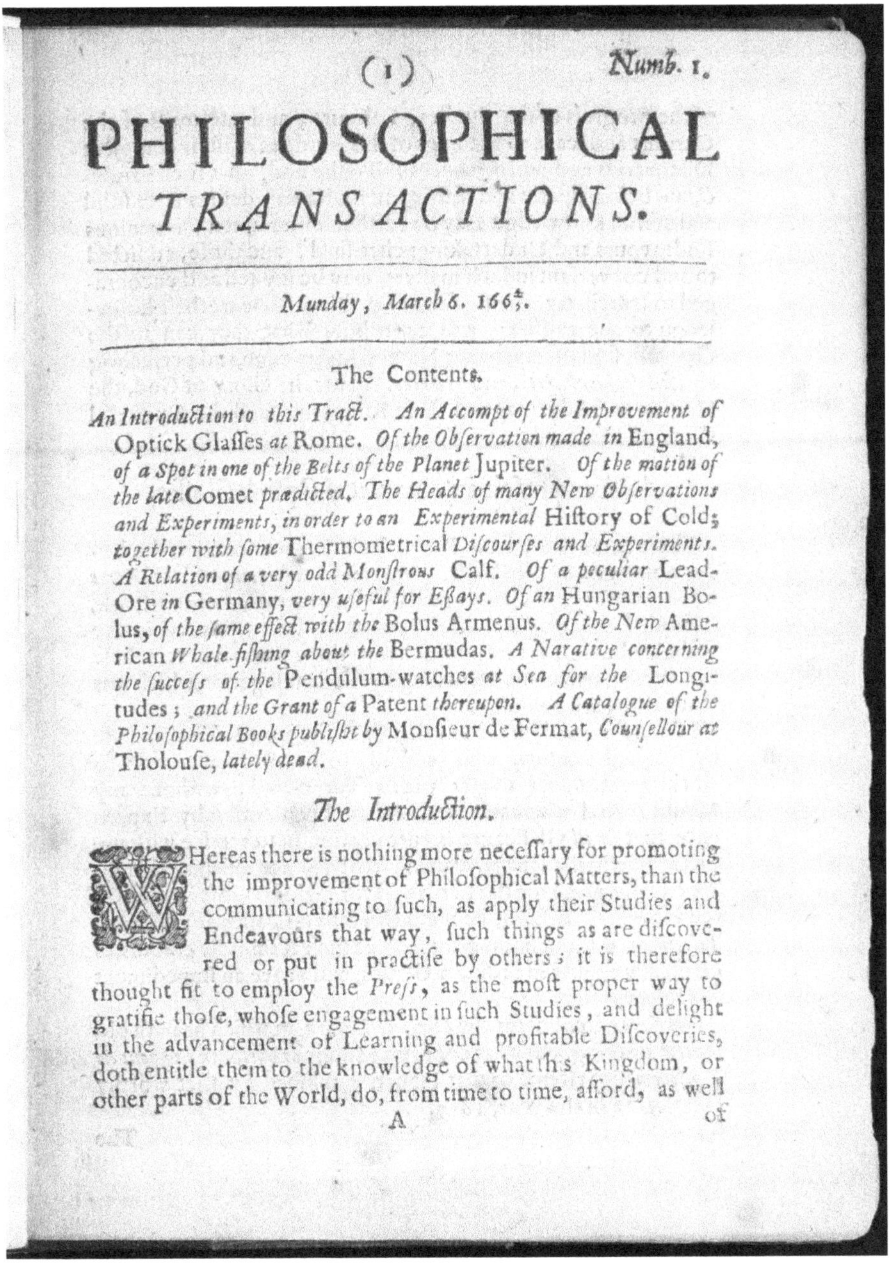

Figure 1.2 'Introduction' to the first issue of the *Transactions*, 1665 © The Royal Society.

Historians of scientific journals have claimed to detect in this preface 'the first sound, as it were, of a scientific editor's voice', as May Katzen put it.[21] The range of Oldenburg's activities – soliciting, shaping, translating and sequencing as well as contributing material – certainly fits the modern term 'editor'.[22] But Oldenburg's vision (and practice) contrasts strikingly with modern conceptions of a 'scientific editor' whose labour should not draw attention to itself in the finished product, allowing the work presented in it to speak for itself in unvarnished terms. Oldenburg, trying to drum up communications, interest and sales for a new kind of learned publication, gave a performance of editorship that was *intended* to be highly visible. This visible editorial presence supports the contention of those historians who have felt that the *Transactions* began life as an extension of Oldenburg's correspondence, even a labour-saving device for him.[23] These early issues all bear Oldenburg's mark, even if only at the level of précis or translation, and they reflect his *authorial* control in a way that would rarely be true of issues produced under subsequent editors. Oldenburg's reputation rested on his privileged access to information, his trustworthy management of it, and his scholarly connections, not his own research, and it was correspondingly important to play up his own role in obtaining information and highlighting the possibilities inherent in his new genre.

The formal and generic variety of the contents in the early *Transactions* makes it difficult to give an appropriate name to the actual units of communication that made up each issue. 'Item' seems preferable as a general term, because 'article' is anachronistic, and 'paper' and 'essay', though in contemporary usage, imply a unitary authorial communication that does not reflect the extent to which much of Oldenburg's content was extensively abridged, rewritten or unattributed. From early 1666 onwards, large parts of each issue were given over to 'accounts of books': in contrast to 'book reviews' (such as those in the *Journal des Sçavans*), these 'accounts' generally avoided critical judgement. The exploratory quality of the early *Transactions* reminds us that it was, after all, an innovation, and its conductor, readers and licensers were still feeling for the possibilities it might afford.

The variety of items, and their sources, can be seen by examining the very first issue. It was 16 quarto pages long, and its contents are listed in Table 1.1. Two pages of Oldenburg's introduction were followed by 10 items under separate headings, ranging from eight lines to almost six pages. As well as the notably wide remit – from optics and astronomy to natural history and monstrous births – we can see that five of the 10 items were unattributed (and for three of them, their origin remains uncertain).

Table 1.1 List of items in the first issue of the *Transactions*

Heading	Approx. length (in 4° pages)	Researcher	Place of origin	From a printed source?	Mentioned in RS minutes?	Originating in RS activity?
Introduction	1	Henry Oldenburg	London	No	No	No
An Accompt of the Improvement of Optick Glasses	1	Giuseppe Campani	Rome/Paris	Yes	Yes	No
A Spot in One of the Belts of Jupiter	0.25	Robert Hooke	London	No	Yes	Yes
The Motion of the Late Comet Praedicted	4.75	Adrien Auzout	Paris	Yes	No	No
An Experimental History of Cold	1.75	Robert Boyle	Oxford/London	Yes	Yes	No
An Account of a Very Odd Monstrous Calf	0.5	Robert Boyle	Oxford / London / Hampshire	No	Yes	No
Of a Peculiar Lead-Ore of Germany	0.5	Anon.	Freyung	No	No	No
Of an Hungarian Bolus	0.25	Anon.	Hungary	No	No	No
Of the New American Whale-Fishing About the Bermudas	2	Anon.	North America	No	No	No
A Narrative Concerning the Success of Pendulum-Watches at Sea for the Longitudes	2	Robert Holmes / Christiaan Huygens	London / Cape Verde Islands	Partial (JdeS)	Yes	No
The Character […] Of an Eminent Person not Long Since Dead at Tholouse	1.5	Anon.	Paris/Toulouse	Yes (JdeS)	No	No

Five were derived in whole or in part from printed sources (covering 10 pages). Seven originated overseas, including three from France (amounting to almost half the total length of the issue). Two of those were directly lifted from the *Journal des Sçavans* without attribution. It is also notable – given the presumed close links between the Royal Society and the *Transactions* – that only one of the 10 items might be said to have originated at the Society (Robert Hooke's observations of 'a spot in one of the belts of Jupiter'), though five others had been mentioned at its meetings. Oldenburg drew on his own correspondence, and on printed sources. Some of these were linked to the Royal Society, but much came from overseas. This would remain a keynote of Oldenburg's editorial work.

Oldenburg's heavy reliance on printed matter as source material is notably different from the later emphasis on originality as a requirement for contributions to scholarly journals. Like the *Journal des Sçavans*, Oldenburg's *Transactions* surveyed the realm of print: the appearance of particular items of print – or their imminent appearance, in the case of Boyle's *Experimental History of Cold* – constituted items of philosophical news in themselves.[24] Boyle's volume was connected to the *Transactions* in more than one way, since Oldenburg was its publisher and was working on it concurrently with the periodical. His preface to the reader of Boyle's volume is dated 10 March, suggesting that the *Transactions* only narrowly preceded it into circulation, and that Oldenburg was using the periodical to help advertise and promote the sale of a volume of which he would be the beneficiary.[25]

We should also note that much of the overseas material in the *Transactions* was partly filtered through France (often through the *Journal des Sçavans*) even if it did not originate there. The account of Giuseppe Campani's observations in Rome, for instance, relied upon a French original, and Oldenburg would make a habit of exploiting the superior connections between France and Italy for news of the latter for the next several years. The 'Narrative concerning the success of pendulum-watches' was derived partly from Robert Holmes's extravagantly falsified account of their trial at sea, delivered to the Royal Society, and partly from the watch-designer's account (Christiaan Huygens, who, though he was then resident in the Hague, published in the *Journal des Sçavans*).[26]

It seems likely that Oldenburg initially envisaged a mutually beneficial, almost symbiotic relationship between the *Transactions* and the *Journal*, and that this hope influenced his international editorial outlook. Bearing in mind that journeys between Paris and London in the mid-seventeenth century typically took between eight and 12 days, depending on the time of year and the difficulty of crossing the Channel, the two learned periodicals responded to one another very nearly as quickly as was possible.[27] Material from the *Journal* from 23 February NS featured in the first issue of the *Transactions* on 6 March OS (equivalent to 16 March NS); while the *Journal* of 30 March NS carried a warm account of the first *Transactions*. Unsurprisingly, it cast the *Transactions* as a philosophical imitator of the *Journal*. (It also refers to direct translations of the *Journal* in Germany and Italy during the first months of 1665 for which no other evidence has ever been found.)

The *Journal*'s encomium on the *Transactions* also raised the problem of language in international exchange, drawing attention to the limitations of publishing in English:

> Mais parce qu'ils sont la plupart escrits en langue Angloise, on n'a pû iusques à present en rendre compte dans ce Iournal. Mais on a enfin trouvé un interprete Anglois, par le moyen duquel on pourra à l'avenir l'enrichir de tout ce qui se fera de beau en Angleterre.[28]
>
> (But because they [the excellent works coming daily from 'that Company', i.e. the fellows of the Royal Society] are mostly written in English, we have not hitherto been able to give an account of them in this journal. But we have finally found an English interpreter, by whose means we shall be able to enrich it in future with all the best of what shall be produced in England.)

It seems quite likely that the 'English interpreter' referred to was Oldenburg himself, whose services the authors of the *Journal* had already secured. The effect of the *Journal*'s review is to subtly relegate the (English) *Transactions* to a local vernacular rendition of what Oldenburg was already supplying to the *Journal*. Oldenburg, of course, was aware that English was not, at this time, widely read in the Republic of Letters: his *Transactions* was primarily intended to enable scholarly gentlemen in England to learn about natural-philosophical news from the Continent, not the other way round.[29] His willingness to publish items in Latin (and sometimes French, German and Italian), indicates that he expected his Anglophone readers to have some fluency in the major European scholarly languages. However, his early hopes for a quarterly Latin digest, which would have made the *Transactions* comprehensible to a wider European audience, never came to pass in a satisfactory manner. Between 1668 and 1670, various printers, first in Leiden and then in

Amsterdam and Copenhagen, announced plans for pirate Latin editions, much to Oldenburg's annoyance. Such editions not only threatened to spoil his own plans, but might, if carried out carelessly, taint his personal standing in the Republic of Letters. An authorised Latin version was eventually produced in Amsterdam in the 1670s, but Oldenburg ended up with deep misgivings about it, too.[30]

The hopes for the future expressed by the *Journal des Sçavans* envisaged a continuing, reciprocal relationship with the *Transactions*. Yet at exactly that point, the situation changed. The *Journal des Sçavans* had its publication licence abruptly withdrawn, and its review of the *Transactions* was the last item to appear in the *Journal* for over nine months.[31] This meant that the *Transactions* was suddenly unique in the realm of European publishing, and needed to plough its own furrow. An event that might have forced Oldenburg to rely more upon the resources of the Royal Society to provide copy for his periodical, however, had no such effect. Instead, Oldenburg maintained his strong orientation towards European and printed material over the next several issues, and continued to publish very little that actually originated in courses of study or experiments planned or carried out at the Society.

Issue 4 is an instructive example of the significance of print as source material for the *Transactions*, of Oldenburg's reliance on overseas communications for copy, and of his willingness to reconfigure material to his own purpose. In this issue, Oldenburg's account of a recent pamphlet by the astronomer Adrien Auzout is split into four separately headed items. The separation is not apparent from the item headings themselves, although it is clear in the main text, but it allowed Oldenburg to treat as distinct Auzout's own arguments and new observations, his comments on the performance of Italian telescopes, and to interpolate Robert Hooke's response to Auzout's criticisms of him in the appropriate place.[32] One of the possibilities the *Transactions* afforded was to reconfigure already published scientific print as dialogue. Published material could be reprinted *and* simultaneously amplified or contested, all within a very short and responsive timeframe. Oldenburg also famously prodded Hooke to respond in lively terms to particular passages in Auzout's letters, which he passed on to Hooke with his annotations.[33]

Oldenburg's early editorial practice, then, emphasised the provisional, the incomplete, the new, the ongoing and the international. Anonymity of authorship was common, Oldenburg's editorial mediation was often highly (and, we have suggested, intentionally) visible, and much of what made its way into the *Transactions* was neither derived from, nor reported to, the meetings of the Royal Society fellows.

For how long did the contents of the *Transactions* remain so curiously detached from the activities of the Royal Society? In some respects, for a very long time indeed. An examination of the material in Oldenburg's *Transactions* in 1674 (vol. 9) reveals that over half the items (discounting book reports, errata, prefaces and advertisements) had not – as far as the Society's minutes report – been reported or discussed at a Society meeting; and in the two years following (Oldenburg's last full years in charge), the proportion of material from outside the Society was about 60 per cent. The average length of a substantive item had risen considerably: by the mid-1670s, it was four or five pages, more than double the length of the typical item in 1665 and 1666. This suggests that Oldenburg had gradually begun to give his correspondents more scope. Ellen Valle has also suggested that the *Transactions* evolved towards the increasing importance of an authorial voice, meaning that fewer pieces were summarised by Oldenburg and were instead allowed to appear in the author's own words.[34] This did not mean, however, that anonymity vanished from the periodical's pages: two essays on vitriol (sulphuric acid) appeared in issues 102 and 103 without any attribution except that they were 'communicated by a Fellow of the Royal Society, who maketh use of chymistry chiefly as subservient to physiology'.[35] This hint might have been enough to suggest to initiates that the author was Daniel Coxe, whose paper on vitriol had recently been read to the Society; for anyone else, specifying that he was a fellow of the Society advanced a claim about his philosophical status and his credibility.[36]

This did not mean that Oldenburg's editorial mediation became strictly invisible, however. He continued, well into the 1670s, to emphasise the chains of transmission by which information came his way, as in the title given to an extract of a letter from Boyle in 1674 (Figure 1.3). Its ponderousness was intentional, and typical: it carefully specified by whose means Oldenburg had received his information – from the younger van Helmont via Boyle – and these names in turn (especially those of Boyle and the elder van Helmont) informed judgements about how interesting or reliable it was likely to be.[37] Oldenburg also continued to draw attention to his own role in bringing new knowledge to light. In late 1673 or early 1674, for example, he received a letter from Christoph Sand in Hamburg, with information about the formation of pearls. 'The matter of [it] being new, but destitute of proof', Oldenburg explained to the reader, '[I] took the liberty of desiring from the author the favour of

> *An Account of the two Sorts of the* Helmontian Laudanum, *communicated to the Publisher by the Honourable* Robert Boyle, *together with the Way of the Noble Baron* F. M. van Helmont *(Son to the famous* Johannes Baptista,*) of preparing his* Laudanum.
>
> AS for the *Helmontian Laudanum,* you may use your own Liberty in suspecting the Receipts that go about of it. For the name it self seems ambiguous to me, who am well inform'd that there are two sorts of the Helmontian *Laudanum*; the one us'd by the elder *Helmont,* the other by his Son. The former was as a great Secret communicated to me by an expert Chymist, sent by a German Prince to Complement *Johannes Baptista Van Helmont,* some of whose Manuscripts (one of which perish'd in the fire of *London,*) he procured to-

Figure 1.3 The title to this item about laudanum (from *PT* 9 (1674)) illustrates the chain of communication through which Oldenburg received information © The Royal Society.

imparting to [me] what ground he had for this assertion', which Sand duly supplied in a letter of February.[38] (Sand was in regular and slightly awkward communication with Oldenburg as the translator employed by the printers in Amsterdam working on the authorised Latin version of the *Transactions*.)

Oldenburg also continued to push together material on related topics by different authors – issue 105, for example, juxtaposed letters from Christiaan Huygens and Giandomenico Cassini in Paris, responding to Robert Hooke's recently published *An Attempt to Prove the Motion of the Earth from Observations*, which had been summarised in issue 101.[39] Oldenburg also allowed himself occasional editorial comments: a letter from the Danish astronomer Johannes Hevelius in Danzig enclosed another from a Dr Wasmuth, which Oldenburg printed in Latin with the comment, 'Thus far Dr Wasmuth: who we earnestly desire may not fall short of his hope and very large promises'.[40] This was probably intended both as a warning and a courtesy, since Wasmuth's letter was discussed at a meeting of the Society in April at which 'the sense of the members seemed to be, that he had promised too much to answer expectation'. This was not the only instance in which the judgements of fellows on other people's work found their way into the *Transactions*, even if only in softened form. In an account of a book about Evangelisto Torricelli's barometric experiments in the same issue, Oldenburg made a point of referring to the judgement of 'some very learned and able men' (presumably including Boyle) about it, emphasising their praise of its ingenuity but wondering whether it had in fact refuted the theory of 'the weight and spring of the air' as thoroughly as the author believed.[41] Oldenburg's tact in this case was either very well judged or very fortuitous, since the author of the anonymous work criticising Boyle's theory of the air's elasticity was Sir Matthew Hale, at the time the Lord Chief Justice of the King's Bench and the second highest-ranking law official in England.[42]

In 1674, three types of communication predominated in the *Transactions*: longer research communiqués, typically six to 10 pages in length; shorter letter extracts containing philosophical developments from Oldenburg's provincial or Continental contacts, many of whom functioned as local news-gatherers for him; and eight to 10 pages of accounts of books. But this did not mean that more speculative or eccentric items could no longer be published; witness 'Sir Samuel Moreland's undertaking for the raising of water', for example, a short paragraph in which the inventor 'under[took] to demonstrate' how his mechanisms would outperform contemporary water-pumps.[43] The same issue also contained a single-paragraph report of 'some unusual diamonds' recently presented to the King of France from the East Indies.[44] Oldenburg and other fellows also used the *Transactions* as an instrument for outreach and steering the research of others; later in the year a list was published of 'Divers rural and oeconomical Inquiries, recommended to observation and tryal'. 'These [28] and the like queries', Oldenburg observed, 'are raised from the Communications of several of our Correspondents; which we are unwilling to deliver positively, till we hear them asserted and confirmed by observing persons upon their own practice and experience'.[45] Readers, throughout the existence of the *Transactions*, were an epistemic resource as well as a pool of possible contributors and (perhaps) a source of income.

In short, by the mid-1670s, the *Transactions* and its contents had stabilised their form to some degree, but almost none of the myriad forms that items had taken in the early years were altogether abandoned. Oldenburg continued to be willing to freely appropriate both printed and manuscript material to his purposes, and the uses to which he put them highlight the continuing flexibility of his new mode of philosophical print. His further willingness to freely manipulate it can be seen as evidence that he had relatively little concern for the integrity of the individual research communication when translating it into print. He frequently pushed together items by different researchers to produce a single item.

An author's or communicator's language was often framed as reported speech, and the basis on which an author's name was communicated or withheld is generally unclear. There are undoubtedly instances where authors were allowed simply to report their own work in their own words, and Rob Iliffe has persuasively shown that Oldenburg stepped into the editorial shadows in 1672 in order to help carve out a philosophical identity in print for the then relatively unknown Isaac Newton.[46] But such instances were not, at least in the early *Transactions*, wholly normative. Taking into account not just the extent but the sheer *visibility* of so much of Oldenburg's editorial mediation, the idea of 'authorship' within the early *Transactions* remains highly elusive.

What then was the purpose of Oldenburg's editorial agency? It seems to have been deployed mainly to emphasise the currency of the *Transactions*, and an air of provisionality and development around the research and the news he reported – an air which helps produce a feeling of complicity and involvement on the part of the reader. It also reveals, sometimes explicitly and exhaustively, the sheer length of the chains of information Oldenburg controlled, and thus the near impossibility of imagining the *Transactions* without him.[47]

Licensing, printing and money

Oldenburg was an active and independent editor, but he needed to work with partners in the London print trades in order to bring his content to his readers. His connections with the Royal Society were, once again, useful, enabling him to use both the Society's licensing privilege and its printers. The question of how much control or oversight was exercised by the Society in return – and whether it can be seen as the origins of peer review – requires close attention, as does the question of the actual financial arrangements between the parties.

Under the terms of its founding Charter, the Royal Society enjoyed the privilege of licensing books for printing on its own authority, without reference to the civil or ecclesiastical authorities whose job it was to screen printed matter for politically seditious or religiously heterodox material in Restoration London. The licensing regime had been introduced in 1662, a few months before the grant of the Society's Charter. Among other effects, and in conformity with the restored monarchy's wish to control the flow of political information in its dominions, the Licensing Act dramatically curtailed the trade in printed periodical news that had flourished during the Wars of the Three Kingdoms and the Interregnum. In this environment, arranging for the *Transactions* to be licensed for printing by the Royal Society allowed it to appear without the intense scrutiny that periodicals normally attracted due to their overwhelming association with political news.[48] Seeking the Society's licence also made sense at a personal level: Oldenburg's position as secretary gave him access to the publishing privilege and helped him turn his duty to the institution to his personal benefit.

The initial licensing order granted for the *Transactions* is relatively complex compared with those for other publications licensed by the Society, because it contains details about the production and examination of future issues. The need to grant a new licence for each new issue of the *Transactions* suggests that the Society had not had periodical publication in mind when its licensing privilege was first established.

> Ordered, that the *Philosophical Transactions*, to be composed by Mr. Oldenburg, be printed the first Monday of every month, if he have sufficient matter for it; and that that tract be licensed by the council of the society, being first reviewed by some of the members of the same; and that the president be desired now to license the first papers thereof, being written in four sheets in folio, to be printed by John Martyn and James Allestry, printers to the Society.[49]

Certain aspects of this order were definitely not followed in practice: for instance, 'four sheets folio' had transmuted to two sheets quarto by the time of going to press. This smaller format was more typically associated with newsbooks, pamphlets and cheap print in general.[50] Equally, while the requirement that each issue be 'reviewed' by the Society's Council might look like a form of censorship or early peer review the oversight exercised by Council was in practice little more than notional, as we will see. If the Council's instructions were intended to establish normal procedures for the Society's involvement in the *Transactions*, the evidence strongly suggests that these were not in fact carried out.

Using a licence from the Royal Society determined Oldenburg's choice of printers: as the order stated, the *Transactions* was to be printed by John Martyn and James Allestry, who had been appointed printers to the Royal Society in 1663. They were not actually printers, but booksellers based in St Paul's Churchyard, the heart of the London book trade. In early modern usage, there was no consistent distinction between 'printers' and 'booksellers': they were all members of the Stationers' Company, and could practise each trade 'promiscuously'.[51] We know very little about the relationship between Martyn and Allestry, or their relationships with the printers to whom they sub-contracted the early issues of the *Transactions* (identified only as 'T.N.' and 'T.R.'); but we do know that their relationship with the Royal Society, and with Oldenburg, was a complex and sometimes contentious affair.[52] In principle, the Society retained the right to appoint other (or additional) printers if it chose, but this appears in practice to have been a threat to hold over the printers' heads if their work proved unsatisfactory, or a way of getting around any unwillingness to undertake a licensed work.[53]

The contract for any particular work licensed by the Society was a matter for negotiation between the author and the booksellers, and the same was true of the *Transactions*. Few details survive of its financial arrangements, and most of what is known from the Oldenburg period comes from the very early years. It is clear that the Royal Society had no financial responsibility or interest in the project, but both Oldenburg and the printer-booksellers hoped to benefit financially, and they shared responsibility for some of the costs. The balance between benefit and responsibility was a key point of contention.

The surviving evidence comes from the second half of 1665, only a few months into the history of the *Transactions*. Plague had struck London, and by July those who could had left for the provinces. Oldenburg stayed, but his correspondence with the Society's vice-president Robert Moray (in Oxford) in October 1665 reveals both the financial arrangements he had initially made, and those he was trying to negotiate.[54] In contrast to the publishers of modern academic journals, Martyn and Allestry had initially agreed to pay Oldenburg outright for the copy he delivered, at the rate of three pounds per sheet. This guaranteed Oldenburg some income regardless of actual sales: in principle, if he managed to deliver 12 issues a year of at least two sheets, this would have earned him over £70 annually. It was not a fortune, but undoubtedly a tidy sum. (It was more, for instance, than was earned by the endowed lecturers at Gresham College.) The correspondence with Moray (of which only Moray's side survives) also reveals that Oldenburg had to pay half the costs of engraving any illustrations, which may explain why the *Transactions* was unillustrated until its fifth issue, and why around half of all the issues produced by Oldenburg had no illustrations.

Martyn and Allestry's willingness to pay Oldenburg for copy suggests that they saw enough commercial potential in the *Transactions* that they expected to be able to recoup those payments, as well as the costs of composition, presswork and paper, and still make some profit. Production costs for early modern print are usually estimated to have been half a penny per sheet, and single-sheet newsbooks often retailed for one penny. According to Robert Hooke's diary, the individual issues of the *Transactions* varied in price between sixpence and one shilling.[55] This suggests that the *Transactions* was much more expensive (and perhaps had a larger profit margin built into its price) than comparable print products. We can make another estimate from an inscription by Martyn on a surviving copy of the first volume, covering 1665–6, in which he records having received 18 shillings from Oldenburg for the volume.[56] Given the length of the volume, this implies a retail price of slightly more than four pence a sheet. Oldenburg's periodical was aimed at a book-buying public that did not mind paying far above the market rate for printed matter; but this high price point may have been necessary to generate income for all the concerned parties.

The dispersal of many fellows of the Royal Society (and members of the royal court) to Oxford during the plague months presented two problems for Oldenburg and the *Transactions*. First, the Royal Society's senior fellows considered that they could not formally meet as a corporate body outside London, and this meant that Oldenburg was suddenly at a loss for a licence to publish.[57] And second, the plague hit the book trade particularly hard. It was most virulent in the closely packed, densely populated houses of the City of London, where most of the booksellers and printers lived and had their shops; and it drove away much of the book-buying public. Oldenburg complained to Moray that Martyn and Allestry had taken advantage of the situation to try to renegotiate their agreement. In the subsequent correspondence, Moray helped Oldenburg make new arrangements with the Oxford printer Richard Davis; and it was in this context that details of his arrangements with Martyn and Allestry were rehearsed.[58]

Moray reassured Oldenburg that he would 'get a licence though the Council meet not', and he eventually obtained one on Oldenburg's behalf from the University authorities: three issues (numbers 6–8) were thus licensed by the University of Oxford rather than by the Royal Society.[59] This had the consequence of freeing Oldenburg from necessarily working with the Royal Society's printers, but it is not clear whether this was what upset Martyn and Allestry, or whether they had already expressed a wish to renegotiate – perhaps to pause publication or to obtain more favourable terms – due to the difficult conditions imposed by the plague. Oldenburg, for his part, was evidently anxious not to lose the momentum of his new periodical, which had only reached its fifth issue when the Society broke up in July.

Moray's assistance suggests that at least some senior figures in the Royal Society were eager to have the *Transactions* continue. In this prolonged period of suspension, the Society risked seeing its relevance and reputation fade, and the interest of its members drift. Under such conditions, the fellows perhaps saw advantage in the perceived association between the Society and the *Transactions*; or were at least grateful for anything that continued to draw the eyes of the learned community towards the natural-philosophical activities of London (even if temporarily in exile).

The arrangement with the Oxford printer, Davis, which was apparently modelled on the existing arrangement with Martyn and Allestry, tells us that the print run was to be 1,250 copies. This suggests an ambition to reach an audience significantly wider than the 150 or so Royal Society fellows who were the most likely customers, and was broadly comparable to the edition size of mid-seventeenth-century newsbooks (even though, as we have seen, the price point was much higher). It was, however, notably higher than the print runs for the *Transactions* in the eighteenth century, and was perhaps tainted by the optimism of youth.

We also learn something about the allocation of those copies: 50 copies were reserved for Oldenburg personally, 'half for yourself, the other half for presents etc.'; 200 copies were for the bookseller; and the 'bargain made upon 1000'.[60] Exactly what these latter two provisions meant is hard to determine. Adrian Johns has suggested that the 200 copies for the bookseller were intended as an acknowledgement of the impossibility of effectually preventing the practice of supernumerary printing, a common form of early modern print piracy whereby the publisher secretly produced and sold more copies of an edition than had been agreed with the author, pocketing the profit from illicit sales of the work while spoiling the remaining market for it.[61] Johns also speculates about the possibility of a profit-sharing agreement between Oldenburg and the booksellers, or – more probably, as he thinks – of a penalty clause in case the edition failed to sell, perhaps in the form of a reduced rate of payment for copy.[62] Moray's letters acknowledged that possibility; he had agreed with Davis that if the copies 'go not off consideration shall be had of the loss'.[63] Yet if the 200 copies were intended simply to ward off a possible piracy, and there was no profit-sharing agreement, there was little reason for Oldenburg to assent to their production; he could not gain from them, and since the enlarged edition drove up both production and warehousing costs, it might also put his copy-payments at risk. On that reckoning a profit-sharing agreement seems likely.

On the other hand, we have the evidence from Davis that he had just broken even with sales of 300 copies for issue 6. This is plausible: 300 sales at 1s per copy would have recouped £15; and our estimated costs for three sheets of print and 1,250 copies for the entire edition, would suggest production costs of around £7 16s., which, combined with the £9 owed to Oldenburg for copy, gives a total expense of about £16 16s.[64] But for Davis to have broken even on 300 sales would have required him to have received, and kept, all the sales income from those copies. It seems possible, therefore, that the 200 copies reserved for the bookseller were intended to (partly) defray the costs of paper, typesetting and printing, which the bookseller bore alone (the costs of engraving were shared). After the first 200 copies, on the sales of the remaining 1,000 copies, there was a 'bargain'; that is, a profit-sharing agreement, on top of Oldenburg's fee for copy.

Sadly for Oldenburg, this arrangement proved never to be as lucrative as it promised. Once printing returned to London in February 1666, there is evidence of Martyn and his successors making every effort to renegotiate the bargain in their favour: for instance, Martyn refused to pay a fee for copy after the Great Fire in September 1666.[65] And Oldenburg complained in 1667 that the *Transactions* had never yet yielded him more than £40 per annum, far less than might have been expected from these 1665 arrangements; it was, as he observed to Robert Boyle, 'little more than my house rent' in Piccadilly.[66] Fortunately, his marriage the following year relieved his financial woes. After 1668, however, we know nothing more about the *Transactions*' sales or (lack of) profitability, during the Oldenburg years.

The Royal Society and the *Transactions*

In one sense, the relationship between the Royal Society and the *Transactions* was a simple one. The Society was merely the licensing authority, performing the same function as the civil and ecclesiastical authorities whose scrutiny and imprimatur were necessary before any book could be published in Restoration London. Licensers had no particular rights in the titles they authorised, and their award of an imprimatur did not signify an endorsement of a book's content – only that it contained nothing actually seditious or heretical. The Society contributed no money towards the publication of the *Transactions*, supplied very little of the early content, and imposed no evident constraints upon Oldenburg.

Yet, the terms of the order for licensing the first issue of the *Transactions* in 1665 suggest an intention of regular Society oversight of the new periodical: future issues were to be 'licensed by the council of the society, being first reviewed by some of the members of the same'.[67] This has suggested to some commentators, then and now, that the Royal Society was in some sense responsible for the *Transactions*. It has also been used to suggest that peer review originated in 1665, even though more nuanced historical accounts have pointed both to other possible origins, and to the significant changes between then and now.[68] Mario Biagioli has posited links with the practices of early modern book censorship, by virtue of the licensing privileges held by the new scientific societies of England and France, while others have pointed to book reviews and the role of personal correspondence.[69] We have argued elsewhere that historical forms of scholarly review had far more diverse purposes than commentators focused upon modern 'peer review' would expect.[70] More broadly, and despite the wording of the 1665 order, we would argue that there is insufficient evidence of regular or rigorous prepublication scrutiny by anyone other than Oldenburg to justify the use of the term 'peer review' in relation to the early *Transactions*.

By May 1666 there was clearly a sufficiently widespread belief that the *Transactions* was put out by the Royal Society that Oldenburg felt the need to respond. He explained in print that the assumption of Royal Society responsibility was 'a meer mistake', and that he composed and published the *Transactions* 'upon his *Private* account ... as a Well-wisher to the advancement of usefull knowledge'. He admitted that he did occasionally use material from Society meetings in the periodical, but only such as 'he knows he may mention without offending them, or transgressing their Orders'.[71] His phrasing suggests either that the Society issued interdictions on the publication of particular papers, a process for which there is only spotty evidence; or that there was customary practice, understood by all parties, governing which papers or experiments heard in the Society might appear in the *Transactions*.

Surviving evidence of actual Society intervention in the editorial process of the early *Transactions* is strikingly thin. We have found only three unambiguous instances of the Society issuing editorial 'orders' during Oldenburg's lifetime, and all stand out as rare and exceptional, rather than normal practice. In spring 1665, the Society's president, Lord Brouncker, declared that an account of William Petty's trials of a double-hulled ship design should not be published until the king had been acquainted with the results. It is possible that the subject was felt to be militarily sensitive, since England was then at war with the Dutch Republic.[72] In 1672, the printing of Robert Hooke's response to Isaac Newton's famous paper on the colours of the visible spectrum was ordered to be delayed. This was explicitly intended as a courtesy to Newton, to prevent the appearance of 'disrespect, in printing so sudden a refutation of a discourse of his, which had met with so much applause at the Society but a few days before'.[73] And in 1677, Council explicitly ordered the publication of a paper by Henry Howard, the Duke of Norfolk. This was a show of special consideration given to the head of an ancient aristocratic family who had been a conspicuous benefactor of the Society, particularly after the Great Fire.[74] On this evidence, the occasional institutional input into the early editorial process of the *Transactions* was variously a function of social deference, scientific civility or the risks of appearing to infringe the crown's prerogative. It was not until 1686, when the Society attempted to take more direct editorial control over the *Transactions*, that it adopted the practice of routinely 'ordering' for papers for publication (see Chapter 2).

The importance of civility and social relations in early modern knowledge-making meant that, even without direct instructions on what to publish, or not, Oldenburg did not have an entirely free hand.[75] As the Society's secretary, he was expected to conform to norms of gentlemanly scholarly etiquette, and the unwritten nature of these norms (not to mention the uncertainty about how they applied to the new printed periodical) meant that some correspondents complained of their treatment. Oldenburg's efforts to manage delicate transnational priority claims with justice to all parties were often made more complicated by the fact that all parties could be fellows of the Society.

One such fellow was Christiaan Huygens, now relocated in Paris as a member of Louis XIV's Académie Royale des Sciences. In March 1669, Huygens complained that Oldenburg and the Royal Society had solicited his expression of the laws of motion at the same time as the subject was being discussed by the Oxford mathematician John Wallis and the architect, astronomer and mathematician Christopher Wren. Huygens' views had turned out to be very similar to Wren's, but he had arrived at them independently. By publishing Wren's theory without mentioning Huygens, Oldenburg had, as Huygens put it, 'to all intents and purpose anticipated me in the publication of those rules, although you had made me hope for the contrary when you asked me to open relations with these gentlemen on the question of motion'.[76] Oldenburg replied that he had assumed that if Huygens had wished to appear in print he would have availed himself of the *Journal des Sçavans*.[77] He added that he did not presume to publish communications without permission and that Huygens' contribution to the debate had been duly recorded in the Society's register book, so there could be no suggestion of plagiarism on either Huygens' or Wren's part. Huygens took Oldenburg's advice and published his account in the *Journal*, whereupon Oldenburg reprinted it with some account of the context in the next issue of the *Transactions*.[78]

The episode testifies to the deepening complexity of managing the competing claims of researchers in a sphere of scholarly discourse that included correspondence, in-person meetings and a pair of periodicals best described as cooperative rivals. Huygens' objection to the non-appearance of his views on the laws of motion in the *Transactions* contrasts with Robert Hooke's complaints that Oldenburg was sometimes too free in sharing information whose author had intended it to be treated confidentially, or at least as not for dissemination outside the semi-private meetings of the Society.[79] After Oldenburg's death, Hooke, newly elected as secretary, would comb his predecessor's papers for evidence of his perfidy in failing first to properly record Hooke's contributions to the Society, and second in disclosing them to his correspondents.[80] Hooke's suspicions point to the importance of Oldenburg's probity in maintaining the social and epistemic integrity of early modern learned communication.

The enterprise of the *Transactions* has often been thought of as an extension of the practices of record-keeping and correspondence undertaken by Oldenburg for the Society. The delicate interaction between manuscript and print explains some of the difficulties of effectually distinguishing between the institution and what appeared in the periodical. When Oldenburg reassured Huygens that his claims had been recorded in the Society's register books, even though they had not appeared in print, he was insisting that the manuscript registers were an adequate public record of priority. Huygens' reply did not mention the registers, but it was clear that he did not think they were public in the same way as the *Transactions*.

In fact, Oldenburg himself evidently shared Huygens' view in contexts where it suited him. In 1669, he had to defend the Society's record-keeping from accusations made in print by Scottish natural philosopher George Sinclair. Sinclair, too, alleged that his ideas had been inadequately recorded: in his case, they had not even been included in the registers. Far from apologising, Oldenburg scrambled to his own defence, issuing a swift rebuttal of Sinclair's accusations and counter-charging that Sinclair had plagiarised from Boyle.[81] The contrasting treatments meted out to Sinclair and Huygens were effects of their different social and intellectual standing. Huygens was the son of an immensely influential Dutch statesman, and, like Oldenburg's mentor Boyle, had the wealth and connections to pursue and publish his researches largely on his own terms.[82] Publishing the work of such a person without his permission ran a real risk of offending. Sinclair, on the other hand, had been professor of philosophy at the University of Glasgow until forced to resign in 1666 by the imposition of religious tests under the restored monarchy; he subsequently worked as a mining surveyor, school teacher and public lecturer in Edinburgh. Furthermore, he had the temerity to oppose Boyle, and thus found the resources of the Royal Society mobilised very rapidly and effectively against him.

In each of the instances we have considered, the already delicate considerations of propriety were made more complex by the Royal Society's being implicated in, but not straightforwardly responsible for, the *Transactions*. Even when their functions plainly overlapped, the Society and the *Transactions* refused to identify fully with one another. The reasons for that refusal were bound up with questions of propriety, but also with the Society's apparent sense that the *Transactions* was not a proper place in which to advance the Society's own claims to have produced experimental natural knowledge.

This concern with the epistemological status of periodical print was never made explicit but the relation between the Society and the *Transactions* is hard to explain without it. The early Royal Society consistently nurtured ambitions to devise, prosecute and bear witness to experimental performances on its own account. Its actual productivity in this respect was uneven, and a wide variety of forms and inducements were tried at different times to encourage this kind of activity and to disseminate its results. They included collaboratively authored tracts bound into a single volume (such as John Evelyn's *Sylva* (1664)); sets of experimental observations and demonstrations performed partly at the Society's behest (such as Robert Hooke's *Micrographia* (1665)); courses of experiments on particular topics sponsored by the Society (such as Nehemiah Grew's *The Anatomy of Vegetables Begun* (1672)); and a string of individually authored pamphlets by senior fellows in the mid-1670s.[83] A selection of early experimental performances produced at the Society's request or demonstrated before its fellows, of the kind that might have been expected to find a home in the *Transactions*, appeared instead in the apologia the Society commissioned Thomas Sprat to write, the *History of the Royal Society of London* (1667). Later attempts to revive the Society's experimental programme in the late 1670s carefully distinguished between the *Transactions* and the mode of publication proposed for the research the Society itself commissioned, conducted and witnessed.[84]

The Society had plenty of research more or less at its command that could have appeared in the periodical. From the facts that it did not, and that when the Society involved itself in the journal's editorial process that interference drew attention to itself as exceptional, we conclude that there was perceived to be a difference between the research in which the Society felt a proprietorial concern, and the rest of the content of the periodical. Much of this content, as we have shown, took the form of scraps of news, summaries of recent discoveries, or excerpts or recapitulations of recently published matter, and by definition much of it was second-hand. This was an early established tendency in the *Transactions* and the emphasis on the provisional, the second-hand and the merely reported may have compromised it in the Society's eyes as a site for establishing credible and substantive claims to knowledge.

The problem of securing the agreement of members of the early modern natural-philosophical community to observations or results they could not personally claim to have witnessed has been much analysed, most influentially by Steven Shapin and Simon Schaffer. They proposed that Robert Boyle's strategies for winning the assent of readers to his experimental conclusions (and in particular to the performance of his pneumatic apparatus) were designed to get around precisely this problem, and christened this cluster of techniques – or technologies, to use their preferred term – 'virtual witnessing'.[85] Oldenburg, particularly when excerpting or summarising material from overseas, made no attempt to establish the claims of his correspondents or the authors he drew upon as matters of fact. In decency, he could do little else, since he was not personally in a position to verify the claims that were communicated to him. Instead, he summarised or framed claims to knowledge as reported speech from their authors.

The Royal Society had a ready solution to the problem of procuring assent to knowledge-claims at a distance. It was in a position to multiply credible witnesses to experiments and observations simply by gathering a significant number of well-credentialled people in the same room every week. But by preferring what was primarily witnessed over what was merely reported, it introduced an epistemic hierarchy, one that was built into the Society's constitution from the outset and which, we suggest, severely constrained its confidence in the *Transactions* as a genre. Various factors could have played into this, including the periodical's confined length, its relatively rapid schedule of publication, and the frequent appearance in its pages of what was merely reported, rumoured or incomplete, which might have tarred the Society's efforts with the same brush. It is vital to recognise that the early *Transactions* made no claims to epistemic authority. The idea that a contribution to a scientific periodical represents an original, adequately described and methodologically sound claim to knowledge, so essential to the modern conception of scientific literature, was simply not current in London in the 1660s. But this did not mean that the periodical was an inherently debased form. Rather, it represented the translation into print of an earlier, more provisional stage in the knowledge-making process.

In an important sense, the story of the emergence of the periodical as the dominant mode of scientific communication is the story of how this epistemic hierarchy was overcome, and of the development of publicly understood protocols for establishing the good faith and methodological soundness of a periodical's contents. Those protocols were intended to create trust and understanding between readers and authors unknown to one another, and thus to enable reliable communication over large physical distances as the scientific community grew. The Royal Society was slow to recognise the potential need for such protocols, because its very existence – as a group of people who met together every week – was an alternative solution to the same problem. Even once the Society recognised the possibilities Oldenburg's periodical held out for extending the natural-philosophical community, for as long as the institution aspired to produce natural knowledge on its own account its favoured technique for doing so restrained it from extending the same degree of credence to anything beyond the scope of that technique – that is to say primary witnessing. Much of the content of the *Transactions* could not meet that standard and the Society was evidently reluctant to allow the experimental natural knowledge produced within the institution to appear alongside it – an attitude that does a lot to explain the periodical's early focus on news from beyond the Society's walls.

———

Oldenburg's *Transactions* carried few of the hallmarks we now associate with scientific periodical publication. There was no systematic review by anyone other than Oldenburg, nor did it necessarily register priority of discovery, a role that was undertaken by the Royal Society's register books. It did act as a channel of communication; but the material it communicated was different both from that in a modern academic journal, and from that contained in contemporary books and treatises. Oldenburg's issues contained a miscellany of items, not all original, not all written by the attributed 'authors' themselves, and often reprinted from other printed sources. It was an experimental format.

It was a commercial enterprise for Oldenburg and the printers, and was run accordingly; a point understood at least as well by senior fellows of the Society as by Oldenburg himself. Its survival was always a precarious matter and depended strongly upon Oldenburg's personal management and reputation. This was why he worried about the possible reputational damage of a sloppy pirate Latin edition, and why he was unhappy about a stopgap number of the *Transactions* that appeared while he was imprisoned in the Tower of London in the summer of 1667. This issue, untitled but paginated and numbered continuously with the rest of the series, has been attributed to the good intentions of Oldenburg's fellow-secretary, John Wilkins, but it appeared to endorse a French claim to priority in blood transfusion experiments. Oldenburg – perhaps newly sensitised to the need to patriotically assert English claims by his recent experience – felt compelled to disavow it upon his release.[86]

It is notable that the early *Transactions* was far less intimately linked to the Royal Society than has usually been assumed. Relatively little of its content was derived from the experimental activities of the Society, nor even from the Society's official correspondence. Nor did the Society direct or oversee the content of the *Transactions*. And yet, in summer 1665, Robert Moray had helped to keep it going; and by the 1670s, the Society found it increasingly difficult to imagine life without the *Transactions*. Oldenburg's death in September 1677 meant this prospect suddenly had to be contemplated.

Notes

1. Marie Boas Hall, *Promoting Experimental Learning: Experiment and the Royal Society, 1660–1727* (Cambridge: Cambridge University Press, 1993); Steven Shapin, '"O Henry", review of *The Correspondence of Henry Oldenburg*', *Isis* 78, no. 3 (1987): 417–24; Michael Hunter, *Establishing the New Science: The experience of the early Royal Society* (Woodbridge: Boydell & Brewer, 1989).
2. Oldenburg to Francis Willughby, 20 January 1670, in A. Rupert Hall and Marie Boas Hall, eds, *The Correspondence of Henry Oldenburg*, 13 vols (Madison: University of Wisconsin Press, 1965–86) vol. 6, 437–8.
3. Henry George Lyons, ed., *The Record of the Royal Society of London*, 4th edn (London: The Royal Society, 1940; reprint, 1992).
4. Iordan Avramov, 'An apprenticeship in scientific communication: The early correspondence of Henry Oldenburg (1656–63)', *N&R* 53.2 (1999): 187–201.
5. 'L'Imprimeur au lecteur', in *Journal des Sçavans* 1, 7 January 1665 NS [author's translation].
6. Jean-Pierre Vittu, 'La formation d'une institution scientifique: le Journal des Savants de 1665 à 1714 [premier article]' *Journal des Savants* (2002): 179–203; Jacob Soll, *The Information Master: Jean-Baptiste Colbert's secret state intelligence system* (Ann Arbor: University of Michigan Press, 2009), 100–1.
7. It is not clear who approached Oldenburg on the French side. The editors of the *Oldenburg Correspondence* suggest Adrien Auzout, an astronomer and one of Oldenburg's most assiduous correspondents during the first few years of the *Transactions*' publication. No letters survive between Oldenburg and Denis de Sallo, the founding editor of the *Journal des Sçavans*, suggesting that Oldenburg's contributions were probably solicited and managed through an intermediary. Hall and Hall, *Oldenburg Correspondence*, vol. 2, 319–20 and 324 (n4).
8. Oldenburg to Robert Boyle, 24 November 1664, RS/EL OB/26.
9. Oldenburg to Boyle, 22 August 1664, Hall and Hall, *Oldenburg Correspondence*, vol. 2, 210.
10. Hall, *Promoting Experimental Learning*, 79.
11. Thomas Birch, *History of the Royal Society of London for Improving of Natural Knowledge*, 4 vols (London: A. Millar, 1756–7) vol. 2, 6 (re 11 January 1665). Until 1752, England operated on the Julian calendar; whereas France had already changed to the Gregorian calendar. Dates in the two calendars differed by 10 days until 1700, and then by 11 days.

In 1665, 11 January in England would have been dated 21 January in France. Also, in the Julian (Old Style, or OS) calendar, the year began on 25 March (rather than 1 January).

12 Robert Moray to Christiaan Huygens, 3 February 1665 OS, in *Oeuvres Completes de Christiaan Huygens*, 22 vols (The Hague: Martinus Nijhoff, 1888–1950), vol. 5, 234–5 (French original; author's translation).

13 Moray to Huygens, 3 February 1665 OS, in *Oeuvres Completes de Christiaan Huygens*, vol. 5, 234–5.

14 Henry Oldenburg, 'Preface', *PT* 1 (1665–6), 2.

15 The work of Adrian Johns makes this clear, particularly with respect to Oldenburg's engagements with the London and Continental print trades; see Adrian Johns, *The Nature of the Book: Print and knowledge in the making* (Chicago, IL: University of Chicago Press, 1998); and Johns, 'Miscellaneous methods: Authors, societies and journals in early modern England', *British Journal for the History of Science* 33, no. 2 (2000): 159–86.

16 Alan G. Gross, Joseph E. Harmon, and Michael S. Reidy, *Communicating Science: The scientific article from the seventeenth century to the present* (New York: Oxford University Press, 2002), 1.

17 Oldenburg found himself imprisoned following the successful Dutch raid on English shipping in the Medway in June 1667, apparently for a letter, addressed to a French colleague, lamenting the ineffectual English response. The episode is mentioned in Samuel Pepys's *Diary*, and discussed in Hall, *Promoting Experimental Learning*, 115–19.

18 There is only one surviving letter from Oldenburg between 10 December and 11 March 1665 (perhaps because he and Boyle were both in London, and thus had less need to correspond), and there is no explicit allusion to *Transactions* in any letter until that summer (Oldenburg to Boyle, 10 August 1665; Hall and Hall, *Oldenburg Correspondence*, vol. 2, 458–9). On Boyle's activities in these months, see the calendar of Boyle's correspondence in *Early Modern Letters Online*, http://emlo.bodleian.ox.ac.uk/forms/advanced?start=500&mail_recipient-person=http://localhost/person/38df1cfc-fa19-421a-9a34-d47cfae8bf4e&people=Robert%20Boyle&start=550; and Birch, *History of the Royal Society*, vol. 2, 448–65 (7 January to 29 April 1665).

19 Henry Oldenburg, 'Preface', *PT* 1 (1665–6): 1–2.

20 For a fuller version of this argument see Noah Moxham, 'Authors, editors and newsmongers: Form and genre in the Philosophical Transactions under Henry Oldenburg', in *News Networks in Early Modern Europe*, ed. Joad Raymond and Noah Moxham (Leiden: Brill, 2016), 463–92.

21 May F. Katzen, 'The changing appearance of research journals in science and technology', in *The Development of Science Publishing in Europe*, ed. A.J. Meadows (Amsterdam: Elsevier, 1980), 177–214 at 193. See also Charles Bazerman, *Shaping Written Knowledge: The genre and activity of the experimental article in science* (Madison: University of Wisconsin Press, 1988), 131.

22 On scientific editorship, see Aileen Fyfe and Anna Gielas, 'Introduction: Editorship and the editing of scientific journals, 1750–1950', *Centaurus* 62, no. 1 (2020): 5–20.

23 Among others, Hunter, *Establishing the New Science*, 248; and David A. Kronick, *A History of Scientific & Technical Periodicals: The origins and development of the scientific and technical press, 1665–1790* (Metuchen, NJ: Scarecrow Press, 1976), 55–6. Hall, *Promoting Experimental Learning* is more circumspect, recognising a connection between the correspondence and the *Transactions* – particularly as sources of income – but, intriguingly, hints at the possibility that Oldenburg's foreign correspondence represented an income stream in itself – though how that could have been is uncertain, see pp. 104–5, 125–6 and esp. 273.

24 Boyle's book was actually in press as Oldenburg was putting together the first issue of the *Transactions*; see Hunter and Davis, 'Introductory Notes', in Michael Hunter and Edward B. Davis, eds, *The Works of Robert Boyle*, 14 vols (London: Pickering & Chatto, 1999–2000), vol. 4, xx–xxii.

25 Oldenburg, 'The Publisher to the Ingenious Reader', in Robert Boyle, *New Experiments Touching Cold* (London: John Crooke, 1665), sigs. a2^r–a4^v.

26 Lisa Jardine, 'Never trust a pirate: Christiaan Huygens's longitude clocks', in Lisa Jardine, *Temptation in the Archives: Essays in Golden Age Dutch Culture* (London: UCL Press, 2015), 33–44.

27 See T.C.W. Blanning, *The Pursuit of Glory: Europe, 1648–1815* (London: Viking, 2007), ch. 1; Mark Brayshay, Philip Harrison, and Brian Chalkley, 'Knowledge, nationhood and governance: The speed of the Royal post in early-modern England', *Journal of Historical Geography* 24, no. 3 (1998): 265–88; and Nikolaus Schobesberger et al., 'European

postal networks', in *News Networks in Early Modern Europe*, ed. Joad Raymond and Noah Moxham (Leiden: Brill, 2016), 19–63, esp. 28–30 and 58–9.

28 *JdeS* 1 (1665): 156.

29 For a valuable discussion of the possibilities and limitations of Latin as a functional universal scientific language in the scientific revolution and after, see Michael D. Gordin, *Scientific Babel: How science was done before and after global English* (Chicago, IL: University of Chicago Press, 2015), 39–49.

30 See Johns, *Nature of the Book*, 516–20; and Pablo Toribio, 'The Latin translation of *Philosophical Transactions* (1671–1681)', in *Translation in Knowledge, Knowledge in Translation*, ed. Rocío G. Sumillera, Jan Surman and Katharina Kühn, Benjamins Translation Library (Amsterdam: John Benjamin, 2020), 123–44.

31 For the circumstances of the *Journal*'s suspension – essentially for its over-enthusiastic endorsement of a Gallican theological tract that prompted a complaint from the papal envoy in Paris – see Vittu, 'La formation d'une institution scientifique'.

32 Adrien Auzout, 'Considerations ... upon Mr Hook's new instrument ...', *PT* 1.4 (1665): 56–62; 'Monsieur Auzout's judgement', *PT* 1.4 (1665): 55–6; 'A means to illuminate an object ...', *PT* 1.4 (1665): 68–9; and 'Touching Signor Campani's book ...', *PT* 1.4 (1665): 69–73.

33 See Hall and Hall, *Oldenburg Correspondence*, vol. 2, 474 for Oldenburg's annotations.

34 Ellen Valle, '"Reporting the doings of the curious": Authors and editors in the Philosophical Transactions of the Royal Society of London', in *News Discourse in Early Modern Britain: Selected papers of CHINED 2004*, ed. Nicholas Brownlees (Bern: Peter Lang, 2006), 71–90.

35 [Daniel Coxe], 'Some observations and experiments', *PT* 9.102 (1674): 41–7.

36 For Coxe's paper at the Royal Society, see Birch, *History of the Royal Society*, vol. 3, 135.

37 *PT* 9.107 (1674): 147–9.

38 Christoph Sand, 'Extracts of two letters ...', *PT* 9.101 (1674): 11–12.

39 Christiaan Huygens and Giandomenico Cassini, 'Letter ... touching his thoughts ...', *PT* 9.105 (1674): 90–1. For the summary of Hooke's work, see *PT* 9.101 (1674): 12–20.

40 Johannes Hevelius and Dr Wasmuth, 'An extract of a letter ...', *PT* 9.104 (1674): 74–7.

41 'An accompt of some books ...', *PT* 9.104 (1674): 78–88.

42 Alan Cromartie, *Sir Matthew Hale, 1609–1676: Law, religion and natural philosophy* (Cambridge: Cambridge University Press, 1995).

43 *PT* 9.102 (1674): 25.

44 *PT* 9.102 (1674): 26.

45 *PT* 9.111 (1674): 240–2. On calls for readers to supply information, see also Daniel Carey, 'Compiling nature's history: Travellers and travel narratives in the early Royal Society', *Annals of Science* 54 (1997): 269–92.

46 Rob Iliffe, 'Author-mongering', in *The Consumption of Culture 1600–1800: Image, object, text*, ed. Ann Bermingham and John Brewer (London: Routledge, 1995), 166–92.

47 For example, 'Some Observations Concerning Iapan, made by an Ingenious person, that hath many years resided in that country; as they were communicated in French by M. I.; whence they are thus English'd by the Publisher; who some months agoe occasion'd this Accompt by some Queries, sent to that Traveller', *PT* 4.49 (1669): 983–6.

48 For discussions of the wider significance of the licence, see Mario Biagioli, 'From book censorship to academic peer review', *Emergences: Journal for the Study of Media & Composite Cultures* 12, no. 1 (2002): 11–45 and Noah Moxham, 'The uses of licensing: Publishing strategy and the imprimatur at the early Royal Society', in *Institutionalisation of Sciences in Early Modern Europe*, ed. Mordechai Feingold and Giulia Giannini (Leiden: Brill, 2019), 266–91. For the wider censorship regime of Restoration Britain, see Peter Hinds, *'The Horrid Popish Plot': Roger L'Estrange and the circulation of political discourse in late seventeenth-century London* (Oxford: Oxford University Press, 2010).

49 Birch, *History of the Royal Society*, vol. 2, 18.

50 Joad Raymond, *Pamphlets and Pamphleteering in Early Modern Britain* (Cambridge: Cambridge University Press, 2003); and *The Oxford History of Popular Print Culture* (Oxford: Oxford University Press, 2011), 4–12.

51 Birch, *History of the Royal Society*, vol. 1, 321; Charles A. Rivington, 'Early printers to the Royal Society 1663–1708', *N&R* 39.1 (1984): 1–27.

52 On the actual printers of the early *Transactions*, see Rivington, 'Early printers to the Royal Society'. The initial printers may have been Thomas Newcombe and Thomas Roycroft, see British Book Trade Index, http://bbti.bodleian.ox.ac.uk/.

53 See Birch, *History of the Royal Society*, vol. 1, 324, for the language of the commission to which Martyn and Allestry originally agreed.
54 Hall and Hall, *Oldenburg Correspondence*, vol. 2, 563.
55 See Johns, 'Miscellaneous methods', 170.
56 Inscription on copy of Volume I exhibited at the Museum of the History of Science, Oxford in 2010, from the collection of Marcus Cavalier, inscribed 'Rec. October 18th 1669 from Mr Oldenburgh Eighteen shillings for this voll: of Transactions by me John Martyn', accessed 20 January 2017.
57 In response to the problem of being unable to meet officially outside London, the Society altered its Charter to allow it to hold its Assemblies anywhere, subject to the institution's own rules of quorum. See 'Translation of Third Charter (1669)', in J.S. Rowlinson and Norman H. Robinson, *The Record of the Royal Society: Supplement to the fourth edition, for the years 1940–1989* (London: The Royal Society, 1992), 282.
58 Hall and Hall, *Oldenburg Correspondence*, vol. 2, 563.
59 See the colophons to issues 6–8 of *Transactions*: *PT* 1 (1665): 118, 130, 146.
60 Hall and Hall, *Oldenburg Correspondence*, vol. 2, 563.
61 Johns, 'Miscellaneous methods', 167.
62 Johns, 'Miscellaneous methods', 170.
63 Hall and Hall, *Oldenburg Correspondence*, vol. 2, 582–3.
64 Based on representative production costs of ½d. per sheet.
65 For Martyn's refusal to resume printing or to pay for copy following the damage done to his business by the Great Fire of September 1666, see Hall and Hall, *Oldenburg Correspondence*, vol. 2.
66 Hall and Hall, *Oldenburg Correspondence*, vol. 4, 58–9.
67 Birch, *History of the Royal Society*, vol. 2, 18.
68 For the 1665 myth in scholarly communications debates, see, for example, Robert Campbell, Ed Pentz, and Ian Borthwick, eds, *Academic and Professional Publishing* (Oxford: Elsevier Science, 2012). The myth probably derives from the use of Oldenburg in the opening sections of the pioneering analysis of peer review, see Harriet Zuckerman and Robert K. Merton, 'Patterns of evaluation in science: Institutionalisation, structure and functions of the referee system', *Minerva* 9, no. 1 (1971): 66–100. For more nuanced accounts, see J.C. Burnham, 'The evolution of editorial peer review', *Journal of American Medical Association* 10, no. 263 (1990): 1323–9; and Noah Moxham and Aileen Fyfe, 'The Royal Society and the prehistory of peer review, 1665–1965', *Historical Journal* 61, no. 4 (2018): 863–89.
69 Biagioli, 'From book censorship'; Mark Hooper, 'Scholarly review, old and new', *Journal of Scholarly Publishing* 51, no. 1 (2019): 63–75.
70 Moxham and Fyfe, 'The Royal Society and the prehistory of peer review'.
71 Oldenburg, 'Advertisement', *PT* 1:12 (1666): 213–14 (emphasis in original).
72 Birch, *History of the Royal Society*, vol. 2, 40, 42, 47 (re 26 April and 10 May 1665). Two years earlier, the Society had refused to give a collective opinion of Petty's designs on the grounds that navigation was 'a state concern … not proper to be managed by the Royal Society'; Birch, *History of the Royal Society*, vol. 1, 250 (27 May 1663).
73 Birch, *History of the Royal Society*, vol. 3, 10 (15 February 1671/2).
74 Birch, *History of the Royal Society*, vol. 3, 336 (22 March 1677). The Howard family had given the Society quarters in its London residence, Arundel House, when it was displaced from its former home in Gresham College following the Great Fire of 1666, and given them the Arundelian library of rare books and manuscripts.
75 On this, see pre-eminently Steven Shapin, *A Social History of Truth* (Chicago, IL: University of Chicago Press, 1994).
76 Huygens to Oldenburg, 30 March 1669 NS, in Hall and Hall, *Oldenburg Correspondence*, vol. 2, 452–3 (editors' translation).
77 Oldenburg to Huygens, 29 March 1668 OS, in Hall and Hall, *Oldenburg Correspondence*, vol. 2, 462–4.
78 *PT* 4.46 (1669): 925–8, from *JdeS* (18 March 1669 NS).
79 For detailed accounts of this 1675 case, involving Huygens and the design of balance-spring watches, see Rob Iliffe, '"In the warehouse": Privacy, property and priority in the early Royal Society', *History of Science* 30, no. 1 (1992): 29–68; Lisa Jardine, *The Curious Life of Robert Hooke: The man who measured London* (London: HarperCollins, 2003), 198–202.

80 See Richard Yeo, *Notebooks, English Virtuosi, and Early Modern Science* (Chicago, IL: University of Chicago Press, 2014), 237. These notes are to be found in the Hooke Folio, RS MS 847, rediscovered in 2007; for the story of its re-emergence, see Robyn Adams and Lisa Jardine, 'The return of the Hooke folio', *N&R* 60.3 (2006): 235–9.

81 Johns, *Nature of the Book*, 502–4; see also Alex D.D. Craik, 'The hydrostatical works of George Sinclair (c. 1630–1696): Their neglect and criticism', *N&R* 72.3 (2018): 239–73. Sinclair claimed that the Society had violated its own practices by failing to register his communications. Oldenburg responded in print by explaining that they had not been registered because Robert Moray, the fellow to whom they had been entrusted, thought them insufficiently interesting or original to warrant the Society's attention. Craik shows that Sinclair had some justification for his grievance.

82 On the reputation, wealth and influence of the Huygens family, see Lisa Jardine, *Going Dutch: How England plundered Holland's glory* (London: Harper Perennial, 2008).

83 See Hunter, *Establishing the New Science*.

84 On the Society's experimental programme during this period, see Marie Boas Hall, *Promoting Experimental Learning*.

85 Steven Shapin and Simon Schaffer, *Leviathan and the Air-Pump: Hobbes, Boyle and the experimental life* (Princeton, NJ: Princeton University Press, 1985), 60–5 (for the description of the technique of virtual witnessing); and Steven Shapin, 'Pump and circumstance: Robert Boyle's literary technology', *Social Studies of Science* 14 (1984): 481–520.

86 *PT* 2 (1667): 27; Hall, *Promoting Experimental Learning*, 120–1.

2
Repeated reinventions, 1677–1696
Noah Moxham and Aileen Fyfe

Henry Oldenburg's death in 1677 caught the Royal Society unawares. For some, notably Robert Hooke and his allies, it represented an opportunity, as their long suspicion of Oldenburg's practices had latterly devolved into open hostility. For most of the fellows, however, it represented a problem.[1] The Society had invested far more in its own experimental enterprises than it had in Oldenburg's communicative efforts, and had endeavoured to keep them separate, but there could be no denying that the *Transactions* had been the more reliable and publicly visible aspect of the Society's early undertakings. Oldenburg had largely managed to maintain the appearance of regularity and continuity for the *Transactions* despite plague, war, fire and the vagaries of the London and Continental print trades, whereas the shape and sustainability of the Society's experimental projects had varied considerably. Thus, while the *Transactions* had remained officially independent, it had nevertheless become vital to the Society's international visibility and prestige.

Much that had made the *Transactions* possible died with Oldenburg: his extraordinary facility for languages; his connections and his epistolary relationships, carefully established over several decades; and his editorial experience. His death represented what we might call the first routine crisis of the Society's existence: the need to replace foundational expertise that is faced by every institution that outlives those who set it up. It confronted the Society with the question of whether, and in what form, to continue his legacy. The ultimate long-term stability of the *Transactions* has meant that the question of why, and in what form, the Royal Society in 1677 would wish to see the periodical continue has never been properly addressed. Nor, even more fundamentally, has the complex matter of how the Royal Society came to assume a practical moral ownership over the *Transactions*, given Oldenburg's clear statement that it was his independent literary property.

As we have seen, Oldenburg had gained valuable moral and intellectual support for his editorship due to his association with the Royal Society. His successors as editor benefited in the same way, for they also were closely connected to the Society. However, in the post-Oldenburg period, we also see other senior fellows of the Society being actively involved in keeping the *Transactions* going. Most notably, they repeatedly persuaded one of their number to take on the editorial role, and, for a while in the 1680s, they offered some direct support. Yet the fact that the *Transactions* would have no fewer than five different editors or editorial teams in less than 20 years suggests that sustaining a natural-philosophical periodical was more challenging than Oldenburg had made it seem. We have grouped these editors into three phases.

Nehemiah Grew was the first to take up the editorial mantle, swiftly followed, in 1679, by Robert Hooke. Both men were, like Oldenburg, secretaries to the Royal Society, though they were operating in the context of a shake-up of the Society's official leadership.[2] The clear determination to continue the *Transactions* was in tension with the desire from some parties (especially from Hooke) to reconfigure it by supplementing its focus on recent information communicated from afar, with material generated under the auspices of the Royal Society (some of which dated back to the earliest years of the Society's existence). For three years under Hooke's somewhat erratic editorship, the periodical occupying the niche of the *Philosophical Transactions* was titled the *Philosophical Collections*.

When Hooke was ousted as secretary to the Society in late 1682, the new secretaries, Francis Aston and Robert Plot, were persuaded to become co-editors and to revive the *Transactions* title. The Society offered some financial support for the first time, by promising to purchase a certain number of copies. Due to Plot's residence in Oxford, this collaborative editorship facilitated a fruitful dialogue between the Society in London and a group of like-minded gentlemen in Oxford. After Plot stepped down, Aston continued the collaborative editorship with William Musgrave for a year, until relations between the London and Oxford groups deteriorated in 1686.

After this, the Royal Society tried to find a way to exert more control over the *Transactions*, not just to ensure that it continued, but also that it would not reflect poorly on the Society. The periodical was still legally and financially an independent undertaking by its editor (or editors), but the Society's fellows were acquiring a more proprietary interest in it. In 1686, Edmond Halley was appointed to the new salaried position of 'clerk', and the Society made the editorship of the *Transactions* part of his job. This attempt to make the editor directly answerable to the Society's Council did not, however, prove notably effective, and by 1692, one of the secretaries, Richard Waller, took over responsibility for the *Transactions*. Rather than editing alone, however, he appears to have worked with the assistance of several other figures in the Society, including Halley and the other secretary, Hans Sloane. The Society's (limited) financial support ceased at this point, but the editorial involvement of three senior figures helped align the *Transactions* with the Society.

The rapid succession of editorial regimes reveals considerable diversity of opinion in this period about what the *Transactions* should be. This was not just about how it should be edited, or its periodicity – though monthly, fortnightly, weekly and quarterly were all proposed at different times – but what it should contain, and thus what its function in the natural-philosophical community should be. Should the *Transactions* be a clearing-house for up-to-date news from across the scholarly world, or focus on observations and discoveries gathered by the fellows and their correspondents? Should it showcase the Society's recent experimental investigations, or be a means of bringing unpublished material from its archives to light? There was still considerable uncertainty about whether a periodical was an appropriate format in which to present substantial new experimental investigations. Even those, like Hooke, who were most determined to overturn Oldenburg's legacy were not necessarily anxious to see their own experimental work in a periodical, particularly one with a wide circulation. And yet, as we will see, by the 1690s, the contents of the *Transactions* would have come to be far more closely associated with the activities of the Society than had been the case in Oldenburg's day. This was partly due to a growing acceptance of the periodical format, and partly to changes in the Society itself, as its meetings increasingly came to involve the reading and discussion of news from correspondents, rather than experiments.

Contested visions: Grew and Hooke

Henry Oldenburg had died with many of the Society's working papers in his possession (as well as books and papers belonging to Robert Boyle personally). Their efforts to recover this property were complicated by the death of Oldenburg's widow, Dora Katherina, just a fortnight later. Sorting out Oldenburg's possessions took several months of negotiation, and was entangled with the need to make arrangements for the care of the Oldenburg's two young children. It would be almost the end of

1677 before the Society regained access to its paperwork.[3] The urgency expressed by the Society during those months suggests the extent to which its day-to-day operations had depended on Oldenburg's record-keeping and correspondence; while the difficulty of retrieving them raises questions about what parts of Oldenburg's legacy might belong to the Society and which to his family – including the *Transactions*.

As far as it is possible to tell, the question of who had the legal or moral right to continue the *Transactions* was not mooted at the time, but it was potentially significant and unusually complex. In this case there were at least three potentially interested parties: John Martyn, the Royal Society's printer-bookseller; Oldenburg's surviving family; and the Royal Society itself. Usually, it would be the first two who might be expected to contest the ownership of the deceased's literary property, but as Oldenburg's orphaned daughter and son were both under the age of five, they would not have had strong voices in any such contestation.

There was, as yet, no law of copyright in England. Ownership of the 'copy' in an early modern title was normally vested, in perpetuity, in the member of the Stationers' Company (that is, a printer-bookseller) who had entered it in the Company's Register; it could then pass to his heirs or be bought by other printer-booksellers.[4] However, because the Royal Society had the power to appoint its own printers, works licensed by it were not usually entered in the Stationers' Company Register. An exception had occurred in 1670, when one of the Society's original printers, James Allestry, died and the Register recorded the subsequent acquisition of some of his literary property (including some of the Society's works) by John Martyn, the other Society printer.[5] Martyn might have believed that the terms of his appointment as printer to the Society gave him de facto ownership of the copy of all works licensed by the Society, and he might thus have felt entitled to take action to find a new editor for the *Transactions*.

Martyn did continue to print the *Transactions* until his own death in 1680, but it is clear that the *Transactions* was continued under the Royal Society's aegis, despite the fact that Oldenburg had repeatedly insisted, in letters and in print, that the *Transactions* was his enterprise and his property and nothing to do with the Society as such. Within a few months, the Society had asked one of its secretaries to take responsibility for compiling the *Transactions*, and had continued to license its publication. This practical solution reflects the fact that it was the printer-bookseller and the Society (not the family) who were in a position to keep the *Transactions* going, and who felt they had something to gain by doing so. This did not, however, settle the legal question of ownership in the title, which would be raised again in the future. Nor did the Society immediately choose, on the face of it, to be any more involved in the editorial and financial aspects of the continued *Transactions* than it had under Oldenburg's editorship.

Securing the *Transactions* took place against a backdrop of significant change in the organisation and management of the Royal Society. The president of the past 15 years was displaced after an acrimonious campaign; and the Society's two curators of experiments, Robert Hooke and the physician and botanist Nehemiah Grew, were installed as the new secretaries.[6] Both men would try their hands at the editorial role in the following years. The initial division of labour, to Hooke's displeasure, allotted responsibility for the Society's record-keeping to him, and that for communication – including the correspondence and the *Transactions* – to Grew. He noted in his diary that, 'It seemed they would have me still Curator, Grew Secretary'.[7] This suggests that, in 1678, the *Transactions* was still understood as fundamentally linked to the activity of the Society's chief correspondent, rather than being a record of its own activities.

Despite its role in continuing the *Transactions*, the Society had been considering some new ideas for publishing under its licence even before Oldenburg's death. In December 1676, a committee had been convened to scour the 'Registers' – ledgers in which experimental demonstrations performed by fellows for the benefit of the Society were recorded – for publishable material, since very little of what was in them had seen the light of day. Oldenburg was to be a member of this committee. We do not know whether it met – certainly it seems not to have officially reported – prior to Oldenburg's death, nor what form of publication of archived material had been envisaged. On 2 January 1678, the committee was revived, with a resolution to revitalise the Society by pursuing systematic research in given fields and publishing the results annually. While both were worthwhile projects, neither described the *Transactions* as Oldenburg had published it.

What actually appeared under Grew's editorship during 1678 combined Oldenburg's focus on recent communications with rediscovered material from the Society's archive. Most of this older material was from the very earliest days of the Society, some of it from before the grant of the 1662 Charter, and most was by prominent fellows who were since deceased.[8] Grew's first issue, covering the months of January and February 1678, passed over in silence the change of editorship, and the lapse of several months since Oldenburg's last issue. Neither did Grew draw attention to the mixture of old and new material in his actual issues, though it would have been clear to readers.

The general pattern of contributions during Grew's year in charge was roughly equal parts communications from correspondents, both overseas and in the British Isles, and old material drawn from the early registers. For instance, recent material in his final issue (number 142, in early 1679) included communications from the physicians William Cole and Edward Tyson, dealing with salt-manufacturing in Worcestershire and morbid anatomy, respectively, and observations of semen from the Delft microscopist Antoni van Leeuwenhoek.[9] Although the issue supposedly covered the months of December 1678 and January and February 1678/9, some of the material (from Leeuwenhoek) was older – and some was actually newer. The letter forwarded by William Cole was dated in March 1678/9, and would not be presented to the Society until late May 1679.[10] This suggests both that the printing of the *Transactions* had begun to fall behind schedule, and that Grew did not take a standard approach to whether, or when, to report correspondence to the Society. The contributions from Leeuwenhoek had not been recorded at all in the Society's archive, even though Hooke had been specially deputised to correspond with him.

Grew's final issue also contained an account of the manufacture of malt in Scotland originally sent to the Society in January 1663 by Robert Moray (d. 1672); an account of the cultivation of maize in New England sent in the same month by John Winthrop (d. 1676); and an account of mining and manufacturing processes sent in 1661 by Daniel Colwall.[11] The origin of some of this material lay in the Society's early History of Trades project, an attempt to survey the state of knowledge in various skilled and manufacturing trades and to suggest areas for improvement. (Moray, for example, was assigned to conduct field research into and to write an account of mining and quarrying in England and Wales.[12]) The use of this material 15 years later indicates not only that the Society felt free to appropriate the work of deceased fellows from its own archives, but that its hope of publishing synthetic overviews of artisan knowledge from its accumulated records had finally been abandoned.

The use of so much archival material had some obvious advantages for Grew as editor. It diminished the need to rely on voluntary external communications to fill the *Transactions*, which relieved the pressure on Grew, who was in no position to match Oldenburg's formidable network of correspondents and contributors. The fact that Grew's issues were, on average, much the same length as Oldenburg's had been, but came out less frequently (only every two months on average), also helped to reduce the quantity of recent communications needed to fill his pages. Digging into the Society's archive also disposed uncontroversially of material that was otherwise unlikely to see the light of day. Thus, Grew's *Transactions* conformed both to the Society's specific orders to dig into the early registers for publishable material; and to the general principle that the *Transactions* was considered a suitable home for external communications, and for older-but-unpublished material in the Society's possession; but not for the Society's own current efforts at knowledge production.

Those efforts had been the focus of the other half of the Council's order of January 1678, which stipulated that 'there be prepared once a year a collection of all such matters, as have been handled that year, concerning four, five or more subjects, which have been well prosecuted, and completed; which may be printed in the name of the Society against the anniversary election-day'.[13] On the face of it, the Society never produced any such published collections; but in fact, a set of tracts put out by Robert Hooke during Grew's editorship conform remarkably closely to this proposal.

These three thematically organised gatherings of papers and lectures, by Hooke and by others, carried separate titles – *Cometa*, *Microscopium*, and *De Potentia Restitutiva* [On the power of elasticity] – though the first two were signed and paginated as one volume and published under the joint title of *Lectures and Collections* (Figure 2.1). They gathered together recent work on related themes, using Hooke's own discoveries and 'lectures' as a core around which 'collections' of contributions from others – including Robert Boyle, Christopher Wren, Antoni van Leeuwenhoek, and the astronomers Johannes Hevelius and Giandomenico Cassini – could be woven. Hooke took the opportunity to vindicate himself against foreign authors – in particular a French book on microscopy which Hooke claimed to have 'met with casually' while 'this discourse was printing' and who had criticised some of Hooke's earlier microscopical work. Not all of the 'collections' were as tightly linked: *Microscopium* also included a recent communication from a naval surgeon in Plymouth, about a man who inhaled a pistol-ball that lodged in his lungs with eventually fatal results, which had little directly to do with microscopy.[14] And Hooke's elasticity lectures, partly the result of demonstrations made to the Royal Society over several weeks in the summer of 1678, were accompanied by contributions from other researchers that had little to do with elasticity.[15]

Hooke's notion appears to have been to combine the function of the *Transactions* as edited by Oldenburg, reporting news and recent external communications of research, with the Society's directive to produce more substantive discourses on particular subjects.[16] His

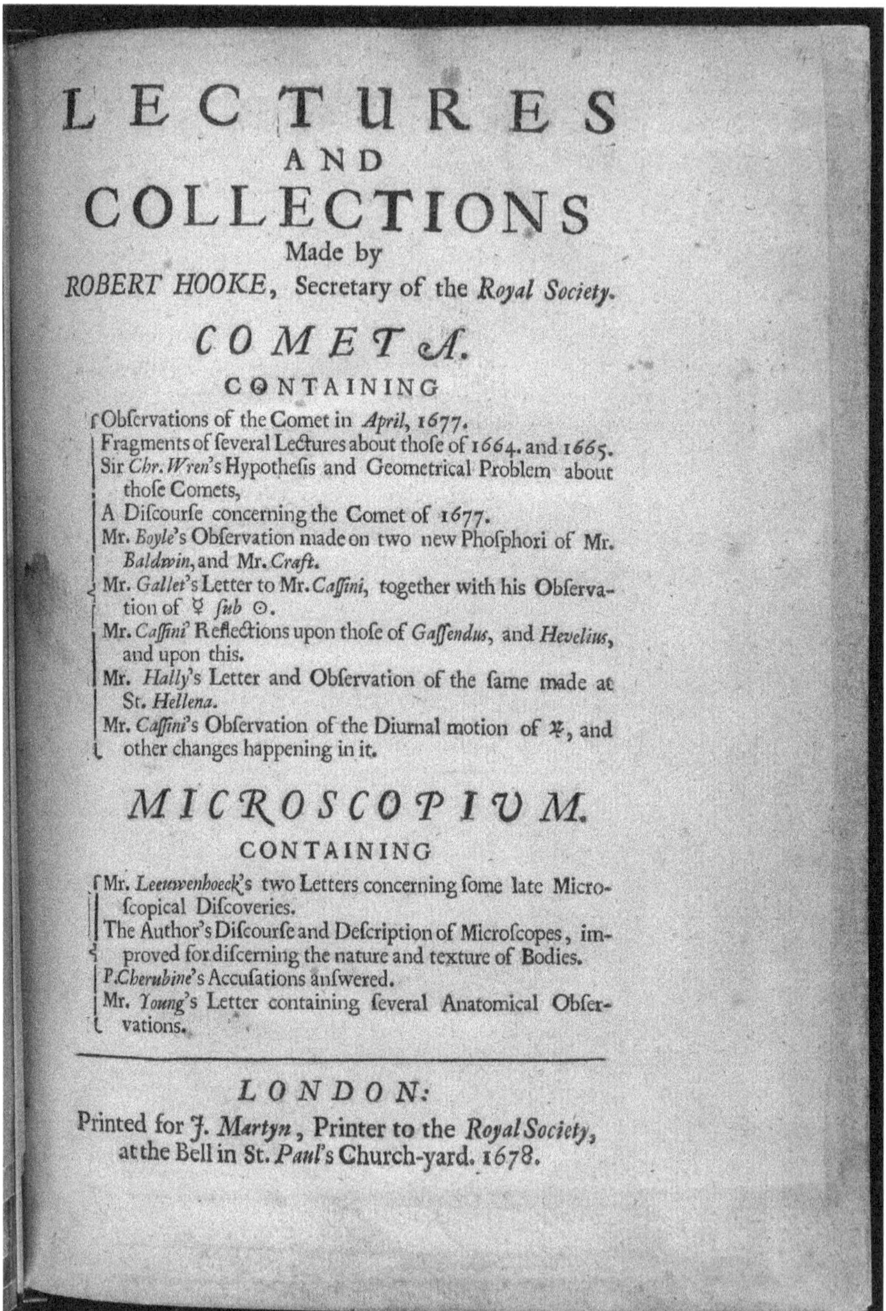

LECTURES
AND
COLLECTIONS
Made by
ROBERT HOOKE, Secretary of the *Royal Society*.

COMETA.
CONTAINING

- Obſervations of the Comet in *April*, 1677.
- Fragments of ſeveral Lectures about thoſe of 1664. and 1665.
- Sir *Chr. Wren*'s Hypotheſis and Geometrical Problem about thoſe Comets,
- A Diſcourſe concerning the Comet of 1677.
- Mr. *Boyle*'s Obſervation made on two new Phoſphori of Mr. *Baldwin*, and Mr. *Craft*.
- Mr. *Gallet*'s Letter to Mr. *Caſſini*, together with his Obſervation of ☿ *ſub* ☉.
- Mr. *Caſſini*' Reflections upon thoſe of *Gaſſendus*, and *Hevelius*, and upon this.
- Mr. *Hally*'s Letter and Obſervation of the ſame made at St. *Hellena*.
- Mr. *Caſſini*'s Obſervation of the Diurnal motion of ♃, and other changes happening in it.

MICROSCOPIUM.
CONTAINING

- Mr. *Leeuwenhoeck*'s two Letters concerning ſome late Microſcopical Diſcoveries.
- The Author's Diſcourſe and Deſcription of Microſcopes, improved for diſcerning the nature and texture of Bodies.
- *P.Cherubine*'s Accuſations anſwered.
- Mr. *Young*'s Letter containing ſeveral Anatomical Obſervations.

LONDON:
Printed for *J. Martyn*, Printer to the *Royal Society*, at the Bell in St. *Paul*'s Church-yard. 1678.

Figure 2.1 Title page of Robert Hooke's *Lectures and Collections*, 1678, showing its thematic arrangement © The Royal Society.

'lectures' and 'collections' maintained the distinction between externally communicated knowledge-claims and those produced within the Royal Society, as was apparent in their titles.

Whether Hooke intended to supplement or supplant Grew's efforts was not made explicit, but Hooke's eagerness to gain control of the Society's communication mechanisms, including the *Transactions*, was evident. These 'collections' of 1678 were the forerunners of his more direct attempt to replace the *Transactions*. On 26 December 1678 the Council (at last) agreed to put Hooke in charge of the Society's correspondence, adding that 'the same shall be continued by the help of a small Journal of some particulars read to the Society'. These, it was ordered,

> shall not be sent or sold to any person but members of the Society, and to such as correspond with Mr Hooke by the Society's directions, and make considerable returns to him for the Society's use; all which returns shall be constantly brought into the Society, and read before them at the very next meeting after the receit thereof.[17]

Grew compiled his final issue of the *Transactions* in spring 1679; and Hooke began issuing a periodical titled *Philosophical Collections*, which appeared erratically and infrequently until 1682. The published issues do not seem sufficiently different in their style, appearance or remit from Oldenburg's periodical to justify the change of name, yet Hooke did apparently intend them to function differently.

The wording of the Council resolution reveals Hooke's desire to limit the circulation of his periodical to the Society's 200 or so fellows, and selected correspondents. This was a very different vision from Oldenburg's initial ambition of 1,250 copies on public sale. The resolution implies that the Society's Council was content with this plan, which suggests that, whatever reputational benefit accrued to the Society from the publication of a periodical at this time, it was thought to derive from the learned community, not the public at large. We can find no evidence that the circulation of the *Philosophical Collections* was, in fact, as semi-private as the resolution suggests, but the wording reflects Hooke's distrust of open systems of dissemination, including Oldenburg's *Transactions*. It also points to a conception of the learned community as a fundamentally closed group, in which knowledge-claims should circulate in private or, at most, semi-public, and on a reciprocal basis. In Hooke's vision, only those able to make substantive contributions of their own were to be initiated into the knowledge community of the Royal Society; and readers of the *Transactions* were also contributors. The question of how to pay for the production of such a periodical was never addressed. Given its very limited pool of possible purchasers, any answer must have involved either the Society's financial support, or asking reader-contributors to pay a high price or a special subscription.

Within just two years of Oldenburg's death, therefore, a surprising number of different models for a learned periodical had been mooted or actually attempted by those involved with the Royal Society. Grew's *Transactions* had combined the publication of communications recently received from outside the Society with the publication of work by deceased fellows that had originally been intended for other purposes. Hooke's *Lectures and Collections*, on the other hand, had attempted to present recently received communications in thematic bundles organised around his own research interests and activities. Both these models had been envisioned as 'public', whereas the plan for Hooke's *Philosophical Collections* reimagined the relevant public as a limited community of initiates who proved their worth by communicating valuable discoveries to it and the Society. The prospects of limited sales may be why John Martyn apparently expressed some unwillingness to publish what he called 'the next *Transaction*' in September 1679.[18] It is also striking that, throughout these discussions with Grew and Hooke, the Society showed no interest in taking what, in retrospect, would seem obvious steps towards the stability and continuity of its periodical, such as funding its production, or taking over editorial responsibility.

Throughout, the suitability of the printed periodical as a site for publishing sustained, systematic investigations remained doubtful. As in Oldenburg's time, this seems to have been a matter of epistemic and social propriety as well as formal limitations. For example, Hooke was reminded in summer 1679 to seek permission from both the author and the Council before using any material recorded in the Society's registers. This implies a working policy of joint ownership, or more precisely a joint power of veto, over work contained in the registers, shared by the author and the Society.[19] It also indicates the continued distinction between knowledge-claims recorded in the registers and external communications which the Society neither warranted nor claimed.

Despite all the editorial changes, three things remained constant in these immediate post-Oldenburg years: the Royal Society remained concerned that some form of periodical should appear under its aegis; that it should be recognisable as the *Transactions* or its direct inheritor; and that the right, or the responsibility, to publish the periodical was associated with responsibility for the Society's correspondence.

Bilateral transactioneering: London and Oxford

By 1682, Robert Hooke's position as secretary of the Royal Society was in peril. He was not doing notably well at revitalising its experimental activity or at restoring its correspondence to Oldenburg-era levels; and only seven issues of *Philosophical Collections* had appeared since 1679 (and four of them were the result of reminders earlier in 1682). Hooke was duly displaced in November 1682 in favour of the Oxford-based physician, chemist and compiler of county natural histories Robert Plot. Plot's appointment ushered in a novel collaboration between the Royal Society in London and a similar grouping of natural philosophers in Oxford.[20] This would affect both the Society's activities and the running of the periodical.

One of Plot's first moves, with his co-secretary Francis Aston, was to propose the revival of the *Transactions*, under that name. The proposal gives no particular hint of their plans and no indication of whether or how the relaunched periodical would differ from its predecessors. As it turned out, the form they adopted mirrored Oldenburg's very closely: as well as keeping his title, the relaunched journal resumed numbering issues where Oldenburg and Grew had left off and adopted much the same format and layout. However, the management of the periodical was significantly different in two ways: first, the editorial work would be done by both secretaries, rather than one alone; and second, Plot and Aston managed to secure financial support from the Society. Up to this point, the Society had spent vanishingly little of its limited income on publishing of any sort, but it now promised to purchase 60 copies of every issue of the *Transactions* printed, thus guaranteeing a minimum income for its conductors.[21] The content published by Aston and Plot was also notably different: it was mostly current, and it became far more British in origin (only around 20 per cent of material was foreign, compared to almost 50 per cent in Oldenburg's *Transactions*). This is partly because Plot and Aston used the *Transactions* as a tool for forging a link between the Royal Society in London and the Philosophical Society in Oxford. It would, however, be one of the key reasons for the eventual collapse of this tentatively glimpsed reorganisation of natural science in Britain in the 1680s.

The credit for this relaunch has usually been given to Plot, whose appointment appears to have precipitated it. Yet Aston was closely involved throughout, and would continue to be involved for a year after Plot resigned as secretary in November 1684. Little is known about Aston, beyond the fact that he was a fellow of Trinity College, Cambridge, to whom Isaac Newton had once sent a letter of advice about foreign travel (even though, according to all the available evidence, Newton had never travelled abroad).[22] Aston was a far more regular participant in Royal Society meetings than Plot, who was most often in Oxford and also had commitments in Kent, where he spent much time gathering material for the natural history of the county he proposed to write. There is no evidence to suggest Aston had a close prior relationship with Plot, but Plot was well connected in the learned communities of Britain and Ireland.[23] The two men worked well together from the outset, tabling their joint proposal and bringing out their first issue within weeks of Plot's election.

The collaboration between Aston and Plot is an invaluable resource for historians, since Plot's absence from London produced a bilateral correspondence chiefly preoccupied with the practicalities of publishing the *Transactions*. Complaints about printers, authors, estimated lengths of articles, how many illustrations to include and when to expect them, arrangements for distribution, profit-sharing, and many other details abound which are simply lacking for every other early modern editorship – including Oldenburg's (whose extensive surviving correspondence was much less preoccupied with the details of producing his periodical).

During the 1680s, the association between the work of the Society and the content of the *Transactions* became closer: an increasing proportion of the material published in the periodical had also been recorded in the Society's minutes. For instance, most of the materials for Aston and Plot's first number, no. 143 for January 1683, came directly from the recent proceedings of the Society: a letter from Sir Robert Southwell describing an underground cave in Gloucestershire, a letter from Johannes Hevelius read on 15 November, and Martin Lister's experiments on the movement and colouring of chyle in the lactea, plus John Flamsteed's London tide-tables.[24] These men were all established figures within the Royal Society, and Hevelius and Lister, in particular, had been regular contributors to Oldenburg's *Transactions*. Southwell was a prominent senior fellow (and future president), Lister's experiments were specifically intended to answer criticisms of his work put by the Society, and Hevelius had been addressing himself to the Society and publishing in the *Transactions* since the mid-1660s.[25] The increasing closeness between the Society and the contents of the *Transactions* reduced the ambiguity about whether correspondents wrote to an editor-secretary in his private or official capacities.

In a notable innovation for the *Transactions*, Aston and Plot also published the results of research originating with, and paid for by, the

Royal Society. In January 1683, the Society had appointed two new 'curators of experiments' in an attempt to revitalise its programme of experimentation. These curators – the anatomist Edward Tyson and the chemist Frederick Slare – had specific instructions to produce experimental demonstrations in their particular fields, and to arrange between them that there be at least one such performance at every weekly meeting. This effectively built the Society's resolution of five years previously, to have courses of experiment on particular subjects systematically prosecuted and published, into the structure of the organisation. And, in contrast to earlier regimes, some of this material now appeared in the *Transactions*. For instance, almost all of Aston and Plot's second issue was devoted to the description and extensive illustrations (Figure 2.2) of the rattlesnake dissected by Tyson; at 34 pages (plus images) it was the longest paper published in the *Transactions* so far.[26]

Aston and Plot tended to be far less editorially obtrusive in the journal than Oldenburg had been; lacking his Europe-wide reputation, their visible involvement would not have added value to the enterprise in the same way. What they could do instead was to effectively recreate the periodical as a collaborative enterprise between overlapping groups in different cities. They would be most successful in linking London and Oxford, but they also tried to drum up support and contributions from elsewhere. For instance, Aston wrote to Edmé Mariotte at the Académie Royale des Sciences in Paris, and to Isaac Newton in Cambridge. But in a subsequent letter to Plot, in which Aston thanked him for 'most acceptable news' of 'correspondence and good assistance' from Oxford, he reported sadly, 'I wish I could tell you of any help from Cambridge'.[27] Acting, as far as we can tell, on their own initiative, Aston and Plot were trying to establish durable correspondences for the Royal Society with intellectual centres in Britain, France and (later) Ireland: possible fringe benefits for the *Transactions* might have included new readers, a source of contributions, or even arrangements for wider distribution.

The strongest link was undoubtedly with Oxford. Plot was based there, where he was part of a group of gentlemen interested in natural philosophy. The Oxford group appears to have been meeting informally from at least 1681, but it was not until Plot's appointment that the Royal Society seems to have taken any formal notice of it.[28] The weekly correspondence between Aston and Plot discussed business arrangements for the *Transactions*, but it rapidly also became a means of sharing natural-philosophical news, experiments, samples and specimens between the natural-philosophical communities in London and Oxford. By

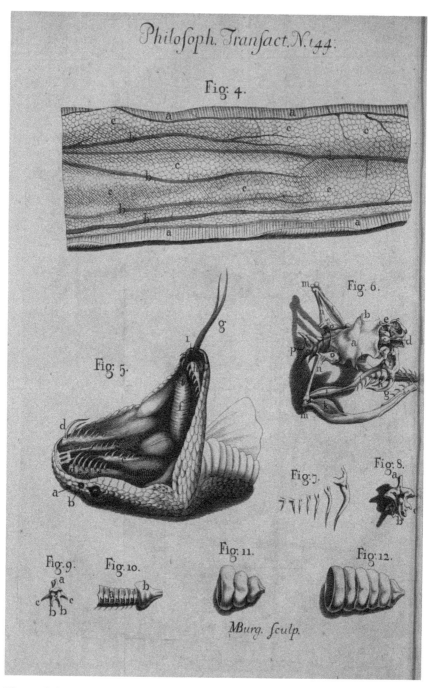

Figure 2.2 Tyson's dissection of a rattlesnake, from *PT* 13.144, 1683, plate 2 © The Royal Society.

October 1683, this had helped formalise the hitherto irregular assembly into the Oxford Philosophical Society.

The London–Oxford link was strengthened by the decision to move the printing of the *Transactions* to the university town. Aston and Plot's experience with the Society's London printer, for their first issue, had not been satisfactory. He had taken days over the composition of Flamsteed's tables, badly mangled Hevelius's letter, and, by the time of the second issue, was already trying to renegotiate his initial agreement to pay Aston and Plot 18 pence a sheet for delivering copy (which was already half what Oldenburg had received from Martyn and Allestry in 1665). By late January 1683, Aston and Plot had resolved to transfer printing to Oxford.[29]

Plot and his Oxford colleagues once again arranged to use the University's printing privilege to get the *Transactions* licensed for publication. This freed the editors from having to work through the Royal Society printer, and that indicates that the Royal Society did not regard its control of the licensing procedure – with its minimal scrutiny – as essential to the identity of the *Transactions*. Aston arranged for the delivery of paper by river from London and sent copy to Plot for publication, as well as cash from retail sales. Plot supervised the press and engravings. Distribution was handled by a coterie of London booksellers, though copies could also be bought from the Royal Society's apartments in Gresham College, managed on commission by Henry Hunt, the Society's 'operator' (and sometime draughtsman).[30] Aston and Plot also had a silent partner in their editorial enterprise, identified in their letters only as 'G', a degree of caution perhaps explained by Aston and Plot's awareness that their letters would sometimes be shown to other members of the London and Oxford groups. The level of G's actual involvement is unclear, other than to induce irritation in Aston at his mistakes, and no more is heard of him after March 1683.[31]

Aston and Plot were clearly closely involved in the practical aspects of printing and distributing a periodical, yet they appeared wholly uninterested in any commercial potential. Their *Transactions* was, in this respect, quite unlike Oldenburg's notion of a natural-philosophical news-sheet appealing to a paying public. Aston and Plot made no attempt to restrict its sale, as Hooke had intended, but their use of it to establish cooperation between the London and Oxford groups shared some attributes with Hooke's vision of reciprocity within a limited learned community.

It is not clear what Aston and Plot were paid for copy by the Oxford printer, nor what share they may have received of the sales income (including that guaranteed by the Royal Society); but there are hints that they did indeed share in the sales income. For instance, their offer of a commission on sales to the Society's operator suggests that they had control of some part of the sales income. A further hint comes from a letter of January 1682, in which Aston suggested that they could fill an entire issue of the *Transactions* with the description of the ostrich then being dissected by Edward Tyson for the Royal Society. (He told Plot that 'Tho the Rattlesnake be hardly finisht yet, wee have bought an Ostrich and Dr Tison is at work upon it'.) Aston mused to Plot:

> I'm considering that if you and I can make a better advantage to him by printing it ourselves as Transaction (tho of 2s price), then he by selling it a bookseller, it were better for ye Doctor; and I think it would be credit for ye Society to have it come out that way ... But Ile think again of this, for it seems but reasonable that Dr Tison giving us ye greatest part of a Transaction, he should have something either by way of present or shares.[32]

Even though Tyson's ostrich dissection did not follow his rattlesnake into the *Transactions*, this proposal sheds intriguing light on Aston and Plot's commercial arrangements.[33] The suggestion that Tyson should receive 'something either by way of present or shares' hints at a financial as well as a reputational dimension to what might be 'better' for the author, and again suggests that Aston and Plot received income that they could potentially share. This seems, as far as we can tell, to have been the only time any *Transactions* editor ever offered to remunerate an author.[34]

We also learn something about the pricing of the *Transactions*. Tyson expected his description to run to seven or eight sheets: this represented an entire double-length issue by Aston and Plot's standards (and close to three times the average length of an Oldenburg issue). The suggested retail price of two shillings hints at a price of a shilling or less for a more usual issue of the *Transactions*. This is significantly higher than was typical of other contemporary periodicals and pamphlets of comparable length; the typical retail price for a stitched work of three or four sheets was somewhere in the region of 2*d*. or 3*d*.[35] This suggests that the audience for the *Transactions* continued to be confined to the most affluent of purchasers of print.

While the printing was at Oxford, there was a marked increase in the frequency and quality of illustration in the *Transactions*. The 142 issues printed by Oldenburg and Grew between 1665 and 1679 had featured just 61 engraved plates, but there would be 43 plates in the 36 issues published at Oxford between 1683 and 1686. Many of these were engraved by the Dutch-born Michael Burghers (including Figure 2.2); some were illustrated by Susanna Lister, daughter of physician and Royal Society fellow Martin Lister.[36] The paucity of illustrations in the early *Transactions* may have been because Oldenburg had to bear half the cost of engraving, or it may have been due to the practical challenges of combining intaglio printing (for plates) with letterpress (relief) printing in the context of the single issue of a periodical.[37] We do not know whether Aston and Plot had to pay part of the cost of engraving, as Oldenburg did. There is some evidence that engraving was cheaper in Oxford than in London, but the enthusiasm for illustrations may be another indication of their more relaxed attitude to the potential profitability of their enterprise.

During 1683 and 1684, Aston and Plot's correspondence about the *Transactions* and about Royal Society business enabled a fluid, responsive and open-ended collaboration between London and Oxford. Experiments were replicated. Findings were discussed in two overlapping forums (since the memberships of the groups overlapped). Samples of earths, minerals and chemicals were sent back and forth for testing.[38] Anatomical specimens and illustrations, to make those discussions intelligible, were a key part of this; and a particular drawing (such as William Gould's drawing of a polypus taken out of a man's heart) might originate in Oxford, be shown to the society there, travel to London to be discussed again, and then be sent back for engraving and inclusion in the *Transactions*.[39] Publication might happen quite a lot later, but the initial response times were rapid.

This collaborative enterprise was evidently felt to be fruitful for the Royal Society, for when Plot resigned as secretary in November 1684, he was replaced by William Musgrave, recently appointed secretary of the Oxford Society. Musgrave had become a fellow of the Royal Society barely a week previously, so this was an unprecedentedly rapid climb to influence that reflects a desire on the part of both organisations to continue the partnership that had developed over the previous two years. The Royal Society would propose a further joint effort with Oxford in 1685: the publication of the ichthyological research by the recently deceased Francis Willughby and the Essex naturalist John Ray.[40] This marked the Royal Society's first serious foray into publishing as an undertaker and not just as a sponsor of research or a licenser, and the Society would invest considerably in it.

The *Transactions* continued to appear in a regular and timely fashion under Aston and Musgrave. It included substantive experimental performances produced and cross-checked between London and Oxford, and apparently proceeded without the need for detailed orders or official goading from the Council. From early 1685, a third society joined the circuit, as natural philosophers in Dublin established a new Philosophical Society there, and exchanges between it and the Oxford Society were routinely funnelled through London.

At the end of 1685, however, the complex arrangements between London and Oxford fell messily apart, as the collaborative publishing projects dissolved into acrimony. Though Aston was re-elected in November for the fifth year running, Musgrave did not continue in his role as secretary of the London society, thus ending the three-year London–Oxford co-secretaryship and co-editorship. Whether this was Musgrave's choice, or the Royal Society's, is not apparent; but we do know that both the *Transactions* and the *Historia Piscium* were causing tension between the two societies, with the secretaries effectively trapped in the middle. Matters were compounded when both Aston and his new co-secretary, Tancred Robinson, abruptly resigned just a fortnight after their election.

The tensions around the *Transactions* arose from the growing influence of the Oxford group over the contents of a periodical that the London society was coming to regard as, in some sense, theirs. Mere weeks before the Royal Society elections, the November issue of the *Transactions* had caused uproar. The fact that it was now being licensed and printed in Oxford meant that there had been nothing to prevent John Wallis, the president of the Oxford group, from publishing an unprecedentedly long account of a short book by the Danzig astronomer Johannes Hevelius, in which Wallis recapitulated in meticulous and insulting detail Hevelius's many criticisms of Robert Hooke (who had, 11 years earlier, criticised Hevelius's refusal to use telescopic sights on his instruments).[41] The review was published anonymously, but the publication in the *Transactions* of an attack on one fellow of the Society (Hooke) by another (Wallis) in the guise of a sympathetic account of a book by a third (Hevelius) presented huge problems within the Royal Society's codes of propriety. It had only been possible because the *Transactions* had, in effect, temporarily passed beyond even the fairly notional oversight usually exercised by the Council. The licensers of the Oxford University Press were unlikely to be concerned with the potential for institutional controversy in Wallis's account. By contrast, it was fairly well understood that the Royal Society would refuse to licence a work containing explicit attacks on the character of a fellow. For instance, when Hooke had attacked Oldenburg's personal and intellectual probity in print in 1676, he had worked with John Martyn but went outside the Society to get it licensed.[42] The Society's remonstrance to Martyn, even though he had not technically violated the terms of his appointment as printer to the Society, reveals its disapproval of printed critiques of one of its fellows.[43] In November 1685, Hooke unsurprisingly protested about Wallis's review of Hevelius. Aston, as the more experienced secretary-editor, was made to feel responsible for the lapse, and this almost certainly contributed to his abrupt resignation in December.

The situation was exacerbated by the realisation of just how extraordinarily expensive the *Historia Piscium* would be to produce.[44] Edmond Halley speculated uncharitably that the resignations of the secretaries had been motivated by a desire to obtain 'better terms of reward' from the Society. There is no other evidence to substantiate this, though since the secretaries volunteered their labour it would hardly be surprising if they decided that their unsalaried posts had ceased to be worth the aggravation. (Halley's report that the Society was 'surprised' by the resignations may, however, be taken as evidence that Aston and Robinson at least were not forced out.)

In December 1685, therefore, the *Transactions* found itself orphaned once again, this time between institutions as well as individuals. The Royal Society had, according to one possible reading of the situation, lost control of the key publication associated with it in a manner that caused discreditable controversy within the Society and with key allies and correspondents; and almost simultaneously lost control of its finances. A system of friendly material and epistemic cooperation with an allied institution had collapsed into recrimination and rivalry, and the Royal Society suddenly found itself without secretaries to organise its correspondence, or editors to publish the periodical that would display its activity.

Keeping the *Transactions* going: Halley, and Waller et al.

The Royal Society's determination to find a way to keep the *Transactions* coming out in a timely fashion, and in a manner that enhanced (rather than diminished) the Society's reputation, would become a long-standing challenge. In 1686, it tried assigning the editorship to one of its employees, but by 1691 it was proposing the creation of an editorial team or committee. Neither model worked long-term (at this point in history), but the attempts confirm the Society's growing acceptance that its public reputation was tied to the *Transactions*, and that it had a moral (if not legal) responsibility for the periodical.

The Royal Society responded to the resignation of its secretaries in 1685 with a sharp change of tack. A new paid role of clerk to the

Society was created, with responsibility for drafting the Society's correspondence, taking minutes, and keeping the Society's records up to date. Shifting these administrative duties from the secretaries (fellows acting in a voluntary capacity) to a salaried employee 'accountable to the council' was an attempt to ensure reliable service.[45] Responsibility for 'drawing up' the *Transactions* was added to the list of the clerk's duties in March 1686.[46] The exact significance of this phrase is hard to parse – Oldenburg had been said to 'compose' the periodical, for comparison – but it seems to imply that the initial selection of material and, where applicable, the drafting of the *Transactions* would fall to the clerk. The intention was clearly to establish the Society's institutional authority over the *Transactions* for what was effectively the first time. Assigning the *Transactions* to an employee carried the possibility, if not necessarily the explicit intent, of giving the Council a greater measure of editorial control than it had previously enjoyed, including the power to order (or prevent) the publication of particular papers.

The distinction between the positions of clerk and secretary was significant. The secretaries were fellows of the Society, and when they acted as editors of the *Transactions*, they did so as independent gentlemen. If the aim was tighter institutional control, it would in principle have been possible to continue with a secretary as editor, while imposing a more stringent licensing procedure – with closer pre-publication scrutiny in Council – to forestall incidents like the Wallis review. It is a reasonable inference that the Society was unwilling to impose such a regime on the secretaries because it would be understood as an impermissible constraint on what Adrian Johns calls 'the freedom of action of a gentleman'.[47] The clerk, on the other hand, was a subordinate figure whose draft of the *Transactions* was (supposed to be) 'perused and approved' by one of the secretaries.[48]

If assigning the editorship of the *Transactions* to a Royal Society employee was intended to provide some much needed editorial and publishing continuity, it failed. Edmond Halley was elected to the clerkship (and duly resigned his fellowship). Halley was 'completely seen in the mathematics and experimental philosophy', as stipulated by the Council; he also had a European reputation in astronomy and mathematics; and had travelled extensively on the Continent. It surely also helped that he was on friendly or at least civil terms with all the principals in the recent controversy, including Hooke, Wallis and Hevelius (whom he had visited and observed with in Danzig in 1679). Given that Halley failed to meet several of the original stipulations for the clerkship (in particular, those dealing with residence and marital status), Council's choice indicated that the intellectual specifications of the job were deemed most important. Yet Halley's editorship cannot be counted a success. Only 16 issues appeared over the next six years, and none at all between the end of 1688 and January 1691. It is difficult to determine Halley's reasons for acting as he did, as his papers and correspondence have survived only sparsely – surprisingly so, for a well-connected mathematician and astronomer who was active for over seven decades and was intimately associated with several long-lasting institutions. (Besides his roles at the Royal Society, Halley was from 1703 the Savilian Professor of Geometry at Oxford, and Astronomer Royal at Greenwich from 1719.)

The Society seems to have had remarkably little hold over Halley as editor, even though it took advantage of his reduced status as clerk to order him to publish certain papers: the first such paper was one of his own, and this practice became more frequent after November 1686.[49] He went on to contribute a large number of his own papers to the *Transactions*, but clearly felt free to set the periodical aside in favour of other projects (including the admittedly taxing and drawn-out process of coaxing the text of *Philosophiae Naturalis Principia Mathematica* out of Isaac Newton, and seeing it through the press in 1686–7).[50] The situation was probably not helped by Robert Hooke, who continued to campaign against the *Transactions* in 1688–9. The fact that the Society paid Halley a salary – albeit often hopelessly in arrears – was not enough to effectually compel him to keep up any regular publication schedule.[51] In early 1691, Halley would attribute his lapse in publication to 'the unsettled posture of Publick Affairs', namely the transition from James II to William and Mary, which 'did divert the thoughts of the curious towards matters of more immediate Concern' than natural philosophy or mechanics.[52] But Halley's publication schedule would be little better in the following years.

The Society's moral case for making any commands about the *Transactions* was weakened by the fact that it still did not take financial responsibility for the periodical. No payments to or from printers connected with the *Transactions* are recorded during this period. The Society did continue its practice of purchasing 60 copies of each issue, at a slight discount, and those payments were made to Halley, suggesting that he was, in some sense, the owner of the *Transactions*.[53] Halley's arrangements with the printers are unknown: he may have been paid by the printers per sheet of copy supplied (like Oldenburg, and Aston and Plot); he may have been paid in printed copies without a share of the profits; or he may have borne part of the cost of publishing and shared the profits. There is not enough surviving evidence to decide the question, but it is reasonable to assume from the extremely lax frequency of publication either that Halley was not being paid for copy, or that he did not particularly need the money. (There is no surviving evidence from the first 90 years of the *Transactions* of any printer or bookseller urgently demanding copy, which, in the absence of other evidence, might have given an indication of its market appeal.)

By January 1691, as the gap between issues got longer, the Council sought to improve matters:

> There being a full Councell it was resolved that there shall be Transactions printed, and that the Society will consider of means for effectually doing it. And Dr [Edward] Tyson, Dr [Frederick] Slare, Dr [Hans] Slone, Mr [Richard] Waller and Mr [Robert] Hook were desired to be assistant to E. Halley in compiling and drawing up the Transactions.
>
> Ordered that Mr Boyle be desired to continue his designe of communicating his small Tracts, to be published in the Philosophicall Transactions.[54]

We notice, again, the Society's determination that the *Transactions* continue to appear, and the acknowledgement that the system currently in place was inadequate. The list of people who were asked to help Halley locate and arrange copy included the Society's curators of experiments, one of its secretaries and several other fellows. This is the first suggestion of something that might be called an editorial committee, a model that would become very common among learned institutions throughout Europe in the eighteenth century.[55] The *Transactions* did begin to appear again, but there is no indication of whether Halley was receiving the proposed support. The collective editorial plan would, however, resurface in a few years' time.

Under Halley, the *Transactions* displayed a distinct bias towards the physical and mathematical side of the Society's remit of improving natural knowledge. In his prefatory 'advertisement' in early 1691, he explained his determination to focus 'for the most part' on '*Physical* and *Mathematical* Enquiries', and to exclude many types of material 'wherewith the forein Journalist usually supply their monthly Tracts'.[56] Halley here referred to the editors of those Continental periodicals that actually went by the title of *Journal* or *Giornale*: these tended to be learned reviews on the model of *Journal des Sçavans* in Paris and *Acta Eruditorum* (Leipzig, f. 1682), with remits much broader than just the natural sciences. However, the *Transactions* continued to include numerous medical and natural-historical papers, despite Halley's stated preference for the physical sciences.

Halley's advertisement-preface also tells us:

> that for the future the *Royal Society* has commanded them [the *Transactions*] to be Published as formerly, and if possibly Monthly. And all lovers of so good a Work are desired to contribute their Discoveries in Art or Nature, addressing them as formerly to Mr *H. Hunt* at *Gresham* College, and they shall be inserted herein, according as the Authors shall direct.[57]

The use of Gresham College (and the Society's operator, Henry Hunt) as a collection point serves to consolidate the link between the *Transactions* and the Royal Society, as does the emphasis on the periodical being revived at the Society's command.

In December 1692, Council was still worried about the irregular issue of the *Transactions*, and instructed the secretaries, Thomas Gale and Richard Waller, to trawl the Society's records in search of material suitable for publication. At the same time, Halley offered, 'that if it shall be undertaken to print a book of philosophicall matters such as the Transactions used to be that he would undertake to furnish de proprio five sheets in twenty'.[58] Halley's wording admits that things were not as they 'used to be', but his offer to supply content for just five sheets in 20 would have been a significant reduction in his editorial responsibility. Halley plainly wished somebody else to assume editorial and financial responsibility for the periodical; he did not make clear whether this was to be the Society or a private individual, only that 'the undertaking' would not be his. The determination to displace the responsibility is a further indication that, at the very least, the rewards of publishing the *Transactions* privately were not worth the trouble taken over them, regardless of whether Halley saw the 15 papers from his own research that appeared in volume 16 as a burden on him as an author, or good use of his privileged access to a publication outlet.[59]

By February 1693 the Council was actively courting alternative editors. They tried to persuade Robert Plot to take it on again, and got as far as agreeing terms. The arrangement ultimately fell through, but its terms offer a reasonable indication of the best deal that could be reached for the *Transactions* with the bookseller in the early 1690s, and may also be an indication of the arrangement between Halley and the bookseller.

> It was resolved that Dr Plott shall print the Transactions and that for his encouragement therin he have the 60 Books agreed by the bookseller to be allowed for the Copy and that the Society will make

up to the Dr what the value of the said books shall fall short of 40 li per annum and they will take the said 60 books as formerly and allow him 12 of each sort to present to his chief Correspondents, to which resolve the Dr agreed.[60]

According to this arrangement, the bookseller was now paying the editor for copy in kind, rather than in cash. The Society agreed to pay for the full allocation of copies to the editor, presumably to ensure that the bookseller and editor were not competing for the same readers; and it agreed to make up the editor's income, if necessary, to £40 a year. For comparison, as clerk Halley had been paid £50 a year; and the income from 60 copies of the *Transactions* would have generated £36 if it had come out monthly and retailed at 1s.[61] This guaranteed income makes it unlikely that there was any further profit- (or, by the same token, cost-) sharing agreement in place.

During Halley's editorship, the link between Society activity and the content of the periodical had continued to consolidate, but this did not diminish the *Transactions*' dependence on external correspondence. Rather, the decline of experimentation at the Royal Society after 1685 (until Newton's efforts to revive it) had meant that the Society's ordinary meetings came to rely heavily on correspondence received. Though the Society's leadership was supposed to exercise more detailed oversight of the *Transactions* under Halley there is no evidence of this happening within Council meetings; if it was happening at all, it was happening informally. At the same time the Society assumed the authority to recommend explicitly particular papers for publications, a change of approach made possible by Halley's status as a paid employee rather than an honorary officer. This was an important, if little-noticed, change, because it normalised the institution's editorial power over the *Transactions*; so much so that the practice was to continue even when the management of the periodical reverted to the long-established model of independent editor-secretaries in the mid-1690s. However, the Society had once again had to intervene to urge continuing publication, proposing a variety of solutions – including a committee of editorial advisers and contributors, and outright subsidy to the tune of up to £40 a year – to Halley's apparent inability or unwillingness to bring out the *Transactions* with any regularity.

By the early 1690s, it was clear that assigning responsibility for the *Transactions* to an employee would not necessarily have a stabilising effect. Though Halley's tenure was not dogged by controversy, it simply failed to sustain production of the periodical at the level that members of

Council felt was necessary, and this model of editorship was abandoned after seven fairly lean years. As someone with an independent reputation, active research interests and intellectual connections of his own, as well as sufficient means to not have to depend on his clerk's salary, Halley was never at the Society's disposal in the way they had hoped. The other problem was that the Society never actually grasped the nettle and exercised its authority in the obvious way – by assuming financial responsibility for the periodical. The reluctance to do so in 1686 probably had less to do with any sense that some other person or institution had a credible claim to the title than with the embarrassed state of the Society's accounts at that time. The reluctance to do so again in 1693, when the institution was not in such immediate crisis, may perhaps be read as part of the mounting circumstantial evidence that publishing the *Transactions* had come to be seen as a reliably money-losing proposition.

In the apparent absence of any other options, the task of compiling the *Transactions* fell back upon the secretaries of the Royal Society. And here, the Society got comparatively lucky. Richard Waller, a gifted linguist, translator and illustrator had been elected secretary in 1687 and, in November 1693, he would be joined by Hans Sloane, a botanist, physician and collector. Both had exceptionally long secretarial tenures, serving for 27 years and 20 years, respectively, mostly together, and they brought stability to the Society. It helped that, in contrast to Oldenburg, Waller enjoyed a significant independent income,[62] and Sloane was well on his way to establishing one of the largest fortunes in Britain. As we will see in the next chapter, Sloane would become the longest-serving editor since Oldenburg, and is routinely (and justifiably) credited with resuscitating and stabilising the *Transactions*. There has, however, been some confusion among scholars regarding the point at which Sloane's editorship actually commenced.[63] Our examination of the surviving correspondence and the contents of the *Transactions* suggests that this ambiguity has arisen because the solution to Halley's editorial struggles was to try a form of collaborative editorship, something along the lines that had been suggested by Council in 1691; this operated from c. 1692 until 1696. Sloane was part of this editorial team, but not initially in sole charge.

Waller, as senior secretary, seems to have taken over principal responsibility for the *Transactions* from Halley in late 1692. Halley appears to have still been assisting, however. For instance, volumes 17 and 18 of the *Transactions* (covering 1693 and 1694) were published under Waller's supervision, but the editorial preface signed by Waller appears in the *second* issue (no. 196) of volume 17, while the first issue (no. 195, Oct. 1692) seems to have been the work of Halley, as two out of its three papers are by him; and further material from Halley appeared later in volume 17.[64] During 1693, Waller was also drawing upon material provided by Sloane and, by 1694, Sloane was helping to shape papers for publication, adding comments and, on occasion, supervising the press when Waller was out of town.[65] As T. Christopher Bond has noted, editorship in 1693–4 was 'more-or-less by a Committee' under Waller's leadership; but these arrangements continued beyond 1694.

By late 1694, Waller was trying to resign the secretaryship. Apologising for 'living so very much in ye Country, of late', in order to look after the estates that provided his living, Waller begged Sloane and others to think of 'a fitter person to serve the Society in my place'. He begged to be taken seriously, adding that he was 'real in this, tis not *nolo episcopari*'.[66] The volume covering 1695–97 (vol. 19) has a dedication signed by Sloane as publisher, but this indicates only that Sloane had taken over by the time the dedication was printed in 1697. And indeed, it was only in June 1697 that Sloane's frequent correspondent, John Ray, noted, as if newly aware, that Sloane was 'now concerned with the Philosophical Transactions'.[67] In fact, it seems that Sloane took over in autumn 1696, because that was the moment when both Halley and Waller definitively withdrew.

By October 1696, Halley had left London to take up a post as supervisor of the Mint at Chester.[68] Until then, he had continued to supply material to the *Transactions*. For instance, the first half of volume 19 (that is, the issues covering 1695 and early 1696) included five papers by Halley himself, as well as several by his mathematical correspondents and collaborators (for instance, John Wallis in Oxford), and several more compiled by Halley on the antiquities of Palmyra. But by autumn of 1696, and only from that point, there was a shift in the pattern of editorial preoccupation towards the medical, natural-historical and antiquarian subjects that interested Sloane, and an increasing roster of contributions from members of Sloane's early intellectual network (for instance, the Yorkshire naturalist and antiquary Martin Lister; John Ray in Essex; and James Petiver in London).

Evidence of Waller's ongoing involvement comes from a letter of September 1696, when Waller told Sloane that he could no longer 'looke after the Printing of them at this distance', thus implying that he had still been doing so. And it was at this point that Waller made arrangements to obtain access for Sloane to Society papers that had been in his and Halley's custody, suggesting a transfer of responsibility.[69] Sloane's editorial preface referred to an interruption in the *Transactions*, and claimed that the Society had 'commanded' him to 'take care to continue them' [that is, the *Transactions*], but there is no documentary confirmation of such a command.[70] It was perhaps fortunate that Sloane's *Catalogue of Jamaican Plants* was finally published in 1696.[71] Sloane may thus have had the time to devote to the *Transactions* when Waller and Halley left him, somewhat reluctantly, holding the editorial reins.

The *Transactions* of the late seventeenth century was repeatedly reinvented to reflect its editors' particular interests and concerns. The title survived, but the periodical to which it referred was constantly changing. Its focus shifted between recent correspondence and material recovered from the archive; between an open or a closed vision of its public; and its subject matter – whether in natural history and comparative anatomy or astronomy and natural philosophy – varied with its editors. Under Aston and Plot, the *Transactions* for a while reflected grand ambitions for collaborative research between London and Oxford. The different visions, and repeated reinventions, suggest that the concept of what a printed periodical of natural knowledge should be and do was still subject to a considerable amount of interpretive flexibility.[72]

The number, and brevity, of editorial tenures in the late seventeenth century points to an instability and uncertainty of purpose – yet that was, in fact, entirely normal for printed periodicals at the time. Newspapers and journals came and went with startling rapidity. If we look beyond the editorial handovers, the fact that the *Transactions* title did resume after Hooke's *Collections*, and did continue thereafter, marks it out as an unusual print product. And here, the role of the Royal Society is surely key. It did not provide much direct editorial or financial support, but its fellows provided moral support to the editors, and its meetings, archives and correspondence provided copy for the pages of the *Transactions*. It is not at all clear that Grew, or Plot, or Halley would have taken on the apparently thankless task of compiling a periodical without at least that level of encouragement and in-kind support.

A process of convergence between the Society and the *Transactions* was, by 1700, occurring, and would establish significant precedents for the first half of the eighteenth century. The normal activity of Society meetings had become the reading and discussion of papers communicated from outside, and the Society's experimental culture remained in abeyance during the 1690s. The precedent established during Halley's editorship survived, and the Society continued to earmark particular papers for publication and recommended them to the editors for that purpose.

Neither Waller nor Sloane was especially anxious about the cost of publishing the periodical. The Society, for its part, saw the *Transactions* appear regularly without having to incur the expense of publication, without which it had struggled to exercise meaningful control over its continuation even where it apparently wanted to. The return, with Sloane, to a model of independent editorship would become a source of friction, coinciding, as it did, with the increasing significance of the *Transactions* to the overall profile of the Society's activity. The more the periodical mattered to the Society, the less acceptable the idiosyncrasies and predilections of particular editors were likely to prove. Moreover, with Newton's ascension to the presidency in 1703 there would be an evident tension between the new president's notable attempts to promote physics and mathematics within the Society, and Sloane's natural-historical bias.

Notes

1 The breakdown in relations between Hooke and Oldenburg – in effect between the Society's experimenter- and communicator-in-chief – is amply detailed in Robert Iliffe, 'In the warehouse: Privacy, property and priority in the early Royal Society', *History of Science* 30, no. 1 (1992): 29–68, from the point of view of systems of registration and priority, and in Adrian Johns, *The Nature of the Book: Print and knowledge in the making* (Chicago, IL: University of Chicago Press, 1998), ch. 7, from the perspective of engagement with the print trades. Also, Thomas Birch, *The History of the Royal Society of London for Improving of Natural Knowledge*, 4 vols (London: A. Millar, 1756–7), vol. 3, 353; H.W. Robinson and W. Adams, *The Diary of Robert Hooke 1672–1680* (London: Taylor & Francis, 1935), 315.
2 On the efforts of Hooke, in particular, to displace Brouncker as President, see Lisa Jardine, *The Curious Life of Robert Hooke: The man who measured London* (London: HarperCollins, 2003), 210–12.
3 Birch, *History of the Royal Society*, vol. 3, 342–3 and 352. Boyle agreed to be financially responsible for the children until the case was settled; they were looked after by Margaret Lowden, a gentleman's daughter.
4 John Feather, 'From rights in copies to copyright: The recognition of authors' rights in English law and practice in the sixteenth and seventeenth centuries', in *The Construction of Authorship: Textual appropriation in law and literature*, ed. Martha Woodmansee and Peter Jaszi (Durham, NC: Duke University Press, 1994), 191–209; Ronan Deazley, *On the Origin of the Right to Copy: Charting the movement of copyright law in eighteenth century Britain (1695–1775)* (Oxford: Bloomsbury Publishing, 2004).
5 Stationers' Company of London and George Edward Briscoe Eyre, *A Transcript of the Registers of the Worshipful Company of Stationers: From 1640–1708 A.D.* 3 vols. (London: Privately printed, 1913), vol. 2, 451–2.
6 On the elections of Hooke and Grew, see Birch, *History of the Royal Society*, vol. 3, 352–3. On Grew's employment as a specialist curator, and the Society's raising subscriptions to pay for it, see Michael Hunter, *Establishing the New Science: The experience of the early Royal Society* (Woodbridge: Boydell & Brewer, 1989), 261–78.
7 Robinson and Adams, *Diary of Robert Hooke*.
8 Issue 137 for January and February 1678, the first assembled by Grew, featured at least three articles drawn from the early Registers by Sir Robert Moray, Christopher Merrett and Jonathan Goddard, all at least 10 years old.
9 Tyson was elected to the fellowship on 1 December 1679, Leeuwenhoek on 29 January 1680, William Cole never. See Birch, *History of the Royal Society*, vol. 3, 512 and vol. 4, 6.
10 Birch, *History of the Royal Society*, vol. 3, 479 (29 May 1679).

11 Colwall's papers on the manufacture of green copperas and alum were dated June and September 1661, see RS Register Books RBO/1/17 and 18.
12 Moray's letters to Oldenburg in summer and autumn 1665 frequently mention his preparations to go into Wales to examine mines and quarries. For instance, A. Rupert Hall and Marie Boas Hall, eds, *The Correspondence of Henry Oldenburg*, 13 vols (Madison: University of Wisconsin Press, 1965–86), vols 2 and 3, *passim*.
13 Birch, *History of the Royal Society*, vol. 3, 369–70.
14 Robert Hooke, *Lectures and Collections* (London: John Martyn, 1678), 101 (critique of French book) and 105–12 (pistol ball in lungs).
15 Hooke, *Lectures: De Potentia Restitutiva* (London: John Martyn, 1678).
16 See Noah Moxham, 'Fit for print: Developing an institutional model of scientific periodical publishing in England, 1665–ca. 1714', *N&R* 69.3 (2015): 241–60, at 248–50.
17 Birch, *History of the Royal Society*, vol. 3, 450–1.
18 Birch, *History of the Royal Society*, vol. 3, 514. In December that year, orders were issued to publish a journal 'such as the *Transactions* were under Mr. Oldenburg, and under that title', but nothing came of this.
19 See Birch, *History of the Royal Society*, vol. 3, 451, 490, 501 (7 July and 9 August 1679).
20 Noah Moxham, 'Making it official: Experiments in institutionality and the employees and publications of the Royal Society, 1675–1705' (PhD, Queen Mary University of London, 2011).
21 Birch, *History of the Royal Society*, vol. 4, 171 (13 December 1682).
22 I. Newton to F. Aston, 18 May 1669, https://www.newtonproject.ox.ac.uk/view/texts/diplomatic/NATP00227. See also Richard S. Westfall, *Never at Rest: A biography of Isaac Newton* (Cambridge: Cambridge University Press, 1980).
23 On the networks of natural historians in Britain and Ireland, and particularly those working in the topographical tradition, see Elizabeth Yale, *Sociable Knowledge: Natural history and the nation in early modern Britain* (Philadelphia: University of Pennsylvania Press, 2016).
24 For the appearance of these items (from *PT* 13.143) at the Society's meetings, see Birch, *History of the Royal Society*, vol. 4, 163, 165 and 177, and RS LBO/8/246.
25 Lister was responding to criticisms put by Nehemiah Grew and comments by Edward Tyson; see Birch, *History of the Royal Society*, vol. 4, 169, 172. Members of the Oxford Society attempted to replicate Lister's experiments on at least three occasions; see Birch, *History of the Royal Society*, vol. 4, 181, 183–4, 186.
26 Edward Tyson, 'The anatomy of a rattle-snake …', *PT* 13.144 (1683): 25–46, and Westfall, *Never at Rest*, 15.
27 Robert T. Gunther, *Early Science in Oxford*, 14 vols (Oxford: Clarendon Press, 1923–45), vol. 12, 14, 16–17. On Newton's later (1685) attempts to organise a philosophical society in Cambridge, see Westfall, *Never at Rest*, 557–8.
28 Evidence for the coherence of the Oxford group comes from Edward Tyson's letters to Plot at Oxford, in which – and seemingly picking up Plot's own use of the term, though Plot's letters do not survive – he refers to the group as a club, from April 1681. From the content of the letters, it was evidently a club with philosophical interests, and with plans to issue communications of its own. Tyson to Plot, 25 April 1681, in Gunther, *Early Science in Oxford*, vol. 12, 5.
29 Aston to Plot, 20 January 1683, in Gunther, *Early Science in Oxford*, vol. 12, 13.
30 On Hunt, see Sachiko Kusukawa, 'Picturing knowledge in the early Royal Society: The examples of Richard Waller and Henry Hunt', *N&R* 65.3 (2011): 273–94.
31 Aston to Plot, 31 January 1682/3, and passim to April 1683, in Gunther, *Early Science in Oxford*, vol. 17, 22. Gunther identifies this person as William Gould, an Oxford physician, but this is entirely unsubstantiated, see *Early Science in Oxford*, vol. 12, 15.
32 Aston to Plot, 27 January 1682/3, in Gunther, *Early Science in Oxford*, vol. 12, 15.
33 Anatomical drawings of an ostrich survive in the Tyson Folio (MS 681) at the Royal College of Physicians.
34 Though not elsewhere: there are instances of paid contributions in the German lands in the eighteenth century in Jeanne Peiffer, Maria Conforti, and Patrizia Delpiano, eds, *Les journaux savants dans l'Europe des XVIIe et XVIIIe siècles / Communication et construction des savoirs / Scholarly Journals in Early Modern Europe. Communication and the Construction*

of Knowledge, vol. 63, Archives Internationales d'Histoire des Sciences (Turnhout: Brepols, 2013).

35 See Jason Peacey, 'Pamphlets', in *Oxford History of Popular Print Culture*, ed. Joad Raymond (Oxford: Oxford University Press, 2011), 457.

36 Anna Marie Roos, *Web of Nature: Martin Lister (1639–1712), the First Arachnologist* (Leiden: Brill, 2011), 302. Anthony Griffiths, 'Michael Burghers', in *ODNB*. On images at the early Royal Society, see Sachiko Kusukawa et al., 'Science made visible: Drawings, prints, objects' (London: The Royal Society, 2018), https://issuu.com/crassh/docs/des5543_1_science_made_visible_exhi.

37 We are indebted for this insight to Meghan Doherty's unpublished work on the variants of the early *Transactions* held in Oxford colleges.

38 See Gunther, *Early Science in Oxford*, vol. 12, 21–32.

39 On Gould's drawing, see Birch, *History of the Royal Society*, vol. 4, 180–1, and eventually published as Gould, 'An account of a polypus …', *PT* 14.157 (1684): 537–48.

40 For the publishing history of Ray and Willughby's work, see Sachiko Kusukawa, 'The *Historia Piscium* (1686)', *N&R* 54.2 (2000): 179–97.

41 Wallis's review of Hevelius's *Annus Climactericus* (Gdansk, 1685) appeared in *PT* 15.175 (1685): 1162–83. Hooke's critique of Hevelius had appeared in his *Animadversions on the first part of the Machina Coelestis* (London: John Martyn, 1674).

42 In Robert Hooke, *Lampas* (London: John Martyn, 1677 [1676]).

43 The episode is discussed in Johns, *Nature of the Book*, 527–9.

44 Kusukawa, 'The *Historia Piscium*'.

45 The discussion of the clerk's responsibilities and competences, and subsequent election, took place at successive meetings of the Council between 16 December 1685 and 3 February 1686. Birch, *History of the Royal Society*, vol. 4, 451–5 (especially the meeting of 27 January, at 453).

46 Birch, *History of the Royal Society*, vol. 4, 462 (re meeting of 3 March 1686).

47 Johns, *Nature of the Book*, 540.

48 Birch, *History of the Royal Society*, vol. 4, 462.

49 See, for example, Birch, *History of the Royal Society*, vol. 4, 486, 511, 516, 521.

50 Westfall, *Never at Rest*; Alan Cook, *Edmond Halley: Charting the heavens and the seas* (Oxford: Clarendon Press, 1997).

51 The question of what made Halley take up the clerkship in the first place is obscure. His father had been a well-to-do London soap-boiler and merchant, owning extensive property. Alan Cook, Halley's biographer, suggests that Halley was desperate for an income while his father's estate was in dispute; but a number of Halley's actions, including particularly his assumption of the financial risk of publishing the *Principia*, and probably also of the *Transactions*, are hardly consistent with this explanation.

52 Halley, 'Advertisement', *PT* 17.192 (1693): 452. This issue appeared in January or February 1690/1 OS.

53 RS AB/1/1/3, entries for years 1686–8.

54 RS CMO/02, 28 January 1691, 82–3.

55 On editorial committees, see David A. Kronick, 'Authorship and authority in the scientific periodicals of the seventeenth and eighteenth centuries', *Library Quarterly* 48, no. 3 (1978): 225–75; and Aileen Fyfe and Anna Gielas, 'Introduction: Editorship and the editing of scientific journals, 1750–1950', *Centaurus* 62, no. 1 (2020): 5–20.

56 Halley, 'Advertisement'.

57 Halley, 'Advertisement'.

58 RS CMO/02, 7 December 1692, 87.

59 Halley's 15 papers amounted to 135 out of the 578 pages in volume 16. Some eighteenth-century editors used their journals as a means of publishing their own work; see Anna M. Gielas, 'Early sole editorship in the Holy Roman Empire and Britain, 1770s–1830s' (PhD, University of St Andrews, 2019), ch. 1.

60 RS CMO/02, 15 February 1693, 88.

61 Halley's salary is mentioned in Birch, *History of the Royal Society*, vol. 4, 45.

62 Lotte Mulligan, 'Waller, Richard (c. 1660–1715), natural philosopher and translator', *ODNB*.

63 Maarten Ultee has Sloane taking over the *Transactions* in 1693, when he became secretary; T. Christopher Bond opts for the end of 1694; James Delbourgo does not commit himself;

and one of the present authors has previously given 1695 (incorrectly, as he now thinks). See Maarten Ultee, 'Sir Hans Sloane, scientist', *British Library Journal* 14 (1988): 1–21, 2; T. Christopher Bond, 'Keeping up with the latest transactions: the literary critique of scientific writing in the Hans Sloane years', *Eighteenth-Century Life* 22 (1998): 1017, 2; James Delbourgo, *Collecting the World: Hans Sloane and the origins of the British Museum* (London: Allen Lane, 2017); Moxham, 'Fit for print', 246.
64 Seventeen papers in volume 17.
65 See RS EL/S2 f. 12, for letters shaped by Sloane into a piece for the *Transactions*, and Alvarez de Toledo, Christopher Love Morley, and Hans Sloane, 'Several accounts of the earthquakes in Peru …', *PT* 18.209 (1694): 78–100. For a paper with additional comments by Sloane, see T.M. and Hans Sloane, 'Concerning the strange effects from the eating dog mercury …', *PT* 17.203 (1693): 875–7.
66 Waller to Sloane, 26 November 1694, BL Sloane MS 4036 f. 194. *Nolo episcopari* meant 'I don't want to be made a bishop', and is proverbial as a refusal in form only.
67 There are 18 extant letters from Ray to Sloane between December 1694 and June 1697 (BL Sloane MS 4036); Sloane's replies do not survive.
68 Halley to Sloane, 12 October 1696, RS EL/H3/48, is dated at Chester.
69 BL Sloane MS 4036 f. 266.
70 'The Preface', *PT* 19 (1695–7): [unnumbered page in volume front matter]. The interruption may refer to the two issues that eventually covered March to August 1696, since the content of those issues carries the imprint of Halley rather less distinctly than the preceding five and may have been pieced together by Waller or (retrospectively) by Sloane.
71 D. Brown and Hans Sloane, *Catalogus Plantarum Quae in Insula Jamaica* (London: D. Brown, 1696).
72 'Interpretive flexibility' is a concept used in the sociology of science and technology to describe the way in which an invention may be imagined or used by different people in very different ways; it most often happens in the early phase, before a widely held and stable meaning for the invention is developed. See Wiebe E. Bijker, *Of Bicycles, Bakelites and Bulbs: Toward a theory of sociotechnical change* (Cambridge, MA: MIT Press, 1995).

3
Stabilising the *Transactions*, 1696–1752

Noah Moxham and Aileen Fyfe

As it entered its fourth decade, the *Transactions* was already unusual among British periodicals for its longevity. Only the official government newsletter, the *London Gazette*, stood comparison, perhaps testifying to the significance of institutional will, if not precisely institutional support, in keeping periodicals going over long spans.[1] That said, the repeated and short-lived cycles of experimentation discussed in Chapter 2 suggested that its continuation could not be taken for granted. Hans Sloane would become its longest-serving editor so far and has been widely credited with rescuing and stabilising the *Transactions* for the future. Dramatic shifts in its form and function did, indeed, become a thing of the past. But, as we will see, the early eighteenth-century *Transactions* was still a very different type of periodical from what it would become in the late eighteenth century.

The *Transactions* was also distinct from a new model of institutional scholarly publishing that was emerging in Paris. The first volume of the *Histoire de l'Académie Royale des Sciences ..., avec les Mémoires de Mathématique et de Physique* covered the year 1699.[2] Its large, ponderous volumes thereafter appeared at annual (or even longer) intervals, and carried lengthy essays by named contributors that had passed the scrutiny of the Paris academicians, and were printed and circulated at the expense of its royal patron. This model of institutional publishing would be widely imitated across Europe. The *Transactions*, on the other hand, maintained a more frequent periodicity, carried shorter contributions, was compiled by an independent editor and made no claim to contain material approved by a parent institution. The paradox of this period of the history of the *Transactions* is that, even while the ongoing tradition of independent personal editorship was criticised for *resisting* institutionalisation (as we will see in Chapter 4), it did in fact stabilise the periodical and, in many ways, helped it become a more institutional product than it had been.

There was no official change in the relationship between the *Transactions* and the Royal Society in the first half of the eighteenth century: the Society's editorial input remained minimal, and its financial support was next to non-existent. And yet, in significant contrast to the previous 30 years, the *Transactions* no longer seemed in regular danger of ceasing publication. Sloane edited and published it until 1713, and he was followed by Edmond Halley (again), James Jurin, William Rutty and Cromwell Mortimer (whose 22-year tenure surpassed even Sloane's). As Sloane had been, they were all secretaries of the Royal Society (Halley had been restored to the fellowship after resigning as clerk in 1699). The periodical seemed to have reverted to being the secretary-editor's personal fiefdom; but at the same time, there is clear evidence that senior fellows of the Society continued to be concerned to keep the *Transactions* going – and, in the case of certain presidents, to exert influence over it.

The story of this period of the *Transactions*' history can be seen as a series of presidential rivalries: Sloane would be edged out of the secretary-editorship by Isaac Newton as president, who then chose Halley and later Jurin; but when Sloane became president after Newton's death in 1727, Jurin resigned, and Sloane chose Rutty and Mortimer. The rise of presidential influence over the *Transactions* in this period is undoubtedly significant, and, as we will see, it shaped not only the choice of secretary-editor, but also the contents of the *Transactions*. But despite the apparent tensions between Sloane the natural historian and Newton the natural philosopher (and their acolytes), the *Transactions* was supported in this period by a core group of senior fellows of the Society who contributed as authors, communicators and editorial assistants, regardless of who was currently in charge of it.[3] The group includes Halley, Sloane and Jurin, as well as regular contributors John Theophilus Desaguliers, Antoni van Leeuwenhoek and the Molyneux family of Anglo-Irish natural historians and mathematicians. We can start to see a sense in which key Royal Society fellows shared a group commitment to the *Transactions*, that would ultimately be formalised in 1752.

This chapter and Chapter 4 both deal with the first half of the eighteenth century. This chapter focuses on the *Transactions* within the Royal Society, and examines how it developed a more durable identity both as the chief publication of the Society, and as the mainstay of its activity. Chapter 4 will examine its cultural and intellectual engagement with worlds beyond the walls of the Royal Society, including the challenges of operating in an increasingly varied, competitive and geographically extended publishing environment.

We begin with an evaluation of Sloane and his editorial practices. Sloane may have been a reluctant editor, but he proved to be an effective one, drawing upon his extensive correspondence networks to increase the amount of *Transactions* material coming from beyond the fellowship, particularly in natural history. We then contrast Sloane's editorial vision and practice with those of his immediate successors, Halley and Jurin. Halley became the first secretary-editor to pragmatically acknowledge the connection between the Society and the *Transactions*. Halley implicitly rebuked Sloane for being insufficiently attentive to the selection and presentation of papers in the periodical, but, as had happened before, he was himself unable to keep up a schedule of regular issues. His replacement, Jurin, was much more attentive to the practical minutiae of editing a periodical, and his detailed surviving correspondence is full of valuable information about the day-to-day management of the *Transactions*. He was an outward-looking editor who gathered information from across Europe to develop his interests in meteorology and in the effectiveness of inoculation against smallpox.

Finally, we consider the tenures of Sloane's hand-picked secretary-editors, Rutty and Mortimer. Mortimer continued after Sloane resigned the presidency in 1741, and died in post in 1751 during the presidency of Martin Folkes. The efforts by Sloane and Mortimer to reorganise the Society's finances and its archives helped to change the relationship between the *Transactions* and the Society.

Hans Sloane's *Transactions*

In the mid-1690s, Hans Sloane was known as a physician, a botanist and a traveller. His burgeoning reputation as a collector of natural and artificial specimens and rarities gave him strong reasons for sustaining a broad network of correspondents in Europe, North America and the Caribbean, and Asia; and the same network made him well-placed to bring in contributions for the *Transactions*. He was also well-placed to take financial responsibility for the *Transactions*. Between his growing metropolitan medical practice, and his advantageous marriage to Elizabeth Langley Rose (who had inherited sugar estates worked by enslaved people in Jamaica from her deceased husband), Sloane was well on the way to establishing a significant (and eventually immense) personal fortune.[4] There is no positive evidence that the Royal Society took any of this into account when asking Sloane to take sole responsibility for the *Transactions* when Waller and Halley backed out in 1696, but since the basic facts of Sloane's situation were widely known, the Council had every reason to hope that the man in charge of the *Transactions* would have the social, intellectual and financial resources to sustain its publication in a way that would demand very little from the institution.

In one sense, this was exactly what happened. Sloane edited the *Transactions* for 17 years, and would later claim to have spent large amounts of his own money supporting it. As editor, he drew both upon his correspondence networks and upon meetings of the Royal Society, though there was a very indistinct boundary between those two categories. For instance, in the collection of Sloane's personal papers now held in the British Museum, there are many letters (and two bound volumes of manuscript scientific papers) that had clearly been communicated to Sloane in his role as secretary-editor, with the hope that they would be shared with the Society or printed in the *Transactions*. Coupled with the evidence of the periodical's content, his incoming correspondence and the material traces of publication left on the manuscript letters help us to partially reconstruct Sloane's editorial style.

As with Oldenburg's correspondence, Sloane's letters were a mix of the personal, editorial and secretarial, often in the same letter. In addition, many of Sloane's scientific correspondents were also his patients, or had family members under his care. The Essex naturalist John Ray, for instance, was one of Sloane's most frequent correspondents at this time. Certain that he was about to die, he kept up a running commentary on his many afflictions, as well as exchanging opinions on new botanical books and answering Sloane's requests for expert advice about the latter's forthcoming *Catalogue of Jamaican Plants*.[5] Since Ray was an eminent fellow of the Royal Society, his communications were routinely read to its assemblies, and he relied on Sloane to judge which parts of his letters could be made public and what should be kept private. Sloane's editorial intervention was usually limited to striking out the obviously personal or confidential parts of letters that he intended for publication.

The shifting social dynamics of these layered exchanges, and the economies of obligation and exchange, affected Sloane's editorial practice. The exchanges with Ray in the mid-1690s, for example, were coloured by the fact that Ray was much the older man and enjoyed the higher reputation. Sloane frequently had to ask Ray for expert opinion and assistance with the publishing projects that were to form the basis of his own reputation as a naturalist, and was thus, in effect, a supplicant. In April 1697, we find Ray acknowledging the gift of a box of sugar from Sloane as well as plant specimens; conceivably the gift was from

Elizabeth's Jamaican estates, which suggests Sloane's use of those assets as part of the economy of botanical communication and exchange.[6] By the 1710s, however, when Sloane's authority and eminence had dramatically increased, he was far less likely to find himself in the supplicant position. Equally, some of Sloane's correspondents were quick to spot that there might be other advantages to be gained from communicating their papers to him than the simple prestige of having them appear in the *Transactions*. In 1698, for instance, Samuel Dale, a neighbour of Ray's in Essex and a fellow naturalist and medical man, offered Sloane a paper on 'the prodigious internal uses of Cantharides to Cowes' and in the same breath asked him if he happened to have any Jamaican shells to spare from his collections.[7]

Dale's offer was unusual for having been originally solicited by the Society's bookseller, rather than by Sloane himself. It is the only positive evidence we have found for members of the print trade getting directly involved in sourcing copy for the *Transactions*. The bookseller was Samuel Smith, of St Paul's Churchyard, and he had been printing the *Transactions* since 1691 (latterly, in partnership with Benjamin Walford). He was also publishing books for John Ray and Samuel Dale, and he acted as a conduit for their correspondence with each other and with Sloane, as they exchanged samples, comment and advice.[8] It is possible that Smith's suggestion to Dale arose from general instructions from Sloane, rather than soliciting contributions on his own initiative.

In other respects, there is not much to be inferred about the relationship between Sloane and the printers and sellers of *Transactions*, except for the implications of the eye-catching estimate (which dates from more than a decade after Sloane ceased to be editor, when he was running for president of the Society) that publishing the periodical had cost Sloane personally £1,500.[9] For Sloane to have *lost* money on this scale – just under £90 per annum – we must assume that he bore the costs of production personally. In this, he differed from Oldenburg and his immediate successors, who had been remunerated either directly, by the publishers for supplying copy; or indirectly, by the Society's commitment to purchase 60 copies of every issue. That commitment from the Society had lapsed by 1693,[10] and by Sloane's time, the financial risk of the *Transactions* had shifted decisively from the printers and booksellers to the editors and publishers. Sloane's financial losses are a strong indication that sales were relatively poor. Given that the early eighteenth-century Royal Society typically ended its financial year with only about £30 cash in hand, it is clear that Sloane's capacity (and willingness) to underwrite the publication of the *Transactions* was a significant asset to all those interested in its continuation. It also removed any incentive for the bookseller to offer better terms.

James Delbourgo suggests that Sloane 'revitalized the *Transactions* in line with his predilection for natural history, making the journal a clearing-house that knitted together reports from around the British Isles as well as Britain's empire'.[11] It is broadly true that there was far more natural-historical and medical content – and less mathematics and physics – than there had been during Halley's first editorship. But these were not the only changes under Sloane.[12]

Unlike Halley, Sloane managed to sustain a fairly regular periodical issue of the *Transactions*. He initially tried to publish an issue every month, but that pattern faded after 1699. Thereafter, he produced bimonthly issues, and (until 1710) biennial volumes. He seems initially to have managed to keep reasonably close to the publication schedule implied by the dates on individual issues, but had fallen well behind by the 1710s.[13] Volume 28, for 1713, is not divided into issues at all, suggesting that it was compiled entirely after the fact, and indeed after Sloane had left office. As Sloane's issues appeared at an increasingly more leisurely schedule, so too the average length of an issue increased gradually, from about 35 pages in the mid-1690s to about 45 pages in the early 1710s. The mean length of an item in Sloane's *Transactions* was about half a dozen pages; slightly shorter in his early years, and slightly longer in the later years of his tenure. Sloane did publish some long, fully articulated, discursive essays running to 10 or more pages (such as those by Leeuwenhoek), but the most typical contribution was a short empirical observation or case history running to only two or three pages, of the sort Sloane hoped others would build on.

Early in his tenure, Sloane was eclectic in acquiring and organising copy for the *Transactions*. He was willing to include material that was not particularly recent, such as the string of papers relating to the activity of the Oxford Philosophical Society in the 1680s that he printed from mid-1697 onwards.[14] He also experimented with new functions: for instance, during the later months of 1698, he tried a regular short section listing books recently printed on scientific subjects by Continental publishers; and in the same year, and again in 1701, he tried a section for scraps of news from recent letters too short for a separate communication. Neither feature proved durable.

Sloane also published extracts and translations from new foreign periodicals, such as Jean Brunet's *Progrès de la Médicine* in 1695, or the Paris Académie's *Histoire et Mémoires* from 1702. But he did not do so systematically (as Oldenburg had done) and did not make much use of either the French *Journal des Sçavans* or the Leipzig *Acta Eruditorum*, possibly because they were well established and accessible to interested readers (provided they could read French and Latin). He also devoted markedly less space to accounts of books than had his predecessors. Only around 10 per cent of Sloane's pages contained book notices, whereas Oldenburg had filled about a third of the pages of his *Transactions* with accounts of natural-philosophical print. In fact, by the 1710s, Sloane became increasingly selective about noticing books – often just a handful per volume – but those he did notice received significant attention. For instance, the fourth and fifth instalments of the catalogue of natural-historical rarities *Gazophylacium Naturae et Artis*, by Sloane's friend James Petiver, received an 11-page description in 1710.[15] Overall, Sloane's reliance on printed material, whether in the form of extracts, summaries or translations, diminished as time went on.

It was in this period that the fellowship of the Royal Society became far more important in providing copy. Almost half of the items printed by Sloane were letters or extracts of letters, and many were written by fellows. In the mid-1690s, fellows had authored only around a third of items in the *Transactions*, but that had risen to over 70 per cent by 1709 (a level roughly maintained during Halley's second editorship).[16] On the one hand, this means that fellows increasingly saw the *Transactions* as a suitable outlet for their own activities. On the other, it meant that contributions from outsiders became less common, and that the *Transactions* was less effective at extending the Society's networks. Some new correspondents – perhaps a quarter or a third – might later become fellows, but by the second half of Sloane's tenure, there were fewer of these. It is striking that when Halley was succeeded in turn by James Jurin, a notably younger and more energetic secretary, the proportion of contributions from outsiders returned to 1690s' levels. It appears that younger editors were more interested in looking actively beyond the Society for correspondents and contributors than were older men with their own established networks.

The high proportion of content by fellows helped consolidate the existing perception of a strong link between the Society and the *Transactions*. So too did a stream of papers by Francis Hauksbee the Elder, a London instrument-maker and researcher in physics who was introduced to the Society by Isaac Newton (elected president in November 1703). Hauksbee specialised in demonstrations of electricity, magnetism, capillarity and other physical phenomena. Hauksbee's fields of research coincided with Newton's interests, and his experimental and instrument-making prowess with Newton's desire to revive experimental demonstrations at the Society's weekly meetings.[17] Hauksbee's experiments were written up for the *Transactions*: the first to appear was published in issue 292 (for July and August 1704) and it was followed by another 37 in little more than five years.[18] Institutional ambivalence about the periodical format was beginning to be overcome.

The contributions might come to Sloane from fellows, but they were just as likely to report the observations of a third-party as of the fellow himself. Thus, as well as becoming increasingly dependent on material provided by the fellowship of the Royal Society, Sloane's *Transactions* was simultaneously far more concerned with the wider world than had been the case under any editor since Oldenburg. During Sloane's tenure, between 35 and 40 per cent of the material either was sent from, or was concerned with, lands beyond the island of Great Britain. And, as we might expect from someone with Sloane's transatlantic networks, this did not simply mean 'Europe'.

Much of Sloane's overseas material derived from regular and repeated contributors. They included the Delft microscopist, Antoni van Leeuwenhoek; Georg Joseph Kamel, a Jesuit in the Philippines; and Samuel Browne, an East India Company surgeon in Madras. Leeuwenhoek had, of course, been sending his long and illustrated essays to the *Transactions* for decades (and Sloane was more careful than Halley had been to cultivate the correspondence).[19] The contributions by Kamel and Browne were handled and shaped for publication by Sloane's close associate and fellow collector, James Petiver.[20] After Browne's death, Petiver parcelled out his collections and catalogues of plants, and published them in the *Transactions*. Petiver seems to have been significantly responsible for coordinating and publishing large parts of a wider overseas network on Sloane's behalf as well as his own.

Despite carrying far more material from or about the world beyond Britain, Sloane's *Transactions* was not intended to be read by that world. English was still not a widely read language on the Continent, and the fields in which Sloane had the strongest interests – natural history, especially botany, and medicine – had strong Latin traditions.[21] Sloane's decision to render so much of his material in English suggests that he assumed no responsibility for making it accessible to a European learned audience. His successors, Halley and Jurin, would feature less material from overseas in their pages, but they would both print about a third of their pages in Latin. Sloane, however, only put a sixth of his material in Latin.[22]

Anonymous contributions declined during Sloane's tenure. Fewer than 3 per cent of the contributions in Sloane's final years as editor were unattributed, compared with about 15 per cent in the mid-1690s. Part of the reason was the decline in notices of new books, but more generally, the *Transactions* appears to have established an identity as a periodical which expected signed contributions. This contrasted with normal practice at literary or news journals of the period, but fits with the culture at the Royal Society, where it was still assumed that the identity of the author or witness was an important part of judging the plausibility of a truth-claim.[23] In the early years of his editorship, Sloane had claimed that it was up to readers to distinguish between matters of fact or mere hypothesis, and to assist them, promised to supply the names, addresses and 'circumstances' behind the 'various relations that come to my hands'.[24]

Attribution did not necessarily operate as we would now expect. The *Transactions* continued to print many communications that were essentially second-hand; that is, the person whose letter reported the event, observation or experiment, and who was named in the item's title, was often not the person who had actually witnessed or carried it out. (This is something that modern digitised databases struggle to represent accurately.) In one notable instance, this practice has obscured the earliest instance of a woman contributing to the *Transactions*: Ann Savile's letter, testifying to the great age of Henry Jenkins of Yorkshire, appeared in number 221, for summer 1696. The piece included her signature, but was framed as an item by the physician and naturalist Tancred Robinson who had apparently exhorted Savile to communicate what she knew of the case in writing.[25]

Under Sloane's editorship, the number of contributions that could broadly be labelled as natural history or medicine outnumbered those that related to the physical sciences by about three to two.[26] Moreover, the natural history contributions tended to be longer (averaging nine pages apiece, rather than the more typical two or three pages), which meant there were about twice as many pages dealing with natural history or medicine as with the physical sciences during Sloane's editorship. Sloane's research interests, his professional activities and the network of friends, agents and correspondents that sustained them decisively skewed the periodical's contents. As we saw in Chapter 2, the shift towards natural history during 1696 suggests the timing of the editorial handover to Sloane, and it intensified during his early years in charge: by 1701, around three-quarters of the pages contained natural-historical or medical content. The critiques of Sloane's editorship in early 1700 made no immediate change, but in the years from 1704 onwards, there was a resurgence of physics and chemistry papers in the *Transactions*.

Compared to just 6 per cent of pages in the volume for 1700–1, the volume for 1708–9 would have physics, chemistry or related material on around 40 per cent of its pages. Much of this shift can be attributed to a sudden stream of contributions from Hauksbee.

There is no evidence to decide whether Sloane particularly welcomed this slew of papers in physics, or whether he would have preferred to maintain the distinctive slant towards natural history and medicine that the *Transactions* had developed under him. Nevertheless, the presence of that material can certainly be attributed, directly or indirectly, to the election of Newton as president. This inaugurated a new and, as it proved, enduring strand of influence over the content of the periodical. The final decade of Sloane's editorial tenure was spent under Newton's presidency, and there were tensions between the interests of the two men. Sloane would eventually be edged out in 1713 and replaced by Newton's chosen secretary-editors, first Edmond Halley and then James Jurin.

Visions of knowledge-making

When Edmond Halley succeeded Sloane, his first act was to issue a preface that laid out his editorial principles, and which was full of veiled criticisms of Sloane.[27] (In other respects his editorship is even more thinly documented than Sloane's, since for the most part not even letters to him survive.[28]) Halley framed his editorial vision with the remark that the *Transactions* needed no introduction, being 'always acceptable to the Learned', so long as 'due care has been taken in the choice of the Collections so recommended to the Inquisitive and Intelligent Reader'.[29] This swipe at Sloane's editorial practice stands in contrast to the apparent warmth that existed between the two men when they assisted Richard Waller with the *Transactions* in the mid-1690s, and that had continued into the new century.[30] It is undoubtedly true that Sloane's editorship had attracted hostility and criticism, both from outsiders with little connection to the enterprise of promoting natural knowledge, and from fellows who shared Sloane's general interests but who reacted against what they saw as his idiosyncrasies and prejudices against their views. Halley's veiled attack suggests that frustration with Sloane's conduct of the periodical extended to those whose interests lay on the physical and natural-philosophical side of early eighteenth-century science.

It had been in response to early criticisms of his editorship that Sloane laid out his only explicit editorial vision for the *Transactions*. As we will see in Chapter 4, in 1700, the anonymous *Transactioneer* pamphlet had critiqued Sloane, and Sloane's preface to volume 21 of the *Transactions* must be read as a response.[31] Sloane described the papers published therein as 'a few of such as have come last Year to the Royal Society'. He apologised for the 'mistakes' that happen in printing, noting that 'there will always be some'; while his references to 'my own Weakness' and to 'others better qualified than my self' represented the editor as a humble and self-effacing intermediary between author and reader. Sloane claimed not to have 'abridged or chang'd any thing' in the contributions he received, a statement that was both a gesture of trust in his correspondents and a denial of editorial responsibility. He insisted it was up to readers (not the editor) to evaluate, scrutinise and accept or reject the claims printed in the *Transactions*.

Sloane expected his 'discerning' readers to distinguish between two types of material: what is 'Matter of Fact, Experiment, or Observation' (which 'must always be useful') and 'what is Hypothesis' (which 'may be pass'd over by such as dislike them'). Sloane presented the *Transactions* as a repository of matters of fact, experiments and observations ('of which all these papers contain, some') rather than hypotheses that might be subsequently proven false. As an example of the harm that could be caused by hypotheses, he cited the long-standing opposition of traditional medical theories to the use of 'the Jesuits Bark' (cinchona): despite a successful demonstration in 1638, when 'a poor Indian' cured the wife of the governor of Peru of an ague, hundreds of volumes of medical treatises continued to argue against the use of a treatment which did not fit 'their Hypotheses'. Sloane hoped his readers would be more open to new observations and matters of fact concerning the treatment of disease in the forthcoming issues of the *Transactions*. And to help readers make that judgement, Sloane promised to provide the names of the correspondents who had sent (and received) the information, and to explain 'the Circumstances of the several Relations that came to my Hands'. Again, Sloane abrogated editorial responsibility for evaluating claims, but left his readers to decide whether authors' accounts could be 'relied on, convicted of falsehood, or further inquired into'.[32]

In contrast to this self-effacing editorial stance, Halley set out to be a more interventionist editor. His 1714 preface invited 'all real Lovers of Knowledge' to assist him by communicating their 'Observations, Discoveries, or Inventions'.[33] He promised to give correspondents 'due Acknowledgement', but he also announced his intention to unilaterally excise all the 'useless parts' of letters, such as the 'Preambles and Conclusions'. He also requested correspondents to 'omit all Personal Reflections', for aspersions on the character of other people would not be conducive to the 'Candor, Respect and Friendship' which Halley felt ought to characterise civil philosophical discourse. Any such aspersions would not be printed, he made clear – although in articulating his standards of correct behaviour among philosophers he could not avoid at least an implicit criticism of his predecessor.[34]

Halley explained in his preface that he intended to publish four types of item in the *Transactions*: short tracts, extracts of letters, an account of experiments at the Royal Society, and notices of books. His emphasis on the *Transactions* as a valuable means of preserving short tracts for 'posterity' was a familiar one,[35] and with the *Transactions* approaching its 50th anniversary, he could make a strong argument that its periodical format no longer implied ephemerality. In reintroducing accounts of books, he held out an expectation that these accounts should be undertaken in a critical or at least a neutral spirit, by promising that, if authors were permitted to write notices of their own works, it would be explicitly noted. And Halley expressed his intention to publish material 'relating to the Improvement of Natural Philosophy, Mathematicks and Mechanicks', a phrasing that invites contrast with the preponderance of medical and natural-historical publishing under Sloane.

Halley's preface also drew attention to the relationship between the Royal Society and the *Transactions*. He intended to notice books that related 'to the Ends of the Royal Society's Institution' and to give accounts of experiments made before 'the Illustrious Royal Society, as they shall please to order or permit the Publication of'. Sloane had also acknowledged that the Society had 'given leave' for materials presented to the Society to appear in the *Transactions*, but Halley's phrasing admitted that the Society had ordered the publication of certain papers or experiments. Thus, Halley became the first person to acknowledge explicitly the involvement of the Society in the editorial process. Sloane had insisted that the *Transactions* was a private venture and that the editor had an irreducible right to publish what he pleased, but Halley alluded to wider norms and standards, and made plain the direct involvement of the Royal Society.

This represented Halley's second stint as editor of the *Transactions*, not counting his involvement in the collaborative editorial team under Richard Waller. It was certainly more successful than his first, in the strictly limited sense that he did not abandon the project within 18 months. Halley also appears to have followed his own precepts. His stated intention of shaping contributions meant that far fewer of the printed items were framed as letters: less than 15 per cent, compared to about half under Sloane. Astronomical and mathematical publishing increased dramatically: on average, Halley published five times more pages of astronomy and mathematics annually than Sloane. Natural-historical and medical material declined almost in proportion: they had filled over half of Sloane's *Transactions*, but only around a fifth of the pages in Halley's *Transactions*. Halley's lack of interest in these topics verged on neglect: for the third time he failed to maintain correspondence with Antoni van Leeuwenhoek in Delft, none of whose letters were published during Halley's 13 years supervising the periodical's content.[36] Halley continued to rely on the fellowship of the Royal Society for contributions (fellows contributed almost three-quarters of the items), but far fewer of these came from, or concerned, the world beyond Britain. At the same time, Halley's *Transactions* more obviously addressed themselves to an international audience, insofar as he published a much greater volume of Latin material than Sloane had done. Halley's editorial practice reinforces the sense, from his preface, that his editorship was framed as intentionally redressing, perhaps even rebuking, some of Sloane's practices.

In 1719, however, Halley was appointed to the position of Astronomer Royal following the death of the first incumbent, John Flamsteed. This called for pretty constant attendance at Greenwich and did not, as Flamsteed had found to his frustration, lend itself particularly well to active participation in the meetings and social world of the Royal Society.[37] Halley already had additional responsibilities as Savilian Professor of Geometry at Oxford, and although he remained as secretary of the Royal Society he found those duties increasingly difficult to fulfil. Something had to give, and – as in the late 1680s – it proved to be the *Transactions*. By November 1721 no issues of the periodical had appeared for two years, and Halley was replaced as secretary and editor by James Jurin.

Jurin was a young physician from the north-east of England looking to build his medical practice in London. He had also previously been a schoolmaster, had given courses of lectures in the provinces on Newtonian physics, and had been taught at Cambridge by Roger Cotes, one of the most talented English mathematicians of the period and Newton's collaborator on the second edition (1713) of *Principia*. Jurin was, in short, 'an ardent Newtonian', taught by Newtonians, and trained in mathematics, physics and medicine.[38] He would prove to be a conscientious secretary to the Society and an industrious editor of the *Transactions*, until he was forced out after supporting the losing side in the presidential election following the death of Newton in 1727.

Jurin came to his editorial role with personal experience of the difficulties of managing the issues of propriety surrounding meetings, conversations and print. In 1719, he had been the author of a paper published in Halley's *Transactions* on the specific gravity of blood. He claimed to have been writing in vindication of John Woodward's view of the matter, and took umbrage at learning that Woodward had not merely objected to some passages in it, but had done so publicly in a coffee-house. Woodward claimed only to have done so 'quietly', to two or three common friends, but Jurin believed he had spoken in a manner audible to strangers.[39] Woodward, for his part, objected to the publication of what he saw as imputations about his experimental methodology, without having had the opportunity to respond or clarify in person. Jurin pointed out that, since the paper had been read in a meeting of the Royal Society, Woodward had had an opportunity of raising his objections then. Woodward pointed out that there was no way for him to know this in advance of the meeting and Jurin eventually took refuge in the argument that the paper had not been published by him but by order of the Society.

Several telling details emerge from the exchange: first, that fellows of the Society were not apparently notified in advance of what the next week's meeting would contain; second, that each man felt the other had transgressed against the proper way of conducting natural-philosophical disagreements; and third, that the publication of Jurin's paper had been ordered by the Society. This is evidence of the Society's input into the choice of material for the *Transactions* – although the claim that he had no power over its appearance is a touch disingenuous, since he would probably have been permitted to withdraw it or to make small alterations. This episode helped form Jurin's sense that good editorial management was necessary to avoid misunderstanding and pointless controversy in natural-philosophical publication.

Back in 1714, amid the contestations between Isaac Newton and Gottfried Leibniz, Jurin had written to a correspondent about his wish for a natural-philosophical adjudicator who could settle disputations between rivals. He referred to his dislike of hypothesis, intellectual overreach and pointless contestation, deploring the 'spirit of Ambition and Contention' that led disputants to 'pinch & wrest the Phaenomena' to fit their hypotheses or to 'magnify' presumed difficulties 'beyond all bounds'. He then imagined a natural-philosophical peace negotiation, in which an adjudicator divided up the disputed regions according to 'a Treaty of Partition for dividing the Mineral World among the several contending powers, in such proportions as to me appears to be doing Justice to their several Pretensions'.[40] This fantasy adjudication is a shrewd appraisal of the characters involved, and manages to affirm Jurin's Newtonian credentials in passing, along with a swipe at

Leibniz's intellectual ambition, construed here as avarice. It is a wry and amused description, and a telling position for a future editor of a natural-philosophical periodical to adopt at the age of 30.

As editor, Jurin was generally sceptical of theorising and hypotheses, even in his own field of medicine. In 1723, for instance, Antoine Deidier, a Paris physician who was fishing for election to the Royal Society, sent two medical treatises in manuscript to Jurin, one on tumours and one on venereal disease. Deidier's idea of the *Transactions* was perhaps informed by the lengthy dissertations published in the *Mémoires* of the Paris Académie, and Jurin attempted to explain that the *Transactions* was different, for it was 'intended rather for scattered pages reporting experiments and observations which otherwise would easily be lost'.[41] Jurin explained that he had, instead, arranged for the treatises to be published independently: 'We have referred them by our joint decision to a certain London bookseller, printer to the Royal Society, to be published as soon as possible; and in this matter we trust we have done nothing contrary to your wishes.'[42]

Jurin's objection to publishing these particular treatises by Deidier in the *Transactions* seems to have been partly their length and partly their hypothetical biases, for Deidier's earlier observational work on plaque had already been published in the *Transactions*. A few years later, Jurin was more explicit, telling Deidier that 'singular Observations and Experiments' were welcome, but in general dissertations on medical subjects were 'not very acceptable', especially when they related to the 'Theory of Medicine'.[43]

Jurin was conscious of a growing prejudice against medical papers in the Royal Society, warning one author that although his piece 'certainly deserves ye light', that 'many Gentlemen of ye Society object against printing much of Physick or Chirurgery in them, as not being so properly ye business of ye Royal Society'. In this case, Jurin thought it might be best to insert an extract and that 'ye Historical part' – that is, the part based on empirical observations – would be least liable to objections. He went on to lament the lack of a dedicated medical journal in early eighteenth-century London:

> It were indeed much to be wish'd, that we had something printed under ye title of Acta Medica or some such like, where this sort of Papers might properly come in & be constantly publish'd, for I am under a necessity of suppressing many valuable Papers of this kind for ye reason above mention'd.[44]

This affords further evidence for the Royal Society's influence over the editing of the *Transactions*, and reinforces the notion that senior figures in the Society felt a strong degree of proprietary interest in the *Transactions* even quite early in the eighteenth century, and believed that it should reflect the Society's purpose and interests; and more, that disciplinary boundaries should be maintained around the *Transactions* even in the absence of an alternative periodical for medical research.

As a doctor himself, Jurin did not share the prejudice against medical research, but when he gave it space in the *Transactions* he maintained a professed commitment to reports of observed phenomena, not hypotheses. He was able to use the periodical to further two of his own data-gathering projects, one on the spread of smallpox and the efficacy of inoculation as a treatment, and the other to gather standardised meteorological data from locations across Britain, Europe and North America. The results of both can be seen in the balance of subject areas that appeared in Jurin's *Transactions*. Jurin's appeal to medical men for local statistical information about the spread and treatment of smallpox in 1722–3 resulted in medicine taking up more page space than any other major disciplinary grouping for the first time in the history of the *Transactions*. And in 1726–7, there was a significant spike in meteorological papers in the *Transactions* in response to Jurin's requests two years earlier for standardised meteorological data: he distributed standardised tables and instruments to his correspondents, and received data from as far afield as St Petersburg (see Figure 3.1). Generally, the proportion of material from the physical and mathematical sciences remained significant through Jurin's editorship (20–40 per cent), reflecting Jurin's understanding of Newton's wishes. Natural history and medicine together accounted for around a third of Jurin's output, but natural history rose as medicine declined (to barely 4 per cent by the end of Jurin's tenure).

As editor, Jurin's correspondence is full of the minutiae of network-building, proof-correction and careful and timely guidance of authors through the process of publishing in the *Transactions*. He worked hard to recruit new contributors to the *Transactions* (some of whom became new members for the Royal Society), as well as maintaining existing relationships. Of all the early eighteenth-century editors, Jurin included the greatest proportion of papers from outside the Society's networks (that is, by correspondents who never became fellows). This was in part a consequence of his own data-gathering projects, which were addressed to a broad potential audience, including overseas correspondents as well

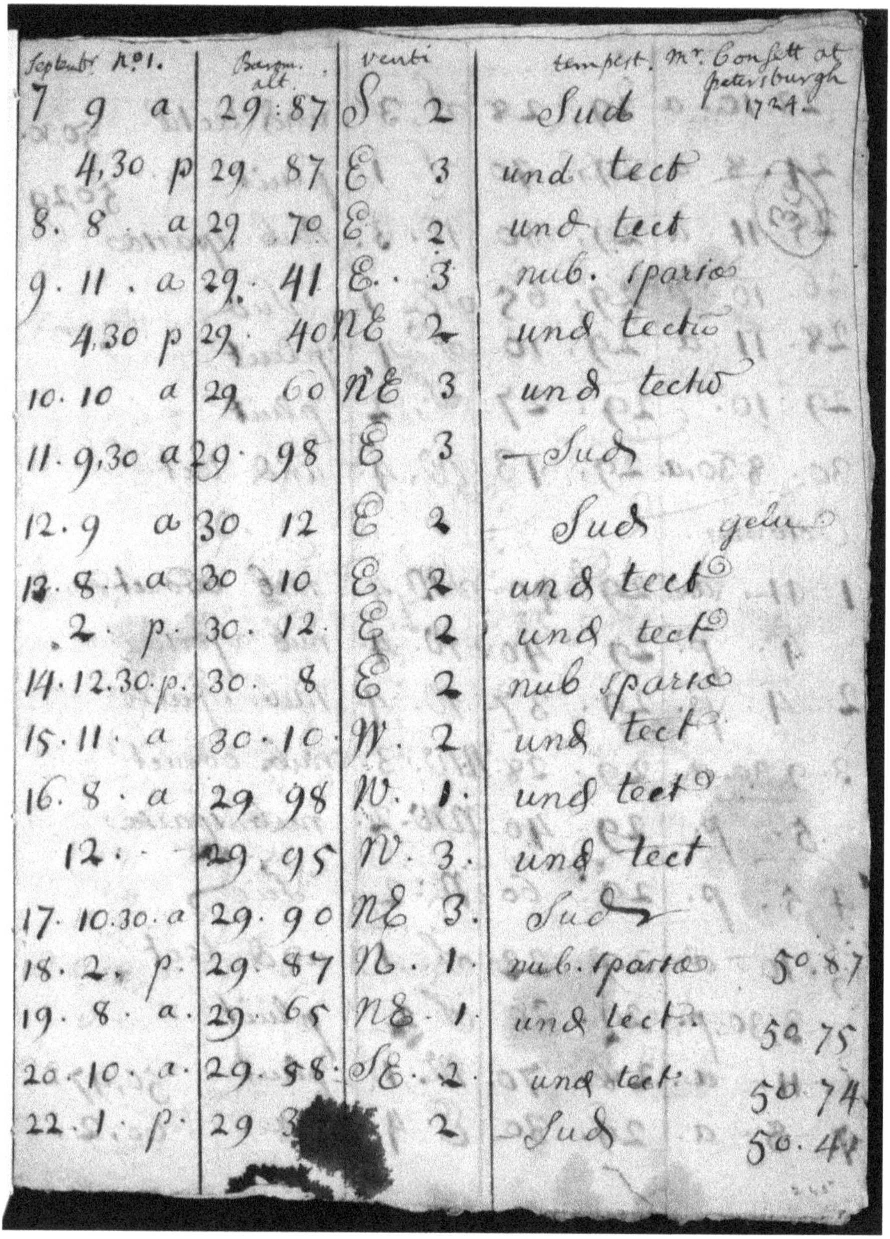

Septemb. No 1.	Barom. alt.	venti	tempest.	Mr. Consett at Petersburgh 1724.
7 9 a	29:87	S 2	Sud	
4,30 p	29 87	E 3	und tect	
8. 8 a	29 70	E 2	und tect	
9. 11 a	29 41	E 3	nub. sparso	
4,30 p	29 40	NE 2	und tecto	
10. 10 a	29 60	NE 3	und tecto	
11. 9,30 a	29 98	E 3	Sud	
12. 9 a	30 12	E 2	Sud gelu	
13. 8 a	30 10	E 2	und tect	
2 p	30 12	E 2	und tect	
14. 12.30 p	30 8	E 2	nub sparso	
15. 11 a	30 10	W 2	und tect	
16. 8 a	29 98	W 1	und tect	
12	29 95	W 3	und tect	
17. 10.30 a	29 90	NE 3	Sud	
18. 2 p	29 87	N 1	nub. sparso	50.87
19. 8 a	29 65	NE 1	und tect	50.75
20. 10 a	29 58	SE 2	und tect	50.74
22. 1 p	29 [illegible]	2	Sud	50.41

Figure 3.1 Meteorological observations from Petersburgh, sent to James Jurin by Thomas Consett, 1724 © The Royal Society.

as those in Britain. Indeed, Jurin's orientation of the *Transactions* towards Continental Europe can be seen in the consistently high level of foreign contributions (hovering around 40 per cent; higher than Sloane's) and the highest amount of Latin content since Oldenburg.

It is clear that Jurin was an outward-looking editor, and had substantial independence that enabled him to use the periodical for his own empirical projects. He occasionally asked other fellows for their opinions or for assistance in correcting proofs in specialist areas, but this should not be seen as a process of peer evaluation prior to publication. His co-secretary at the Royal Society was the astronomer and mathematician John Machin, and Jurin relied heavily on Machin for the selection and press-correction of mathematical papers. Machin's involvement helped Jurin maintain a recognisably Newtonian emphasis in the physical and mathematical contents of the periodical. In contrast, when Jurin forwarded some of Deidier's long treatises to Hans Sloane in 1723, it was because Deidier – not Jurin – wished to learn Sloane's opinion. Jurin's letter to Sloane also indicates that they were on cordial terms, despite the Newtonian politics within the Society: indeed, Jurin was at the time treating Sloane's wife in Tunbridge Wells, and the letter also updated him on her condition.[45]

Indeed, it is one of the striking things about the early eighteenth-century *Transactions* that so many men maintained such very long connections with it, despite the tensions arising from Newton's efforts to shape the Royal Society's activities. One might imagine from the bare bones of the story – which includes Sloane taking over from Halley in the mid-1690s; Sloane being forced out as editor-secretary in 1713 in favour of Halley; and Jurin resigning after not supporting Sloane in the presidential election of 1727 – that these men would not be on speaking terms. Yet Sloane channelled at least 17 papers to the *Transactions* during Halley's and Jurin's tenures (and would become more closely involved again after becoming president). Halley contributed or communicated at least two dozen papers to the *Transactions* when he was not involved as editor or compiler. Jurin continued to communicate papers after 1727, and his meteorological data-gathering network continued to send results that found their way into the *Transactions* deep into the 1730s.

And it was not just former-editors who remained involved behind the scenes of *Transactions*. Figures such as John Machin and John Theophilus Desaguliers were never in control of the *Transactions*, yet maintained 30-year associations with it, helping to vet and comment on, respectively, mathematical and physical papers (and, in Desaguliers' case, also contributing a vast number of papers). Martin Folkes, who had contested the presidency acrimoniously with Sloane in 1727 and would succeed him in 1741, wrote or communicated 10 papers within our sample, and would communicate many more after becoming president. In other words, during the early eighteenth century, there emerged a pool of active, long-lived communicators, contributors and assistants who supported the *Transactions* for decades, and maintained lasting connections with it regardless of who was then its editor. We suggest that it was this, more than any individual strategic effort, which helped bring about the *Transactions*' long-term stability.

The *Transactions* undoubtedly changed in form and content to reflect the personal interests, allegiances and agendas of Sloane, Halley and Jurin. Yet at the same time, the *Transactions* became more formally and generically stable. By the 1730s, papers had become longer and fewer; anonymity had declined almost to vanishing; the reliance on the fellowship as a pool of contributors had generally increased; and the link to the Society's activity was increasingly acknowledged. The shared commitment to the *Transactions* from a group of core figures in the Royal Society helped to make true the widely held assumption that the *Transactions* was an organ of the Society. These processes were gradual and cannot be easily attributed to the acts or intentions of any single individual, but they would be consolidated under the presidency of Hans Sloane.

Institutionalising the *Transactions*

After the death of Isaac Newton in February 1727, both Hans Sloane and the mathematically trained antiquarian Martin Folkes sought election as president. James Jurin supported Folkes, who had been a vice-president under Newton; but it was Sloane who won, though Folkes would succeed him in 1741. Jurin's co-secretary, John Machin, managed to retain his role under Sloane (and under Folkes), but Jurin's explicit support for Folkes in 1727 meant he was more or less compelled to resign, despite having been an efficient and active editor.

Just as Newton had, Sloane wished to have a trusted man in the role of secretary-editor. Sloane's initial choice was William Rutty, a Cambridge-trained physician, who had been a fellow since 1720 and had recently dedicated his *Treatise of the Urinary Passages* (1726) to Sloane. The editorial handover was not graceful. Sloane allowed a letter critical of Jurin to be read to a meeting of the fellows. Jurin immediately defended himself, and sought an apology, which was, in turn, read out at a meeting.[46] But then Jurin, in his last act as editor, dedicated his final volume of the *Transactions* to Folkes, quoting Newton to the effect that

'Natural history might furnish materials for natural philosophy; however, natural history was not natural philosophy'.[47] The reflection on Sloane was, as has been noted by others, unmistakable.[48] However, William Rutty did not prove a successful editor: he struggled to maintain timely publication of the *Transactions* and then, despite being only in his early 40s, died suddenly in June 1730.

Sloane's second choice was a man with whom he had a close professional relationship, and who lived nearby. This would enable Sloane to exercise substantial influence over the *Transactions* in the 1730s. Cromwell Mortimer was also a physician, and his recent election to the fellowship had been supported by Sloane. In 1729, he had moved to live near Sloane in Bloomsbury Square, where he assisted him in prescribing for his patients. In 1730, after barely two years in the Society, Mortimer became its secretary and the editor of the *Transactions*, roles he would retain for the next 20 years. His climb to office was notably quick and speaks to the power of presidential influence.

As neighbours and as colleagues, Sloane and Mortimer had ample opportunity to discuss the selection of material for the Society's meetings and for the *Transactions*. Although their face-to-face discussions have not left much in the way of archival traces, we know that subsequent close relationships between presidents and secretaries (most notably, Joseph Banks and Charles Blagden, in the 1780s and 1790s; see Chapter 6) could give presidents a great deal of behind-the-scenes influence on the *Transactions*. It is also not impossible that the very wealthy Sloane once more helped to sustain the *Transactions* financially in this period.

Sloane certainly remained an influential conduit for papers to the *Transactions*. Our sample suggests that just over a tenth of material in the *Transactions* in the 1730s and 1740s had been explicitly written by, communicated by, or addressed to Rutty or Mortimer, whereas almost a sixth was written by or channelled through Sloane. Given how few papers Newton had communicated to the *Transactions* during his presidency, Sloane's record suggests both the reach and activity of his networks (which were certainly far more extensive than those of the younger Rutty and Mortimer), but also Sloane's ongoing interest in, and commitment to, the continuation of the *Transactions*.

The early years of the Sloane presidency were characterised by efforts to rationalise and organise the Society's affairs. Sloane sought to clear dead wood from the membership, imposed stricter admission criteria, concentrated power in the president and Council, and sought to create a more robust and flexible financial basis for the Society. These reforms, coupled with the clear personal interest that Sloane took in it, helped create the conditions that made the *Transactions* a much more distinctively institutional product than it had been at the beginning of the century.

Sloane gave Council members more control over who could be proposed and elected as new fellows. Intended as a way to prevent the size of the Society growing beyond the capacity of its meeting rooms, it had the lasting effect of concentrating institutional power in the hands of the governing Council (rather than the fellowship at large). This may have been Sloane's response to the bruising election experience.[49] Sloane also tasked the secretaries, Machin and Rutty, to remedy deficiencies in the keeping of the Society's duplicate minute books and the registers of experiments performed and papers read. Sloane appears to have felt that things had been poorly organised since he himself resigned the secretaryship in 1713.

The Society's finances were also in a state of confusion. This was not helped by the death of the treasurer, which revealed that, just as in the 1670s (Chapter 2), the Society could face challenges reclaiming its property from deceased officers. Steps were taken to ensure that the Society's assets, particularly its investment certificates (amounting to £900 of bonds in the South Sea Company), would in future remain in the Society's custody, secure in an iron chest.[50] Two months later, Sloane instigated serious efforts to chase members who had defaulted on their promised membership fees. Lists of defaulters were prepared, and the fellows who had sponsored their membership were deputed to approach them to settle accounts.[51] These efforts proved broadly successful: the Society's income during the presidencies of Sloane and Folkes would average £980 per annum, compared with around £530 per annum at the start of the century. In addition, its average end-of-year balance increased from around £180 to £244. This meant that the Society could spend more each year, and also build up its reserves. The new treasurer was given discretionary authority to invest the Society's surplus cash, as and when it was available. (He was initially limited to South Sea bonds, but by the 1740s, the Society would also be investing in the East India Company.[52]) This was the basis of an investment portfolio which grew slowly over the rest of the eighteenth century, and more rapidly in the nineteenth; but it was also an indispensable condition for any potential institutional takeover of the *Transactions*. Back in the 1690s, when Sloane first became involved with the *Transactions*, the Society's finances had been so shaky that taking on the financial risk of publishing it would have been unimaginable.

There is no evidence from the 1730s and 1740s to suggest any plan or intention to make the *Transactions* an official Society publication, but the conditions were being created that might finally make it thinkable. Sloane's reforms were important steps on this road. So too was the renewed dependence upon the fellowship for contributions to the *Transactions*. Their contributions had filled over 70 per cent of the pages in Sloane and Halley's editorships, but this had declined somewhat under Jurin, due to his data-gathering networks. Under Rutty and Mortimer, the fellowship became again the dominant source of material, supplying almost 80 per cent by the late 1730s. Finally, the development of new administrative procedures formally integrated the *Transactions* into the Society's mechanisms for organising and preserving papers communicated to it.

William Rutty and Cromwell Mortimer left only modest editorial correspondences, but their surviving letters do give us some insight into processes of European distribution and exchange, and the ongoing importance of reciprocity and propriety in natural-philosophical print.[53] They also offer some evidence for fellows' involvement in the decisions about what to publish in the *Transactions*, and early proposals for an editorial committee.

For instance, Rutty continued Jurin's correspondence with Sir Thomas Dereham, an English traveller and Catholic gentleman resident in Italy, who served as fixer to the Stuart court in exile. Dereham funnelled correspondence, natural-philosophical communications, books and periodicals from Italy to the Royal Society via the Grand Duke of Tuscany's London secretary. In August 1728, Rutty wrote to thank Dereham:

> I must now inform you that out of regard to these & other communications transmitted by yr means to the Socicty I have this day sent you by their order the following small present of philosophical Books lately printed here sch. Our learned President has undertaken to convey to you Viz. (1.) Tables of ye antient Coins (2) Woodward's of Fossils (3) Newton's System of ye World (4) His Optical Lectures (5) Leadbeater's System of Astronomy in 2 Vol. (6) Philosophical transactions No. 400. 401. 402. all wch. they desire yr. kind acceptance of.[54]

This letter demonstrates that Rutty was not yet dreadfully far behind with the publication of the *Transactions*, since issue 402, nominally covering April to June 1728, was clearly in print by 20 August at the latest. It also reveals the way in which publications, including the *Transactions*, were used as tokens in a system of gift exchange. Dereham had transmitted to the Society new observations from the astronomer Francesco Bianchini and the physician Nicola Cirillo, and in return, he was sent books and periodicals. Yet this was more than just thanks to an individual, for these gifts would enable Dereham to share the Society's natural-philosophical activities with his Italian network of scholars.

Rutty had little opportunity to develop his editorial practices before his death, but his letters reveal that, as Jurin had done, he sometimes sought the opinion of other fellows on papers in fields outside his competence as a physician – particularly in mathematics and astronomy – and he sometimes took account of comments made by fellows at meetings where papers from correspondents were read out. For instance, in 1729, two papers by Marten Triewald, a Swedish-born engineer who had spent some time working in Northumbrian coal-fields, were read at Society meetings. One was about the cohesion of lead balls, the other about the behaviour of gold leaf when subjected to static electricity. The first was published, but Desaguliers made lengthy objections to the second. Rutty summarised for Triewald the 'difficulties wch still remain wth us wth regard to yr ingenious hypothesis', but suggested that Triewald would be able 'to satisfy these doubts'. The passage reads like an invitation to revise the paper, which, added Rutty, would then 'be very agreeable to the whole society'.[55]

In this letter, Rutty carefully attributed the criticisms to the Society, and not to Desaguliers personally. Given the Society's early reputation for avoiding issuing corporate judgements wherever possible, the deliberate assumption of corporate responsibility for the sceptical appraisal of Triewald's paper is striking. Keeping confidential the identity of the specific individual making the comments created a mechanism by which one fellow could critique another fellow with propriety, something that, as we have seen, was a long-running challenge for the Royal Society and the *Transactions*. The use of confidentiality anticipates what would become the Society's standard way of generating judgements on papers read before it.

We have fewer of Mortimer's letters to guide our understanding of his editorial practice, even though he held the post of editor for 20 years altogether – the longest tenure during the period of private editorship of the *Transactions*. We do, however, find occasional evidence that Mortimer continued the practice of seeking occasional assistance from fellows, or taking the views of the Society into account, when it came to judging particular papers. In 1741, for instance, he sought Desaguliers' opinion upon two papers – one entitled 'a Physico-mathematical Description of the Impossibility of Vortices', and the other, a 'Contrivance for a Coach, Chariot or Chaise to go over uneven ground without overturning'. Mechanical inventions were rarely published in the *Transactions*, and this one was no exception. On the vortices treatise, Desaguliers wrote a brief description, and reported that it was unoriginal but perhaps more readily comprehensible to a general readership than Newton's work on the same subject, and therefore deserved to be published – as it duly was, appearing in volume 41.[56] The following year, Mortimer asked John Machin (who still held the other secretaryship) to revise the proofs of a paper on pendulums by the Padua mathematics professor Giovanni Poleni. Machin complied, though he pretended to believe that the paper had been written 'for no end as I can see but to be a plague and a torture to me'. He did ask Mortimer to 'keep this Letter that it may be produced in my justification hereafter if needful; and to shew that though I revised the sheet I did not however subscribe to the opinion of the Author'.[57] The seeking of second opinions continued to be occasional and unsystematic; and the decision to publish a paper did not yet carry the social or epistemological weight that the development of more formal peer-review processes would later give it (see Chapter 8).

Despite Mortimer's willingness to publish a paper that Desaguliers believed to be unoriginal, originality was generally coming to be more prized in the *Transactions*, and increasingly had to be acknowledged and accounted for. Thus, when publishing Roger Pickering's work on the 'seeds' (spores) of mushrooms, Mortimer followed Sloane's practice of adding occasional editorial notes, in this case explaining that he 'thought proper to print' the work, even though it was not strictly original, because the work of the Italian botanist and mycologist Pier Antonio Micheli was little known in England. He suggested that, since Micheli's book was 'printed at Florence, [it] is not in many people's hands here, & as that is in Latin I thought it would not be disagreeable to our Gardiners to have an account of this Discovery in English'. Mortimer also suggested that Pickering's paper was a legitimate case of independent discovery, and printing it was 'but doing justice to Mr. Pickering's Diligence in searching into the works of Nature, since he was so fortunate as to succeed in a discovery wch had eluded many curious Botanists, & that without having taken any hint from Micheli'.[58] The idea that the printing of Pickering's paper was specifically for the benefit of English 'Gardiners', who Mortimer presumed read no Latin, points intriguingly to the hope that the *Transactions* might find an audience among skilled artisans and tradesman and not just among the traditionally learned. Latin papers continued to appear in the *Transactions*, however, although it had become normal practice to translate papers from French and Latin into English before reading them at a meeting of the Society. This suggests a shift in even a learned audience's capacity to follow Latin when spoken aloud.

Mortimer also used editorial notes to direct readers to papers already published in the *Transactions* on a similar subject and, occasionally, to disagree with an author. For instance, in 1744, he printed a description (by surgeon George Aylett) of a spine malformation that had caused the death of a five-day-old baby girl. He added a note pointing out that, 'Dr Rutty late Secr. R. S. has communicated a case like this see these Transact. n. 366 p. 98'. But he also wrote a legend to accompany the engraved image of the lower spine, Aylett having apparently not supplied one. One of these image labels ran to a full paragraph, in which Mortimer – himself a physician – argued that Aylett's description of the case was wrong. Mortimer believed that the illustration showed an opening 'quite into the Canal of the Vertebrae', which had 'been mistaken for a Parting of the spinal Processes into Two Rows ... and so have given rise to the Notion of a Spina bifida; which Case I doubt whether it ever exists'. He then went on to define 'a perfect Spina bifida' and signed off 'C.M.'.[59]

The surviving manuscripts of papers submitted, and their proof corrections, help us examine Mortimer's editing practices. They typically carry various annotations – titles and numbers added by Mortimer, printers' brackets and signatures to indicate page breaks in composition, and various small changes and deletions (Figure 3.2 and Figure 3.3). Mortimer often struck out salutations and valedictions from letters, as well as excesses in his authors' prose – including jokes and elaborate self-deprecation.[60] In at least one other case he deleted a large section of a book review, which happened to quote a passage in which the French author took a lightly disrespectful tone with Newton.[61] Even 15 years after Newton's death, the editors of the *Transactions* would intervene to protect his reputation.

Mortimer was also involved in a very important structural change to the *Transactions*. Despite Sloane's efforts early in his presidency, the state of the Society's record-keeping and archiving practices remained a matter of concern. In July 1742, early in Folkes's presidency, Mortimer was part of a committee that recommended stringent new procedures and the formation of an annual 'Committee of Papers'. It would meet each autumn to consider a list of the titles of all papers 'read or shew'd to the Society', organised in 'the respective order' by date.[62] The committee's task would be to decide what, if anything, needed to be done to preserve the observations and findings presented in those papers. If a paper had already been printed – whether in the *Transactions* or elsewhere – no further action was needed, although a note should be made of the details of publication. But if it had not been printed, the committee needed to decide whether the abstract written by the secretary for the Society's minute books was an adequate record, or whether the full paper should

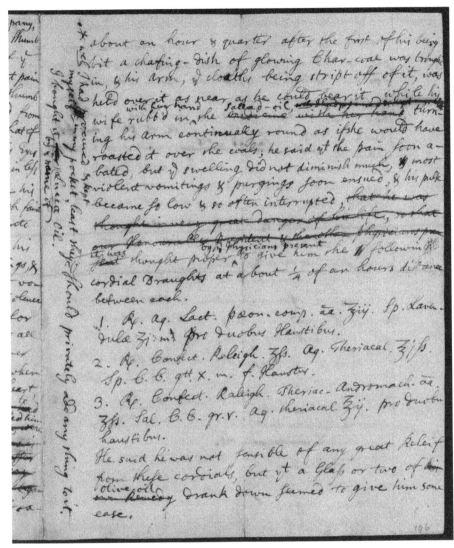

Figure 3.2 Editorial interventions (including additions, deletions and annotations to the printer) on Cromwell Mortimer's own 1734 paper 'Experiments to show the effects of viper bites on a man' © The Royal Society.

be copied into the Society's register books for future consultation. These recommendations formally recognised the *Transactions* as part of the Society's record-keeping for the first time. They also proposed a committee that would judge the worth of papers; something that would, in time, be applied to the periodical as well as the archive.

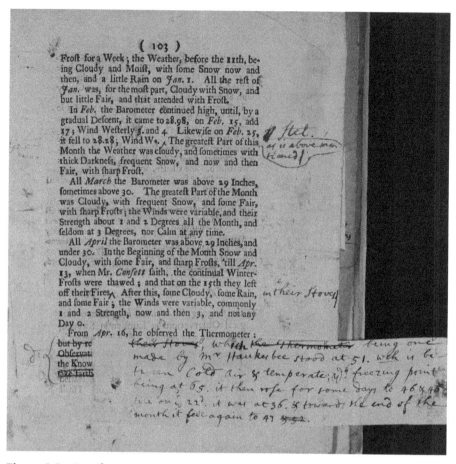

Figure 3.3 Proof corrections to William Derham's 1733 'Abstract of meteorological diaries . . .', which included the observations sent to Jurin by Consett 10 years earlier. © The Royal Society.

In the end, the system was adopted in a modified form. The proposed hierarchy of abstracts, registered copies and published versions was discarded; instead all papers had their abstracts minuted, and all originals were kept (in the order in which they were read, and annotated if published) in a new series of guard-books known as 'Letters and Papers'. This created a single series that functioned as the repository of the Society's scientific activity, to which the minute books would then serve as a guide.

After 1742, issues of the *Transactions* became much longer, typically containing 20 or more papers, and they appeared much less frequently.

The *Transactions* was still formally the domain of an independent editor, albeit an editor who drew upon input from the Society's president, senior fellows and acknowledged experts, but it now seemed as though each issue emerged from a kind of periodic stock-taking of Society activity, as had been envisaged in the proposed reforms. Also from 1742, the papers printed in the *Transactions* routinely reported not only the name of the author, but also the date on which they had first been read to a meeting of the Royal Society, and, if applicable, the name of the fellow who had communicated the paper for an author who was not a fellow. These practices tacitly acknowledged the expectation that everything now published in the periodical had first been read before the Society, thus definitively and publicly linking the Society's meetings and the periodical. Submitting a paper to the Royal Society had become the sole route to publication in the *Transactions* rather than, as had been the case in earlier decades, one of several possible routes.

The reforms of 1742 were, in one view, a set of tweaks to record-keeping practices rather than a radical overhaul. In another they were rather more significant. They effaced any remaining distinction between the research produced at the Society and what was merely communicated to it, and removed the epistemic hierarchy implied by that distinction. They brought the *Transactions* into the Society's archiving practices and turned it into the version of record, replacing the manuscript registers and letter books of previous administrations. The *Transactions* became, de facto, the record of the Society's best work, with the Society's work now defined almost exclusively as what was communicated to it. The reforms thus acknowledged the *Transactions*' status as a Royal Society product, 10 years before any move to take it over officially.

By the 1740s, then, the *Transactions* had become a more decisively institutional product. With its more leisurely rhythm of publication, and a greater reliance on the work and credibility of individual authors (increasingly fellows of the Society), it had come to more closely resemble the memoirs and transactions of other learned institutions that proliferated in the German lands, in Sweden, in France and in Russia during the first half of the eighteenth century. In contrast to those academies that received direct funding and/or in-kind support from their royal patrons, the Royal Society's status as a voluntary society of fee-paying members meant that its finances were far shakier. Thus, the improvement in the Society's financial situation was another important element of the story of the *Transactions*; without financial security, it would have been impossible to assume responsibility for the *Transactions* without the associated risk of wrecking the Society.

The harmonisation of the functions of the *Transactions* and the Society emerged gradually, from a ferment of contrasting editorial styles and priorities. When Hans Sloane had reluctantly assumed control of the periodical in the autumn of 1696, he had initially been very unsure of how best to run it, allowing himself to rely heavily on short, discrete, miscellaneous communications, on scraps of older material newly brought to light, and on the parcelling out of catalogues and descriptions of collections of specimens sent to or acquired by himself and his associates. He used the periodical to address and supply an English audience and was, he proclaimed, unwilling to advance his own epistemic judgements either as an author or editorially. Indeed, his early vision of editorship was essentially one of non-intervention, allowing the periodical to function primarily as a repository of empirical observations. This struck a number of people, including his two immediate successors, as inadequate. Halley and Jurin were both concerned to assert the primacy of natural philosophy over natural history, and Halley expressed dissatisfaction with Sloane's *laissez-faire* editorial policy.

By the time Mortimer became editor, during Sloane's presidency, the meaning and practice of editing the *Transactions* had stabilised. Halley, Jurin and Mortimer were all far more interventionist editors than Sloane had been, feeling able to shape the contributions that they printed. Book notices had largely disappeared, and contributions took broadly similar forms. Whereas Sloane had sought copy for his *Transactions* from a variety of sources, Mortimer drew entirely upon material communicated to the Royal Society. By the 1740s, the *Transactions* had become more closely associated with the Society and its processes than ever before.

Despite obvious areas of friction between editor-secretaries and other senior fellows within the Society, and despite differences in disciplinary interests from regime to regime, one of the most telling features of the history of the *Transactions* in the first half of the eighteenth century was the emergence of long-term commitments to it by a group of senior fellows of the Royal Society. It is not just that individual editors served longer tenures than their late seventeenth-century predecessors, or that past editors remained involved for decades after, though that is true. It is also the evidence of other fellows being committed contributors and communicators over periods of decades. John Machin served intermittently as a proof-reader and evaluator of mathematical papers throughout the three decades of his secretaryship. Francis Hauksbee and John Theophilus Desaguliers had begun sending their contributions on physical and mathematical topics during Newton's presidency; but they continued doing so for a decade and a half after his death. Halley maintained an active connection with the *Transactions* over a period of nearly 50 years, though his role as a compiler of astronomical observations from disparate sources seems to have been taken over in later years by the watchmaker George Graham and later by John Bevis. Some of Jurin's data-gatherers, including William Derham in Essex and Giovanni Poleni in Italy, continued to send material to the *Transactions* a full decade after he had left office.

The combined effect of the long-term commitment of these men suggests their recognition of the value of the *Transactions* to them individually, and also collectively. This helped to gradually consolidate the Society's sense of what the *Transactions* was, and what role it served in the Society. By the early 1740s, any notion that the periodical was an ephemeral format, useful for collecting snippets of second-hand news but unsuitable for publishing substantial new knowledge-claims, had gone. At the same time, the decline (again) of an active experimental programme meant that meetings of the Royal Society had come to focus on observations and experiments reported from elsewhere, and the overlap between material presented at meetings and printed in the *Transactions* became closer than ever.

The editor was still, in principle, acting independently, rather than on behalf of – or on the orders of – the Royal Society, but there was undoubtedly a close association between the Society and the *Transactions* in both public perception and in practice. The idea that the editor's independence was largely a fiction, and moreover undesirable for the Society, would become the principal bone of contention in the early 1750s. Before we continue the story of how the Society came to assume control of the *Transactions*, however, we should reflect on the role the *Transactions* came to play in the wider world – in London, Britain and beyond. This, too, was different than it had been in the late seventeenth century.

Notes

1 The *Gazette* began publication in Oxford in the summer of 1665; printing transferred to London when the Court returned there after the plague outbreak of that year receded.

2 The volume for 1699 did not actually appear until 1701, a delay that quickly became routine. For a discussion of the Académie's eighteenth-century publishing practices, see Anne-Sylvie Guénon, 'Les publications de l'Académie des Sciences', in *Histoire et mémoire de l'Académie des Sciences: guide de recherches*, ed. Académie des Sciences, Éric Brian, and Christiane Demeulenaere-Douyère (Paris: Tec & DOC Lavoisier, 1996), 113–28; and James McClellan III, 'Specialist control: The Publications Committee of the Académie Royale des Sciences (Paris), 1700–1793', *Transactions of the American Philosophical Society* 93, no. 3 (2003): 1–134.

3 For the tensions between Sloane and Newton, see Mordechai Feingold, 'Mathematicians and naturalists: Sir Isaac Newton and the Royal Society', in *Isaac Newton's Natural Philosophy*, ed. Jed Z. Buchwald and I. Bernard Cohen (Cambridge, MA: MIT Press, 2001), 77–104.

4 On the sources of Sloane's wealth, see James Delbourgo, *Collecting the World: Hans Sloane and the origins of the British Museum* (London: Allen Lane, 2017), 151–2, 185–90, 192–4.

5 See BL MS Sloane 4036 ff. 160 (31 January 1692/3), 238 (23 June 1696), 260 (7 September 1696), for instance. In fact, Ray lived on until 1705.

6 BL MS Sloane 4036 f. 298, 2 April 1697.

7 BL MS Sloane 4037 f. 52, 12 April 1698. That paper was never published, but a previous one – on a 'very large eel' – was published in *PT* 20.238 (1698): 90–7. The original is BL MS Sloane 4037 f. 37, 9 March 1698.

8 Smith published Dales's *Pharmacologia* (1693) and various works by Ray, including revised editions of *The Wisdom of God Manifested in the Works of Creation* (1692, 1701).

9 According to Johann Jacob Scheuchzer, writing to an unknown aristocratic correspondent, 1727, in BL Sloane MS 4026 ff. 270–1.

10 See RS AB/1/1/3, 1693 ff., concerning Richard Waller's secretary-editorship.

11 Delbourgo, *Collecting the World*, 160.

12 The following discussion is based upon our analysis of samples from Sloane's editorship: the first three volumes (from 1695 to 1699 inclusive) and every second volume thereafter. For more, see A. Fyfe (and N. Moxham) 'The *Transactions* in the early eighteenth century' (7 February 2022), https://arts.st-andrews.ac.uk/philosophicaltransactions/the-transactions-in-the-early-eighteenth-century.

13 For this we have the circumstantial evidence of Leibniz taking a long time to realise he had been attacked in the *Transactions*, and the admittedly hostile testimony of John Woodward, who alluded to a two-year backlog in 1711. See Chapter 4 for a fuller discussion.

14 The papers were communicated by William Musgrave, who was by this time practising medicine in Exeter. Why he chose this moment to start communicating older material from the Oxford Society is unclear, but there may be a connection to the recent death (April 1696) of Robert Plot, the Oxford Society's long-time secretary.

15 'An account of a book, entitled, Gazophylacii Naturae & Artis …', *PT* 27.331 (1710): 342–52.

16 The mid-1690s' figure would have been slightly higher if Halley's papers during the time when he was clerk (and thus, not a fellow) were included.

17 On Hauksbee, see Stephen Pumfrey, 'Hauksbee, Francis (bap. 1660, d. 1713)', in *ODNB*. On Newton's role in bringing him to the Society, see John L. Heilbron, *Physics at the Royal Society during Newton's Presidency* (Los Angeles: University of California, 1983). On the resurgence of experimentation at the Society under Newton, see also Marie Boas Hall, *Promoting Experimental Learning: Experiment and the Royal Society, 1660–1727* (Cambridge: Cambridge University Press, 1993), 116–23 and 136–7.

18 Francis Hauksbee the Elder, 'An experiment, to show the cause of the descent of the mercury …', *PT* 24.292 (1705): 1629–30.

19 In terms of space, Leeuwenhoek's 41 long papers represent 11 per cent of everything published in the *Transactions* during Sloane's tenure. Their engraved illustrations also represent a significant financial investment.

20 On Kamel, see Sebastian Kroupa, '*Ex epistulis Philippinensibus*: Georg Joseph Kamel SJ (1661–1706) and his correspondence network', *Centaurus* 57, no. 4 (2015): 229–59.

21 Sietske Fransen, 'Latin in a time of change: The choice of language as signifier of a new science?', *Isis* 108, no. 3 (2017): 629–35; Sietske Fransen, Niall Hodson, and Karl A.E. Enenkel, eds, *Translating Early Modern Science* (Leiden: Brill, 2017).

22 On average, 16 per cent of Sloane's material in our sample was in Latin, compared to 34 per cent for Halley and 34 per cent for Jurin.

23 See, for example, Steven Shapin, *A Social History of Truth* (Chicago, IL: University of Chicago Press, 1994), ch. 6, esp. 266–91. Contrast with Robert J. Griffin, 'Anonymity and authorship', *New Literary History* 30 (1999): 877–95.

24 Hans Sloane, 'The Preface', *PT* 21 (1699): [i–ii].

25 Ann Savile and Tancred Robinson, 'An account of one Henry Jenkins …', *PT* 19.221 (1695): 265–8.

26 From 1695 to 1709, our sample contains 294 natural history/medicine articles, and 183 in physical sciences. See also Aileen Fyfe (and Noah Moxham) 'The *Transactions* in the early eighteenth century' (7 February 2022), https://arts.st-andrews.ac.uk/philosophicaltransactions/the-transactions-in-the-early-eighteenth-century.
27 Edmond Halley, 'The Preface', *PT* 29 (1714): 3–4.
28 The published selection of Halley's surviving correspondence, constituting most of what survives, amounts to fewer than 100 letters from a research career spanning seven decades. Eugene Fairfield MacPike, ed., *Correspondence and Papers of Edmond Halley* (Oxford: Oxford University Press, 1932).
29 Halley, 'The Preface', *PT* 29 (1714): 3–4.
30 James Delbourgo cites a friendly letter from Halley to Sloane as evidence of a social ease that ran counter to Sloane's general reputation; see Delbourgo, *Collecting the World*, 160.
31 Sloane, 'The Preface', *PT* 21 (1699): [i–ii]. Volume 21 nominally covered 1699, but it was completed in 1700.
32 Sloane, 'The Preface', *PT* 21 (1699): [ii].
33 Halley, 'The Preface', *PT* 29 (1714): 3–4.
34 Halley, 'The Preface', *PT* 29 (1714): 3–4.
35 See, for example, Richard Waller, 'The Preface', *PT* 17 (1693): 581–2; and [Francis Aston and Robert Plot], 'The Preface', *PT* 13 (1683): 2. Oldenburg, by contrast, tended rather to amplify the importance of the *Transactions* and its contribution to natural knowledge in his annual prefaces, especially from 1667 onwards.
36 Sietske Fransen, 'Antoni van Leeuwenhoek, his images and draughtsmen', *Perspectives on Science* 27, no. 3 (2019): 485–544.
37 Adrian Johns, 'Flamsteed's optics and the identity of the astronomical observer', in *Flamsteed's Stars: New perspectives on the life and work of the first Astronomer Royal*, ed. Frances Willmoth (Woodbridge: Boydell & Brewer, 1997): 77–106.
38 Andrea Rusnock, *The Correspondence of James Jurin* (Amsterdam: Rodopi, 1996).
39 Jurin, 'Some experiments ... of human blood', *PT* 30.361 (1719): 1000–14; for the exchanges between Jurin and Woodward, 13–15 February 1719, see Rusnock, *Jurin Correspondence*, 78–81.
40 Jurin to William Nicholson, 14 October 1714, in Rusnock, *Jurin Correspondence*, 69–74.
41 Jurin to Antoine Deidier, 30 December 1723, in Rusnock, *Jurin Correspondence*, 215–16. Deidier had been elected a fellow on 14 November that year.
42 Jurin to Deidier, 30 December 1723, in Rusnock, *Jurin Correspondence*, 215–16.
43 Jurin to Deidier, 3 April 1725, in Rusnock, *Jurin Correspondence*, 293.
44 Jurin to John Huxham, 3 August 1727, in Rusnock, *Jurin Correspondence*, 364.
45 Jurin to Hans Sloane, 12 August 1723, in Rusnock, *Jurin Correspondence*, 198.
46 Jurin to Dereham, 7 August 1727, in Rusnock, *Jurin Correspondence*, 365; Jurin to Dereham, 12 February 1728, in Rusnock, *Jurin Correspondence*, 371–2; Dereham to Jurin, 27 March 1728, in Rusnock, *Jurin Correspondence*, 373–4.
47 [Jurin], 'Dedication', *PT* 34.392 (1726–27): [ii].
48 Feingold, 'Mathematicians and naturalists'.
49 RS CMO/03, 9 January 1728/9, 21.
50 Treasurer Alexander Pitfeild had died in late October 1728. It was not until the following March that the new treasurer received the Society's cash and bonds from his widow. See RS CMO/03, 17 March 1728/9 and 8 May 1729.
51 RS CMO/03, 20 August 1729.
52 RS CMO/03, 8 May 1729, and 22 February 1741.
53 There are about 40 letters in RS EL/R/2 ff. 1–38.
54 RS EL/R/2 f. 14, 20 August 1728.
55 RS EL/R/2 f. 20, 25 March 1729. For Desaguliers' objections, see RS JBO/14, 28 February 1729, p. 305.
56 RS L&P/1/1 ff. 74 and 75. M. de Sigorgne, 'A physico-mathematical demonstration of the impossibility and insufficiency of vortices', *PT* 41.457 (1739–41): 409–35.
57 John Machin to Mortimer, 21 March 1742 OS, RS MM/20 f. 11. Published as: Johannis Marchionis Poleni, 'De novis quibusdam Cogitationibus ... num pendula ...', *PT* 42.468 (1742–3): 298–306.

58 Roger Pickering, 'A letter ... concerning the seeds of mushrooms', *PT* 42.471 (1742–3): 593–8 (Mortimer's note is at 598). The original paper is RS L&P/1/5 f. 224, dated 30 October 1743, read to RS on 10 November 1743.
59 George Aylett, 'An observation of a spina bifida, so called', *PT* 43.472 (1744–5): 10–11, at 11. Original paper at RS L&P/1/6 f. 256, read to RS 2 February 1743/4 OS.
60 See, for instance, paper by Robert Campbell of Kernan, RS L&P/1/3 f. 140.
61 RS L&P/1/3 f. 144.
62 'Proposal concerning the papers of the Royal Society', in RS CMO/03, 12 July 1742.

4
The *Transactions* and the wider world, c. 1700-1750

Noah Moxham and Aileen Fyfe

The years around 1700 were a time of significant change in the publishing environment in England. In 1695, the legislation that had controlled the publication of printed matter – the 1662 Licensing Act – was not renewed by parliament.[1] This removed the value of the Royal Society's licensing privilege; and more generally, it enabled an efflorescence of political and cultural comment in print, particularly in periodical format, that would be only partially calmed by the 1710 Act for the Encouragement of Learning (the UK's first copyright Act). The more open publishing environment of the early eighteenth century meant that both the *Transactions* itself, and the kinds of material that it carried, would become objects of discussion, scrutiny and critique in a broader periodical press. In short, there was an increasingly large number of forms that a learned periodical might take, and greater scope for competition and criticism.[2]

There was still nothing quite like the *Transactions* in print within Britain or Europe. This was, however, increasingly a matter of fine distinctions. The *Transactions* differed from the *Journal des Sçavans*, and its various Italian analogues, in having a more narrowly natural-philosophical remit and a close association with a learned society. The Leipzig *Acta Eruditorum* (established in 1682) and Pierre Bayle's *Nouvelles de la République des Lettres*, published in the Netherlands from 1684, were similarly wider ranging than the *Transactions*. A closer analogue might be the *Miscellanea Curiosa*, issued in the German lands from 1670 which was exclusively dedicated to science and medicine, mainly the latter, and was linked to a learned body, the Academia Naturae Curiosorum. The major difference was that the Academia had no corporeal manifestation similar to the Royal Society's meetings, and this significantly influenced the paths followed by the two periodicals. In London, there was a variety of (usually short-lived) periodicals dedicated to useful knowledge and the mechanical arts, and to reviews of literature.[3] And by the 1740s, there would be more learned periodicals in Britain, most notably Alexander Monro's *Medical Essays and Observations*, published in Edinburgh from 1731, and the occasional publications of the Edinburgh Philosophical Society.[4]

In the previous chapter, we showed that the *Transactions* was shaped by the vagaries of the Royal Society's internal politics, and by the whims of its individual independent editors and the presidents who chose them. But it was also part of a wider and increasingly diverse intellectual landscape of print in the early eighteenth century. This, too, shaped the way the *Transactions* was edited, managed and circulated. Thus, in this chapter, we consider four instances of the varied interactions between the *Transactions* and the wider world, ranging from London to Naples. In each case, we see a contrast between the way the *Transactions* was regarded within the Royal Society, and how it was perceived outside that community.

We start by examining the anonymous *Transactioneer* pamphlet of 1700. It was the criticisms in this pamphlet that forced Hans Sloane to defend his vision of editorship in a preface to the *Transactions*, as we saw in the previous chapter. Here, we consider what the pamphlet – and the Royal Society's reaction to it – tells us about the place of the *Transactions* in wider print culture. Next, we re-examine Isaac Newton's determination to choose his own secretary-editor in 1714 and 1720, which established a precedent for presidential influence over the *Transactions*. We suggest that Newton's approach to the editorship of the *Transactions* had much to do with his efforts to manage the wide-ranging and ferocious controversy over the discovery of calculus. The *Transactions* became one of the forms of printed ammunition utilised by Newton.

Our third section uses James Jurin's correspondence to examine how an active editor could use the *Transactions* as a tool to gather information from fellows and correspondents in Britain, Europe and across the Atlantic. The *Transactions* was embedded in networks of correspondence, and Jurin, in contrast to certain of his predecessors, was adept at managing his distant correspondents and gathering copy in a timely manner. The story also shows how correspondents understood the *Transactions* in relation to other modes of print, such as newspapers.

Our final section turns to the question of how the *Transactions* itself travelled over distance and through time. While readers in early eighteenth-century London could have encountered the latest issues of the *Transactions* as soon as they appeared, most readers further afield would have had a different experience. They might eventually encounter the bulky annual volumes that bound together a whole year's issues, but they were just as likely to encounter material from the *Transactions* excerpted, summarised or translated in the pages of other periodicals and books, possibly years after its original publication. The longevity of the *Transactions* had turned its back-run into a potentially important scholarly resource, but complete sets were rare, expensive and hard to acquire. Entrepreneurial printers and editors sought commercial success by reprinting, abridging and reorganising the *Transactions* to bring material from its back-run to new generations of scholarly readers.

Critique, satire and reputation in London

In the early months of 1700, an anonymous pamphlet titled the *Transactioneer* was published in London. A preface made clear that its target was the *Philosophical Transactions* and the man who edited it. Hans Sloane was not named, but the brutally personal attack on someone who had 'slipped into the post of Secretary', and was now author and editor of the *Transactions*, was impossible to mistake.[5] The *Transactioneer* was a short, satirical dialogue mocking the style, pretensions and preoccupations of editor and publication alike. It relied on extensive (and radically decontextualised) quotation to portray Sloane's work, and that of his correspondents, as relentlessly trivial, self-evident, ill-written, unoriginal, uncritical and ridden with solecisms.

The anonymous author denied any ill-intention towards the Royal Society, claiming rather to be animated by a concern for the Society's reputation. He noted that 'Learned Men abroad have ever had a very just esteem for the English Society', but claimed to worry that this esteem was 'now like to decline; they having no other way of judging of it but by the *Philosophical Transactions*'. He acknowledged the formal independence of the *Transactions* from the Society, it having been 'begun by Mr. *Oldenburg*, who all along declar'd the R. Society were not concern'd in those *Transactions*'. But despite there being 'no Ground for that Opinion', the *Transactioneer* was confident that, 'the World everywhere looks on [the *Transactions*] as a kind of Journal of the R. Society'.[6]

Unsurprisingly, Sloane perceived the anonymous critique as an attack on his management of the *Transactions*, and assumed it came from a fellow of the Society who was dissatisfied with his editorship. In fact, the attack came from outside the fellowship, and was intended to discredit the Royal Society's entire project. The *Transactions* was made to stand both for the Society's day-to-day activity and its wider purpose, and the editor was assumed to be responsible for all of it, including the absurdity of seeking to publish on so noble and important a subject as natural knowledge in a form so trivial and debased as a periodical. The episode, therefore, speaks both to the internal currents within the Royal Society, and to wider cultural reactions to the Society and its projects, of which the *Transactions* was assumed to be one.

Sloane had recently had significant disagreements over the origin of fossils with the physician and antiquary John Woodward, and with his friend and associate, the cleric, lecturer on science and author John Harris. Fresh from those disagreements, Sloane was apparently confident enough in his suspicions about the *Transactioneer* to 'insinuat[e] about Town' that Woodward and Harris were responsible.[7] These insinuations reached Harris, who wrote to the Society's vice-president vehemently denying responsibility and calling for either an apology from Sloane or for permission from the Society's Council to vindicate himself in print.

Despite characterising the *Transactioneer* as a 'Trifling Pamphlett', Harris agreed with part of its argument, thus confirming Sloane's sense that some in the Society were not happy with his editorship. Harris claimed that the *Transactions* had of late 'been very much censured and complain'd of, both by our own Countreymen and Forreigners', adding that 'I have often my self been forced to alledge in defence of ye. R.S. (for whom I have justly a very great Respect) yt. the Transactions are not Their's'. It was, however, 'impossible to convince men of this, while Papers Read before ye Society are published there, while Letters to Members of it are there inserted, & while Experiments made before them are there printed'. He joined the *Transactioneer* in blaming Sloane for the lapse in quality, and called upon the Council either to 'publickly disown' the *Transactions* or else subject them to the oversight of a committee before publication.[8]

Harris's letter was addressed to the Council, so within the context of the Society it represented a very public and no doubt embarrassing attack upon Sloane. The substance of Harris's and Woodward's grievances with Sloane lay precisely in the arbitrariness of his editorial authority, which then took on, in the eyes of the public, the weight of a collective verdict from the Society. For instance, Harris claimed that Sloane had suppressed a recent paper on fossils in defiance of the Council's recommendation 'because [the]re was something in it which confirmed Dr Woodward's notions' about their formation.[9]

The Council's response to the *Transactioneer* was to resolve upon the expulsion of any fellow found to have been involved in it; to declare a collective belief in Harris's and Woodward's protestations that they were not its authors; and (virtually in the same breath) to declare themselves satisfied with Sloane's conduct as secretary and his management of the *Transactions* in particular.[10] This whitewash preserved proprieties, but did little to mend the fault lines within the Society.

Sloane had been accused of using the *Transactions* to advance particular scientific ideas and to suppress rival theories. The fact that a group of fellows protested against the notion of complete editorial autonomy represents a crucial point in the development of the *Transactions*. Their protest indicates worries about the nature of independent editorship. By implication, the critics held some conception of a higher institutional authority that would be fairer and less arbitrary than the editor's whim.[11] There was, of course, nothing preventing the aggrieved parties from publishing their ideas and observations elsewhere, so their desire to be able to publish in the *Transactions* implies that these fellows believed that appearing in its pages did, and should, carry the suggestion of institutional approval. They (unsuccessfully) pressured Council to create a formal basis for that institutional approval, by removing the *Transactions* from the control of individual editor-secretaries.

Sloane thought he was dealing with an internal critique, albeit externalised through publication, and he sought to use internal institutional mechanisms to suppress it. Part of what is telling about this episode is precisely that he was wrong: *Transactioneer* was actually an attempt to enlist the Society in a much larger cultural debate.

The author of *Transactioneer* was, in fact, William King, a lawyer and literary hack, with no apparent interest in natural knowledge. *Transactioneer* was not his only attack on Sloane: in 1709, he would publish parodies of Sloane's *Voyage to Jamaica* as part of a series of three issues of mock-*Transactions*.[12] We know that these attracted Sloane's attention – a manuscript copy of the title page of the second survives in his papers – but they elicited no official response from the Society.[13] Shortly after King's third mock-*Transactions* was published, Jonathan Swift helped him get the job of editing the London *Gazette*.[14]

Swift's patronage links King (and the *Transactioneer*) to the fierce debates over the respective literary merits of classical and contemporary authors that Swift christened 'the Battle of the Books'. Politically, Swift and King were High Church Tories, firmly on the side of the Ancients against the Moderns. Swift would later compose his own satire on the Royal

Society: the Academy at Lagado, from the third book of *Gulliver's Travels* (1726), is widely understood as a caricature of the futility and pretension of the fellows' activities in general, and of the reductive absurdity (as Swift saw it) of their plans for language reform in particular.[15] As T. Christopher Bond has pointed out, in his *Tale of a Tub* (composed in the mid-1690s; published in 1704), Swift was one of those who criticised the idea of using miscellaneous tracts as a form of scholarly communication.[16] A good deal of King's attack on the *Transactions* is similarly focused, deriding the periodical's style at least as much as its content. This is a revealing emphasis; the fellows and their published works were drawn into the Battle of the Books as extreme adherents of modernity. The implied argument of Swift and King was to link the degradation of the written style of modern natural philosophy with the barrenness of the fellows' intellectual pursuits in a vicious circle.

Swift and King had no particular personal reasons to resent Sloane or the *Transactions* and their attacks were to some extent opportunistic. Sloane and the *Transactions* were simply the most visible manifestation of the Royal Society's activity in print, and were targeted in order to taint the wider enterprise of modern philosophy, especially in its deliberate epistemic and stylistic departures from ancient learning. The episode also demonstrates the difficulty of remaining above the fray. Though neither Sloane nor the *Transactions* directly engaged in controversy on these points, Swift in particular appears to have been concerned that the Society should not be allowed to maintain a posture of aloofness in the Battle of the Books, however much it might have wished to.[17] He compared the *Transactions* unfavourably with John Dunton's *Athenian Mercury*, a periodical that ran between 1691 and 1697. Dunton and his small group of co-authors had dealt with many of the same topics as the *Transactions* but made more of a point of engaging with readers (partly through the medium of an advice column). They also represented the periodical as the print manifestation of the meetings of an actual association, though in fact no such society ever existed.[18] The *Athenian Mercury* drew upon an image of what the Royal Society and the *Transactions* were widely presumed to be, while being more visibly and densely engaged with other authors and periodicals. The *Transactions* was both a model to be imitated and (for some) an example of what to avoid; its posture of disengagement from the rough-and-tumble of Grub Street literary criticism and feuding itself drew the attention of Grub Street. Even if Sloane and his successors had wished to do so, it was impossible to detach the *Transactions* from the emergent forms of printed discourse, the revival of political newsprint and the emergence of monthly magazines, essay-periodicals and reviews that characterised the worlds of authorship and publishing in London in the early eighteenth century.[19]

King's *Transactioneer* had made clear the risks of Sloane's style of editorship, with its heavy reliance on correspondents, and on the interest and fidelity of what they related. The plea of his fictional editor that, 'I rely so much on the sincerity of my correspondents, that I cannot tell how to disbelieve [them]', was resonant because it was plausible.[20] The flow of material to the *Transactions* from private individuals reporting from outside London depended upon a presumption of their good faith; naive credulousness was the associated risk. Nonetheless, such considerations of public credit, and the desirability of editorial accountability, were outweighed by the value of Sloane's resources, namely, his extensive international correspondence, his substantial personal wealth and his willingness to put both at the Society's disposal. Woodward and Harris continued to critique Sloane and, in 1709, they successfully campaigned to get Harris elected as one of the secretaries – but it was Richard Waller, not Sloane, that he replaced. In less than a year, Harris resigned (in favour of the returning Waller), unable to work effectively with Sloane after spending a decade attacking his conduct and credentials.[21] In 1713, Sloane's credentials as the editor of the *Transactions* – in terms of his reputation, connections and wealth – were stronger than ever; and yet that would be the year he was replaced.

The *Transactions* as Newtonian propaganda in the 1710s

We have little direct testimony about the events that led to the end of Sloane's tenure as secretary and editor. It is quite possible that he was tired of being criticised, or weary of throwing good money after bad. But there is also circumstantial evidence for Isaac Newton intentionally taking control of the *Transactions* through a change of editor, and a clear motive for him to do so, arising between 1711 and 1713. The wider context was the long-running and increasingly vituperative priority dispute between Isaac Newton in London and Gottfried Wilhelm Leibniz in Hanover concerning the discovery of calculus. The quarrel developed in correspondence, and in periodicals and pamphlets across Europe. The pages of the *Transactions* became weapons for Newton and his followers, a development that reinforced the widespread perception that the *Transactions* spoke for the Royal Society and its president.

In contrast to the literary aspersions cast on the Royal Society's pursuit of natural knowledge during the *Transactioneer* critiques, the subject matter of the calculus debates lay squarely within the

Transactions' usual sphere: the participants were scholars engaged in similar pursuits and governed, in theory, by shared codes of civility; and the debates were conducted in analogous forms of print.[22] The episode shows how the *Transactions* could be deployed instrumentally to press the claims of particular people, who happened to be the president of the Royal Society, his allies and acolytes. The actions of the Society's Newtonian leadership during the 1710s inaugurated a form of presidential influence over the periodical that would remain significant for more than a century. This episode also gives an idea of the European scope of learned interactions in print and the position of the *Transactions*, as well as the potential complexity of navigating these even when the periodical was operating within its more usual sphere.

The calculus dispute became fully public when the Scottish mathematician John Keill published a letter to Edmond Halley asserting that the 'now very famous method of arithmetic fluxions' had been first discovered by Newton. This appeared in issue 317 of the *Transactions*, which in principle carried material relating to September and October 1708. Keill's letter also made a passing swipe at Leibniz for having 'subsequently published the same [method] in the *Acta Eruditorum*, having changed only the name and notation'.[23] Public critique by one fellow of another fellow was unusual and, unsurprisingly, Leibniz composed a forceful response, demanding a retraction and an apology. Sloane, as secretary-editor, received this demand in March 1711, suggesting either that it took a long time for copies of the *Transactions* to reach Hanover or, more probably, that the issues were once more appearing some months after the dates shown on their title pages.[24] Richard Westfall has shown that, in response to Leibniz's letter, Newton organised a further letter by Keill to appear in the *Transactions*. An advance copy of the letter was sent to Leibniz, so that his response could be included.[25] Leibniz was clearly in no doubt about who was behind the letter, and responded directly to the Society and to Newton, rather than to Keill (as the supposed author of the letter) or to Sloane (as editor).[26]

So far, the dispute had taken place through private correspondence made intentionally public in the pages of the *Transactions*. But things were rapidly escalating. The next instalment – Newton's response to Leibniz's second letter – took the form of a privately printed book, rather than appearing in the *Transactions*. This was the *Commercium Epistolicum*, printed at the Society's expense in January 1713. It drew upon the correspondences of John Collins and Henry Oldenburg both to assert Newton's priority, and to insinuate that Leibniz's discovery had not been independent but had relied on Newton's reported conversation and the sight of some of his letters during visits to London in 1673 and 1676. The book was in fact marshalled and drafted by Newton, but was presented as notionally the work of a committee of the Society packed with Newton's close associates and, most unusually, featuring neither of the secretaries. Sloane and the *Transactions* were, for the time being, excluded from the exchanges, and we have Woodward's testimony that Sloane and Newton had by this time 'become bitter enemies to each other'.[27] In the summer of 1713, both Newton and Leibniz published versions of their case in a new French-language journal recently launched in the Low Countries, the *Journal Literaire*, whose editors apparently had an instinct for lively copy.[28] It was in November that Sloane was replaced as secretary of the Society.

Contemporaries, including Woodward, testify that Sloane was in effect forced out by Newton, leaving the president a free hand 'to transact all as he pleases'.[29] Woodward's phrasing is significant, because it suggests Newton's desire to obtain control of the *Transactions*. Once again, it was assumed that editorship of the periodical went hand-in-hand with the secretaryship. It is also notable that Sloane appeared to have no claim on the past or future issues of the *Transactions* despite having supported it financially for almost two decades. It is, of course, possible that Sloane no longer had any desire to assert such a claim, in view of the delays to publication attested to by Woodward and the heavy costs of production (see Chapter 3).

We do not know the precise cause of the antagonism between Sloane and Newton reported by Woodward, or what role it played in Sloane's departure. We can consider the broader position, however. Sloane was financially independent and had built up his own correspondence networks. His interests in natural history and antiquarianism were not wedded to Newtonianism, he had no need for Newton's patronage, and he had many times insisted on the editorial independence of the *Transactions*. Such a person could not be relied upon to conduct the *Transactions* as Newton wished. Throughout the calculus disputes, Newton preferred to remain in the background and act through proxies (as did Leibniz), and this may explain why – though he was certainly wealthy enough to do so – he did not take on the ownership of the *Transactions* himself. He also did not seek to have the Royal Society take formal control of the *Transactions*, perhaps because the Society itself could not have afforded to do so. Instead, Newton made no formal shift in policy but replaced Sloane with a string of biddable allies. Those allies gave Newton absolute control over the *Transactions* on those occasions when he particularly wanted it; as well as reorienting the periodical and the Society towards his own priorities in, and conception of, natural philosophy.

The next eight years saw a rapid turnover of secretaries, all of them friends, collaborators, acolytes or pupils of Newton's: Edmond Halley (replacing Sloane) in 1713; the mathematicians Brook Taylor (replacing Waller) in 1714 and John Machin (replacing Taylor) in 1718; and James Jurin (replacing Halley) in 1721 (who remained in the post until 1727, the year of Newton's death). Newton had been president of the Society for a decade, but it was from 1713 – just as the quarrel with Leibniz was escalating – that he chose to consolidate his hold over the Society's Council, and its mechanisms of publication in particular.

Newton took continued care to mask his direct involvement. For instance, in 1715, the *Transactions* carried a lengthy review of *Commercium Epistolicum* that had been authored and placed there by Newton himself, but anonymously.[30] During the later phase of the dispute, contributions by Newton's acolytes John Keill and Brook Taylor appeared in the *Transactions*, while those of Leibniz's supporters, including Johann Bernoulli, appeared in the *Acta Eruditorum* of Leipzig. The *Acta Eruditorum* was widely, and correctly, assumed to be a mouthpiece for Leibniz.[31] Keill complained to Newton in 1719 about the way the editors of the *Acta Eruditorum* had turned even its index into fodder for controversy (by creating separate entries for insulting criticisms of Newton and Keill), and referred to them as the 'Lipsick [Leipzig] Rogues'.[32] And despite Newton's efforts to remain in the background, the *Transactions* was widely assumed to represent both his views, and the approbation of the Royal Society over which he presided. Brook Taylor wrote against Bernoulli in 1719, and told a correspondent that:

> the place that [my] piece has in the Philosophical Transactions, will shew the World that what I say has the approbation of the Royal Society ... I am very sensible that Sir Isaac Newton is very glad that any thing should be publisht which affects Bernoulli, & believe that is the only reason which makes him willing this piece should appear in the Transactions.[33]

These words from one of the Society's secretaries demonstrate that it was now not just outsiders who believed in a tight relationship between the *Transactions* and the Royal Society, as well as the personal influence of the president on the periodical, despite the alleged independence of the editor. More broadly, the calculus disputes illustrated both the European reach of the *Transactions* and the variety of publications with which it was in dialogue – including privately printed books, anonymous pamphlets and separates, Continental learned journals, and commercial reviews.

The dispute itself counts as perhaps the most spectacular instance of an author exploiting the ambiguities of the *Transactions*' position to his own ends. This was possible because of the author's unique position; he was the most celebrated natural philosopher in Europe, and the president of the Society. During the calculus dispute, the Society's leadership endeavoured to maintain an appearance of editorial neutrality, but equally made certain that the *Transactions* was absolutely at their disposal when occasion required. This undoubtedly consolidated the tendency, both within and without the Society, to view the *Transactions* as an extension of the Royal Society, regardless of its formal status.

Newton's ambition to influence the Society's agenda (including the *Transactions*) from the president's chair was unique in its history up to that point, but the model of presidential governance he effectively instituted would prove lastingly influential. When Sloane became president after Newton, he immediately emulated his approach to the *Transactions* by appointing a close colleague as secretary-editor. When the *Transactions* was again publicly critiqued, around 1750, it was assumed that the president – then Martin Folkes – was responsible for the intellectual direction of both the Society and the periodical. Later still, Joseph Banks's consistently expressed desire to influence both Council and the *Transactions* sparked concerns about the extent and legitimacy of such control both at the start and end of his long presidency. Nonetheless, the basic fact of presidential control remained normative, if periodically controversial, for over a century.

Periodicity, timeliness and smallpox in the 1720s

The slow pace at which the calculus dispute unfolded over the course of the 1710s speaks both to the practical logistics of circulating scholarly print around learned Europe (which will be discussed further in the next section) and to the editorial challenges of running a periodical in a timely manner. For editors working in the commercial marketplace, maintaining a fairly regular periodicity of issue for newspapers and magazines was crucial to keeping the loyalty of readers. The *Transactions*' failure to keep to a regular schedule (which would continue right into the twentieth century) indicates how far removed it was from those sorts of commercial pressures, despite Oldenburg's original ambitions. That said, while Continental learned academies routinely accepted long delays in the appearance of their publications, the Royal Society had tended to replace secretary-editors who did not keep the *Transactions* running at least vaguely regularly. Edmond Halley, for instance, was replaced as secretary-editor for these (and other) reasons both in the early 1690s and again in 1721, by which point no issues had been published since 1719. Prolonged inactivity was evidently more than the Council was prepared to accept.

For James Jurin, therefore, taking on the editorship of the *Transactions* in 1721 meant, first, clearing the backlog of editorial work, and second, establishing editorial practices that would generate a steady flow of copy to enable regular publication of material that was not too out of date. Jurin proved to be an excellent choice of secretary and editor. He brought the *Transactions* up to date in commendably quick time, publishing the backlog concurrently with the contemporary issues. By about 1723, Jurin was managing to issue the *Transactions* every two months, a periodicity that was rapid by the standards of the official memoirs of European learned societies, though still slow compared to most news publications and some commercial literary periodicals. Jurin was a civil, prompt and efficient correspondent, business-like in his dealings and unwilling to get bogged down in controversy. His substantial surviving correspondence reveals how an editor in the 1720s developed a correspondence network, negotiated with authors, dealt with editing and revision, and managed time-sensitive material.[34]

A few months into his editorship, Jurin apologised to a colleague for having taken so long to reply to his letter, and explained that:

> ye business of ye Society is so much in arrear, there having been no Transactions printed for ye two last years, & ye Materials for them being to be collected out of a great many hands, they lie dispers'd in, that this affair alone takes up at present a great part of my time.[35]

Among the arrears were letters from Antoni van Leeuwenhoek, who had for four decades kept up a faithful stream of communications to the periodical that had first published his work, but now complained that his recent letters had (again) neither received a response nor appeared in the *Transactions*.[36] Jurin diplomatically blamed 'ye difficulty of finding proper Persons to translate them' for the delay, and arranged to have the letters read and published. He was subsequently careful to keep up a regular correspondence with Leeuwenhoek, and to send him batches of recent issues of the *Transactions*.[37]

Even with correspondents within Britain, distance was a challenge for Jurin's desire for speed and timeliness. For instance, in spring 1723, he published a letter from a doctor in south Wales, subsequently apologising that, 'I own I ought not to have taken this Step, without first asking your leave for doing it, but considering ye distance of place, I hope you will excuse it'.[38] As he had warned his correspondent, speed was more important than precise detail:

> As my account [of the incidence of smallpox] will be printed in ye Phil. Transactions in about a fortnight or three Weeks, I must desire you will favour me with an answer before that time. I had rather it should be imperfect as to ye number, than come too late.[39]

This letter of 22 January 1723 related to an article in issue 374. Given that this issue nominally covered November and December 1722, Jurin's expectation that it would be in print in 'a fortnight or three Weeks' shows that he had cleared Halley's backlog, and was putting out numbers of the *Transactions* with a delay of only a few weeks from what they advertised on the title page. Compared to his recent predecessors at the *Transactions*, who were routinely a year or more behind, this counted as very timely publication indeed.

Jurin had noted that the materials for the *Transactions* backlog needed to be 'collected out of a great many hands'. He would become adept at managing this process, not merely 'collecting' but also soliciting and shaping contributions from correspondents across Britain, and beyond. In the early 1720s, there was significant medical interest in the possibility of smallpox inoculation. Two of King George I's granddaughters would be inoculated in April 1722 (following a widely reported trial on six criminals convicted of capital crimes, who were offered the possibility of a pardon should they survive the treatment). Jurin used his correspondence networks to make the *Transactions* a forum for gathering case studies and statistical data on the efficacy of inoculation. One of those correspondents was Thomas Nettleton, a medical practitioner from Halifax, Yorkshire (who was not a fellow of the Society). Their correspondence reveals the detail with which Jurin influenced when, how and where Nettleton's contributions would appear.

Jurin included two accounts of smallpox inoculation in issue 370 of the *Transactions*, one from New England and one from Yorkshire. The Yorkshire account was from Nettleton, and the bulk of it was a letter, dated 3 April, to William Whitaker, a London doctor, detailing Nettleton's experience of inoculating at least 40 people.[40] Even before Nettleton's letter had been read to the Royal Society, Jurin had written to suggest publishing it in the *Transactions*, and asking for further details. On 5 May, Nettleton thanked Jurin for the offer to publish, saying that, 'This Favour I had some thoughts of being so bold as to request from you but you have prevented me by your most oblidging offer, which shall be very thankfully comply'd with'.[41]

Nettleton also mentioned that his knowledge of the technique of inoculation 'was entirely from the Philosophical Transactions'.[42] Jurin was quick to capitalise on an opportunity to demonstrate the usefulness of the *Transactions* and, by implication, the Royal Society. He told Nettleton that, 'The obliging acknowledgement you are pleased to make of ye Information you received in this affair from ye Phil. Transactions, was very agreable to ye Society', and hinted that it would be even more so 'if you should think proper to mention it in ye account which is to be made publick'. He went on to suggest that this might be accomplished by Nettleton writing a follow-up letter to Jurin, 'as in answer to one from me enquiring about your farther progress in Inoculation, since your letter to Dr Whitaker'.[43] This could then be published alongside the original letter. Nettleton's ingenuous praise of the Society and its work was an opportunity, if taken up quickly, to have the Society's utility trumpeted in print: a valuable one, since it was precisely for its failure to adequately produce or promote useful knowledge that the Society frequently found itself critiqued. Jurin was skilfully managing a new correspondent to generate extra copy and praise of the Society, and was helping shape the specific form of the exchange so that it would look natural in print.

Jurin's letter to Nettleton on 2 June tells us a great deal about the practical scheduling of getting papers into print. Jurin told Nettleton that his original letter (to Whitaker) 'may be printed first, just as it was before, or with what alterations you think proper' while the additional letter 'may follow it, containing ye additional Observations'. Jurin outlined his schedule:

> On Monday next I think to send to ye Press ye Transactions for ye three first months of ye present Year, & shall reserve room for your Letter, which I suppose may make up about a sheet & half when printed. If you judge it will make considerably more, I beg you will let me know so much by ye next Post, but if not you need not give your self ye trouble of writing, till you send ye Paper itself; which I shall expect, if you please, in a fortnights time.[44]

Jurin's reference to the possibility of alterations, and his uncertainty over the length of the finished contribution, implies that he had returned the manuscript letter to Nettleton after it had been read to the Society. This is in keeping with Sloane's claim (if not his practice) not to have 'abridged or chang'd anything' himself, but 'when it was possible' to have the papers 'corrected by those who communicated them'.[45] Jurin's estimate of the finished length was broadly accurate: as printed, the original paper occupies a little over 13 pages, marginally more than the sheet-and-a-half of printed quarto he had anticipated; the follow-up letter, dated 16 June, is a further three-and-a-half pages. The two items appeared in the same issue, but the typesetting – particularly the blank space at the end of the first item, with the extra letter starting on a new page – suggests that they arrived at the printers separately.[46] The original paper may even have been on its way back to London when Jurin wrote.

The exchange reveals, with unusual accuracy for the eighteenth century, the exact date on which the copy for a particular issue of the *Transactions* (no. 370) was sent to the printer. Since Jurin wrote on a Saturday to a correspondent 200 miles away, his reference to 'Monday next' most likely meant 11 June. On the one hand, this gives us a good indication of the time in which an efficient secretary-editor could turn around a paper with a hitherto unknown correspondent: from early April to mid-June was about 10 weeks from composition to communication to acceptance and finally to print. On the other hand, since this issue nominally covered 'ye three first month of ye present Year', other papers in the same issue had taken almost six months to find their way into print, reminding us that Jurin was still in his first year of editorship, and not yet up to full speed. Jurin also gives a schedule here for typesetting – up to five days for the bulk of it, given that he intended sending the issue to the press on the 11th while reserving room for Nettleton's contribution, which he expected on the 16th. The possibility of a 10-week turnaround (including revisions) in the pre-digital age is impressive, and vastly better than the typical time taken to publish in the late nineteenth and early twentieth centuries.

In October 1723, about 18 months into his correspondence with Nettleton, Jurin received a letter which brought the *Transactions* into contact with the wider world of news periodicals, in a manner that threw the issue of timeliness (and audience) into sharp relief. Nettleton wrote:

> I was very much surprised to find it inserted in severall of the News Papers of this last Post, as a Current Report, that I had been lately call'd to a Patient who was Inoculated about a year ago, &

who is since dead of the natural Small Pox, & I cou'd not omit this first opportunity to assure you that the said report is entirely false, none who have been Inoculated having dyed of the natural Small Pox afterwards, neither have we any reason to think worse of the Practice here.[47]

The London papers were, at this time, tracking the progress of an outbreak of smallpox in France, publishing reports of the infection's arrival in particular districts, and the death or infection of notables. Patent remedies were widely advertised in *Mist's Weekly Journal*, the *Daily Courant* and the *Daily Journal*. The claim that an inoculated patient had subsequently died of smallpox was a serious threat to those arguing for the efficacy of inoculation. Nettleton told Jurin he had written in haste: 'I did not see the News Papers till this Evening & the Post going out early in the Morning I cannot write so much at large as I cou'd wish, but this matter being of so great concern I wou'd not defer giving you a true account of it.'[48]

Nettleton admitted that one child he had inoculated had died during a smallpox outbreak – though he maintained that there were grounds for supposing the child had in fact died of something else – but he strenuously denied that another child inoculated previously had died of the natural outbreak of smallpox.[49] It was one thing to admit that inoculation carried a risk, but the imputation that it might be ineffective as well as dangerous had to be totally refuted. Nettleton offered Jurin true accounts of the case from credible witnesses, and asked for advice in challenging the newspaper report.

Jurin's response is missing, but on 11 November Nettleton sent the promised account (Figure 4.1), with certified testimonies. Jurin reported on the 19th that:

> Upon talking with Dr Whitaker about your Letter, we both agreed upon publishing it immediately together with ye Certificates in one of ye Newspapers. The trouble of this ye Dr took upon himself, but could not get it inserted in any other Paper, at least not without a very extravagant consideration, than ye daily Journal, in which it was yesterday published.[50]

Nettleton's defence thus appeared in the *Daily Journal* for Monday 18 November 1723. The choice of a newspaper demonstrates the limitations of a periodical like the *Transactions*: Jurin was managing to issue the *Transactions* roughly every two months, but this was slow indeed

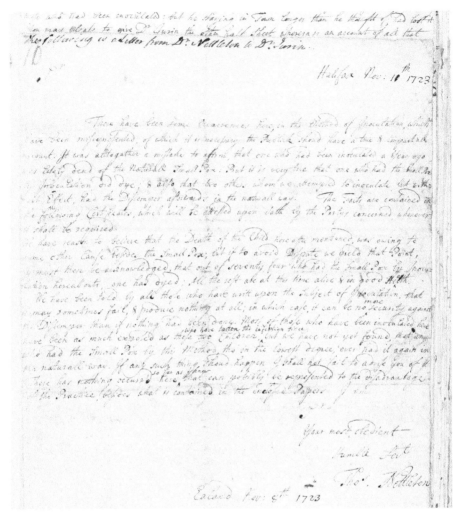

Figure 4.1 William Nettleton's letter about inoculation, to James Jurin, 11 November 1723 © The Royal Society.

compared to a daily newspaper. Michael Harris estimates the typical circulation of the London dailies of the 1720s at perhaps 1,000 copies.[51] For the *Transactions*, on the other hand, there is good circumstantial evidence that its print run in the late 1710s and early 1720s was around 500 copies, which is more than the fellowship of the Royal Society (about 300 fellows), but rather fewer than the newspapers.[52] The price of each issue of the *Transactions* varied with its length – which was usually somewhere between 40 to 60 pages – and the scanty surviving evidence from the early eighteenth century hints at issue prices such as 10*d*., 1*s*. or (presumably for a long issue) 1*s*. 7*d*.[53] It was still priced for the affluent end of the print market.

Nettleton's immediate intuition – and Jurin agreed – was that the *Transactions* appeared too slowly, and circulated too narrowly, to be an effective way of stemming the tide of public opinion if it started to flow against inoculation. A problem of public perception had to be addressed in a public forum. The underlying assumption is that the *Transactions* was a place where the learned spoke to one another, not a periodical for the general reader. Jurin had significantly improved the timeliness and regularity of the *Transactions* but it could not compete with newsprint for speed and range.

Repackaging the *Transactions*: abridgements and translations

Jurin and his successors continued to publish the latest *Transactions* content in separate periodical issues, but for many readers, an encounter with the *Transactions* was far more likely to involve bulky bound volumes that circulated only slowly through the commercial book trade, or alternative channels. A number of learned societies, most notably the Académie Royale in Paris and the Academia Naturae Curiosorum in the German lands, did in fact issue their 'periodicals' only as annual or biennial volumes (and these were often some years behind the times). This partly reflected a lingering sense that books were a more appropriate form of publication for the knowledge produced or sponsored by corporate bodies, especially those with royal patrons.[54] The slow publication schedule also reduced some of the pressure on editors.

Even learned publications that were issued periodically – including the Leipzig *Acta Eruditorum* and the *Philosophical Transactions* – were also issued as more substantial bound volumes covering a whole year. And it was in this format that readers at a distance were most likely to encounter learned periodicals. Thus, in 1723, Glasgow mathematician Robert Simson told Jurin that, 'The Act. Lips. [i.e. the Leipzig *Acta Eruditorum*] for 1722 have not as yet come to this place'.[55] The fact that something published in Leipzig in 1722 had not reached Glasgow by spring 1723 reminds us that the circulation of these learned periodicals was generally slow and uneven. Jurin, in London, *had* seen the 1722 *Acta*, which is why he declined to publish Simson's paper, on the grounds that his claimed discoveries had been anticipated elsewhere.[56] This desire that contributions should be original, and not previously published elsewhere, marks a significant change in the *Transactions* since Oldenburg's time. Establishing originality depended on good access to print networks.

This section discusses two other ways in which the transformation of the periodical *Transactions* into book format enabled it to reach wider audiences. Abridgements served those wishing to gain access to material in the back-run, which was expensive and difficult to obtain. Translations (which were often also abridgements) served overseas scholars who wanted access to more recent material, but faced the double challenge of inefficient distribution channels and the language barrier. It was the commercial print trade, not the Royal Society, that filled the demand for wider access to the *Transactions*. The success of these retrospective abridgements and translations shows that there was a significant group of scholarly readers in the eighteenth century who were willing to purchase the *Transactions* in a book format. For these readers, the *Transactions* did not represent novelty and newness, but a repository or archive of durable, and valuable, claims to knowledge.

Abridgements

Despite deprecatory comments from its early editors about the ephemerality of what it contained, the back-run of the *Transactions* had become a significant repository of research and knowledge-claims. By 1700, its 21 volumes contained letters and contributions by philosophers, naturalists and mathematicians of European reputation – including Halley, Leeuwenhoek and Martin Lister – that had never been published separately. These older issues were, however, difficult and expensive to obtain, whether for personal or institutional libraries seeking to complete their holdings, or for younger scholars wishing to consult particular papers.

There simply were not many copies of the original numbers in existence, let alone in circulation and available for purchase (especially if we assume that 500 is a more realistic estimate of the print run than the ambitious 1,250 of early 1665). Sets of volumes occasionally came onto the market via auctions of the libraries of deceased fellows. An undated letter in the Sloane papers, probably written by Waller or Sloane, informed a foreign correspondent that a bookseller had been found who could provide a complete set of the *Transactions* down to the (uncertain) present for the sum of £12 10s.[57] This was a huge amount for a private individual to spend on a single work: it was more than eight times the retail price of Nicholas Rowe's 1709 edition of Shakespeare, and represents a purchasing power equivalent to well over £1,500 today.[58]

In 1723, Jurin directed a correspondent to William Innys, the Society's bookseller, as 'the only man that can furnish you' with the past 20 volumes of the *Transactions*.[59] This reveals that it was the bookseller – not the Society, and not the editor – who owned any unsold stock; and since Innys had only been involved in distributing the *Transactions* for 10 years, it also implies that whenever there was a change of bookseller – which after 1695 was at the editor's discretion – the incoming bookseller also handled any remaining stock.

Even if back issues of the *Transactions* could be acquired, they were not easy to navigate or search, because the internal organisation was virtually random. Most, but not all, of the individual volumes were indexed, but there was no general index for the entire series (although an index had been produced in 1678 for the first 12 volumes).[60] Thus, by the early eighteenth century, the difficulties of acquiring and navigating the back-run of the *Transactions* created a potential opportunity for entrepreneurs. In the following decades, several found commercial success by abridging the back-run to a more affordable size, and rearranging the contents for access. The existence of these projects, however, raises once more the question of who had the rights to reprint, reuse and modify material from the *Transactions*.[61]

The Royal Society appears to have first learned of a plan for an abridgement in early 1700, at the height of the controversy over the *Transactioneer*. It is unclear who was behind this plan, and the Council passed a resolution forbidding any fellow from undertaking such an enterprise without its permission and that of the authors.[62] As so often with the *Transactions*, it is unclear what (if any) legal authority the Society could have used to enforce this resolution. Its licensing privilege had lapsed in 1695, there was as yet no copyright, and, as the Society had insisted in response to the *Transactioneer*, the *Transactions* was formally the property of its editor. Yet this resolution appeared to infringe the interests of Sloane and his editorial predecessors, since it seemed to forbid them from making an abridgement of their own periodical. More likely, perhaps, is that the Council intended to protect Sloane by assuming authority over any other fellow who might have wished to undertake an abridgement. We should also note the limits of this position, however; namely that the Society appeared to recognise that it would not be able to prevent non-fellows from undertaking the project.

Any person making such an abridgement would have considerable power to shape the public representation of the *Transactions*, its editor and the Society. On this reading, the stipulation that permission would be needed from the various contributors to the *Transactions* looks more like an effort to throw a logistical obstacle into the path of the would-be abridger, than a serious assertion of the moral rights of authors.

No more was heard of that plan for an abridgement, but a few years later, John Lowthorp proposed to Council the production of a systematic abridgement of the *Transactions* from 1665 to 1700. Lowthorp was librarian to the Duke of Chandos, and with his patronage had recently been elected to the fellowship.[63] It seems highly unlikely that Lowthorp acquired permission from all the authors whose work he abridged, but his plan did receive the Society's official endorsement in 1703 and imprimatur in 1705.[64] The Society offered no financial backing, and nor did Lowthorp work with the Society's regular printers.[65]

Lowthorp's abridgement was published as *The Philosophical Transactions and Collections to the end of the year MDCC Abridged, and Disposed under General Heads* (1705–8). In his preface, he envisaged two classes of reader: the generally curious who read for pleasure and instruction, and those scholars who wished to use it to 'write something of their own'. It was apparently for the sake of readers in the first category that he regrouped the material by subject ('under General Heads'), while adding marginal references to the original articles for the benefit of the second group. These marginal references mean that it is impossible to tell whether eighteenth-century scholars referring to pre-1700 issues of the *Transactions* were working from the original, or the abridgement.

Lowthorp managed to compress the 21 volumes of the original into just three (Figure 4.2). He omitted book notices and papers that had already been made available elsewhere (in collected editions of an author's works, for example). In reorganising the remaining material by subject matter, he prioritised Newtonian preoccupations in mathematics, mechanics and optics. As editor, Lowthorp was perfectly willing to combine separate contributions when they referred to the same subject (with a marginal note to give attribution and reference correctly); or to split a single communication between two separate volumes, if it treated two very different subjects. Lowthorp's abridgement amounted to a significant transformation of the historic content of the *Transactions*, making it more accessible, physically and financially, to readers, as well as easier to navigate.

Lowthorp's abridgement appears to have been a striking success. It clearly met a demand, for it ran to five editions by 1749. The fifth edition appears, from the evidence of the Bowyer printing ledgers, to have had a print run of 500 copies.[66] Even if the earlier editions were no larger, this implies that there were at least twice as many copies of

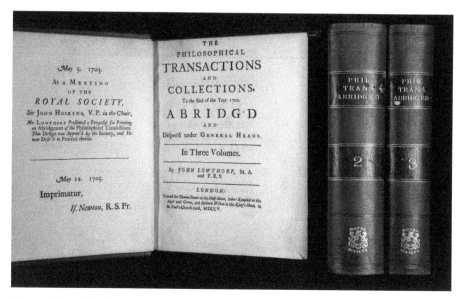

Figure 4.2 John Lowthorp's *Abridgement* (1705–8) condensed the back-run of the *Transactions* into just three volumes © The Royal Society.

the abridgement in existence than there were of the original issues (and probably more).

The concept of an abridged and reorganised version of the *Transactions* content was sufficiently successful that other entrepreneurs would extend it to the post-1700 volumes. Two separate publisher-editors extended Lowthorp's abridgement from 1700 to 1720, both, strangely, with explicit permission from the Royal Society. Again, the edition sizes point to the market demand for these volumes: Benjamin Motte's extension was printed in 1,000 copies, as was the first edition of its rival from Jones et al. The Jones extension went into a second edition of 750 copies.[67] A further extension, by Andrew Reid, John Gray and John Martyn, took the coverage up to 1749, by which time 47 volumes of the *Transactions* had been compressed into 10 volumes of abridgement. A rival abridgement, edited by Benjamin Baddam, also appeared, this one preserving the sequence of the original papers instead of regrouping them by subject matter.[68] It, too, ran to a second edition. Besides these there was also a subject-specific collection of *Medical Essays*, edited by S. Mihles and running to two volumes, and a selection published by the Essex naturalist William Derham alongside other Royal Society material under the title of, confusingly enough, *Miscellanea Curiosa*.[69]

These abridgements were all undertaken as commercial ventures. Lowthorp's three volumes were offered at an advance subscription price of 35s. or for retail at 50s., making them vastly more affordable than the £12 10s. quoted earlier for a complete second-hand set of unabridged volumes of the *Transactions*.[70] The sale of 2,500 sets of these volumes would thus have generated at least £4,300 income and probably more. We know nothing about the cost of production, but the rule of thumb for this period is that retail prices were set at roughly twice the cost of production; we might estimate that half of the income was profit. We have no information about Lowthorp's financial arrangement with his publishers, but it is clear that the Royal Society never saw any of this money and there is no evidence that any of the periodical's current or former editors did either (or that they ever sought it). When we remember that Hans Sloane was at this time subsidising the current issues of the *Transactions*, the commercial success of the abridged back issues is striking. The abridgement probably *made* as much (or more) for its undertakers as the editor of the *Transactions lost* publishing new numbers of the periodical over a similar period.

This state of affairs would persist through the eighteenth century. In the later part of the century, we know that the Society would routinely be left with surplus stock of the current numbers (even though two-fifths of the 750 copies then printed was distributed to the fellows); and yet the abridgement covering 1665–1749 went through two further editions, and a new abridgement to 1800 would find commercial success in the early nineteenth century (see Chapter 7). The overwhelming impression is that, contrary to perceptions of the scientific journal as a tool for rapid communication, in the eighteenth century, there was considerably more demand for the abridged, reorganised and retrospective version of the *Transactions* than for the fresh, original content in the current numbers.

Translations

Repackaged and abridged volumes of *Transactions* content were also important on the Continent, not least because they made translation more practical. In Continental Europe, as far as we can tell, the original *Transactions* circulated only inefficiently or not at all. It is reported to have been available for consultation in the public libraries of Paris in 1728, though it is not clear how it got there.[71] Some copies reached Italy via the London agent of the Grand Duke of Tuscany, who forwarded the issues to Sir Thomas Dereham, a roving diplomat and agent for the Stuart court in exile.[72] Failures in this chain of communication exacerbated the breach between Jurin and Sloane in late 1727, when Sloane blamed Jurin for Signor Pucci's failure to forward packages to Dereham.[73] Commercial book trade channels seemed equally unreliable, and there was as yet relatively little in the way of systematic exchanges between institutions. The Royal Society was no longer supporting the *Transactions* by subscribing for significant numbers of copies, but it did purchase a small number of copies each year to be sent to its very best correspondents. In the Jurin years, these included Leeuwenhoek in Delft and the Swedish theologian Eric Benzel, who was involved in Jurin's meteorological data-gathering.[74] Overall, the best way for foreign scholars to receive the *Transactions* was to contribute to it regularly.

Even if it was physically accessible, the *Transactions* was predominantly in English, a language not widely spoken outside the British Isles in the late seventeenth and early eighteenth centuries. Its contents were, therefore, quite likely to be unintelligible to many Continental scholars. There do not appear to have been any Latin editions since the efforts that dissatisfied Henry Oldenburg, and by the eighteenth century scholarly readers seem to have preferred vernacular translations. Summaries or extracts from the *Transactions* commonly appeared in the *Journal des Sçavans* and *Acta Eruditorum*, and brief notices of its contents had sometimes appeared in Pierre Bayle's *Nouvelles de la République des Lettres*, a monthly French-language review of European learned publishing printed in Amsterdam until the 1710s.[75] Undertaking a full translation of a current periodical was difficult for translators to sustain in a regular and timely manner. Working with retrospective content was arguably a more achievable task. Such translations were often also abridgements.

In France, the young scholar François de Bremond began his year-by-year translation with the 1732 volume, and it appeared in print from 1738. It took his successor until 1760 to complete the translation up to 1744.[76] The *Collection Academique* (1754–87), an attempt to collate and summarise all the material from obscure learned periodicals in Europe, also translated some of the content from the earliest *Transactions*. As Lowthorp had done, it omitted material that readers could access by other means: in this case, papers that had been translated from – or already reproduced in – French journals (principally, the *Journal des Sçavans*), and book reviews. Together, these translation projects meant that a fair amount of the material from the first 70 years of the *Transactions* was available in French.

This patchwork of French translations covering various periods, and following different rules of selection, was eventually superseded by a 10-volume abridgement undertaken in the 1780s by the physician and librarian Jacques Gibelin.[77] Gibelin appears to have worked from the original volumes, doing both abridgement and translation; whereas, a few years earlier, Nathaniel Leske had produced a German translation based heavily upon the existing English-language abridgement by Lowthorp and his successors.[78] Like Lowthorp, Gibelin rearranged the *Transactions* material by subject area, but unlike Lowthorp, he chose to exclude the mathematical and astronomical papers which were beyond his competence and which, he claimed, would have swelled the work to unreasonable dimensions. Gibelin's introductory note offers two further hints about the circulation of the *Transactions*: he observed that, by the 1730s and 1740s, half a year or more's worth of issues would be released at a time (in other words, that Cromwell Mortimer did not manage to maintain Jurin's bi-monthly schedule), and that, in the 1780s, it was very difficult to find any Paris library other than the Bibliothèque du Roi with a complete set of the original *Transactions*.[79] This was why Gibelin's abridgement would be as welcome to Francophone scholars as Lowthorp's abridgement had been in Britain.

Italian access to the *Transactions* was mediated by Thomas Dereham. Dereham's interest in the Royal Society's activities, coupled with his contacts among Italian virtuosi, meant that he was an influential conduit for promoting the *Transactions* and its contents, and the Society's secretaries were well aware of this. For instance, in 1728/9, William Rutty sent Dereham – in the Grand Duke of Tuscany's diplomatic bag – the last of six issues of the *Transactions*, along with the title page and index to allow him to have them made up as volume 35. Rutty was also careful to instruct Dereham on using the *Transactions* to bolster the Society's reputation abroad. For instance, he drew attention to an astronomical paper by James Bradley, which he described as 'a very considerable discovery in Astronomy', shedding 'great Light & confirmation' on various 'grand doctrines' including the motion of the earth and the immense distance of the stars. He added, 'I don't doubt but that you are pleased, that this is owing to one of ye Fellows of our own Society'.[80] It is a heavy hint about what the Society wished Dereham to promote to his Italian interlocutors. But Dereham did not only circulate copies of the English issues. He also undertook an Italian translation of Lowthorp's abridgement. Dereham seems to have done the translation around 1726, and then spent some time trying to convince printers in various Italian cities to publish it. The five-volume translation, covering the *Transactions* from 1665 to 1720, was printed in Naples between 1729 and 1734.[81]

The protection of the Society's licensing privilege had enabled the *Transactions* to be one of very few periodicals able to function in the highly regulated, even oppressive environment of the late seventeenth-century print trades. The lapse of the Licensing Act in 1695 had brought about a fresh explosion of periodicals of many kinds – price-currents, newspapers, essays and reviews, many of them jostling for readers and snapping at one another. In consequence, the *Transactions* was operating in a much busier print landscape, in which it found itself borrowed from, criticised, imitated and lampooned. It was ill-equipped to match the combative tone and scurrility of the Grub Street hacks and satirists, and yet found itself embroiled with them. The Society was so ill-prepared for this possibility in 1700 that Hans Sloane and his allies assumed that so pointed a challenge to his credentials could only have come from interested parties within the Society.

Two decades later, Jurin's use of the London newspapers to refute allegations about the dangers of smallpox inoculation would demonstrate a growing awareness of different niches within the print trades, and of the *Transactions'* place in that ecosystem. Jurin recognised that the *Transactions* could be a valuable forum through which his network of provincial and Continental correspondents could collate data on smallpox and meteorology, or a way of prosecuting disputes among scholars, as Newton had done. But Jurin also recognised that the *Transactions* had neither the reach nor the rapidity that would have enabled it usefully to address public fears or panics.

The *Transactions* was now part of a print ecosystem that included scholarly periodicals, books and pamphlets produced in various centres of learning across Europe, as well as a flourishing array of newspapers and reviews closer to home. Its longevity was already a distinctive feature: other periodicals dealing with medicine, the practical arts and the sciences did exist, but only a few had the editorial and financial resources to keep going long term.[82]

By the middle of the eighteenth century, the *Transactions* had acquired a curious parallel existence, and one that has come to characterise research journals in the long term: it provided current content, and also acted as a long-term repository. On the one hand, those involved in producing it – its editors and their supporters among the fellows of the

Royal Society – focused upon gathering original, previously unpublished, material for the upcoming issues. Jurin was particularly adept at doing this, and at maintaining an efficient and timely publishing cycle. The periodicity may have slowed under Mortimer in the 1730s and 1740s, but the *Transactions* continued to focus upon original observations and results reported by members of the Royal Society's networks. For readers of the current issues – who were likely to be members of those networks – the *Transactions* provided access to the up-to-date contributions from other members, in natural philosophy, antiquities and natural history.

On the other hand, the content of the *Transactions* was far more widely available in book form than in unbound, periodical-issue form in the eighteenth century. The challenges faced by those outside the Royal Society's networks in accessing current issues, and by everyone seeking back issues, had created an opportunity for entrepreneurial printers, abridgers and translators to create derivative print products. Those published beyond British shores were beyond its influence, but nonetheless helped form the Royal Society's international scholarly reputation. The Society's attempt to control potential British abridgements in 1700 appears to have had no legal basis, and yet its claim on the *Transactions* was sufficiently credible that, over the following decades, several entrepreneurial editors and printers felt obliged to seek the Society's permission for their projects. In the 1750s, the Royal Society's assumption of editorial and financial responsibility for the *Transactions* would finally establish its ownership of current and future issues. It also stabilised a slow and stately periodicity – six-monthly – that kept the *Transactions* quite apart from the flourishing periodical culture of Georgian newspapers, reviews and magazines.

Notes

1. R. Astbury, 'The renewal of the Licensing Act in 1693 and its lapse in 1695', *The Library* 5th series, 33 (1978): 296–322.
2. On the involvement of the Royal Society and its fellows in the early eighteenth-century culture wars known as 'the Battle of the Books', see Joseph M. Levine, *Dr. Woodward's Shield: History, science, and satire in Augustan England* (Ithaca, NY: Cornell University Press, 1991); and Gregory Lynall, *Swift and Science: The satire, politics, and theology of natural knowledge, 1690–1730* (Basingstoke: Palgrave Macmillan, 2012).
3. Adrian Johns, *The Nature of the Book: Print and knowledge in the making* (Chicago, IL: University of Chicago Press, 1998), 457.
4. David A. Kronick, 'Medical "publishing societies" in eighteenth-century Britain', *Bulletin of the Medical Library Association* 82, no. 3 (1994): 277–82; James Tierney, 'Periodicals and the trade, 1695–1780', in Michael Suarez and Michael Turner, eds, *The Cambridge History of the Book in Britain, vol. V, 1695–1830* (Cambridge: Cambridge University Press, 2009).
5. [William King], *The Transactioneer With some of his Philosophical Fancies: in Two Dialogues* (London: no imprint, 1700), sigs. A2r-v. See T. Christopher Bond, 'Keeping up with the

6 [King], *The Transactioneer*, sigs. A3r-v.

7 The disagreement turned on Woodward's theories concerning the origins of fossils. See Levine, *Dr. Woodward's Shield*, esp. 24–32 and 82–92. For Sloane's spreading the rumour that Harris and Woodward were responsible, see Harris's reply, BL Sloane MSS 4026 ff. 253–4 (also discussed below), and RS CMO/02, 28 February 1699/1700, 109.

8 BL Sloane MS 4026 ff. 253–4; John Harris to Sir John Hoskyns, 27 February 1699/1700. Harris's letter was endorsed 'to be communicated to the Councill'.

9 In fact, the Society's Council minutes for this period show no reference to this paper by Roger Morton. RS CMO/02, 1698–1700 *passim*.

10 RS CMO/02, 28 February 1699/1700.

11 On individual versus collective editorship, see Aileen Fyfe and Anna Gielas, 'Introduction: Editorship and the editing of scientific journals, 1750–1950', *Centaurus* 62, no. 1 (2020): 5–20.

12 *Useful Transactions in Philosophy*, 3 issues, January–September 1709, printed in London by Bernard Lintott. Sloane's manuscript transcription is in BL Sloane MSS 4026 f. 211. The first volume of Sloane's West Indian natural history and travelogue, *A Voyage to Madera, Barbados, Nieves, S. Christophers and Jamaica*, had been published in 1707.

13 BL Sloane MSS 4026 f. 211.

14 See letter to Archbishop King of Dublin, 8 January 1712, in Harold Williams, ed., *The Correspondence of Jonathan Swift*, 5 vols (Oxford: Clarendon Press, 1963), vol. 1, 286. No better explanation of King's motives has ever been offered than that he was acting in consultation with Swift, if not actually on his instructions. King did not last long in his editorial post at the *Gazette*.

15 See also Levine, *Dr. Woodward's Shield*, 247–50; Joseph Levine, *Between the Ancients and the Moderns: Baroque culture in Restoration England* (New Haven, CT: Yale University Press, 1999), 29–30; Lynall, *Swift and Science*, 20–1.

16 Bond, 'Keeping up with the latest transactions'.

17 Bond argues that Swift believed the Society *ought* to have been able to maintain its dignity but had failed to, and that his satirical treatment of it was a suitable punishment for its failure in print to live up to its lofty ambitions. See Bond, 'Keeping up with the latest transactions'.

18 On Dunton and the *Athenian Mercury*, and Swift's praise of it, see Johns, *Nature of the Book*, 457.

19 See Pat Rogers, *Hacks and Dunces: Pope, Swift and Grub Street*, abridged ed. (London: Methuen, 1980), especially 1–55 on the topography of the area meant by the fictive 'Grub Street' and its character as a literary and cultural milieu. More generally, see the essays in 'Section C: Serial Publication and the Trade', in Suarez and Turner, eds, *History of the Book in Britain 5*.

20 [King], *The Transactioneer*, 54–5; and for a further discussion of King's specific criticisms see Delbourgo, *Collecting the World*.

21 Woodward's continuing attacks on Sloane, resulting in a reprimand and then expulsion from the Council in 1706 and 1710, respectively, are documented in Sloane's notes in BL Sloane MS 4026 ff. 266, 195–6, and 297–9; these include an undated draft of a form of words for a mutual apology. Waller, for his part, had been endeavouring to resign the secretaryship for almost a decade; see BL Sloane MS 4036 f. 194, and RS CMO/2 for Harris's election.

22 On scientific civility more broadly, see above all Anne Goldgar, *Impolite Learning: Conduct and Community in the Republic of Letters, 1680–1750* (New Haven, CT: Yale University Press, 1995) and Steven Shapin, *A Social History of Truth* (Chicago, IL: University of Chicago Press, 1994).

23 'Epistola … de legibus virium centripetarum', *PT* 26.317 (1708–9): 174–88, at 185 (our translation).

24 RS LBO/7/114, received 4 March 1711. On the timeliness of the *Transactions*, Woodward asserted in May 1711 that no issues had appeared for almost two years; Levine, *Dr. Woodward's Shield*, 112, citing letters to Johann Jakob Scheuchzer in Zurich.

25 See Richard S. Westfall, *Never at Rest: A biography of Isaac Newton* (Cambridge: Cambridge University Press, 1980), 715–16, 720–2.

26 Leibniz's response is in RS LBO/7/115, 29 December 1711; received 31 January 1712.

27 Woodward to Scheuchzer, 12 September 1712, cited in Levine, *Dr. Woodward's Shield*, 112.

28 'Extrait d'une letter de Londres', in *Journal literaire* 1 (May–June 1713): 206–14; 'Remarques sur le different entre M. de Leibnitz, & M. Newton', *Journal literaire* 2 (November–December 1713): 445–53. On this phase of the dispute, see Westfall, *Never at Rest*, 761–4.
29 Levine, *Dr. Woodward's Shield*, 112.
30 'An account of … *Commercium Epistolicum* …', *PT* 29.342 (1714–16): 173–224. Issue 342 covered January and February 1715. Over 50 pages long, this was easily the longest account of any book to appear in the periodical.
31 For evidence of the first editor of the *Acta*, Otto Mencke, manipulating publishing procedures to favour Leibniz, see Uwe Mayer, '"Kein tummelplatz, darauff gelehrte leut Kugeln wechseln": Principles and practice of Mencke's editorship of the Acta eruditorum in the light of mathematical controversies', *Archives internationales d'histoires des sciences* 63, nos. 170–1 (2013): 49–59.
32 Keill to Newton, 24 June 1719 in A. Rupert Hall and Laura Tilling, *The Correspondence of Isaac Newton*, paperback edn, 7 vols (Cambridge: Cambridge University Press, for the Royal Society, 2008), vol. 7, 48–9.
33 Brook Taylor to Edward Jones, 5 May 1719, in Hall and Tilling, *The Correspondence of Isaac Newton*, vol. 7, 39. Taylor's piece was 'Apologia D. Brook Taylor …', *PT* 30.360 (1719): 955–63.
34 Andrea Rusnock, *The Correspondence of James Jurin* (Amsterdam: Rodopi, 1996); and her 'Correspondence networks and the Royal Society, 1700–1750', *British Journal for the History of Science* 32, no. 2 (1999): 155–69.
35 Jurin to Langwith, 17 February 1721/2, in Rusnock, *Jurin Correspondence*, 85.
36 Sietske Fransen, 'Antoni Van Leeuwenhoek, his images and draughtsmen', *Perspectives on Science* 27, no. 3 (2019): 485–544.
37 Jurin to Leeuwenhoek, 22 May 1722, in Rusnock, *Jurin Correspondence*, 99–100. The particular difficulty with Leeuwenhoek's letters, as Jurin observed (and modern scholars agree), was finding someone who could not only read Low Dutch but who had a good understanding of the subject matter.
38 Jurin to Perrott Williams, 7 March 1722/3, in Rusnock, *Jurin Correspondence*, 138.
39 Jurin to Perrott Williams, 22 January 1722/3, in Rusnock, *Jurin Correspondence*, 124.
40 Thomas Nettleton, 'A letter … concerning the inoculation of the small pox', *PT* 32.370 (1722–3): 35–48. On Nettleton, see Arthur Boylston, 'Thomas Nettleton and the dawn of quantitative assessments of the effects of medical interventions', *Journal of the Royal Society of Medicine* 103 (2010): 335–9. Also, Rusnock, *Jurin Correspondence*, 22–7.
41 Nettleton to Jurin, 5 May 1722, in Rusnock, *Jurin Correspondence*, 98.
42 Nettleton to Jurin, 5 May 1722, in Rusnock, *Jurin Correspondence*, 98.
43 Jurin to Nettleton, 2 June 1722, in Rusnock, *Jurin Correspondence*, 104.
44 Jurin to Nettleton, 2 June 1722, in Rusnock, *Jurin Correspondence*, 104.
45 Sloane, 'The Preface', *PT* 21 (1699–1700): [i–ii].
46 Nettleton, 'A letter … concerning his farther progress in inoculating the small pox', *PT* 32.370 (1722–3): 49–52.
47 Nettleton to Jurin, 28 October 1723, in Rusnock, *Jurin Correspondence*, 203–5. We have been unable to track down the report Nettleton had seen.
48 Nettleton to Jurin, 28 October 1723, in Rusnock, *Jurin Correspondence*, 203–5.
49 *Daily Journal*, 18 November 1723 (issue 885).
50 Rusnock, *Jurin Correspondence*, 210.
51 Michael Harris, *London Newspapers in the Age of Walpole: A study of the origins of the modern English press* (Madison, NJ: Fairleigh Dickinson University Press, 1987), 57.
52 The print run estimate comes from Benjamin Motte, *A Reply to the Preface publish'd by Mr Henry Jones* (London: J. Jones, 1722), 12. Motte had no inside knowledge, but was printer to an unofficial abridgement of the *Transactions*.
53 In the 1720s, the Society seems to have paid around 10*d.* or 1*s.* for the copies it sent to Leeuwenhoek, see RS AB/1/1/3. For the mid-1740s, there is a hint of a price in Birch Papers, BL Add. MSS 4441.
54 On late seventeenth-century preferences for book publication versus periodicals, see Noah Moxham, 'The uses of licensing: Publishing strategy and the imprimatur at the early Royal Society', in *Institutionalisation of Sciences in Early Modern Europe*, ed. Mordechai Feingold and Giulia Giannini (Leiden: Brill, 2019), 266–91; and Anita Guerrini, *The Courtiers'*

Anatomists: Animals and humans in Louis XIV's Paris (Chicago, IL: University of Chicago Press, 2015), 147–64.
55 Robert Simson to Jurin, 22 March 1723, in Rusnock, *Jurin Correspondence*, 140.
56 Jurin to Simson, 5 March 1723, in Rusnock, *Jurin Correspondence*, 135.
57 BL Sloane MS 4026 f. 209 – an undated Latin letter, unknown author and unknown recipient.
58 On the price of Shakespeare, see Robert D. Hume, 'The value of money in eighteenth-century England: Incomes, prices, buying power – and some problems in cultural economics', *Huntington Library Quarterly* 77, no. 4 (2014): 373–416, 374–5. MeasuringWorth.com estimates £12 10s. as the equivalent of between £1,483 and £26,480 depending on the basis of comparison.
59 Jurin to Henry Newman, 22 April 1723, in Rusnock, *Jurin Correspondence*, 96.
60 'A general index or alphabetical table to all the Philosophical Transactions, from the beginning to July 1677' (London: Printed by J. M. for John Martyn, Printer to the Royal Society, 1678). It is unknown who originated this project. There was, however, discussion in spring 1678 (after Oldenburg's death) of various attempts to index the Society's books, including the Council minutes and the register books. See Thomas Birch, *The History of the Royal Society of London for Improving of Natural Knowledge*, 4 vols (London: A. Millar, 1756–7), vol. 3, 395–6.
61 These issues are discussed more generally in Aileen Fyfe, 'The production, circulation, consumption and ownership of scientific knowledge: historical perspectives', in *CREATe Working Paper* 20/4 (Glasgow: CREATe: UK Copyright and Creative Economy Centre, 2020), https://doi.org/10.5281/zenodo.3859493.
62 RS CMO/02, 28 February and 20 March 1699/1700, 110–11.
63 Lowthorp was elected on 30 November 1702.
64 John Lowthorp, *The Philosophical Transactions Abridged, and Disposed under General Heads*, 3 vols (for Thomas Bennet, Robert Knaplock, Richard Wilkin, 1703–1708), imprimatur leaf (dated 12 May 1705). The imprimatur leaf also records the Society's original approval of the project, on 5 May 1703. There was no Council meeting on the date of Newton's actual grant of the imprimatur, and we must assume he acted on his own authority in doing so. On imprimatur, see Moxham, 'The uses of licensing'.
65 For the printers of the *Abridgement* see note 64. The *Transactions* was printed by Samuel Smith and Benjamin Walford in 1705.
66 Keith Maslen and John Lancaster, eds, *The Bowyer Ledgers: The printing accounts of William Bowyer father and son*, 2017 edn (London: Bibliographical Society, 1991), entry 3586.
67 Motte, *A Reply to the Preface* ..., 12, in which Motte claims that the size of each of the rival continuations' first edition was 1,000 copies; Maslen and Lancaster, *Bowyer Ledgers*, entries 837, 1691.
68 Benjamin Baddam, *Memoirs of the Royal Society: Being a New Abridgment of the Philosophical Transactions ... 1665 to ... 1735 ... the Whole Carefully Abridg'd from the Originals, and the Order of Time Regularly Observ'd*, 10 vols (London, 1738–41).
69 David A. Kronick, 'Notes on the printing history of the early "Philosophical Transactions"', *Libraries & Culture* 25, no. 2 (1990): 243–68.
70 For the subscription price, see Kronick, 'Notes on the printing history'; the retail price is mentioned in Eric Gray Forbes, Lesley Murdin, and Frances Willmoth, eds, *The Correspondence of John Flamsteed, the First Astronomer Royal*, 3 vols (Boca Raton, FL: CRC Press, 1995–2001).
71 William Rutty to John Woolhouse at Paris, 10 and 20 August 1728, RS EL/R/2 ff. 3 and 14.
72 Jurin's correspondence with Dereham is discussed in Rusnock, *Jurin Correspondence*, 22–31.
73 Jurin was accused of failing to maintain a proper correspondence with Dereham, but it was Pucci (the Tuscan agent in London) who had been negligent, not Jurin. See letters between Jurin and Dereham (31 March 1726; 7 August 1727; 12 February 1728; 27 March 1728) all in Rusnock, *Jurin Correspondence*, 333–74. On the breach with Sloane, see also Richard Parkinson, ed., *The Private Journal and Literary Remains of John Byrom*, 2 vols (Manchester: Chetham Society, 1854), vol. 1, 278–9.
74 See RS AB/1/1/3, particularly the 1720s. On Benzel, see Rusnock, *Jurin Correspondence*, 29.

75 See, for instance, the issue for August 1700, which contains the titles of papers published in the *Transactions* for December 1699.
76 For an overview of the eighteenth-century translations of the *Transactions*, see Kronick, 'Notes on the printing history', 257 ff.
77 Gibelin's French translation-abridgement was published by Buisson in Paris in 1787–91. Its 14 volumes covered the *Transactions* from 1665 to the recent present – at least as far as 1783.
78 Leske's German translation-abridgement was published by Donatius in Lübeck and Leipzig in 1774–8. Its five volumes covered the *Transactions* from 1699 to 1720.
79 Jacques Gibelin, *Abrégé des transactions philosophiques de la Société royale de Londres* (Paris: Buisson, 1787–91), vol. 1, xviii.
80 Rutty to Dereham, 4 March 1729, RS EL/R/2 f. 16. The paper referred to is James Bradley, 'An account of a new discovered motion of the fix'd stars', *PT* 35.406 (1728–9): 637–61.
81 *Saggio delle Transazioni Filosophiche della Societa Regio, compendiate da Giovanni Lowthorp & Benianmino Mottes, tradotte dall'Inglese nell'idioma Toscana dal Tommaso Dereham*, 5 vols (Naples: Moscheni, 1729–34).
82 David A. Kronick, *Scientific and Technical Periodicals of the Seventeenth and Eighteenth Centuries: A guide* (Metuchen, NJ: Scarecrow Press, 1991). Ch. 5 identified 34 journals carrying substantive scientific or technical content published (at some point) in Britain between 1665 and 1790; most appeared after 1750.

Part II
Maturity and institutionalisation, 1750–1820

5
For the use and benefit of the Society, 1750–1770

Noah Moxham and Aileen Fyfe

By 1750, Cromwell Mortimer had been editor of the *Philosophical Transactions* (and secretary to the Royal Society) for two decades.[1] He had held the role even longer than his patron Hans Sloane, and in contrast to several earlier secretary-editors, Mortimer managed to keep his position after Sloane's presidency ended in 1741. In fact, Mortimer appears to have managed to rub along comfortably enough with Sloane's successor, and erstwhile rival, the antiquarian Martin Folkes. Historians have tended to characterise the Royal Society in this post-Newtonian period as leisurely, complacent and lacking in scientific dynamism.[2] The Society's meetings were overwhelmingly taken up with reading papers communicated by fellows, foreign members and strangers. Experimental demonstration had largely ceased, as had efforts to devise and conduct systematic research through the Society's limited institutional means.

This complacent inactivity could equally be seen, however, as evidence of an institutional stability that the Society had lacked in its earlier days. The financial situation was much improved, as investment and membership income grew. There were fewer abortive enterprises, and the danger of imminent collapse seemed to have been averted more or less indefinitely. The same was true of the *Transactions*: its contents were closely linked to the activity of the Society, and drawn from the material presented at meetings. Under Mortimer, it had settled into a publication rhythm of (roughly) quarterly issues. Every two years, these were also presented as bound volumes of eight issues. At the halfway point of the eighteenth century, the *Transactions* had all the appearance of a stable publication.

In September 1751, Martin Folkes suffered a stroke that left him partly paralysed; he was unable to attend Society meetings, and would soon be replaced as president. Meanwhile, in early January 1752, Cromwell Mortimer died. It was in the context of this power vacuum that the Society's relationship to the *Transactions* was transformed. After 87 years of (theoretically) independent existence, the *Transactions* was brought formally under the editorial and financial control of the Royal Society. The initial proposal, on 23 January 1752, was that the *Transactions* should henceforth be run 'for the sole use and benefit of this Society', rather than being privately controlled by one of the secretaries for his (admittedly notional) benefit.[3] On paper, this was a significant change of direction for a Society that had long disclaimed responsibility for the *Transactions*; yet in practice, there was little sense of disruption. It generated procedures that insulated the *Transactions* from changes of secretary-editor, and kept it following the same broad template for the next 80 years.

Mortimer's death created the opportunity for change, but the Society's willingness to embrace that opportunity was only partly due to perceived inadequacies in his editorial practice. During the previous 18 months, the Society, its president and the *Transactions* had all been subjected to very public critique and satire. It was not, of course, the first time that public critique had been directed at the Society or the *Transactions*, but the response would be notably different. Back in 1700, the Society's reaction to the *Transactioneer*'s critiques had been to reiterate its official lack of responsibility for the *Transactions*, its satisfaction with Sloane as secretary, and its particular gratitude to him for his work as editor (see Chapter 4).[4] In contrast, the reaction in 1752 was to implicitly acknowledge that the Society had indeed been responsible for the *Transactions*, and to take steps to ensure that, in the future, the Society would be able to control it more effectively. This involved the creation of collective editorial processes that replaced the judgement of an individual editor. These collective processes distributed responsibility among the fellows of the Society, and provided a shield that protected individuals.[5] They also created an impression of corporate approbation for what was published, something the Society initially sought to downplay.

Control of its reputation, via control of the *Transactions*' editorial processes, was clearly central to the Society's ambition to get the 'sole use and benefit' from the periodical after 1752. This chapter explores various possible 'benefits' to the Society, but it will become clear that money was not among them. The reformers were well aware that taking on the *Transactions* would mean taking on responsibility for funding its production. Indeed, it is only because the Society's financial position and general stability had improved so much over the early eighteenth century that something unimaginable in 1700 had come to seem possible by 1752.

Public critiques

Even before its president suffered a stroke, the start of the 1750s were proving difficult for the Royal Society, as the conduct of its leaders and fellows, and its (alleged) management of the *Transactions*, were brought firmly into the public eye. The sober accounts of new issues of the *Transactions* that usually appeared in the literary monthlies were joined by responses to the wild accusations levied against the Society in two anonymous pamphlets and one stout quarto book.[6]

The first pamphlet, *Lucina sine concubitu* (1750), caught the attention of the London public by describing how an unnamed fellow of the Royal Society had allegedly endeavoured to pass off the pregnancy of a chambermaid in his household as a case of immaculate conception. This scurrilous parody of the sorts of topics discussed at Royal Society meetings became a notable commercial success, reprinted at least seven times in London and once in Dublin, as well as being translated into French.[7] The second pamphlet, *A Dissertation upon Royal Societies* (1750), purported to be a series of letters from a foreign nobleman contrasting the meetings of the Society unfavourably with those of the Académie Royale in Paris. The following year, *A Review of the Works of the Royal Society* (1751) mercilessly exposed the weakness of some of the papers published in the *Transactions*. Its accusations were noticed in most of the contemporary literary monthlies, including the *London Magazine*, the *Universal Magazine* and the *Scots Magazine*.[8] Although only the *Review* carried the name of its author, by early 1751, literary London seems to have been aware that all three publications were the work of the same author.[9]

John Hill was an actor, botanist, author and apothecary. In 1746 and 1747, he had papers read to the Royal Society and published in the *Transactions*, and he initially had sufficient support among his reasonably wide acquaintance among the fellowship to hope to be admitted to the fellowship himself.[10] Yet this support evaporated, and his name was not even put forward as a candidate for the fellowship in 1748. Hill's subsequent satirical attacks on the Royal Society are usually interpreted as the simple manifestation of his disappointment and outrage. Tensions between him and the Society clearly existed by June 1750, when Hill was refused guest admission to meetings of the Society: *Lucina* had appeared by then, but it was not mentioned as the reason (and Hill had also criticised the Society in the *British Magazine*).[11]

The precise nature of the objection to Hill's election is not clear, although his biographer points reasonably to Hill's treatment of his friend Emanuel Mendes da Costa (which had involved Hill racing into print a book project that anticipated a similar one that da Costa had discussed with Hill), as well as a more general failure to observe the niceties of rank and reputation in his dealings with other scholars.[12] It is also not impossible that Hill's journalistic activity may have counted against him: he was heavily involved with both the *British Magazine* and the *London Magazine,* and later with Ralph Griffiths's *Monthly Review.* Writing for Grub Street periodicals in the eighteenth century was sometimes regarded uneasily by the fellows, although in this matter (as in so many others) the Society was scarcely consistent. For instance, in 1751, the Society elected Matthew Maty, the editor of the *Journal Britannique.* Maty's reputation as a literary reviewer was for largely uncritical positivity, and he was known to the Society in particular for increasing awareness of British publications (including the *Transactions*) on the Continent, both of which may have been an advantage.[13]

Hill was far more critical. He lambasted the conduct of the Society's meetings, the behaviour of its president and secretaries, and the quality of the papers it heard at its meetings and published in the *Transactions.* His strategy was most developed in the *Review,* and had three essential elements. First, to tarnish the Society's reputation by making it appear to be responsible for a periodical full of embarrassing trivialities; second, to attribute this to the corruption and incompetence of the Society's leadership; and third, to imply that the author's dissatisfaction with the Society's leadership and conduct was shared by some of the fellows themselves. Hill also elaborately sought to repudiate his own prior involvement with the Society.

In the 1750 pamphlets, Hill represented the Royal Society as a divided, ill-mannered, ignorant, inarticulate, arbitrary, corrupt, tedious, idle, self-regarding and nepotistic organisation, in which the president deliberately selected 'dissertation[s] upon nothing' to be read at meetings, which he would then sleep through, purely as a form of petty tyranny over the fellows.[14] The meetings themselves are a vision of pandemonium: crowded, noisy and incompetently run, at which papers of a lamentable standard were read inaudibly.[15] In the *Review,* Hill turned his attention to the *Transactions,* where he rode roughshod over any distinction the Society might wish to draw between itself and the *Transactions.* He conflated the *Transactions* entirely with the work of the Society, and suggested in unmistakable terms that the *Transactions* had been publishing worthless work since its inception, and that this was the fault of the Royal Society.

Hill's criticisms began at the top, protesting against the abuse of power (or incompetence) of the president, Martin Folkes. We actually know very little about Folkes's involvement with the *Transactions*, but, as we have seen, previous presidents – notably Isaac Newton and Hans Sloane – certainly did exercise personal influence over the periodical, and Folkes may have followed suit. But regardless of how much Folkes actually influenced Mortimer's editorial work, Hill rightly saw him as responsible for setting the tone for the Society's meetings, including the standards of scholarly behaviour and the acceptability of various topics.

Hill opened the *Review* with a kind of inverted dedication to Folkes, claiming that Folkes's bad character, ignorance and favouritism made it necessary for Hill to write his *Review*. This mock dedication made it clear that Hill's resentment was aimed at Folkes personally, not just at the president *ex officio*. Though the precise details of the affair are impossible to reconstruct, Hill accused Folkes of being two-faced, complaining that he had 'made me much more than I deserved' in conversation with Hill, while 'representing me to a noble friend' in an entirely different character.[16]

Hill also attempted to make Folkes personally responsible for the alleged collapse in standards at the *Transactions*, suggesting that 'if any body, except your great Self, had been in the high Office you so worthily fill at present [...]', such trivial papers would never have been published under the Society's imprimatur.[17] The accusation conflated the inherent corruption of patronage with intellectual corruption and the degradation of standards at the Society. George Rousseau has speculated that Hill was not so much sincerely concerned about the exercise of presidential influence, as aggrieved at the failure of those mechanisms to work for him personally.[18] This was a legitimate grievance, since the standards required for election to the fellowship remained ill-defined (despite attempts by Hans Sloane to tighten them) and many gentlemen had been elected who had made appreciably fewer contributions to the Society and the *Transactions* than Hill. In the *Review*, Hill was necessarily circumspect about his earlier ardent campaign to join an organisation he now excoriated, using the coincidence of the Society's recent election of another John Hill (a Northamptonshire MP) to claim that he was not in fact the gentleman who had been seeking election.[19]

Despite Hill's undoubted disgruntlement with the current leadership and operation of the Society, his critique of the *Transactions* began at the beginning, ridiculing work dating to its very foundation. His main text opens with a contemptuous account of a paper from Oldenburg's third issue suggesting that the scent of pennyroyal was fatal to rattlesnakes.[20] Animadversions on other papers from the late seventeenth century were scattered through the *Review*. Hill took it for granted that the Society should be seen as responsible for the contents of the *Transactions*, thereby insinuating that the Society's penchant for trivial and unsubstantiated papers was a tradition extending back virtually to its foundation.

Hill did not attack the Society's activities indiscriminately and some sections contained serious scientific reflection and criticism. As Rousseau has noted, the brunt of Hill's attack focused on papers dealing with natural history and antiquities.[21] It was not that Hill disdained natural history and antiquities: rather, these were precisely the areas in which Hill was most interested and in which he hoped to distinguish himself. His treatment of a 1702 paper on 'a Way to catch Wild Ducks' illustrates his efforts to deflect any criticism of himself, or any possible defence by the Society, back to the discredit of the Society. Hill gave an essentially fair summary of the actual paper, describing a duck-hunting technique practised in Ceylon that involved the hunter wearing an earthenware pot with eyeholes, venturing into a body of water where ducks congregate, and then advancing slowly on the ducks with only the jar showing above the water and pulling them down briskly by the legs in turn. To this, Hill proposed an alternative method, advocating the tying of a piece of fat bacon to a long string. The ducks would eat and then excrete the bacon, each in turn, so that 'in the Morning a whole String of Ducks will be found ready catched, and there needs only the drawing in the String to take them all up'.[22] Hill claimed that the Society had supressed the account of this alternative technique, not because of any absurdity, 'but because it contradicted the former'. Thus, Hill not only brought to light the triviality of certain papers in the *Transactions* and (in passing) linked the Society to editorial decision-making, but also accused the Society of ignorance and intellectual dishonesty.

As regards the more recent *Transactions*, Hill's critique was directed at Folkes in general (for allowing, or encouraging, the publication of substandard papers), and at two unusually prolific fellows whom he particularly despised: the Norwich-based naturalist William Arderon and the microscopist Henry Baker, in London. In the anonymous *Dissertation*, he had lampooned them as Hardyrun and Bokur, and accused them of being chiefly responsible for 'render[ing] the *Philosophical Transctions* of late so ridiculous throughout Europe'.[23] Baker's name certainly appeared frequently in the *Transactions*: he published 24 papers of his own between 1739 and 1766, and communicated a further 87 papers by other people. One of those others was Arderon, who had 20 papers in the *Transactions* between 1744 and 1750 alone, almost one per issue on average. Arderon and Baker had thus been highly visible within the pages of the *Transactions* in the 1740s, and their prolific output made them a convenient target. So too did the nature of Arderon's work on local natural history, which Hill stigmatised as relentlessly trivial. In the *Dissertation*, Hill insinuated that Folkes would intercede to ensure Arderon's papers were read as soon as received, implying either that the Society was bereft of worthwhile material or that its protocols were manipulated in favour of the president's cronies.[24] It is perhaps unsurprising that Arderon decorated his own copy of Hill's *Review* with a mock title page describing it as 'a lying and abusive Representation' (Figure 5.1).

The fact that Hill had himself published in the *Transactions* might have been seen as diminishing the power of his critique, yet he adroitly turned an acknowledgement of his own contributions into a further accusation against the Society. He referred to those contributions as the record of observations made at the philosophical meetings at his house, that had been 'somewhat too pompously recorded in their [the Society's] Transactions'.[25] This phrasing distanced Hill from responsibility for his own paper, both for putting it into print and for its literary style. Editorial marks on surviving manuscripts from this time show that the most typical editorial intervention was simply to strike out the personal elements of a letter.[26] Substantive alterations – such as those implied by Hill – were rare. However, it was true that the published version of Hill's letter on the seeding of mosses carried a note specifying that it had been read (and published) 'with alterations'.[27] Those changes were not enough in themselves to justify the attempt to blame the Society for their stylistic defects, but the hint that the Society habitually mangled the work of authors in bringing it to print was unmistakable.

For all his critiques of the Society, its leadership and its *Transactions*, Hill was careful not to simply rage against the extent to which social position and clubbability counted within the Society. He placed the blame, not on the fellowship as a whole, but on the 'busy and ignorant Persons' who 'thrust themselves into Employment', that is, the president and secretaries. These officers were allegedly wrecking the Society, failing to carry out their responsibilities and, by their personal characteristics, discouraging new members, for 'those who are able to do the Body Credit' would 'refuse to join their Labours to those of such unworthy Associates'. In the same breath he implied that there were others within the fellowship, 'Men great in all Senses of the Word', who

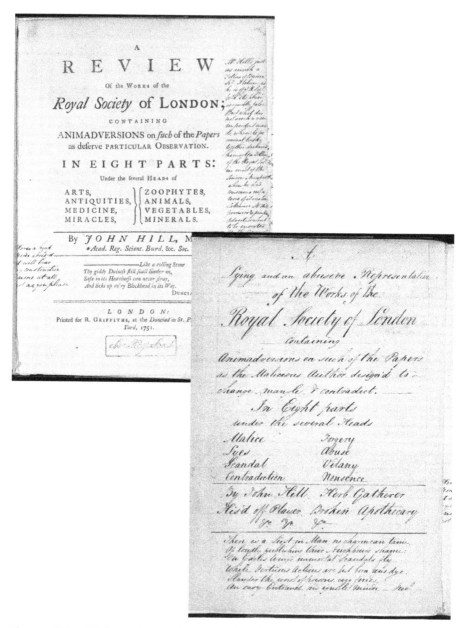

Figure 5.1 William Arderon's copy of John Hill's *Review*, 1751, with his handwritten alternative title page for a work he considered 'A lying and abusive Representation' © The Royal Society.

'agree with the author in his opinions', and were determined to use Hill's work as 'the Basis of a Reformation' of the Society.[28] As would become clear after Cromwell Mortimer's death, there certainly were reform-minded individuals in the fellowship, but it is not clear whether they were the men imagined by Hill.

The *Review* received extensive coverage in both the *Gentleman's Magazine* and the *Monthly Review* in February 1751, though it is likely that neither came from a neutral party. The *Gentleman's Magazine* featured an extract from the *Review*, a straight-faced discussion and refutation of a paper on fossil plants from Oldenburg's *Transactions*. Such extracts were a normal aspect of noticing and reviewing books at the time. The intriguing thing is how this extract was framed. The previous issue of the *Gentleman's Magazine* had appeared just before Hill's *Review* was published, and had contained a short paper on Derbyshire fossils, signed 'L.C.', and a letter on Derbyshire minerals, signed 'R. Roe'.[29] In February, 'L.C.' wrote again to the editor of the *Magazine* explaining that, having now seen Hill's *Review*, he realised that his earlier contribution – based on material from the *Transactions* – was 'exploded, by the above ingenious writer' (namely, Hill), and suggesting that the editor should share with his readers 'a specimen of his manner of treating the *Society* and his subject' (that is, the extract from the *Review*).[30]

The elements of this exchange line up suspiciously conveniently for Hill: a paper that might have been calculated to give the *Review* a chance to shine appeared in the same month as Hill's *Review*; and the unknown author then graciously acknowledged his mistake, and showcased Hill's research in the same gesture. It is possible that Hill was both 'Roe' and 'L.C.' ('Roe' was a conventional pseudonym of the period, and Hill had previously adopted it in connection with *Lucina*). Hill was certainly sufficiently experienced and well-connected in the world of London literary periodicals to have orchestrated such a sequence. If this was the case, Hill was continuing to manage the reception and impact of his work, and was partly responsible for actively stoking the affair in public. In the swirl of elaborate fictions and false identities, the mixture of truth and falsehood, the counterpoint between legitimate critique and savage partiality, it is possible to detect a self-delighting as well as self-protecting and polemical element. Hill's possible involvement in the reaction to his own work makes it hard to analyse the impact of his critiques.

The other extended treatment of Hill's *Review* is also potentially linkable to Hill, for it appeared in the other major literary periodical of the day, the *Monthly Review*, edited by Ralph Griffiths. Hill had been a regular contributor to the *Monthly* from its inception in 1749; and Griffiths was the publisher of Hill's *Review*. Perhaps unsurprisingly, the *Monthly* devoted 15 pages to its notice of the *Review*. The anonymous reviewer began by gently reproving Hill's tone for a couple of paragraphs, but for the next 14 pages recapitulated, in Hill's own 'merry' tones, his criticisms of the Society and its officers:

> [Readers] are to know then, that, through the whole of this work, the author acts the petulantly humourous critic, in order to turn the *royal society* and their writings into ridicule. He is particularly bitter against the president, and a few others; and takes an ill-natured pleasure in treating them with the utmost contempt and severity, not to say, ill-manners.[31]

The reviewer also reproached Hill with throwing the baby out with the bathwater: 'To argue, as he does, that because some unphilosophical papers are to be found among the society's writings, therefore the whole must be one mass of absurdity, is altogether unfair.' With that said, however, the *Review* 'will be attended with considerable advantages to the public', in three respects: forcing the Society to impose stricter criteria of merit for publication, teaching a credulous public not to take on trust everything stamped with the Society's imprimatur, and correcting a number of errors in natural history.[32]

The concession of the utility and justice of some of Hill's critique (while reproving its excesses) is particularly suggestive because authorship of this notice has been attributed to John Ward.[33] Ward was an established fellow of the Royal Society, and 18 months after this notice appeared, he would become part of the new governing clique within the Society that, among other things, took formal control of the *Transactions*. If it was indeed Ward, this implies that there were, as Hill had claimed, a group of influential fellows within the Society who shared at least some of his concerns. There is no positive evidence that the reformers alluded to by Hill were the same as those who did indeed reform various aspects of the Society in 1752; and the discussions within the Society in early 1752 made no public acknowledgement of Hill's attacks. That said, contemporaries had no difficulty in seeing a general link between those attacks and the move to take over the *Transactions*. When the *Gentleman's Magazine* took note of the reform proposals in March 1752, it reported that the plans were made for the Society's 'honour and reputation', which had been 'much injured by an enemy to that body of which he attempted, but in vain, to become a member': that is, unmistakably, John Hill.[34]

The 1752 reforms

In an 'advertisement' printed in the *Transactions* afterwards, the Royal Society attributed its decision to take on the management of the *Transactions* to the difficulties faced by an individual editor coping with growth: the fellowship was 'of late years greatly inlarged', and 'their communications' were 'more numerous'.[35] The burden placed on secretaries by the growth in the fellowship, and in the quantity of material being published, would become a constant theme in the Society's administration for the next 250 years. However, the records of the Council meetings make clear – even without mentioning John Hill – that it was the public reputation of the Society, not the editorial workload, that really motivated the change.

The key figure in the reform was George Parker, the Earl of Macclesfield and one of the 21 members of the Society's ruling Council; he would replace the paralysed Martin Folkes as president later in 1752.[36] He was supported by the vice-presidents Sir Hugh Willoughby and Lord Charles Cavendish, and they were all part of the 'Hardwicke Circle', a clique with overlapping literary and scientific interests and shared Whig allegiances. Philip Yorke, later the second Earl of Hardwicke, was one of the Council members who supported Macclesfield's plans.[37]

The death of Cromwell Mortimer, in January 1752, gave the reformers their opportunity. Just a week after Mortimer's death, Macclesfield was urging his fellow Council members to consider the Society's role in the future publication of the *Transactions*, and in particular, to consider 'in what manner the Papers, which have been Communicated to the Society have afterwards been introduced to the Publick'. As Macclesfield pointed out, the *Transactions* intimately affected the public reputation of the Society, since 'in the general estimation of the publick and more especially of foreigners' the periodical was perceived as being 'ushered into the world under the aprobation, or at least under the inspection of the Society'.[38]

Detailed proposals for the future management of the *Transactions* by the Society were drawn up in the form of questions for the members of Council, who approved the substantive points at their next meeting on 15 February. By late May, the Council had developed a series of rules that would govern the future editing, production and distribution of the *Transactions*.[39] The first volume printed under the new dispensation appeared in late spring or early summer 1753: this was volume 47, covering 1751 and part of 1752.[40] Thus, the *Transactions* came to be the Society's business (and from this point on, there is more surviving archival evidence about its operations).

The speed with which Macclesfield and his colleagues put forward their comprehensive, carefully justified and well-supported proposals suggests that they had been waiting for an opportunity. Macclesfield admitted that he and his comrades were taking advantage of 'this Season of a Vacancy in the Secretary's Office' to push their proposals through. He insisted that no criticism of any person, living or dead, was intended, even though Hill had vilified Folkes personally, and even though Mortimer, as the long-serving editor of the *Transactions*, had to be an implicit target of the reforms. However, the reformers scrupulously avoided the risk of derailing their own objectives by getting embroiled in quarrels about personalities and individual competence. Mortimer's death allowed his editorial practices to be buried without public consequence, and Folkes's illness removed him from the scene. The reformers also sought to avoid casting aspersions on the competency of any of the prospective candidates to replace Mortimer as secretary, claiming that, since they 'could never be suspected of any want of iudiciousness in choosing the proper pieces to be made public', they could not possibly be offended by this move to dramatically limit the scope of the office they sought.[41]

In contrast to earlier occasions when the Royal Society's reputation had seemed threatened by the *Transactions*, the idea of simply reiterating the Society's traditional official distance from the *Transactions* did not arise. It would have been more difficult to be convincing now that the printed *Transactions* papers carried explicit evidence of their association with the Society's meetings and fellows: since 1742, the date of the meeting at which papers were read and, if applicable, the name of the fellow communicating the paper appeared on the published papers. But it may also have been the case that none of the current Council members were entirely certain as to the precise history of the relationship between the Society and the *Transactions*. Indeed, two of the more historically minded members, John Ward and Thomas Birch, were tasked to search the Society's archives and the pronouncements of past editors in an effort to clarify the situation.[42] But even before they had reported back, it was apparent that plans were afoot to bring the *Transactions* into the institutional fold.

There was probably no serious alternative. The *Transactions* enabled the Society to appear more than the mere learned discussion club that some critics believed it to be. But as the sole public manifestation of the Society's activity, it became the basis for public judgement on the Society, as Hill's critiques had so forcefully pointed out.[43] John Harris had made the same point during the *Transactioneer* spat in 1700, but this time, the Council accepted the implications. By conceding the reputational link between the Society and the *Transactions*, some measure of formal institutional control became inevitable. Seen in this light, the Society's decision finally to assume responsibility for the *Transactions* is relatively unsurprising and its wider logic straightforward. More intriguing are the questions of why and how the reformers sought to create collective editorial practices, rather than rely on an individual editor.

The 1752 reforms created a 'Committee of Papers' which took on the responsibility of selecting papers for publication in the *Transactions*. The value of spreading the editorial workload among a group of people had previously been tested by the co-editorships of the early 1680s and the editorial team of the early 1690s (see Chapter 2), but since then, Sloane, Jurin and Mortimer had performed their solo editorial roles with apparent success. In contrast to the late seventeenth-century use of co-editorship to share workload, the focus of the 1752 reforms would be on creating a committee that shared decision-making and responsibility among a group. From this point onwards, the *Transactions* had, in theory, no editor other than the Committee of Papers, which would remain the formal editorial decision-making body for the Society's periodical(s) until the twentieth century. The Committee of Papers succeeded in sustaining the regular, if less frequent, production of the *Transactions*. In the 1750s, it issued one 'part' each year (amounting to a volume every two years); and from 1763, it issued 'parts' every spring and autumn, and bound them as annual volumes. The *Transactions* would retain this (roughly) six-monthly periodicity until the late nineteenth century.

Before we look more closely at how the Committee of Papers actually operated, it is worth considering how it differed from John Hill's call for 'a Committee of Inspection of Papers'. The key differences concerned its position within the Society's processes, and the composition of its membership. (It also differed from an internal 1742 proposal for a 'Committee of Papers', which would not have been an editorial committee at all, but a group of fellows charged with undertaking a retrospective survey of the fate of all papers presented to the Society over the past year, and to determine what action, if any, should be taken to preserve those papers that had *not* been published in the *Transactions*.[44])

The procedures drawn up in 1752 formalised two elements that would later come to be seen as distinctive features of journals published by learned societies, as distinct from those managed by individual editors. The first was the use of a committee to make editorial decisions. The second was the continuation of the 1742 practice that only papers previously read at meetings of the Society could (potentially) appear in the *Transactions*. This meant that oral presentation to the fellowship functioned as an initial step in the editorial process for publication. This model was used at other learned societies; and it continued in use at the Royal Society until the 1890s, when the expansion of research finally overwhelmed the capacity of a one-hour weekly meeting.[45] The 1752 arrangements gave the appearance of fulfilling John Hill's desire that there be an opportunity for the 'Sense of the whole Society' to be heard upon the papers presented for possible publication.[46] However, there would be no formal mechanism to enable the views of the listening fellows to be taken into account by the Committee of Papers (and under some later presidents, notably Joseph Banks, there was in fact no discussion of papers during the formal part of the meetings).

John Hill's proposal for 'a Committee of Inspection of Papers' involved a more radical overhaul of the Society's procedures than was in fact implemented in 1752. He wanted to improve the quality of the weekly meetings, as well as of the *Transactions*, and proposed a committee that would intervene at a much earlier stage, as Figure 5.2 illustrates. Hill's hypothetical committee would have scrutinised the papers submitted to the Society, and would have been charged with selecting those 'worthy to be read' to the fellows at a meeting. This role was theoretically performed by the president and secretaries as they prepared the agenda for meetings, but, as Hill had pointed out, the current level of scrutiny seemed minimal, with almost anything submitted by a fellow appearing on the agenda. The discussion at the meeting of fellows provided Hill's next stage of scrutiny, and then, he proposed, 'let them pass a second Examination [before the committee] before they are ordered to be printed'.[47] Hill's editorial committee would thus have functioned as an initial gate-keeper, as well as a later court of pre-publication scrutiny. The actual 1752 Committee of Papers performed only the latter role, and left control of the meeting agenda in the hands of the president and secretaries (where it remained until the 1890s).

Hill's proposals would have added rigour and some transparency to the pre-publication processes of scrutiny, and, among other things, would have meant that ordinary members of the Society would no longer be able to take it for granted that their papers would be read, let alone published. By long-established custom, the formal communication of a letter or paper to the Royal Society required the intercession of one of its fellows, and this practice continued after 1752, even though it was not mentioned in the new procedures. By the 1780s, the principle of accepting papers only when communicated by a fellow would be so

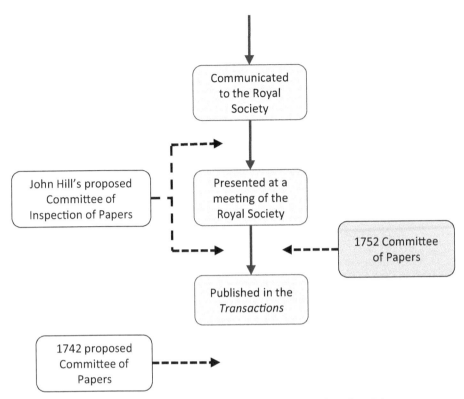

Figure 5.2 The role of the 1752 Committee of Papers in the editorial process.

solidly entrenched in the Royal Society's procedures that it could be invoked as a rule by the president, even though it was not actually written down anywhere.[48] Allowing fellows the de facto right to have their own work read before the Society (and considered for the *Transactions*) gave them a privileged position in the publishing system. And allowing them to put forward the work of their friends, relations and associates created a system in which the Society's ordinary membership acted as gate-keepers controlling the access of outsiders. It meant that someone like Hill, who had made himself a pariah to the institution, could be irrevocably excluded from both the Society and its publications. The fellows' privilege of communicating papers would remain a crucial, if occasionally contested, element of the Society's editorial process until the late twentieth century.

Macclesfield and his reforming colleagues agreed in principle that the Society's editorial judgements needed a broader basis, but their actions showed them to be far less interested in curbing existing forms of privilege and authority than Hill would have liked. Hill wanted to significantly neutralise the traditional influence exercised by the president, the secretaries and their coterie, and he wanted to transfer decision-making power to a committee of ordinary fellows. He proposed a committee of at least three fellows, selected to ensure the committee had knowledge of each of 'the several Sciences that relate to the Business of the Society', and he proposed to exclude the president and secretaries.[49]

Hill's argument implied that the senior officers of the Society were not competent to judge. Unsurprisingly, as members of Council themselves, this was something that Macclesfield and his colleagues did not accept. Thus, their 1752 proposals had the president, vice-president and secretaries as *ex officio* members of the new editorial committee, and insisted further that, in addition to requiring a quorum of five members, the Committee of Papers could not be convened without the president or his deputy. This was a clear assertion, *contra* Hill, that the officers who had influenced the *Transactions* hitherto should retain an editorial voice within it, even if it was now diminished.

Further, the new regulations insisted on the scientific competence of the Council. In contrast to Hill's idea of a separate editorial committee, with its own distinct membership, the reformers declared the existing Council to be (also) the new 'Committee of Papers'. In other words, although the Council and the editorial committee would meet separately, with their own agendas and minute books, both meetings were attended by the same individuals. This proposal, so opposed to Hill's ideas, was put forward by Philip Yorke.[50] As Yorke was a key ally of Macclesfield, it seems highly likely that the proposal was fronted by him to avoid the appearance of a naked power grab by Macclesfield and his associates on Council. That said, it was made in the context of a plenary meeting of all fellows that had been called in response to the reading of the Council's proposals, at which ordinary fellows were simultaneously notified of the Council's intended reforms, and informed that their approval was not actually required.[51]

During the spring months of 1752, Macclesfield and his associates developed a detailed set of procedures for the operation of their new Committee of Papers. A striking feature was that decision-making was to be done by secret ballot, without 'discussion or deliberation'. The justification for the use of a secret ballot was to enable each committee member to be 'more at liberty to declare freely his Opinion, in favour of or against the Question', a phrasing that suggests a desire to avoid allegations of undue influence by certain senior members of the committee.[52] The secret ballot had the effect of generating a collective decision that left no recorded trace of individual opinions, thus providing a mechanism for one fellow of the Society to disagree with another fellow without generating rancour that might disrupt the norms of gentlemanly politeness.

This approach to collective decision-making contrasts with that taken by other learned institutions that recognised a value in discussion and consensus-formation in the evaluation of knowledge. For instance, the Paris Académie Royale already had an editorial committee overseeing its *Histoire et Mémoires*. James McClellan has shown that, on at least some occasions, sub-groups of the committee worked together to produce consensual reports on papers submitted to the Académie, a process that necessarily involved discussion.[53] And when the Royal Society of Edinburgh came to consider the materials for the first volume of its own *Transactions* in 1784, its Committee of Publication divided itself into subject-specific subcommittees, whose members were to circulate the papers between themselves in advance of a meeting at which 'a conversation may be held on the different papers' to determine which were 'fit for publication'.[54] The final decision took place by a ballot, but only after opportunities to build consensus through discussion. The London society's determination to avoid discussion or deliberation may have been intended to prevent senior figures exerting undue influence over the votes of other committee members, but it is difficult to see how the meetings could have functioned without at least some discussion. For instance, there is later evidence that the committee was able to agree to defer consideration of certain papers to future meetings.

The contrast with Paris illuminates another aspect of the London practices. In Paris, at least by the later eighteenth century, the academicians scrutinised the observations submitted to them by outsiders in great detail. This process could take a year or more, and could involve checking the experiments or asking for additional experiments, as well as requesting (or making) revisions to the paper.[55] In London, the Royal Society's committee operated quite differently. First, the focus at the committee meetings was not usually the full text of the paper, but the 'minute' of each paper recorded by the secretary in the Society's 'Journal-book' as a record of the Society's meetings. These summaries were read to the committee, in order, ahead of each vote. Any committee member could request that the entire paper be read out, but the norm was theoretically a vote based on the secretary's summary plus any personal recollection of hearing the paper at an ordinary Society meeting sometime in the previous month.

Second, the use of a ballot meant that the committee was constructed to make yes/no decisions about potential publication. It had no mechanism for discussing the adequacy, or not, of experimental procedures, or the validity of the conclusions drawn from them, let alone requesting further work or revisions. There is evidence from subsequent practice that committee votes could be conditional upon some revision – usually, straightforward deletion of certain sentences or paragraphs – and also that revisions of this type were sometimes requested (or made) by the secretaries after the committee meeting, but the point remains that the London committee did not at this time engage in sustained examination or revision of papers in anything like the way that its Paris equivalent did. This is one of the reasons that papers could appear in the *Transactions* within six months of submission, whereas mid-eighteenth-century volumes of *Histoire et Mémoires* generally appeared two or three years after the year actually printed on the title page.

The 1752 procedures included a provision for the committee to call for additional assistance in evaluating papers needing expertise not adequately represented on the committee. The committee could invite the participation of other fellows 'who are knowing and well Skilled, in any particular branch of Science that shall happen to be the subject matter of any Paper which shall be then to come under their deliberation'. Any such fellow who was invited to attend a specific meeting of the Committee of Papers would be permitted to vote on all papers considered at that meeting.[56] There is no archival trace of these mechanisms being extensively used in the 1750s or 1760s.[57] However, this provision is significant because it later became the basis of the practice of 'referring' the full text of papers to a fellow with appropriate expertise, whose report (initially oral, later written) would be received by the committee ahead of the vote (see Chapter 8). The inclusion of this provision in the 1752 procedures also raises substantial, and unanswerable, questions, about what was really expected to happen at meetings: how could a committee that was not allowed to discuss papers make the decision to call in external expertise? And how could a committee where each man's vote was supposed to count equally, give additional weight to the opinion of an expert member of the committee?

As well as practical questions about the operation of the Committee of Papers, the 1752 reforms also raised an epistemological question about the meaning and significance of publication decisions made by the committee. Despite having created a mechanism for making collective decisions, the Royal Society explicitly insisted that it never, as a body, officially endorsed anything submitted to its judgement. This position was presented in the 'advertisement' printed at the beginning of volume 47, the first that appeared under the new management. This 'advertisement'

(Figure 5.3) insisted that 'it is an established rule of the Society, to which they will always adhere, never to give their opinion, as a body, upon any subject, either of nature or art, that comes before them'.[58] It would be reprinted in every subsequent volume of the *Transactions* until 1959.

The 'advertisement' specifically warned that the Society never endorsed the claims of inventors or 'projectors', and thus, if any such person should claim that their invention or scheme enjoyed the support of the Society, that claim was by definition fraudulent. A surviving draft of the 'advertisement' placed even greater emphasis on the potential claims of 'projectors', suggesting that this issue – rather than the epistemological status of the papers in the *Transactions* – may have had greater prominence in the minds of the reformers.[59] Again, the Society's position stands in striking contrast with that of the Paris Académie Royale. The Académie had a formal duty to adjudicate the claims of inventors (in order to award patents) on behalf of the king, and this is one reason why its eighteenth-century editorial approach involved rigorously examining and, where possible, re-testing the claims of what outsiders communicated to it.[60] The Royal Society had no such authority, and was extremely reluctant to allow its imprimatur to be translated into public endorsement.[61]

In Paris, the imprimatur of the Académie Royale could be seen as the stamp of approbation. In London, in contrast, the voluntary association of learned gentlemen that was the Royal Society sought to explain their reasons for selecting some papers over others without claiming to use truth or certainty as a criterion. The 1752 'advertisement' explained that:

> The grounds of their choice are, and will continue to be, the importance or singularity of the subjects, or the advantageous manner of treating them; without pretending to answer for the certainty of the facts, or propriety of the reasonings contained in the several papers so published, which must still rest on the credit or judgement of their respective authors.[62]

Thus, the Society disclaimed all responsibility for the certainty or reliability of the claims to knowledge advanced in the papers it chose to publish in the *Transactions*. This position would be emulated by the Manchester Literary and Philosophical Society, three decades later, which copied the London model yet admitted in the preface to the first volume of its *Memoirs* that 'a majority of votes, delivered by ballot, is not an infallible test of excellence, in literary or philosophical productions'. It therefore also copied London by insisting that all 'responsibility

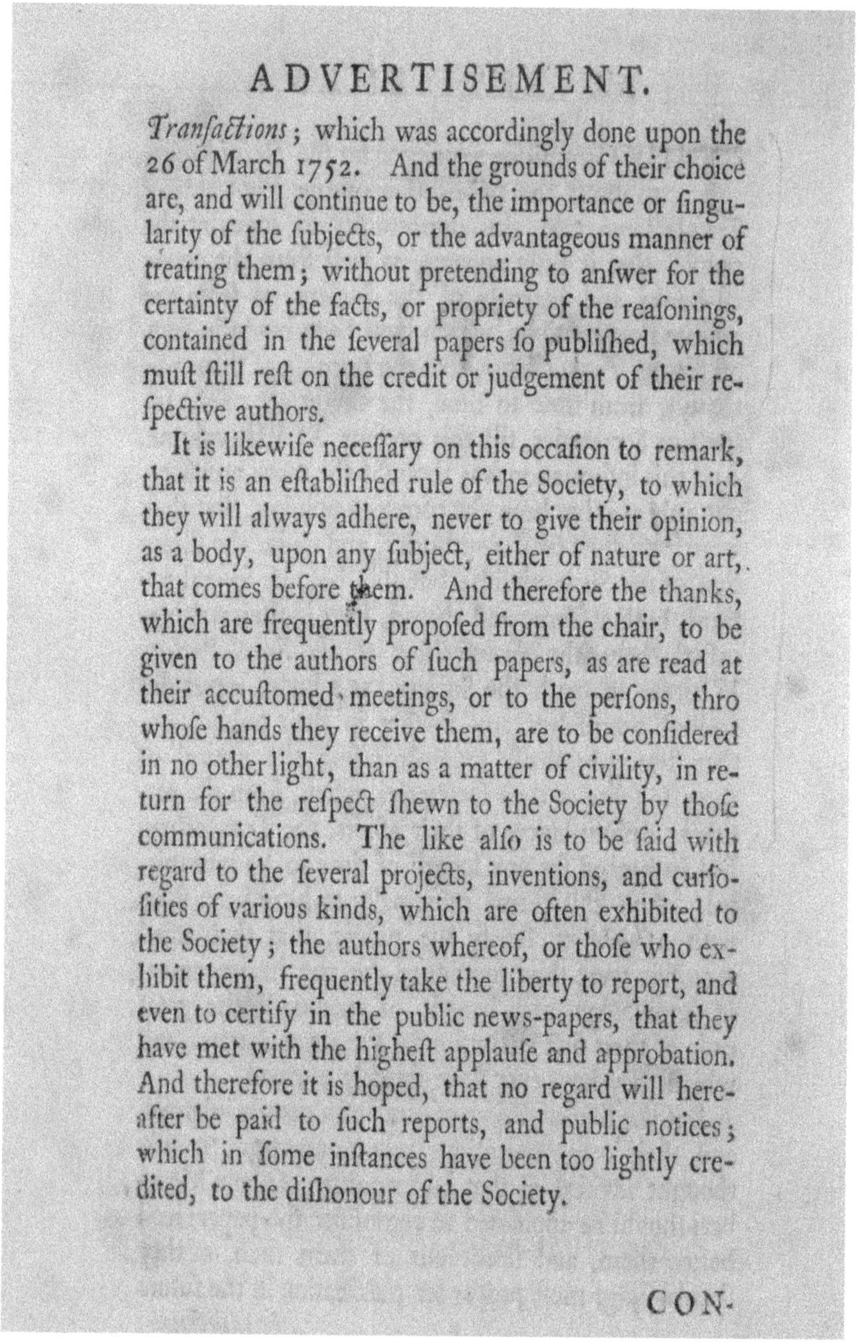

ADVERTISEMENT.

Transactions; which was accordingly done upon the 26 of March 1752. And the grounds of their choice are, and will continue to be, the importance or singularity of the subjects, or the advantageous manner of treating them; without pretending to answer for the certainty of the facts, or propriety of the reasonings, contained in the several papers so published, which must still rest on the credit or judgement of their respective authors.

It is likewise necessary on this occasion to remark, that it is an established rule of the Society, to which they will always adhere, never to give their opinion, as a body, upon any subject, either of nature or art, that comes before them. And therefore the thanks, which are frequently proposed from the chair, to be given to the authors of such papers, as are read at their accustomed meetings, or to the persons, thro whose hands they receive them, are to be considered in no other light, than as a matter of civility, in return for the respect shewn to the Society by those communications. The like also is to be said with regard to the several projects, inventions, and curiosities of various kinds, which are often exhibited to the Society; the authors whereof, or those who exhibit them, frequently take the liberty to report, and even to certify in the public news-papers, that they have met with the highest applause and approbation. And therefore it is hoped, that no regard will hereafter be paid to such reports, and public notices; which in some instances have been too lightly credited, to the dishonour of the Society.

CON-

Figure 5.3 The 'advertisement' prefaced to volume 47 of the *Transactions* © The Royal Society.

concerning the truth of the facts, the soundness of reasoning, or the accuracy of calculation' must lie with the authors, not the society.[63]

The collective decision-making processes created by the 1752 reformers – namely, the Committee of Papers – were supposed to evaluate 'importance', 'singularity' and literary style, not certainty or truth. This was a fine line to tread, and the epistemological status of the committee's editorial decisions would often be misunderstood, particularly once papers in the Society's periodicals came to play a role in the reward and recognition systems of nineteenth- and twentieth-century scholarly life.

The public response to the new editorial regime was muted. In 1754, the *Monthly Review* – which had consistently antagonised the *Transactions* – glossed the 'advertisement' as meaning that the works published under the Society's auspices should be considered as incremental contributions to the progress of natural knowledge. Despite what the Parisian academicians sought to do, the *Review* assumed that 'it is impossible, in the nature of things, that the importance' of new claims or observations should 'appear at once'. Rather:

> The hints of one year, may the next be carried on to experiments; and those experiments gradually open either a new, or an improved field of natural knowledge. The design of the [Royal] society is to incite the learned, in all parts of the world, to improve upon their labours, to correct them where necessary.[64]

This suggests the persistence of the late seventeenth-century sense that periodicals were not the place for definitive, authoritative statements. It would be repeated in the *Critical Review* in 1756, whose condemnation of the unevenness of the material in the first part of volume 49 of the *Transactions* compared it to 'a collection of rude drawings in a school of young painters, among which we, now and then, meet with the sketches of a master'.[65] This language implies that even the best work in the *Transactions* was seen as preparatory 'sketches' rather than definitive; but it had something in common with the *Monthly Review*'s vision of the pieces appearing in the *Transactions* as incremental building blocks, or valuable steps on the way to something more authoritative.

Overall, the 1752 reforms had created protocols for the evaluation of papers that were intended to be more equitable and less arbitrary than the former reliance on an individual secretary-editor. The new system appeared to represent the collective voice of the fellows, rather than just one of the secretaries (who might be more or less heavily influenced by the president), but the group of fellows involved was far more limited than John Hill had wanted. The procedures also left unclear who would actually manage the new editorial system, from setting the agenda for Committee of Papers meetings, to acting on its decisions. There remained, therefore, ample opportunity for the senior officers of the Society to exercise hidden influence in the interstices and around the margins of the written rules.

It is difficult to say how the new mechanisms worked in practice, because the minutes of the Committee of Papers prior to 1780 have unfortunately not survived (other than a few largely uninformative draft lists of papers considered from 1753 and 1754). It is clear from later evidence, however, that the Society's leaders did retain significant ability to determine which papers would be published. For instance, in 1774, the then-president felt able confidently to predict – ahead of the committee meeting – that a paper by Joseph Banks would be published.[66] That said, the judgements of the Society's leaders were now subsumed into the collective verdict of a committee, which shielded those individuals from the imputations of individual negligence or corruption that had been levelled at Hans Sloane, Martin Folkes or Cromwell Mortimer.

The reformers did not adopt Hill's suggestion for the pre-screening of all papers, and the privileged status of fellows as contributors to the *Transactions* was maintained. For the *Critical Review*, this meant that the Society continued to have 'crude and trivial essays' (or 'fungous excrescences'!) forced upon it by 'the impertinence of frivolous correspondents', as well as more 'valuable fruit'.[67] The *Critical Review*'s complaints that the *Transactions* should have 'been better weeded before it was presented to the public' suggest that the new editorial regime did not immediately transform the public reputation of the *Transactions*.[68] Nor does Richard Sorrenson's analysis of the range of topics treated in the *Transactions* show any significant change after 1752, despite the theoretical removal of any bias towards secretarial-presidential interests.[69]

Thus, the public reviews suggest that the effect of the editorial reforms on the content of the *Transactions* was perhaps not very noticeable to contemporaries. Furthermore, the evidence of the manuscript papers in the Society's archives suggests there was no distinctive change in editing practice at the level of individual detail. The significance of the 1752 reforms lies in the Royal Society's formal assumption of responsibility for the *Transactions*, after many decades of remaining at arm's length, and in creating the conditions that would later lead to editorial practices that became central to academic recognition and reward.

Costs and benefits

As well as editorial control, in 1752, the Royal Society assumed financial responsibility for the *Transactions*. This turned out to be a project that would not show an annual profit for two centuries, and which would, by the 1890s, come to burden the Society with costs it struggled to control or recoup. However, the commercial language of profit and loss is not the only framework for understanding the finances of the *Transactions* in the late eighteenth and nineteenth centuries. The Royal Society's willingness to support the costs of publishing and circulating the *Transactions* came to be part of its mission for scholarship, in which material quality and the widespread circulation of knowledge would be worth paying for.[70]

Financial arrangements were very much the second order of business for the 1752 reformers, after the editorial protocols. Nobody appears to have tried to estimate the likely cost or possible profit of publishing the *Transactions* before taking it over, though the notion that it might turn a profit was at least countenanced. John Hill had insinuated that Cromwell Mortimer was a busy seeker after income-generating 'employment', even though there is no evidence that Mortimer (or any of the secretaries since Oldenburg) had made any money at all from the *Transactions* (in contrast to Hill, who certainly did make money from his publishing activities).[71]

The initial resolution of Council implied the possibility of profit when it proposed that, 'for the future, the Philosophical Transactions shall not be printed and published for the benefit and Advantage of the Secretaries or of either of them, or of any Printer or Bookseller to be appointed by them or either of them'. However, any reference to pecuniary 'advantage' was eliminated from the formal statute that approved, which simply specified that the periodical should henceforth be run 'at the Sole charge, and for the Sole use and benefit of the Society, and the Fellows thereof'.[72] Receiving a copy of the *Transactions* became a benefit of membership to every fellow who paid their membership fees; and any financial advantage arising from sales to the wider learned public would accrue to the funds of the Society, not to the bookseller.

The Society's bookseller in 1752 was Charles Davis (1693–1755), a well-established publisher of literature, poetry and other things, who had been organising the printing and distribution of the *Transactions* for Cromwell Mortimer since 1736.[73] He was not himself a printer, and outsourced the printing. On his retirement in 1754, he was succeeded by his nephew Lockyer Davis, in partnership with Charles Reymers. The Society's new agreement with Davis was hammered out by Thomas Birch and John Ward in May 1752, no doubt informed by Ward's recent experience scouring the archives to investigate the historical relationship between the periodical and the Society. Their 'Scheme for Printing and Publishing the Philosophical Transactions of the Royal Society at their own Expence' made the secretary responsible for editorial and production decisions (without specifying *which* secretary; it would in fact be Birch), while the Society's paid clerk managed the distribution of copies to the fellowship.[74]

The secretary had discretion over the choice of paper, types and format, as well as the print run; was to negotiate prices for composition and printing; to supervise the proof corrections; and to agree the 'shop price' of copies to be sold by the bookseller. He was also responsible for paying the bills for the printing and, separately, for the engraving and printing of copper plates for illustrations. Meanwhile, as a bookseller, Charles Davis was to take custody of all the printed sheets and illustrative plates (issuing a receipt to the secretary); to organise for them to be folded and collated into parts, ready for distribution; to deliver some copies to the Society's clerk at its Crane Court premises; to put other copies, stitched 'in blue Paper' covers, on sale; and to advertise their availability in 'the Public Papers' (newspapers and periodicals). As well as advertising the copies for sale, this was a convenient way of letting the fellowship know that their copies were ready for collection from the Society's clerk (who had to keep a list of who had received their copies). The proportion of the print run available to be sold was not specified: it was whatever remained after the bookseller had delivered to Crane Court 'such Numbers as the Secretary shall direct by an Order Written with his own hand'.[75] Elsewhere, the Council minutes record that Davis was agreeable to taking 'the 200 Copies or such other Number' as were left after the 480 or so fellows had received their copies. With this in mind, Council ordered a print run of 750 copies of volume 47.[76]

Financially, these arrangements meant that the Society was liable for all the costs of paper, printing and illustrations, and would receive, in return, copies of the *Transactions* for all its fellows plus 80 per cent of the retail price on any sales effected by the bookseller. The engraver, printer and paper merchant were all paid directly for their labour and materials, but Davis the bookseller was paid via a 20 per cent commission on income from sales (which had to cover his advertising expenses). He was also paid a 20 per cent commission on the copies earmarked for the

Society's fellows, in recognition both of his handling costs and the fact that this distribution substantially reduced the likely market for Davis's sales. The Society retained ownership of all the unsold or uncirculated copies of the *Transactions* unless Davis chose to purchase some or all of those copies at a 25 per cent discount once the Society had taken delivery of the copies intended for the fellows.

In July 1753, the Society's clerk, Francis Hauksbee the Younger, tried to persuade Council that it would lose money under the new arrangement.[77] There was no doubt that Hauksbee himself would lose out, because, under the previous regime, he had received a commission on the copies he distributed for Mortimer, whereas distribution of the *Transactions* was now encompassed within his employment by the Society. In response, John Ward sought to calculate 'the gain or loss to the Society by printing the Transactions'. His calculations were not recorded in the minutes, which merely noted, in June 1754, that 'they are gainers by the new Established method of printing them'.[78] His calculations do, however, survive elsewhere, and they show that Ward's estimated 'gain' on volume 47 was marginal, of the order of £10. They also show that Charles Davis, the bookseller, could expect an income of almost £65 (from which his unknown costs would have to be deducted).[79]

Ward's surviving calculations also reveal that he included the joining fees paid by new members of the Society as income for the *Transactions*. The joining fee had been substantially increased, from two to five guineas, on the grounds that it was 'but reasonable' that those entitled to gratis copies of the *Transactions* 'should contribute in some measure towards defraying the said extraordinary Expence'.[80] Without this membership income, Ward's calculations would have shown volume 47 of the *Transactions* to have cost the Society around £34, rather than gaining £10. Putting some fraction of the membership income towards the *Transactions* was, however, a way of demonstrating the value that the periodical now brought to the Society: it was a perquisite for members, and copies could also be used as gifts to those the Society wished to acknowledge or reciprocate.

Ward's calculations were also based on the optimistic assumption that 450 copies would be sold, which is rather more than the 270 or so copies that could entered the retail trade if all the fellows claimed their copies. Ward's assumption may reflect his knowledge of how many fellows had in fact claimed their copies of volume 47, or of the number of fellows who were ineligible to claim because their membership fees were in arrears. The evidence from later decades suggests that only around two-thirds of fellows claimed their copies, and the main reason was the practical difficulty of claiming in person for those not resident in London.

In recognition of this, in the late eighteenth and nineteenth centuries, Council would routinely allow late claims, if they were no more than five years delayed.[81]

Ward's assumption that the entire print run would either be claimed by fellows or sold through the trade was to prove wishful thinking. In 1765, there would still be 55 copies of volume 47, unsold, in the booksellers' hands, and the treasurers' accounts suggest that Davis's successors, the booksellers Davis & Reymers, were managing to sell only around 20 copies a year (at best) of the *Transactions* in the 1760s.[82]

In 1765, the Society undertook a reckoning of the *Transactions*' finances. This revealed that, over 12 years, it had spent £2,209 on production costs, and recouped £964 in sales income.[83] On average, that was an expenditure of roughly £184 and an income of just £80 per year. Including the fees from new members (of whom there were in fact more per year than Ward had anticipated) improves the picture slightly, by adding a further income of £63 per year.[84] But with sales so much lower than Ward had anticipated, this still left the Society at a loss. That said, the Society's finances were healthy enough to support the *Transactions* to the tune of around £40 a year.

The significant financial difficulties that had formerly faced the Society had been resolved by the mid-eighteenth century, and during the 1750s, the treasurer could afford to purchase new investments to add to the portfolio created from bequests.[85] Modern accountants tend to look at annual income/expenditure, but the measure the Society itself used to assess its financial health appears to have been the balance of cash in hand remaining at the end of the year. This had averaged £266 in the 1740s, but had risen to £407 in the decade after 1752.[86] The Society's income at this time derived from members' fees (of various sorts), and from rent on the properties it had been bequeathed. There were relatively few calls upon those funds: the Society employed a clerk, and a small domestic staff to oversee its premises in Crane Court.[87] It had its weekly meetings, but there were no longer any elaborate demonstrations involved. In other words, the mid-eighteenth-century Royal Society was well able to support the *Transactions* despite what appears, to modern eyes, as disappointing sales figures. Indeed, supporting the cost of publishing the *Transactions* may well have seemed like an excellent use of the Society's funds.

Owning the *Transactions* enabled the Society to provide a benefit to its members; but it also enabled the strategic use of the printed volumes as gifts. Such gifts might flatter an eminent patron, acknowledge a gift received by the Society, or be a means of spreading awareness of the Society and its activities. The idea of using the *Transactions* as a gift was not new. In 1724, for instance, Antoni von Leeuwenhoek's widow had been sent 'a plain silver Bowl' and 'the two last Volumes of the Philosophical Transactions ... for a Present' in acknowledgement of her husband's contributions to those volumes.[88] But until 1752, the Society had had to purchase copies to use in this way. Thus, when, in 1750, the Paris Académie Royale sent the Royal Society a complete set of its *Mémoires*, the Londoners found it embarrassingly difficult to reciprocate. Despite consulting with booksellers, the Society's Council was unable to acquire a spare complete set of the *Transactions*. A diplomatic faux pas was averted only once Macclesfield became president and donated his own set.[89] Gifts could also be a fruitful way of widening the circulation of the *Transactions*. In 1753, for instance, the Jesuit missionaries in China who had sent observations to the Society were rewarded with some volumes of the *Transactions*, and in the same year, volumes were donated to the Royal College of Physicians.[90]

A careful hierarchy of gifts developed, ranging from a single copy of the current volume to a set of volumes. The significance of 1752 in enabling this can be seen in the wording of the gift to the Royal Stockholm Academy in 1757: it was to include all the volumes that had appeared from the time 'the Society took the publication of them into their own hands'.[91] By 1765, the Society had set up regular arrangements to gift each new volume of the *Transactions* to about a dozen recipients (Figure 5.4). These included the universities of Oxford and Cambridge, the British Museum and (surprisingly belatedly) the king.[92] Other recipients included the Royal College of Physicians and the Society of Antiquaries; the royal academies in Paris, Madrid, Stockholm and Berlin; the imperial academy in St Petersburg; and the Leopoldina of the Holy Roman Empire.[93] Unlike the ancient universities or the king, these learned institutions might respond to a gift of the *Transactions* in kind. But such arrangements could be erratic or sporadic. In the 1770s and 1780s, for instance, the American Philosophical Society did not receive the London *Transactions* regularly, but only when it had sent a volume of its own; and the slower publication schedule of the American periodical meant that Philadelphia's collection of the *Transactions* was full of gaps.[94]

As more academies and learned institutions emerged across Europe and the world in the later eighteenth and nineteenth centuries, the Royal Society developed dozens of ongoing arrangements for reciprocal gift-giving of publications. These proved enormously important for circulating knowledge outside the commercial book trade. They enabled the libraries of learned institutions to provide their members with access to publications from across the scholarly world for little visible cost,

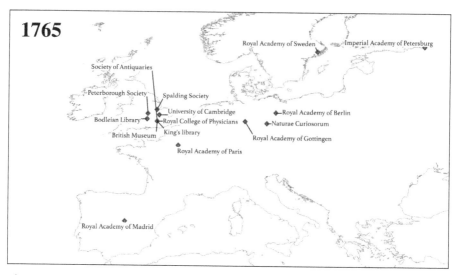

Figure 5.4 Locations of institutions that were sent gifts of the *Transactions* in 1765.

since the costs of printing and shipping these copies were absorbed in the publication expenses or general office expenses of the sending institution. These are often referred to as 'exchange' arrangements, but the Royal Society's continued use of the language of 'presents' ensured that a kernel of eighteenth-century patronage remained at the heart of its nineteenth-century schemes for circulating knowledge.[95]

From 1752 on, the Royal Society clearly owned the current and future *Transactions*, with a legal claim on the physical copies it had paid to produce, and a moral claim on the intellectual property. In early 1757, its claim to the pre-1752 volumes would be tested when John Innys sought to sell at auction the stock and copyrights of his late father. William Innys had been the Society's bookseller from 1714 to 1736, and had managed the printing and sale of the *Transactions* for the various independent secretary-editors of that period. The 1757 auction catalogue offered the sale of remaining stock and the rights in all the back numbers of the *Transactions* up to number 422 (that is, up to 1730), and a half share in issues 423 to 443 (because Innys had taken a partner from 1730 to 1736).[96] In common with almost all booksellers in London at the time, John Innys was still working with a common law notion of perpetual property rights in 'copies', rather than the limited-term concept of 'copyright' that had been on the statute books since 1710.[97] In addition to the volumes actually issued by William Innys, John Innys claimed that Edmond Halley and Hans Sloane had sold his father their 'copies' – that is, their rights in the back numbers published by them.

The Royal Society's Council indignantly summoned John Innys to a meeting, for his actions threatened the Society's ability to control the way in which its past activities appeared in print. It admitted Innys's right to sell the copper plates engraved to illustrate the numbers published by his father, but denied he had any other rights to the intellectual content of the *Transactions*. Despite having reiterated its claim about the independence of past *Transactions* editors in the 1752 'advertisement', the Council was now insisting that those past secretary-editors had had no personal rights in the issues they had produced. It argued that the responsibility of editing the *Transactions* and the right to enjoy the profit (if any) were limited to their period of editorship, after which the rights reverted to the office of secretary, and thus customarily passed from one secretary to the next. Such a claim had never been formalised, and its legal basis must have been shaky in the extreme.[98] But the episode makes clear that the Society believed it (now) possessed the moral rights to the entire back-run of the *Transactions*, and also that it had sufficient power and influence in the mid-eighteenth-century London book trade to get its way.[99]

The reform of the *Transactions* in 1752 had been a balancing act, a simultaneous consolidation of established trends and a significant change of course designed to deflect savage public criticism. By assuming collective editorial responsibility, the Society officially accepted that the periodical was the most significant projection of its own identity and activity. It was meant to add the appearance of probity to an editorial procedure that had been attacked for its arbitrariness and lack of transparency, but also to shield individual actors (notably the president and secretaries) from being held responsible for editorial decisions. It was a way to reconcile the Society's longstanding aversion to accepting responsibility for any claims advanced in the periodical, with a new model of collective editorship that would, in increasingly complex iterations, underpin the publication of the Society's periodicals for the next two centuries.

Financially, the Society proved surprisingly well able to weather the cost of supporting a periodical whose public sales can only be described as modest. It is a pointed reminder that the 'use and benefit' that Macclesfield and his associates anticipated from the Society's ownership of the *Transactions* was not principally financial. The Society benefited by owning the printed copies of the *Transactions*, which it could use as a perquisite for its fellows; as gifts to patrons, contributors and learned institutions; and as tokens in a gift exchange with the growing number of similar societies and academies emerging throughout Europe and across the Atlantic. Such exchanges ultimately helped stock the Society's library, and extended the Society's reputation among international scholarly elites. Taking ownership and responsibility for the *Transactions* turned out to be not simply a means of protecting the Society's reputation against criticisms by the likes of John Hill, but a way of enhancing it.

But in the 1760s and 1770s, the evidence for the success of these reforms was still ambiguous at best. The international gifting and exchange of publications was in its infancy. Creating systems of collective editorial control had only enacted what much of the periodical's audience believed had already been happening, and could even be seen as cloaking editorial decisions in secrecy. Clarifying the protocols for submission explicitly favoured the fellows over outsiders. And the power of cliques, and of presidential cliques in particular, would soon return.

Notes

1 W.P. Courtney, 'Mortimer, Cromwell (c. 1693–1752), physician and antiquary', *ODNB*; also RS JBO/22, 27–8.
2 For conventionally dismissive accounts of the Royal Society in the early eighteenth century, see John Heilbron, *Physics at the Royal Society during Newton's Presidency* (Los Angeles: University of California, 1983); Joseph M. Levine, *Dr. Woodward's Shield: History, science, and satire in Augustan England* (Ithaca, NY: Cornell University Press, 1991); and for useful correctives, Andrea Rusnock, 'Correspondence networks and the Royal Society, 1700–1750', *British Journal for the History of Science* 32, no. 2 (1999): 155–69; Richard Sorrenson, *Perfect Mechanics: Instrument makers at the Royal Society of London in the eighteenth century* (Boston, MA: Docent Press, 2013); and Palmira Fontes da Costa, *The Singular and the Making of Knowledge at the Royal Society of London in the Eighteenth Century* (Newcastle: Cambridge Scholars Publishing, 2009).
3 RS JBO/22, 23 January 1752, 22–3.
4 RS CMO/02, 28 February 1699/1700, 109–10.
5 On distributed editorial responsibility, see Aileen Fyfe, 'Editors, referees and committees: Distributing editorial work at the Royal Society journals in the late nineteenth and twentieth centuries', *Centaurus* 62, no. 1 (2020): 125–40.
6 For instance, in *The London Magazine: or, Gentleman's Monthly Intelligencer* (January 1751), the latest number of the *Transactions* – number 492 – appears in the 'Monthly Catalogue' of new books for sale under the heading 'Philosophy, Physick &c', while the *Review* is listed under 'Miscellaneous'. Number 496 of the *Transactions* (the last to be printed under the independent editorial regime) would be listed in the December issue (p. 576), along with an extract (p. 562).
7 It also became an anthology piece, appearing in a much-reprinted collection, *Fugitive Pieces on Several Subjects, by Various Authors* later in the eighteenth century.
8 The *Review* was noted by the *London Magazine* (January 1751): 48; and *Universal Magazine* (January 1751): 47.
9 The *Scots Magazine* was published in Edinburgh, but drew a connection between the *Review* and the *Dissertation*, thus indicating that Hill's authorship of at least one of the anonymous satires of the previous year had become public knowledge; see *Scots Magazine* (January 1751): 55–6.

10 Hill's papers were 'A letter ... concerning Windsor loam', *PT* 44.483 (1746–7): 458–63, originally read 19 March 1746/7; and 'A letter ... concerning the manner of the seeding of mosses', *PT* 44.478 (1746–7): 60–6, read 13 February 1745/6. Hill's campaign is discussed in George S. Rousseau, *The Notorious Sir John Hill: The man destroyed by ambition in the era of celebrity* (Bethlehem, PA: Lehigh University Press, 2012), 40–50.
11 Rousseau, *The Notorious Sir John Hill* draws upon the Stukeley papers in the Bodleian Library and to his diary for this information.
12 Rousseau, *The Notorious Sir John Hill*, 46–8.
13 See P.R. Harris, 'Maty, Matthew (1718–1776)', *ODNB*. For a later contrast, see Iain Watts, '"We want no authors": William Nicholson and the contested role of the scientific journal in Britain, 1797–1813', *British Journal for the History of Science* 47, no. 3 (2014): 397–419, 405 (and n.34 for the suspicion of the journalistic activities of Alexander Tilloch and Joseph Des Barres).
14 [John Hill], *A Dissertation upon Royal Societies: In three letters from a nobleman on his travels, to a person of distinction in Sclavonia* (London: printed for John Doughty, 1750), 21 (sig. C3r-v).
15 [Hill], *Dissertation upon Royal Societies*, 16–31.
16 John Hill, *A Review of the Works of the Royal Society* (London: R. Griffiths, 1751), sig. A2v.
17 Hill, *A Review of the Works of the Royal Society*, sig. A2r-v.
18 Rousseau, *The Notorious Sir John Hill*, 78.
19 See Hill, *A Review of the Works of the Royal Society*, 2. For the election of the other John Hill, in 1748, see RS EC/1748/10.
20 [Silas Taylor], 'A way of Killing Ratle-Snakes', *PT* 1.3 (1665): 43.
21 Rousseau, *The Notorious Sir John Hill*, 74–6.
22 Hill, *A Review of the Works of the Royal Society*, 12.
23 [Hill], *Dissertation upon Royal Societies*, 28 (sig. D2v.)
24 [Hill], *Dissertation upon Royal Societies*, 22–3.
25 Hill, *A Review of the Works of the Royal Society*, 5.
26 RS L&P series, *passim*; also BL Sloane MS 4025–6.
27 Hill, 'Seeding of mosses', 60. Hill's manuscript, in RS L&P/10/450, does in fact bear the mark of substantive alteration by another hand; certain words have been replaced throughout ('half' for 'cell', for instance), several sentences and short passages are so thoroughly scored out as to be illegible, and Hill's characterisation of his own observations as 'diligent and faithfully related' has been deleted.
28 Hill, *A Review of the Works of the Royal Society*, 3–4.
29 *The Gentleman's Magazine* (January 1751): 21.
30 *The Gentleman's Magazine* (February 1751): 68–71.
31 *Monthly Review* 75 (February 1751): 279.
32 *Monthly Review* 75 (February 1751): 280.
33 George S. Rousseau, ed., *The Letters and Papers of Sir John Hill 1714–1775* (New York: AMS Press, 1982), 43; Kevin J. Fraser, 'John Hill and the Royal Society in the eighteenth century', *N&R* 48.1 (1994): 43–67, 52.
34 *The Gentleman's Magazine* 22 (1752): 138.
35 'Advertisement', *PT* 47 (1751–2): sig. a1r-v.
36 Contemporary attestations make clear that Macclesfield's elevation to the Society's presidency in 1752, after Folkes had signalled his intention to resign, was more of a coronation than an election; but it is not clear whether it was done with the latter's blessing. BL Add MSS 35398, ff. 104–8.
37 On the Hardwicke Circle, see David Philip Miller, 'The "Hardwicke circle": The Whig supremacy and its demise in the 18th-century Royal Society', *N&R* 52.1 (1998): 73–91.
38 RS JBO/22, pp. 22–3.
39 RS JBO/22, meetings between 15 February and 28 May 1752.
40 It was advertised and reviewed in the general periodical press from July 1753, for example, *Monthly Review* 8 (July 1753): 37 ff.; *Gentleman's Magazine* 23 (1753).
41 RS JBO/22, pp. 29–30. The prospective candidates were probably, on the evidence of a letter of Emanuel Mendes da Costa to William Borlase, Thomas Birch and Gowan Knight. It was Birch who was actually appointed, with, as da Costa believed, the backing of Macclesfield and his socially eminent party within the Society. BL Add MSS 28535 ff. 70–3, 27 February 1751/2.

42 Their report was entered into the Council Minutes on 20 February 1751/2, see RS CMO/04, 55–66.
43 On the need for public judgement on public bodies, see Alex Csiszar, *The Scientific Journal: Authorship and the politics of knowledge in the nineteenth century* (Chicago, IL: University of Chicago Press, 2018), ch. 2.
44 'Proposal Concerning the Papers of the Royal Society', in RS CMO/03, 12 July 1742.
45 On the relationship between meetings and publications, see Aileen Fyfe and Noah Moxham, 'Making public ahead of print: Meetings and publications at the Royal Society, 1752–1892', *N&R* 70.4 (2016): 361–79.
46 [Hill], *Dissertation upon Royal Societies*, 39.
47 [Hill], *Dissertation upon Royal Societies*, 39.
48 On Banks's correspondence with communicators, see Noah Moxham, '"*Accoucheur* of literature": Joseph Banks and the *Philosophical Transactions*, 1778–1820', *Centaurus* 62, no. 1 (2020): 21–37, 30.
49 [Hill], *Dissertation upon Royal Societies*, 39.
50 RS JBO/22, 27 February 1751/2, pp. 68–9.
51 The reforms were enacted as 'statutes', which were a matter for Council, not the fellowship at large. RS JBO/22, 27 February 1751/2, pp. 68–9.
52 The procedures included rules for tied votes, and for deferring votes. RS CMO/04, 15 February 1751/2, 48–55.
53 James McClellan III, 'Specialist control: The publications committee of the Académie Royale des Sciences (Paris), 1700–1793', *Transactions of the American Philosophical Society* 93, no. 3 (2003): 1–134, for example ch. 7.
54 Minutes of the Royal Society of Edinburgh, 17 July 1784, National Library of Scotland, archive of the Royal Society of Edinburgh, Acc.10000/1.
55 McClellan III, 'Specialist control', see chs 4–7. McClellan's examples range over the whole eighteenth century, but most of the detailed examples of close scrutiny appear to be from the 1760s and after.
56 RS CMO/04, 15 February 1751/2, 48–55.
57 Noah Moxham and Aileen Fyfe, 'The Royal Society and the prehistory of peer review, 1665–1965', *Historical Journal* 61, no. 4 (2018): 863–89.
58 'Advertisement', *PT* 47 (1751–2): sig. a1r-v
59 BL Add MS 4441 f. 20. This draft, apparently by the vice-president James Burrow, survives in the Birch Papers and uses rather more confrontational language than the final version.
60 McClellan III, 'Specialist control'; Roger Hahn, *The Anatomy of a Scientific Institution: The Paris Academy of Sciences, 1666–1803* (Berkeley: University of California Press, 1971), ch. 1; Alice Stroup, *A Company of Scientists* (Berkeley: University of California Press, 1990). On the Royal Society and patents, see for instance Christine MacLeod, *Inventing the Industrial Revolution: The English patent system 1660–1800* (Cambridge: Cambridge University Press, 1988), 27; and the Royal Society's own patent application of 1665, BL Add MS 4441 ff. 101–2.
61 The Royal Society had initially hoped to serve a similar patent-inspecting function in England, see Noah Moxham, 'Natural Knowledge, Inc.: The Royal Society as a metropolitan corporation', *The British Journal for the History of Science* 52, no. 2 (2019): 249–71, 255–7 and 270–1. The possibility was mooted again by the Crown as late as 1713. See BL Sloane MS 4026 f. 240, and RS CMO/02, 29 January 1712/13 and JBO/11 22 January 1712/13.
62 'Advertisement', *PT* 47 (1751–2): sig. a1r-v
63 'Preface', *Memoirs of the Manchester Literary and Philosophical Society* 1 (1785): viii–ix.
64 *Monthly Review* 11 (September 1754): 211–12 [referring to *PT* 48 (1753–4)].
65 *The Critical Review, or Annals of Literature* 1 (July 1756): 528–9 [referring to *PT* 49 (1755–6), Part 1].
66 Pringle to Banks, [January] 1774, in Neil Chambers, *The Scientific Correspondence of Sir Joseph Banks, 1765–1820*, 6 vols (London: Pickering & Chatto, 2007), vol. 1, 56.
67 *The Critical Review* 4 (August 1757): 130–1 [referring to *PT* 49 (1755–6), Part 2]. D.P. Miller attributes authorship of the *Critical*'s notices of the *Transactions* to Tobias Smollett, its founding editor: see Miller, 'The "Hardwicke circle"', 84.
68 *The Critical Review* 1 (July 1756): 528–9 [referring to *PT* 49 (1755–6), Part 1].
69 Sorrenson, *Perfect Mechanics*, 163. Note that Sorrenson's tables use the number of articles and not a page count, so no regard is paid to the actual amount of space given over to

each subject area. Furthermore, the apparent spike in natural-historical articles is almost entirely due to 1750 alone, in which 81 articles were published in that category; 52 more than in any other year during the entire half century under review.

70 See Aileen Fyfe, 'Journals, learned societies and money: *Philosophical Transactions, ca.* 1750–1900', *N&R* 69.3 (2015): 277–99; 'The Royal Society and the noncommercial circulation of knowledge', in *Reassembling Scholarly Communications: Histories, infrastructures, and global politics of open access*, ed. Martin Paul Eve and Jonathan Gray (Cambridge, MA: MIT Press, 2020), 147–60.
71 Hill, *Review of the Works of the Royal Society*, 3–4. On Hill's success as an author, see Rousseau, *The Notorious Sir John Hill*.
72 RS CMO/04, 19 March 1751/2, ff. 71–82.
73 On Davis, see James Raven, 'Location, size and succession: The book shops of Paternoster Row before 1800', in *The London Book Trade*, ed. Robin Myers, Michael Harris, and Giles Mandelbrote (London: Oak Knoll Press & the British Library, 2003), 89–126.
74 RS CMO/04, 25 May 1752, ff. 85–8.
75 RS CMO/04, 25 May 1752, ff. 85–8.
76 RS CMO/04, 27 March 1752, ff. 83–5 (Davis and 200 copies); 25 May 1752, ff. 85–8.
77 BL Add MSS 6180, ff. 242–5; RS AB/1/2/3, Clerk's Accounts.
78 RS CMO/04, 20 June 1754, ff. 138–40.
79 We have found Ward's detailed calculations in BL Add MSS 6180 f. 239.
80 RS CMO/04, 27 February 1751/2, ff. 66–71.
81 The hard limit of five years was in place by 1773, see RS CMO/06, 20 December 1773; it was retained in the 1847 statutes.
82 RS DM/1/102. It is unclear whether these were still half-yearly or annual accounts; half-yearly seems plausible, both in view of the Society's stated preference, the rhythm of *Transactions* publication, and the level of stock in hand compared with the average of copies sold. See also R.K. Bluhm, 'Remarks on the Royal Society's finances, 1660–1768', *N&R* 13.2 (1958): 82–103, table 7.
83 Strictly speaking, 13 years; but the accounting only covers entries for 12 years, from 1754 to 1765, so the averages have been arrived at by dividing by 12. RS DM/1/105, partially duplicated in RS AB/1/1/11 (Clerk's Accounts).
84 This is our estimate, based on manual counts of elections recorded in the Society's Council minutes. In the mid-1750s, between 14 and 29 new members were admitted each year, around 20, on average. It is, however, possible that some new fellows may never have taken up their places, and others might have gone unrecorded.
85 Henry Lyons, 'The Society's finances Part I – 1662–1830', *N&R* 1.2 (1938): 73–87, 80; and R.K. Bluhm, 'Remarks on the Royal Society's finances', 91.
86 The trend was not consistently upward: the average cash in hand had reached £523 by the 1770s, but had been just £174 in the 1760s.
87 Henry William Robinson, 'The administrative staff of the Royal Society 1665–1861', *N&R* 4.2 (1946): 193–205.
88 RS JBO/13, 21 May 1724, p. 391.
89 RS CMO/04, 4 July 1750 and 18 January 1753.
90 RS CMO/04, 17 November 1753.
91 RS CMO/04, 17 November 1757.
92 RS CMO/04, 25 June 1761 (BM, and the universities) and 12 December 1765 (the king). On earlier gifts to the king, see Adrian Johns, *The Nature of the Book: Print and knowledge in the making* (Chicago, IL: University of Chicago Press, 1998), 493.
93 In 1816, a list of all recipients was compiled, including the date on which the gift commenced, RS CMO/09, 14 March 1816. The original decisions were usually recorded in RS CMO/04-5.
94 George L. Sioussat, 'The "Philosophical Transactions" of the Royal Society in the libraries of William Byrd of Westover, Benjamin Franklin, and the American Philosophical Society', *Proceedings of the American Philosophical Society* 93, no. 2 (1949): 99–113.
95 Aileen Fyfe, 'The Royal Society and the noncommercial circulation of knowledge', 147–60; Jenny Beckman, 'Editors, librarians, and publication exchange: The Royal Swedish Academy of Sciences, 1813–1903', *Centaurus* 62, no. 1 (2020): 98–110.
96 *A Catalogue of Books in quires, and copies, being part of the stock of the late Mr William Innys*, (London, 1757).

97 Ronan Deazley, *On the Origin of the Right to Copy: Charting the movement of copyright law in eighteenth century Britain (1695–1775)* (Oxford: Bloomsbury Publishing, 2004).
98 RS CMO/04, 24 February 1757, 189–90.
99 Aileen Fyfe, Julie McDougall-Waters, and Noah Moxham, 'Credit, copyright, and the circulation of scientific knowledge: The Royal Society in the long nineteenth century', *Victorian Periodicals Review* 51, no. 4 (2018): 597–615.

6
Sociability and gatekeeping, 1770–1800

Noah Moxham and Aileen Fyfe

In 1776, the *Philosophical Transactions* was renamed. Oldenburg's ponderous subtitle, with its claim to cover the 'Present Undertakings ... of the Ingenious, in many Considerable parts of the World', was replaced by the simple declaration that the *Philosophical Transactions* was 'of the Royal Society of London'. As well as belatedly acknowledging the institutional takeover, the new subtitle dropped any pretensions to European (or wider) coverage. The Society did still receive material from overseas scholars: for instance, between 1779 and 1781, the Society's foreign secretary arranged for about a dozen papers received in French, Latin, Swedish and Italian to be translated for presentation at meetings.[1] The *Transactions* occasionally printed papers in French or Latin, but most papers now appeared in English translations: the last Latin paper would appear in 1785.[2] Over the next century, the *Transactions* would become principally a site for the publication of work in English, by and for British natural philosophers and savants.

By the late eighteenth century, the *Transactions* had long ceased to be unique in the scholarly world. The academies of science in Paris, Berlin, Uppsala and St Petersburg, for instance, issued their own 'memoirs' or 'transactions', using the more widely understood languages of French or Latin. And within the English-speaking world, the *Transactions* was joined by the publications of the American Philosophical Society (from 1771), the Manchester Literary and Philosophical Society (from 1783), the Royal Society of Edinburgh (from 1783), and the Royal Irish Academy (from 1787). This proliferation of learned institutions and their periodicals meant that many scholars in the late eighteenth century were now able to address an informed community of colleagues within their own national, regional or linguistic boundaries.

The new subtitle for the *Transactions* emphasised its one remaining unique attribute: its connection to the Royal Society. The *Transactions* was 'of the Royal Society' both because it showcased the activities of the Society's fellows, and because access to its pages was controlled by the fellows. Twenty-one of those fellows served on the Committee of Papers that made the final decisions to publish (or not), but all 520 or so fellows had the privilege of being able to communicate their own findings to the Society, and of being trusted to forward observations from suitable friends, relatives or associates. In theory, the 1752 reforms had ensured that the voices of the president and secretaries carried no greater weight than those of other members of the editorial committee, but in practice, as we shall see, their influence remained strong.

From 1778, the president of the Royal Society was Joseph Banks. The heir to a large landowning Lincolnshire family, Banks had privately financed an entourage of botanical artists and assistants and accompanied James Cook's voyage aboard the *Endeavour* (1769–71), returning with a fabulous collection of largely unknown flora, insects and fauna from Tahiti, New Zealand and the east coast of the Australian continent. At the time of his election as president, he was a genuine scientific celebrity whose reputation as an amateur botanist sat alongside powerful political and social clout.[3] That clout enabled him to survive what has been called 'the largest row the Society had ever seen', when his attempt to replace the Society's foreign secretary united a disparate group of variously dissatisfied fellows into a rebellion against the Banks presidency.[4] The 'dissensions' lasted for several months over the winter of 1783 to 1784, but Banks survived and would continue as president until 1820, the longest tenure on record.

The society that Banks had taken over in 1778 was in reasonably robust financial health – in the broad sense that its outgoings rarely outstripped its income – and his lengthy tenure provided organisational stability for both the Society and the *Transactions*. Though he published no papers of his own in the *Transactions*, he communicated hundreds and passed judgement on hundreds more. Nominally one voice among many in the running of the periodical, he was in practice as powerful an influence on the *Transactions* as any figure since Oldenburg. This influence was grounded in his central position in the social world, and the social codes, of science in late eighteenth-century Britain. Banks was supported by his trusted associate Charles Blagden, especially but not only during the period when Blagden served as secretary, from 1784 to 1797.

To an outsider, the editorial process for the *Transactions* appeared to be codified in the 1752 procedures for the Committee of Papers but, in practice, those procedures described only part of the evaluation, selection and revision processes that shaped the *Transactions*. The Committee of Papers was part of a much wider system of scrutiny, much of which operated without written rules. Banks and his circle were able to develop flexible mechanisms for controlling the *Transactions*, and were able to achieve most of what they desired in ways that lay outside the scope of official oversight by the Committee of Papers.

This chapter and the next cover broadly the same time period, but Chapter 7 will explore the circulation of the *Transactions*, and its place in wider print culture. The current chapter draws upon the minute books of the Committee of Papers, and the correspondence between Joseph Banks and Charles Blagden, to reveal the ways in which social status and relations influenced who could participate in the 'transactions' of the Royal Society. The Society's editorial processes were embedded in the sociability of its formal and informal activities.[5] Social dynamics were particularly significant in those aspects of editorial process that were *not* explicitly covered by the 1752 procedures.

Collective editing in practice

As laid down by the 1752 procedures, the first step in the editorial process for the *Transactions* was the reading of a paper at a meeting of the Royal Society (Figure 6.1). This meant that the late eighteenth-century *Transactions* was tightly tied to the rhythms of Society activity. Each year, the Society's session ran from the beginning of November to the end of June, and during that period the fellows met every Thursday evening in the Society's rooms, which, from 1780, were in Somerset House, on the Strand.[6] The lack of meetings in the summer break meant that no papers could be presented during that period. At the meetings, papers were read aloud by one of the Society's secretaries, not by their author (even if he were present). Selected papers presented from November through to about February would appear in Part 1 of that session's *Transactions* in the spring; while papers presented in the second half of the session (up to June) would appear in Part 2 in the autumn.

The management of the weekly meetings affected when and how a paper was presented to the fellows, as well as their opportunity to comment upon it ahead of publication. One of Banks's first initiatives on becoming president had been to move the time of the Thursday meetings from six to eight o'clock in the evening, and to restrict them to one hour. This strict time limit not only restricted the number of papers that could be presented each week, but changed the nature of the meetings by removing

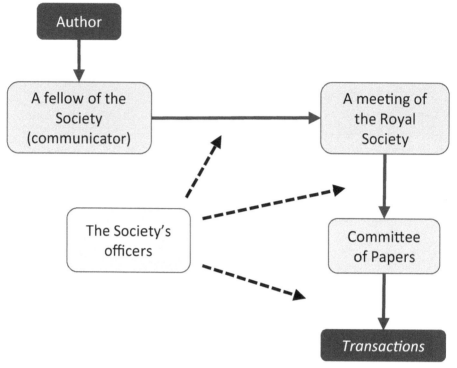

Figure 6.1 The formal editorial process for the *Transactions* in the late eighteenth century, and the informal influence of the Society's officers.

the opportunity for discussion of the papers. This format appears to have been to Banks's taste, since he remarked in 1783 that he would scorn to hold the presidency if it amounted to being 'the moderater in any Shape whatever of a Debating Club'.[7] Without discussion, the president and his allies could control the conduct and content of meetings, and there was little scope for the ordinary members to speak up.

The compression and formalisation of Thursday evening meetings had in fact been driven by a desire for improved sociability outside the context of formal meetings: it enabled fellows both to dine together at a fashionable hour, and to attend the meetings of the neighbouring Society of Antiquaries, if they wished.[8] Scientific discussion was by no means discontinued, but was increasingly pushed to other sites, as Charles Blagden's diary illustrates. He regularly discussed papers with the crowd who repaired to the Crown & Anchor Tavern after the Thursday evening meetings, and he was a frequent guest at the *conversazioni* and scientific breakfasts held in Banks's Soho Square house.[9] Separating sociable from official activity went hand in hand with the concentration of power in the person of the president, who personally provided some of the most important social spaces orbiting the Society.

After a paper had been read at a meeting, the next step towards publication was the monthly meetings of the Committee of Papers. Minute books survive from the beginning of the Banks presidency onwards, and the entries record the short titles of papers, their authors' names, the dates of the meetings at which they had been read, and the committee's verdict (for instance, Figure 6.2). Since the committee members would, in principle, have already heard the papers read aloud, their meeting did not usually consider the original manuscripts, but relied on the substantial abstracts, of perhaps 500 to 2,000 words, prepared by the secretaries for the Society's 'journal book'. The minute books of the Committee of Papers offer no evidence of discussion ahead of the vote by ballot, suggesting that

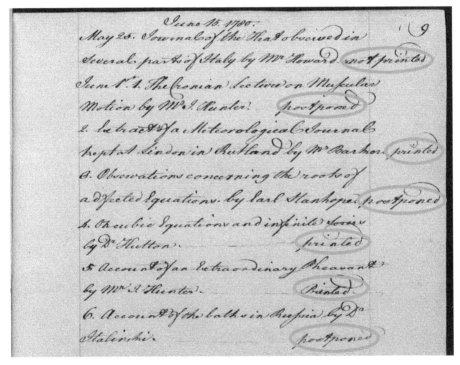

Figure 6.2 Decisions of the Committee of Papers, 15 June 1780: this section of the minutes records that three papers were approved for printing, one was declined ('not printed'), and decisions on three more were 'postponed' to a future meeting © The Royal Society.

the secretaries were aware of the 1752 ban on discussion. However, they also reveal that votes were sometimes deferred to a future meeting, to allow members to consult the paper in full, which suggests that some form of conversation was possible at meetings. Votes were only very rarely deferred to enable the consultation of other fellows 'who are knowing and well Skilled, in any particular branch of Science': if the minute books are to be believed, this provision for acquiring additional expertise was only rarely invoked, as there were just four recorded instances in the first decade of Banks's tenure, and none at all in the remainder of his presidency.[10]

The minute books also reveal that around three-quarters of the papers considered by the Committee of Papers were approved for printing or, to put it another way, that around a quarter of the papers read at meetings were, for some reason or another, not deemed appropriate for the *Transactions*.[11] This statistic was not publicly visible, as, unlike the Berlin and Paris academies, the Royal Society did not, at this time, publish abstracts of unpublished papers or reports of its meetings: all that was publicly known about its activities came from the *Transactions*. One of its secretaries, Joseph Planta, did propose in 1783 that a full record of meetings should be published annually, but the idea was thoroughly quashed in Council. Planta had argued that the proceedings of the Society's meetings were already known to the fellows (and their guests), and accused the Committee of Papers of cowardice in refusing to have its deliberations made public.[12] Banks's objections to the idea of letting it be known whose papers had been declined for publication were evasively expressed, but likely arose from a sense that such 'naming and shaming' would have been a gross breach of gentlemanly etiquette.

Banks would also have been aware that there were good reasons for some papers to have been read but not published. For instance, just a few years earlier, Banks himself had chosen not to pursue publication of his one and only contribution to the Society. His account of two types of 'Labradore stone' had been read in spring 1774, and the then-president, the military physician John Pringle, had suggested he recast it for publication in the form of a letter to Pringle or one of the secretaries. Instead, Banks withdrew the paper before it could be put to the ballot.[13] His reasons are obscure, but it is notable that Banks was elected to the Society's Council shortly afterwards: presenting a paper could have been a carefully managed act to enhance Banks's position within the Society, and for that purpose, having it published for a wider audience would be unnecessary. As far as it is possible to tell, he never sought to publish another paper in the *Transactions*.[14]

There were other cases where the Society's officers allowed papers to be read at a meeting without any intention or expectation of later publishing them. For instance, in December 1795, Charles Blagden lamented the necessity of reading a paper on light from Edinburgh that he felt was 'long, tedious and not well drawn up'. He claimed that, 'if it had not been for the scarcity of matter for reading, I should have desired the author to take it back & alter it'.[15] In this case, the Society's secretary desperately needed material to fill in time at the meeting. In another case, in October 1786, Blagden agreed to the reading of a paper on hygrometry from Brussels, even though he described it as 'perfectly foolish'. It had, however, 'a certain degree of obscurity & pomp which will make it pass, without detection, in a public reading, which probably we must be obliged to allow it'.[16] Blagden's assumption that the Society was obliged to hear the paper illustrates the importance of international diplomacy: the paper was by Abbé Theodore Mann, recently appointed perpetual secretary of the Royal Academy of Brussels.

Scholarly diplomacy could also be more overtly political. Benjamin Franklin had been a fellow since his residence in Britain before the American Revolution. In November 1783, Banks wrote to Franklin to tell him that he had opened the first meeting of the Society since the establishment of a peace between Britain and its former colonies, 'by reading to them your two Communications upon the subject of the Aerostatique Machines lately executed in France'. Reading papers by the American philosopher carried clear political symbolism, and Banks was careful to assure Franklin that his observations had been received with 'an Evident pleasure'. Yet Banks appears to have felt that reading was enough: he was not enthusiastic about hot air balloons in any case, and rather than sending the papers directly to the Committee of Papers, he instead asked Franklin whether 'you would chuse to have these Essays printed', and hinted that 'some more general detail' would need to be added to the 'short' papers.[17] This politely but effectively diverted Franklin's papers from the editorial process, their public reading having already served its purpose.

The minute books suggest that about 5 per cent of papers were 'withdrawn', voluntarily or otherwise, from the Committee of Papers ballot. A further 20 per cent were rejected in the ballot, which raises the question of what the Committee of Papers was looking for in a paper suitable for the *Transactions*.[18] Further examination of the unpublished papers surviving in the archive might reveal more details, but there is one criterion that consistently emerges from Banks's correspondence with disappointed authors: material published by the Royal Society should not have previously appeared in print, either by the author or by anyone else. This expectation of originality would, of course, come to be an enduring feature of scholarly publishing.

The requirement for original material was by no means an obvious or necessary position for a learned periodical in the late eighteenth century. Reprinting was fundamental to the practices of newspaper journalism at the time, enabling editors with limited news-gathering resources to bring their readers news from far and wide.[19] For similar reasons, many of the entrepreneurs who founded scientific journals around this time relied at least partly on reprinting and excerpting from other publications, to supply natural-philosophical news from across Europe.[20] Lorenz Crell's *Chemisches Journal* (1778) or William Nicholson's *Journal of Natural Philosophy* (1797) could be seen as reinventing one of the features of Oldenburg's *Transactions* that had been lost (or left to the general literary periodical press) in the intervening decades. But where those editors aimed to produce journals that readers might wish to purchase, the function of the *Transactions* had become that of providing a printed, public representation of the activities of the fellows of the Royal Society and their acquaintances. Insisting on originality in the papers submitted to it was a way of ensuring that the Royal Society was not (seen to be) wasting its resources by discussing or printing material that was already familiar to scholars.

The contemporary strangeness of the Society's stance can be seen from Lorenz Crell's difficulties in navigating its unwritten code of publication ethics. Crell was professor of medicine at the University of Helmstedt, in the German principality of Brunswick-Wolfenbüttel, but made his international reputation as an editor of journals after starting the *Chemisches Journal* in 1778; the subsequent *Chemische Annalen* would be his longest-running journal. Crell assiduously cultivated correspondents across Europe who could keep him abreast of the latest chemical discoveries, several of whom had connections to the Royal Society. His own first paper in the *Transactions* had appeared in 1777, but his efforts to have more papers published there foundered on his lack of familiarity with the Society's codes of civility, especially as they related to the etiquette of publishing.[21]

For instance, in 1785, the Society discovered that papers sent by Crell to London, and under consideration by the Committee of Papers, had already appeared in the *Chemische Annalen*. Banks conveyed the Society's 'displeasure' that the Committee of Papers might have approved the publication of material that was already in print. Crell's response was a profuse apology, combined with a two-part defence. First, he suggested, with some justification, that it was not his fault that the London editorial process had taken so long. He had expected the German publication to come out after, not before, the *Transactions*. Unfortunately, his papers had been 'lost' in London and 'read but 2 years, & longer, after they were sent'. Second, Crell argued that the Royal Society's claim to absolute priority in the printing of papers communicated to it was highly unusual. Crell assured Banks that, 'Count Sickingen has published his paper on the Platina, read before the French R Academie, first in German' and that the Swedish chemist Carl Wilhelm Scheele 'sends me always a description of his discoveries; before they are published in Swedish'. He went on to name several other academies in the German lands, including the Royal Academy of Berlin, to support his claim that other learned institutions did not have the Royal Society's distaste for work that had been previously published, particularly if permission were sought in advance.[22] Banks was unmoved.

Crell's relationship with the Royal Society recovered sufficiently for him to be elected to the fellowship in 1788, but he still struggled to grasp the Society's understanding of original publication. In 1797, ever ambitious, he aspired to the Society's Copley Medal, which was awarded to papers published in the *Transactions*. Crell's problem was that some of his research was already in print elsewhere. Thus, he asked Banks whether he could rewrite it in an 'other & more lucid order', and introduce additional material, so that the result might 'be looked upon, as a new memoir, at least, a new corrected edition *of it*, – if, I say, on this case it might not be inserted into the Phil. Transactions?'[23] Unfortunately for Crell, the Society's position remained that prior appearance in print anywhere, in any language, ruled out publication in the *Transactions*.

For Crell, it was his own prior publications that made his work ineligible for the *Transactions*, but Banks's correspondence reveals that papers were also rejected if their claims had been pre-empted or disproven by other scholars. In July 1790, for instance, Banks informed the American astronomer and natural philosopher Samuel Williams that his paper had been turned down by the Committee of Papers because some of its findings had already been disproven by research that Banks politely supposed had not yet reached the other side of the Atlantic.[24] Rejected authors did not automatically receive a reason, but in this case, Banks had felt obliged to justify why he was declining to communicate any further papers from a correspondent of apparently good standing: Williams had, until recently, been a professor at Harvard College. Scholars who were distant from the major centres of scholarly publishing were at a disadvantage in a system which presupposed everyone was equally well-informed about all the latest publications, but Banks's action suggests that he was not entirely convinced by the polite fiction of delays in the circulation of transatlantic knowledge.

The procedures of the Committee of Papers may have been written in statute, but its evaluation criteria were tacit. The expectation of original publication is an example of a criterion that emerged gradually, from experience. The committee's decision-making, however, was only one aspect of the evaluation of papers: a significant fraction of the pre-publication evaluation was performed by ordinary fellows of the Society when they decided whether, or not, to forward submissions from non-fellows. This evaluation was as much about the personality of the author as it was about the intellectual content, and unwritten codes of politeness and sociability were central.

Assessing authors: identity and reputation

Before a paper was ever considered by the Committee of Papers, it had to be presented at a meeting. The processes for presenting or communicating papers to a meeting therefore performed a gatekeeping function, in which evaluation of the author's identity and reputation was key. Regardless of their contents, only papers by known authors of good character would be evaluated for their suitability for publication. All fellows of the Society passed this test automatically: they were known quantities, whose social and philosophical attributes had already been assessed in the process of election to the fellowship. Even though the majority of fellows never made any contributions to the *Transactions*, around 60 per cent of papers published in the late eighteenth-century *Transactions* were authored by fellows.[25] The suitability of authors from beyond the fellowship was ensured by requiring that papers could only be communicated to the Society by one of its fellows. This was a privilege for fellows, but also a responsibility. The question of how well the communicating fellow needed to know an author, or their work, was ambiguous.

Submissions to learned society periodicals in the eighteenth and nineteenth centuries were never judged anonymously. This ban on anonymity protected the rights of authors in their discoveries and claims to knowledge, and ensured that the Society knew to whom it was lending its prestige. It also ensured that responsibility for the contents of a paper – and particularly any errors or embarrassment – could be laid on the author, as per the disclaimer in the 1752 'advertisement'. In 1795, Blagden wrote to Banks about a paper where the authors wished to remain unknown, explaining his concern that 'at present, no one is responsible to the Society for the truth of the paper; it has in appearance no father'.[26] A public acceptance of authorship was seen as necessary to protect the Society's reputation.

The fellows' privilege of communicating papers to the Society implied that anything a fellow wished to share would be worthy of (at least some) attention from other fellows. It would certainly be read, in some form, at a meeting. There was, however, no right to have that paper published in the *Transactions*. Papers by fellows enjoyed no automatic privilege at the Committee of Papers. That said, it is clear that some fellows with high scientific reputations, or with close connections to Banks and his circle, were routinely given more leeway than others. For instance, William Herschel's papers, eagerly anticipated by the Society and the astronomical community at large, were often afforded special treatment. Thus, when some questions arose in 1782 about Herschel's claims for the magnifying power of his telescopes, he was given the benefit of the doubt: rather than declining his paper, Banks wrote to urge him to give a clearer account. Furthermore, the Society's processes were adapted to cope with Herschel's tendency to send in early results as soon as he had them and then confirm or extend them later. His papers were often intentionally expedited or deferred on their way through the Committee of Papers.[27] When Herschel discovered the sixth satellite of Saturn in August 1789, he was able to announce it in a single sentence affixed to the end of his catalogue of nebulae already in the press for the *Transactions* before anyone at the Society had had the opportunity to verify it; it was only subsequently written up as a full paper.[28] As a trusted insider, Herschel was allowed far more flexibility than others.

There were risks, however, even in relying on trusted insiders. Another regular contributor to the *Transactions* was the London-based Scottish surgeon John Hunter. In 1785, he believed he had identified a fossil bone as belonging to a Roc, the giant bird of Arabian mythology. This startling discovery would have appeared in the *Transactions*, had not the visiting Dutch anatomist Peter Camper pointed out that the bone actually belonged to a tortoise. Blagden's report to Banks carried the clear assumption that, in the normal course of events, a paper by Hunter would have appeared in the *Transactions* as a matter of course, but Blagden feared that, while 'Mr. Hunter's name would have proved an ample shield' for a small error, in a matter 'so much calculated to make people Stare as this of the Roc', there would have been 'a sufficient Stock of ridicule both for him and us'.[29] Blagden's expectation that the Society would have shared in the embarrassment of publishing such a ridiculous claim hints at the practical limits of the Society's formal disclaimer of responsibility for 'the certainty of the facts, or propriety of the reasonings'.[30] The Society's officers hoped to bask in the reflected glory of publishing important contributions, but the 1752 transfer of editorial power meant that they could not avoid some portion of the blame for publishing ridiculous or trivial material.

The 40 per cent or so of papers in the late eighteenth-century *Transactions* that were not written by fellows represent a group of authors who were, to varying extents, outsiders to the Society, but who had successfully navigated the invisible and unwritten protocols. The requirement that papers be communicated by a fellow meant that potential authors who had personal connections to the fellowship, such as friendship or a family relationship, were at an advantage over those who did not. For instance, in 1795, a precocious 17-year-old, Henry Brougham, wrote to Blagden from Edinburgh, apologising that he had not 'the honor of being known to yourself or any of the other members' but asking advice on whether, and how, he could submit a paper on optics to the Society; Blagden agreed to communicate the paper for him.[31]

The act of communication did not necessarily entail any comment on the quality or content of the paper, but it did act as a check on the identity and character of the author. This was as much a check of social standing as of scientific capabilities or interest. Fellows often communicated papers for their friends, neighbours or relatives. For instance, the Midland physician and member of the Birmingham Lunar Society Erasmus Darwin forwarded a paper by his son Robert in 1786; and astronomer William Herschel was first introduced to the Royal Society through his local connections in Bath. His 1781 discovery of a new comet – that turned out to be the planet Uranus – reached the Society through letters from William Watson, a physician in Bath, to his father, also William Watson, in London.[32] Both Watsons were already fellows of the Society, and they would be the first two signatures on Herschel's own election certificate shortly afterwards.[33]

Family connections to the upper social and scientific echelons of the Royal Society could even, if rarely, overcome the conventions which excluded women. In 1786, Caroline Herschel wrote directly to Blagden, invoking his friendship with her brother William as her excuse for doing so: 'In consequence of the friendship which I know to exist between you and my Brother, I venture to trouble you in his absence with the following imperfect account of a comet.' Blagden arranged for her discovery to be read to the Society, and it was later published (accompanied, in print, by an explanatory note from William).[34] The circumstances were admittedly unusual: had William been in England rather than Germany, he would surely have been the one to write to Blagden, but a new comet was time-sensitive news which could not wait until his return. The episode demonstrates the considerable power of having the right connections. (The second woman to have a paper published in the *Transactions* under her own name would be Mary Somerville, in 1826; her husband was a fellow.[35]) Caroline Herschel's paper helped to establish her reputation as an astronomer, and several of her subsequent discoveries also appeared in the *Transactions*, though some were subsumed under her brother's name.[36]

The expectation that a communicator would vouch both for the identity and character of an outside author was made explicit in an exchange of correspondence in 1791. John Latham, a London physician, forwarded a paper about gunpowder without sharing the author's name. Banks explained that 'the R.S. never read a Paper sent to them without a name'.[37] Yet, as the subsequent correspondence revealed, the name alone was not enough. Banks asked Latham to provide 'not the bare signature but such additions Local & Professional as may Lead any of us at once to a knowledge of the Person intended by it'.[38]

For Banks, an author's identity, character and place in the world were legitimate objects of the Society's scrutiny. Ten years earlier, Blagden had reflected on Joseph Priestley's slightly spotted reputation when considering whether to find space for his paper in a meeting. Priestley was a chemist, dissenter and radical, and Blagden noted that '[s]o many things of the same kind have fallen upon this latter Gentleman, that one is led to suspect he goes as near the boundary line as possible, if he does not sometimes tread over it'.[39] Those evaluating the allocation of time in Society meetings and space in Society publications weighed the characters of those involved, as well as the research claims themselves.

Identity and reputation were part of the process of evaluating scholarship for publication, but the reverse was also coming to be true: publication was becoming an increasingly conventional route to scholarly reputation, including the sort of reputation that would secure election to the Royal Society. And as it became conventional, so conventions arose around the process itself. This was the tenor of Banks's advice to Thomas Wedgwood in 1792. Wedgwood was sufficiently well connected: he was the son of Josiah Wedgwood, the Midland pottery manufacturer and chemist who had been a fellow of the Society for almost 10 years. But Banks recommended that Thomas defer putting himself forward as a candidate for election until two papers of his 'respecting the production of Light & heat from various bodies' had appeared in the *Transactions*. The papers had already been read, but Banks suggested that 'you will gain much upon the Good opinion of the members when they read your Papers at their Leisure & Consequently are able to understand them more fully than can be done by hearing them read'.[40] It is notable that Banks was suggesting that, even within the fellowship, publication in the *Transactions* would be a more effective form of communication than being read aloud at a meeting.

Erasmus Darwin seems to have been well aware of the value of a *Transactions* paper for those seeking election to the fellowship. His grandson Charles Darwin would later claim that the paper Erasmus communicated for Charles's father, Robert, had in fact been substantially written by Erasmus himself. According to Charles, the paper was submitted under Robert's name as part of a successful campaign to get Robert elected a fellow of the Society (despite Robert's near-total lack of interest in science).[41]

Banks gave his clearest statement of his understanding of the role of scholarship among the criteria for election to the Society in a letter in the early 1790s: 'dignity of station and wealth we look upon as ornaments to our body', he explained, but 'learning we look upon as a sure passport' – unless it was accompanied by 'gross deficiency of moral character'.[42] As Banks's own trajectory showed, 'learning' could be demonstrated in various ways, but it was now most clearly and flatteringly demonstrated by submitting good papers for publication to the Society's *Transactions* prior to seeking election.

Gatekeeping the flow of papers into the Society's meetings and publications through an insistence on communication by a member was not unique to the Royal Society. It was common to many voluntary learned societies in the eighteenth century, and, in later decades it would come to be a feature that distinguished their periodicals from those unaffiliated with formal communities of scholars. For those fellows who introduced outsiders to the Society's meetings and publications, acting as a gatekeeper entailed a responsibility not to waste the fellows' time or to embarrass the Society's reputation. This pre-submission evaluation of authors is a prime example of how unwritten rules grounded in social and cultural norms shaped the published content of scientific knowledge. We can use Banks's correspondence to explore how he conceived this responsibility.

The first element was a duty to know enough about the topic of the paper to be able to evaluate it, at least to some extent. For instance, Banks refused to put forward a chemical paper by Robert Harrington in 1792, claiming that 'I do not Find myself capable [of] Comprehending their scope or their force sufficiently to feel Convincd that the Establishd theories of Modern Philosophy are subverted by them'.[43] Harrington felt ill-used, and demanded to know why Banks had not consulted a specialist in chemistry. However, the idea of evaluation by someone with subject-specific expertise was not (yet) a usual part of the Society's processes. As we will see, Banks did sometimes consult other people, but he was under no obligation to do so, nor to put forward papers whose conclusions he claimed not to understand. Harrington was an unusually difficult and manipulative customer – the testimonial he proffered for his work in chemistry was a book written by himself under a pseudonym.[44] During a subsequent exchange about another paper, Banks pointed out to Harrington that if could find another fellow willing to communicate the paper he, Banks, would not be able to prevent it. This may have been true in principle, but there is no known instance of an author successfully finding another point of entry to the Society's meetings after being turned down by Banks.

Another responsibility implicitly laid on communicators is revealed by an episode from 1783. The Astronomer Royal, Nevil Maskelyne, had communicated an astronomical paper by Patrick Wilson, of Glasgow, which was approved for the *Transactions*. Blagden was subsequently 'much mortified' to find that its principal conclusions had been anticipated, and critiqued, in print by the French astronomer Jérôme Lalande. Blagden felt that Maskelyne surely 'ought to have been acquainted with this, &, if he was, ought to have prevented the publication of Wilson's paper, at least in its present form; but it was he that communicated it'.[45] Taxing him on the subject (over dinner with Henry Cavendish), Blagden found Maskelyne evasive. He told Banks that he suspected that Maskelyne had not known of Lalande's work, but that Wilson might well have heard of it, 'considering the great connexion which the Scots always keep up with France'.[46] In other words, there was a double problem: it was bad enough that Wilson's conclusions were not original to him; but also, Maskelyne, as the communicator, had failed to properly scrutinise Wilson's work.

Communicators had a personal interest in being careful about what they communicated to the Society because, as Banks explained to a fellow in 1789, if the paper were to be read at a meeting, 'it must … be introduced to the fellows in your name'.[47] There was a risk of semi-public embarrassment for the communicator of a weak paper. This was balanced by the chance of public credit for communicating a paper approved for the *Transactions*, since the name of the communicator as well as the author would appear in print.

Banks's position as president made him an obvious point of contact for potential authors with no other connection to the Society. Between 1780 and 1799, for instance, 136 of the 455 papers published in the *Transactions* (30 per cent) were either communicated by Banks or explicitly took the form of letters addressed to him.[48] The same was true, to a lesser extent, of the two secretaries. Thus, a high proportion of what appeared in the *Transactions* was communicated through one of these official representatives, rather than by ordinary fellows. This opportunity to communicate – or not communicate – papers by outside authors was one of several ways in which the president and secretaries retained substantial influence over the editorial process. Despite the efforts of the 1752 reforms to constrain the power of individuals and cliques over the *Transactions*, effective control of the periodical in the late eighteenth century was almost as firmly vested in the hands of the president and secretaries as it had been during the era of private editorship.

The invisible influence of the presidential coterie

As well as acting as gatekeepers, the president and secretaries controlled the agenda for the Society's weekly meetings and for the Committee of Papers meetings, and they oversaw the progress of approved papers through the press. These spaces were not explicitly covered by the written statutes, but should nonetheless be seen as part of the editorial process, broadly understood. In these spaces, there were substantial opportunities to shape the length, framing or content of the papers that would appear in the *Transactions*. Furthermore, presidential influence over the elections of officers and Council members ensured that many of the members of the Committee of Papers owed allegiance to the president.

Joseph Banks was hardly the first president to exert his influence over the publications – after all, he had himself received advice from his predecessor John Pringle on his paper on Labrador stone – but Banks was notably not self-effacing. His autocratic approach to the presidency was widely recognised at the time, and after. One of the accusations he faced during the 1783–4 dissensions was that of 'tyrannical overbearing conduct', alongside mismanagement of the reading of papers at meetings and a dislike of mathematics.[49] However, Banks did not wield power alone. He relied on the assistance of a coterie of trusted colleagues and allies, the most important of whom was Charles Blagden.

The relationship between Blagden and Banks was crucial to the smooth running and social organisation of the Society in the late eighteenth century. Blagden was, as John Gascoigne has put it, Banks's chief 'scientific lieutenant', an interlocutor and confidante.[50] They kept each other informed of goings-on in London when one was absent, and constantly exchanged fragments of scientific news and opinions on particular papers. Their friendship, and their working relationship, was interrupted by a quarrel in 1788 which turned upon social questions: Blagden felt that Banks had neglected to push Blagden's social advancement, and their reconciliation in 1792 seems to have been brought about partly by Blagden's being given a knighthood. Even when Blagden resigned the secretaryship in 1797, claiming that his eyesight was no longer up to the task of reading papers at the Society's candlelit meetings, he continued to be closely involved in the affairs of Banks and the Society.[51]

Banks's apparent lack of interest in publishing himself did nothing to diminish his involvement in the publications of others, and he played pivotal roles in the production of numerous scientific works and travelogues of the late eighteenth and nineteenth centuries, including the accounts of the voyages of James Cook and William Bligh, Lord Macartney's China embassy in the 1790s, and William Roxburgh's *Plants of the Coast of Coromandel* (1793–1813) among others. His role usually fell somewhere between that of a patron and a fixer, sometimes acting as financial sponsor, and at other times as a broker between author, bookseller and sponsoring institution.[52] He frequently advised colleagues on where and when to publish, supplied information and specimens, allowed researchers access to his excellent botanical library, liaised with printers and booksellers, solicited subscriptions, or supplied them with assistants.

Unsurprisingly, then, Banks regarded the *Transactions* as a large part of his responsibility as president of the Royal Society. His perception of his own role can be seen in a letter to Blagden in summer 1782. Banks reported that he had relatively little to do in the summer, because the fellows were all busy working on papers that he would be called upon to 'produce to the Light' when the Society's meetings resumed in November (and perhaps in the *Transactions* the following spring). Jokingly, he referred to himself as 'an *Accoucheur* merely of literature'; in other words, as its midwife.[53] He was at once disclaiming the role of an author, while claiming principal responsibility for actually bringing the work of the fellows and their acquaintances to public notice.

Banks believed that management of the *Transactions* was part of the reason for the protests against his power over the Society's activities in 1783–4. He pointed out to Blagden that his opponents 'have every one of them had their /publications// papers/ repulsd & probably wish to print them in the R.S. Transactions'.[54] Rumours that the dissenters would launch a rival periodical, variously referred to by Blagden as the 'seceding' or 'mathematical' transactions, were still circulating in 1785.[55] Some of those involved were undoubtedly bothered by the relatively poor mathematical coverage in the Society's meetings and the *Transactions*, though Banks suspected others simply wished to be able to assert some sort of control over the publication of their own work.

When Blagden became co-secretary in 1784, he was junior to Joseph Planta, who served from 1776 to 1804. Despite this, Blagden appears to have taken immediate charge of the managerial aspects of the *Transactions*, including liaising with printers and engravers, and checking the proofs. His personal association with Banks meant that he was also closely involved in the selection and management of papers for the Society's meetings, and Banks appears to have trusted Blagden to evaluate papers in fields such as physics and chemistry, which were more remote from Banks's own interests. These two areas – the agenda for meetings and managing papers for the press – are where the invisible influence of the president and his friends was most effective.

Drawing up the agenda

The rule that authors hoping to publish in the *Transactions* should either be fellows or have the support of a fellow was an explicit and well-understood requirement that authors could learn to navigate. The next stage – of being added to the agenda for a weekly meeting of fellows – was shrouded in comparative mystery and guided by no written rules. Banks and Blagden drew up the running order for each meeting, which gave them the opportunity to scrutinise all the incoming papers, even those by fellows who were, in principle, fully entitled to have their papers read without question. This was another reason why fellows frequently sent their contributions directly to Banks, but even papers communicated to Blagden were likely to be seen by Banks when the two men discussed the upcoming meetings.

Banks claimed to have no special privilege of scrutinising papers ahead of meetings, telling a correspondent that 'I am not in the habit of preferring my own opinion to those of my friends'.[56] But this stance was undermined by his willingness to politely suggest that fellows might perhaps change their minds about certain papers they had proposed to communicate. Thus, having read a paper forwarded by Richard Gough in 1789, and discovering himself to be 'in some instance unable to admit the *data* the author has assumed, and in other to follow the reasoning he has grounded upon them to the establishment of the conclusions he wishes to draw', he wrote to Gough hinting of the embarrassment he might incur if he proceeded to communicate the paper. Banks stopped short of actually prohibiting it, but exerted clear moral pressure on the communicator to change his mind. By hinting that Gough might have been misled since 'the course of your studies has not directed your attention' to the matter, Banks offered a polite retreat.[57] Banks's intervention is particularly noteworthy as Gough had originally communicated the paper to Blagden: Banks was intervening where he had not been directly consulted.

The pressure Banks applied to Gough not to insist on exercising the privilege of communicating his friend's paper indicates that there were additional stages of evaluation that went on among a circle of trusted friends and colleagues or acknowledged experts before papers were accepted for reading at a meeting. Control of the agenda gave Banks and Blagden advance sight of papers, and a window of opportunity within which to exercise their influence if they thought the paper weak, unoriginal, unpalatable or better diverted elsewhere.

The presidential power over meetings was not simply a matter of selection, but also of practical scheduling. The constraint of one hour per week had implications for the form and length in which papers were presented to meetings. A letter from Blagden in February 1792 illustrates how he and Banks scheduled a meeting:

> I calculated the remainder of Schmeisser's paper, to take near quarter of an hour, & Schröter's will occupy about the same time. I think it should be read; and, as Mr Planta's minute of Topping's paper cannot be long, I should suppose there will be time to read a bit of Hunter's, if you take the Chair at 10 past 8.[58]

As this reveals, some papers were split between meetings (those by Johan Gottfried Schmeisser and John Hunter), and others (including Michael Topping's trigonometrical survey of the Coromandel coast) were read in abstract rather than in full.[59]

Banks was also willing to make cuts, as can be seen in the case of Jan Ingen-Housz, a Dutch-born chemist. In May 1782 Ingen-Housz wrote to Banks from Vienna pleading with him to read a hastily written paper of his before the Society's summer recess. A fellow of 13 years' standing, who had spent time in London a decade earlier, Ingen-Housz did not need Banks's support to get his paper read, but he understood the power the president exerted over the prioritisation of papers for reading at meetings.[60]

Ingen-Housz wanted to defend his priority in the discovery that light, rather than heat, was responsible for the production of dephlogisticated air (that is, oxygen) by plants. His claim had been effectively usurped in print by his former collaborator, the English chemist Joseph Priestley, whose criticisms of Ingen-Housz had been recapitulated in the general periodical press.⁶¹ Ingen-Housz thus wanted his defence of his own work read before the premier natural-philosophical assembly in England, as soon as possible. He hoped that Banks's 'friendly sentiments for me' would encourage him to 'get it read' before the summer recess, and then into print in a timely fashion. As he explained to Banks, he wanted it to 'be known among the philosophers that I can defend my cause'.⁶² The issue was not whether Ingen-Housz's paper would be read to the Society, but whether it could be read *soon*.

Banks's reply to Ingen-Housz reveals that, at the end of May, there were only three meetings remaining before the Society's recess, and at least two and a half of those were already spoken for. Banks explained that the decision about what to include in the remaining 30 minutes of reading time would depend on 'the importance of the paper & the priority of delivery'.⁶³ It hardly needed to be stated that decisions about importance, originality and urgency would be made by Banks himself.

In this case, Banks managed to meet Ingen-Housz's desires by unilaterally editing and shortening his paper. The cuts were necessary not simply to fit into the time available, but to avert the spectacle of one fellow critiquing another under the auspices of the Society. Casting aspersions on the character of other fellows was as ungratifying in the 1780s as it had been 70 years earlier when Edmond Halley had urged his correspondents to display 'Candor, Respect and Friendship' for each other (Chapter 3).⁶⁴ Banks told Blagden that Ingen-Housz had sent 'some good experiments mixed with no small abuse' directed at Priestley. Banks thus 'cut out the abuse and read the Experiments to the Soc', adding that the committee 'will probably order them to be printed'.⁶⁵ The Committee of Papers would consider the paper as presented at the meeting, which meant that Banks's interventions shaped the form of the published paper. (Despite Banks's and Blagden's willingness to make cuts, the average length of a *Transactions* paper was growing consistently: it had been less than 10 pages in the 1760s, but would be over 20 pages by 1800.)

Banks did not rely entirely on his own judgement to evaluate the papers submitted to the Society. He often consulted Blagden, and his friend, the chemist Henry Cavendish. This sometimes happened at Banks's breakfasts, where it is clear that a good deal of Royal Society business, as well as more general scientific conversation, was carried on. Blagden's diary recorded 'settled with Engraver' and 'conversation with [Sir J.B.] about gout, papers', as well as 'settled the Council', and 'pushing forwardness of Rennell about the Paper', among other things.⁶⁶ Part of the work of organising the schedule of upcoming meetings could include seeking opinions on proposed papers from other members of their social circle.

Sometimes these reports were made orally, but there is surviving evidence for written evaluations too. For instance, in early 1800, Blagden wrote to Banks from Bath, giving his opinion on a paper on luminescence in fish by London physician Nathaniel Hulme. Blagden was no longer secretary, but this letter illustrates his approach to evaluating a paper after almost two decades of editorial experience with Royal Society authors. Blagden saw merit in Hulme's paper, telling Banks that, if revised, it would be 'a valuable memoir for the Transactions' and, in a later letter, that 'The experiments on the extinction & revival of the light open in some respects a new field of inquiry'.[67] However, he had serious qualms about its current state, and offered three paragraphs of critique as well as extensive annotations on the original manuscript (as can be seen in Figure 6.3).

Blagden took issue with the phrasing of the paper's title; with the author's understanding of the physical nature of the luminescence; and with an 'unfounded hypothesis' based on 'a misunderstood experiment'. Blagden referred to a recent essay by Benjamin Thompson, Count Rumford, as justification for his claim that Hulme misunderstood 'the propagation of heat in fluids'; and this misunderstanding had led Hulme into a series of deductions that Blagden felt 'must be struck out of the paper, as surely the Society cannot give its sanction, under any terms, to such an opinion'.[68] Blagden's clear concern that publishing such opinions in the *Transactions* could be understood as approval by the Royal Society is further evidence that the disclaimers in the 1752 'advertisement' were already seen as an ineffective fig leaf.

The solution was revisions. Blagden told Banks, 'I have struck out, with pencil a large part of the paper; probable [sic] more than the Author will choose to sacrifice'. He had also indicated some sections which could be made fit for publication 'if the Author will write them over anew, leaving out the exceptionable reasonings & expressions', and added 'a few observations in the margin, suggesting some alterations in the order, as well as in certain terms; together with some reflexions for the author's consideration'.[69] It is clear that Blagden was not at all sure the author would be willing to make such extensive revisions, and that Banks would need to exercise diplomatic persuasion.[70] Banks appears to have succeeded, for Hulme's paper was read in February, and later published in the *Transactions*.[71]

Banks did not only rely for advice on his inner circle of associates, and it is evident that he was especially likely to call on outside help when the subject of the paper lay beyond his immediate expertise, or if the paper was of special importance. For instance, Banks sought the

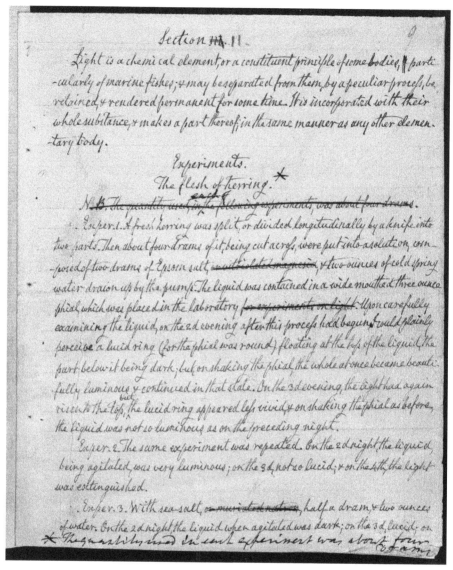

Figure 6.3 Charles Blagden's annotations on an 1800 paper on luminescence in fish, by London physician Nathaniel Hulme, included the removal of an entire section (and consequent re-numbering), as well as deleting phrases © The Royal Society.

opinion of Oxford astronomer Thomas Hornsby on William Herschel's 1781 paper.[72] This paper was being considered for the Society's Copley Medal, but there were doubts about what it was Herschel had actually discovered: he had initially characterised it as a new comet, but some members of the Society, notably Maskelyne, had begun to realise that it might be a planet. (Herschel did indeed win the medal, and it was indeed a planet.)

Banks also sought advice on a 1796 paper by Edward Jenner outlining his theory of vaccination against smallpox. He received at least one positive report, but the London physician Everard Home expressed some doubts about the commensurability of cowpox and smallpox, and the small number of actual cases presented in Jenner's paper.[73] Banks, therefore, declined to read the paper to the Society; and thus the Committee of Papers never had the chance to decide whether to publish Jenner's investigations (later published as *An Inquiry into the Causes and Effects of the Variolæ Vaccinæ*, 1798). Banks's seeking of advice was not systematic and followed no formal procedure.

Banks was willing to pass on some hints of the comments he received from his advisers. Just as he had passed on his own doubts about the paper communicated by Gough in 1789, so he informed Matthew Boulton in 1795 that the paper he had communicated for civil engineer William Chapman 'does not Prove satisfactory Either in Point of the theory he Endeavours to Establish or in the Conclusions he has drawn'.[74] He sent the original manuscript back to Boulton, to encourage the author to revise it, and enclosed the letter of advice he had received (presumably, with the critic's signature). He hoped it would explain to Boulton how problematic the mathematical treatment was. He told Boulton it was sent 'in full Confidence however that you will not Communicate it to any one'. He was welcome to tell Chapman 'any of the Objections Provided you do not tell him from what Quarter it is they are made', for 'in all such cases it Leads to an immediate Controversy which we Cannot maintain'.[75] Just to be sure, Banks asked for the letter to be returned once Boulton had read it. Banks's insistence on keeping the identity of his advisers confidential stands in contrast to his insistence on knowing the identity and character of authors. This asymmetry would be perpetuated in the procedures the Society developed for the refereeing of papers in the nineteenth century.

Revisions and corrections

By the time papers reached the Committee of Papers, they had already been scrutinised, evaluated and possibly altered by the president and secretaries. Those that were approved for publication were then entrusted to the secretaries to see through the press. Rather than waiting for all the papers which would make up the next 'part' of the *Transactions*, the secretaries transmitted copy to the printers as it was ready. This enabled the printers to spread the work of typesetting, proofing and printing across the year, striking off the papers separately and only later compiling them into parts and volumes. (Authors' purchase of separate copies will be discussed in Chapter 7.) The secretaries' oversight of the print production process created another window of opportunity for Banks and Blagden to shape the final printed form of the papers.

In 1776, the Society had introduced a statute that 'the original copy' of every paper read at a Society meeting 'shall be considered as the property of the Society'.[76] Banks usually offered to return manuscripts that were considered unworthy of being read out in the first place; but for papers that were read, the Society's claim was fiercely enforced. The physical manuscripts were deposited in the Society's library, where they could be consulted by fellows and approved guests. These included any papers not ultimately approved for publication, and thus 'archived' became a euphemism within the Society for 'rejected'. This arrangement sometimes proved awkward for authors who had sent in their original text and drawings without making a copy, though the Society routinely granted requests by such authors to make copies (or, as in the case of bee expert John Debraw, in 1783, to pay for them to be copied).[77]

The Society appears to have understood its common law ownership of the 'copy' of the paper as giving it the exclusive right to publish (or not).[78] In 1783, Joseph Planta noted that, 'When an Author gives in a paper to the Society, he delivers up his whole right to it', and that Council was 'peremptory' in asserting its unrestricted right not just to publish the papers it received, but also to alter them as it saw fit.[79] This could be done before sending the manuscript to the press, or in proof corrections.

Many of the original manuscripts surviving in the Royal Society archives bear annotations indicating alterations that were to be made ahead of printing, as Figure 6.3 showed.[80] Sometimes they also indicate where and by whom those alterations were to be made: this was most often to be done by the author at the author's house, but there are also numerous instances of editorial work taking place at Banks's house. For instance, in 1787, the visiting Swedish botanist Olof Swartz revised his paper there (to include the genus name in its title), as the annotation in Figure 6.4 makes clear.[81] Sometimes, the editorial work at Banks's house was done by Blagden, or one of Banks's scientific assistants-cum-librarians, or (less frequently) Banks himself. This had the definite advantage of limiting the delays as manuscripts and proofs traversed the English countryside. Blagden was particularly vexed by William Herschel's attempt to save money by using a private coach service, rather than the Royal Mail, writing in 1787 that 'your letter which ought to have been with me monday, did not arrive till yesterday, that is, two days later than it ought to have been deliverd, & a day later than it wo[ul]d. have

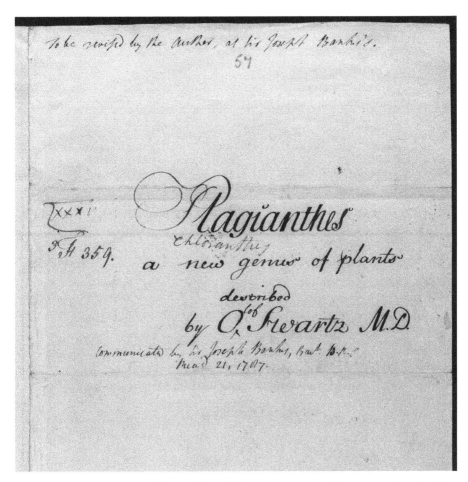

Figure 6.4 Manuscript of paper by Olof Swartz, 1787, with annotation showing that the author was to revise it at Banks's house © The Royal Society.

come had you waitd [sic] for the post'. He pleaded with Herschel 'always to trust the post' in future.[82]

We have already seen how Banks and Blagden negotiated with Nathaniel Hulme about revisions to his paper: as a London-based fellow, communication with Hulme was relatively easy. With Jan Ingen-Housz, in Vienna, correspondence would have been slow and thus, knowing Ingen-Housz's concerns for speed, Banks made the revisions himself. Some authors were willing to defer to Banks's judgement about the necessary revisions, as we see from a 1790 instance, where another spat between fellows threatened to disrupt the Society's surface politeness.

The problem arose when instrument-maker Jesse Ramsden laid eyes on the draft of William Roy's report on the trigonometric measurement of the distance between Dover and Calais, an effort to connect the surveys of Britain and France.[83] Roy had become so exasperated by delays to the project that he documented his grievances in the report, and accused Ramsden of 'inexcusable negligence'.[84] Ramsden complained to the Council, arguing that 'a Tradesman or Mechanic' could not 'suffer his professional character in particular to be publicly traduced', particularly by an organisation to which he himself belonged.[85] Ramsden was one of a very small group of instrument-makers and designers whose admission to the fellowship depended upon their profession, rather than their gentility, and he was justifiably concerned for his livelihood if Roy's paper were to be printed without revision.[86]

Urgency was less pressing in this case than in that of Ingen-Housz, for the project had been ongoing since 1783, and so Banks wrote to Roy (in Lisbon), urging him to tone down his language. He sought permission to moderate 'the epithets', and suggested Roy might 'give up the inexcusable on condition of retaining the negligence'.[87] Roy wrote back to argue that the Society and the public ought to be warned of Ramsden's unprofessional behaviour, but nonetheless gave Banks permission 'to smooth down Mr. Ramsden's back, by removing the whole or any part of the asperity of my mode of expression, in the manner you may think best. – You have my full leave to do it.'[88]

Proof correction was another area where the president and secretary exerted hidden influence over the final form of papers in the *Transactions*. Blagden's letters reveal that his usual editorial practice was to entrust the proof-reading of the main text to the author where possible, but that he would check it himself as well. He also usually took responsibility for checking the pagination, catch-words and running titles, though on one occasion when both he and Planta were going to be out of town, he had to ask William Herschel to take care of the paratextual proof-reading for his paper then going through the press.[89]

Roy's 1790 paper illustrates the challenges facing proof-readers: it was over 200 pages long and included pages upon pages of numerical tables. These were particularly difficult to check, for errors could be introduced in calculation, transcription or typesetting, and yet would be invisible to the casual glance. Blagden told Banks,

> I find it totally impossible by myself to collate the printed sheets with the MS, especially in figures, so as to be sure that no blunder

shall escape: all other persons, with whom I have conversed on the subject, acknowledge the same inability; indeed it is evidently a sort of business which cannot be done properly but by two persons one of whom shall read aloud, whilst the other attentively looks at the copy.[90]

Proofs were usually dispatched to authors for correction, but not always. For instance, in 1787, the engraved plate accompanying Cambridge professor Samuel Vince's paper on the precession of the equinoxes was sent to Nevil Maskelyne for proof correction. The correspondence makes clear that Maskelyne was also in possession of the original manuscript, so he may also have been checking the proofs for the text of the paper. The following year, a paper by Edward Waring was, in turn, sent 'to be corrected by the Revd. Mr Vince', perhaps for his expertise, or perhaps because it would be quicker than sending it to Waring near Shrewsbury.[91] The practice of asking someone other than the author or the secretary to check proofs seems to have been more common with mathematical or astronomical papers, and it is possible that Blagden did not have complete faith in his own ability to read the proofs.

Charles Blagden and Joseph Banks were not the first secretary and president to collaborate closely together on the editorship of the *Transactions*, but their surviving correspondence means that we know far more about their approach to editorship than we do about, for instance, Cromwell Mortimer and Hans Sloane. It is also important to note that Blagden and Banks operated within a different organisational framework: they were managing the *Transactions* as a formally acknowledged part of their roles within the Society, whereas Mortimer had been, in theory, an independent editor. Blagden and Banks had to work within the codified procedures that the 1752 reformers had laid down to govern the decision-making of the Committee of Papers, but, as we have seen, those written rules governed only one part of a much broader editorial process.

Evaluation of papers began with an evaluation of their authors, whose social status and character were assessed by the Society's tight gatekeeping processes before the contents of their papers could even be considered. As potential communicators, all fellows of the Society had a nominal role in the editorial process, even though few of them exercised it. Fellows also had the privilege of communicating their own papers, but, in contrast to the two-track system for 'academicians' and 'strangers' used at the Paris Académie Royale, papers authored by fellows went through the same formal process as those by outsiders. The advantage fellows had, as authors, lay in their better ability to recognise and navigate the tacit parts of the Society's processes, something that remained true long into the nineteenth century.

Fellows clearly had privileges over non-fellows, but it is also clear that members of Banks's inner circle were privileged and trusted over ordinary fellows, both as authors and as participants in the editorial process. In particular, the Society's secretaries and president wielded substantially more editorial power than anybody else. Control of the agenda for meetings gave Banks and Blagden substantial hidden power over the entry point to the editorial process. They were able to divert papers that they, or their advisers, regarded as weak or undesirable, but they were also able to decide when, at what length, and in what shape papers would be read by suggesting or imposing alterations to the text. This power of cutting and shaping papers vested considerable additional influence in the Society's officers, above and beyond simple gatekeeping. This hidden power held by the individuals who managed the supposedly collective editorial processes remained important into the twentieth century, though after Banks it would come to rest more with secretaries than with presidents.

It was during Banks's tenure that publication came to be viewed as an indispensable marker of scientific reputation. In this respect, Banks himself was something of a throwback: his own reputation in natural history did not rest on anything he had published, and by the time of his death this was enough of a peculiarity that at least one of his obituarists felt compelled to explain it away.[92] Furthermore, as president of the Royal Society, he was willing to admit those with genteel or aristocratic backgrounds chiefly on grounds of social acceptability rather than scientific accomplishment. Yet Banks advised other prospective fellows that a publication – or at least, submitting a paper to the Society – would help their chances of election. By the end of his tenure, having members who demonstrated their accomplishment by contributing papers to the *Transactions* had come to be seen as enhancing the reputation of the Society.[93] The link between publication and scientific reputation predates the professionalisation of scholarship and the development of academic career paths: it emerged in the context of a voluntary learned society principally composed of learned amateurs.

By the time Banks died in 1820, some of the younger fellows would be desperate to reform the Society for the new century. They wanted to replace the Banksian *dilettantes* with men of proven scientific accomplishment, with publication records. Yet even after those reforms, many of the social practices of scientific authorship, editorship and publishing that had become firmly embedded during the Banks regime remained pervasive. These include the traditions that scientific papers are not published anonymously; that they should be original, and not previously published; and that authors 'present' their papers to the scholarly community, with both gift and reward being understood in a non-financial economy of prestige or reputation. The processes for selecting and evaluating papers (and their authors) for publication were utterly embedded in the social dynamics of an eighteenth-century English gentlemanly learned society, with all that implies about class and gender. The *Transactions* was now firmly acknowledged as being 'of the Royal Society of London', and this would give it a distinct identity within the increasingly diverse landscape of scientific periodical publishing that emerged after 1800.

Notes

1 Benjamin Wardhaugh, 'Charles Hutton and the "dissensions" of 1783–84: scientific networking and its failures', *N&R* 71.1 (2017): 41–59, 47.
2 On the procedures for printing of foreign language material, and translations, see RS CMO/06, 20 May 1773 (the use of smaller type for the foreign language) and RS CMO/07, 20 January 1780 (translations to go before Committee of Papers, and be printed at end of volume).
3 On Banks and science, see David Philip Miller, 'Joseph Banks, empire and "centres of calculation" in late Hanoverian London', in *Visions of Empire: Voyages, botany, and representations of nature*, ed. David Philip Miller and Peter Hans Reill (Cambridge: Cambridge University Press, 1996), 21–37; John Gascoigne, *Science in the Service of Empire: Joseph Banks, the British state and the uses of science in the age of revolution* (Cambridge: Cambridge University Press, 1998); Marie Boas Hall, *All Scientists Now: The Royal Society in the nineteenth century* (Cambridge: Cambridge University Press, 2002), ch. 1; John Gascoigne, *Joseph Banks and the English Enlightenment* (Cambridge: Cambridge University Press, 2003); Simon Werrett, 'Introduction: Rethinking Joseph Banks', *N&R* 73.4 (2019): 425–9. On his celebrity, see, for example, Peter Simon Pallas's letter of congratulation to Banks, 13 March 1779 NS, Letter 156 in Neil Chambers, *The Scientific Correspondence of Sir Joseph Banks, 1765–1820*, 6 vols (London: Pickering & Chatto, 2007), vol. 1, 196–7.
4 Wardhaugh, 'Hutton and the "dissensions"', at 41. Wardhaugh recasts earlier accounts of the 'dissensions', such as Hall, *All Scientists Now*, ch. 1; John Heilbron, 'A mathematicians' mutiny with morals', in *World Changes: Thomas Kuhn and the nature of science*, ed. Paul Horwich (Cambridge, MA: MIT Press, 1993), 81–129.
5 For broader discussions of the role of sociability in the making of knowledge, see Steven Shapin, *A Social History of Truth* (Chicago, IL: University of Chicago Press, 1994); Anne Goldgar, *Impolite Learning: Conduct and community in the Republic of Letters, 1680–1750* (New Haven, CT: Yale University Press, 1995); Elizabeth Yale, *Sociable Knowledge: Natural history and the nation in early modern Britain* (Philadelphia: University of Pennsylvania Press, 2016).
6 D.C. Martin, 'Former homes of the Royal Society', *N&R* 22.1/2 (1967): 12–19.
7 Banks to Blagden, 27 December 1783, Letter 456 in Chambers, *Banks Scientific Correspondence*, vol. 2, 243–4.

8 See RS CMO/07, 25 May and 15 June 1780, and Samuel Wegg to Banks, 14 October 1780, Letter 191 in Chambers, *Banks Scientific Correspondence*, vol. 1, 241–2. For the frequent attendance of fellows at both, see Charles Blagden's diary, RS CB/3/3, *passim*.
9 Blagden's diary, RS CB/3/3 f. 8r, 3 July 1794; f. 68r, 7 September 1795; f. 77v 17 November 1795.
10 Provision for expert advice: RS CMO/04, 15 February 1751/2. Four papers are characterised as having been 'referred' in this way: two each on 25 May 1780 and 29 June 1786, see RS CMB/90/2. See also Noah Moxham and Aileen Fyfe, 'The Royal Society and the prehistory of peer review, 1665–1965', *Historical Journal* 61, no. 4 (2018): 863–89.
11 Based on a manual count of verdicts recorded in the minutes of the Committee of Papers; RS CMB/90/2, 1780–1828.
12 It was very sketchily recorded in the Council Minutes, CMO/07, 3 April 1783. See correspondence between Planta and Banks in April 1783, Letters 323–8 in Chambers, *Banks Scientific Correspondence*, vol. 2, 64–71. On public scrutiny of learned institutions in a slightly later period, see Alex Csiszar, *The Scientific Journal: Authorship and the politics of knowledge in the nineteenth century* (Chicago, IL: University of Chicago Press, 2018), ch. 2.
13 Pringle to Banks, [January] 1774, Letter 49 in Chambers, *Banks Scientific Correspondence*, vol. 1, 56. Banks's paper survives in RS L&P/52/44, with a manuscript annotation noting that it has been 'withdrawn from Ballot'.
14 He did, however, publish widely in other spheres, notably on issues of agricultural and land reform. See Julian Hoppit, 'Sir Joseph Banks's provincial turn', *Historical Journal* 61, no. 2 (2017): 403–29.
15 Blagden to Banks, 23 October 1795, Letter 1334 in Chambers, *Banks Scientific Correspondence*, vol. 4, 390–1.
16 Blagden to Banks, 22 October 1786, Letter 694 in Chambers, *Banks Scientific Correspondence*, vol. 3, 229–30. Mann's paper did not appear in the *Transactions*.
17 Banks to Franklin, 7 November 1783, Letter 429 in Chambers, *Banks Scientific Correspondence*, vol. 2, 209.
18 Manual count of minutes of the Committee of Papers, RS CMB/90/2, 1780–1820. Some 20.3 per cent of papers considered by Committee were not printed by the Committee's decision.
19 Will Slauter, 'The paragraph as information technology: How news traveled in the eighteenth-century Atlantic world' [Le paragraphe mobile], *Annales. Histoire, Sciences Sociales* 67, no. 2 (2012): 253–78; and 'Upright piracy: Understanding the lack of copyright for journalism in eighteenth-century Britain', *Book History* 16, no. 1 (2013): 34–61.
20 On the wave of new scientific journals, see Jonathan R. Topham, 'Anthologizing the book of nature: The circulation of knowledge and the origins of the scientific journal in late Georgian Britain', in *The Circulation of Knowledge between Britain, India, and China*, ed. Bernard Lightman and Gordon McOuat (Boston, MA: Brill, 2013), 119–52; and 'The scientific, the literary and the popular: Commerce and the reimagining of the scientific journal in Britain, 1813–1825', *N&R* 70 (2016): 305–24. Also Alex Csiszar, 'Proceedings and the public: How a commercial genre transformed science', in *Science Periodicals in Nineteenth-Century Britain: Constructing scientific communities*, ed. Gowan Dawson et al. (Chicago, IL: University of Chicago Press, 2020), 103–34; and Anna M. Gielas, 'Early sole editorship in the Holy Roman Empire and Britain, 1770s–1830s' (PhD, University of St Andrews, 2019).
21 On Crell's journals, including his connections to the Royal Society, see Gielas, 'Early sole editorship', ch. 2.
22 Crell (in Helmstädt) to Banks, 10 March 1785, Letter 566 in Chambers, *Banks Scientific Correspondence*, vol. 3, 34–5.
23 Crell to Banks, 27 April 1797, Letter 1416 in Chambers, *Banks Scientific Correspondence*, vol. 4, 481–3.
24 Banks to Samuel Williams, 20 July 1790, Letter 1002 in Chambers, *Banks Scientific Correspondence*, vol. 4, 2.
25 A manual count of papers published in the *Transactions* in the five years 1780–84 shows that 62 per cent were published by fellows. In 1827, just 109 of 714 fellows had ever published in the *Transactions*, see Charles Babbage, *Reflections on the Decline of Science in England* (London: B. Fellowes, 1830), 155.
26 Blagden to Banks, 5 October 1795, Letter 1330 in Chambers, *Banks Scientific Correspondence*, vol. 4, 386–7. The problem here was not that Blagden did not know the

credentials of the author(s), but that the authors were disinclined to be publicly seen to critique a deceased but eminent colleague. The paper was published as Edward Williams, William Mudge and Isaac Dalby, 'An account of the trigonometrical survey ...', *PT* 85 (1795): 414–591.

27 Banks to Herschel, 15 March 1782, Letter 243 in Chambers, *Banks Scientific Correspondence*, vol. 1, 311; Banks to James Edward Smith, 11 May 1787, Letter 734 in Chambers, *Banks Scientific Correspondence*, vol. 3, 276.

28 Herschel to Banks, 30 August 1789, Letter 940 in Chambers, *Banks Scientific Correspondence*, vol. 3, 503; Postscript to William Herschel, 'Catalogue of a second thousand of new nebulae ...', *PT* 79 (1789): 255.

29 Blagden to Banks, 16 October 1785, Letter 611 in Chambers, *Banks Scientific Correspondence*, vol. 3, 104.

30 'Advertisement', *PT* 47 (1751–2): sig. a1r-v.

31 Henry Brougham to Blagden, 26 July 1795, in RS CB/1/2/191.

32 Watson's role in the early communication of the news of the 'comet' is mentioned in Thomas Curtis to Blagden, 31 March 1781, in Blagden Papers, RS CB/1/3/121. See also 'Account of a Comet. By Mr. Herschel; communicated by Dr. Watson, Jun. of Bath F.R.S.', *PT* 71 (1781): 492–501.

33 Election certificate for William Herschel, RS EC/1781/19.

34 C. Herschel, 'An account of a new comet', *PT* 77 (1787): 1–3; and W. Herschel, 'Remarks on the new comet', *PT* 77 (1787): 4–5. See also Emily Winterburn, 'Learned modesty and the first lady's comet: A commentary on Caroline Herschel (1787): An account of a new comet', *PTA* 373.2039 (2015).

35 M. Somerville, 'On the magnetizing power of the more refrangible solar rays', *PT* 116 (1826): 132–9. See also Kathryn A Neeley, *Mary Somerville: Science, illumination, and the female mind* (Cambridge: Cambridge University Press, 2001); and James A. Secord, *Visions of Science: Books and readers at the dawn of the Victorian age* (Oxford: Oxford University Press, 2015), ch. 4.

36 W. Herschel [and Caroline Herschel], 'Observations on a comet', *PT* 79 (1789): 151–3, and 'Miscellaneous observations' *PT* 82 (1792): 23–7. See also Claire Brock, *The Comet Sweeper: Caroline Herschel's Astronomical Ambition* (Icon Books, 2017); and Michael Hoskin, *Discoverers of the Universe: William and Caroline Herschel* (Princeton, NJ: Princeton University Press, 2011).

37 Banks to John Latham, 18 December 1791, Letter 1078 in Chambers, *Banks Scientific Correspondence*, vol. 4, 89.

38 Banks to John Latham, 28 December 1791, Letter 1083 in Chambers, *Banks Scientific Correspondence*, vol. 4, 93–4.

39 Blagden to Banks, 30 June 1782, Letter 262 in Chambers, *Banks Scientific Correspondence*, vol. 1, 333.

40 Banks to Thomas Wedgwood, 17 March 1792, Letter 1103 in Chambers, *Banks Scientific Correspondence*, vol. 4, 113.

41 Blagden to E. Darwin, 14 September 1786, CB/2/34; and Charles Darwin, *Life of Erasmus Darwin*, ed. Desmond King-Hele (Cambridge: Cambridge University Press, 2003), 14.

42 Banks to John Eardley Wilmot, 11 March 1793, Letter 1175 in Chambers, *Banks Scientific Correspondence*, vol. 4, 204.

43 Banks to Harrington, [March] 1792, Letter 1102 in Chambers, *Banks Scientific Correspondence*, vol. 4, 111–12.

44 Georges Cuvier, 'Eloge historique de Sir Joseph Banks, lu le 2 avril 1821', *Memoires de l'academie des science de l'institut Francais* 5 (1827): 204–30. The possibility that Bewley might be a pseudonym was circulating early, and is mentioned in a review of Bewley's (that is, Harrington's) work in the *Gentleman's Magazine* 70 (July 1791): 640.

45 Blagden to Banks, 16 October 1783, Letter 410 in Chambers, *Banks Scientific Correspondence*, vol. 2, 178–9.

46 Blagden to Banks, 16 October 1783, Letter 411 in Chambers, *Banks Scientific Correspondence*, vol. 2, 180–1.

47 Banks to Richard Gough, 23 February 1789, Letter 898 in Chambers, *Banks Scientific Correspondence*, vol. 3, 461.

48 Data derived from manual counts of papers published in the *Transactions* for the period, and those considered for publication by the Committee of Papers (RS CMB/90/2,

Committee of Papers minute-book 1780–1828). Almost exactly two-thirds of all papers considered during this period were published, leaving a further 50 or so that were read to the Society but rejected for publication.

49 Notes on 'Firepower' by Charles Hutton (c. 1783), quoted in Wardhaugh, 'Hutton and the "dissensions"', at 48.

50 Gascoigne, *Science in the Service of Empire*, 41. See also Hannah Wills, 'Charles Blagden's diary: Information management and British science in the eighteenth century', *N&R* 73.1 (2019): 61–81; and 'Joseph Banks and Charles Blagden: Cultures of advancement in the scientific worlds of late eighteenth-century London and Paris', *N&R* 73.4 (2019): 477–97.

51 Blagden to Banks, 25 April 1797, Letter 1407 in Chambers, *Banks Scientific Correspondence*, vol. 4, 475.

52 Edwin D. Rose, 'Specimens, slips and systems: Daniel Solander and the classification of nature at the world's first public museum, 1753–1768', *British Journal for the History of Science* 51, no. 2 (2018): 205–37; 'From the South Seas to Soho Square: Joseph Banks's library, collection and kingdom of natural history', *N&R* 73.4 (2019): 499–526; and 'Publishing nature in the age of revolutions: Joseph Banks, Georg Forster, and the plants of the Pacific', *Historical Journal* (2020): 1–28.

53 Banks to Blagden, 19 August 1782, Letter 279 in Chambers, *Banks Scientific Correspondence*, vol. 2, 7. See also Moxham, '*Accoucheur* of literature'.

54 Banks to Blagden, December 27 [1783], Letter 456 in Chambers, *Banks Scientific Correspondence*, vol. 2, 243.

55 Wardhaugh, 'Hutton and the "dissensions"', 51.

56 Banks to Gough, 23 February 1789, Letter 898 in Chambers, *Banks Scientific Correspondence*, vol. 3, 461.

57 Banks to Gough, 23 February 1789, Letter 898 in Chambers, *Banks Scientific Correspondence*, vol. 3, 461.

58 For example, Blagden to Banks, 23 February 1792, Letter 1096 in Chambers, *Banks Scientific Correspondence*, vol. 4, 108.

59 For other instances of the partial reading of mathematical papers, see Wardhaugh, 'Hutton and the "dissensions"'.

60 This episode is more fully discussed in Aileen Fyfe and Noah Moxham, 'Making public ahead of print: Meetings and publications at the Royal Society, 1752–1892', *N&R* 70.4 (2016): 361–79.

61 Joseph Priestley, *Experiments and Observations Relating to Various Branches of Natural Philosophy, with a Continuation of the Experiments Upon Air. The Second Volume* (London: Birmingham, Pearson and Rollason for J. Johnson, 1781), 180–91. For the press coverage, see *Critical Review* 52 (July 1781): 133–41 and 176–84. For wider context, see Robert E. Schofield, *The Enlightened Joseph Priestley: A study of his life and work from 1773 to 1804* (University Park: Pennsylvania State University Press, 2004), 154–6; Norman Beale and Elaine Beale, *Echoes of Ingen Housz: The long lost story of the genius who rescued the Habsburgs from smallpox and became the father of photosynthesis* (Salisbury: The Hobnob Press, 2011), 310–19.

62 Jan Ingen-Housz to Banks, 6 May 1782, Letter 253 in Chambers, *Banks Scientific Correspondence*, vol. 1, 321–2.

63 Banks to Ingen-Housz, 31 May 1782, Letter 256 in Chambers, *Banks Scientific Correspondence*, vol. 1, 326–7.

64 Halley, 'The Preface', *PT* 29 (1714): 3–4.

65 Banks to Blagden, 16 June 1782, Letter 259 in Chambers, *Banks Scientific Correspondence*, vol. 1, 330.

66 Blagden's diary, RS CB/3/3 f. 42v, f. 50r, f. 78v, f. 77v.

67 Blagden to Banks, 5 and 17 January 1800, Letters 1527 and 1532 in Chambers, *Banks Scientific Correspondence*, vol. 5, 1–3 and 7–8.

68 Blagden to Banks, 5 January 1800, Letter 1527 in Chambers, *Banks Scientific Correspondence*, vol. 5, 1–3.

69 Blagden to Banks, 5 January 1800, Letter 1527 in Chambers, *Banks Scientific Correspondence*, vol. 5, 1–3.

70 See also Blagden to Banks, 17 January 1800, Letter 1532 in Chambers, *Banks Scientific Correspondence*, vol. 5, 7–8.

71 Nathaniel Hulme, 'Experiments and observations on the light which is spontaneously emitted ...', *PT* 90 (1800): 161–87.
72 Thomas Hornsby (in Oxford) to Banks, 23 November 1781, Letter 232 in Chambers, *Banks Scientific Correspondence*, vol. 1, 294–5.
73 Everard Home to Banks, 21 April 1797, Letter 1406 in Chambers, *Banks Scientific Correspondence*, vol. 4, 474–5.
74 Banks to Boulton, 25 May 1795, Letter 1310 in Chambers, *Banks Scientific Correspondence*, vol. 4, 366–7.
75 Banks to Boulton, 11 June 1795, Letter 1314 in Chambers, *Banks Scientific Correspondence*, vol. 4, 370. See also David Philip Miller, 'The usefulness of natural philosophy: The Royal Society and the culture of practical utility in the later eighteenth century', *British Journal for the History of Science* 32 (1999): 185–201, 189.
76 Statutes of 1776, described in *The Record of the Royal Society of London*, 3rd ed. (London: The Royal Society, 1912), 168.
77 RS CMO/07, 24 February 1783. Debraw's 1777 paper had been published (*PT* 67), but this request appears to refer to a later, unpublished, paper.
78 Feather, 'From rights in copies to copyright: The recognition of authors' rights in English law and practice in the sixteenth and seventeenth centuries', in *The Construction of Authorship: Textual appropriation in law and literature*, ed. Martha Woodmansee and Peter Jaszi (Durham, NC: Duke University Press, 1994), 191–209. For the Society's approach to the relatively new concept of 'copyright', see Aileen Fyfe, Julie McDougall-Waters, and Noah Moxham, 'Credit, copyright, and the circulation of scientific knowledge: The Royal Society in the long nineteenth century', *Victorian Periodicals Review* 51, no. 4 (2018): 597–615.
79 Joseph Planta to Banks, 5 April 1783, Letter 323 in Chambers, *Banks Scientific Correspondence*, vol. 3, 62–3. In practice, authors were often, but not always, consulted about such changes.
80 See the 'Letters & Papers' series, at the RS archive.
81 The manuscript survives at L&P/9/57/1; it was published as O. Swartz, 'Chloranthus, a new genus of plants', in *PT* 77 (1787). See also Neil Chambers, *Joseph Banks and the British Museum: The world of collecting, 1770–1830* (Abingdon: Pickering & Chatto, 2007), vol. 3, 31.
82 Blagden to William Herschel, 17 May 1787, RS CB/2/54.
83 The survey had originally been proposed by the French in 1783. See Banks to Blagden, 13 October 1783, Letter 405 in Chambers, *Banks Scientific Correspondence*, vol. 2, 171–2. See also Rachel Hewitt, *Map of a Nation: A biography of the Ordnance Survey* (London: Granta, 2010), ch. 3. The published version of Roy's report appeared in *PT* 80 (1790).
84 See Banks to William Roy, 30 January 1790, Letter 971 in Chambers, *Banks Scientific Correspondence*, vol. 3, 532–4; Roy's paper survives at RS L&P/9/168.
85 Ramsden to RS Council, 13 May 1790, Royal Society Miscellaneous Manuscripts, MM/3/30, 1–9.
86 On Ramsden's background, see Anita McConnell, *Jesse Ramsden (1735–1800): London's leading scientific instrument maker* (Aldershot: Ashgate, 2007).
87 Banks to Roy, 30 January 1790, Letter 971 in Chambers, *Banks Scientific Correspondence*, vol. 3, 532–4.
88 Roy to Banks, 6 March 1790, Letter 974 in Chambers, *Banks Scientific Correspondence*, vol. 3, 536–8.
89 Blagden to William Herschel, 15 September 1789, in Blagden Letter-book, RS CB/2, f. 297.
90 Blagden to Banks, 23 September 1790, Letter 1008 in Chambers, *Banks Scientific Correspondence*, vol. 4, 8–9.
91 For Vince's paper ('On the precession of the Equinoxes', *PT* 77 (1787): 363–7, see RS L&P Decade IX Vol. 83, f. 58; for Waring's paper revised by Vince ('Some properties of the sum of the divisors of numbers', *PT* 78 (1788): 388–94), see L&P Decade IX Vol. 84, f. 95.
92 See Cuvier, 'Eloge historique'. Hannah Wills has usefully pointed to Cuvier's remark as evidence of the different significance of publishing in building scientific careers in Britain and France at this time, in Wills, 'Banks and Blagden'. On Banks's reputation, see James A. Secord, 'How scientific conversation became shop talk', in *Science in the Marketplace: Nineteenth-century sites and experiences*, ed. Aileen Fyfe and Bernard Lightman (Chicago, IL: University of Chicago Press, 2007), 23–59.

93 Charles Babbage used publication in the *Transactions* as a proxy for evaluating the scientific activity of the fellowship in Babbage, *Reflections on the Decline of Science in England* (London: Fellowes, 1830). See also David Philip Miller, 'The Royal Society of London, 1800–1835: A study in the cultural politics of scientific organization' (PhD, University of Pennsylvania, 1981); Hall, *All Scientists Now*, ch. 3. On the consolidation of the practice of using publication lists to evaluate research, from the mid-nineteenth century, see Alex Csiszar, 'How lives became lists and scientific papers became data: Cataloguing authorship during the nineteenth century', *British Journal for the History of Science* 50 (2017): 23–60.

7
Circulating knowledge, c. 1780–1820
Noah Moxham and Aileen Fyfe

In 1802, Joseph Banks, president of the Royal Society, assured the journal editor William Nicholson that the Society's aim was 'to disperse as widely as possible the knowledge communicated to them by their members'.[1] The printed copies of the *Transactions* were the main way in which the approved version of the 'knowledge communicated' to the Society could circulate beyond the people who attended its meetings. Yet, as Nicholson had pointed out just a few years earlier, the circulation of 'academical Transactions' – such as those of the Royal Society – was 'very limited', and that 'even the best memoirs they contain must continue unknown' to many scholars.[2]

Nicholson was the founding editor of the *Journal of Natural Philosophy* (f. 1797), one of a new breed of periodicals devoted to the sciences. For Nicholson, originality was less important than 'utility', and he aimed to give his readers access to the latest developments in the sciences throughout Britain and Europe.[3] This type of periodical, combining original contributions with reprinted excerpts, summaries and reportage, was familiar on the Continent through such titles as François Rozier's *Observations sur la physique* (f. 1771) and Lorenz Crell's *Chemische Annalen* (f. 1784), but it was still new in Britain in the 1790s. These periodicals were regarded by contemporaries as a distinctly different genre from the *Transactions* and *Mémoires* produced by royal societies and academies. New learned societies, such as the 1788 Linnean Society and the 1807 Geological Society of London, would (at least initially) emulate the publishing style of the older learned institutions; but independent editors took a different route.

The distinction was made clear in 1813 by the London-based Scottish chemist Thomas Thomson. He had recently been elected a fellow of the

Royal Society, but was also in the process of launching his own monthly periodical, the *Annals of Philosophy*. In a preface surveying the history of scientific periodicals in Britain, he acknowledged the *Transactions* as 'the first periodical work of science', but, as Jon Topham has noted, he argued that the way it had been transformed from Oldenburg's news-sheet into the '*Transactions* of the Royal Society' meant that, for him, it no longer qualified as 'a periodical philosophical journal'.[4] His down-playing of the contemporary *Transactions* was undoubtedly part of a pitch to promote his own *Annals*, but it emphasises how different the Royal Society's gentlemanly, institutional model of publishing was from that being adopted by the new periodicals that were being labelled 'journals'.[5]

For Thomson, part of the reason the *Transactions* could not be seen as a true 'journal' was its erratic and infrequent periodicity: it appeared in chunky 'parts' twice a year, usually but not necessarily in June and November. But the real difference was one of scope. Thomson saw a journal as something that reached beyond the activity of one organisation and its members, with a willingness to draw in material from a variety of sources to keep readers up to date with goings-on in the world of scientific scholarship. It was a vision not unlike that of Oldenburg but, as Thomson pointed out, the contemporary *Transactions* no longer took 'notice of the discoveries made by foreigners, nor of the scientific books which have made their appearance in different countries', but focused exclusively on original papers by fellows and their acquaintances.[6] This did not mean that nothing foreign was ever reported in the *Transactions* and its ilk – a steady flow of overseas communications was maintained and eminent Continental and American men of science were elected to the Society as foreign members – but rather that readers could not use it for an overview of scientific developments in the broad sense. Thomson promised his own prospective readers not just original papers but also reprinted foreign material, book notices, 'scientific intelligence', and reports 'of the proceedings of Philosophical Societies', including the Royal, Linnean and Geological societies.[7]

The appearance of the 'journals' edited by Nicholson, Thomson and others points to the emergence of a British scientific public interested in syntheses of recent work as well as original research papers, and eager for news of discoveries from abroad (but without the financial means or linguistic ability to acquire it directly). Nicholson's vision of his audience imagined 'a very large class of men of science' beyond those 'extreme few' who, like Joseph Banks and friends, were 'so fortunate as to have access to all the expanded sources of philosophical intelligence'.[8] The new journals have often been described by historians as 'commercial', to emphasise an engagement with the British book trade that was notably closer than that of the Royal Society and similar institutions.[9] We prefer to think of them as 'independent' journals, a term which makes no assumptions about the editor's economic aims, can be used beyond the context of early nineteenth-century Britain, and encourages us to think about other ways in which the journals differed from learned society transactions. The independent journals not only had a different business model, but also different editorial visions and intended audiences.[10]

The independent journals should not be seen as competing directly with the society transactions, but as serving different purposes. Particularly in their early years, the new journals depended heavily on the learned societies for content to report or reprint; but it can also be argued that societies, including the Royal Society, came to depend on the 'literary replication' and reportage provided by the independent journals as a means of expanding the circulation of their material (or at least, increasing awareness of its existence).[11] Banks and his associates were quicker to recognise the value of this in the European context, where distribution of the *Transactions* itself was slow and erratic, and comprehension was limited by linguistic barriers; but they also came to appreciate its value within Britain, as long as it was carried out in a manner that did not pre-empt or diminish the *Transactions*.

Changes in the production arrangements of the *Transactions* in the 1790s would give the Royal Society an opportunity to improve its physical appearance. By the 1810s, therefore, the *Transactions* was a handsomely produced publication, whose generous quarto format and copper-plate engravings set it visually apart from something like Thomson's more sparsely illustrated *Annals* with its smaller octavo format.[12] This was only possible because the Society funded the production of the *Transactions* as a perquisite for its members and for use as a gift to other learned institutions: it chose to fund high-quality production for the honour and reputation of the Royal Society. Independent editors whose business model depended on income from public sales had to make different choices.

The independent journals were not the only new element in scientific print culture in these decades. The *Transactions* had been joined by equivalent institutional periodicals produced by academies and learned societies across Europe and North America.[13] Together with the independent journals, they helped to create communities of scholars defined by their shared readership of certain periodicals, and separated by language, geography and scientific specialisation. The development of national or regional scholarly communities had clear benefits, but also made it trickier to participate in a transnational scholarly community. The Swedish chemist Jöns Jakob Berzelius, for instance, argued for the importance of Swedish-language publication for the honour of Sweden and the Royal Swedish Academy of Sciences, while also arranging for his own papers to appear in the Paris-based *Annales de chimie* and various German journals.[14] Transnational scholarship had ceased to function as a 'republic of letters', and become a patchwork of fragmented republics that managed to sustain some inter-communication.[15]

One of the motivations for the editors of the new journals was to transcend national publishing cultures by bringing transnational news to their readers. For similar reasons, learned institutions sought to position themselves as key agents in the circulation of transnational scientific information, presenting their own publications to cognate institutions and seeking to acquire reciprocal gifts with which to stock their own libraries. Joseph Banks sought to use these gifts as a benevolent patronage that would establish a clear role for the Royal Society and its *Transactions* in an increasingly crowded landscape. In 1804, for instance, he assured an American correspondent that the Society was 'very desirous of giving every aid in their power to bodies, who like them are busied in the promotion of Science'.[16] He cast the Royal Society as the senior institution, willing and able to assist its younger siblings or offspring; in this case, it meant the promise of a regular gift of *Philosophical Transactions* to the American Philosophical Society. Banks was similarly gracious in accepting gifts from learned institutions that honoured the Royal Society by sending their own *Transactions* or *Memoirs* to London for the interest and edification of the fellows there.

This chapter will explore the ways in which the *Transactions*, and the papers printed therein, circulated in the decades around 1800. As well as the practical logistics of domestic and overseas distribution, it investigates the relationship of the *Transactions* to the increasingly diverse array of other forms of scientific print at home and overseas.

The transnational circulation of knowledge

Most copies of the *Transactions* circulated within Britain, but the officers of the Royal Society were well aware of its role in spreading the name and reputation of the Society overseas. As we will see in the next section, discontent with its print quality was stimulated by the fear that 'foreigners' would think it 'worse printed than the publications of most of the other learned Societies of Europe'.[17] In the early 1790s, it was hoped that a change of printer would help the *Transactions* to 'do credit to the British

Press in Foreign Parts'.[18] This did depend, however, on copies of the *Transactions* managing to reach 'foreign parts'. The circulation of news and scientific discoveries was slow and erratic at the best of times, and made worse for most of the 1790s and 1800s by the years of war between Britain and France. The surviving correspondence between Joseph Banks and Charles Blagden illuminates the ways in which scientific news travelled to and from Britain. Personal contacts, correspondence and private travel remained a key element of that inter-communication: Blagden's trips to Paris, and his friendships there, were invaluable. But their correspondence also reveals how much they depended on printed material – books, pamphlets and, especially, periodicals – for knowledge of scientific developments beyond their personal networks. Along with details of the Society's meetings and publications, they would routinely pass on snippets from their reading, and direct each other's attention to interesting articles.

When Blagden was in Paris, the commissions he undertook for Banks and for the Royal Society reveal some of the challenges involved in the international acquisition of scientific periodicals and books. In 1783, for instance, we learn that Banks already had an arrangement with a London bookseller to acquire Rozier's *Observations*, but he asked Blagden to source other periodicals directly from Paris. Blagden reported that he had made arrangements for the *Journal des sçavans*, the *Journal de Medicine* and *L'Esprit des Journaux*.[19] A few years later, these arrangements were still in place: Banks was getting most of his French journals via an agent in Paris, but Rozier's journal was apparently supplied by 'a french Bookseller in Greek Street'.[20]

For his information on developments in the German and northern European lands, Banks relied on a succession of Scandinavian assistants and librarians in the 1780s and 1790s. The Swedish botanist Jonas Dryander managed the smooth flow of letters, printed copies and specimens between Banks's London residence, his Lincolnshire estate and his many Continental correspondents. His letter to Banks include such comments as, 'I learnt it from the Intelligenz-blatt to the Jena Journal', which suggest that he was extracting information from the German periodical literature for Banks.[21] On another occasion, he told Banks that: 'From the Gottingische Anzeigen which came to day, I saw that Crell has also printed his Memoire ... in the Volume of Nova Acta Acad. Naturae Curiosorum, which is just come out. There was also a review of a new volume of the Berlin Society's transactions.'[22] Thus, Dryander's reading of one German-language journal – the *Göttingische Gelehrte Anzeigen* – alerted Banks to potentially interesting material in two other German periodicals, enabling him to make the decision about whether to acquire the originals.[23]

In 1788, Blagden told Banks that Horace-Bénédict de Saussure's narrative of his ascent of Mont Blanc was scheduled to appear in the weekly newspaper *Journal de Genève*. Blagden was travelling in Switzerland at the time, and considered acquiring a copy for Banks, but then realised that, 'as such things are immediately copied into the Paris Journal, I did not attempt to [send] it you, from the conviction that you would get [it] sooner in your ordinary course'.[24] In this case, the existence of a reprinted account removed the need to acquire the (more obscure) original.

Acquiring news from overseas, whether in correspondence or in print, depended upon postal services that often connected poorly.[25] Royal or governmental mail services usually only carried letters and were expensive, so larger packets were typically sent by private carrier services. Lorenz Crell, for instance, had sent his letters to Banks and Blagden in packages that also contained copies of his journal. The packages kept going missing in Amsterdam, leading Blagden to complain to Banks: 'Unless we can stop him from sending the letters with his journals, they will never be delivered.'[26] He told Crell that his letters ought to 'go by the same post as carries the letters', rather than 'by a sort of diligence, or wagen' which delivered them to 'some office in Amsterdam, different from ... the post-office, where they lie neglected, no one taking charge of them'.[27]

Transnational scientific exchange often depended on the ingenious use of personal contacts. It was common practice to send letters or other material destined for several people to a single trusted point of contact, who would distribute them further. This was one of the functions that Dryander performed for Banks: for instance, in 1791, he reported to the absent Banks that, 'An other batch of Kirwan's paper came [from Dublin], directed to you. It contained a copy for Mr. Cavendish, which I have sent to him; the rest are for different people abroad'.[28] In 1804, the American naturalist Benjamin Smith Barton, a vice-president of the American Philosophical Society, asked Banks to 'excuse the liberty' of sending multiple copies of the American *Transactions* 'under cover to you' though intended 'for some other Academies'.[29] Three months later, Banks was able to reassure Barton that the copies 'have reached my hands Safe & have been distributed to the Several persons & Societies for which they were intended'.[30]

For few and small items, the most secure option was to ask a traveller to convey the items in their luggage, though it was sometimes difficult to find an acquaintance travelling at the right moment. In early 1815, Banks was trying to send copies of the *Transactions* and Thomson's

Annals to Paris. On 14 February, he told Blagden that he had 'delayed this Letter too Long in the hopes of finding means to Forward to you by a Private hand', but had finally sent it by the public 'diligence'. As luck would have it, just six days later, Banks learned that 'a Friend of Sir E Home' (Everard Home, the surgeon) was travelling to Paris, and could take the recent issue of the *Edinburgh Medical Journal*.[31] On 2 March, he was able to send another parcel 'by Mrs Damer', and on 8 March, another letter was carried 'by the hands of Mr William the Proprietor of one of the rolling Printing Presses'.[32]

Judging the reliability of travellers who were merely acquaintances of acquaintances could be difficult. In 1820, Banks was 'uneasy' that a recent part of the *Transactions* sent to the French physicist Jean-Baptiste Biot had gone astray. He investigated the history of its travels, and was able to confirm that it had been taken by 'Gould the Porter of the R.S.' to the home of Dr Roche on Carey Street, whose brother was intending 'to set out for Paris that Evening'. Gould confirmed that the *Transactions* had been delivered before the traveller left; but Banks could track it no further. (He sent a replacement copy, just in case.[33])

Those scholars unable or unwilling to rely on personal acquaintances to transmit the latest issues of periodicals of interest, could instead engage the services of a bookseller with international networks. This was how Banks had acquired his French journals, but it did depend on finding the right bookseller. For instance, when the chemist Richard Kirwan returned to Dublin in 1787, he relied on the Royal Society's bookseller to send him the latest periodicals. In 1789, Kirwan told Banks that he hoped soon to see the new *Annales de chimie*, and an article in a 'late Volume of the Academy of Turin'. But he was not optimistic: the bookseller still had not sent 'the Memoirs of Paris for 1785' that he had requested; nor 'the Memoirs of Berlin for 1784'; nor 'the first part of our transactions [*Philosophical Transactions*] for 1788, tho' he has the last part'. Kirwan ended in despair, 'Thus am I used by him, nor does he even answer my letters'.[34] Despite the plaintive tone, this letter demonstrates that Kirwan was clearly aware of what was being published in the European world of science (even though he could not get hold of it). It also reminds us of the time – potentially measured in years – it could take for publications to travel long distances.

One of the things Blagden did while in Paris in early 1803 was try to track down copies of the *Transactions* that had apparently been sent through booksellers' networks. They had not arrived at the Institut National (into which the former Académie Royale had been incorporated), and the suspicion was that the missing issues were sitting in an unknown bookshop somewhere in Paris. Blagden told Banks, 'It is a pity that you did not mention to me who are the correspondents of [the London booksellers] Payne & MacKinlay: they are not known either at the Institut or to me'.[35] The following year, Banks told a Philadelphia correspondent that the Royal Society had 'found by long experience, that they cannot trust their Bookseller for remittg. the Books to their very numerous correspondents to whom they are Sent, with any degree of regularity'.[36] Banks politely imputed this habitual failure to the scale of the task rather than the negligence of the bookseller, but nonetheless advised his correspondent about alternative channels.

The erratic and unreliable nature of communications with Continental Europe was not entirely the booksellers' fault. Following the French Revolution, Britain was continuously at war with France between 1793 and 1814, except for the brief respite of the Peace of Amiens from 1802 to 1803. Travel and communication were disrupted between the warring nations, and between other countries whose usual communication links lay through France.[37] Correspondents at a distance – such as the president of the Academy of Sciences in St Petersburg – could not be sure whether the fact that 'we have not Received the Transactions of your Society for several years' was due to incompetence or to the vagaries of war. Nor could they be sure whether their own dispatches had got through: 'I hope you have received ours.'[38]

Letters, books and periodicals did cross the Continental blockade, but it required creativity, determination and, ideally, useful political contacts. Military and diplomatic individuals were among those who could travel across enemy lines, and Banks and Blagden made use of them when they could. Blagden's letters often travelled with (neutral) American diplomatic couriers, while Banks arranged for copies of the *Transactions* to be taken to the French Institut by the Neapolitan general Prince Francesco Pignatelli, and by the Spanish minister of war General O'Farrell.[39] There was, however, a limit to what an individual traveller could reasonably carry. When the American minister to France, Robert Livingston, travelled from Paris to London to negotiate the Louisiana Purchase in 1803, he carried letters and papers to the Society from the American-born (but resident in Paris) inventor Benjamin Thompson, Count Rumford; but the following year, Banks apologised to Rumford that 'it does not prove convenient to Mr. Living[s]ton to take with him the Societie's Transactions' when he returned to Paris (though he did agree to take a few books).[40]

As well as taking advantage of military-diplomatic travellers, Banks tried other routes. In 1796, while negotiating for South Pacific botanical specimens seized by the Royal Navy to be returned to their French collectors, he made a useful contact in Jean Charretié, the French Commissioner in London for the exchange of prisoners of war.[41] He then sought Charretié's assistance in return, explaining the difficulties of 'Obtaining from Paris the Scientific Journals publishd there Especialy those that give An Account of the Proceedings of the Institute Nationale', and hoped that 'by your intercession Leave *might* be granted for these Journals to be sent to me'.[42] Banks routinely mixed his personal interests with his official ones as president of the Royal Society. When Charretié appeared willing to act as a scientific conduit in both directions, Banks sought a meeting with Prime Minister William Pitt, to check whether there were any political or military objections to 'opening a Communication with Paris'. To Pitt, he explained his aim to ensure both the transit to London of the papers of the Institut National, 'Some of which are highly interesting to the Royal Society' and 'For Sending in Return the Philosophical transactions, the Greenwich Observations &c which have not been Sent for Some years Past'.[43] When Charettié's term of office ended, Banks expressed his 'warm Gratitude' for his aid in 'Keeping open the *Small* Communication I have lately had with my Scientific friends at Paris', and hoping that 'your successor may Feel the Same Liberality of Sentiment'.[44]

As Banks's meeting with the prime minister demonstrates, there could be political risks in communicating (or being seen to communicate) with people or institutions in an enemy country, and distrust remained, even during the Peace of Amiens. Banks was taken brutally to task in the British press by William Cobbett for the warm terms in which he accepted his election to the Institut de France in that period.[45] And, some months after his visit to Paris during the peace, Blagden was disturbed to learn that Napoleon believed that he had in fact been acting as 'espion du Gouvernement Anglois' [a spy of the English Government]. Blagden protested that Napoleon 'never spoke to me alone', and insisted, 'I need scarcely add to you, that I never sent any thing from France, directly or indirectly, to the English newspapers'.[46] The episode temporarily damaged Blagden's scientific friendships in Paris.[47]

Banks and Blagden continued to use military-diplomatic channels to get scientific publications to and from Paris, but their success depended very much on the phases of the war. For instance, in 1808, Blagden told the mathematical secretary of the Institut National that the copies of its *Mémoires* that should have crossed the Channel with returning prisoners of war were probably still in port in Brittany: he explained that the last British ship that had tried to exchange French prisoners had not been 'suffered to land them nor any thing else, nor to receive anything from the shore'. He promised that if prisoner exchanges resumed, 'we will make use of the opportunity, to send you the Phil. Trans, &c.'.[48]

Once the war was finally over, international travel and communication became somewhat easier. But even in peace time, transnational scientific communication still involved negotiating the varied opportunities of fragmented and poorly connecting postal systems, freight carriers and private travellers. It was still slow and unreliable over long distances. Despite these difficulties, Banks and Blagden clearly managed to see – or to see reports of – scientific news from across Europe and, to some extent, North America. Private travellers, correspondence networks and the reports in the periodical press were valuable resources for scholars seeking both to find out what had been published, and to find ways to get hold of it. Those seeking to find out what had been published in the *Philosophical Transactions* had two main options: acquire a copy of the *Transactions* itself, through commercial or non-commercial channels, or rely on the literary replication of its contents through the reprinting, extracting, translating and reportage activities of independent editors.

Commercial and non-commercial circulation

In 1791, the London bookseller Lockyer Davis died. He had held the role of 'bookseller' to the Society since taking over from his uncle Charles Davis in 1755. This involved organising the printing as well as the sales of the *Transactions*. Davis had initially contracted the printing to William Bowyer, and then, from 1777, to Bowyer's successor, John Nichols. Nichols was well regarded in the London literary scene: he also printed for the Society of Antiquaries, and was the printer and editor of the *Gentleman's Magazine*; but, by the late 1780s, the Royal Society had become unhappy with Nichols's services.[49] Davis's death provided an opportunity to act. This appears to be the first time that the Society's officers had explicitly chosen a printer for the *Transactions*, rather than relying on its bookseller's arrangements. The considerations involved in that choice show an overriding concern with the quality of printing that trumped mere cost.[50]

The unhappiness with Nichols's work first surfaced in 1786, when Council refused to pay his bill until all the printing work was delivered. Charles Blagden, who, as secretary to the Society, was most directly involved in the printing of the *Transactions*, was particularly unhappy. He reported that some of the recent parts had been found to be missing several of their printed sheets, and complained of 'great negligence & irregularity on the part of Mr Nichols the Printer'. Nichols was ordered to 'make compensation' for these errors 'occasioned by his neglect'.[51] The Society also began to suspect that Nichols was using cheaper paper than it had requested, and carefully retained a sample so as to be able to check the work delivered.[52] Two years later, Blagden again complained of 'great irregularity', because the printers had 'mislaid' some sheets of corrected proofs. Nichols was summoned before a special committee. He was not only tasked with delays and uncorrected proofs, but informed that the committee considered his printing of the *Transactions* to be 'ill executed, and the ink foul and of a bad colour'. They even described some of the copies as being executed in a 'slovenly manner'. Nichols apologised profusely and promised 'his utmost endeavours' for the future.[53]

The Society's willingness to change printers was doubtless influenced by its ongoing frustrations with Nichols, but the final decision was officially grounded in a desire to improve the quality of the printing, or, as Banks told a correspondent, 'to Augment the Beauty' of the *Transactions*.[54] The new printer was to be William Bulmer, who was then working on a lavishly illustrated edition of the works of Shakespeare, and was noted for what the Royal Society Council referred to as 'the avowed superiority' of his printing.[55] Banks himself described Bulmer's printing as 'acknowledgd as the best in the Country', and he believed the improved quality of the *Transactions* would be 'for the honor of the Society', particularly 'in Foreign Parts'.[56] The Society's representatives might also have been reassured by Bulmer's promise to look after their business 'with all the accuracy & unremitting attention in his power'.[57] The move to Bulmer in early 1792 was accompanied by a decision to increase the print run from 850 copies to 1,000 copies.[58] This suggests that Banks and the Society had ambitions for circulating the handsome *Transactions* more widely. The timing was, however, unfortunate: barely a year later, France and Britain would be at war.

Since 1778, some of the copies of the *Transactions* had been printed on larger sheets of paper, so that the text was surrounded by broader margins.[59] The numbers printed by Nichols in the 1780s – 250 copies 'with broad Margins and 500 with narrow Margins' – suggest that these more expensive copies were intended either for customers willing to pay for a handsome publication or for presents to people or institutions that the Society wished to impress. The smaller format would have been used for distribution to fellows, of whom there were around 540.[60] When appointing Bulmer in 1792, however, the Society ordered him to shift towards the more handsome format, and to print 750 copies on large paper, and only 250 on 'small paper'.[61] (The difference between the formats can be seen in the new digitised edition of the *Transactions*, by comparing a page printed by Nichols in 1777 with one printed by Bulmer in 1794.[62]) In 1792, the Society also agreed to pay for the paper to be 'hot-pressed', to give the printed page a more polished appearance, in keeping with Banks's ambition 'to Augment the Beauty' of the *Transactions*.[63] As things transpired, the increased print run of 1,000 proved too optimistic; but the subsequent reductions were made by cutting the 'small paper' part of the run, until it was entirely discontinued in 1808. Thereafter, Bulmer was printing only 750 copies, but all were on large paper with broad margins.[64] (In the 1830s, the Society would find cost-savings by sacrificing those margins to fit more text onto the page.)

The Society's discontent with Nichols and its decision to move to Bulmer has left traces in the archival record. Its involvement with other members of the print trades appears to have been less contentious, and thus its paper suppliers, engravers and stitchers (who stitched the printed sheets into parts) appear only as names whose bills were to be paid from time to time, as in Table 7.1, from November 1806. These bills reveal that paper, printing and illustrations each accounted for about a third of the total production costs of the *Transactions*.[65]

The Society's Council discussed the supply of paper only rarely, and the choice of engraver even less often. For the engravers, that was because the Society worked with the same family for three generations

Table 7.1 Costs of producing the *Transactions* in 1806

Payee	Amount	Proportion of total costs
Messrs Bowles & Gardiner, stationers	£253 17s 6d	31%
Mr Bulmer, printer	£265 2s 6d	32%
Mr Basire, engraver	£166 19s 0d	34%
Mesrs Cox & Co, copper plate printers	£117 8s 0d	
Mr Sacheverell, for sewing Transactions	£25 16s 0d	3%
Total	£829 3s 0d	

(and almost a century). James Basire had been appointed engraver to the Society in 1771, after the death of the previous incumbent, James Mynde, but in 1806, it was his son, James Basire II, who was in charge of the family firm. For Basire, the well-illustrated *Transactions* provided a welcome source of regular income (though the neighbouring Society of Antiquaries was an even more lucrative client in the early 1800s). Printing from copper plates required a different technique – and a rolling press – than the letter-press printing done by Bulmer, and the Basires regularly worked with the printer Daniel Cox and his son John. The two families were joined by marriage in 1795, enabling the next generation, James Basire III, to combine both businesses and become engraver and plate printer to the Royal Society in the 1820s. In contrast to the Society's occasional dissensions with its printers, its relationship with the Basire family continued until the death of James Basire III in 1869.[66]

During the Banks presidency, there were between 750 and 1,000 copies printed of each part of the *Transactions*, but what happened to them? The first call on the print run was 'for the use and benefit' of the Society and its fellows, as mandated by the 1752 reforms. All the fellows were entitled to free copies of the *Transactions*, as long as they had paid their annual fees, and this potentially accounted for at least half the print run. This mode of distribution ensured that the *Transactions* reached many of the people most likely to be interested in its contents, in other words, people who were already members of a society dedicated to improving natural knowledge. It did not, however, include those whose socio-economic status, gender or geography excluded them from the Royal Society.

The fellows did not all receive their copies, or at least, not necessarily in a timely manner. To avoid the difficulties and expense of shipping the *Transactions* to hundreds of individual addresses, the Society's regulations required fellows to claim their copies in person from the Society's apartments, where they had to sign a book to confirm receipt (Figure 7.1). An advertisement in the *St James's Chronicle* notified fellows when their copies were available, and they were supposed to claim within a year or else forfeit their claim.[67] This posed obvious difficulties for those who were neither resident in London, nor regular visitors to the capital. It was possible to appoint a friend or relative to collect on one's behalf; but even so, the minutes of almost every Council meeting included requests from fellows seeking to claim their copies beyond the initial year. In 1773, Council had imposed a five-year limit after which it would refuse such claims, but it still made exceptions.[68] For instance, in 1801, Sir William

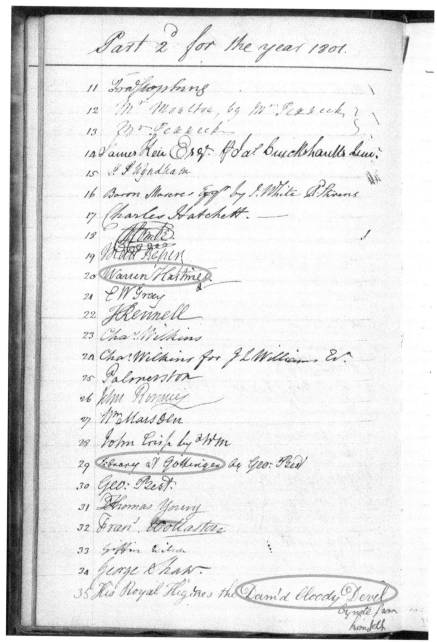

Figure 7.1 Signatures of those who had claimed copies of Part 2 of the *Transactions* for 1801, including Warren Hastings (the former Governor-General of Bengal, newly elected to the fellowship in 1801); the university of Göttingen; and (apparently) 'His Royal Highness the Dam'd bloody Devil' © The Royal Society.

Hamilton was allowed to claim all the volumes 'he has not yet received', including those beyond 'the usual limitation', on the grounds of 'his long absence on His majesty's Service' as a diplomat in Naples.[69] And in 1804, David Pitcairn was granted the volume from 1795: as a London physician, he was usually able to collect his copies, but had missed that one due to absence and 'extreme ill health'.[70]

The Society's sense of obligation to its fellows drove it to retain copies that might possibly be claimed. These accumulated copies were stored in a rented warehouse along with the copper plates, and some instruments owned by the Board of Longitude.[71] This could be seen in business terms as wasted resources: the Society's funds had been spent on paper, printing and engraving that was now lying in a warehouse costing rent. But this policy of retaining surplus stock enabled the Society to react with generosity to belated claims from fellows, and from other learned institutions.

In 1804, the Society dealt with two institutional requests that shed light on the extent of the surplus holdings. The Imperial Academy of St Petersburg (belatedly) reported that its copy of the second part of the *Transactions* from 1779 had never been received; the Society was able to provide a replacement from its warehouse.[72] Harvard College, Massachusetts, also discovered that its library had never received the regular gift of the *Transactions* offered by the Royal Society in 1782; the Society was able to make up a 20-year run that was missing only five of its parts.[73] The unavailable parts were from the 1780s, the low stock of which may relate to a 1790 decision that keeping just 10 copies of older material (over five years) would be sufficient to meet any future 'extraordinary demands'.[74] The demand for particular parts was highly variable. The Society's effort to fulfil an earlier request reveals that two of the parts that could not be supplied to Harvard had already been in low stock back in 1794. However, it also reveals that at least 10 copies of 1785 Part 2 (which included papers by Joseph Priestley, Henry Cavendish and William Roy) had still been available at that point, although they would all be gone by the time Harvard wanted them.[75]

Council had the statutory right to dispose of excess stock, but it rarely did so until space in the warehouse became critical. For instance, in 1824, the Society's bookseller was authorised to sell all remaining volumes prior to 1810 'at one third the original price to Members of the Society, and at one half price to the Public'.[76] In 1828, this was extended to volumes prior to 1813, though 20 copies of each volume were to be retained.[77] We do not know how many surplus copies were sold (or disposed of) in this way, but it is clear that the number of copies of the

Transactions in circulation was less than we might imagine from the print run.

Making presents to learned institutions in Russia or Massachusetts was part of the Royal Society's efforts to ensure that its *Transactions* – and thus the Society's name and the knowledge produced under its auspices – was available to scholars across the learned world. As well as sporadic gifts of single parts or volumes to particular individuals or institutions, the list of institutions to which the Society's Council had agreed to make regular 'presents' of the *Transactions* was growing: the dozen institutions of 1765 had become 32 institutions by 1816 (see Figure 7.2). The additions included Harvard and the Boston Philosophical Society, both added in late 1782 once British military action against the former colonies had ceased; the Royal Society of Edinburgh and the Royal Irish Academy in 1790; the Asiatic Society of Bengal in 1806; the École des Mines de France in 1815; and the Royal Academy of Sciences in Lisbon in 1816. Closer to home, copies were also regularly presented to the Linnean Society, the Royal Institution and the Geological Society of London.[78]

There appears to have been little plan or policy: national academies of sciences across Europe were likely to be granted the *Transactions*. So, too, were several voluntary societies of scholars committed to the natural sciences, but apparently only if they were overseas, for none of the provincial British literary and philosophical societies were honoured. The timing of the grants does not correlate neatly with the foundation dates of the institutions, which suggests that the Royal Society's decisions to gift the *Transactions* were reactive rather than proactive. A newly established institution that announced its existence to the Royal Society by sending a gift of its own *Transactions* or *Mémoires* was very likely to receive a gift of the *Philosophical Transactions* in return. As Banks told a correspondent in 1804, 'I know of no one that has sent their books to the Rl Sy. to which in return they have not ordered theirs to be delivered'.[79] The regular, reciprocal gifting of periodicals looks like an exchange scheme that benefited both institutional libraries, but the Royal Society's language remained that of giving and receiving gifts.[80] In the nineteenth century, a list of institutions that were entitled to receive a copy of the *Transactions* would be printed annually; separately, the publications received by the Society's library were acknowledged under the heading 'presents'.[81]

The Royal Society's decision to make a regular and ongoing gift to another learned institution did not guarantee that the *Transactions* did appear in the other institution's library in a timely fashion, or at all. As we have seen, Banks and Blagden clearly took some pains to get copies of the *Transactions* to the Paris Institut during the war years, but, as with

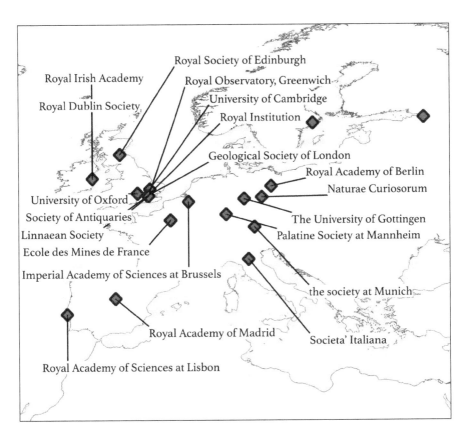

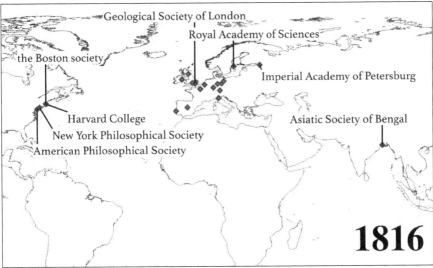

Figure 7.2 Institutions to which presents of the *Transactions* were sent in 1816.

fellows, the burden of claiming copies usually fell on the recipient. When the American Philosophical Society, in Philadelphia, complained that 'the Boston academy' was receiving the *Transactions* more regularly than it was, Banks insisted that no slight by the Royal Society was intended: the problem was entirely about logistics.[82] Banks explained that the Society could not expect any bookseller to undertake 'a business so multifarious' and, implicitly, unprofitable, as sending out free copies of the *Transactions* to international recipients. Instead, 'each Society or Person to whom the transactions are to be delivered', whether fellows or overseas institutions, should 'authorise Some agent in London to receive & forward them'. Banks explained that these agents should be asked to 'call at the apartments of the R.S. in Somerset place, & enquire for Mr Gilpin their Clerk', who would then 'without hesitation or delay' hand over the *Transactions*, asking the agent only to sign 'in their receipt Book'.[83] This arrangement saved the Royal Society the effort and cost of arranging international shipping, but Harvard's experience demonstrates its disadvantages as an effective distribution system.

The Society's non-commercial distribution of the *Transactions* was directed to the gentlemen who were fellows of the Society, and to the members of other national and international scholarly institutions. Getting the *Transactions* onto the shelves of personal and, especially, institutional libraries made it accessible to more readers than the mere number of copies would suggest, and it did so into posterity. Gifted copies became the basis of the runs of the *Transactions* held in research libraries across the world, where many of them still remain. Institutions or individuals who did not receive the *Transactions* in this way could seek to buy the new issues as they appeared in the commercial book trade, or scour the second-hand trade – and the private auctions that sold the libraries of deceased scholars – for the back issues.[84]

Those who wished to purchase the current *Transactions* could do so from the Society's apartments, or from its booksellers. Since 1780, the Society had appointed Peter Elmsley as well as Lockyer Davis to that role. They were granted a commission of 15 per cent on the retail price of the copies they sold.[85] There is no indication that Davis and Elmsley were business partners: rather, Elmsley's reputation as an importer of foreign books, and his contacts with the French book trade, suggests that the two men served in complementary roles.[86] Although Kirwan had complained about his poor service, Elmsley would continue as the sole bookseller to the Society after Davis's death. The colophon of the *Transactions* now read 'Sold by Peter Elmsly, Printer to the Royal Society;

from the Press of W Bulmer & Co', a phrasing that reminds us of the confusing inter-changeability of the terms 'printer' and 'bookseller' in contemporary usage.

It is, however, likely that at least some of the distribution of the *Transactions* in the 1790s was managed by George Nicol. Nicol was the royal bookseller, but he was also in partnership with Bulmer, and sold books printed by him. In 1801, Nicol sounded out Banks about the possibility of a formal appointment as bookseller to the Royal Society. He described the Society's payments for the printing of the *Transactions* as so liberal 'as might have made me blush at any further application' for patronage, but, while deprecating himself as a 'coscomb' [sic] who 'preferr'd Fame to Fortune', he admitted that he sought public recognition.[87] Nicol's importunity paid off six months later, when Elmsley tendered his resignation 'of the Office of Bookseller to the Society', shortly before he died. Without any discussion, Banks presidentially informed the Council that he had appointed Nicol to the vacant position.[88] Nicol's son, William, would become Bulmer's successor as printer of the *Transactions* in the 1820s.

George Nicol was clearly keen to acquire the role of bookseller to the Royal Society, but it is far from clear that much 'fortune' would accompany the 'fame' of the appointment. There are no records in this period of the funds remitted to the Society from its bookseller, nor of the commission he received, but we can make some estimates. In 1800, for instance, from a print run of 900 copies, only about 350 copies would be available for public sale after the copies for fellows and institutional presents had been set aside. That is the same number as Davis is known to have been allocated back in 1779, when both the print run and number of free copies needed were lower.[89] The price of the *Transactions* varied with length, but 13s. 6d. was typical for one part in the first decade of the nineteenth century. If Nicol sold 350 copies at that price, and received the same 15 per cent commission as his predecessors, he would potentially have earned a modest £71 a year.[90]

From the Society's perspective, subsidy was built into the business model. In 1799, for instance, the Council minutes reveal that the retail price for the *Transactions* was based 'upon a calculation of the produce of 600 Copies defraying the expence of an impression of 900 Copies'.[91] In theory, this meant that the sale of 600 copies would fund 300 'free' copies. But since the Society was actually using about 550 'free' copies, and selling perhaps 350 copies, most of the 'free' copies were really funded from the Society's membership and endowment income.

In fact, it seems unlikely that Nicol was managing to sell 350 copies. In 1808, the print run was reduced to 750 copies. The Society's own needs had not diminished, so the cut affected the number of copies on public sale: there were now perhaps only 200 copies of the *Transactions* available for purchase. Such a figure stands in sharp contrast with the new independent journals, whose print run was almost entirely available for sale: Nicholson's *Journal of Natural Philosophy* achieved sales of around 750 copies in the late 1790s, while the *Philosophical Magazine* had a print run of 1,250 copies in 1803.[92] Each issue of these monthly journals was much cheaper than a part of the *Transactions*, though a year's supply of Nicholson's *Journal* actually cost around the same price – about 30s. – as a year's supply of the *Transactions*.[93] But the small number of copies of the *Transactions* even available for public sale suggests that Nicholson had been entirely justified in claiming that its circulation was 'very limited'.[94]

One reader who did receive the *Transactions*, and did so with delight, was Elizabeth Grey, of Northumberland. In the 1780s and 1790s, she received copies as gifts from Charles Blagden, though it is not clear whether Blagden was personally paying for those copies, or using his position as secretary of the Society to send presentation copies. Blagden was able to use Grey's male family members to convey the *Transactions* to her. In the mid-1780s, it was her military officer husband Charles Grey (later Earl Grey) who carried them north; by the late 1790s, the courier was her son Charles Grey, then developing a career in London as a Whig politician. Elizabeth Grey's letters of thanks to Blagden demonstrate the keen enthusiasm with which scientific news might be greeted in a part of Northumberland 'without one Philosophical Neighbour'. The *Transactions* supplied her with 'intellectual food sufficient to live upon for some time'. Grey modestly claimed that much of the details 'will be out of my reach no doubt' but was sure that it would nonetheless 'both instruct and entertain me much'.[95] The letters reveal that, for instance, during 1784, Grey read Blagden's own paper on mercury; commented on papers by Henry Cavendish; struggled to understand one of William Herschel's dissertations because she had mislaid the two separate sheets of figures that accompanied them; and set out what she understood to be 'the difference between Mr Schiele, Priestley & Mr Lavoisier in regard to fixed Air or vitiated Air being produced by respiration'.[96] Fifteen years later, she was still thanking Blagden for sending the *Transactions* to her (and for taking the time 'to point out the Papers most worthy of notice, with your own judgement').[97]

Reprinting, abstracting and timeliness

The six-monthly parts received by Elizabeth Grey were not the only ways in which the *Transactions* and its contents circulated. Separate printed copies of individual papers circulated through the authors' private correspondence networks, often ahead of the formal publication of the *Transactions*. Excerpts and summaries of papers were printed in newspapers and periodicals in Britain and overseas, based sometimes on reports from those who attended the weekly meetings, and sometimes on one of the printed versions. And, for those who wanted access to the older content of the *Transactions*, retrospective abridgements provided a slimmer and more conveniently organised version. These forms of circulation offered different levels of detail – from the author's full text to a third-party summary – and reflected different requirements for timeliness. Abridgements were retrospective, but offprints and reportage could circulate in advance of the parts. This raised difficult questions for the Royal Society about whether such advance circulation aided the circulation of knowledge, or undermined the role of the *Transactions*.

Earliest access to new papers was granted to the fellows and their guests who attended the Royal Society's Thursday evening meetings. Whether they acquired a full understanding of the arguments or details depended on the secretary's oral delivery style, the quantity of mathematics, and the extent to which the paper had been excerpted to fit into the available time. For this reason, Thomas Thomson apologised that 'mistakes and inaccuracies' would 'unavoidably occur' in the reports in his journal, because they were written 'merely from hearing the papers read'.[98]

For the full printed version of the paper, it was the authors who had fastest access. They had the option of paying for separate printed copies of their own papers, which they could then circulate to friends and correspondents. One of the earliest mentions of this practice – later known as 'offprints' – is from 1781, when Blagden (not yet secretary) mentioned to Banks that 'I wish to have some separate copies struck off for my own use', of his forthcoming paper on the Gulf Stream.[99] This was, as he later told a correspondent, 'a private transaction between the Author & the Printer', in which the Society's only involvement was to insist on a limit to 'prevent the printer from striking off more than 100 copies'.[100] The existence of that limit suggested a perceived risk to the *Transactions* if too many separate copies were in circulation, though, as we shall see, it was more likely to be reputational than financial.

Authors who wanted separate copies needed to make their request in good time. The secretaries tended to send each approved manuscript to the printer as soon as possible, to allow him to avoid a last-minute rush. Once copies of each paper had been printed, the loose type would be dispersed and used for other projects. In 1786, Blagden had told Banks that he would have to apologise to the Dutch anatomist Peter Camper, 'because the sheets of his paper are now all worked off, & the forms broken up, so that it is impossible to gratify him in the article of separate Copies'.[101] If separate copies were wanted, they had to be printed while the type was still set.

A consequence of this workflow was that many papers were printed months before they were assembled into a 'part' for official publication. Most of the copies would remain in the printer's workshop awaiting collation, but if the author had paid for 'separates', he was allowed to receive them when they were ready. Separate copies could thus be in private circulation weeks or months ahead of the official publication of the *Transactions*.

It would be misleading to imagine that there were 100 separate copies of every *Transactions* paper in advance circulation in this period, for not all authors requested separate copies, and not all wanted many copies. In 1779, a scholar in Brussels had estimated that he would not need more than half a dozen separate copies to supply all his English-speaking acquaintances, while in 1783, Blagden told Banks that another author 'wanted 12 copies only for himself'.[102] On the other hand, Council was also willing to break its own rule, as it did in 1797 when it allowed Robert Marsham to purchase 150 copies of his updated data on tree growth.[103] Some authors, including physician Charles Pears and naval architect Robert Seppings, requested copies only of the illustrations to their articles.[104]

Editors of scientific journals were among those who were keen to get hold of separate copies of *Transactions* papers, so they could accurately excerpt, summarise or translate them for their readers. Lorenz Crell, for instance, repeatedly asked Banks and Blagden to write with the latest news, or, preferably, to send 'some of the last physico-chemical papers, contained in the Phil. Transact. & printed separately'.[105] He explained that receiving separate papers enabled him to get them 'sooner, & communicate them to my countrymen' in the pages of his *Chemische Annalen*, than if he had to wait until 'the entire Volume comes to Germany'. Crell suggested using a Hanoverian diplomat in London as an intermediary.[106]

The risk of this advance circulation of separate copies was that publication in the *Transactions* might be pre-empted, and thus undermine the Society's insistence that the material published in the *Transactions* must be original. In September 1785, for instance, Jonas Dryander had told Banks that Henry Cavendish's experiments on air, which had been read to the Society in June, was already being translated for publication in Rozier's journal in Paris. It seemed likely 'that it will be published there, before the Transactions come out'.[107] A separate copy of Cavendish's paper had reached Paris, perhaps through Blagden, who reflected that pre-emptive publication in Rozier's journal was 'a circumstance which would not well be foreseen, as there was no idea that any one would think of translating the Paper'. He added, 'For the future, however, such an accident shall be prevented'.[108] Thus, when Blagden sent another paper to the chemist Claude-Louis Berthollet the following year, he explicitly asked that he 'not communicate it' (or any future papers) to a journal editor in any way that would 'anticipate its publication in the Transactions, which will probably happen in June & November of each year'.[109] Preventing such 'accidents' would, however, be a long-standing problem in which the Society's desire to promote the circulation of knowledge sat in tension with its insistence that it held the moral right to first publication of any paper communicated to it.

A partial solution was reached in 1787, when the Society ordered that, in the future, 'all titles of the Copies of Papers worked off for the Authors, specify that the said papers are parts of the Phil: Transactions'.[110] Printed copies circulating ahead of publication would now identify themselves specifically with the *Transactions*, ensuring that any reprinting could at least have appropriate attribution to the Royal Society. It did not, however, clarify whether – or when – other periodicals could excerpt, reprint or translate such papers. Copyright law was of little relevance here: British legislation had no jurisdiction beyond Britain; and even within Britain, it was not until 1842 that copyright law explicitly applied to periodicals. Reprinting and excerpting was common practice among periodical editors, and norms such as attribution and 'fair use' were still developing.[111] The Royal Society appears to have resigned itself to accepting reprinting, translating and excerpting by foreign editors as a valuable service to scholars overseas; but it was less happy when such reprinting occurred in the English-language scientific journals in Britain.

Matters came to a head in 1802, when the journal editor William Nicholson sought clarification from Joseph Banks about '*whether or not I may publish the papers of the Royal society in my Journal as soon as they separately appear?*'[112] Nicholson was not a fellow of the Society, but he and Banks knew each other well, and were close neighbours in

Soho Square.[113] Between March and July 1802, Nicholson and Banks exchanged views, in letters and in conversation at Banks's house, about reprinting papers from the *Transactions*. Throughout, Nicholson insisted that he was not trying to harm the Society's reputation or finances, but trying to spread the knowledge of its papers to an even wider audience. But his position was weakened by the fact that on 1 June, the *Journal of Natural Philosophy* reprinted the first sections of papers by Thomas Young and Charles Hatchett that were about to appear in the next *Transactions*, later in June.[114] For Nicholson, it was an inadvertent mistake in timing; for Banks, it seemed an attempt to diminish the *Transactions*.

June 1802 was not the first time Nicholson had reprinted from the *Transactions* – he had, for instance, reprinted another paper by Young the previous year[115] – and he was aware of the Society's desire that such reprinting should not occur before the public appearance of the *Transactions*. As he explained to Banks, he had tried to check the publication schedule of the *Transactions*. Young's and Hatchett's papers had been read at Society meetings in November 1801, and were clearly in print as separate copies by mid-April. Nicholson had already seen them, and was considering reprinting them in his May issue. Jonas Dryander, as Banks's personal secretary, had told him that the *Transactions* would not be out in time for Nicholson's May issue, but since there was 'only one proof waited for', he thought it was highly likely that the *Transactions* would be out before June.[116] As things turned out, Dryander was wrong: proof corrections must have taken rather longer than expected, and thus Nicholson's June issue had inadvertently pre-empted the *Transactions*.

The fact that the proof corrections for one paper had delayed the issue of the entire part is another indication of the differences between the *Transactions* and Nicholson's *Journal*. Nicholson had to keep to strict deadlines in order to release a new issue on the first of every month: a delayed paper would have to wait until his next issue. The Royal Society's approach to timeliness revolved around the date of the meetings at which a paper had been read aloud, rather than the date of print publication. Volumes of the *Transactions* reported papers from a particular year of meetings, but they did not have to be published on any particular date. The actual dates on which its parts became available varied by weeks and sometimes months. It is a reminder that, while the *Transactions* was certainly issued periodically, it did not maintain the sort of regular periodicity that readers were coming to expect from the monthly reviews, magazines and journals.

Nicholson defended his June issue by arguing that Young's and Hatchett's papers had in fact been public – and thus available for reprinting – ever since the separate copies were put in circulation. He argued that the Royal Society had implicitly accepted this definition by allowing authors to distribute their separate copies 'to the learned in all parts of Europe before the whole Book is ready', thus enabling 'Journalists in all foreign parts' to reprint them 'immediately'.[117] He interpreted the inclusion of a clear attribution to the *Transactions* on each separate copy – intended 'to prevent mistake in foreign Journals' – as an implicit acceptance 'that these single Papers should be the Instruments of diffusing knowledge, by those Journals'.[118] And he argued that it was unfair that 'Journalists within the Realm should be put in a less favored Situation than foreign philosophers' in Paris or Helmstedt.[119]

Banks's response acknowledged that there were different ways of defining the moment at which the *Transactions* should count as 'published'. He was, he said, 'well aware' that in libel cases, a paper would count as 'published' as soon as it was 'put into the hands of a third Person'. But he argued that the Royal Society was 'a body who wish to Receive back a Portion at Least of the expence they incur by Printing', and so the only valid definition must be the moment 'when their Volume is Publishd for Sale', that is, available from the bookseller's shop.[120] This effort to cast the Royal Society as a commercial publisher whose sales might be harmed by advance reprinting was tendentious, for it was reputational, rather than financial, harm that actually bothered Banks. As discussed in Chapter 6, originality had come to be a necessary feature of papers considered for the *Transactions*, and advance reprinting undermined that. The papers by Young and Hatchett had not been in print when the Committee of Papers approved them for publication in the *Transactions*, but by the time they were published in the *Transactions*, the start of each paper had already appeared in Nicholson's *Journal*. For those who did not understand the Royal Society's processes, it might look as though the *Transactions* was reprinting material that already existed elsewhere, and doing so in a more expensive format.[121] And if that were so, it could raise questions about the purpose and prestige of the *Transactions*, and thus about the return to the Society on its expenditure on publishing.

Banks also defended the Society's different attitudes to reprinting by British journal editors and foreign editors. In Continental Europe, the limited number of readers of English meant that awareness of *Transactions* papers depended heavily upon their literary replication in the translations and summaries carried out by foreign journal editors.

Thus, wrote Banks, 'it appears to me wise ... to throw no Obstacle in the way of a Foreign Journalist who wishes to republish their papers as Early as possible'. But the situation within Britain was different, because British readers could read the *Transactions* itself. That said, Banks accepted that the reprints, excerpts or summaries in other journals performed a valuable service for those who could not afford or acquire the *Transactions* itself. The Society did not object to reprinting 'by English Journalists instantly after the volume is Put into Circulation', but that it should be *afterwards* was the point on which they 'made their Stand'. Banks felt that permitting instant reprinting – in contrast, for instance, to the 14-year period of copyright that applied to books – was already generous. 'If the British Journalists are not Satisfied' with this 'Liberality', he wrote, 'I confess I Consider them as ungratefull for a Favor'.[122]

During his exchange with Nicholson, Banks wrote as if the Royal Society had the absolute right to withdraw the 'favour' of allowing British journal editors instant reprinting. Such a right depended upon the social and cultural power exerted by the Royal Society within London's scientific community, arising both from its age as an institution and from the social status and connections of the gentlemen associated with it.[123] Nicholson may have disagreed with Banks's definition of the moment of publication, but he did accept the Society's claim on *Transactions* material. He wrote that, 'I am well aware that the property of the Copies here discussed is invested in the Corporation',[124] and later, that 'I neither desire nor claim nor intend to exercise any other power over the Royal Society's Copy than what they may think fit to grant'.[125] He further accepted that this was not just a matter of the Society's right to first publication of that material, but 'an undoubted and unlimited power to allow or refuse the reprinting of the Transactions', and that the Society could legitimately regard a person who contravened its desires 'as the violator of their property and liable to the loss or disgrace that may ensue'.[126] The implication is that Nicholson believed that the Society's influence could harm the social or professional reputation of scientific editors and publishers.

By the end of June 1802, Nicholson had accepted that the Society did intend to treat British journal editors differently from foreign editors. He promised that in future, he would not reprint from separate copies, but only from the *Transactions* once it 'shall have been sold by order of the Society'.[127] This does not, however, appear to have satisfied Banks and his associates. In July, Council agreed to add a new declaration to the covers of the separate copies, adjuring authors 'to use their endeavour to prevent them from being reprinted till one Month after the publication of the Part of the Volume of the Philos: Transactions in which they are inserted'. The Society would continue to 'indulge' authors by permitting them to receive separate copies in advance of formal publication, but it asked authors to exercise care in how they circulated those copies.[128] (Nicholson was notified of this new rule personally.[129])

This compromise suggests Banks recognised the value of advance circulation within limited scholarly circles. Some years later, he had personal experience of the inconvenience arising from the different practice at the Society of Antiquaries, who 'do not allow their Private Copy to be deliverd till the volume is Publishd'. He wanted to see a copy of a paper on the Rosetta stone, but complained to Blagden, 'it may be years before I am able to Obtain it. I Shall however Try'.[130] Rather than preventing the advance circulation of separate copies, the Royal Society chose to shift the perception of the appropriate timing of reprinting, from instant to one month after publication. This would remain the Royal Society's default position until the end of the century, when speedy circulation of knowledge had become a greater priority. The fact that journal editors appear to have largely respected the Society's wishes is testament to the Society's influence.

Nicholson had been reprinting the entire text of *Transactions* papers. Other than splitting papers over two issues, his only other change was to add marginal annotations summarising the key points of each paragraph. Other editors and publishers transformed the *Transactions* by excerpting or summarising its contents. For the many British readers who were not closely associated with the Royal Society, this literary replication would have been their most likely encounter with the *Transactions*, as they came across synopses – and sometimes lengthy excerpts – of new issues in such general periodicals as the *Monthly Review*, the *Critical Review* or the *European Magazine*, as well as in science-focused periodicals such as the *Philosophical Magazine*.[131]

Summarised or abridged versions of the *Transactions* were especially important overseas, as the task of undertaking a translation of the entire *Transactions* on an ongoing basis was just as challenging in the early nineteenth century as it had been in the late seventeenth century. Rozier and Crell translated or summarised only a handful of *Transactions* papers in their journals, for instance. For Francophone readers, the completion of Jacques Gibelin's 14-volume abridgement in 1791 made *Transactions* material far more easily accessible, but it included only the back-run to 1785, not current material, and it excluded astronomy or mathematics (see also Chapter 4).[132]

Until 1809, Anglophone readers' access to the accumulated material in the back-run depended on the abridgement by John Lowthorp and its subsequent extensions: these provided coverage from 1665 to 1749. In 1803, mathematician Charles Hutton announced 'his intention to undertake the care of arranging and printing a new abridgement', covering the entire run to 1800. He informed Banks that he trusted 'that the President and Council would please to countenance this undertaking'.[133] This was a lot to take on trust. First, although Hutton and Banks had known each other for over 30 years, their relations had not been cordial since Banks had ousted Hutton as the Society's foreign secretary, thus sparking the 'dissensions' against his presidency back in 1783. Furthermore, Hutton's proposal might be seen as an entrepreneurial attempt to exploit publications subsidised by the Society for his own gain.[134] After all, Lowthorp's abridgement had found sufficient customers to justify five editions and three successive expansions, so a new abridgement would have been an attractive commercial proposition for Hutton, his co-editors and his publishers. Indeed, later accounts would claim that Hutton was paid £6,000 for his work on the abridgement (though this is likely to have included payments to assistants).[135] Despite all this, and despite Banks's claim to Nicholson the previous year that the Society was keen to protect its financial return from the *Transactions*, the Society appears to have made no response to Hutton's 1803 note. But six years later, Hutton's publishers, Charles and Robert Baldwin, did receive permission to dedicate the resulting 18-volume work to the 'President, Council and Fellows of the Royal Society'.[136]

The retrospective abridgements undertaken by Gibelin and Hutton enabled material from the *Transactions* to circulate more widely – and in a more accessible format – than the original publication. Both abridgements were very clear about their debt to the *Transactions*: they did not threaten its role as place of first publication, but enhanced its status as what would now be termed the 'version of record'. Like the activities of foreign journal editors, the work done by the abridgers could be appreciated as a valuable service to scholarship, that would further Banks's objective 'to disperse as widely as possible the knowledge communicated' to the Royal Society.[137] Reprinting only seemed a threat if it pre-empted the originality of the *Transactions* among its own English-language readership.

Over the course of the eighteenth century, the Royal Society and its officers had welcomed the creation of royal and imperial academies across Europe, and of philosophical societies in North America. They had seen these new learned institutions as kindred spirits to be supported through correspondence and gifts of publications. The flood of local 'literary and philosophical societies' that were established across Britain in the second half of the eighteenth century did not, however, receive the same recognition, suggesting the seeds of a scholarly hierarchy that privileged the national over the local or provincial.

Another tension among learned institutions would emerge from a new breed of societies with more specialist interests. The Society of Antiquaries, founded in 1751, and the Linnean Society, founded in 1788, both catered for gentlemen with interests in a sub-set of the topics covered by the Royal Society, specifically antiquities (a meld of archaeology and history) and botany. They would be followed by many other specialist learned societies in the nineteenth century that, thanks to their London location, would ultimately come to be recognised as the national bodies for their field of enquiry.

The Royal Society initially welcomed these societies, and many of its fellows were (and would continue to be) members of one or more of the specialist societies, but by the early nineteenth century, it was becoming clear that their existence affected the meetings and publications of the Royal Society. Papers on antiquities, numismatics or linguistics had largely disappeared from the meetings and publications of the Royal Society by 1800, having found a new home in the Society of Antiquaries. There was a similar story with botanical papers: an 1830 analysis of the contents of the *Transactions* revealed that no botanical paper had appeared in it since (at least) 1800.[138] By the early nineteenth century, Joseph Banks had become more careful about supporting new organisations that might divert people and papers away from the Royal Society.

Recognising this, in 1808, the organisers of a proposed 'Society for Animal Chemistry' offered it as an 'assistant society', claiming their aim was not to narrow the remit of the Royal Society, but to exert 'Every means in their power to be usefull to that very respectable body of which the greater part of them are already members'. They suggested their new group could be something like a special interest group within the Royal Society for those interested in questions that would now be labelled organic chemistry or physiology. They asked Banks 'to induce the Royal society to take this Society under its Protection as an assistant Society', and proposed that the best of the papers heard by the Society for Animal Chemistry would be 'presented to the Royal Society for the purpose of being read & if thought worthy of that distinction Printed in the Philosophical Transactions'.[139] In spring 1809, the Royal Society's Council approved the constitution of the Society for Animal Chemistry 'as an Assistant Society to the Royal Society', but little was heard of it subsequently.[140]

This vision of specialist 'assistant societies' subservient to the Royal Society and feeding into its meetings and publications processes fitted Banks's vision of centralised, institutional science, and would have kept control among the gentleman amateurs (referred to by critics as '*dilettantes*') that Banks was proud of elevating to the Royal Society's Council. However, some of those involved in the founding of the new organisations were acting specifically to free themselves from the influence of Banks and the Royal Society. The founders of the Geological Society of London, in 1807, hoped to encourage a wider variety of people and practices than were then welcomed at the Royal Society, and wanted to be able to publish under their own auspices.[141] Banks initially joined the new society, but changed his mind two years later.[142]

Banks claimed that he found he 'Could not duely fulfill my duties to the royal & to the Geological Society at the Same time without an interference between them'.[143] But he also claimed that there had been a 'misunderstanding' between himself and the officers of the new society about its role.[144] His decision to break with the Geological Society occurred in the same month as the Royal Society accepted the proposed Society for Animal Chemistry, and the point at issue appears to have been the geologists' unwillingness to accept a similar arrangement. In early March 1809, the Geological Society rejected a proposal to become an 'assistant society'.[145] Although the geologists did subsequently offer Banks the opportunity to see if any of their papers were 'worthy the honor' of being read at the Royal Society 'or of being published in their Transactions', Banks insisted such an arrangement was impossible for 'a Society intirely unconnected with the R.S.'.[146]

The problem for Banks was that he could not realistically prevent other groups from establishing themselves as societies. The Geological Society would be followed by the Astronomical Society (1820), the Zoological Society (1826), the Entomological Society (1833) and the Chemical Society (1841), among others. All of these societies established periodicals to publish the papers presented to them. As the new societies and their periodicals became more established, they did divert papers and attention away from the Royal Society. If an astronomer could find a convivial gathering of gentlemen with shared interests at the Astronomical Society, why go to the Royal Society's meetings? And if the natural home for a research paper on chemistry was now the *Memoirs* of the Chemical Society (or, later, its *Proceedings* or its *Journal*), what sort of chemistry paper – if any – could or should still appear in the

Philosophical Transactions? By the twentieth century, the increasing specialisation – or fragmentation – of knowledge, and of scholarly life, would be posing major challenges to the relevance of the broad-remit Royal Society and its periodicals.

For scholars in nineteenth-century Britain, the existence of both the independent journals and the specialised learned societies generated more reading material to keep abreast of, and offered more choices for communicating new observations and discoveries. For the Royal Society, the existence of these other options meant that the fellows who acted as communicators, and as members of the Committee of Papers, had to reassess what it meant for a paper to be 'suitable' for the Royal Society. In the medium-term, the periodicals of the specialist societies posed a bigger threat to the *Transactions* than did the independent journals, because they largely shared its scholarly mission. Like the *Transactions*, they too were grounded in scholarly communities, emphasised originality, and developed committee-based editorial processes to enact collective responsibility to the parent organisation. As scholarship was increasingly turned into a career for professional academics, being published in the periodicals of the learned societies carried reputational value. Whether being published by the Royal Society carried the greatest weight was, however, no longer certain.

Thus, even though the editorial model of the *Transactions* remained largely unchanged during the long years of Joseph Banks's presidency, its function changed because the landscape of the sciences and of scientific print changed around it. In that landscape, the *Transactions* came to appear determinedly generalist in an age of increasing specialisation, slow and ponderous, and distinctly British-focused (as well as handsomely produced). In the decades following the death of Banks, questions of how the *Transactions* could meet the new expectations of speed, expert judgement and intellectual quality would become increasingly pressing.

Notes

1 Banks to William Nicholson, 12 March 1802, Letter 1637 in Neil Chambers, *The Scientific Correspondence of Sir Joseph Banks, 1765–1820*, 6 vols (London: Pickering & Chatto, 2007), vol. 5, 148–9.
2 [William Nicholson], 'Preface', *Journal of Natural Philosophy* 1 (1797–8): iv.
3 Iain Watts, '"We want no authors": William Nicholson and the contested role of the scientific journal in Britain, 1797–1813', *British Journal for the History of Science* 47, no. 3 (2014): 397–419; William Nicholson, *The Life of William Nicholson, 1753–1815: A memoir of enlightenment, commerce, politics, arts and science*, ed. Sue Durrell (London: Peter Owen, 2018); Anna Gielas, 'Turning tradition into an instrument of research: The editorship of William Nicholson (1753–1815)', *Centaurus* 62, no. 1 (2020): 38–53.
4 Thomas Thomson, 'Preface', *Annals of Philosophy* 1 (1813): 2. On Thomson, see Jack B. Morrell, 'Thomas Thomson: Professor of chemistry and university reformer', *British*

Journal for the History of Science 4 (1969): 245–65. On his journal, see Jonathan R. Topham, 'The scientific, the literary and the popular: Commerce and the reimagining of the scientific journal in Britain, 1813–1825', *N&R* 70 (2016): 305–24.

5 On the emergence of scientific 'journals', as against scientific periodicals more generally, see Jonathan R. Topham, 'Anthologizing the book of nature: The circulation of knowledge and the origins of the scientific journal in late Georgian Britain', in *The Circulation of Knowledge between Britain, India, and China*, ed. Bernard Lightman and Gordon McOuat (Boston, MA: Brill, 2013), 119–52; Gowan Dawson and Jonathan R. Topham, 'Scientific, medical, and technical periodicals in nineteenth-century Britain: New formats for new readers', in *Science Periodicals in Nineteenth-Century Britain: Constructing scientific communities*, ed. Gowan Dawson, et al. (Chicago, IL: University of Chicago Press, 2020), 35–64. Also Watts, 'We want no authors'; and Alex Csiszar, *The Scientific Journal: Authorship and the politics of knowledge in the nineteenth century* (Chicago, IL: University of Chicago Press, 2018).

6 Thomson, 'Preface', *Annals of Philosophy* 1 (1813): 2.

7 Thomson, 'Advertisement', *Annals of Philosophy* 1 (1813): iii.

8 [Nicholson], 'Preface', *Journal of Natural Philosophy* 1 (1797–98): iii–iv.

9 William H. Brock, 'The development of commercial science journals in Victorian Britain', in *The Development of Science Publishing in Europe*, ed. A.J. Meadows (Amsterdam: Elsevier, 1980), 95–122; Bill Jenkins, 'Commercial scientific journals and their editors in Edinburgh, 1819–1832', *Centaurus* 62, no. 1 (2020): 69–81. See also Topham, 'The scientific, the literary and the popular'.

10 On independent editors, see Aileen Fyfe and Anna Gielas, 'Introduction: Editorship and the editing of scientific journals, 1750–1950', *Centaurus* 62, no. 1 (2020): 5–20.

11 On 'literary replication', see James A. Secord, *Victorian Sensation: The extraordinary publication, reception and secret authorship of* Vestiges of the Natural History of Creation (Chicago, IL: University of Chicago Press, 2000), 126.

12 Compare, for instance, the appearance of the *Transactions* in 1815 (https://royalsocietypublishing.org/doi/pdf/10.1098/rstl.1815.0002), and the *Annals* in the same year (https://www.biodiversitylibrary.org/page/15907001).

13 On scientific academies of the eighteenth century, see James McClellan III, *Science Reorganized: Scientific societies in the eighteenth century* (New York: Columbia University Press, 1985), and histories of individual institutions.

14 Jenny Beckman, 'The publication strategies of Jöns Jacob Berzelius (1779–1848): Negotiating national and linguistic boundaries in chemistry', *Annals of Science* 73, no. 2 (2016): 195–207.

15 We feel that McClellan overplays the extent to which transnational European communication continued into the late eighteenth and (especially) the nineteenth century. Cf. McClellan, *Science Reorganized*.

16 Banks to Benjamin Smith Barton, Philadelphia, 10 September 1804, Letter 1793 in Chambers, *Banks Scientific Correspondence*, vol. 5, 374.

17 RS CMO/07, 9 July 1789.

18 Banks to Josiah Wedgwood, 28 December 1791, Letter 1085 in Chambers, *Banks Scientific Correspondence*, vol. 4, 94–5.

19 Blagden to Banks, 23 July 1783, Letter 362 in Chambers, *Banks Scientific Correspondence*, vol. 2, 109–10. His order for Rozier's journal was to 'continue on the old footing'.

20 The Paris agent was a M. Pissot, who is mentioned in several letters in the Banks correspondence, for example, Lumisden to Banks, 20 February 1785, Letter 565 in Chambers, *Banks Scientific Correspondence*, vol. 3, 32–3. Sourcing Rozier's journal is also mentioned in Kirwan to Banks, 10 January 1789, Letter 891 in Chambers, *Banks Scientific Correspondence*, vol. 3, 456–7.

21 Jonas Dryander to Banks, 5 September 1791, Letter 1056 in Chambers, *Banks Scientific Correspondence*, vol. 4, 63–4.

22 Dryander to Banks, 6 October 1783, Letter 399 in Chambers, *Banks Scientific Correspondence*, vol. 2, 160.

23 On the German literary review magazines, see Thomas Broman, 'Criticism and the circulation of news: The scholarly press in the late seventeenth century', *History of Science* 51 (2013): 125–50; and 'The profits and perils of publicity: Allgemeine Literatur-Zeitung, the Thurn und Taxis Post, and the periodical trade at the end of the eighteenth century', *N&R* 69.3 (2015): 261–76.

24 Blagden (in Bern) to Banks, 14 September 1788, Letter 869 in Chambers, *Banks Scientific Correspondence*, vol. 3, 429–30.
25 For a thorough discussion of the early modern challenges of news and the post, see Joad Raymond and Noah Moxham, eds, *News Networks in Early Modern Europe* (Leiden: Brill, 2016).
26 Blagden to Banks, 16 October 1785, Letter 611 in Chambers, *Banks Scientific Correspondence*, vol. 3, 104.
27 Blagden to Lorenz Crell, 13 October 1786, in RS CB/2/45.
28 Dryander to Banks, 5 September 1791, Letter 1056 in Chambers, *Banks Scientific Correspondence*, vol. 4, 63–4.
29 Barton to Banks, 6 June 1804, Letter 1779 in Chambers, *Banks Scientific Correspondence*, vol. 5, 354.
30 Banks to Barton, 10 September 1804, Letter 1793 in Chambers, *Banks Scientific Correspondence*, vol. 5, 374.
31 Banks to Blagden, 14 February, 20 February and 22 February 1815, Letters 2031–3 in Chambers, *Banks Scientific Correspondence*, vol. 6, 165–8.
32 Banks to Blagden, 8 March 1815, Letter 2036 in Chambers, *Banks Scientific Correspondence*, vol. 6, 172–3.
33 Banks to Blagden, 1 January and 18 January 1820, Letters 2207–8 in Chambers, *Banks Scientific Correspondence*, vol. 6, 377–9.
34 Kirwan to Banks, 19 April 1789, Letter 916 in Chambers, *Banks Scientific Correspondence*, vol. 3, 483. The bookseller in question was Peter Elmsley.
35 Blagden to Banks, 29 January 1803, Letter 1715 in Chambers, *Banks Scientific Correspondence*, vol. 5, 267–8.
36 Banks to Barton, 10 September 1804, Letter 1793 in Chambers, *Banks Scientific Correspondence*, vol. 5, 374.
37 For wider context, see Gavin De Beer, *The Sciences Were Never at War* (London: Nelson, 1960); and Iain Watts, 'Philosophical intelligence: Letters, print and experiment during Napoleon's continental blockade', *Isis* 106, no. 4 (2015): 749–70.
38 Nikolay Nikolayevich Novosiltzev (in St Petersburg) to Banks, 7 July 1803, Letter 1743 in Chambers, *Banks Scientific Correspondence*, vol. 5, 307–8.
39 On Blagden's letters, see Watts, 'Philosophical intelligence', 754–5. The eminent couriers are mentioned in Banks to Delambre, [4 January 1805], Letter 1809 in Chambers, *Banks Scientific Correspondence*, vol. 5, 405–6.
40 Banks to Rumford, 6 June 1804, Letter 1780 in Chambers, *Banks Scientific Correspondence*, vol. 5, 355–6.
41 This was Labillardière's collection of South Pacific specimens; see Elise S. Lipkowitz, 'Seized natural-history collections and the redefinition of scientific cosmopolitanism in the era of the French Revolution', *British Journal for the History of Science* 47, no. 1 (2014): 15–41.
42 Banks to Jean Charretié, 4 February 1797, Letter 1391 in Chambers, *Banks Scientific Correspondence*, vol. 4, 458.
43 Banks to William Pitt, 17 March 1797, Letter 1397 in Chambers, *Banks Scientific Correspondence*, vol. 4, 464–5. In subsequent stages of the war, it is clear that Napoleon approved this two-way scientific communication, see Watts, 'Philosophical intelligence', 758.
44 Banks to Jean Charretié, [January 1798], Letter 1457 in Chambers, *Banks Scientific Correspondence*, vol. 4, 528–9.
45 The pseudonymus 'Misogallus' criticised Banks's election in *Cobbett's Weekly Political Register* 1 (2 April 1802): 327–30; and attacked him again later that year, *Cobbett's Weekly Political Register* 2 (3 to 10 December 1802): 509–11. Cobbett also carried a letter signed 'by a Fellow of the Royal Society' urging the fellows not to re-elect Banks as president, *Cobbett's Weekly Political Register* 2 (6 to 13 November 1802): 395–400.
46 Blagden to Banks, 6 October 1803, Letter 1750 in Chambers, *Banks Scientific Correspondence*, vol. 5, 320–1.
47 His relationships recovered. See, for instance, Blagden to Pierre-Simon de Laplace, 6 April 1811 and 23 November 1809, RS CB/1/4/L20d and L20e.
48 Blagden to Delambre, 1 August 1808, RS CB/1/3/D21a. It is not perfectly clear what 'Convention' refers to here; the Berlin Decree of 1806 had forbidden all trade and contact with Britain by France and her allies.

49 On Nichols, see Julian Pooley, '"A Laborious and Truly Useful Gentleman": Mapping the networks of John Nichols (1745–1826), printer, antiquary and biographer', *Journal for Eighteenth-Century Studies* 38, no. 4 (2015): 497–509. Julian Pooley manages the Nichols Archive Database, see https://www2.le.ac.uk/departments/history/people/staff-pages/previous-staff/jpooley/julian-pooley.

50 And this was true even though costs were rising, see Aileen Fyfe, 'Journals, learned societies and money: *Philosophical Transactions, ca.* 1750–1900', *N&R* 69.3 (2015): 277–99.

51 RS CMO/07, 16 November 1786 and 25 January 1787.

52 RS CMO/07, 20 December 1787; cf. RS CMO/07, 16 December 1779.

53 RS CMO/07, 9 July 1789.

54 Banks to Josiah Wedgwood, 28 December 1791, Letter 1085 in Chambers, *Banks Scientific Correspondence*, vol. 4, 94–5.

55 RS CMO/08, 22 December 1791.

56 Banks to Wedgwood, 28 December 1791, Letter 1085 in Chambers, *Banks Scientific Correspondence*, vol. 4, 94–5.

57 Bulmer to Banks, [attributed December 1792, but probably 1791], Letter 1151 in Chambers, *Banks Scientific Correspondence*, vol. 4, 172.

58 RS CMO/08, 22 December 1791. The print run had been increased from 750 copies to 850 copies in 1788, see RS CMO/07, 20 December 1787.

59 RS CMO/06, 9 April 1778. Since libraries usually hold only single copies of the *Transactions*, the concurrent existence of two different formats for almost 30 years has not been recognised. The *Transactions* is sometimes said to have shifted to the larger quarto format in 1792, see *Record of the Royal Society of London* (3rd edn) (London: The Royal Society, 1912), 274.

60 RS CMO/07, 25 November 1779. This print run was the result of an investigation into the uptake of the formats among fellows, see RS CMO/07, 18 November 1779.

61 RS CMO/08, 22 December 1791 (order to Bulmer). It is not clear who decided which fellows received which format.

62 Compare Alexander Mackenzie, 'Account of a woman ...' *PT* 67 (1777): 1–11, https://royalsocietypublishing.org/doi/pdf/10.1098/rstl.1777.0002 with William Herschel, 'Account of ... the late eclipse', *PT* 84 (1794): 39–42, https://royalsocietypublishing.org/doi/pdf/10.1098/rstl.1794.0008.

63 Banks to Josiah Wedgwood, 28 December 1791, Letter 1085 in Chambers, *Banks Scientific Correspondence*, vol. 4, 94–5.

64 RS CMO/08, 15 November 1798 (reduced to 900 copies); 22 November 1798 (only 50 small paper); and 11 February 1808 (reduced to 750 copies); 24 March 1808 (small paper discontinued).

65 From the bills paid, RS CMO/08, 13 November 1806.

66 On the Basire family, see Richard Goddard, *'Drawing on Copper': The Basire family of copper-plate engravers and their works* (Maastricht: Maastricht University Press, 2016). For their work with the Royal Society, see particularly pp. 282–8; for the relationship to the Cox family of printers, see p. 297.

67 For instance, the first part of the *Transactions* for 1800 was announced as 'ready to be delivered to the Members of the Royal Society, at their Apartments in Somerset-Place, on Thursday next, the 29th Instant', *St James's Chronicle* (24–27 May 1800), Issue 6621.

68 RS CMO/06, 20 December 1773.

69 RS CMO/08, 15 January 1801.

70 RS CMO/08, 22 March 1804.

71 For Banks's involvement in renting the warehouse (and getting the Board of Longitude to pay part of the rent), see Meetings of 13 July and 7 December 1782, Confirmed Minutes of the Board of Longitude, 1780–1801, Papers of the Board of Longitude, Cambridge Digital Library, https://cudl.lib.cam.ac.uk/view/MS-RGO-00014-00006/181 and https://cudl.lib.cam.ac.uk/view/MS-RGO-00014-00006/185. Also Rebekah Higgitt, 'Instruments and relics: The history and use of the Royal Society's object collections c. 1850–1950', *Journal of the History of Collections* 31, no. 3 (2018): 469–85, 22. For the Society's copper plates, see RS CMO/07, 12 August 1785.

72 RS CMO/08, 17 May 1804.

73 RS CMO/08, 28 June 1804.

74 RS CMO/07, 22 April 1790.
75 RS CMO/08, 23 January 1794. Mrs Watt requested to be allowed to purchase parts of the *Transactions* dating from 1781 to 1785, and was permitted to purchase some but not all of the parts requested.
76 RS CMO/10, 6 May 1824.
77 See RS CMO/10, 24 April 1828 and RS DM/1/90 [undated; c. 1824–8].
78 A list of institutions, with dates for the commencement of 'presents', appears in RS CMO/09, 14 March 1816.
79 Banks to Barton (Philadelphia), 10 September 1804, Letter 1793 in Chambers, *Banks Scientific Correspondence*, vol. 5, 374.
80 On exchange schemes, see Jenny Beckman, 'Editors, librarians, and publication exchange: The Royal Swedish Academy of Sciences, 1813–1903', *Centaurus* 62, no. 1 (2020): 98–110.
81 The list of institutions entitled to copies appeared annually in the *Proceedings* after its creation in the 1830s; it was later moved to the *Year Book* in the late 1890s. For the list of 'presents' received, see for example 'Presents, January 10, 1878', *Proc* 27 (1878): 141.
82 Barton to Banks, 6 June 1804, Letter 1779 in Chambers, *Banks Scientific Correspondence*, vol. 5, 354.
83 Banks to Barton, 10 September 1804, Letter 1793 in Chambers, *Banks Scientific Correspondence*, vol. 5, 374.
84 For the American Philosophical Society's attempts to complete its back-run, see George L. Sioussat, 'The "Philosophical Transactions" of the Royal Society in the libraries of William Byrd of Westover, Benjamin Franklin, and the American Philosophical Society', *Proceedings of the American Philosophical Society* 93, no. 2 (1949): 99–113.
85 RS CMO/07, 24 February 1780.
86 O.M. Brack, 'Peter Elmsley, 1735/6-1802, bookseller' in *ODNB*.
87 George Nicol to Banks, 27 August 1801, Letter 1603 in Chambers, *Banks Scientific Correspondence*, vol. 5, 108–9.
88 RS CMO/08, 4 February 1802.
89 RS CMO/07, 11 March 1779.
90 Prices of parts issued between 1801 and 1811 were listed in an undated document from the later 1820s (when surplus stock was sold off), see RS DM/1/90. Calculation assumes two parts priced at 13s. 6d. per year; 350 copies of each; and 15 per cent commission.
91 RS CMO/08, 23 May 1799.
92 Topham, 'Anthologizing the book of nature', 137 and 141.
93 For a rich description of the contrast between the periodicals, see Watts, 'We want no authors'. The 1802 *Transactions* cost 11s. for Part 1 and 17s. 6d. for Part 2. Nicholson's *Journal* cost 2s. 6d. in 1806 (thus, 30s. a year), according to Watts, p. 13.
94 [Nicholson], 'Preface', *Journal of Natural Philosophy* 1 (1797–98): iv.
95 Elizabeth Grey to Charles Blagden, 30 June 1785 in RS CB/1/4/G43.
96 Grey to Blagden, 8 June 1784 in RS CB/1/4/G37 (Blagden on mercury); 15 July 1784 in CB/1/4/G38 (Cavendish); 31 August 1784 in CB/1/4/G39 (Blagden on mercury again); 24 November 1784 in CB/1/4/G40 (Herschel, Cavendish, Scheele, etc.); 26 November 1784 in CB/1/4/41 (missing figures for Herschel).
97 Grey to Blagden, 29 May 1799 in RS CB/1/4/G44.
98 [Thomson], 'Advertisement', *Annals of Philosophy* 1 (1813): iii. On journalists at society meetings, see also Csiszar, *The Scientific Journal*, chs 1 and 2.
99 Blagden to Banks, 28 September 1781, Letter 223 in Chambers, *Banks Scientific Correspondence*, vol. 1, 292–3.
100 Blagden to Dr [Erasmus] Darwin (Derby), 14 September 1786, RS CB/2/34.
101 Blagden to Banks, 19 October 1786, Letter 691 in Chambers, *Banks Scientific Correspondence*, vol. 3, 223–5.
102 Mann to Banks, 31 December 1779, Letter 177 in Chambers, *Banks Scientific Correspondence*, vol. 1, 221–2; Blagden to Banks, 7 October 1783, Letter 400 in Chambers, *Banks Scientific Correspondence*, vol. 2, 161 (re Thomas Hutchins' paper on freezing mercury).
103 RS CMO/08, 6 April 1797.
104 RS CMO/08, 20 February 1806 (Pears); and CMO/09, 17 November 1814 (Seppings).
105 Crell to Banks, [May 1790], Letter 998 in Chambers, *Banks Scientific Correspondence*, vol. 3, 590–1.

106 Crell to Banks, 20 January 1796, Letter 1351 in Chambers, *Banks Scientific Correspondence*, vol. 4, 412–14.
107 Dryander to Banks, 1 September 1785, Letter 595 in Chambers, *Banks Scientific Correspondence*, vol. 3, 78. The paper was H. Cavendish, 'Experiments on air', *PT* 75 (1785): 372–84. It was being translated by Bertrand Pelletier.
108 Blagden to Banks, 3 September 1785, Letter 596 in Chambers, *Banks Scientific Correspondence*, vol. 3, 79–80.
109 RS CB/2 f.29, Blagden to Berthollet, 13 September 1786.
110 RS CMO/07, 29 June 1787.
111 Aileen Fyfe, Julie McDougall-Waters, and Noah Moxham, 'Credit, copyright, and the circulation of scientific knowledge: The Royal Society in the long nineteenth century', *Victorian Periodicals Review* 51, no. 4 (2018): 597–615; and more generally, Will Slauter, 'Introduction: Copying and copyright, publishing practice and the law', *Victorian Periodicals Review* 51, no. 4 (2018): 583–96.
112 Nicholson to Banks, 12 March 1802, Letter 1636 in Chambers, *Banks Scientific Correspondence*, vol. 5, 147–8 (emphasis in original).
113 This episode has been discussed by Watts, 'We want no authors', who focuses on Nicholson's radical journalism; and by Gielas, 'Turning tradition into an instrument of research', who situates it in the context of their longer relationship. See also Nicholson, *Life of William Nicholson*.
114 Nicholson admitted his error in Nicholson to Banks, 28 June 1802, Letter 1675 in Chambers, *Banks Scientific Correspondence*, vol. 5, 206–8. The second parts of the papers were reprinted in Nicholson's July issue.
115 Young's 'Outlines of experiments and enquiries respecting sound and light' had appeared in the *Transactions* in 1800, and was reprinted in Nicholson's *Journal* between May and July 1801.
116 Nicholson to Banks, 28 June 1802, Letter 1675 in Chambers, *Banks Scientific Correspondence*, vol. 5, 206–8.
117 Nicholson to Banks, 12 March 1802, Letter 1636 in Chambers, *Banks Scientific Correspondence*, vol. 5, 147–8.
118 Nicholson to Banks, 28 June 1802, Letter 1675 in Chambers, *Banks Scientific Correspondence*, vol. 5, 206–8.
119 Nicholson to Banks, 12 March 1802, Letter 1636 in Chambers, *Banks Scientific Correspondence*, vol. 5, 147–8.
120 Banks to Nicholson, 12 March 1802, Letter 1637 in Chambers, *Banks Scientific Correspondence*, vol. 5, 148–9.
121 Attentive readers would not be misled, for the *Transactions* paper included the date it had been read at a meeting, and the *Journal of Natural Philosophy* version included an attribution to the (then-forthcoming) *Transactions*.
122 Banks to Nicholson, 12 March 1802, Letter 1637 in Chambers, *Banks Scientific Correspondence*, vol. 5, 148–9.
123 Fyfe et al., 'Credit, copyright'.
124 Nicholson to Banks, 12 March 1802, Letter 1636 in Chambers, *Banks Scientific Correspondence*, vol. 5, 147–8.
125 Nicholson to Banks, 28 June 1802, Letter 1675 in Chambers, *Banks Scientific Correspondence*, vol. 5, 206–8.
126 Nicholson to Banks, 28 June 1802, Letter 1675 in Chambers, *Banks Scientific Correspondence*, vol. 5, 206–8.
127 Nicholson to Banks, 28 June 1802, Letter 1675 in Chambers, *Banks Scientific Correspondence*, vol. 5, 206–8.
128 RS CMO/08, 15 July 1802.
129 Banks to Nicholson, [n.d.] July 1802, Letter 1676 in Chambers, *Banks Scientific Correspondence*, vol. 5, 209.
130 Banks to Blagden, 2 January 1814, Letter 1997 in Chambers, *Banks Scientific Correspondence*, vol. 6, 118.
131 Topham, 'Anthologizing the book of nature', 123–4.
132 Jacques Gibelin, *Abrégé des Transactions Philosophiques de la Société royale de Londres* (Paris: Buisson, 1787–91).

133 RS CMO/8, 31 March 1803. See also the editorial notes to Letter 89 in Benjamin Wardhaugh, *The Correspondence of Charles Hutton: Mathematical networks in Georgian Britain* (Oxford: Oxford University Press, 2017).
134 See Marie Boas Hall, *All Scientists Now: The Royal Society in the nineteenth century* (Cambridge: Cambridge University Press, 2002), ch. 1; and Benjamin Wardhaugh, 'Hutton and the "dissensions" of 1783–84: Scientific networking and its failures', *N&R* 71.1 (2017): 41–59.
135 Eneas Mackenzie, *A Descriptive and Historical Account of the Town and County of Newcastle-upon-Tyne*, 2 vols (Newcastle-upon-Tyne: Mackenzie and Dent, 1827), vol. 1, 559.
136 RS CMO/8, 27 April 1809.
137 Banks to Nicholson, 12 March 1802, Letter 1637 in Chambers, *Banks Scientific Correspondence*, vol. 5, 148–9.
138 [Augustus Bozzi Granville], *Science Without a Head; or, The Royal Society dissected. By one of the 687 F.R.S.* (London: T. Ridgway, 1830), 86.
139 William Thomas Brande to Banks, 15 October 1808, Letter 1885 in Chambers, *Banks Scientific Correspondence*, vol. 5, 499–500.
140 RS CMO/10, 27 April 1809. See also N.G. Coley, 'The Animal Chemistry Club; assistant society to the Royal Society', *N&R* 22.1 (1967): 173–85.
141 Cherry Lewis and Simon Knell, *The Making of the Geological Society of London* (London: Geological Society of London, 2009).
142 On Banks's relationship with the Geological Society, see David Philip Miller, 'The Royal Society of London, 1800–1835: A study in the cultural politics of scientific organization' (PhD, University of Pennsylvania, 1981), 141–56; and Cherry Lewis, 'Doctoring geology: The medical origins of the Geological Society', in *The Making of the Geological Society of London*, ed. Cherry Lewis and Simon Knell (London: Geological Society of London, 2009), 49–92, at 78–80.
143 Banks to James Laird, 4 April 1809, Letter 1904 in Chambers, *Banks Scientific Correspondence*, vol. 6, 3.
144 Banks to G.B. Greenough, [1 March 1809], Letter 1897 in Chambers, *Banks Scientific Correspondence*, vol. 5, 513.
145 Laird to Banks, 13 March 1809, Letter 1901 in Chambers, *Banks Scientific Correspondence*, vol. 6, 1.
146 Greenough to Banks, 18 February 1810, Letter 1926 in Chambers, *Banks Scientific Correspondence*, vol. 6, 26–7; Banks to Greenough, 19 February 1810, Letter 1927 in Chambers, *Banks Scientific Correspondence*, vol. 6, 27.

Part III
The professionalisation of science, 1820–1890

8
Reforms, referees and the *Proceedings*, 1820–1850

Noah Moxham and Aileen Fyfe

Both the *Transactions* and the Royal Society had enjoyed a high degree of stability under Joseph Banks and his trusted associates. But to a younger generation, this could seem like inflexibility. In the years after Banks's death, every election of a new president – in 1820, 1827, 1830, 1838 and 1848 – raised hopes of reform. One fellow later described the Society in these years as having been 'in a species of suspended animation', as its members tried to decide what sort of president would be best suited to lead the Society and its *Transactions* in the nineteenth century.[1]

The period covered in this chapter is a familiar element in histories of nineteenth-century science. Calls for the reform of British science, and of the Royal Society in particular, coincided with movements for political reform, such as Catholic emancipation, the abolition of slavery and the expansion of the electoral franchise. The would-be reformers at the Royal Society sought to transform it from an association dominated by scientific *dilettantes*, with just a handful of noted researchers, into a more elite organisation that recognised and supported those who were actively contributing to the sciences. The attempted reforms of 1830 are often seen, along with the foundation of the British Association for the Advancement of Science in 1831, as the first steps in the professionalisation of science in Britain.[2]

The first president after Banks was Humphry Davy, whose reputation was built on his fame as a public lecturer at the Royal Institution, and on an impeccable record of publication that included over 30 papers in the *Transactions*.[3] But, despite his humble, provincial origins, he did not distance himself as clearly from the aristocratic party within the Society as the reformers would have liked, and he was tainted by his association with the Banks regime, in which he had served as secretary from 1807 to 1812. The hoped-for reorientation of the Society did not materialise during Davy's tenure, nor during the brief tenure of the 'indecisive and irresolute' Davies Gilbert.[4] They would be followed by a series of aristocratic presidents – the Duke of Sussex, the Marquess of Northampton and the Earl of Rosse – and it was in their tenures that reforms were gradually made.

The year 1830 was particularly significant in Royal Society history because the election for president was contested (which it usually was not), and because two of its fellows – mathematician Charles Babbage and physician-midwife Augustus Bozzi Granville – criticised it in public. Babbage's *Reflections on the Decline of Science in England* (1830) is the better-known, partly because of the debates it generated and partly because of Babbage's enduring fame as the inventor of a mechanical computer.[5] Babbage had a track-record of interest in the reform of science dating from his student days at Cambridge, when he and the astronomer-to-be John Herschel had sought to reform the obsolete teaching in mathematics then offered in the university. He and Herschel had later been among the founders of the Astronomical Society in 1820, promoting a different approach to astronomy from that encouraged at Banks's Royal Society.[6] And then, during Davy's final illness in 1827, they were both part of a Royal Society committee that raised the question of 'limiting … admission of members into the Society', and being less 'indiscriminate' in accepting new members.[7] Their vision for reform included reform of the *Transactions*, for, just as in Banks's day, it was acknowledged that the Society's 'character abroad can only be appreciated by the nature and value of its Printed Transactions'.[8]

In *Reflections on the Decline of Science in England*, Babbage's criticisms of the Royal Society were embedded in his wider concerns about the state of science in Britain compared (in particular) to France, including the administration of scientific patronage, and the lack of state support for science. He used statistics prepared for the 1827 committee, on the number of papers published by fellows in the *Transactions*, to argue that there had been a decline in the Society's contribution to science since the mid-eighteenth century. The headline figure that only 109 of the current 714 fellows had ever published in the *Transactions* appeared to support his argument that far too many of the current fellows were contributing nothing to science.[9] The Royal Society responded to Babbage's criticisms with its long-perfected but fundamentally inadequate formula of dignified outward silence.

Babbage and most of the scientifically active fellowship supported John Herschel for president in 1830. Augustus Granville, in contrast, supported Prince Augustus Frederick, the Duke of Sussex and younger brother of King George IV. Granville's anonymous *Science Without a Head*

(1830) was intended 'to secure the election of the royal duke', so that the Society would have 'a head at last'.[10] Granville believed that Sussex's social eminence would be more useful to the Society than any specific scientific accomplishments, and argued that, being 'equally the patron of every science', Sussex's election might defuse rivalries between the different disciplinary communities within the fellowship.[11]

Granville may have differed from Babbage in his presidential choice (and on whether science in Britain was actually 'in decline'), but he shared the desire to reform the organisation that they both saw as the natural home of scientific authority in Britain. *Science Without a Head* focused directly on the 'state of confusion and disorganization' at the Royal Society.[12] The fact that its harsh criticism was based on a detailed examination of the Society's own archival records, by an insider, made it particularly damaging. Granville complained about 'absurd and unsatisfactory' election procedures, the lack of meaningful discussion at meetings, the lack of financial oversight, and the mode of evaluating papers for reading and publication in the *Transactions*.[13] Like Babbage, he understood the authorship of original research papers in the *Transactions* as a marker of a contribution to science.

The Duke of Sussex's narrow victory in 1830 has usually been seen as delaying the reform of the Royal Society: it was not until 1847 that new statutes, including changes to the processes for admission of new members, were finally approved.[14] However, significant changes were in fact made to the Society's operations in the intervening decades, including the management of the finances, library and publications. The *Transactions* was entangled with the wider changes for two reasons. First, the proposed reduction in the admission of new fellows would significantly reduce the Society's income, thus potentially affecting its ability to fund its publications. Second, with papers published in the *Transactions* increasingly being seen both as a marker of eligibility for admission to the fellowship, and a demonstration of contribution once in the fellowship, editorial decision-making mattered more than ever.

In its efforts to demonstrate the scientific uselessness of much of the recent intake of fellows, the 1827 committee investigated the contributions made by every fellow: the award of the Copley or Rumford Medals, service on Council, and number of papers in the *Transactions*.[15] They recommended that future printed lists of fellows should carry an appendix listing all living members who had contributed at least two papers to the *Transactions* or who had been awarded one of the medals in the Royal Society's gift.[16] Had this been approved, it would have created two classes of fellow – the scientifically active, and the others – and could have been seen as an admission that most of the fellows were unworthy of their places. Indeed, by Babbage's subsequent account, this was precisely the intention.[17] In 1830, Granville took the analysis further, and 'dissected' the contributions of the fellowship by rank and profession. His tables showed that physicians had authored more papers in the *Transactions* than clergymen, but that the 63 noblemen in the fellowship had made precisely '000 contributions towards improving natural knowledge'.[18] These are early instances of an effort to create a metric to measure research output, with publication in the *Transactions* being seen as a contribution to the Society, which in turn was seen as a proxy for scientific eminence.[19]

During this period, the *Transactions* changed very little on the surface: it retained its stately eighteenth-century periodicity and format; its print run was not high enough for its production to shift to the new steam-powered printing technologies then being adopted by newspapers and high-circulation magazines; and most of its illustrations continued to be engraved on copper plates, though there was some use of the newer (cheaper) technique of lithography. On the editorial side, the rules that papers could only be communicated by a fellow and must be read at meetings before being considered for publication, remained in force. The Society's officers continued to exercise significant invisible power over the agenda for meetings, although the aristocratic presidents of the 1830s and 1840s tended to delegate that power to the secretaries, of whom the longest serving were Peter Mark Roget (1827–48) and Samuel Hunter Christie (1837–54).

There were, however, significant changes behind the scenes. The years around 1830 would see the first competitive tender for the role of printer to the Royal Society; the creation of a new periodical, the *Proceedings of the Royal Society*; and the adoption of practices which may be seen as the origin of modern 'peer review'. By mid-century, the Society would be operating a two-tier publishing system in which the boundary between the *Transactions* and *Proceedings* was policed by the expert scrutiny of fellows acting as referees. These changes consolidated the emerging function of a memoir in the *Transactions* as a token in the (somewhat) meritocratic economy of scientific reputation-building.

Printers and the *Proceedings*

Tendering for a printer: Richard Taylor

When the Society's printer, William Bulmer, retired from business in 1821, its printing and bookselling needs were seamlessly transferred to his partner George Nicol. But when William Nicol became the principal in G. & W. Nicol after his father's retirement, things went less smoothly. In January 1828, he was summoned to appear before a Royal Society committee chaired by the new president, Davies Gilbert, and was informed that his printing had been found, 'on comparison with other works containing analogous matter, greatly inferior; so much so, as to call for some decided change'. He was informed that the Society would be seeking competitive tenders. Rather than participate, Nicol chose to resign.[20]

This competitive tender process contrasted with previous appointments: Joseph Banks had simply informed the Council of his choices. The 1828 committee included Herschel (then vice-president) and both the secretaries. They issued a circular to 'some of the principal printers of the Metropolis', announcing their desire to find the printer 'most likely to execute the printing of the Philosophical Transactions in the manner most creditable to the Society'.[21] They sent specimen pages from a recent part of the *Transactions* to demonstrate the variety of material in its pages, and requested samples of work from each printer, as well as an estimated cost for typesetting and printing 1,000 copies of the *Transactions*, paper not included. The request for samples reflects the ongoing concern with print quality, but the committee also showed an awareness of the practicalities of running a printing business. Each printer was asked: 'What number of sheets could you furnish per week?' and 'What number of sheets could you allow to be set up at one and the same time?'[22] Such questions suggest an interest in reliability and efficiency, as well as quality and price.

The committee contacted five printers, the most notable of whom were William Clowes and Richard Taylor.[23] Clowes had built his business on government printing contracts, and had been using steam-powered printing machines since 1823; by the mid-1830s, his printing works would be the largest in the world.[24] Taylor's business, on the other hand, was smaller but grounded in London's scientific and scholarly communities. He had served his apprenticeship in the late 1790s in the workshop that printed both the *Transactions* of the Linnean Society and the new *Philosophical Magazine*; Taylor took over and extended this business. By the late 1820s, he was printing for the Geological Society, the Zoological Society and the Astronomical Society, as well as the Antiquaries; and he had even been elected to the fellowship of several learned societies. Yet as well as providing print-related services to the learned societies and their members, he was a key player in the publication of independent scientific journals. Since 1825, he had been editor and owner (as well as printer) of the *Philosophical Magazine*, whose role as the key monthly journal of science in Britain was consolidated by its acquisition and merger of former rivals, including Nicholson's *Journal of Natural Philosophy* (in 1814), Thomson's *Annals of Philosophy* (in 1827) and Brewster's *Edinburgh Journal of Science* (in 1832).[25]

In mid-February 1828, the Royal Society's printing committee considered the submissions from the four printers who had responded. All the estimates received were cheaper than the £120 the Society had actually paid for printing the last part of the *Transactions*: Clowes's estimate worked out at £90, and that of Taylor at £96.[26] The decision to appoint Taylor rather than Clowes suggests that cheapness was not an overriding priority. Taylor had a track-record of high-quality scholarly typesetting, and, although the print run of the *Transactions* was raised (back) to 1,000 copies in 1828, this was not nearly high enough – nor its periodicity frequent enough – to need the steam-printing capability that Clowes could offer. (The print run would remain at 1,000 until 1898.[27])

Taylor was already well established in the world of scientific printing prior to 1828, but winning the Royal Society contract was a feather in his cap. It was also a valuable contract: for instance, in 1832, he billed the Royal Society for £890 of work, compared to just over £57 for the Geological Society.[28] The work done in 1832 included 500 copies of 'Regulations for the Library'; 750 copies of the 'List of Fellows'; 450 copies of the list of potential candidates for election to the fellowship; and 1,500 copies of a list of the Society's collection of portraits.[29] At other times, there were sets of draft minutes 'for the Use of the Council only'; the annual accounts; and notices to fellows who were overdue with their subscriptions. None of this printing had the public prestige of the *Transactions*, but it provided plenty of paid work for Taylor's firm, and it reminds us that the Society's relationship with its printer was more wide-ranging than the *Transactions* alone.[30]

Taylor was appointed specifically as the Society's printer, not its bookseller, and, in 1829, the Council formally agreed that the *Transactions* would be 'sold at the Apartments of the Society, instead of a Bookseller being employed to sell them'.[31] This implies that the Society did not consider the *Transactions* as a product whose circulation might need professional expertise. The Society's tiny staff could sell copies to individual callers at Somerset House (as well as issuing copies to fellows and the agents of institutions on the Society's list for gifts), but they did not try to market the *Transactions*.

Nonetheless, it is clear that Taylor did, in fact, manage some of the marketing and sales of the *Transactions*. He was responsible for the newspaper advertising that announced each new part: in December 1832, for instance, he placed advertisements in five London newspapers, the weekly *Athenaeum*, and the monthly trade journal *Bent's Literary Advertiser*.[32] These advertisements also served as announcements to fellows and learned institutions that their copies were available for collection. Most of the sales took place in the months immediately after publication, though there was a steady trickle of requests for older parts. For instance, a surviving income/expenditure book in the Taylor & Francis archive reveals that, between 1 and 14 August 1835, Taylor sold 19 copies of the recently released Part 1 of volume 125, as well as a further 14 copies of various parts from the previous four years. All the sales were in ones or twos.[33] There are no records of the purchasers, but these records probably represent the fulfilment of internal book trade orders rather than sales to individual readers. Any London bookseller whose client requested a current or recent copy of the *Transactions* would know to apply to Richard Taylor, in Red Lion Court, off Fleet Street.

The *Proceedings*

Taylor's appointment would be a crucial element in the creation of the *Proceedings of the Royal Society*. This had its roots in the meshing of Taylor's interest – as editor of the *Philosophical Magazine* – in reporting the proceedings of learned society meetings, and the interests of the officers of learned societies in making public fair and accurate accounts of the papers presented at their meetings, in a timely manner. The Royal Society was just one of many learned institutions, in London and across Europe, that created *Proceedings*-type periodicals in the years around 1830. Such periodicals allowed learned institutions to complement their traditional, stately (and slow) *Transactions* or *Mémoires* with briefer and more rapid accounts. The most notable example was the *Comptes rendues*, produced by the French Académie des sciences from 1835, though its weekly periodicity was rarely imitated.[34]

However, the creation of the *Proceedings of the Royal Society* was also entwined with a more backward-looking project. Access to the historic content in the back-issues of the *Transactions* was a long-standing challenge for scholars, to which the multi-volume abridgements undertaken by John Lowthorp and Charles Hutton had offered a solution. These abridgements had been compiled without the Society's involvement, yet the Society did have an in-house resource that could serve a similar purpose. The minute books kept by its secretaries contained summaries of all the papers read at the meetings, usually running to several hundred words. Back in the 1780s, a suggestion to use these minutes as the basis of an annual report on the meetings of the Society had been thoroughly quashed (see Chapter 6).[35]

Public or semi-public access to the minute books of learned societies remained highly contentious in the late 1820s: it was sought by journal editors to improve their reportage of meetings, and, in the case of the Royal Society in particular, by critics seeking to build the case for reform. Societies that already worked with Richard Taylor found that providing him with the secretaries' abstracts for printing in the *Philosophical Magazine* improved the speed and accuracy of the reportage of their meetings. The secretaries of both the Geological and Astronomical societies had been collaborating with Taylor in this way since 1825.[36] The Royal Society, on the other hand, had an uneasy relationship with the *Quarterly Journal of Science*, whose editor, William Brande, was secretary to the Society as well as an employee of the Royal Institution. Charles Babbage noted the 'inconvenience' that Brande's multiple roles had created for the Society.[37] By 1828, Brande was no longer secretary, and mere days after appointing Taylor, Council gave the secretaries authority to oversee the process of communicating 'to the public … accounts of the proceedings of the Society', along the same lines as the other learned societies.[38] (Brande was explicitly refused further access to the minute books.[39])

As well as printing the abstracts in the pages of the *Philosophical Magazine*, Taylor offered to print separate copies for his learned society clients, for circulation to their members. By reusing the types already set for the *Magazine*, Taylor could charge the societies only for paper, presswork and the setting of any additional material (such as presidential addresses, annual accounts or other society business) not reported in the *Magazine*. By 1827, the Geological and Astronomical societies both had these separately printed *Proceedings*, which helped the societies amplify the debate and discussion around claims to knowledge, and to present themselves as open to public scrutiny.[40] As Alex Csiszar has pointed out, the overlap between the reports of 'proceedings' in the *Philosophical Magazine* and the *Proceedings* of the learned societies shows there could be constructive collaboration between learned societies and the independent publishers who are often considered their rivals.[41]

Back at the Royal Society, barely a fortnight after Sussex's victory in the presidential election, the new Council of the Royal Society resolved in December 1830, that henceforth 'the Abstracts of the Papers read at the Society's Meetings and entered upon the Minutes of their proceedings, be printed for the use of the Fellows'.[42] This led both to the creation of the

Proceedings of the Royal Society, containing abstracts of the papers read at recent meetings, and to the publication of two retrospective volumes, containing the abstracts from 1800 to 1830. Both projects would prove important for reaching audiences beyond the fellowship.

Taylor printed 750 copies of the first number of the 'proceedings' in late February 1831: it started with a report of the meetings held in November 1830 (see Figure 8.1).[43] The project followed the same pattern as with his other clients: it drew upon the secretaries' minute books; it reused the type set for the *Philosophical Magazine*; and it was initially intended primarily for internal circulation among the 660 or so fellows. Taylor's quotation made the overlap with the *Philosophical Magazine* clear, assuring the Society that, 'The price of the composition of such part of the Proceedings as can be used in the Philosophical Magazine & Annals – not to be charged; but only the Presswork & Paper'.[44] The phrase 'Types partly standing from Phil Mag' would be repeated in Taylor's records of printing the *Proceedings* throughout the 1830s, with the use of 'partly' reflecting the fact that the *Proceedings* contained a more complete account of Royal Society meetings and business than appeared in the pages of the *Philosophical Magazine*.

Interest in the *Proceedings* from beyond the fellowship emerged in just a few weeks. In March 1831, the Society's assistant secretary requested six extra copies for distribution to friends; in April, the secretary of the nearby Athenaeum Club asked for a copy, and Francis Beaufort of the Admiralty wanted half a dozen 'for the use of the Gentlemen connected with him in the Hydrographical Office', as well as six more for the former Royal Astronomer of Ireland; in May, the Society's treasurer asked for 12 copies for his friends.[45] A little over a year later, in May 1832, the Council resolved to print title pages and indexes for what was now regarded as 'Part 1, 1830–31' of a new periodical, thus effectively acknowledging its public character.[46]

The public demand for the printed abstracts can be seen in the print runs: 250 copies of the first issue had to be reprinted in May 1832, while in March 1833, an extra 750 copies were reprinted of all 10 issues to date. From then on, 1,500 copies became the standard run.[47] Compared to the 1,000 copies printed of the *Transactions*, this indicates the interest in faster access to news of the latest findings, even if only in brief. Like the *Transactions*, the *Proceedings* came out erratically and was tied to the Society's season. In the 1830s, there were usually several issues in the spring months of the year, nothing over the summer, and one or two issues in the winter. Material could appear in print more quickly in the *Proceedings* than with the *Transactions*, but the lack of regular periodicity

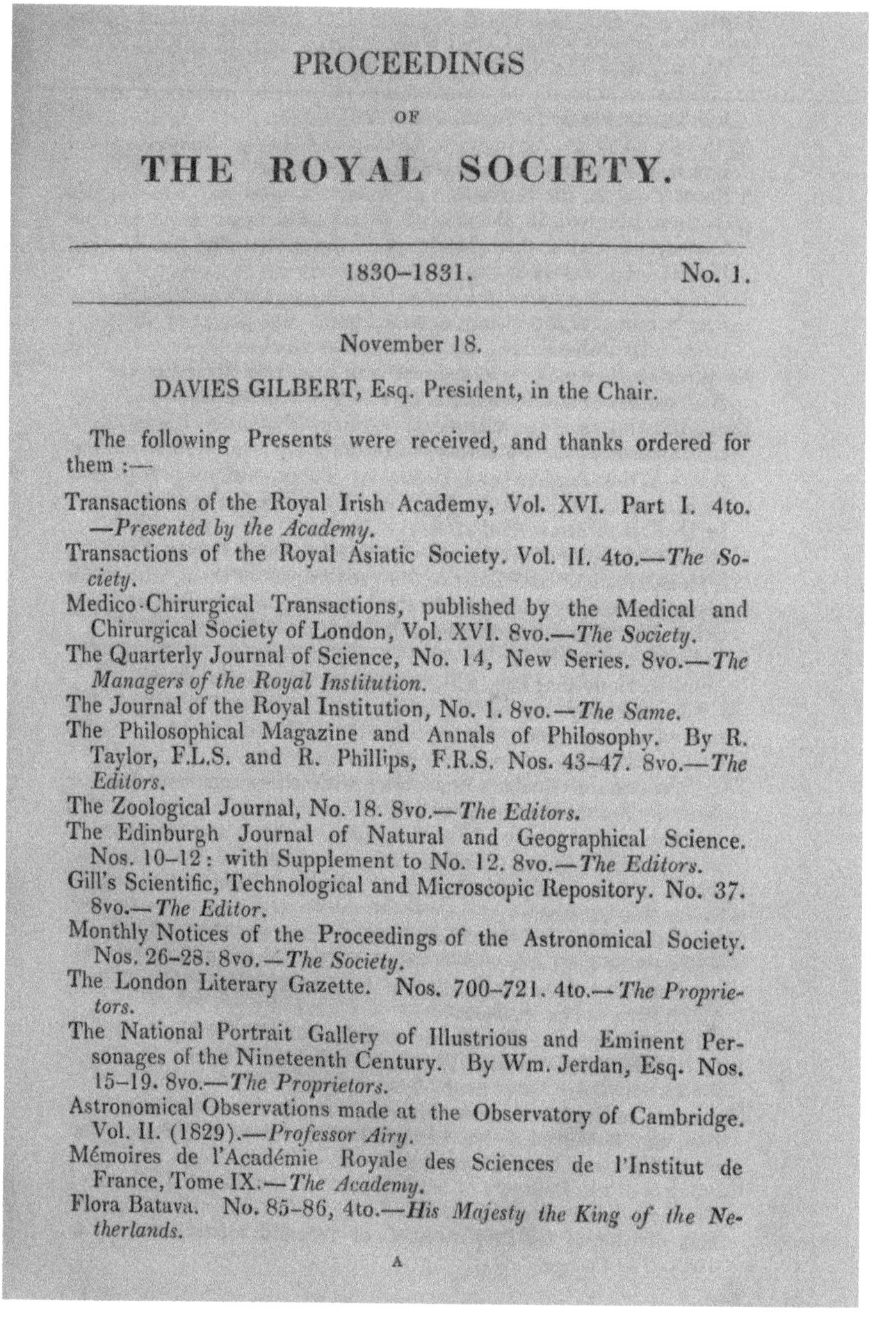

PROCEEDINGS

OF

THE ROYAL SOCIETY.

1830–1831. No. 1.

November 18.

DAVIES GILBERT, Esq. President, in the Chair.

The following Presents were received, and thanks ordered for them :—

Transactions of the Royal Irish Academy, Vol. XVI. Part I. 4to. —*Presented by the Academy.*

Transactions of the Royal Asiatic Society. Vol. II. 4to.—*The Society.*

Medico-Chirurgical Transactions, published by the Medical and Chirurgical Society of London, Vol. XVI. 8vo.—*The Society.*

The Quarterly Journal of Science, No. 14, New Series. 8vo.—*The Managers of the Royal Institution.*

The Journal of the Royal Institution, No. 1. 8vo.—*The Same.*

The Philosophical Magazine and Annals of Philosophy. By R. Taylor, F.L.S. and R. Phillips, F.R.S. Nos. 43–47. 8vo.—*The Editors.*

The Zoological Journal, No. 18. 8vo.—*The Editors.*

The Edinburgh Journal of Natural and Geographical Science. Nos. 10–12: with Supplement to No. 12. 8vo.—*The Editors.*

Gill's Scientific, Technological and Microscopic Repository. No. 37. 8vo.—*The Editor.*

Monthly Notices of the Proceedings of the Astronomical Society. Nos. 26–28. 8vo.—*The Society.*

The London Literary Gazette. Nos. 700–721. 4to.—*The Proprietors.*

The National Portrait Gallery of Illustrious and Eminent Personages of the Nineteenth Century. By Wm. Jerdan, Esq. Nos. 15–19. 8vo.—*The Proprietors.*

Astronomical Observations made at the Observatory of Cambridge. Vol. II. (1829).—*Professor Airy.*

Mémoires de l'Académie Royale des Sciences de l'Institut de France, Tome IX.—*The Academy.*

Flora Batava. No. 85–86, 4to.—*His Majesty the King of the Netherlands.*

A

Figure 8.1 The first issue of the *Proceedings*, printed in February 1831, but reporting the meeting of 18 November 1830. The first item of business was to acknowledge the 'presents' received for the Society's library © The Royal Society.

for the *Proceedings* would be a point of frequent criticism in decades to come.

These printed abstracts of papers were referred to as 'Proceedings' from their earliest days: this was the title used by Taylor when he quoted for the work, and it appeared in the header from the very first issue. However, the Society created the grounds for subsequent confusion when, in 1837, it gathered all the issues to date into a single volume with a title page that described it not as the *first* volume of the *Proceedings* but as the *third* volume of *Abstracts of the Papers Printed in the Philosophical Transactions* (even though it also contained abstracts of papers not printed in the *Transactions*). When the title of the volumes was finally changed to the *Proceedings* in the 1850s, the misleading volume numbering would be retained.

The *Abstracts of the Papers Printed in the Philosophical Transactions* was the other outcome of the Society's new willingness to make public 'the Abstracts made by the Secretaries'. In 1830, Babbage had suggested that 'the knowledge of the many valuable papers' accumulated in the 'extremely bulky' back-run of the *Transactions* could be 'much spread, by publishing the abstracts of them.... Perhaps two or three volumes octavo, would contain all that has been done in this way during the last century.'[48] During 1832, the secretaries and treasurer gathered the abstracts for all the papers in the *Transactions* between 1800 and 1830, which 'possess in themselves much intrinsic value'. They did not try to go back a full century, but to create 'an useful sequel to the Abridgement of the Philosophical Transactions of which the public is already in possession'.[49] The key difference from Lowthorp's and Hutton's abridgements was that the editorial work and financial responsibility was now undertaken by the Society itself. Taylor printed the two thick octavo volumes in 1832 and 1833, '*uniformly* with those [abstracts] which are now published for the present year' (Figure 8.2).[50] The choice of title page and volume number for the first volume of the *Proceedings* indicates that the publication of the current and retrospective abstracts were seen as part of a combined project.

The origins of the *Proceedings* are thus entangled in two separate discourses. It could be seen as opening up the Society's meetings to public scrutiny, in response to the calls for reform and a desire to improve the reliable and timely circulation of information about papers presented to Society meetings. It was from this perspective that the *Mechanics' Magazine* was one of the formerly critical voices that welcomed the 'spectacle' of 'so stiff and unbending a Society' giving up 'a good portion of its haughty spirit of exclusiveness'.[51] But the use of the official abstracts meant that it was also part of an effort to tighten (not relax) the Society's

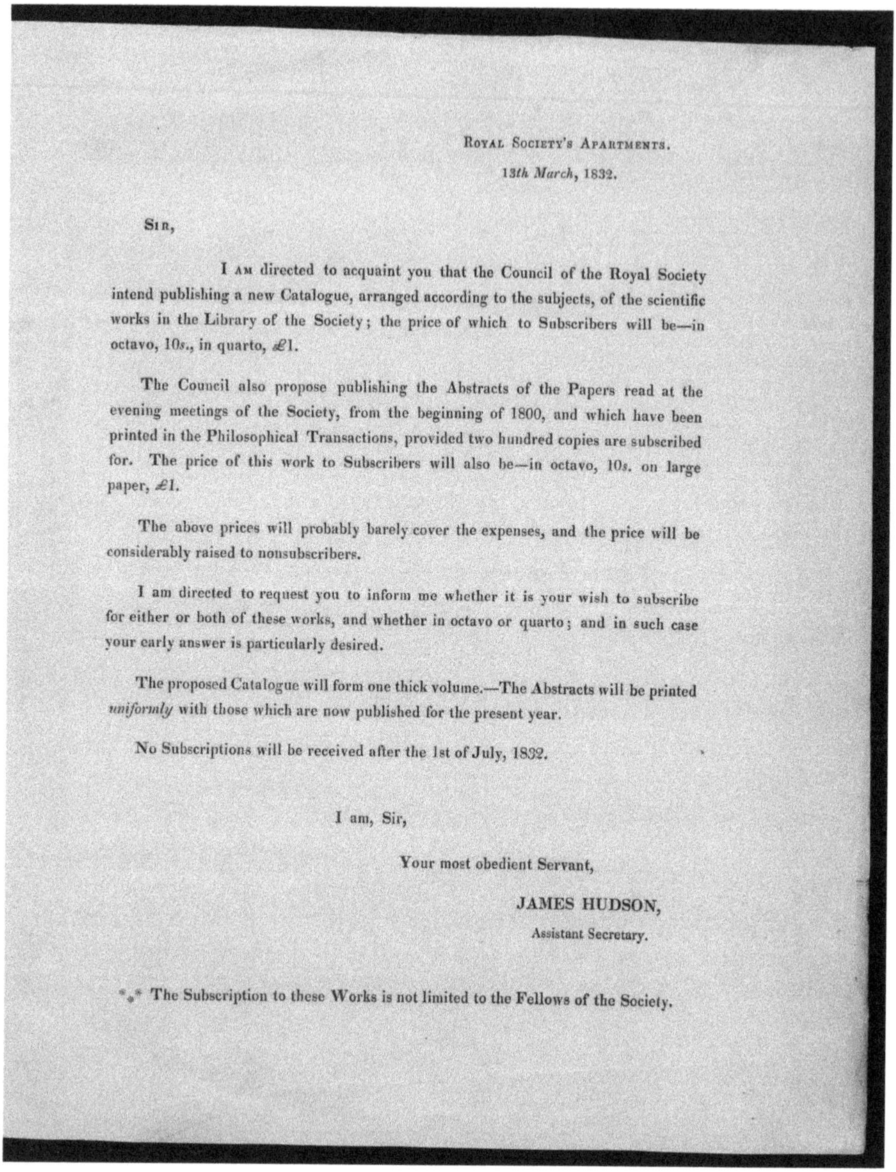

ROYAL SOCIETY'S APARTMENTS.

13th March, 1832.

SIR,

I AM directed to acquaint you that the Council of the Royal Society intend publishing a new Catalogue, arranged according to the subjects, of the scientific works in the Library of the Society; the price of which to Subscribers will be—in octavo, 10*s*., in quarto, £1.

The Council also propose publishing the Abstracts of the Papers read at the evening meetings of the Society, from the beginning of 1800, and which have been printed in the Philosophical Transactions, provided two hundred copies are subscribed for. The price of this work to Subscribers will also be—in octavo, 10*s*. on large paper, £1.

The above prices will probably barely cover the expenses, and the price will be considerably raised to nonsubscribers.

I am directed to request you to inform me whether it is your wish to subscribe for either or both of these works, and whether in octavo or quarto; and in such case your early answer is particularly desired.

The proposed Catalogue will form one thick volume.—The Abstracts will be printed *uniformly* with those which are now published for the present year.

No Subscriptions will be received after the 1st of July, 1832.

I am, Sir,

Your most obedient Servant,

JAMES HUDSON,

Assistant Secretary.

*** The Subscription to these Works is not limited to the Fellows of the Society.

Figure 8.2 Printed flyer, 13 March 1832, inviting advance subscriptions to the retrospective *Abstracts*, 1800–30 'be printed *uniformly* with those which are now published for the present year' [i.e. the *Proceedings*] © The Royal Society.

control over the circulation of knowledge about its meetings and the papers presented there. In 1828, for instance, at the same time as the secretaries were authorised to share extracts from their minute books with Taylor, they were urged to use 'their discretion' to prevent others from copying from the minute books.[52]

The retrospective volumes contained only abstracts of those papers that had been approved by the Committee of Papers and appeared in the *Transactions*. In contrast, the *Proceedings* quickly became a fuller record of papers read at meetings. This meant that it included reports presented to the meetings (including annual reports by the treasurer or president), as well as abstracts of papers that had not been approved for publication in the *Transactions*. Abstracts would already be in press (and sometimes in print) while referees and committees were still considering whether or not to publish the full paper. In some cases, these abstracts ran to several pages, and by the 1840s, the *Proceedings* also began to carry the full versions of papers that were short enough to fit among the long abstracts.

By the 1840s, as Csiszar has shown, the success of *Proceedings*-style periodicals led some societies to consign the *Transactions* format to the past. For instance, the new Geographical Society (f. 1830) decided to create a *Journal* rather than a *Transactions*, and the Horticultural Society abandoned its *Transactions* in 1845. Other societies kept their *Transactions*, but developed their *Proceedings* into something more like the journals produced by independent publishers: accepting independent articles, rather than abstracts of longer memoirs, and introducing material other than research papers, such as book reviews, news and reportage. The Geological Society, for instance, turned its *Proceedings* into a *Quarterly Journal* and moved it to the publishers Longman, whose financial interest in its success motivated a more active approach to advertising and marketing than was usual with society periodicals.[53] The Royal Society, however, did neither of these things. It remained committed to its *Transactions*, to which the *Proceedings* would remain firmly subsidiary for many decades, and there was no visible effort to market or commercialise either periodical.

Printing and publishing in the 1830s and 1840s

The costs of publishing and distributing the *Transactions* and the *Proceedings* were the largest item of expenditure for the Royal Society and came under correspondingly close scrutiny as the Society's officers planned for the consequences of potential membership reforms. In 1830,

Table 8.1 Average 'ordinary' income and expenditure of the Royal Society 1828–33, as reported by the treasurer in 1833

Expenditure		Income	
Salaries	£645	Rents	£284
Lighting and coal	£120	Dividends on stock	£501
Charwoman and servant	£42	Fellows' (weekly and quarterly) contributions	£270
Miscellaneous (including postage, tax, insurance)	£302	Fellows' admission fees and compounded annual contributions	£1,100
Sub-total	£1,109	Sub-total	£2,155
Transactions – printing	£350	Sales of *Transactions*	£350
Transactions – paper	£259		
Transactions – engraving	£285		
Sub-total	£894	Sub-total	£350
Total	£2,003	Total	£2,505

John Lubbock had become the Society's treasurer. He served with the Duke of Sussex, but his strong links to the reformers (especially through shared membership of the Astronomical Society) helped him to reconcile some of the divisions caused by the contested election.[54] He was a member of a banking family, and his financial innovations at the Society included the creation of a standing Finance Committee.[55] This would be the body with oversight of all printing-related decisions for the next hundred years, and its records provide more detailed insight into the costs and income of the Society's publications than exist for earlier periods.

In 1833, Lubbock compiled a report that situated the publication finances in the wider finances of the Society during the first years of Taylor's tenure. Lubbock's figures are summarised in Table 8.1, and show that the publications (listed here as the *Transactions*, but including the early issues of the *Proceedings*) had accounted for a striking 45 per cent of the Society's entire 'ordinary' expenditure over the previous five years.[56]

Lubbock's analysis suggests that the Society's periodicals were a net cost of around £540 a year. In the context of the possible reduction in membership income, this was sufficiently worrying for the Finance Committee to investigate possible cost savings. Top of the list was a change of format: the *Transactions* would remain a quarto publication, but Taylor suggested a layout that would save the Society money on paper, trimming the margins and allowing him to fit 'twelve additional lines in each page'.[57] Council also agreed to recommend more economy in the use of illustrations, but their resolve was undermined by their desire to avoid 'injuring the character' and handsome appearance of the *Transactions*.[58] Analyses by subsequent treasurers suggest that the net cost of publishing remained at somewhat over £500 a year until the mid-1840s.[59]

Lubbock's figures show the Society was receiving an average of £350 a year as income from publication sales in the early 1830s. The price of the parts of the *Transactions* still varied with their length: in 1827, both parts had been sold at 18s., but in 1832, Part 1 retailed for 15s. 9d. while Part 2 cost £1. 10s. 0d.[60] At those prices, and without allowing for the cut taken by the agent, an income of £350 might represent sales of around 150 to 190 copies. It is also notable that the sales income Lubbock reported in the early 1830s was marginally less than it had been in the mid-1820s,[61] and that, by the late 1830s and mid-1840s, subsequent reports would show total sales income of only £270 or £280 a year.[62] This declining sales income, despite the higher print run and the addition of the *Proceedings*, may be a consequence of the 1828 decision not to appoint a bookseller, and reinforces the sense that the Royal Society did not see its publishing operations as a business.

Circulation figures (rather than sales income) are very rare in the Society's archive, but in early 1846, the Finance Committee did investigate the actual numbers of copies of the *Transactions* circulated through different channels over the preceding 10 years (see Table 8.2).[63] The figures make clear that the *Transactions* had a very small paid-for circulation: its annual sales varied between just 130 and 160 copies. Even with an expansion of the number of learned institutions and others on

Table 8.2 Annual circulation of the *Transactions*, on average for the years 1835–44

Sales	140
To learned institutions	64
Other gratis copies	20
Claimed by fellows	476
Claimed by foreign members	48
Total circulation	748
Copies remaining on hand	*251*

the list of recipients of 'presents', the vast majority of copies still went to fellows. Non-commercial distribution remained the dominant model of circulation.[64]

That said, only about half of the now-750 fellows claimed their copies within two years, rising to about two-thirds within five years. As a result, around a quarter of the print run was actually accumulating as surplus stock in the warehouse, but there was no discussion of reducing the print run.[65] There is no equivalent breakdown for the circulation of the *Proceedings*. We know that its print run was reduced from 1,500 to 1,250 in 1844, but this still suggests a wider circulation, though not necessarily wider sales, than the *Transactions*.[66]

In mid-December 1846, the reform of the Society's statutes, including a reduction of the membership, was finally in progress, and Council wondered 'whether any diminution could properly be effected' in the Society's publication costs.[67] The Finance Committee organised another competitive tender, this time including paper supply as well as printing. There do not appear to have been questions hanging over the quality or efficiency of Richard Taylor's work, just its price. Taylor managed to submit an estimate that was more financially competitive than the others, particularly in the typesetting of mathematical work (see Figure 8.3).[68] The committee swiftly declared Taylor's bid 'the most advantageous', and reappointed him.[69] Five paper makers were also asked to quote for the supply of 150 reams of paper annually, of a quality 'equal to the specimen sent herewith'; again, and on Taylor's advice, the committee decided to remain with their existing supplier, Bowles & Gardiner.[70]

The other major expense associated with the publications was the illustrations, but the Finance Committee felt unable to run a tender process for these, remarking that 'so much difficulty attends the giving of estimates' when the illustrative requirements of individual papers could vary so much.[71] The illustrations for the *Transactions* were still almost entirely supplied by the Basire family, though they had been using lithography as well as copper engraving since the early 1830s. The Finance Committee decided to develop a pool of trusted suppliers, and to seek competitive quotations for each batch of work.[72] Thus, although the Basire family continued to work for the Royal Society, they lost their decades-old monopoly to become one supplier among half a dozen or so.[73]

The re-tendering process may not have resulted in significant changes, but it did persuade the Society's paper suppliers and printer to lower their prices. Coupled with figures showing that publication costs had fallen during the early 1840s, the Finance Committee concluded in early 1847 that print costs were under control. In less than five years,

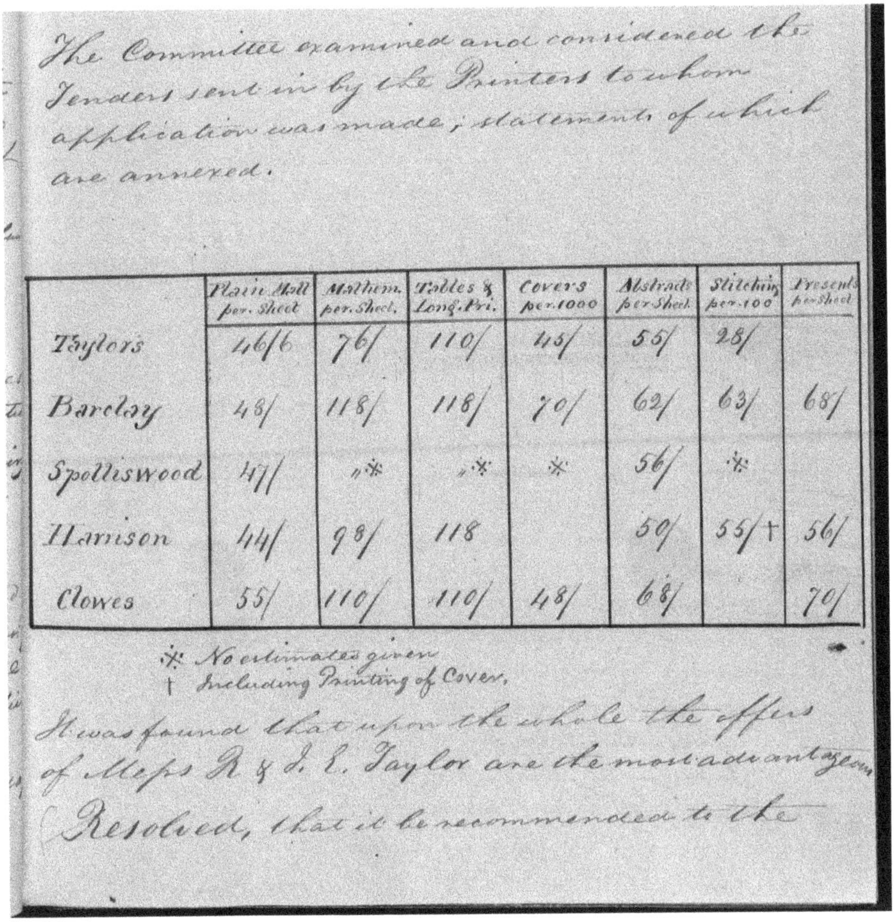

The Committee examined and considered the Tenders sent in by the Printers to whom application was made; statements of which are annexed.

	Plain Matt per Sheet	Mathem. per Sheet.	Tables & Long. Pri.	Covers per 1000	Abstracts per Sheet	Stitching per 100	Presents per Sheet
Taylors	46/6	76/	110/	45/	55/	28/	
Barclay	48/	118/	118/	70/	62/	63/	68/
Spottiswood	47/	"*	"*	*	56/	*	
Harrison	44/	98/	118		50/	55/†	56/
Clowes	55/	110/	110/	48/	68/		70/

* No estimates given
† Including Printing of Cover.

It was found that upon the whole the offers of Messrs R & J. E. Taylor are the most advantageous

Resolved, that it be recommended to the

Figure 8.3 Estimates of printing costs received by the Finance Committee, December 1846 © The Royal Society.

another treasurer would show that this confidence had been mistaken. But he would also show that the root of the problem of spiralling costs lay not in the printer's workshop or stationer's shop, but in the prolix writing habits of the Society's authors, and an editorial process that considered each submission on its own merits.[74]

Editorial reforms: scrutiny and expertise

During the 1830s and 1840s, the well-established procedures of the Committee of Papers would be rendered increasingly complicated by the introduction of written refereeing, the creation of specialist sub-committees, and the creation of the *Proceedings*. The first two were attempts to introduce closer scrutiny and more expert judgement, while the latter shifted the implicit criteria by which a contribution was deemed 'worthy' of the *Transactions* by providing an alternative publication venue.

Augustus Granville had pointed out the problems with the existing editorial system in 1830. He recognised merit in the argument that a committee 'conversant with all the infinite varieties of scientific subjects' might be the 'fairest' method of making decisions, since it prevented any individual judge exercising undue influence, but, from his close examination of the minute books, he pointed out the problems in practice. Some members of the committee had 'not the smallest pretension to any knowledge whatever of the subject under consideration'; and in some cases, thanks to Banks's willingness to embrace *dilettantes*, they lacked knowledge even of 'science in general'. The uneven availability of relevant competence was exacerbated by patchy attendance: meetings were 'sometimes' well-attended, but 'at other times, very few only are present'.[75] Unlike John Hill, 80 years earlier, Granville did not seek to ridicule the papers that had in fact been published in the *Transactions*, but he did claim to be 'lost in astonishment' that the effects of so 'clumsy' a system had 'not been more injurious, more ridiculous or more frequent'.[76]

Granville knew that the Committee of Papers already had the power to call upon additional expertise, and he wrote that, 'every communication *is supposed* to have been previously put into the hands and referred to the judgement of some competent member who reports his opinion'. The basis for the claim that refereeing should have been happening regularly is not entirely clear, but Granville's point was that it was not happening, and that 'the fate of a paper' was 'much oftener … committed to the chances of the mere yea-or-nay box' than referred to 'a competent judge'.[77]

During 1830, as Granville was carrying out his investigations, the Society's officers did, in fact, seek informed views on a handful of papers. For instance, in March, the Astronomer Royal recommended that an unnamed paper should be withdrawn and shortened,[78] and two months later, the director of the Royal Institution laboratory, Michael Faraday, had 'hastily looked' at some papers, and reported that one 'was worth nothing and the other is not I think fit for the *Transactions*' (Figure 8.4).[79] Faraday would look at another paper the following April.[80] These are the first recorded instances of refereeing since the 1780s, but it was still very sporadic and far from routine. It contrasted, for instance, with the procedures adopted by the Geological Society in 1817, under which all

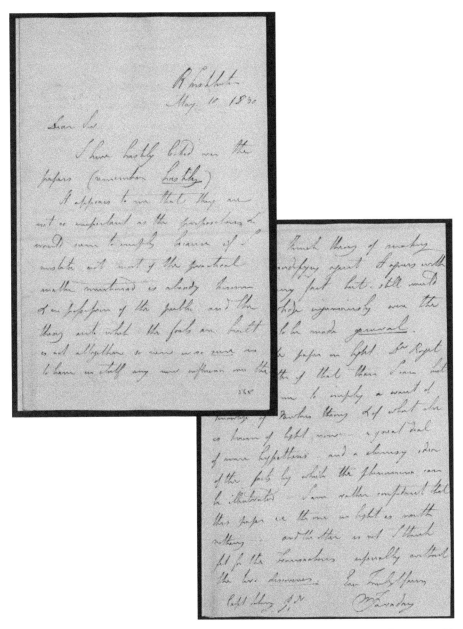

Figure 8.4 Michael Faraday conveyed his opinion on two unnamed papers, 10 May 1830 © The Royal Society.

papers were to be referred for written reports on their merits.[81] Granville himself thought that relying on the opinions of individuals could be 'vastly objectionable', and he suggested that the Royal Society should instead create standing committees for each broad area of science that, among other things, would possess the relevant subject expertise to make an informed recommendation to the Committee of Papers.[82]

Another possible mechanism for refereeing had been proposed by Babbage, Herschel and the other members of the 1827 committee, when they addressed the 'important task of deciding on what papers shall be published'.[83] They suggested that each paper should be 'referred to a separate Committee', apparently meaning the sort of small ad hoc committees that had been used by the Paris Académie since the eighteenth century, rather than a standing committee in the appropriate field. The Astronomical Society had adopted a version of this system in 1821, as Herschel would have known, but in practice its reports could simply amount to an oral expression of approval (or not) during a meeting of Council.[84] It may be with this in mind that the Royal Society's 1827 committee was explicit that its proposed reporting committees should 'have sufficient time given them to examine it carefully'; and that they should report 'not only their opinion, but the grounds on which that Opinion is formed' to the Committee of Papers. They also suggested that the committee be permitted to communicate directly with the author 'on any doubtful parts'.[85]

In the winter of 1831–2, the Royal Society began to experiment more extensively with refereeing. Referees would indeed be expected to take time to scrutinise papers, and to make 'a written report' of the grounds of their opinion.[86] The model initially adopted was to have jointly written reports, with pairs of Council members acting as ad hoc committees, but it differed from the other proposals by intending the reports for a public audience. The origins of this idea seem to lie in March 1831, when John Lubbock had reached out to Cambridge polymath William Whewell for suggestions for reforming the Society. Whewell had proposed that the Society might take on a more public role in reporting and evaluating recent research. He referred to the reports on the state of knowledge produced by Georges Cuvier and Jöns Jakob Berzelius in their roles as secretaries to the Paris and Stockholm academies, and to the occasional publication of the reports made by the Paris academy's *rapporteurs* on papers submitted, which were, said Whewell, 'often more interesting than the memoirs themselves'.[87]

The public evaluation of contributions would have implied a fundamental reconfiguration of the position that the Society never vouched for the truth or certainty of papers appearing in the *Transactions*. Only a few years earlier, John Herschel had described the idea of the Society's officers 'publicly commenting on the contents of papers' as 'something like indelicacy'.[88] Herschel was perfectly happy to communicate his thoughts on a paper by Faraday to the author himself, in a 'purely conversational' manner, but he felt that publishing those comments could 'establish a precedent that might prove very inconvenient in future'.[89] He was conscious of being a serving secretary of an organisation which had been insisting since 1752 that it did not make public assessments of value. In 1832, however, there was no evidence of such qualms among the Society's officers.

The first joint public reports were commissioned by the Committee of Papers in December 1831.[90] In his presidential address the following November, the Duke of Sussex would report that, 'for the greatest part of the last year', Council had adopted a plan 'to increase the usefulness and to uphold the credit of the Royal Society'. They had decided to 'allow no Paper to be printed in the *Transactions* ..., unless a written Report of its fitness shall have been previously made by one or more Members of the Council, to whom it shall have been especially referred for examination'.[91] Sussex shared Whewell's admiration for the value of the reports written by the Paris academicians, but also noted that refereeing helped distinguish 'the genuine cultivators' of science, whose contributions would be 'properly examined and appreciated by those who are most competent to judge', from those who had 'pretensions' to appear as men of science without just cause.[92] This experiment with joint public refereeing took place in the very same months that also saw the launch of the *Proceedings*. No clear link between the two projects is visible in the surviving record, but, as Sussex noted, the *Proceedings* provided a mechanism for publishing the reports, after they had been read at a Society meeting.

The intention to rely on members of Council to write the reports is significant. Sussex described the academicians who wrote the equivalent reports in Paris as 'veterans' of 'European reputation', who possessed 'an authority sufficient to establish at once the full importance of a discovery, to fix its relation to the existing mass of knowledge, and to define its probable effect upon the future progress of science'. He also claimed their position put them above 'personal feelings of rivalry or petty jealousy'.[93] For the Royal Society, drawing on Council members was an implicit vote of confidence in the scientific competence and authority of that Council. This form of refereeing would indeed provide for careful scrutiny of each paper, and a written evaluation of its merits, but it would not ensure that papers would be scrutinised by someone with specialist – rather than general – knowledge of its content.

Despite Sussex's hope that written refereeing would 'become a permanent law of the Council', only a handful of papers received joint-authored, public reports. One problem was, as Sussex acknowledged, that writing these reports increased 'the labours and responsibility' of Council members. He had hoped that a sense of duty to the 'scientific character' of the Society – and the country – would encourage them to make 'the occasional sacrifice both of time and labour'.[94] Yet Council only had 21 members, and every year, about 30 papers were accepted for the *Transactions* (not to mention the ones that were not approved). Moreover, the secretaries were already responsible for preparing abstracts of all the papers, and it had just been agreed that they would be made public in the *Proceedings*. Those abstracts had originated as summaries for the minute books, without the intention to contextualise and evaluate, but there was clear overlap between the two projects.

The idea of public evaluations also sat in potential tension with the Society's long-established tradition of gentlemanly politeness, particularly between fellows. As Sussex had acknowledged, it was only reports 'of a favourable nature' that could be printed.[95] Yet, if refereeing was also to keep down the pretensions of the undeserving, and protect the Society from publishing ridiculous or trivial papers, then frankness was more important than politely worded reports for public consumption. If refereeing was to encourage frank criticism, then its reports would need to be confidential.

A third problem lay in the pragmatics of evaluating and writing collaboratively. In early 1832, two refereeing teams were looking at papers: John Lubbock and William Whewell tried to evaluate a paper by George Airy, while secretary Samuel Hunter Christie and vice-president John Bostock considered one by Michael Faraday. Christie and Bostock found themselves in agreement that Faraday's demonstration that electricity could be 'excited' by magnetism was 'so important'.[96] As Csiszar has shown, however, Lubbock and Whewell found it more difficult to agree. Lubbock was uncomfortable with the way that Airy had applied the mathematics of Laplace and Lagrange, but Whewell saw this as little more than a difference of opinion that did not affect the overall merits of the paper. It took several rounds of correspondence before Lubbock agreed to put his name to a revised version of Whewell's initial report, acknowledging 'some peculiarities' in Airy's mathematical methods but agreeing that it was a 'valuable' contribution that demonstrated 'care ... in the numerical calculations' and 'sagacity'.[97] The difficulties of reaching consensus were a clear warning of just how much of a burden a full implementation of this vision for public reporting would have placed on Council members.

After 1832, written refereeing at the Royal Society became confidential, was usually done individually rather than collaboratively, and came to involve the wider fellowship not just Council.[98] The creation of a series of new discipline-specific committees, similar to those proposed by Granville, was key to widening the pool of referees. The first subject-specific scientific committees had been created in 1833 with the specific remit of advising on the award of the Royal Medals.[99] These medals were awarded annually, but as the field of science being recognised changed each year, the committee members were initially refreshed each year. From 1838, 'permanent Committees in each department of science' were created, covering the fields of mathematics, physics, chemistry, astronomy, meteorology, mineralogy and geology, botany and vegetable physiology, and zoology and animal physiology. They were known as the 'scientific committees' at the time; in a later incarnation, they would be known as 'sectional committees'. The revised editorial process – including referees and committees – is represented in Figure 8.5.

These scientific committees gave Council access to expert advice from 'such Members of the Society as have most devoted their attention to particular branches' of science, and provided a collective voice for different sub-communities within the Society.[100] As well as medal nominations, the committees acted as groups of subject experts 'to whom papers on that subject should be referred'.[101] This gave the Committee of Papers access to appropriate expertise and spread the burden of refereeing more widely. It also made it somewhat more difficult for authors to guess who was likely to have refereed their paper, though the pool of experts on any particular subject remained small. For instance, Sloane Despeaux has pointed out that there were just 21 mathematicians in the Royal Society in 1830.[102]

A set of resolutions from April 1839 reveal how this system was expected to operate. Printed copies of the secretaries' abstracts were to be circulated 'by the twopenny post [that is, within London] to all the Members of the Council, and of the Scientific Committees to which the Secretaries may judge the subject to belong, within reach of the said post'. The package of abstracts was to be accompanied by 'a circular letter' asking scientific committee members to give their opinions 'upon the propriety of printing' each paper at the next meeting of their committee. As well as reading the abstracts, committee members could visit the Society's rooms in Somerset House to consult the original manuscripts. The committee was also 'empowered to transmit' those manuscripts 'to a competent

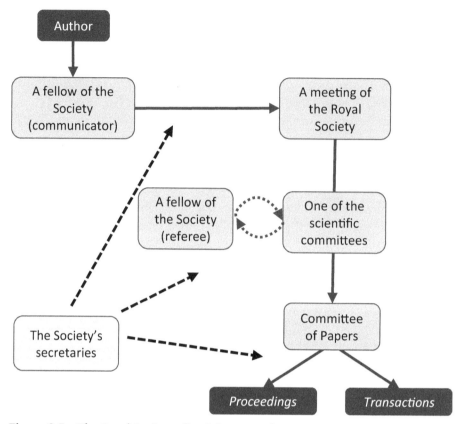

Figure 8.5 The Royal Society editorial system after 1838.

person … for the purpose of reference', which opened up the possibility of involvement by non-London fellows. The committees were to make recommendations to the Committee of Papers from some unspecified combination of their own opinions and the reports of referees.[103] It was the remit of the Society's secretaries to manage and coordinate all of this: sending titles and abstracts to the committee members; sending manuscripts to referees; receiving referee reports and sending them to the committee chairs; and conveying the recommendations to the Committee of Papers. Most of the scientific committees met in person only once or twice a year, but their chairmen became useful advisers to the secretaries, and the committee members often served as referees.

Both referees and scientific committees were incorporated into the Society's editorial processes in the 1830s and 1840s without any revision to the statutes: their role could be seen as a liberal interpretation of the existing right of the Committee of Papers to call for expert help when required. Provision was now made for close scrutiny by competent judges, and discussion of the merits of a paper, but this all happened outside (and before) the meeting of the Committee of Papers, which could, therefore, continue to operate in its traditional way, and remained the final authority.

Early refereeing in practice

During the 1830s, refereeing developed from an experimental practice into a familiar aspect of scholarly life for all those affiliated with the Royal Society, or similar learned societies. Two volumes of handwritten reports from referees survive in the Royal Society archives for the period 1830 to 1848 (Figure 8.6). They contain almost 300 referees' reports, with opinions on around 230 papers. The reports vary in length from a single paragraph to eight sides of letter paper, and occasionally more. About half of them recommend publication in the *Transactions*.

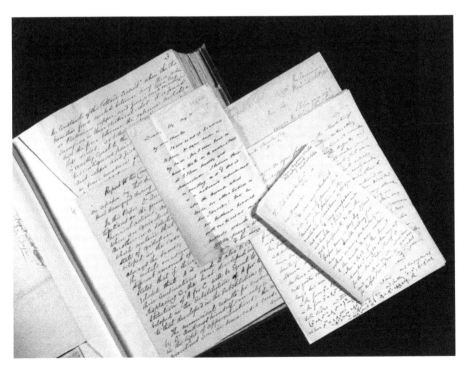

Figure 8.6 The volumes of surviving referees' reports from the 1830s and 1840s © The Royal Society.

Table 8.3 Most frequent referees in the period 1830–48

Name	Date elected to RS (+ service to RS in this period)	Occupation	Reports written
George B. Airy	1836	Professor of Astronomy, University of Cambridge	27
George Peacock	1818	Professor of Astronomy and Geometry, University of Cambridge	17
Richard Owen	1824	Conservator, Hunterian Museum of Royal College of Surgeons	14
Samuel H. Christie	1826 Secretary 1837–54	Professor of Mathematics, Royal Military Academy, Woolwich	11
John Bostock	1829	Lecturer in Chemistry, Guy's Hospital	10
Edward Sabine	1818 Secretary 1827–30; Vice-President 1839–41; Foreign Secretary 1845–50	Army officer	9
William H. Pepys	1808	Scientific instrument maker	8
John Herschel	1813	Independent astronomer	8
Thomas Graham	1836 Vice-President 1847	Professor of Chemistry, University College London	8
Robert B. Todd	1838	Professor of Physiology and Morbid Anatomy, King's College London	8

Even once refereeing spread beyond Council, most fellows were rarely involved. Just 15 men were responsible for writing a full half of the reports that survive (see Table 8.3).[104] This list of core and active referees includes no *dilettantes*, and is dominated by men who held professorships. Their areas of interest – particularly the dominance of mathematical physics, chemistry and physiology or anatomy – reflect the submissions received by the Society, and hint of a growing tendency for scholars writing on botany, zoology, geology or astronomy to communicate their work elsewhere, perhaps to the specialist societies in those fields. Notably, this group of referees was not entirely the 'veteran' authorities that the Duke of Sussex had anticipated: Peacock, Sabine, Pepys and Herschel did have seniority within the Society, but Graham and Todd were very recent recruits. So too was Airy, though his 1835 appointment as Astronomer Royal undoubtedly secured his authority.

There was a clear preponderance of London and Cambridge men among the most active referees. This reflected the dominant social networks within the Royal Society, but was also pragmatic. Bulky manuscripts, with their illustrations or appendices of data, could not at this time be transmitted with correspondence through the Royal Mail, so they either had to be collected in person – while attending a meeting of Council, for instance – or dispatched by some other mode. In 1843, for instance, John Phillips (a professor at King's College London) returned two papers 'by Mail Train ... this evening', having apparently taken them to his home in Yorkshire.[105] The separate transmission of correspondence and papers could lead to confusion: Lubbock complained, 'I have received a paper, from the clouds for aught I know, but I conjecture that it may have been sent to me ... for a Report.'[106] It is perhaps surprising that few papers appear to have been lost in transit, though there was an embarrassing month in 1848, when two different authors had to be asked to supply (if possible) a replacement copy of their paper, as the original had been 'mislaid' by 'the Gentlemen to whom it was referred'.[107]

These referees wrote their reports individually, but not necessarily independently. The practicalities of postal systems meant that, if opinions were requested from two referees, the first referee would send the paper directly to the second referee. In 1839, for instance, John Bostock sent a paper on the circulation of the blood on to the London comparative anatomist Richard Owen, and enclosed his own (critical) report along with an explicit request for Owen's opinion.[108] In such cases, a second referee might simply agree with the first report, or he might attempt to persuade his colleague. Thus, in 1842, Oxford mathematics professor Baden Powell saw merit in the 'general character' of a paper by S.M. Drach, but he deferred to the 'superior acquaintance with the details' possessed by Edinburgh natural philosophy professor James David Forbes, who thought some of the arguments were 'faulty'.[109] The case is unusual not only because of the dialogue between referees, but because both were outside London. The dates of the surviving reports – Powell: 20 April, Forbes: 23 May, and Powell: 8 June – give us a sense of how long it took for manuscripts to travel round the country and, probably more pertinently, for busy professors to find time to scrutinise them.

The Society's long-standing insistence on attaching a name to all papers read at its meetings was transferred into the refereeing system, and there was no attempt to conceal the identities of authors from referees. In contrast, referees' names were known only to those involved in the decision-making process, and for most of this period, were not even recorded in the minute books. In the mid-1840s, the Society's editorial processes would be attacked in the public press. Critics would point out that this one-sided secrecy left authors at the mercy of referees who might be 'full of envy, hatred, malice and all uncharitableness', and unable to 'soar above all personal feelings'. And with no way of knowing whether referees had appropriate expertise or not, or of assessing what would now be called conflicts of interest, how could secret judgements be trusted? After all, 'If the judgment be a righteous one, why not give us the clearest evidence of it? – why seek to shroud yourselves in the privacy of the Star-chamber, or the darkness of the Inquisition?'[110] This call for more openness in editorial decision-making would not be heeded. Confidential refereeing enabled frankness while hiding from the public the spectacle of fellows criticising the judgement of other fellows (whether as authors or communicators of papers). It came to be seen as a mechanism that allowed individual experts to speak on behalf of the wider community.[111]

Behind the scenes, refereeing was not always as confidential as it appeared and the secretaries would become an important conduit for conveying suggestions from referees to authors. Some referees saw their role primarily as judges, and wrote very brief recommendations. Some made recommendations for cuts that could be imposed by the Committee of Papers before printing. And others seem to have seen their role as somewhat more akin to Herschel's 'conversational manner': as an opportunity to engage with the content of the paper. Their reports could run to several sides of paper with their reactions, comments or suggested alterations, and took on the form of an implicit dialogue with the author, in which the secretaries acted as intermediaries to preserve the referee's anonymity.

There was a tricky balance to be found, however; revisions might improve the logic, clarity or fluency of the argument, but there was a risk that the paper would be so changed that it would no longer be a true representation of the oral paper presented to the Royal Society (and whose date was attached to the printed version). Thus, as a relatively new referee in 1838, George Airy was tentative in his suggested revisions, wondering whether the Committee of Papers would be willing 'to make representations to the author … regarding the addition or the removal of particular parts'.[112] Two years later, John Herschel recommended the publication of a paper on the photographic process, but noted that, 'I think the paper would be improved by the revision by the author of a few passages which I have marked'.[113] By 1843, with a dozen reports under his belt, Airy was making quite specific suggestions for revision, recommending, for instance, that Edward Sabine should express all the forces in his latest paper on terrestrial magnetism in the same units.[114] As Herschel's annotations imply, authors might be allowed to have their original manuscripts returned 'for the purpose of making various emendations in it', but they were supposed to return it to the Society afterwards.[115] This allowed the secretaries to check that no substantial changes – such as new observations or results – had been introduced and spuriously attributed to the original date of reading.

Referees were usually limited to evaluating the paper as presented by the author. Mathematical proofs could be checked, but referees who were unconvinced by experimental results or observations could do little more than express their unease and recommend against publication.[116] As Faraday had noted in 1831, 'I should be obliged to make experiments before I should feel satisfied. This I have not time to do nor should it be expected. The paper ought to tell the story clearly.'[117] Richard Owen did report on one occasion that he had taken the trouble to examine 'most of the [anatomical] preparations and parts described by the Author', to satisfy himself of 'the accuracy of his descriptions and figures', but this was unusual.[118] In contrast to the thorough investigation of truth claims that had been undertaken by special commissions of the Paris Académie in the late eighteenth century, refereeing was a paper-based practice of scrutiny.[119]

Refereeing initially operated without any formal evaluation criteria, and it relied heavily on tacit norms. As fellows of the Royal Society, all referees shared certain characteristics: they were, in this period, all male, white and sufficiently affluent to be able to pay the fees. There was some variety in their social backgrounds, but most of the core referees socialised together to some extent: for instance, they routinely met at the meetings of the various London scientific societies.[120] But the extent to which they agreed on what a publishable article might look like is far from clear.

The vocabulary of the surviving reports gives us some indication of what these early referees believed they were looking for. Papers were routinely recommended for publication in the *Transactions* on the grounds that they were 'valuable', 'important', 'worthy' and (occasionally) 'ingenious'. Such phrases as 'a real contribution' or 'a valuable addition' (or even 'an acceptable addition') indicate a sense that *Transactions* papers should add to existing knowledge. The search for a distinct contribution to knowledge can also be seen in the steady stream of complaints from referees about papers that contained nothing new: in 1830, Faraday had reported that, 'if I mistake not most of the practical matter mentioned is already known & in possession of the public'.[121] Sixteen years later, another referee would complain that a paper communicated by Faraday for its Swiss author also contained 'nothing which has not already reached the public through other channels'.[122] As another referee put it, a paper for the *Transactions* needed to 'present some novelty in method or result'.[123] And discovering that an author's ideas, however valuable, had been anticipated by another scholar was reason enough for a referee to change his mind.[124] Originality mattered just as much as it had to Joseph Banks (see Chapter 6).

Referees could be damning about papers that they regarded as poorly written, particularly if they were 'of great and unnecessary length'.[125] They condemned 'vague discussion',[126] and disliked papers that were 'redundant and pompous'.[127] Even papers that were recommended for acceptance could be improved by trimming excessive quotation or superfluous introductions. Perhaps the best early example is Adam Sedgwick's 1839 report on Charles Darwin's paper on the 'parallel roads' of Glen Roy, which went into finely detailed recommendations for improving a paper of which he essentially wholly approved.[128]

One of the most scathing critiques of prose style came from Peter Mark Roget, who waxed lyrical about the faults to be found in an 1848 paper on the movements of the intercostal muscles:

> His style is marked throughout by want of perspicuity & precision: his meaning is often obscured by a diffuseness & laxity of expression, rendering it difficult to follow the course of his argument. He employs terms in senses quite different from their common acceptation ... His orthography also is in many instances grossly incorrect.[129]

Roget's critique of the style, spelling and vocabulary of this paper is the more striking as this was a paper that had already been revised and abridged by its author.

Looking back from the 1870s, Granville would argue that the introduction of refereeing had led to a visible improvement in the quality of papers published in the *Transactions*: 'Who can deny that since the adoption of this plan ..., the value of that renowned collection of scientific memoirs has risen a hundred-fold in the estimation of all the continental savants and academies ..., as well as in that of the English lovers of science?'[130] At least part of this improvement was literary and rhetorical. Scholars have shown that, by the later decades of the nineteenth century, both the style of scientific writing and the structure of scientific research papers had come to be much more standardised than they had been in the days of Banks and Blagden.[131] The suggestions for revisions put forward by referees, coupled with authors' anticipation of likely scrutiny, were a powerful mechanism for this transformation.

Written refereeing had certainly become familiar and commonplace by the end of the 1840s, but it was far from universal. Between 1832 and 1848, 386 papers were published in full in the *Transactions*, yet the surviving referees' reports offer verdicts on only around 230 papers. Furthermore, of the surviving reports that make a clear recommendation, barely more than half recommended publication in the *Transactions*. It is entirely likely that many early reports have not survived, but these figures strongly suggest both that papers were not always sent to referees, and that papers could still be published in the *Transactions* without a referee report (despite what the Duke of Sussex had said in 1832).

The significance of refereeing in the 1830s and 1840s should not be overstated: it was only one element in the wider editorial system depicted in Figure 8.5. The Committee of Papers drew upon the opinions of the scientific committees and the secretaries, and in many cases, that input was felt to be sufficient to recognise certain papers as being clearly appropriate for either the *Transactions* or the *Proceedings*. Contributions by prominent or trusted authors with a track-record fell into the first category: papers by Richard Owen, for instance, seem to have had a smooth journey to the *Transactions*. And for many other papers, the secretaries who had laboured to produce the official abstracts would have had a good sense of how fully those abstracts represented the original paper. Of the items that ultimately appeared only in the *Proceedings*, some had been sent to referees for consideration for the *Transactions*, but most represent cases where the secretaries and committee chairmen felt that the abstract would be sufficient. It was for the cases when the secretaries were not sure, that referees were most needed.

Referees helped to police the boundary of the Society's new two-tier publishing system, helping to decide which of many adequate contributions already read at meetings and summarised in the *Proceedings* were (or were not) 'sufficiently important' to deserve 'a place in the Society's *Transactions*'.[132] That place needed to be deserved, because a *Transactions* paper represented both a financial commitment from the Society (because these papers were lengthy and well-illustrated), and a mark of prestige for the author and for the Society (because of the glory potentially reflected on the Society for having published important research). The refereeing process could be publicly represented as a mechanism for generating expert evaluation of research, but it also (internally) served to protect the Society's reputation and finances.[133] The growing significance of publication in the *Transactions* was built into the Society's administration: from 1840, authorship of a *Transactions* paper, but not one in the *Proceedings*, was seen as sufficient evidence of scientific contribution to justify a discount on the life membership fee, thus implicitly creating two classes of fellows, much as Babbage had hoped.[134]

In 1847, significant reforms of the Royal Society's statutes were finally enacted under the presidency of Spencer Compton, the Marquess of Northampton.[135] These included new procedures for electing fellows and Council members, and, in their insistence on scientific credentials and their limitation on the number of fellows elected each year, they are usually seen as the belated implementation of the ideas put forward around 1830 to transform the Society from a club of gentlemanly *dilettantes* into a respected group of scholarly men of science. In fact, as we have seen, some reforms had been quietly implemented well before 1847.

The Society's publications were entangled, both financially and editorially, with the wider reforms. The new attention to the Society's finances meant that the expenditure on printing and publishing came under closer scrutiny, while the desire to be able to identify distinguished men of science through their significant contributions to knowledge, lay behind the changes in editorial procedures. The other major influence on the Society's publications was the appointment of Richard Taylor as printer. Taylor created a route for collaboration between the independent periodical press and the learned societies, particularly through the printing of *Proceedings*-style periodicals.

The creation of the *Proceedings* enabled a gradual shift in the role and meaning of the *Transactions*: with the majority of papers recorded and made public as relatively short summaries in the *Proceedings*, the *Transactions* became increasingly selective. In the Banks era, about 75 per cent of papers read to the Society had been printed in the *Transactions*; in the period between 1835 and 1846, only marginally more than 40 per cent appeared there; and by the late 1850s, it was down to around 30 per cent.[136] *Transactions* papers had successfully navigated a complex set of social, intellectual and literary evaluation processes that marked them out from contributions to the *Proceedings* and, equally, to the independent periodical press.

Close scrutiny by referees was at the heart of that evaluation process. After experimenting with different versions of refereeing, a system of confidential written reports by individual referees had become familiar and accepted. By the late 1840s, a form of 'normal' practice emerged from the Society's social practices and tacit shared values. Referees routinely commented on whether the work they reviewed contributed something extra to what was known, and tried to assess the value, worth or importance of that contribution. They tried to spot errors in logic or mathematics. They preferred clarity to vagueness, and brevity to prolixity.

Refereeing did not reliably solve all the problems associated with the Society's publication regime prior to the 1830s: the complaints of secrecy in the Committee of Papers' ballots re-emerged as complaints about the secrecy of referees' reports in the mid-1840s, for instance. Nor did it avoid the danger of concentrating power in particular fields in the hands of one or two individuals, if they should happen to be trusted referees. George Airy and George Peacock held significant influence over the prospects of papers submitted in mathematical physics and astronomy, as did Richard Owen over all matters of anatomy or palaeontology. For a new generation of men of science, particularly those from different academic perspectives or backgrounds, the power exercised by these trusted insiders could be as difficult to navigate as that wielded by Banks in an earlier generation.

In 1852, for instance, the young Thomas Henry Huxley wrote to his sister about his hopes of getting a paper on cephalous molluscs into the *Transactions*. The problem was that it was highly likely to be 'referred to the judgment of my "particular friend"', that is, Owen. Huxley described Owen as having become such a 'great authority' over the previous two decades that he was 'determined not to let either me or any one else rise if he can help it'. Huxley was confident that, 'He won't be able to say a word against it, but he will pooh-pooh it to a dead certainty'. With the advantage of being recently elected to the fellowship himself, Huxley planned to 'manœuvre a little to get my poor memoir kept out of his hands' (and succeeded).[137]

Despite referees' repeated suggestions for shortening papers, the new editorial processes seem to have done little or nothing to address concerns with the cost of publishing. The *Transactions* might be more selective, but, after some lean years in the early 1840s, the total output of the Society was growing. In the 1820s, the Society had published around

460 pages in the *Transactions* each year; in the 1850s, that would become almost 700 pages a year, plus the *Proceedings*. By 1852, the treasurer would report that publishing costs were 'much beyond what is usual' and argued that, 'The remedy is obvious; – the selection of papers for the *Transactions* should have reference to the pecuniary means at the disposal of the Council, as well as to the merits of the several communications.'[138] Finding a mechanism for balancing an evaluation of production costs with intellectual merit would become an ongoing problem for the Society.

One curiosity of the 1847 reforms is that the scientific committees were not formally adopted in the new statutes. The committees were still occasionally meeting at that point, but, judging by their minute books, they seem to have disappeared by late 1849.[139] In their absence, the burden of selecting and managing referees would fall more heavily upon the Society's secretaries than at any point since 1752. The Society's secretaries once more became (de facto) editors, but in an organisational and scientific context very different from that of the early eighteenth-century secretary-editors.

Notes

1 Augustus Bozzi Granville, *Autobiography of A.B. Granville; being eighty-eight years of the life of a physician. Edited with a brief account of the last years of his life*, ed. Paulina Granville, 2 vols (London: H.S. King, 1874), vol. 2, 217.

2 Jack Morrell and Arnold Thackray, *Gentlemen of Science: Early years of the British Association for the Advancement of Science* (Oxford: Oxford University Press, 1981); A.J. Meadows, *Victorian Scientist: The growth of a profession* (London: British Library, 2004).

3 See Jan Golinski, 'Humphry Davy: The experimental self', *Eighteenth-Century Studies* 45, no. 1 (2011): 15–28; Frank A.J.L. James and Sharon Ruston, 'New studies on Humphry Davy: Introduction,' *Ambix* 66, nos. 2–3 (2019): 95–102; Tim Fulford and Sharon Ruston, eds, *The Collected Letters of Sir Humphry Davy* (Oxford: Oxford University Press, 2020), Intro.

4 D.P. Miller, 'Davies Gilbert [formerly Giddy]', in *ODNB*.

5 Charles Babbage, *Reflections on the Decline of Science in England* (London: Fellowes, 1830). On the book, see James A. Secord, *Visions of Science: Books and readers at the dawn of the Victorian age* (Oxford: Oxford University Press, 2015), ch. 2.

6 Jonathan R. Topham, 'A textbook revolution', in *Books and the Sciences in History*, ed. Marina Frasca-Spada and Nicholas Jardine (Cambridge: Cambridge University Press, 2000), 317–37; David Philip Miller, 'Between hostile camps: Davy's presidency of the Royal Society, 1820–1827', *British Journal for the History of Science* 16 (1983): 1–48; William J. Ashworth, 'The calculating eye: Baily, Herschel, Babbage and the business of astronomy', *British Journal for the History of Science* 27, no. 4 (1994): 409–41.

7 RS CMO/10, 1 March 1827.

8 Minutes of the Committee for Limiting Membership, 11 June 1827, RS CMB/1/20/2.

9 Babbage, *Decline of Science in England*, 155.

10 Granville, *Autobiography*, vol. 2, 218. The book was [Granville], *Science Without a Head; or, the Royal Society Dissected. By One of the 687 F.R.S.* (London: T. Ridgway, 1830). A subsequent, expanded edition carried his name: Augustus Bozzi Granville, *The Royal Society in the XIXth Century; being a statistical summary of its labours during the last thirty-five years* (London: Printed for the author; sold by John Churchill, 1836). On Granville, see Ornella Muscucci, 'Augustus Bozzi Granville', in *ODNB*.

11 [Granville], *Science Without a Head*, 116.

12 Granville, *Autobiography*, vol. 2, 218.

13 He summarised his key concerns in Granville, *Autobiography*, vol. 2, 217–20 (quotation on p. 219).
14 On the 1830 campaigns and their immediate aftermath, see Marie Boas Hall, *All Scientists Now: The Royal Society in the nineteenth century* (Cambridge: Cambridge University Press, 2002), 58–66. Also, Roy M. MacLeod, 'Whigs and savants: Reflections on the reform movement in the Royal Society, 1830–48', in *Metropolis and Province: Science in British culture, 1780–1850*, ed. Ian Inkster and J.B. Morrell (London: Hutchinson, 1983).
15 RS CMO/10, 22 March 1827.
16 RS CMB/1/20/1, 24 May 1827.
17 Babbage, *Decline of Science in England*, 155–6.
18 [Granville], *Science Without a Head*, 34–49.
19 Babbage's and Granville's use of authorship as a metric for scientific reputation is discussed in Alex Csiszar, *The Scientific Journal: Authorship and the politics of knowledge in the nineteenth century* (Chicago, IL: University of Chicago Press, 2018), 121–9. On publication lists and reputation, see also Alex Csiszar, 'How lives became lists and scientific papers became data: Cataloguing authorship during the nineteenth century', *British Journal for the History of Science* 50 (2017): 23–60.
20 Committee for Printing Philosophical Transactions, 25 and 28 January 1828, RS CMB/1/24/1 (quotation from 25 January; Nicol attended on 28 January).
21 'Circular to the printers', [n.d., early February 1828], RS DM/1/85.
22 'Memorandum of questions to printers', [n.d., late January 1828], RS DM/1/96 (an abbreviated version is at RS DM/1/91).
23 The other printers were Priestley & Weal, James Moyes and R. Watts. RS DM/1/95 Memorandum of printing estimates; RS DM/1/94 Memorandum of Watts' printing estimates.
24 Alexis Weedon, 'William Clowes (1779–1847), printer', in *ODNB*.
25 William H. Brock and A.J. Meadows, *The Lamp of Learning: Taylor & Francis and the development of science publishing* (London: Taylor & Francis, revised edn 1998).
26 Memo of printing charge estimates, [13 February 1828], RS DM/1/95. This document compares the charges of Clowes, Taylor and Moyes. The estimate from Watt came in late, and is summarised in 'Memorandum of printing estimates', 14 February 1828, RS DM/1/94.
27 Print run when Taylor took over: RS CMO/10, 28 February 1828. Print run reduced to 800: RS CMP/07, 20 January 1898.
28 St Bride Library, Taylor & Francis archive [hereafter T&F], Taylor Journal, 1830–40.
29 T&F Taylor Journal, 1830–40 (these entries are from the Royal Society account, January to June 1832).
30 On the importance of this type of printing in the eighteenth century, see James Raven, *Publishing Business in Eighteenth-Century England* (London: Boydell & Brewer, 2014).
31 RS CMO/11, 14 May 1829.
32 T&F Taylor Journal Book 1830–40, December 1832.
33 'Sale of Phil Trans and Abstracts' for August 1835, in T&F Income and Expense Account, 1835–55.
34 On the rise of *Proceedings*-journals, see Csiszar, *The Scientific Journal*, ch. 2 and 'Proceedings and the public: How a commercial genre transformed science', in *Science Periodicals in Nineteenth-Century Britain: Constructing scientific communities*, ed. Gowan Dawson, Bernard Lightman, Sally Shuttleworth and Jonathan R. Topham (Chicago, IL: University of Chicago Press, 2020), 103–34.
35 RS CMO/07, 3 April 1783.
36 See pp. 108–9 in Alex Csiszar, 'Proceedings and the public'.
37 Babbage, *Decline of Science in England*, 57. On Brande's *Quarterly Journal*, see Anna M. Gielas, 'Early sole editorship in the Holy Roman Empire and Britain, 1770s–1830s' (PhD, University of St Andrews, 2019), ch. 6.
38 RS CMO/10, 28 February 1828.
39 RS CMO/11, 11 March 1830.
40 Csiszar, 'Proceedings and the public', 110.
41 Csiszar, *The Scientific Journal*, 76–9; and 'Proceedings and the public'.
42 RS CMO/11, 16 December 1830.
43 For the printing of the first issue of the *Proceedings*, see entry for 25 February 1831, in T&F Taylor Journal Book 1830–40. Taylor initially quoted for 500 copies, see 'Estimate of the

44 'Estimate of the costs ... from R. Taylor', [c. January 1831], in RS DM/1/97.
45 RS CMO/11, 17 March 1831, 12 April 1831, 5 May 1831.
46 RS CMO/11, 10 May 1832, and the entry for the same date in T&F Taylor Journal 1830–40.
47 For reprinting of issues 1–10, see entry for 21 March 1833 in T&F Taylor Journal 1830–40. For the increased print run, see entry for [undated] April 1833, in T&F Taylor Journal 1830–40.
48 Babbage, *Decline of Science in England*, 197.
49 'Report of the council to the Anniversary Meeting on St. Andrew's Day, 1832', *Abstracts of the Papers Printed in the Philosophical Transactions of the Royal Society of London* [i.e. *Proc*] 3 (1830–37): 155–8, 156.
50 Printed flyer advertising the retrospective abstracts, 13 March 1832, RS DM/1/98.
51 J.C. Robertson, in *Mechanics' Magazine* 21 (1834), quoted in Csiszar, 'Proceedings and the public', 111–12.
52 RS CMO/10, 28 February 1828.
53 Csiszar, 'Proceedings and the public', 117–21.
54 Some of the moderate reforms he was able to promote were reported in 'Report of the council to the Anniversary meeting on St. Andrew's Day, 1831,' *Abstracts of the Papers* [i.e. *Proc*] 3 (1830–7): 85–90. More generally, see Timothy L. Alborn, 'Sir John William Lubbock, third baronet (1803–1865)', in *ODNB*.
55 *The Record of the Royal Society of London* (3rd edn) (London: The Royal Society, 1912), 270. The earliest surviving minutes appear to be from 1834. See also Henry Lyons, 'The Society's finances Part II: 1831–1938', *N&R* 2.1 (1939): 47–67.
56 RS CMP/01, 14 November 1833. 'Ordinary' expenditure excluded one-off projects, such as the retrospective abstracts. There was no separate accounting for the *Proceedings*, which was surely included under Lubbock's heading for the *Transactions* (as it was in Taylor's bills to the Society).
57 Minutes of the Finance Committee, RS CMB/42, 2 January 1834.
58 Council referred the paper quality to an ad hoc 'printing committee', see RS CMP/01, 6 February 1834 and RS CMB/42, 15 February 1834.
59 See 'The last five years', in Report of the Charter Committee, in RS CMP/01, 18 June 1846; and 'The last ten years', in Report of the Finance Committee, in RS CMP/02, 14 January 1847. These are summarised in Fyfe, 'Journals, learned societies and money: *Philosophical Transactions, ca.* 1750–1900', *N&R* 69.3 (2015): 277–99, table 1.
60 1827 prices from RS DM/1/83; 1832 prices from T&F Income & Expense Account 1835 [*sic*].
61 For the mid-1820s sales income, see RS DM/1/83.
62 In the 1830s, sales income was reported in the annual accounts and published in the *Proceedings*; an analysis for 1838–47 was included in a Finance Committee report, in RS CMP/02, 10 December 1846.
63 Undated circulation figures [before 12 February 1846], RS CMB/86/A. The figures cover the years 1835 to 1844, so presumably date from 1845 or early 1846.
64 Aileen Fyfe, 'The Royal Society and the noncommercial circulation of knowledge', in *Reassembling Scholarly Communications: Histories, infrastructures, and global politics of open access*, ed. Martin Paul Eve and Jonathan Gray (Cambridge, MA: MIT Press, 2020), 147–60.
65 Finance Committee Review, RS CMB/86/A [before 12 February 1846].
66 Entry for September 1844 in T&F Taylor Journal 1839–44, fol. 239.
67 First meeting of that committee, 12 December 1846, RS CMB/86/A.
68 RS CMB/86/A, 12 December 1846.
69 RS CMB/86/A, 14 December 1846.
70 RS CMB/86/A, 12 December and 17 December 1846.
71 RS CMB/86/A, 12 December 1846. On changing illustrative technologies in scientific periodicals, see Jonathan R Topham, 'Redrawing the image of science: Technologies of illustration and the audiences for scientific periodicals in Britain, 1790–1840', in *Science Periodicals in Nineteenth-Century Britain: Constructing scientific communities*, ed. Gowan Dawson, Bernard Lightman, Sally Shuttleworth and Jonathan R. Topham (Chicago, IL: University of Chicago Press, 2020), 65–102.

72 RS CMB/86/A, 12 December 1846.
73 On the Basire family and the Royal Society, see Richard Goddard, *'Drawing on Copper': The Basire family of copper-plate engravers and their works* (Maastricht: Maastricht University Press, 2016), 286–7 and 296. The first lithograph in the *Transactions* had appeared in 1832, drawn and printed by Charles Hullmandel's firm, but from 1834 to 1847, James Basire III provided all the *Transactions* lithographs, see *Drawing on Copper*, 287.
74 Sabine to Rosse, recorded in RS CMP/02, 19 February 1852.
75 [Granville], *Science Without a Head*, 53–4
76 [Granville], *Science Without a Head*, 54–5.
77 [Granville], *Science Without a Head*, 54 (emphasis on 'is supposed' added by us).
78 John Pond, on an unnamed paper by Joseph Jackson Lister, 11 March 1830, RS RR/1/152.
79 Michael Faraday to Edward Sabine, on unnamed papers, 10 May 1830, RS RR/1/165.
80 Michael Faraday, on a paper 'On a new combination of chlorine and nitrous gas' by Edmund Davy, 15 April 1831, RS RR/1/49.
81 Csiszar, *The Scientific Journal*, 134–8.
82 [Granville], *Science Without a Head*, 54, 84–8.
83 Report of the Committee for Limiting Membership, 11 June 1827, RS CMB/1/20/2.
84 Csiszar, *The Scientific Journal*, 134–8.
85 Report of the Committee for Limiting Membership, 11 June 1827, RS CMB/1/20/2.
86 By December 1831, the requirement to report in writing was explicit, see RS CMB/90/C, 8 December 1831, 43.
87 Whewell to Roget, 22 March 1831, in RS DM/1/30; also Whewell to Lubbock, 12 March 1831, in RS JWL/41. Whewell's proposal, and its troubled implementation, has been discussed in Alex Csiszar, 'Peer review: Troubled from the start', *Nature* 532, no. 7599 (2016): 306–8 and *The Scientific Journal*, 142–6.
88 Herschel to Faraday, 8 June 1826, in Frank A.J.L. James, ed., *The Correspondence of Michael Faraday*, 6 vols (London: Institution of Electrical Engineers, 1991–2012), vol. 1, Letter 299. The exchange began with Faraday sending the paper to Herschel on 26 May 1826.
89 Herschel to Faraday, 8 June 1826, in James, *Faraday Correspondence*, vol. 1, Letter 299.
90 RS CMB/90/C, 8 December 1831, 43.
91 Frederick Augustus Duke of Sussex, '[Presidential Address 1832]', *Abstracts of the Papers* [i.e. *Proc*] 3 (1830–1837): 140–55, 141.
92 'Presidential Address 1832', 142.
93 'Presidential Address 1832', 142.
94 'Presidential Address 1832', 141–2.
95 'Presidential Address 1832', 141.
96 Hunter and Bostock reported to the Committee of Papers on 9 February 1832; their report was read to a general meeting of the Society on 5 April 1832 and it was then printed in the *Proceedings*, as Samuel Hunter Christie and John Bostock, '[Report on Faraday's] experimental researches in electricity', *Abstracts of the Papers* [i.e. *Proc*] 3 (1830–37): 113–19. See also Ronald Anderson, 'The referees' assessment of Faraday's electromagnetic induction paper of 1831', *N&R* 47.2 (1993): 243–56.
97 William Whewell and John William Lubbock, '[Report on Airy's] On an inequality of long period in the motions of the Earth and Venus', *Abstracts of the Papers* [i.e. *Proc*] 3 (1830–37): 108–13, 111 and 113. See also Alex Csiszar, 'Objectivities in print', in *Objectivity in Science: New perspectives in science and technology studies*, ed. Flavia Padovani, Alan Richardson, and Jonathan Y. Tsou (Dordrecht: Springer, 2015), 145–69, esp. 155–6; 'Peer review'; and *The Scientific Journal*, 143–6.
98 On mathematical refereeing, see Sloan Evans Despeaux, 'Fit to print? Referee reports on mathematics for the nineteenth-century journals of the Royal Society of London', *N&R* 65.3 (2011): 233–52.
99 RS CMP/01, 23 March 1833. On the Royal Medals, see RS CMO/10, 11 November 1826 (award of the first Royal Medals); on their establishment, see Hall, *All Scientists Now*, 28–9.
100 RS CMP/01, 31 May 1838.
101 RS CMP/01, 8 March 1838 (initial proposal). For subsequent amendments, see RS CMP/01, 29 March, 10 May, 31 May, 13 July 1838.
102 Despeaux, 'Fit to print?', 12–13.
103 RS CMP/01, 14 March 1839.

104 This trend – of a small number of very active referees – remained true into the late nineteenth century, see Aileen Fyfe, Flaminio Squazzoni, Didier Torny, and Pierpaolo Dondio, 'Managing the growth of peer review at the Royal Society journals, 1865–1965', *Science, Technology and Human Values* 45, no. 3 (2020): 405–29.

105 Phillips to Weld, [n.d.] June 184[3?], RS RR/1/103. The papers were by S.C. Homersham and J.F. Miller.

106 John Lubbock to Peter Mark Roget, 15 May 1846, in RS RR/1/76. The paper was by the eccentric naturalist and traveller, Charles Waterton, and Lubbock was not impressed.

107 RS CMP/02, 20 July 1848.

108 Bostock, on a paper 'On the motion of the blood' by James Carson, 17 June 1839, RS RR 1/39.

109 Baden Powell, on a paper 'On the diurnal temperature of the Earth's surface …' by S.M. Drach, 8 June 1842, RS RR/1bis/20; and J.D. Forbes, report on the same paper, 23 May 1842, RS RR/1bis/18. (See also the original report by Baden Powell, 20 April 1842, RS RR/1bis/18.)

110 *Wade's London Review* 1 (1845): 583, quoted in Csiszar, *The Scientific Journal*, 153. Csiszar discusses several critiques of the Society's processes in the years 1843–5.

111 Csiszar, *The Scientific Journal*, 152.

112 George Biddle Airy, on a paper 'Of such ellipsoids consisting of homogeneous matter …' by James Ivory, 10 February 1838, RS RR/1/120.

113 John F.W. Herschel, on a paper 'On the influence of iodine in rendering several Argentine compounds, spread on paper, sensitive to light …' by Robert Hunt [1840], RS RR/1/110.

114 George Biddle Airy, on a paper 'Contributions to terrestrial magnetism, No. IV' by Edward Sabine, 26 May 1843, RS RR/1bis/44.

115 Peter Mark Roget, on a paper 'Researches on the function of the intercostal muscles …' by John Hutchinson, 31 July 1848, RS RR/1/111.

116 On mathematical refereeing, see Despeaux, 'Fit to print?'

117 Michael Faraday, on a paper 'On a new combination of chlorine and nitrous gas' by Edmund Davy, 15 April 1831, RS RR/1/49.

118 Richard Owen, on a paper 'On the organ of hearing in crustacea' by Arthur Farre, 16 September 1843, RS RR/1/63.

119 For instance, the investigation into Anton Mesmer's claims, see Arthur Donovan, *Antoine Lavoisier: Science, administration and revolution* (Cambridge: Cambridge University Press, 1993); Robert Darnton, *Mesmerism and the End of the Enlightenment in France* (Cambridge, MA: Harvard University Press, 1968; new edn 1995).

120 On scientific sociability, see Hannah Gay and John W. Gay, 'Brothers in science: Science and fraternal culture in nineteenth-century Britain', *History of Science* 35 (1997): 425–53; James A. Secord, *Victorian Sensation: The extraordinary publication, reception and secret authorship of vestiges of the natural history of creation* (Chicago, IL: University of Chicago Press, 2000), ch. 6.

121 Michael Faraday on an unnamed paper by an unnamed author, 10 May 1830, RS RR/1/165.

122 Thomas Graham, on a paper 'On the production of ozone by chemical means' by C.F. Shoenbein, in a letter to Michael Faraday, 27 October 1846, RS RR/1/222.

123 George Peacock, on a paper 'On factorial expressions …' by [?] Tate, 29 April 1843, RS RR/1bis/48.

124 Letter from Robert Bentley Todd, re a paper 'Researches on the function of the intercostal muscles …' by John Hutchinson, 24 October 1848, RS RR/1/112, in which he withdraws his report of 18 July 1848, RS RR/1/113 [*sic*].

125 John Bishop, on a paper 'Researches on the function of the intercostal muscles …' by John Hutchinson, 25 May 1848, RS RR/1/118.

126 George Biddle Airy, on a paper 'On the construction and use of single achromatic eye-pieces …' by Joseph Bancroft Reade, 25 February 1840, RS RR/1/191.

127 William Henry Smyth, on a paper 'New ideas on electricity, applied to the invention of an earthquake guard' by [?] Mancini, 4 May 1838, RS RR/1/160.

128 Adam Sedgwick, on a paper 'Observations on the parallel roads of Glen Roy …' by Charles Darwin, 26 March 1839, RS RR/1/46.

129 Peter Mark Roget, on a paper 'Researches on the function of the intercostal muscles …' by John Hutchinson, 31 July 1848, RS RR/1/111.

130 Granville, *Autobiography*, vol. 2, 220.
131 Dwight Atkinson, *Scientific Discourse in Sociohistorical: The Philosophical Transactions of the Royal Society of London, 1675–1975* (London: Routledge, 1998); Alan G. Gross, Joseph E. Harmon, and Michael S. Reidy, *Communicating Science: The scientific article from the seventeenth century to the present* (New York: Oxford University Press, 2002), chs 6–7.
132 Robert Bentley Todd, on a paper 'Microscopical examination of the contents the hepatic ducts …' by Thomas Wharton Jones, 29 July 1848, RS RR/1/126.
133 Noah Moxham and Aileen Fyfe, 'Royal Society and the prehistory of peer review, 1665–1965', *Historical Journal* 61, no. 4 (2018): 863–89.
134 Fellows could either pay annual fees, or 'compound' their future fees by paying a hefty £60; this 'compounded' fee was reduced to £40 for those with a *Transactions* paper. See *Record of the Royal Society* (1912), 170. This bias towards the *Transactions* was discontinued in 1887, *Record of the Royal Society* (1912), 275.
135 Hall, *All Scientists Now*, chs 3–4.
136 For the Banks era, the figure comes from our manual count of verdicts of the Committee of Papers, RS CMB/90/2. The publication rate for 1835–46 was calculated by the treasurer in 1852 (because he was concerned about a worrying increase in papers approved for the *Transactions* in the intervening years), see Sabine (treasurer) to Rosse (president), in RS CMP/02, 19 February 1852. Publication rates for the later 1850s come from analysis of our digital database of the 'Register of Papers'. Cf. rates calculated from samples by Atkinson, *Scientific Discourse in Sociohistorical Context*, 33–4, and 39–40.
137 T.H. Huxley to Mrs Elizabeth Scott, 5 March 1852, at *The Huxley File*, at https://mathcs.clarku.edu/huxley/letters/52.html. His paper on cephalous mollusca was refereed by Edward Forbes (27 July 1852) and Thomas Bell (n.d. August 1852), RS RR/2/113–14.
138 Sabine to Rosse, in RS CMP/02, 19 February 1852.
139 The minute books of the Astronomical, Mathematical and Geological Committees all record their last respective meetings around April 1849, see for example RS CMB/283, 284, 288, respectively. Baldwin has suggested that controversy surrounding the award of the Royal Medal may have led to their discontinuance, see Melinda Baldwin, 'Tyndall and Stokes: Correspondence, referee reports and the physical sciences in Victorian Britain', in *The Age of Scientific Naturalism: Tyndall and his contemporaries*, ed. Bernard Lightman and Michael S. Reidy (London: Pickering & Chatto, 2014), 171–86, 176.

9
Editing the journals, 1850s–1870s
Julie McDougall-Waters and Aileen Fyfe

In April 1851, William Thomson asked his friend and fellow mathematical physicist, George Gabriel Stokes, whether he might be tempted to 'take a turn at Editorship?' Thomson was looking for someone to succeed him as editor of the *Cambridge and Dublin Mathematical Journal*. Stokes was one of the group of 'mathematical friends' who assisted Thomson editorially, 'by giving me reports on papers', and, as the Lucasian professor of mathematics, he was firmly based at Cambridge. Thomson assured Stokes that the editorial duties were 'not on the whole onerous', not least because of the help of his friends. Thomson also assured Stokes that the editor bore no 'pecuniary risk', despite the publisher's concern about growing losses. Nonetheless, Stokes declined the offer.[1]

Just a few years later, however, in 1854, George Stokes became one of the secretaries of the Royal Society. He would spend the next 31 years immersed in the editorial responsibilities for the *Philosophical Transactions* and the *Proceedings*. This was, however, a somewhat different matter than editing the *Mathematical Journal*. The Royal Society's vision for the *Transactions* as a valuable means for circulating knowledge meant that there was no pressure to make a profit. And rather than Thomson's informal but 'sufficient *council*' of friends, Stokes would work within formal structures that gave him assistance in making decisions, as well as collective support for those decisions. This was all the more necessary as the *Transactions* ranged over a far wider array of subjects than the *Mathematical Journal*, and Stokes would find himself dealing with papers on comparative anatomy, chemistry and botany as well as mathematical physics. Contemporaries often assumed that Stokes was responsible only for the papers in the physical sciences, and left the biological sciences to his co-secretary, but, as Stokes insisted in 1884, 'I am editor of the *Transactions*'.[2]

Editorial work became central to Stokes's scientific identity. His own research output diminished, but he remained actively involved in developing the work of others through his comments as editor and referee. As his secretarial colleague Michael Foster later remarked, 'It is indeed difficult to say how much science gained through Stokes' secretaryship by his editorial influence on the work of others, stopping that which was not fit to appear, and moulding a crude, imperfect effort into something worthy of being made known.'[3]

Stokes and Thomson were part of the new generation of fellows admitted to the Royal Society after the reforms of 1847. The changes were intended to turn the Society into a community dominated by scientifically active researchers, rather than the wealthy *dilettantes* of an earlier age.[4] From 1847, the number of new fellows admitted annually was limited, and candidates were expected to demonstrate suitable qualifications. A candidate might be described as 'distinguished for his acquaintance with the science of …' or 'eminent as …', but he was more likely to be described as the 'discoverer of …', the 'inventor or improver of …' or, most likely of all, the 'author of …'.[5]

This was the period in which the publication of research became important, not just to the construction and circulation of knowledge, but to the careers of men of science. Charles Babbage had argued that active fellows could be recognised by dint of having published in the *Transactions*, but it is clear from election certificates of the 1850s that the *Transactions* was not the only periodical in which a successful candidate might prove his expertise. In 1851, William Thomson was described as the author of 'several Mathematical Papers in the Cambridge & Dublin Mathematical Journal, of which Journal he is Editor', as well as two papers in the *Transactions*.[6] Stokes was credited on his certificate (see Figure 9.1) with a paper in the *Transactions*, as well as having 'communicated to the *Cambridge Philosophical Transactions* and to the *Cambridge and Dublin Mathematical Journal* various papers, too numerous to specify in detail'. He was also the 'author of a report … and other communications' to the British Association for the Advancement of Science.[7] Other fellows elected in 1851 included the naturalist Thomas Henry Huxley (one paper in the *Transactions* and 'other papers'),[8] the physiologist James Paget (one paper in the *Transactions*, 'numerous papers' in the *Medico-chirurgical Transactions* and 'other publications')[9] and the organic chemist Wilhelm Hofmann

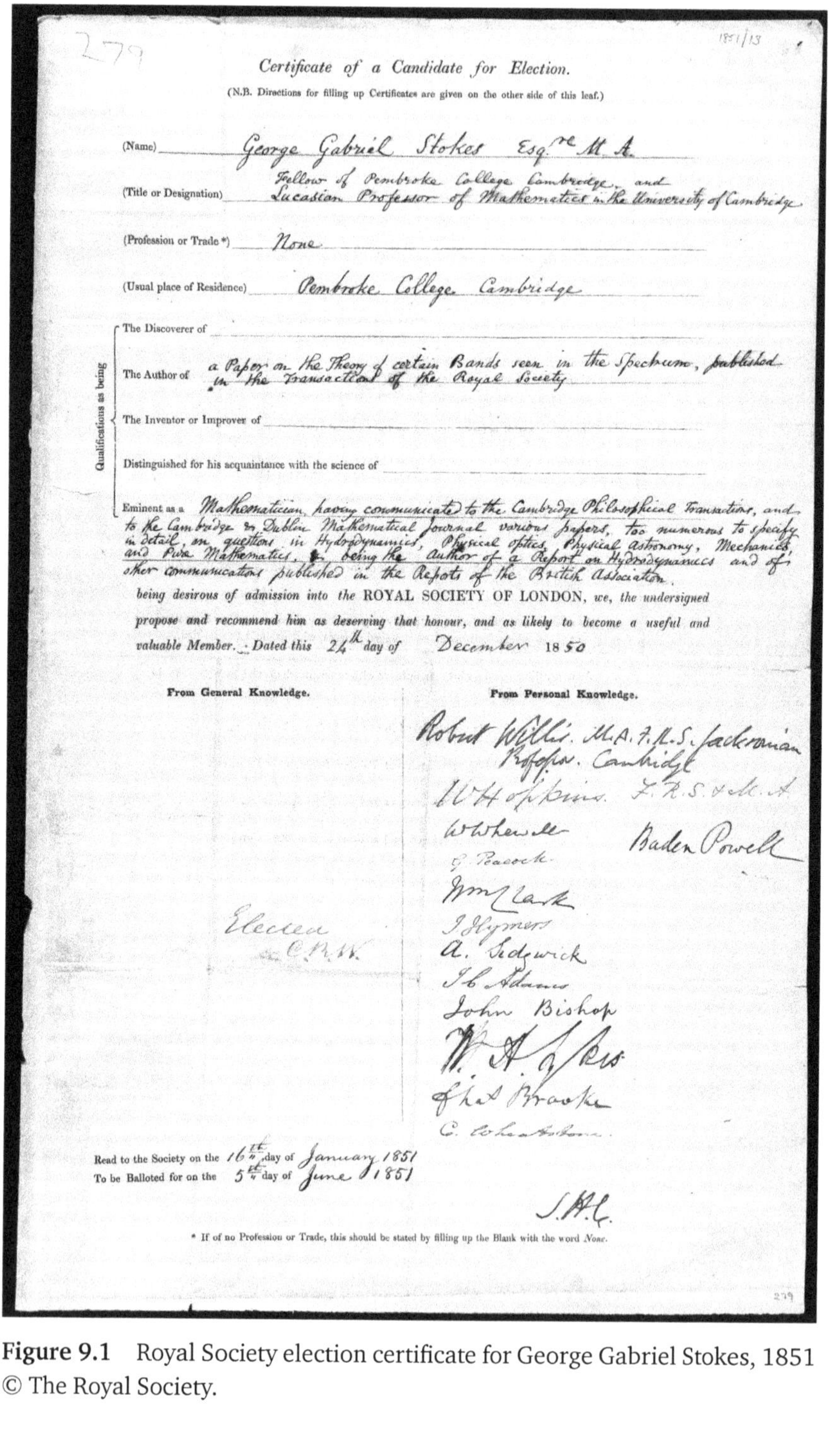

279

1851/13

Certificate of a Candidate for Election.

(N.B. Directions for filling up Certificates are given on the other side of this leaf.)

(Name) George Gabriel Stokes Esq.re M.A.

(Title or Designation) Fellow of Pembroke College Cambridge, and Lucasian Professor of Mathematics in the University of Cambridge

(Profession or Trade *) None

(Usual place of Residence) Pembroke College Cambridge

Qualifications as being

The Discoverer of

The Author of a Paper on the Theory of certain Bands seen in the Spectrum, published in the Transactions of the Royal Society

The Inventor or Improver of

Distinguished for his acquaintance with the science of

Eminent as a Mathematician, having communicated to the Cambridge Philosophical Transactions, and to the Cambridge & Dublin Mathematical Journal various papers, too numerous to specify in detail, on questions in Hydrodynamics, Physical optics, Physical astronomy, Mechanics, and Pure Mathematics, & being the author of a Report on Hydrodynamics and of other communications published in the Reports of the British Association.

being desirous of admission into the ROYAL SOCIETY OF LONDON, *we, the undersigned propose and recommend him as deserving that honour, and as likely to become a useful and valuable Member. Dated this* 24th *day of* December 1850

From General Knowledge.	From Personal Knowledge.
Elected C.R.W.	Robert Willis. M.A. F.R.S. Jacksonian Professor. Cambridge
	W Hopkins. F.R.S. & M.A.
	W Whewell
	Baden Powell
	G. Peacock
	Wm Clark
	J Hymers
	A. Sedgwick
	J C Adams
	John Bishop
	W. H. Sykes
	Chas Brooke
	C. Wheatstone

Read to the Society on the 16th day of January 1851
To be Balloted for on the 5th day of June 1851

S.H.C.

* If of no Profession or Trade, this should be stated by filling up the Blank with the word *None.*

279

Figure 9.1 Royal Society election certificate for George Gabriel Stokes, 1851 © The Royal Society.

(a 'series of memoirs' published variously in the *Transactions*, Liebig's *Annalen* and the journal of the Chemical Society).[10]

One of the other key effects of the 1847 reforms was to increase the power of Council relative to the president. Never again would there be a president as long-serving as Joseph Banks; and the revised regulations for the election and turnover of Council members were intended to limit the president's power to control the Council. Thus, in the 40 years after the reforms, nine presidents held office, some of them for only a few years. The longest serving was Edward Sabine, an army officer and investigator of magnetism, who served from 1861 to 1871. Despite the efforts to reduce the risk of dominance by a single individual, Sabine's Royal Society career demonstrates that it did remain possible for a person to have long-lasting influence. Before becoming president, Sabine had already served the Society for 21 years (not continuously) as secretary, foreign secretary, treasurer and vice-president. His first role was in 1827, and his last ended in 1872. Sabine was presumably a valuable source of institutional memory for the Society, and yet he is far less visible in the Society's public history than Banks had been.[11]

With presidents handing over frequently, a long-serving secretary could come to have significant influence. During his own long tenure at the Society (secretary 1854–85, president 1885–90, vice-president 1890–2), George Stokes does not seem to have engendered the active opposition to his rule that Banks had done. By the 1870s, the members of the X-club might have come to see him as a conservative force, and personally lacking in the leadership qualities that a president would ideally display; but they nonetheless respected his hard work as secretary. He was widely seen as a fair, careful and effective administrator to the Society, and it has been said that he 'quietly managed the business of Victorian science'.[12]

Through his long service, Stokes brought some much-needed stability to the Society's activities and processes. This is particularly clear in the editorial processes, which had seen several dramatic changes in the 1830s and 1840s, relating to the experiments with referees and scientific committees. Under Stokes, things settled down, and in the absence of scientific committees after 1849, he became the person who selected and managed referees. Little was yet codified or formalised, but norms and routines emerged through the interactions between Stokes, his co-secretaries and the fellows who communicated or refereed papers. As well as their shared membership of the Society, many of these men knew each other in person, and some were regular correspondents. They met at meetings of the Society, as well as the other organisations to which they belonged, from the specialist London societies to the annual meetings of the British Association for the Advancement of Science.[13]

Under Stokes, refereeing came to be understood as a two-part task, involving a recommendation to the Committee of Papers, and suggestions or advice for the author. It became standard practice for referees' names and reports to be confidential by default (though some referees intentionally revealed themselves), but conveying the gist of their suggestions to the author became a routine part of the secretary's role. Norms of scholarly etiquette and of appropriate language and structure also emerged during this period. None of this was laid down by Stokes, but it emerged through the social interactions of the community of fellows that he facilitated. The significance of his role as facilitator and institutional memory is surely indicated by the fact that many of these things came to be formalised as 'guidelines' and 'procedures' in the years immediately after Stokes stepped down as president in 1890.

George Stokes as secretary

Why did Stokes find the role of secretary of the Royal Society more attractive than editor of the *Mathematical Journal*? It is not just that becoming secretary to a prestigious and historic institution might be regarded as higher status than becoming editor of a respected but struggling independent journal. The secretarial role was more wide-ranging and would involve working on a wider variety of activities. But there are also two features of Stokes's own situation that helped make the secretarial role attractive to him in 1854: money and sociability.

In 1854, Stokes was 35 years old; he had no independent income to support him; and like so many other men of science of his generation, he was already holding several paid positions simultaneously. As well as his Cambridge professorship, he had just acquired a lectureship at the government School of Mines, in London. The £105 honorarium granted to the secretaries of the Royal Society was a welcome further addition and helped create a sufficiently secure income for him to marry in 1857. Stokes hoped his marriage would help him move beyond the loneliness of intense intellectual effort, and he may have seen similar potential in the opportunities for social contact provided by the secretaryship.[14]

The Society had two honorary secretaries, of whom Stokes became the junior; while a third fellow assisted with foreign correspondence. The definition of the secretaries' duties in the statutes focused on dealing with its correspondence and arranging and keeping the records for Society meetings. One of the secretaries was present at every meeting of the Society, weekly between November and June, which is why the role provided such good opportunities for networking. But the continuing tradition that papers were read aloud by the secretaries, rather than their authors, initially made Stokes feel that he was 'a sort of reading machine' or 'a pair of bellows to be blown for the benefit of the Society'.[15]

In contrast to the close involvement of eighteenth-century presidents in the organisation of meetings and publications – often through a tight relationship with hand-picked secretaries – the shorter-serving Victorian presidents left far more to the secretaries elected alongside them. Thus Sabine (as president) explained, in 1869, that although the arrangement of papers for reading at meetings 'is understood to be subject to the Regulation of the President', it was 'a matter in which I very seldom interfere; and when I do so, I always endeavour to do so with the approval of the Secretaries'.[16]

The formal description of the secretaries' role referred only briefly to 'the charge (under the direction of the Committee of Papers) of printing the *Philosophical Transactions* and correcting the press', thus glossing over the editorial work that was done before the papers were ever sent to the printer. The practical aspects of dealing with printers and distributing the printed *Transactions* was typically the responsibility of the assistant secretary – an employee (not a fellow) working under the supervision of the secretaries – and the secretaries themselves focused on the editorial decision-making. Even though the statutes had been revised in 1847, there was no mention at all of any responsibility for the *Proceedings*.[17] Nonetheless, as the *Proceedings* developed in the 1850s, it took up an increasing amount of editorial time. In 1860, Sabine (as treasurer), noted that the *Proceedings* was now 'found so useful by the Fellows', but had 'added considerably' to the labour of the secretaries; he successfully proposed that the secretaries' honoraria be increased to £200.[18]

Stokes had initially assumed the duties of the secretary of the Royal Society would not be too onerous, and reassured Sabine that 'I know of no reason why I should not undertake them', for his teaching duties in Cambridge were light enough that he could plan to be in London for most of his first year as secretary.[19] Stokes undoubtedly underestimated the impact that being secretary – and de facto editor of the Society's periodicals – would have on his life and research. By the 1880s, his friends and colleagues were suggesting other academic roles that would pay well enough for him to be able to afford to give up his Society duties, and his co-secretary, Michael Foster, commented that 'it has been painful to see how his energy has been wasted in this way'.[20] Yet Foster also recognised how much science had gained from Stokes's work as a facilitator and mentor of others.

Stokes was not the only scientific manager in Victorian Britain, but in contrast to someone like George Biddell Airy at the Royal Greenwich Observatory, or the directors of the Geological Surveys, Stokes and his co-secretary had just one paid assistant. Rather than running a team of staff, they had to persuade a geographically dispersed group of fellows to undertake unpaid work as committee members or referees, for the benefit of the Society to which they all belonged. One of the results was a voluminous correspondence: it is estimated there are at least 30,000 letters surviving to or from Stokes, twice as many as in the better-known correspondence of Charles Darwin. Stokes's correspondence often mixed editorial matters with Society business, and with personal correspondence, for Stokes was friend, colleague or adviser to many men of science. Despite his initial plans to spend time in London, Cambridge became the family home after his marriage, and it was here that a lot of his editorial work was done. His daughter remembered his study crammed with 'as many tables as the room would hold' on which 'papers were piled a foot or more deep', and in which 'he could find nothing'.[21] In 1879, he acquired an early typewriter, to the relief of the many correspondents who found his handwriting difficult.[22]

Stokes began his secretarial term alongside William Sharpey, the professor of anatomy and physiology at the University of London. Sharpey, having taken up office in 1853, was formally designated the 'senior secretary', and remained so until his retirement in 1872. From 1856, they were joined by the Cambridge professor of mineralogy, William Hallowes Miller, as foreign secretary.[23] During the 1850s, this new secretarial team was supported by the experienced assistant secretary (and Society historian) Charles Weld who, unlike them, had been in post since before the reforms. In 1861, he was replaced by Walter White, formerly the Society's sub-librarian, who would remain in post until the end of Stokes's secretaryship in 1885.[24]

After Sharpey's retirement, Stokes was joined first by the naturalist Thomas Henry Huxley and then, from 1881, by the physiologist Michael Foster.[25] Foster recalled that, when he joined Stokes, the secretarial labour was arranged around a division between internal and external activities, rather than (as would become the case in the twentieth century) by their disciplinary interests. According to Foster, Stokes had chosen to take charge of 'internal' activities such as the organisation of the weekly meetings and the editing of the periodicals. He did almost all of the correspondence 'from the receipt of the communication in manuscript until its publication in the *Transactions* or *Proceedings*'. He chose the referees and corresponded with them, and made recommendations to the Committee of Papers. Once a decision was made on a paper, it was usually left to White to deal with the printers, engravers and lithographers. With the editorial work mostly taken care of, Foster, like Huxley before him, took on responsibility for the Society's engagement with other learned societies, with the government and with other outside bodies.[26] The statutes may have described two secretaries sharing responsibilities equally, but it is clear that Stokes – seemingly by choice – controlled and directed the Society's editorial process for three decades.

The editorial community

Foster credited Stokes with great skill in moulding 'a crude, imperfect effort into something worthy of being made known' and published.[27] Stokes and his co-secretaries were undoubtedly highly visible and influential actors in the Royal Society's editorial processes (see Figure 9.2). They received submissions of papers; arranged the agenda for the meetings at which those papers would be read (in whole or in part) to the fellows and their guests; wrote the abstracts that would be printed in the *Proceedings*; decided which papers might be worth considering for the *Transactions* and (after the demise of the scientific committees) selected the referees for them; organised the agenda for the Committee of Papers, often including recommendations on papers; and oversaw authorial revisions and proof-reading. They had, however, much less autonomy than independent journal editors – such as William Crookes at *Chemical News* (f. 1859) or Norman Lockyer at *Nature* (f. 1869) – for they had to work within the editorial system described in the rules and procedures of the Society, in which formal decision-making remained a collective responsibility, vested in the Council sitting as the Committee of Papers.

The Royal Society's *Transactions* and *Proceedings*, like those of other learned societies, were distinguished from independent journals by the community involvement in their editorial work. As well as the secretaries and Council members, substantial numbers of ordinary fellows of the Society were involved in the work and responsibility of editing. As in the days of Joseph Banks, all fellows were potential gatekeepers because only fellows could communicate papers to the Society. As well as their own research (which was almost always automatically read in some form at a meeting), fellows could communicate research by others: about a quarter of all papers submitted during Stokes's time were authored by non-fellows. Since the *Proceedings* now reported at least the titles (and

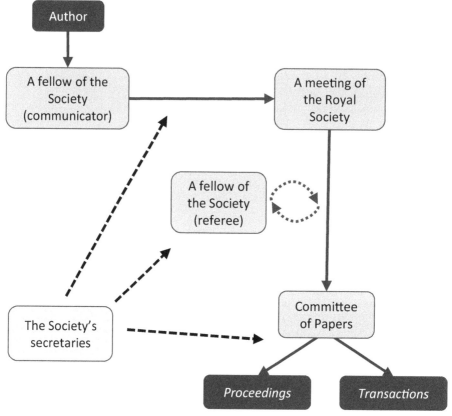

Figure 9.2 The Royal Society editorial process in the 1860s.

usually an abstract) of all papers read at meetings, the ability of fellows to identify appropriate research was open to public scrutiny.

Refereeing was another way in which ordinary fellows could be involved. Acting as referee typically (but not always) involved a closer engagement with the intellectual content of the paper than did acting as communicator. Only a third to a half of submitted papers were sent to referees (of whom there were usually two).[28] This was because most of the short papers that were increasingly appearing in the *Proceedings* were decided in the traditional way, by the Committee of Papers acting on the secretaries' recommendation. When a paper went to referees, the referee was not so much being asked *whether to publish*, but which periodical it should appear in.

Thanks to an innovation made by William Sharpey when he became secretary in 1853, we are able to reconstruct the editorial community

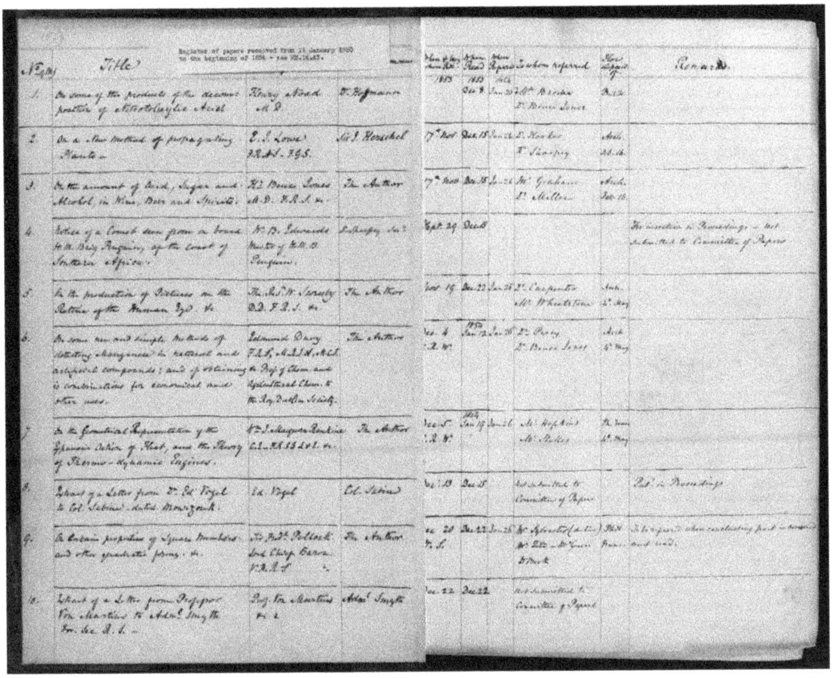

Figure 9.3 First page of the 'Register of Papers', 1853 © The Royal Society.

behind the Royal Society's periodicals. He had a ledger drawn up, to be known as the 'Register of Papers', with columns recording the author's name, title of paper, by whom communicated, when and by whom received, when read to the Society, when referred, to whom referred and how the paper was ultimately 'disposed of' (see Figure 9.3).[29] Since 1850, the secretaries had kept a master list of all papers considered by the Committee of Papers – easier for reference than having to search through the minutes of each meeting – but Sharpey's 'Register of Papers' was a fuller record.[30] Some of the information it contained was later printed on the published paper (the date received, the name of the communicator), but the names of the referees remained confidential, as did the dates on which they were asked and reported, and all the details for papers that were not ultimately published. The Register was maintained by Walter White, and then by his successors as assistant secretary, and continued until 1990.[31] The Register allows us a deep insight into the operation of the refereeing process, and its subsequent development.

The first thing that scrutiny of the 'Register of Papers' reveals is that the extent of participation by ordinary fellows should not be over-stated.

In the 1850s and 1860s, around 40 fellows a year acted as referees, and around 20 or 30 acted as communicators for papers by outsiders. Given some overlap between these two groups, this represents less than 10 per cent of the fellowship. The proportion grew during the second half of the nineteenth century, both because more editorial work was needed to deal with the increased number of submissions, and because the size of the fellowship was falling (from a high of 765 fellows in 1847 to a little over 500 fellows by the 1890s). However, the proportion of fellows involved – in some way – in the editorial process would still be less than a quarter at the start of the twentieth century.

Moreover, of those who were 'active' in editorial work, most were only minimally involved. For instance, over Sharpey and Stokes's co-secretaryship (1854–72), about 140 fellows acted as referees, but 45 of them wrote only one report in those years. And of the 120 or so fellows who communicated a paper by an outsider during this period, barely half a dozen did so more than 10 times. The public impression of collective, Society-wide, editorial structures was based upon a relatively small sub-group of keen and active fellows.

The most active referees and communicators were typically past, current or future members of Council. In other words, fellows who were active in one area of the Society's service tended to be active in other areas. They were, as Ruth Barton puts it, 'reliable committee men'.[32] They were often based in or near London and thus able to attend the Society's weekly meetings regularly; but the Victorian postal service was good enough for fellows to referee at a distance, as William Thomson did from Glasgow. As Table 9.1 shows, the secretaries continued to be among the most active referees. The predominance of referees with mathematical and physical interests reflects the sorts of papers the Royal Society was receiving in this period; while the quantity of reports written (compared to Table 8.3) reflects both an increased use of refereeing and the increase in the number of submissions. It is also notable that Edward Sabine, who had been an active referee in the earlier period, wrote only two reports during this period. On the other hand, Thomas Henry Huxley wrote 23 reports during this period, and would write 35 more after becoming secretary in 1872.

During Stokes's time, surprisingly few fellows exercised the right (and responsibility) of communicating a paper on behalf of a friend, relation, colleague or student. Three men were far and away the most active communicators of papers by non-members, and they all held official positions in the Society's hierarchy. Stokes, Sharpey and Sabine each communicated over 40 papers during the period 1854 to 1872

Table 9.1 Most frequent referees in the period 1854–72

Name	Date elected to RS (+ service to RS in this period)	Occupation	Reports written
George Stokes	1851 Secretary 1854–85	Professor of Mathematics, University of Cambridge	52
William A. Miller	1845 Vice-President 1855–57, 1861–70; Treasurer 1861–70	Professor of Chemistry, King's College London	41
George Busk	1850	Surgeon; later retired	32
Arthur Cayley	1852	lawyer; later Professor of Mathematics, University of Cambridge	31
William Thomson	1851	Professor of Natural Philosophy, University of Glasgow	31
Henry J.S. Smith	1861	Professor of Geometry, University of Oxford	30
William H. Miller	1838 Foreign Secretary 1856–73	Professor of Mineralogy, University of Cambridge	30
William Sharpey	1839 Secretary 1853–72	Professor of Anatomy and Physiology, University of London	29
James Clerk Maxwell	1861	Professor of Natural Philosophy, Marischal College, Aberdeen; later at King's College London; later Professor of Experimental Physics, University of Cambridge	27
Charles Wheatstone	1836	Inventor, sometime Professor of Experimental Philosophy, King's College London	25

(Stokes alone communicated 60). For authors who had no personal connection to the fellowship networks, writing to someone in a publicly visible position of authority was a point of access to the Society, as it had been in Banks's day. The fact that the president and secretaries did so often formally communicate such papers suggests that the Society's meetings and periodicals were not quite such a closed-shop as might at first appear. It also demonstrates that the secretaries and presidents continued to perform a significant role as gatekeepers. (The foreign secretary, Miller, communicated far fewer papers than his colleagues, reflecting the relative paucity of international input to the Society in the nineteenth century.)

The handful of other fellows who made repeated use of their right to communicate papers by others included some of the active referees (notably Thomson, Cayley and Huxley); the chemists Edward Frankland, Henry Roscoe and Wilhelm Hofmann; and the surgeon, microscopist and instrument inventor, Charles Brooke. Several of these men (Frankland, Roscoe, Hofmann and Thomson) were in academic positions where they had a stream of students and junior co-workers who might benefit from being presented to the Royal Society. Mostly, they communicated just one paper by an author: some of those authors went on to become fellows, and others have lapsed into obscurity. But occasionally there was a longer-term relationship between an author and a communicator. For instance, between 1863 and 1868, Thomson communicated six papers for his former student, Joseph D. Everett, who was a professor in Nova Scotia at the time of the first paper; and then worked in Thomson's Glasgow laboratory. Everett went on to gain a professorship in Belfast (1869), but it was not until the late 1870s, after more publications in mathematical journals, that Thomson successfully proposed him for fellowship of the Royal Society.[33] On the other hand, Cayley had communicated four papers on polyhedra by the Reverend Thomas Kirkman in 1855 and 1856, but their relationship foundered after the fourth paper was turned down.[34]

Since editorial roles – whether communicator, referee, committee member or secretary – were all carried out by the fellowship, every person involved in the evaluation of research at this time was male. After the papers by Caroline Herschel and Mary Somerville, we have found no records of any other women submitting papers to the Society until the late 1870s. In 1877, the pioneer doctor Frances Hoggan (née Morgan) would submit two papers co-authored with her doctor husband; they were sent to referees and considered for the *Transactions*, but, like many

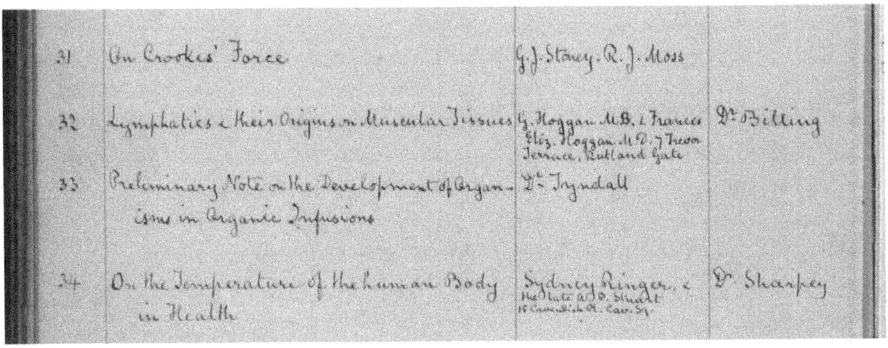

31	On Crookes' Force	G. J. Stoney. R. J. Moss	
32	Lymphatics & their Origins in Muscular Tissues	G. Hoggan M.B. & Frances Eliz. Hoggan M.D. 7 Trevor Terrace, Rutland Gate	Dr Billing
33	Preliminary Note on the Development of Organisms in Organic Infusions	Dr Tyndall	
34	On the Temperature of the Human Body in Health	Sydney Ringer, & the late A. P. Stuart 15 Cavendish Pl. Cav. Sq.	Dr Sharpey

Figure 9.4 The 'Register of Papers' records the submission of a paper by George Hoggan and Frances Elizabeth Hoggan, communicated by Dr Billing, 18 January 1877 © The Royal Society.

papers, were ultimately published only in abstract in the *Proceedings* (Figure 9.4).[35] By the 1880s, there would often be two or three papers a year by women authors or co-authors, but the evaluators continued to be all male.

Trusting the judgement of fellows seems usually to have been very effective at gatekeeping, for the number of papers formally received and read but not subsequently published in either periodical was very low (usually, below 10 per cent). For reasons of gentlemanly etiquette, the Society did not formally 'reject' papers that had, by tradition, been gifted to it and had been vouched for by one of its fellows. A handful of the unpublished papers were withdrawn by their authors for publication in a more appropriate or more rapid venue, and both Stokes and referees occasionally recommended this course of action. Most of the unpublished papers were formally 'archived', meaning that the manuscript was deposited in the Society's library. Such papers were, in principle, available for consultation by suitably qualified individuals, and thus, in 1892, Lord Rayleigh (then secretary) was able to rediscover an 1845 paper which had pre-empted Maxwell's theory of gases.[36] The importance of the archive as historical record was such that the Council was extremely embarrassed to discover, in 1860, that a paper archived seven years earlier, and whose author now wished to copy the illustrations for a new publication, could not be found.[37]

Relying on a fellow's evaluation of a paper – as author, communicator or referee – was not fool-proof, however. Consider the case of Charles Piazzi Smyth, fellow of the Society and Astronomer Royal for Scotland.

In 1870, one of his papers was read at the Society, considered by two referees, and sent to the archive. When, in 1871, he learned that a paper on a similar topic had been accepted for publication by the Society, he complained publicly in a letter to the editor of *Nature*, abusing his referees for having kept the paper 'shut up with themselves for upwards of seven months' only to 'condemn it before the Council'. Piazzi Smyth claimed that his paper had been 'instantly extinguished', and his subsequent complaint seen as 'rebellious opposition to the despotic dictates of the *Secret* Committee'.[38] Relations between Piazzi Smyth and the Society did not improve: another paper in 1871 was read but not published; and the final straw came when, in January 1874, his paper discussing the significance of 'the length of a side of the base of the Great Pyramid' was 'returned to author' without even having been read. Within days, Piazzi Smyth resigned his fellowship.

Piazzi Smyth complained about the length of time that the referees had taken, but the 'Register of Papers' reveals that his manuscript actually went to its first referee, back to the Society, and out to its second referee in less than a fortnight. But it was then seven months until the Committee of Papers came to a decision. It may be that the second referee took much longer to review it than the first had done, but it seems more likely that most of the delay was indecision among the secretaries or the Committee of Papers.[39] The Society did not have any clear guidelines for declining to publish papers that had been received and read, and a case involving a paper by a fellow was diplomatically sensitive. Such cases remind us why a secretary might not wish to be solely (or visibly) responsible for difficult decisions.

Piazzi Smyth's first referee had been remarkably swift, but it is clear that sending papers out to referees slowed down publication decisions. Decisions on the short papers to be published in the *Proceedings* could usually be made within four to six weeks. In contrast, by the end of Stokes's secretaryship, the average time taken to make a decision that involved referees (that is, to publish in the *Transactions*, or not to publish at all) had risen to six or seven months, almost double what it had been when he took on the role. Some sense of what Council's sense of an appropriate timeframe for referees might be can be gauged from its repeated unwillingness to insist on a two-month turn-around.[40] Publication speed was not yet the key performance indicator for journals that it would later become. It was not a metric on which the *Transactions* could have been competitive, given the number of people involved (sometimes, at a substantial distance) in the multiple stages of the Royal Society's distributed editorial process.

Selection criteria

A large proportion of the papers that George Stokes communicated for non-fellows ultimately appeared in the *Proceedings*, not the *Transactions*. He and his co-secretaries had the power to decide whether to send a paper to referees or not. If they thought a paper was neither strong enough to be considered for the *Transactions* nor so weak its author should be asked to withdraw it, then they could send it (or its summary) straight to the printers for the *Proceedings* without further consultation. The destiny of so many of the papers communicated by Stokes suggests that, although he was perfectly willing to present material from outsiders to Society meetings, relatively little of it seemed to him worthy of consideration for the *Transactions*.

The role of the *Proceedings* was changing. In 1863, a review of its arrangements was undertaken, partly with the aim of reducing the costs of publishing the growing amount of research (see also Chapter 10). Some adjustments were made to its subscription price and to its page layout, though the committee's strong recommendation that the *Proceedings* be 'issued regularly at specific times', and that publication should not cease during the Society's summer recess, was not adopted. The committee acknowledged that the *Proceedings* had proved useful 'as a medium of publicity' for authors who hoped to see their results in print, under the Society's auspices, more quickly than the *Transactions* could manage.[41]

The 1863 review report did not discuss the types of papers that should appear in the *Proceedings* rather than the *Transactions*. A few months later, however, the treasurer described the discussions around the review as having included the suggestion that 'it would be desirable to adopt a somewhat higher standard' for papers intended for the *Transactions*.[42] A stricter selection policy for the *Transactions* had been promoted by Sabine (as treasurer) in the 1850s, on the grounds that it would save money, and the argument would be reiterated by treasurers in the 1890s.[43] In the 1860s, however, a corollary of this approach to the *Transactions* was that other papers, deemed to be 'of merit, [but] confined chiefly to details of research', were diverted to the *Proceedings*, where they appeared alongside the abstracts of longer papers under consideration for the *Transactions*.[44] This shifted the function of the *Proceedings* from the advance notification of full papers that would follow later, to a way for the Society to publish the increasing number of submissions it received in a briefer (and cheaper) format.

It also set up a two-tier publication system, with differing (but largely unwritten) criteria for the *Transactions* and the *Proceedings*. It meant that Stokes, Sharpey and their referees could no longer work with a binary distinction between what was appropriate for the Society's publications, and what was not. Publishable research now came in two categories. Stokes's correspondence reveals how referees tried to navigate the implicit criteria.

When considering whether the Society should publish a paper at all, Stokes and his colleagues were very concerned with whether a paper's experiments or observations had been carefully and diligently undertaken, and whether they were accurately represented and logically followed through. Thus, an 1856 referee praised a chemical paper for being 'carefully worked out'; two years later, one of its authors then praised a different paper in similar terms, as having 'been very carefully carried out'.[45] Stokes asked one author for 'some evidence' that his results 'were in fact more accurate' than those of a rival, while James Clerk Maxwell praised an 1873 submission as 'probably far more accurate than any yet made'.[46]

The usual way to deal with doubts about the diligence of the investigation or the accuracy of the results was to ask (the communicator to ask) the author for clarification. But referees occasionally went further. Richard Owen still had a tendency to contact authors and discuss their papers directly. Thus, in 1857, he arranged a meeting with an author so that he could examine for himself the specimens described in a paper on ectopic pregnancy.[47] Undertaking replication or testing of a paper's claims as part of the refereeing task remained vanishingly rare, although, in 1873, Stokes did receive a short note from one of Thomson's assistants in Glasgow, to let him know that their experiments on a voltaic battery had agreed with those reported by the author of a paper under review.[48] But for the most part, referees confined themselves to scrutinising the manuscript text and any accompanying illustrations.

But accuracy alone was not enough: W.H. Miller described an 1856 paper as 'perfectly satisfactory in regard to its accuracy', but it 'does not appear to me to be of sufficient importance' to be worthy of publication in the *Transactions*.[49] In 1860, Stokes advised Faraday that 'negative results' might be included 'incidentally' in a paper containing other results, but would 'scarcely do for an independent communication' to the *Transactions*. Nonetheless, he added: 'I don't think there would be any objection to the paper's appearing in the Proceedings.'[50] Phrases like 'valuable contribution', 'deserving' and 'considerable ingenuity' crop up regularly in referee reports in the 1850s and 1860s, along with their opposites, such as 'perfectly satisfactory in regard to its accuracy, [but] does not appear to me to be of sufficient importance to be worthy' of the *Transactions*.[51] Captain Edward Boxer, of the Woolwich arsenal, was even more damning about an 1858 paper on gunpowder, which he judged to contain nothing likely 'in the smallest degree to further the object of the Society in "improving natural knowledge"'.[52]

The Duke of Sussex's 1832 desire that papers should be of 'importance' continued to influence referees, particularly in discussions of possible *Transactions* papers. In 1858, mathematician Archibald Smith carefully weighed up the characteristics of a paper by Arthur Cayley, himself an experienced referee. He praised Cayley's investigation of the planes passing through an ellipsoid as 'conducted with great skill and completeness' and admitted that 'the results are curious and instructive'. He believed it had 'intrinsic merits' and 'mathematical interest', but he was not sure that it was right for the *Transactions*. He sought guidance from Council on whether a paper that 'as far as I am aware has no interest beyond itself', nor introduced any 'new mode of investigation', could be published – or, 'whether it is not rather better fitted for publication in some exclusively Mathematical Journal'.[53]

This example also reveals that suitability for one of the Royal Society periodicals was increasingly defined in the context of the newer, more specialised societies and journals. The Royal Society sought to retain its privileged position by appealing both to its historic tradition and its unique claim to general coverage of all areas of natural knowledge. It came to be assumed that papers presented to and published by the Royal Society would be of wider, more general interest or significance than those that would appeal to the specialist societies (an assumption that would remain influential in the 1960s; see Chapter 14). Referees increasingly suggested that certain papers might be redirected to a different society. In 1873, for instance, two referees reported themselves unable to find 'anything … of sufficient novelty' to justify the publication of a paper on crystal formations in urine, but they hoped it would 'find a more suitable place' with a medical or surgical society.[54] Charles Wheatstone similarly argued that a paper on voltaic batteries did not contain 'sufficient originality' for the *Transactions*, but might well be suitable 'as a communication to the British Association, or in the Proceedings of the Society of the Telegraphic Engineers'.[55] Stokes himself had told a disappointed author in 1871 that his paper 'was hardly of a kind suited to us', but the description 'would come more suitably before some other Society'.[56] Such opinions suggest an emerging perceived hierarchy among scientific periodicals (and also learned societies). They also imply that authors were now permitted to withdraw their papers for resubmission elsewhere, in contrast to the Society's traditional insistence on retaining the manuscript in the archive.

A final key criterion for publication with the Royal Society, in either periodical, was originality. At its simplest, this meant that papers should not have appeared in print elsewhere, as had been the requirement since Banks's time. Thus, Faraday, in 1837, had advised a German colleague that the Society 'seldom if ever' prints papers on topics that 'have recently been dealt upon' by other societies or journals, unless there are 'some decisive views or some new discovery'.[57] At its strongest, originality could be used to mean creativity and novelty. Papers for the *Transactions* were usually expected to be intellectually original, as well as not having appeared in print before. Judging originality did, of course, assume that the referee himself was keeping up with the literature. The Irish natural philosopher Humphrey Lloyd confessed in 1855 that he could not really form a satisfactory opinion on George Airy's latest paper because 'I have not read any of the other investigations relating to the same subject'.[58] In contrast, the zoologist and physiologist William Carpenter opened his report on a different paper by surveying what was already known to European researchers. Having demonstrated his familiarity with the field, he felt able to state that the memoir was a 'most valuable contribution … well-deserving of publication'.[59]

The ongoing operation of the Society's system of pre-screening for papers by non-members meant that the vast majority (around 90 per cent) of the papers received were felt to be appropriate for publication by the Society. The question, then, was where, and at what length, to print them. The key unwritten rule was length. Three-quarters of papers in the *Proceedings* in the 1860s and 1870s were less than six pages long, while *Transactions* papers were typically 20 to 25 pages, but could be 40 or more.[60]

Understanding the tacit criteria for the *Proceedings* is complicated by the miscellany of types of items appearing in its pages in this period. In 1865, for instance, some items simply appeared under their title; others had the word 'abstract' after their title; and others had titles incorporating phrases such as 'preliminary note on …'. The preliminary notes tended to be shorter (about a page), but 'abstracts' could be just as long as the items not so labelled. The majority of these items appeared in print as a direct consequence of the paper being read at a Society meeting, without much or any additional editorial scrutiny; indeed, the relatively rapid print schedule for the *Proceedings* meant that there was little time for such scrutiny.

Two types of item had, however, received more attention. First, those labelled 'abstracts', a label that implied the existence of a longer version of the paper currently under consideration for full publication in the *Transactions*, but, at the time of printing, with an unknown decision: someone, presumably Stokes, had decided abstracts should be prepared for these papers, while their eventual fate was being determined. (Author-generated abstracts appear to have begun in the 1890s.[61]) Second, the *Proceedings* also contained a handful of papers that appeared out of sequence with the reports of meetings. These were papers for which abstracts had appeared in the appropriate chronological sequence; which had been considered for the *Transactions*, but were now – for whatever reason – being printed instead in the *Proceedings*, but at greater length than the original abstract (perhaps nine pages rather than three).[62] Such items carried a citation back to the abstract; the original abstracts could not carry citations forward to the final place of publication.

When papers were sent out to referees, Stokes included a covering letter. By the 1880s, this had evolved into a pre-printed letter (see Figure 9.5), in which he advised referees that they could recommend papers be 'printed *in extenso* in the *Proceedings*', although he noted that this happened only 'occasionally'. Given that referees were considering the papers after the meeting and, usually, after the 'abstract' had been printed, one simple question was whether there was any need to publish anything more than had already appeared. Thus, in 1865, Thomas Hirst reported that the 'general character' of a paper on plane stigmatics 'will be very well understood from the succinct and lucid abstract which has already been published in the Proceedings', and there was therefore no need for fuller publication.[63] Similarly, a few years later, John Burdon-Sanderson commended the 'real research and labour' and 'interesting observations' of a paper on monocular and binocular vision, but judged that the already-published abstract 'appears to be sufficient'.[64] What we seem to be seeing here is an emerging preference for brevity in the scientific article, and a questioning of the intellectual value of excess verbiage.

Apart from length, however, referees in this period remained uncertain how to distinguish between a *Transactions* paper and what Stokes's letter called those 'deemed worthy of publication in full, but … considered to be of a kind better suited for the *Proceedings* than the *Transactions*'.[65] The 1863 treasurer's criterion that *Proceedings* papers might be those 'confined chiefly to details of research' does not seem to have gained traction.[66] Some referees assumed originality and significance were more necessary for the *Transactions*, but others seem to have felt

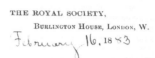

THE ROYAL SOCIETY,
BURLINGTON HOUSE, LONDON, W.

February 16, 1883

Sir,

    I am directed to send you the accompanying Paper,— *On the Affinities of Thylacoleo by Professor Owen.* and to request that you will favour the COMMITTEE OF PAPERS with your opinion as regards its eligibility for publication in the Philosophical Transactions.

    Papers which have already appeared in abstract, and which are deemed worthy of publication in full, but are considered to be of a kind better suited for the Proceedings than the Transactions, are occasionally ordered to be printed *in extenso* in the Proceedings. Should this appear to you the most suitable course to follow in the present instance, it is open to you to recommend its adoption.

    The Committee hope that you will be able to make a sufficient examination of the Paper within a month from the present date; if, however, your engagements prevent, I should feel obliged by an intimation to that effect.

I remain,
Sir,
Your obedient Servant,

for G. G. Stokes
Walter White
Assistant-Secretary R. S.

J. W. Hulke, Esq.
F.R.S.

Figure 9.5 Standard printed letter requesting a referee's report, 1883 © The Royal Society.

that the *Proceedings* was a good venue for short papers of any significance. Thus, a referee wrote apologetically to Stokes in 1873: 'I scarcely know what to say as to which of the two [journals] should be chosen … I do not know what rule is followed in this matter'.[67]

Refereeing as social practice

Refereeing in the second half of the nineteenth century was governed by codes of gentlemanly sociability, just like the meetings of the Society itself. In contrast to the situation that would follow in the early twentieth century, most of the authors whose papers were being evaluated in this period were fellows of the Society, as were all the referees. Referees and authors knew each other by reputation, and often personally; and they shared a loyalty to the Society to which they all belonged. Refereeing was a way of continuing discussions beyond the meeting, and in more detail, and of gaining input from fellows who lived too far away to attend every meeting. This context is important when we consider the growing tendency for referees not merely to make a recommendation, but to suggest revisions and improvements.

Refereeing was far less anonymous, and somewhat less confidential, than we might expect. For one thing, referees routinely knew the name of the author. Even if they had not been present at the meeting when the paper was read, or seen the report in the *Proceedings*, Stokes's covering letter would provide the name. Thus, writing to John Herschel in 1856, Stokes said 'I send you Mr Pole's paper on colour blindness which was referred to you if you were disposed to take it'.[68] (See also Figure 9.5.) Referees often knew the author personally. For instance, when Stokes asked William Thomson to referee a paper by Lord Rayleigh, Thomson responded, 'I saw some of the work in progress and heard many details regarding it from Lord Rayleigh from time to time'.[69] The names of the referees were supposed to be confidential from the authors, but some referees did not try very hard to remain anonymous. As we have seen, Richard Owen sometimes met with authors to view their specimens, and when Charles Darwin was asked to referee a geological paper by his friend Leonard Horner, he submitted a one-sentence report to Stokes, and wrote directly to Horner with more detailed suggestions for 'a little shortening'. He had no qualms about admitting that, 'I may as well say that I have had to Report on your paper'.[70]

The 'Register of Papers' reveals that it was now standard practice for the manuscript to be returned to the Society (with a report) before being posted out again to the next referee, rather than sent directly by one referee to another, and the dates recorded for 'Sent to referee' allow us a glimpse of the speed with which first referees responded. When writing to the second (or subsequent) referees, Stokes sometimes indicated who had already looked at it and what their general thoughts or concerns had been. In certain research fields, the same referees often worked together, whether knowingly or not. This is most striking in physics, where most papers were refereed by at least one (and often two) of Stokes, Thomson or Maxwell. Refereeing pairs in the life sciences were more varied, but Thomas Huxley, George Busk and William Henry Flower were the most dominant group.[71]

In the absence of any formal rubric to guide referees, shared community norms were particularly significant. Such norms were not, however, tightly confined to the Royal Society: several other learned societies had a refereeing process by the 1850s, and men of science might well referee for multiple societies.[72] Darwin, for instance, refereed for (at least) the Linnean, Geological and Royal Societies.[73]

As in the 1830s and 1840s, referee reports in the 1850s and 1860s varied enormously in length and style, ranging from one-sentence recommendations to the half a dozen pages (sometimes more) written by George Busk, John Burdon-Sanderson and George Stokes, among others. While Darwin merely made a recommendation, many other referees gave reasons for their recommendation, often praising experiments or observations that had been 'very carefully carried out'.[74] For instance, referees praised Richard Owen's 'laborious accuracy and acute philosophic induction', while others found originality and significance in the papers under review.[75]

For authors in the biological sciences, Stokes acted as facilitator, passing on the referee's opinion and suggestions with little comment of his own. But for authors in the physical sciences, Stokes routinely inserted his own opinions into the correspondence especially, but not only, when he had also acted as one of the referees. For Victorian physicists, Stokes became an important adviser and sounding-board, offering advice before submission and suggestions for revision afterwards. His own referee reports were often combined with his letter to the author, or enclosed with it, and when he was not the referee, he often added his own opinions nonetheless. This is particularly clear in his correspondence with John Tyndall, which spanned 40 years from the mid-1850s.[76] In 1883, Tyndall apologised for bothering Stokes, but commented that 'in your replies you always say something that is useful to me'.[77] These were often suggestions for more clarity, for changes in wording, or occasionally, for rearrangement of the paper as a whole. These changes were never a condition of publication, but Tyndall was usually willing to take the advice, assuring Stokes in 1864 that 'all the points to which you refer me shall be attended to'.[78]

Stokes and Tyndall were not close personal friends, but they were near contemporaries, both having arrived on the London scientific scene in the mid-1850s (albeit from very different previous occupations). To his contemporaries and his juniors, Stokes appears to have been an excellent mentor. Foster, for instance, praised his ability to 'encourage the worth of the initial efforts of young workers who had not presented their work in the best possible way, and who might have been hardly judged by those who could not distinguish so clearly and patiently as did Stokes, the essential from the passing and the irrelevant.'[79]

When dealing with older men of science, Stokes took a more deferential tone. Barely a year after taking up the secretaryship, he apologised to Michael Faraday, almost 30 years his senior: 'I should not have ventured to write as I did, considering how deeply you have thought over the subject and how little I have attended to it.'[80] Two days earlier, Stokes had written to Faraday to let him know that his latest paper of experimental researches in electricity would be read at the next meeting of the Society; and he had added a lengthy paragraph musing on the behaviour of phosphorus in a magnetic field.[81] Stokes's apology suggests that Faraday had been taken aback by the younger man's intervention; but he must then have encouraged Stokes to share his thoughts, since in this, and several following letters, Stokes set out his thoughts in more detail.[82] The following year, Stokes had another diplomatic stumble, when he ventured to suggest that 'I call you Faraday', and then apologised, 'I suppose I must keep to the "Sir" for I feel it due to your age and standing in the scientific world that you should make the change if any is to be made'.[83] There are few surviving letters from Faraday to Stokes so we do not know what Faraday thought of Stokes's style of dealing with authors, or how it compared with his previous experiences with Royal Society secretaries.

What was the point of Stokes – or any referee – entering into dialogue with the author of a paper? For the physicists who knew each other well, it was possible that their discussion in and around the refereeing process would spark new lines of enquiry for future research. But the correspondence was usually focused on the current paper, and referees appear to have been motivated by a desire to make the paper as good as it could be. This suggests a concern for the author's reputation and for the reputation of the Society that would publish it (as one referee would express it, 'in the interests of the author and the *Transactions*'[84]), as well as a willingness to spend the time and energy on improving someone else's work. Since referees were not named in the final publication – unlike communicators – there was no public credit to be gained by being a conscientious and altruistic referee. Nevertheless, the surviving files of referees' reports in the Royal Society clearly indicate that many Victorian men of science gave up substantial amounts of time and energy to assist their peers.

Referees commented on literary style, sometimes praising 'an excellent exposition', but more often complaining.[85] And the revisions suggested by referees in this period continued to be what was then termed 'literary'; we might say 'stylistic'. They may have helped standardise the structure and style of scientific articles, but they did not involve authors making significant changes to the intellectual content of the paper.[86] The published paper still carried the date at which it had been received by the Society and the date on which it was read at a meeting of the Society, providing evidence by which priority claims could potentially be adjudicated.[87]

An 1848 episode demonstrates just how worried the Royal Society officers were about the possibility of intellectually substantial changes. One of the papers in the editorial process that April was on marine molluscs by the naturalist John Edward Gray. A suspicion arose that Gray's changes in proof had been more than minor, and since the original manuscript had been mislaid, this could not be checked. Given that almost a year had passed since Gray's paper was read, it was not inconceivable that his thoughts had developed in the interim. Council's concern about the registration of priority led to a recommendation that a note be added to Gray's paper to indicate the date of his corrections. In the end, the manuscript was found, and the secretary reported that the changes were in fact 'merely verbal', and thus permissible.[88] The issue did not go away, however, as can be seen from an 1880 Council resolution that 'any substantial changes in the papers or abstracts made between the time of their being read and their being published should be properly dated'.[89]

Given these concerns, what sorts of revisions were Stokes and his referees suggesting? They pointed out errors, mistaken assumptions and lapses of scholarly etiquette. Thus, Darwin wrote to Horner in 1855, 'you seem, (or give the impression) in your Nile Paper to quite believe Russegger's statement … [but] from what I have seen myself … I cannot avoid a total disbelief'.[90] And when, in 1872, the astronomer Norman Lockyer submitted what would become his second paper in the *Transactions*, Stokes dropped a hint about appropriate forms of reference (citation) to the work of previous scholars, and provided the necessary offprint: 'I need only refer to the papers by the late Professor Miller and myself in the *Phil Trans* for 1862. I send you a copy of the latter by book post'.[91] In both these cases, the advice was conveyed directly from the referee to the author, but in most cases, it was sent to Stokes as secretary to pass on as he saw fit. Thus, in 1876, one referee asked Stokes, in a letter accompanying his report on a paper on the expansion of sea water:

> Do you not think the words underlined in pencil on page 1 are rather unnecessarily hard on Muncke who, living at Heidelberg [?] fifty years ago, probably could not get sea-water whenever he liked and no doubt did the best he could under his circumstances? What the authors mean is most likely quite true, but the meaning is a little harshly expressed.[92]

Some referees requested clearer explanations from authors, while others objected to unnecessarily lengthy introductions. William Carpenter wished that one author had 'more fully expressed his reasons for dissenting from the statements of Prof Kollicker' on involuntary muscular fibres, and that he had managed to give 'a clearer elucidation' of the nature of the different types of fibres.[93] On the other hand, both referees recommended that the 'introductory explanations' to an 1865 paper on the brains of marsupials (and some of the illustrations) 'should be curtailed or even omitted'.[94]

Papers were virtually always submitted in English, and the Society's foreign secretary was no longer expected to act as translator. The Society now received relatively few foreign submissions, and when it did, its referees tended to be critical of non-Anglophone authors. John Tyndall communicated many papers for his own German correspondents, and proposed several of them as foreign members of the Society, but he could be a harsh critic. He told Stokes in 1865 that a paper by a German explorer 'seems to have been written somewhat hastily. Its style needs improvement.'[95] The same year, a different referee complained of another German foreign member that his 'English should undergo a careful revision. As it stands, it is in parts almost unintelligible.'[96] (Nonetheless, he won the Society's Copley Medal a year later.)

Despite ongoing concerns about changes made after the date of receipt, revisions became a familiar part of the Society's refereeing process during Stokes's secretaryship. They were, however, almost always couched as suggestions, not requirements. In 1855, for instance,

Stokes informed Tyndall that the Committee of Papers had approved the printing of his paper in the *Transactions*. He also enclosed referees' comments, and asked if Tyndall felt 'disposed to make any material alteration'.[97] An author who chose to make revisions at this stage had to balance the advantages to be gained from presenting his work in the best possible light, against slowing down the printing process.

Occasionally, referees had very clear recommendations that could be presented to the Committee of Papers as part of the approval process, such as the omission of complete sections or illustrations. Thus, in 1857, Wilhelm Hofmann annotated the paper he was reviewing, and recommended 'that the part between B and C be omitted'.[98] Similarly, referees in 1873 recommended printing a paper 'with the exception of the portion pp.18–33 indicated between blue lines, which we do not consider to be sufficiently novel and definite to merit a place in the Transactions'.[99] The Committee accepted this recommendation and ordered the paper to be printed 'with omission of a certain portion as recommended'.[100] The authors in these cases appear to have had no role in, nor choice about, the changes: they were presented with a *fait accompli* when Stokes reported the approval of the revised paper.

Revision by cutting was relatively simple: Stokes could strike out paragraphs or pages just as easily as Banks and Blagden had done. Revision that involved rewriting, rephrasing or clarifying was more tricky, in a purely practical sense. Most authors only had one fair copy of their manuscript, which they sent to the Society. On receipt, the manuscript was treated as the property of the Society and a part of its historical record. Thus, when Stokes returned a paper to Tyndall for revision in 1855, he informed him that he should 'send back the paper in the original state with fresh leaves containing the substitutions in order that we may have the evidence in case of any future dispute relating to priority'.[101] Inserting extra sheets, or pinning on small pieces of paper, along with annotations to guide the typesetters, was a way of enabling authors to make changes without having to recopy the entire manuscript, while retaining the integrity of the original paper. This concern remained valid throughout Stokes's secretaryship, as we find him in 1883 informing the author of a paper deemed too long to publish: 'Please therefore clearly to understand that the drawings and the manuscript are only returned to you on loan, to enable you, if willing, to draw up such an abbreviated account of the substance of the paper.'[102]

The physicality of the materials used in Victorian authorship and editing is nicely illustrated in another episode from the Stokes and Tyndall correspondence. In 1873 and 1874, Tyndall presented work to the

Society based on his recent investigations into the transmission of sound through fog (on behalf of Trinity House, the organisation responsible for lighthouses). The referees felt that revisions to his *Transactions* paper were needed, not least to refer to Tyndall's official report on the matter. Stokes wanted a very substantial restructuring, and tried to persuade Tyndall that his proposal would not be as much trouble as 'it might appear on first sight'. Stokes suggested that Tyndall take 'a couple of waste printed copies of the report to operate on';[103] the rearrangement would then involve 'little writing, being chiefly a matter of brain, scissors and paste'. He suggested using excerpts cut from the printed report, pinned to or interleaved with the manuscript, with suitable 'marks on it to guide the printer'.[104] As well as involving less rewriting, Stokes noted that, 'As this is a case in which questions of priority might arise [over fog signals], it would be well to be more cautious than ever to leave the original MS intact'.[105]

By the 1880s, British scientific print culture was even richer and more varied than it had been when Stokes took up office in the 1850s. The last of the so-called 'taxes on knowledge' (the paper duty) had been repealed in 1861. Coupled with technological innovations in steam-powered printing, paper-making, mechanical typesetting and railway distribution – not to mention improvements in literacy – this made it easier than ever before to make a success of periodical publishing.[106] The 1860s had seen a flourishing of newspapers, especially at local level; and also of magazines, where a huge range of new titles targeted very specific demographics and interest groups. Readers interested in the sciences were among the many who benefited.

The *Geological Magazine* (f. 1864) was hardly the first attempt to create a community of people with geological interests that reached more widely (socially and geographically) than the Geological Society of London, but it was the first to achieve long-term success.[107] Other editors targeted specific occupational groupings: the *Chemist and Druggist* (f. 1859) and *Chemical News* (f. 1862) catered to a readership that included practising pharmacists and apothecaries as well as academic chemists; while a re-launched *Electrician* (f. 1878) thrived on the flood of academic and commercial interest in the technical challenges of electrical telegraphy and other applications of electricity.[108]

These independent commercial journals were very different from the Royal Society's *Transactions* or *Proceedings*: they aimed at a community of readers far beyond the fellows (or potential fellows) of a London society, and they offered those readers a regular miscellany of short articles and up-to-date content. Where the Royal Society periodicals published only more or less lengthy descriptions of well-developed results from original research, the scientific magazines typically included a mixture of book reviews, reports of meetings of scientific societies, abstracts of research articles published elsewhere (including foreign journals), scientific news, letters to the editors, and 'notes and queries' sections that enabled readers to contribute observations that might be useful to other readers. The editors used their journals as tools to create virtual communities of readers with shared passions for geology, entomology or electricity. In contrast, the periodicals of the Royal Society (and other London learned societies) supported and benefited from existing communities of scholars – that is, their members – who provided referees and editorial committee members, as well as many of the periodicals' authors and readers.

Many fellows of the Royal Society read and contributed to these specialist magazines, as well as to general-interest periodicals such as *Macmillan's Magazine* (f. 1859) or the *Fortnightly Review* (f. 1865). And some of them were more closely involved: in 1861, Huxley became a co-editor of the *Natural History Review* (1856–65), and tried to turn it into a 'journal of biological science'.[109] After its demise, he became closely involved with *The Reader* (1863–7), a more wide-ranging periodical whose then-editor, Norman Lockyer, was aiming to improve the way in which 'THE PROGRESS OF SCIENCE, and THE LABOURS AND OPINIONS OF OUR SCIENTIFIC MEN, are recorded in the weekly press'.[110] By the end of the 1860s, these experiences with the hard realities of commercial publishing seem to have dissuaded Huxley from initiating any further journalistic ventures, but he continued to contribute to many journals – including Lockyer's new venture, the weekly *Nature* (f. 1869).

By the early twentieth century, *Nature* would have become the journal to which critics of the Royal Society routinely pointed when they argued that the *Transactions* and the *Proceedings* needed to be reformed to better fit the needs of scientists. They would particularly emphasise the speed of publication that was possible for short articles in a weekly periodical that did not use referees: it undoubtedly did contrast sharply with the many months of waiting for an editorial decision, and the possibility of having to make revisions, that were associated with the Royal Society periodicals. But in the 1870s, it was not yet clear what role *Nature* would take on. Melinda Baldwin highlights its similarities with *Chemical News*, though with a wider disciplinary remit: it carried editorials, book reviews, reports from scientific meetings, correspondence and miscellaneous 'news'.[111] It was, therefore, potentially a useful medium for increasing awareness of the Royal Society and the research it published and it created a venue where scientists could discuss shared concerns about the practice and organisation of science, but it did not seem to be a direct competitor with the Royal Society's periodicals for the publication of substantial original research.

During the Stokes years, the development of refereeing practices (and expectations) meant that getting a paper published in the *Transactions* was a slow and complicated matter. There were many ways in which authors could have got their findings into print more quickly. However, having their work accepted as papers for the *Transactions* meant that the Royal Society paid the printing and (especially) the illustration costs. And publication in the *Transactions* also carried the prestige of association with – and, by implication if not formal statement, of approval by – the Royal Society. The greater prestige of publication in the *Transactions* rather than the *Proceedings* was made explicit in the Society's public presentation of itself: in the official lists of fellows, those who had published at least one paper with the Society were distinguished with a 'P.' (for 'published') against their names – but only if it had appeared in the *Transactions*.[112]

As the editorial processes for the *Transactions* became slower, more complex and more demanding, the *Proceedings* became a more attractive venue for some researchers. For instance, in 1872, astronomer George Airy submitted a paper on the periodicity of magnetism, and explicitly stated that he intended it 'for the Proceedings'.[113] By this time, an article that was less than about six pages, and had been communicated through the traditional Society channels, could be pretty much guaranteed to appear in print in the *Proceedings* within a month or two of submission, without much scrutiny or revision. For researchers who valued speed, and did not need substantial length, the *Proceedings* might serve a more useful function than the *Transactions*. Airy was just one of many voices over the years wishing that the Society would develop the *Proceedings* further. He told Stokes, 'I think we want a weekly publication, not interrupted by the Society's holidays, and not strictly confined to the presentations of its meetings'.[114] The aborted 1863 proposals would not have turned the *Proceedings* into a weekly, but they would have enabled it to fit better into the practices of the commercial book trade and, perhaps, reach more widely through the scholarly scientific community. However, as we shall see, the Royal Society's approach to the distribution of scientific knowledge was still firmly grounded in its self-image as a patron of science, focused on circulating rather than selling.

Notes

1 Thomson to Stokes, 21 April 1851, Letter 80 in David B. Wilson, ed., *The Correspondence between George Gabriel Stokes and Sir William Thomson, Baron Kelvin of Largs*, 2 vols (Cambridge: Cambridge University Press, 1990), vol. 1, 117. See also Tony Crilly, 'The Cambridge Mathematical Journal and its descendants: The linchpin of a research community in the early and mid-Victorian Age', *Historia Mathematica* 31, no. 4 (2004): 455–97.

2 Stokes to Owen, 1 November 1884, CUL Stokes Papers Add. 456 069. Stokes was writing in response to a letter in which Owen blamed co-secretary Foster for a mistake in the labelling of figures, see Owen to Stokes, 10 October 188[4] [misdated 1886], CUL Stokes Papers ADD 456 068.

3 Joseph Larmor, ed., *Memoir and Scientific Correspondence of the Late Sir George Gabriel Stokes*, 2 vols (Cambridge: Cambridge University Press, 1907), vol. 1, 100.

4 Marie Boas Hall, *All Scientists Now: The Royal Society in the nineteenth century* (Cambridge: Cambridge University Press, 2002).

5 See extant Certificates, RS EC collection.

6 Certificate for William Thomson, 4 January 1851, RS EC/1851/15.

7 Certificate for George Stokes, 24 December 1850, RS EC/1851/13.

8 Certificate for Thomas Henry Huxley, 26 February 1851, RS EC/1851/10.

9 Certificate for James Paget, 30 January 1851, RS EC/1851/12.

10 Certificate for August Wilhelm von Hofmann, 5 March 1851, RS EC/1851/09.

11 Gregory A. Good, 'Sir Edward Sabine (1788–1883)', in *ODNB*.

12 Wilson, *Stokes–Kelvin Correspondence*, vol. 1, xvii. Even those who did not particularly like Stokes could appreciate his administrative skills, see Ruth Barton, *The X-Club: Power and authority in Victorian science* (Chicago, IL: University of Chicago Press, 2018), ch. 4.3, compare pp. 273–4 and p. 284. See also Paul Ranford, 'Sir George Gabriel Stokes, Bart (1819–1903): His impact on science and scientists', *PTA* 378.2174 (2020).

13 For many instances, see Barton, *The X-Club*, ch. 4.

14 David B. Wilson, 'Stokes, Sir George Gabriel, first baronet (1819–1903)', in *ODNB*. On the value of the secretarial honorarium to a professor, see Barton, *The X-Club*, 274.

15 Stokes to Mary Stokes (née Robinson), 30 January 1857, in Larmor, *Stokes Correspondence*, 54. Thanks to Graeme Gooday for drawing this letter to our attention, and for discussions about Stokes.

16 Sabine to Tyndall, 16 January [1869], RI MS JT/1/S/29. We are grateful to the editors of the Tyndall Correspondence project for giving us advance access to their correspondence files. The correspondence to 1868 has been published so far, see Bernard Lightman, Michael S. Reidy, and Roland Jackson, eds, *The Correspondence of John Tyndall* (Pittsburgh: University of Pittsburgh Press, 2016–). It is indexed at https://epsilon.ac.uk.

17 1847 statutes, reprinted in *The Record of the Royal Society of London* (3rd edn) (London: The Royal Society, 1912), 138.

18 RS CMP/03, 23 February 1860.

19 Stokes to Sabine, 16 October 1854, in Larmor, *Stokes Correspondence*, vol. 1, 143.

20 Foster to Rayleigh, 1884, in Wilson, *Stokes–Kelvin Correspondence*, vol. 1, xv. On alternative jobs, see Thomson to Stokes, 30 November 1884, in *Stokes–Kelvin Correspondence*, vol. 2, 572.

21 Isabella Humphry (née Stokes) in Larmor, *Stokes Correspondence*, vol. 1, 34–5.

22 Isabella Humphry (née Stokes) in Larmor, *Stokes Correspondence*, vol. 1, 42.

23 W.H. Miller served 1856–73.

24 On Weld and White, see Hall, *All Scientists Now*, 80; and Walter White, *The Journals of Walter White, Assistant Secretary to the Royal Society* (London: Chapman and Hall, 1898).

25 On Huxley's election to the secretaryship, see Barton, *The X-Club*, 272–3.

26 Larmor, *Stokes Correspondence*, vol. 1, 98. See also Barton, *The X-Club*, 277.

27 Larmor, *Stokes Correspondence*, vol. 1, 100.

28 Aileen Fyfe, Flaminio Squazzoni, Didier Torny, and Pierpaolo Dondio, 'Managing the growth of peer review at the Royal Society journals, 1865–1965', *Science, Technology and Human Values* 45, no. 3 (2020): 405–29.
29 'Register of Papers' 1853–85, RS MS/421. There is no direct evidence that this was Sharpey's innovation, but the entries begin with papers read at the meeting of 8 December 1853, the first meeting after Sharpey's election. The analysis that follows (and also in later chapters) is based upon the *Virtual Register of Papers* database we created: it holds editorial data transcribed from the 'Register of Papers' for sample years. The sample years include all years ending -0 or -5, and also certain continuous runs, including 1862–79.
30 The earlier list can be found at RS MM/14/43.
31 On White's role, and the fact that the 'Register' was kept 'private' (rather than available for consultation), see RS CMP/03, 29 March 1860.
32 Barton, *The X-Club*, 284.
33 See C.H. Lees (rev. Graeme J.N. Gooday), 'Everett, Joseph David (1831–1904)', in *ODNB*; and Certificate for Joseph David Everett, 15 March 1878 [*sic*], RS EC/1879/14.
34 See Robin J. Wilson, 'Kirkman, Thomas Penyngton (1806–1895)', in *ODNB*.
35 George Hoggan and Frances Elizabeth Hoggan, 'II. Lymphatics and their origin in muscular tissues', *Proc* 25, no. 171–8 (1877): 550–1; and 'I. On the minute structure and relationships of the lymphatics of the mammalian skin, and on the ultimate distribution of nerves to the epidermis and subepidermic lymphatics', *Proc* 26, no. 179–84 (1878): 289–90. The slow refereeing process for the first paper appears to have led to the reverse order of publication of the abstracts. The papers included dozens of histological drawings, which were returned to the authors. See M.A. Elston, 'Hoggan [née Morgan], Frances Elizabeth (1843–1927)', in *ODNB*.
36 J.J. Waterston and Lord Rayleigh, 'On the physics of media that are composed of free and perfectly elastic molecules in a state of motion', *PTA* 183 (1892): 1–79.
37 The paper was by Dr J. Newton Heale, on the 'Blood-vessels of the lung'. Heale's request to copy illustrations was approved by Council on 22 December 1859; the MS was found to be missing on 23 February 1860 and a committee investigating the loss, and the security of the Society's archive, reported on 29 March 1860, all in RS CMP/03.
38 C. Piazzi Smyth, 'Solar science at the pleasure of *Secret* Referees', *Nature* (13 April 1871): 468–9.
39 C. Piazzi Smyth's 'On supra-annual cycles of temperature in the Earth's surface-crust', received 4 March 1870, sent to Balfour Stewart 6 May 1870, sent to Francis Galton 19 May 1870, considered by Committee of Papers 15 December 1870, 'Register of Papers', RS MS/421.
40 Two months for all reports to be received was proposed (unsuccessfully) both on 20 July 1848 and 14 April 1853, RS CMP/02.
41 RS CMP/03, 5 November 1863. It is not clear why the changes to the pattern of issue were not implemented.
42 Treasurer's report to Council (W.A. Miller), in RS CMP/03, 17 March 1864.
43 Sabine's arguments are discussed in Chapter 10. See RS CMP/02, 19 February 1852 and RS CMP/03, 24 February 1859.
44 Treasurer's report, in RS CMP/03, 17 March 1864.
45 John Stenhouse to Stokes, on a paper 'Researches on the action of sulphuric acid …' by G.B. Buckton and A.W. Hoffman, 2 July 1856, RS RR/3/30; G.B. Buckton to Stokes, on a paper 'On the constitution of the essential oil of rue' by Charles Greville Williams, 16 April 1858, RS RR/3/290.
46 Stokes to S. Houghton, regarding a paper 'On the reflexion of polarized light …' by Samuel Haughton, 24 July 1862, RS RR/5/125; J.C. Maxwell to Stokes, on a paper 'Determination of the number of electrostatic units …' by Dugald M'Kichan, [c. May 1873], RS RR/7/249.
47 Owen to [Stokes], on a paper 'On the existence of the decidua … in four cases of fallopian-tube conception …' by Robert Lee, 12 August 1857, RS RR/3/178.
48 James Bottomley to Stokes, on a paper 'On a standard voltaic battery' by Latimer Clark, 26 March 1873, RS RR/7/231.
49 William H. Miller to Sharpey, on a paper 'Researches on the velocities of currents of air …' by W.D. Chowne, 28 October 1856, RS RR/3/70.
50 Stokes to Faraday, 8 June 1860, Letter 3788 in Frank A.J.L. James, ed., *The Correspondence of Michael Faraday* 6 vols (London: Institution of Electrical Engineers, 1991–2012), vol. 5.

The paper was Faraday's note 'on the possible relation of gravity with electricity or heat'; after Stokes's advice, Faraday did not submit it to the Society.

51 For 'originality', see J. Stenhouse, on a paper 'Researches on the action of sulphuric acid ...' by G.B. Buckton and A.W. Hoffman, 2 July 1856, RS RR/3/30.

52 Edward Boxer, on a paper 'On the nature of the action of fired gunpowder' by Lynall Thomas, 12 December 1858, RS RR/3/260.

53 Archibald Smith, on a paper 'On the envelope of planes through the points of an ellipsoid ...' by Arthur Cayley, 5 May 1858, RS RR/3/60.

54 G.W. Busk to Stokes, and J. Burdon-Sanderson to Stokes, on a paper 'Structural elements of urinary calcale' by Henry John Carter, [n.d.] 1873, RS RR/7/224-225.

55 Charles Wheatstone to Stokes, on a paper 'On a standard voltaic battery' by Latimer Clark, 1 July 1873, RS RR/7/232. The paper was in fact printed.

56 Stokes to J. [Halmes?], 29 June 1871, MC/9/228.

57 Faraday to C.F. Schoenbein, 6 February 1837, Letter 968 in James, *Faraday Correspondence*, vol. 2.

58 Lloyd to C.R. Weld, on a paper 'Discussion of the observed deviations of the compass ...' by George Biddell Airy, 29 October 1855, RS RR/3/1.

59 W.B. Carpenter, on a paper 'On the ultimate arrangement of the biliary ducts ...' by Lionel Beale, 25 August 1855, RS RR/3/20.

60 Median page length for the *Transactions* in 1865 and 1875 is between 20 and 25 pages, but 25 per cent of papers were longer than 40 pages. Another 25 per cent of papers were shorter than 15 pages.

61 Aileen Fyfe, 'Where did the practice of "abstracts" come from?' (8 July 2021), https://arts.st-andrews.ac.uk/philosophicaltransactions/where-did-the-practice-of-abstracts-come-from.

62 See, for instance, William Marcet, 'On a colloid acid, a normal constituent of human urine', *Proc* 14 (1865): 1–9. The paper was read on 16 June 1864, and a three-page 'abstract' had appeared in *Proc* 13 (1864): 314–16. It was refereed by Mr Graham and Dr Bence Jones.

63 T.A. Hirst to Stokes, on a paper 'Introductory memoir on plane stigmatics' by Alexander John Ellis, 18 September 1865, in RS RR/5/69.

64 Burdon-Sanderson to Stokes, on a paper 'Visible direction ... the study of monocular and binocular vision' by James Jago, 12 May 1873, RS RR/7/246. The other referee was Maxwell (RS RR/7/247), who agreed it was not suitable for the *Transactions*.

65 Printed referee report form, 16 February 1883, RS RR/9/171 (asking J.W. Hulke to report on R. Owen).

66 Treasurer's report, in RS CMP/03, 17 March 1864.

67 Henry Clifton Sorby to Stokes, on a paper 'On a new locality of amblygonite ...' by Alfred Louis Olivier Des Cloizeaux, 24 February 1873, RS RR/7/234.

68 Stokes to Herschel, 10 October 1856, RS MM/16/57, re W. Pole, 'On colour-blindness', *PT* 149 (1859): 323–39.

69 Thomson to Stokes, on a paper 'Experiments to determine the value of the British association unit of resistance ...' by John William Strutt, Lord Rayleigh, 23 April 1882, RS RR/9/86.

70 Darwin to Leonard Horner, 18 March 1855, Letter 1649 in *Darwin Correspondence Project*. The Darwin correspondence can be accessed at http://www.darwinproject.ac.uk or in print as Frederick Burckhardt and James A. Secord, *The Correspondence of Charles Darwin*, 30 vols (Cambridge: Cambridge University Press, 1985–2022).

71 These findings come from our analysis of the *Virtual Register of Papers*.

72 Benjamin Newman, 'Authorising geographical knowledge: The development of peer review in The Journal of the Royal Geographical Society, 1830–c.1880', *Journal of Historical Geography* 64 (2019): 85–97; James Mussell, *Science, Time and Space in the Late Nineteenth-Century Periodical Press* (Aldershot: Ashgate, 2007), 129–30.

73 The earliest example in the *Darwin Correspondence* of Darwin acting as referee is for the Geological Society in 1838: Darwin to the Geological Society, 7 September 1838, Letter 426 in *Darwin Correspondence Project*. He also refereed for the Linnean, see Darwin to the Linnean Society, [10] May 1869, Letter 6722 in *Darwin Correspondence Project*.

74 George Bowdler Buckton to Stokes, on a paper 'On the constitution of the essential oil of rue' by Charles Greville Williams, 16 April 1858, RS RR/3/290.

75 Thomas Bell, on a paper 'Description of the skull and teeth of the Placodus laticeps' by Richard Owen, 8 April 1858, RS RR/3/210.
76 Melinda Baldwin, 'Tyndall and Stokes: Correspondence, referee reports and the physical sciences in Victorian Britain', in *The Age of Scientific Naturalism: Tyndall and his contemporaries*, ed. Bernard Lightman and Michael S. Reidy (London: Pickering & Chatto, 2014), 171–86.
77 Tyndall to Stokes, 9 December 1883, Stokes Correspondence, CUL MS Add. 7656.
78 Tyndall to Stokes, 25 June 1864 [Tyndall 2115], Stokes Correspondence, CUL MS Add. 7656.
79 Larmor, *Stokes Correspondence*, vol. 1, 103.
80 Stokes to Faraday, 7 November 1855, Letter 3036 in James, *Faraday Correspondence*, vol. 5.
81 Stokes to Faraday, 5 November 1855, Letter 3034 in James, *Faraday Correspondence*, vol. 5. This was the 30th paper in the series of that name.
82 Letters 3039–3040 in James, *Faraday Correspondence*, vol. 5.
83 Stokes to Faraday, 25 October 1856, Letter 3200 in James, *Faraday Correspondence*, vol. 5.
84 Hoffman to Stokes, on a paper 'On ... the destructive distillation of boghead coal' by Charles Greville Williams, 1 June 1857, RS RR/3/280.
85 George Busk, report on a paper 'Researches on the poison-apparatus in the actiniadae' by Philip Henry Gosse, 15 April 1858, RS RR/3/120.
86 Dwight Atkinson, *Scientific Discourse in Sociohistorical Context: The Philosophical Transactions of the Royal Society of London, 1675–1975* (London: Routledge, 1998); Alan G. Gross, Joseph E. Harmon, and Michael S. Reidy, *Communicating Science: The scientific article from the seventeenth century to the present* (New York: Oxford University Press, 2002).
87 In fact, the Society had no mechanism for adjudicating such claims, as it discovered during a controversy between the chemists Hermann Sprengel and Frederick Abel in 1891–3 (but concerning a paper published in 1874; the 'date read' was not an issue in this case), see documents in RS MM/16.
88 RS CMP/02, 25 May and 6 July 1848.
89 RS CMP/05, 13 May 1880.
90 Darwin to Horner, 18 March 1855, Letter 1649 in *Darwin Correspondence Project*.
91 Stokes to Lockyer, 22 Nov 1872, in Larmor, *Stokes Correspondence*, vol. 1, 398–400. The paper appeared as J.N. Lockyer, 'Researches in spectrum-analysis ...', *PT* 163 (1873): 253–75.
92 George Carey Foster to Stokes, on a paper 'On the expansion of sea-water by heat' by Thomas Edward Thorpe and Arthur William Rucker, 5 February 1876, RS RR/7/489. Foster's formal report is RR/7/490.
93 Carpenter, on a paper 'Researches into ... the urinary bladder' by George Viner Ellis, 18 December 1856, RS RR/3/100.
94 Sharpey, on a paper 'On the ... cerebral hemispheres of the marsupialia and monotremata' by William Henry Flower, [n.d.] 1865, RS RR/5/78. John Marshall's report on the same paper, 18 May 1865, RS RR/5/77.
95 Tyndall to Stokes, on a paper 'Numerical elements of Indian meteorology' by Hermann De Schlagintweit, 13 June 1865, RS RR/5/240. The full paper was not published, but (as with the other two papers in the series) the abstract appeared in the *Proceedings*.
96 H.J.S. Smith, on a paper 'On a new geometry of space' by Julius Plücker, 28 March 1865, RS RR/5/188.
97 Stokes to Tyndall, 26 April 1855 [Tyndall 1076], RI MS JT/1/S/217.
98 Hofmann to Stokes, on a paper 'On ... the destructive distillation of boghead coal' by Charles Greville Williams, 1 June 1857, RS RR/3/280. Hofmann recommended cutting seven pages.
99 E. Frankland and H. Debus, on a paper 'On the union of ammonia nitrate with ammonia' by Edward Divers, 3 April 1873, RR/7/238. Frankland and Debus had initially written individual reports, but were requested to confer on a recommendation.
100 RS CMB/90/5.l, 1 May 1873.
101 Stokes to Tyndall, 26 April 1855 [Tyndall 1076], RI MS JT/1/S/217.
102 Stokes to David Douglas Cunningham, re Cunningham's paper 'On the relations of particular structural features', 7 July 1883, RS RR/9/133.

103 Stokes to Tyndall, 21 March 1874, RI MS JT/1/S/259. The paper became Tyndall, 'On the atmosphere as a vehicle of sound', *PT* 164 (1874): 183–244, with a footnote explaining its origins in a number of separate presentations to the Society.
104 Stokes to Tyndall, 19 March 1874, RI MS JT/1/S/258.
105 Stokes to Tyndall, 21 March 1874, RI MS JT/1/S/259. American researchers did indeed claim priority.
106 Aileen Fyfe, *Steam-Powered Knowledge: William Chambers and the business of publishing, 1820–1860* (Chicago, IL: University of Chicago Press, 2012); Martin Hewitt, *The Dawn of the Cheap Press in Victorian Britain: The end of the 'Taxes on Knowledge', 1849–1869* (London: Bloomsbury, 2013); Andrew Hobbs, *A Fleet Street in Every Town: The provincial press in England, 1855–1900* (Cambridge: Open Book Publishers, 2018).
107 Gowan Dawson, '"An independent publication for geologists": The Geological Society, commercial journals, and the remaking of nineteenth-century geology', in *Science Periodicals in Nineteenth-Century Britain: Constructing scientific communities*, ed. Gowan Dawson, Bernard Lightman, Sally Shuttleworth and Jonathan R. Topham (Chicago, IL: University of Chicago Press, 2020), 137–71.
108 Graeme Gooday, 'Periodical physics in Britain: Institutional and industrial contexts, 1870–1900', in Dawson et al., *Science Periodicals*, 238–73.
109 Barton, *The X-Club*, 177–82; Geoffrey Belknap, 'Natural history periodicals and changing conceptions of the naturalist community, 1828–65', in Dawson et al., *Science Periodicals*, 35–64.
110 Advertisement page of *The Reader* (16 January 1864), quoted and discussed in Barton, *The X-Club*, 206.
111 Melinda Baldwin, *Making 'Nature': The history of a scientific journal* (Chicago, IL: University of Chicago Press, 2015), ch. 1. See also A.J. Meadows, *Science and Controversy: A biography of Sir Norman Lockyer* (London: Macmillan, 1972); Bernard Lightman, '"Knowledge" confronts "nature": Richard Proctor and popular science periodicals', in *Culture and Science in the Nineteenth-Century Media*, ed. Louise Henson, Geoffrey Cantor, Gowan Dawson, Richard Noakes, Sally Shuttleworth and Jonathan R. Topham (Aldershot: Ashgate, 2004), 199–210.
112 This practice stopped in 1887, see *Record of the Royal Society* (1912), 275.
113 Airy to Stokes, 27 March 1872, in Larmor, *Stokes Correspondence*, vol. 2, 165–6.
114 Airy to Stokes, 27 March 1872, in Larmor, *Stokes Correspondence*, vol. 2, 165–6.

10
Scientific publishing as patronage, c. 1860–1890
Julie McDougall-Waters and Aileen Fyfe

The Royal Society's biggest publication project in the late nineteenth century was not the *Transactions* nor the *Proceedings*, but the mammoth effort to create a *Catalogue of Scientific Papers*. Since 1858, a team of staff and volunteers had been working in the Society's library to compile a universal index to the vast and growing periodical literature of science. As Thomas Henry Huxley noted in 1885:

> It has become impossible for any man to keep pace with the progress of the whole of any important branch of science. If he were to attempt to do so his mental faculties would be crushed by the multitude of journals and of voluminous monographs which a too fertile press casts upon him.[1]

In contrast to the Society's earlier involvement with abridgements, indexes and abstracts (see Chapters 4 and 8), the scope of the *Catalogue* reached far beyond its own meetings and publications: it would survey research being published by national academies, subject-based societies and independent editors, whether in Britain, Continental Europe, North America or further afield. At its completion in 1925, it would list the authors and titles of papers published in 1,400 international scientific periodicals between 1800 and 1900.[2]

The task of gathering, copying and sorting the details of papers published in so many journals was not something that could be achieved in anything like real-time, with the techniques available in the late nineteenth century. The first volume, covering papers published between 1800 and 1863, did not appear until 1867. By 1887, the project's coverage had reached 1873, but, as George Stokes told the fellows that year, work on the most recent decade was proving 'more arduous than was expected'. This was partly because the printed output to be surveyed 'very much exceeds in bulk' that of earlier decades. But it was also a question of locating and accessing copies of the periodicals to be surveyed. For the earlier period, much of the work had been done in the Society's own library, but its holdings had not kept pace with the flourishing of specialist journals in Britain and internationally. The determination to make the *Catalogue* 'as complete as possible, and to include scientific serials in all languages' involved members of the team in 'gleaning stray papers from works in other libraries' across London, and sometimes further afield.[3]

The scale of the *Catalogue* project reflected the expansion of scientific research and journals in the late nineteenth century. But the Society's willingness to take on the project – originally suggested by the American physicist Joseph Henry – and bring it, eventually, to a successful conclusion over 40 years later, also exemplifies the Society's vision of itself as a benefactor or patron of the scholarly world. Stokes made this explicit in his 1889 presidential address, when he explained that: 'This work, in the preparation of which the Royal Society has spent a large sum, is for the benefit of the whole civilized world.'[4] This paternalistic benevolence was a guiding theme of the Society's activities, including its publishing, throughout the Stokes years. It can be seen in the increasing range of research projects and publications sponsored by the Society, including the reports of the 1872–6 oceanographic research expedition of HMS *Challenger*, and reports on the 1883 eruption of Krakatoa and the 1886 total solar eclipse.[5] It can be seen in the number and global distribution of the institutions receiving free copies of the Society's *Transactions* and *Proceedings*; and in the Society's lack of interest in trying to use copyright legislation to restrict reprinting of papers it had published. It can also be seen in the generous production values lavished upon papers selected for publication in (particularly) the *Transactions*, and the willingness to accept that large numbers of images were necessary for the sake of scholarship.

The production values and mode of distribution of the Royal Society publications were guided by what the Society's officers believed to be good for scholarship, regardless of what the commercial market would bear. The *Proceedings* and the *Transactions*, therefore, operated within an intellectual and business framework that was quite distinct from that of the independent journals. Such journals as the *Philosophical Magazine*, the *Annals of Natural History* (both owned by Taylor & Francis) and *Nature* (owned by Macmillan) operated with the aim of generating enough income from sales to cover costs (though this aim was not always realised).[6] For editors and proprietors of such journals, finding the right combination of material that readers wanted, at a price they were willing to pay, was crucial. The Royal Society rarely, if ever, defined its intended audience, but it was implicitly a scholar whose needs were assumed to be much like those of the fellows of the Society, and whose needs should not be constrained by mere financial considerations. Much like the university presses, the Society's officers believed that works of scholarship should be published to advance human knowledge, even although they were expensive to produce, and had a limited potential for sales. These were coming to be recognised as the sorts of publications that readers would access in libraries, rather than as personal copies.

The Royal Society's ongoing commitment to generous and gentlemanly standards of scholarly publication was another way in which its periodicals were marked out from the many newer journals and magazines of late nineteenth-century Britain (in addition to the editorial practices discussed in Chapter 9). But, as we shall see, maintaining this commitment was beginning to put the finances of the entire Society under strain. That tension, between the mission for scholarship and the practical difficulties of funding it, would be an ongoing challenge for the Society until the 1950s.

Circulating scientific knowledge

There is little surviving evidence that the officers of the Royal Society paid much attention to the circulation of the *Proceedings* or the *Transactions* in the nineteenth century. The presidents' anniversary addresses routinely reported the number of papers, pages and illustrations published, but not the number of copies printed, sold or otherwise distributed. The income from sales was an item in the annual accounts, but never explained or analysed. The one form of circulation that the Society did celebrate publicly was its extensive programme of gifts. The list of institutions entitled to free copies of the Society's periodicals was printed annually: in 1878, it ran to four pages, detailing 160 recipients of a complete set of the periodicals, plus a further hundred or so receiving only the *Proceedings*.[7] Even in the archive, there are only occasional traces of any attempt to survey the different modes of circulation used by the Society. Part of the reason may be that no one person or committee had oversight: copies for fellows were distributed by the assistant secretary; gifts and exchanges with other institutions were managed by the Library Committee; the treasurer received any income from sales, but the management of sales was left to the assistant secretary and the printer.

Table 10.1 Circulation of the *Proceedings* in 1863

The fellows of the Society	606
Foreign members	50
Privileged libraries	5
Presented to institutions and individuals	158
Gifts	**819**
The Chemical Society	366
The Royal Irish Academy	100
Deeply discounted or exchange	**466**
Non-commercial circulation	*1,285*
Subscribers	56
Occasional sales	12
Commercial circulation	*68*
Copies remaining on hand	147

The most detailed information we have about circulation in the later nineteenth century concerns the *Proceedings*, and comes from 1863. The committee that was reviewing the periodicity and format of the *Proceedings* also investigated its mode of distribution. At the time, Taylor & Francis were printing more copies of the *Proceedings* (1,500 copies) than of their own *Philosophical Magazine* (550 copies) or *Annals of Natural History* (400 copies). The committee's findings are presented in Table 10.1, and dramatically demonstrate both the Society's minimal engagement in the normal channels of the commercial book trade, and the extent of its non-commercial circulation.[8]

The 1863 figures reveal that only 68 of the copies, less than 5 per cent of the print run, had been sold to the public.[9] They also show that the advance subscription had proved an attractive option, perhaps because it was just 5s. a year. In comparison, the *Philosophical Magazine* cost 2s. 6d. per monthly issue.[10] The review committee's recommendation to regularise the issue of the *Proceedings* as a monthly (to make it easier for booksellers to handle) was not accepted, but its recommendation to raise the price to 10s. was implemented.[11]

The Society's annual accounts in the 1860s usually reported a total sales income for all the publications of around £400. By the 1870s, this would rise to around £630 a year. Most of this income must have come from the *Transactions*: given that the cost of a year's worth of *Transactions* papers was about £4 in the 1870s (varying with length), that equates to fewer than 200 copies sold.[12] We have been unable to find an analysis for the *Transactions* equivalent to that for the *Proceedings*. We do know that its print run was still 1,000 copies, and since the number of copies notionally reserved for fellows, foreign members and learned institutions was around 760 (in the 1870s), we can assume that sales would not have been higher than 240 copies. Reports on the state of the Society's warehouse at the turn of the century suggest that substantial quantities of unsold periodicals had accumulated.[13] Again, this points towards a likely commercial circulation for the *Transactions* of 100–200 copies in the 1860s and 1870s.

Another hint at the low priority placed on commercial sale as a form of distribution for the *Transactions* and the *Proceedings* is the Society's lack of an appointed bookseller in this period. In the Banks era, the Society had had both a 'printer' and a 'bookseller', but Richard Taylor had been appointed only as 'printer' (Chapter 8). By the 1870s, some learned societies, including the Linnean and Chemical societies, had appointed a publisher or bookseller to act as their retail agent;[14] but the Royal Society did not follow suit. People wishing to purchase copies of the Society's periodicals could do so directly from the Society's assistant secretary, or from Taylor & Francis, but neither of those agents actively engaged in publicity or marketing. Taylor & Francis did sell copies to individual customers, and fulfilled requests passed on by other members of the trade (presumably including the subscriptions for the *Proceedings*). Surviving cash books from the late 1850s suggest a pattern of perhaps half a dozen or so purchases each month, amounting to sales of between 10 and 20 copies of various parts of the *Transactions* over the course of the month.[15] Taylor & Francis did organise for new parts of the *Transactions* to be advertised in the periodical press. For instance, in early 1873, advertisements were placed in the *Times*, *Morning Star*, *Athenaeum* and *Nature*.[16] But the address listed in the advertisements was the Society's (new) premises in Burlington House, not the printers. These advertisements were aimed at fellows – to advise them that the latest part was now available for collection – as much as the commercial book-buying public.

Institutional gifts

The modest commercial prospects for the *Transactions* and the *Proceedings* were in large part a consequence of the Society's extensive programme of non-commercial distribution.[17] As Table 10.1 showed, 85 per cent of the print run of the *Proceedings* was distributed through learned and scientific institutions: to fellows of the Royal Society and certain other institutions, and to the libraries of universities, societies and scientific institutions. The Society received a modest return from the institutions with whom it had arrangements for bulk distribution: the Royal Irish Academy sent copies of its own *Proceedings* for distribution to the fellows of the Royal Society; while the Chemical Society made a £50 cash payment, thus purchasing copies for its members at about half-price.[18] Nonetheless, the arrangements are best understood as part of the philanthropic scholarly distribution of the journal, rather than commercial sales.

The number of institutions whose libraries were entitled to receive copies of the *Transactions* or the *Proceedings* had grown four-fold over the course of the nineteenth century: by the 1870s, there were more than 270 such institutions.[19] Their geographical spread had extended, as the contrast between Figures 7.2 and 10.1 shows. The majority of copies still went to institutions in Britain, including almost all the universities and university colleges, most of the metropolitan scientific societies and a variety of other national scientific organisations. London had the greatest density of copies of the *Transactions*: in 1878, in addition to the private copies of fellows and the set in the Royal Society library, it was available for consultation in at least 28 locations. These included the libraries of the Chemical and Entomological societies, the War Office and the Admiralty. The Society's generosity was extended to equivalent organisations in Continental Europe, and also to certain libraries, museums and observatories. A handful of colonial institutions benefited, as did a dozen institutions in the United States of America. The content printed in the Society's periodicals was predominantly of British origin in this period, but the patterns of free distribution point to a concern with the global circulation of knowledge, particularly (but not only) in the imperial and Anglophone world.[20]

The Royal Society's library received periodicals as gifts from other learned institutions, and sometimes these came from the same institutions that had been sent the Society's publications. Historians have frequently referred to these as institutional or library 'exchanges', a phrasing that ignores the unreciprocated gifts that the Society also gave (for instance, to university libraries). It also ignores the Society's own ambivalence about whether it was 'exchanging' or gifting. Joseph Banks had certainly thought in terms of 'presents' from a generous Royal Society and, in 1830, Michael Faraday had reported to a Genevan correspondent that 'the RS does not *exchange* with any Society'; rather, 'they *present* Transactions to Royal & National Societies'.[21] By the 1860s, usage was changing, perhaps

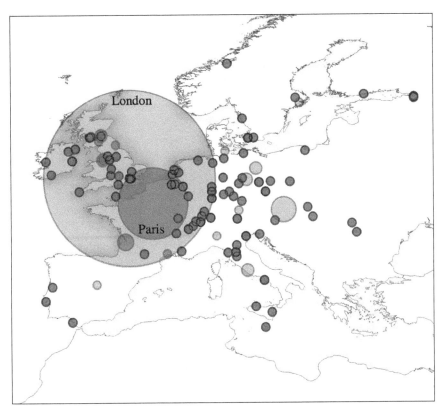

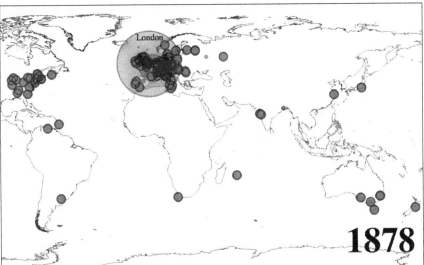

Figure 10.1 Global distribution of free copies of the *Transactions* and/or the *Proceedings* in 1878. The size of the spots is proportional to the number of gifts per city.

due to the influence of a Library Committee that was becoming more alert to the potential value of return gifts. In 1863, for instance, the Council agreed that copies of the *Proceedings* could be granted to the library at the new Royal Victoria military hospital, 'in exchange for the "Reports of the Army Medical Department"'.[22] In the 1930s, the Society would formally acknowledge that the institutions receiving its largesse could be distinguished between a 'free list' and an 'exchange list' (see Chapter 12).

The list of institutions receiving 'presents' was not planned nor actively managed. It evolved organically from the numerous requests received by the Society. Requests were considered several times a year by the Library Committee, whose recommendations were then presented to Council. No rules survive to indicate how the Library Committee was supposed to decide which institutions were eligible, and, if eligible, what they should receive, though some can be inferred. For instance, in May 1874, Council approved the following presents: future and retrospective (from 1830) volumes of the *Transactions* and the *Proceedings* to the Guildhall Library, London, and the library of the Royal Engineers; odd numbers of the *Transactions* and the *Proceedings* to the Société Géologique of Paris and the Société de Physique et d'Histoire Naturelle of Geneva, to enable them to complete their back-runs; and future issues of the *Proceedings* (only) to the Physikalische-medicinische Societat of Erlangen, and the Wisconsin Academy of Sciences of Madison.[23] King's College, Nova Scotia, was allowed a copy of the *Catalogue of Scientific Papers* after Stokes had made enquiries 'as regards the status of the College', but enquiries into the publication activity of the Torquay Natural History Society resulted in a decision against sending the *Proceedings* to Torquay. The request of the Bilston Free Library (Wolverhampton) was refused without any stated reason.[24] An implicit hierarchy of learned institutions was developing: newer, and especially provincial, scientific societies (whether foreign or British) were often granted the (cheaper) *Proceedings*, whereas the older institutions that had been on the list for decades usually received a full set of the *Transactions* and the *Proceedings*.

The Royal Society was willing to send its publications to organisations whose members produced research publications that would be a valuable addition to the Society's library, and to institutions whose libraries seemed desirable locations to place copies of the *Transactions* and the *Proceedings*. Institutions that could offer neither publications (Torquay) nor an appropriate pool of library users (Bilston) were likely to be refused. Jenny Beckman has shown that the Royal Academy of Sciences in Stockholm carefully managed its periodicals to maximise their effectiveness in stocking the Academy's library through return exchanges.[25] The Royal Society, in contrast, appears to have prioritised the distribution of its own publications over the possibility of exchange.

Personal scholarly networks

Virtually all of those who read the *Transactions* or the *Proceedings* during the late nineteenth century probably did so without purchasing their own copies. They might gain access as a member of a university, observatory, government department or research institution that received a set for the library. Or they might be fellows of the Society (or family, friends or colleagues with access to a fellow's book shelves). Free copies of the *Transactions* had been a membership perquisite for fellows since the 1752 settlement, and the *Proceedings* had been added to that without apparent comment. During the nineteenth century, some of the newer learned societies – including the Geological Society – had been financially obliged to adopt a different arrangement, offering their fellows discounted rather than free copies, but the Royal Society never appears to have considered removing its fellows' privilege.

The Society also promoted the circulation of knowledge through non-commercial channels by its continued provision of separate copies (or offprints) to authors. Authors sent them to their correspondents, and they could expect requests to arrive in the post from researchers who wanted a personal copy. John Tyndall's correspondence is full of references to the papers he received from correspondents in Britain and Germany, and his own actions in sending copies of his own papers; he also took copies with him on his summer travels to the Alps.[26] The Royal Society continued to provide separate copies as soon as they were printed, which meant that those who were in the author's scholarly networks might get advance access to the full versions of the latest papers. Even those who received them after publication gained the advantage of a personal copy to peruse or annotate at their leisure, more conveniently than with a library copy.

Separate copies still had to be requested at the right moment, and, by the 1870s, the Society sought to discipline authors by developing a printed form to be attached to the proofs, which authors could return to order offprints. In the 1860s, authors had sometimes been told that they were too late: that the type had already been dispersed, and additional copies could not be printed except at the significant expense of re-setting the type.[27] Incidentally, this indicates that the Society's periodicals were still being printed from loose type, despite the increasingly widespread use of stereotype and electrotype plates in other sectors of the trade.

Another indication of the extent of this private circulation can be judged from the Royal Society's rules about the number of copies available to authors. In the 1850s and 1860s, the standard number of copies allowed at the Society's expense was 100, but authors who had plans for wider distribution could acquire additional copies from the printer, provided they were willing to pay 'the expense of the paper and press-work'.[28] Council tried to save money by reducing the number of copies funded by the Society and, by the 1870s, standard provision had been reduced to 50 free copies. However, it could still be increased to 100 on request, and authors could still choose to pay for up to 150 further copies.[29] In 1875, for instance, the zoologist E. Ray Lankester requested, and was granted, 25 additional copies of his *Transactions* paper on the development of molluscs.[30]

The widespread circulation of offprints through private channels could lead to confusion about where and when a paper had in fact been published, and thus how to find it or cite it. Offprints of *Transactions* papers had been clearly marked as such on their covers since the Banks era, but it was not until 1878 that the volume number and date were added.[31] Having an appropriately identified separate copy did not, of course, always help: Tyndall had to tell a correspondent in 1874 that he did 'not know where Clerk Maxwell's lecture was published – He sent me a separate copy – but I cannot lay my hands on it'.[32]

The separate copies included images as well as text, and this could be particularly valuable to authors who wanted to reuse or repackage their journal papers. In the 1860s and 1870s, for instance, comparative anatomist Richard Owen collated several series of his papers on fossil remains into volumes which he sent to, among other places, 'public libraries of cities and towns in Australia and Tasmania', using the ready-printed images from his separate copies to provide the illustrations.[33] Text could be relatively easily set by any printer, but copying images was a skilled and expensive task; it was far easier to incorporate ready-printed images, if they could be acquired. The scale of the requests received for extra copies of illustrations suggests some substantial ambitions for reuse: for instance, in 1870, one author was permitted to acquire (at his own expense) 500 extra copies of a lithographic plate from his *Transactions* paper; and in 1876, another was allowed 250 extra copies of the illustrations (in addition to the maximum number of separate copies) from his paper on sharks and dogfishes, probably for his forthcoming book.[34]

Owen's actions drew a reprimand from the Society in 1877, because the Council assumed that, in creating his volumes, he had caused additional copies of his images of Australian fossils to be printed without asking permission. Owen explained that he had actually extracted all the image pages from his legitimate supply of separate copies; but he offered to pay for any damage he might have caused to the 'property of the Society', if such damage could be valued. The Council ignored the offer of financial compensation and directed the secretary to convey 'regret' that the Society and the *Transactions* had been inadequately acknowledged and, in particular, 'that the Society's mark at the head of the plates should have been obliterated without express permission'.[35] (The way the bibliographic information should appear can be seen in Figure 10.2, later in this chapter.) The Society's concern was not financial, but reputational. When a new set of guidelines was drawn up in 1878, the Society was careful to limit the possible reuses. The guidelines reiterated that all separate copies could be used 'for gratuitous distribution only'. They insisted that the copies will 'in every case bear the name and date of the volume from which they are extracted' and that '[t]he original paging of the letter-press and numbering of the plates will always be retained'. Authors who wished to use the images in a new volume (such as a book based on several journal articles) could request 'additional paging and numbering', but they could not have traces of the original publication venue removed.[36] The Society's aim was to ensure that the pages and plates remained identifiable as *Transactions* pages.

Until the mid-1870s, only those who were part of the correspondence networks of Royal Society authors could benefit from the private and non-commercial circulation of separate copies of papers. In March 1875, however, the Library Committee agreed to a suggestion from Charles Darwin (communicated via his friend, the then-president Joseph Hooker) that 'the sale of the papers of the Philosophical Transactions in a separate form' would promote 'the interests of science'.[37] Putting separate copies on sale at a bookseller's shop would make them available to people who were not in the author's personal network. Arrangements were made with Messrs Trübner & Co., a well-established Anglo-German publisher and agent. The decision to seek a partner suggests that Taylor & Francis were recognised to lack expertise in sales and distribution, but it is also notable that the arrangement with Trübner concerned only the separate copies of *Transactions* papers, not complete parts or volumes of either the *Transactions* or the *Proceedings*. In other words, the Society's interest in seeking a commercial market for its periodicals remained almost non-existent.

Trübner understood the new arrangement as being intended to make the *Transactions* more 'accessible to students', and he promised to 'send copies of each to my various depots in the principal cities of Europe' and to use his monthly list of new publications to 'keep the whole before the public'.[38] He pushed the Society to agree to various standard trade terms, including: allowing Trübner's imprint to appear, alongside the Society's, on the wrappers; granting Trübner a 10 per cent commission on sales; and allowing him to offer retail booksellers 25 copies for the price of 24.[39] Despite the optimistic vision of sales that this suggested, Trübner expected that the sale would likely be low.[40] Even the Society's Council felt that the Library Committee was being over-optimistic in hoping that 50 copies of each paper might be sold in this way.[41] In fact, Trübner was sent 25 copies of each paper and, even so, in 1883, he returned 'a large quantity of unsold stock', suggesting that in future 'not more than ten copies' of each paper would be sufficient.[42] Despite this disappointment, Trübner (and the successor firm, Kegan Paul) continued to sell separate papers of the *Transactions* for another decade.[43]

Even though the sales of separate copies of *Transactions* papers were exceptionally modest, the decision had far-reaching ramifications. First, the century-old tradition of issuing the *Transactions* in six-monthly parts, ahead of an annual volume, rapidly disappeared. From the 1870s until the mid-twentieth century, the *Transactions* was issued either as a series of individual papers, or as bound annual volumes. It might be better described as a series of memoirs rather than a periodical in the usual sense: the separate papers had no regular periodicity and no fixed cover price, and each was a monograph on a single topic rather than a collection of items by different contributors. Nonetheless, from the Society's perspective, both the *Transactions* and the *Proceedings* continued to be considered as periodicals.

Second, by enabling purchasers to buy only the specific papers that actually interested them, this move undermined the identity of the *Transactions* as a broad generalist periodical. With no more six-monthly parts, fellows and learned institutions also now received their free copies either as separate papers or as annual volumes, and some chose to claim only the papers that were of direct interest to them. It can therefore be seen as a step towards the 1887 decision to issue (and bind) the *Transactions* as two separate series, one for the physical sciences and one for the life sciences.

Third, issuing the *Transactions* as separate papers speeded up the publication of many papers, as they no longer had to wait until all the papers for the current part were ready. And finally, it shifted the Society's official position on the true date of publication. Joseph Banks's argument (see Chapter 7) that separate copies were private, and the

Transactions was only formally published when its volume appeared, was no longer tenable when the separate copies were on public sale in Trübner's shop. This was a significant issue in relation to the question of when the wider periodical press could report or reprint material from the Society's meetings and publications. The officers of the late nineteenth-century Society shared Banks's desire that the Royal Society should be unambiguously recognised as the venue of first publication, but his insistence that 'one month after formal publication' was an appropriate delay before reprinting no longer made sense in a world in which the *Proceedings* had already become the key source of news about Society meetings. For the increasing number of short papers that were being published in the *Proceedings*, there was no other possible moment of publication. But for papers whose abstracts appeared in the *Proceedings*, to be followed some months later by the full version in the *Transactions*, it was difficult to say whether rapid reporting, based on the abstract, should be preferred over the long delay that would be necessary if reporting was to be based on the full published paper. Putting the separate copies of *Transactions* papers on sale helped solve this problem, enabling editors to report, discuss and cite full *Transactions* papers more quickly than before.

Reprinting

The Society's interest in enabling more rapid reporting, as well as its willingness to help authors reuse images in their books, are both examples of the way in which the Society's commitment to circulating knowledge was not limited to the physical distribution of the copies of the *Transactions* and the *Proceedings*. The landscape of scientific periodicals and journalism had been transformed by the appearance of monthly and weekly magazines, and the Society's traditional distance from those venues had been moderated by a widespread recognition of the usefulness of a rich ecosystem of types of periodicals. Oldenburg's *Transactions* had managed to supply a variety of research, reviews and news to its relatively homogeneous community of readers; but since then, the expansion and fragmentation of knowledge, and the expansion and diversification of the communities involved in the sciences, made it impossible for any one periodical to fill all these niches. The fact that so many of the fellows of the Royal Society were involved with other periodicals – as readers, contributors and editors – eased the antagonism that Banks had once felt towards those who ran journals. By the 1890s, the Society would even create a distribution list of editors who should automatically receive advance copies of forthcoming papers.[44]

One way in which we can see a new willingness to engage with the scientific press is in the requests repeatedly recorded in the Council minutes for permission to reuse the illustrations from the *Transactions* and the *Proceedings*. In the 1860s and early 1870s, those wishing to reprint or reuse images had, as we have seen, sought permission to acquire extra printed copies of the images. The Society was, on occasion, willing to grant such permission to journal editors, not just to authors. In 1871, it granted a request from the Admiralty (with the author's permission) for '750 or 1000 copies of the table and diagrams' from a paper by William Thomson that had appeared in the *Proceedings*. These images would be used 'to illustrate a reprint in the "Nautical Magazine", the expense to be defrayed by the conductors of that publication'.[45] By the 1870s and 1880s, however, the requests shifted to permission to reuse the engraved metal plates, lithographic stones or wood-blocks that carried the illustrations. These physical materials were stored in the Society's warehouse, where they represented a substantial investment of capital. Allowing their reuse – always with suitable attribution to the Society – helped circulate the Society's research, and its name, a little further at no extra cost to the Society. And if the recipient was willing to pay, it even generated a little income.

Almost every Council meeting in the 1870s and 1880s included at least one request for images. The requests sometimes came from authors themselves: for example, Tyndall was given leave in 1870 to 'use the woodcut blocks and copper plates of the figures illustrating his paper on diamagnetism' for a reprint of the paper.[46] In 1877, Taylor & Francis sought permission to use Royal Society plates, some of which were 40 years old, for a (posthumous) facsimile reprint of Michael Faraday's *Experimental Researches*.[47] But the requests also came from editors of news magazines wanting images to accompany their reports of the latest research published by the Society. Thus, in 1871, the 'editor of *Nature*' was allowed to have 'electrotype copies' made 'at his own expense, of the cuts illustrating the Report of the [HMS] "Porcupine" expedition'.[48] The following year, the editor of *Popular Science Review* was allowed to borrow 'the Blocks of the two cuts in Dr Carpenter's paper on Deep-Sea Temperatures'.[49] By 1891, the routine nature of such requests can be seen in the way permission was granted to Norman Lockyer 'to have electrotypes [made]' of illustrations in a paper by William Ramsay 'on the usual conditions'.[50]

The Society routinely granted bona fide requests of this type, whether to the original authors or their heirs, or to the editors of journals whose reporting would circulate knowledge further than the *Proceedings* and the *Transactions* could do.[51] Unlike the newspaper proprietors who sought changes to copyright law in the 1890s to prevent competitors from reprinting their exclusive material, the Royal Society generally welcomed efforts to distribute, report, abstract and index its published papers.[52] It still insisted on having the prestige and credit of being the point of first publication for new research; and it insisted on attribution, both as a form of acknowledgement of the Society's generosity and as a tool to aid scholarly literature searches and citations.

This new relationship with the press was part of the Society's growing attention to other ways of circulating knowledge, in addition to its traditional provision of separate copies and gifts to learned institutions. By supporting third-party reporting, commenting and reprinting, it helped some of the Society's publications, and their contents, become known to a wider public.[53] The Society occasionally agreed to send its volumes to public libraries in some of the large industrial cities, but its primary efforts were directed towards a restricted scholarly public. This included a variety of people who were in some way part of a scholarly community, whether through employment in a university or research institute, or through membership in a scientific society or association. But it was not the public at large.

Generous production values

When George Stokes told the fellows that 1,482 pages of letterpress and 76 plates had been prepared for the *Transactions* in 1887, he was detailing the extent of the Society's generosity to its authors. It had paid for the typesetting, printing and paper for all those pages, and covered the costs of all those illustrations.[54] It also insisted on good-quality paper, and the type size and line spacing was a clear sign that the Society was not one of the publishers seeking to reduce costs by cramming as much text as possible onto the page. This applied to both periodicals, but the generosity was far more visible in the *Transactions*, for it carried more illustrations, and had much longer papers. The Society could have saved significant amounts of money by making different decisions about how the *Transactions* should look and feel.

The length at which authors could describe their research in the *Transactions* was of a completely different order to what was possible in either the *Proceedings* or the commercial periodical press. The median length of a *Proceedings* item (whether abstract or independent paper) was four or five pages in the 1870s. An entire issue of *Nature* was just 16 pages, and an 80-page issue of the *Philosophical Magazine* might well contain a dozen different items. In contrast, the median length of a *Transactions* paper in the 1870s was between 18 and 25 pages, and around a quarter of papers ran to more than 40 pages.[55] By the mid-1880s, the median paper length had risen above 30 pages.

Referees did often comment on length. Extreme length could be used as a reason for declining to publish in the *Transactions*, as, for instance, when an 1879 paper was described as 'diffuse and lengthy', and as providing 'no new information';[56] or when an 1880 paper was 'extremely lengthy and poorly communicated'.[57] But it is equally true that referees, and the Committee of Papers, accepted many papers that were long. And sometimes the referees suggested making them longer. An 1877 report on a mathematical paper suggested that the author should add 'an explanatory introduction', work out some of the solutions 'in full, to provide detail and clarity', and incorporate 'a few references that had been omitted'.[58] Quite what referees understood as the appropriate length for a *Transactions* paper is unclear, but the Society placed no limits on the number of pages or illustrations for the *Transactions* until the 1890s (and even then, it did not rigorously enforce the notional limits).

Even after the mechanisation of paper-making, paper remained the most significant cost in publishing. Nor had the removal of the long-standing tax on paper in 1861 reduced prices as much as had been hoped, because demand outstripped the supply of cotton and linen rags used to make it (and this was exacerbated by the disruption of the cotton industry during the American Civil War). Print costs had now been reduced by the adoption of steam-powered, mechanised printing – of which, Richard Taylor had been an early proponent – but typesetting remained a substantial cost, because of the high level of skill needed from compositors who picked out and arranged the individual letters of type. Setting the text for tables and equations required even more skill, and although machines for typesetting would be developed in the late 1880s, the complex scientific material in the Society's periodicals would still have to be set by hand.[59]

The *Transactions* in the late nineteenth century was notable for the number and quality of its illustrations, and this was another area where the Society did not stint its largesse. Commercial periodicals had limited space in each issue and were constrained by their fixed cover price. But for the Royal Society, illustrations seemed essential explanatory tools, whether for describing new species of plants and animals, demonstrating the arrangement of scientific apparatus or reporting a spectrographic analysis. Papers dealing with natural history might easily have a dozen engraved plates. In 1874, Stokes was dealing with the publication of a paper by E. Ray Lankester on molluscs, and jovially remarked to his co-secretary, Huxley, that, 'If the snail knew they were to be honoured with TEN PLATES in the Phil Trans wouldn't they cock up their horns?'[60] (In fact, those particular snails were ultimately honoured with 20 plates.)

Authors could be confident that their papers would be illustrated as required: Albert Günther had felt able to argue that 13 plates were entirely 'necessary for the proper illustration' of his extinct Australian lungfish,[61] and Richard Owen's descriptions of fossil mammals were famously well illustrated. Owen published more than 50 papers with the Society from 1832 to 1886, and is one of the few authors who may have discovered limits to the Society's generosity when he was asked, in 1858, to contribute to the costs of engraving the illustrations of the *Megatherium*.[62] But this was a vanishingly rare instance. The Society's willingness to pay skilled craftsmen to produce the engraved metal plates, lithographic stones or engraved wooden blocks needed for printing such images made it an attractive publication venue for scientific authors who would otherwise have had to choose between no pictures, or paying for them themselves. Furthermore, the Society's willingness to grant authors permission to reuse the images (as printed offprints, or by loaning the plates or blocks) meant that it was often also subsidising an author's subsequent book publication.

The Society had been using engraved metal plates for its images since the eighteenth century, but by the 1870s, several alternative technologies had become widely available.[63] Of those, wood engraving offered the greatest potential cost savings, since its raw material was cheaper, and the blocks could be inserted amid ordinary type and printed on a regular printing press or machine. It came to be widely used in the *Proceedings* but was only used in the *Transactions* for simple line drawings. Lithography was the other option, and had been used in the *Transactions* (alongside metal engraving) since the early 1830s.[64] The Society had been spending around £250 a year on illustrations in the 1840s; by the early 1890s, despite the change in technology, that had risen to around £800, due to the greater number of well-illustrated papers being published.[65]

There was so much illustrative work to be done for the Society's periodicals that the Society worked with a large number of different artists, engravers and lithographers. In 1871, Albert Günther specifically requested that the illustrations for his papers describing extinct bony fishes should be done by G.H. Ford, who had been 'sketching the parts as I proceeded in their examination'. These preliminary sketches would be 'useless in the hand of another artist', but Ford's close work alongside Günther meant that he 'understands now fully the structure'.[66] Ford painted the images onto a lithographic stone, which Mintern Brothers used to print the images (see Figure 10.2). Whenever metal engraving or lithography was used, the plates were printed on separate sheets of paper from the text; they were later collated with the pages of text printed by Taylor & Francis.

Including so many high-quality illustrations in the *Transactions* was not just a cause of expense, but also of delay. When Stokes chastised Huxley in 1864 for having had the proofs of his paper on *Glyptodon* 'in hands for about 3 months', Huxley shifted the blame.[67] He claimed: 'I have not been at all remiss in "getting on with my own" paper—The artist is keeping me waiting & I have stirred him up over & over again— Perhaps you will try your hand now.'[68] Huxley's paper eventually appeared about a year after it was first delivered, with nine lithographic plates.[69]

It was partly for this reason that illustrations in the *Proceedings* were fewer in number, and usually executed as wood engravings. For authors whose illustrations could be executed as line diagrams, wood engraving was entirely adequate. However, the Society could at times be generous about the illustrations in the *Proceedings*, too. In 1872, William Carpenter had shown his awareness of the pros and cons of the Society's periodicals by telling Stokes that he had 'no wish that' his paper reporting hydrographic and zoological surveying activity in the Mediterranean 'should appear elsewhere than in the "Proceedings"'. He wanted it to appear swiftly. However, he was 'anxious that it should have some illustrations of a kind not usually given there'.[70] Since Carpenter's paper ran to 116 pages, it would normally have appeared in the *Transactions* regardless of its illustrations. Yet, Carpenter's paper was published in the *Proceedings*, and it was even accompanied by a set of lithographic graphs and charts depicting sea temperature around the British Isles and in the Mediterranean. The episode reminds us that the lack of hard-and-fast rules about Society publishing gave the officers a certain leeway to act in the manner that seemed most appropriate at the time.

The Royal Society's choices about the length of papers and the number of illustrations had financial repercussions that would become problematic in later decades, but also marked out the *Transactions* as a different class of periodical from the wealth of cheap print that was becoming common in late Victorian Britain. In the *Transactions*, authors could be given the space and illustrations to report and explain their research as they wished, without having to cut it (much) to fit the practical limits of the journal.

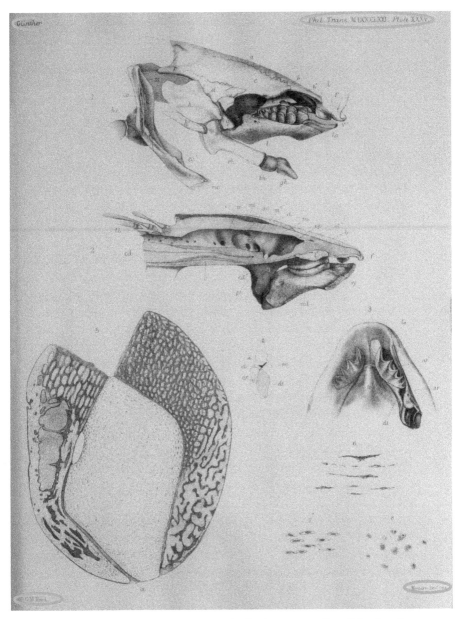

Figure 10.2 Lithographic plate of the skull of an Australian fish, showing names of author, lithographer and printer (as well as the *Transactions*), from A. Günther, 'Description of Ceratodus...', *PT* 161, 1871, plate 35 © The Royal Society.

Financial sustainability

The question of how, or whether, the Royal Society could financially sustain this generosity to its scholarly authors, while simultaneously gifting most of its copies to its readers, was rarely discussed prior to the 1890s. It was becoming an increasingly pertinent question for the Society's treasurers, however. While the Society's presidents proudly reported the number of pages and plates in their anniversary addresses, the Society's treasurers wondered how to fund it all.

Edward Sabine appears to have been the first treasurer to raise serious concerns about the cost of the publications. In 1852, he pointed out that volumes of the *Transactions* were getting longer, and urged the Committee of Papers to exercise a 'greater strictness in selection'. He suggested that most papers could be reported in briefer (cheaper) form as abstracts in the *Proceedings*.[71] Seven years later, he was still making the point, arguing not merely that money would be saved, but also that the 'utility of the Philosophical Transactions' itself would be increased by 'a more discriminating selection' of papers, and 'by the promotion of greater condensation and brevity on the part of authors'.[72]

Sabine's concerns led to the 1863 review of the *Proceedings*, which included an intention to find ways 'to diminish this expense'.[73] They approved a new typesetting layout, which would fit one-third more material onto each page with less overall expense. They also doubled the cost of the annual subscription to the *Proceedings* to a still very modest 10s.[74] But, as William A. Miller, then-treasurer, pointed out the following year, 'It is not, however, certain that this alteration ... will be attended with any great saving of expense', not least because of the growth of output.[75]

Discussions of growth tended to focus on the *Transactions*, the volumes of which kept getting longer. As Figure 10.3 shows, individual volumes ranged enormously in size, but the most common length in the first half of the century had been around 400 pages; by the 1860s and 1870s, it had grown to about 780 pages. But while Sabine seems to have attributed this growth (at least partly) to an increase in the number of papers being published, the number of articles published in the *Transactions* per year hovered around 200 for the entire century. It was the page count of each article that was increasing after 1850, as Figure 10.3 shows. There was an increase in the number of papers published by the Society each year, but most of those papers were appearing in the *Proceedings*. They did contribute to the increased total costs – and especially so once the *Proceedings* began carrying more

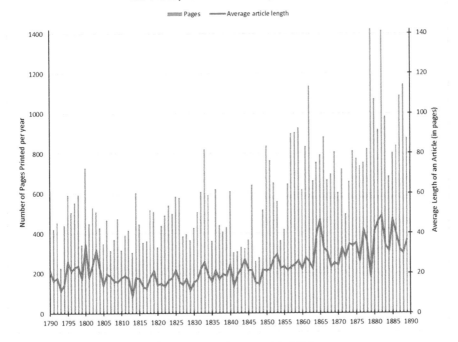

Figure 10.3 The growth of the *Transactions*, 1790–1890.

illustrations in the 1860s – but it was undoubtedly a more cost-effective way of publishing the flood of research than doing so at well-illustrated length in the *Transactions*.

The current state of Royal Society finances was routinely mentioned at the anniversary meeting every November, but not at any length. Throughout the 1870s and 1880s, the remarks on finances were routinely brief and generally upbeat. For instance, in 1874, the treasurer, William Spottiswoode, delivered the anniversary address in the absence of then-president Joseph Hooker, and confidently assured the listening fellows that there was 'no cause for apprehension in respect of the Society's funds or income'. In particular, he noted that, even though the forthcoming volume of the *Transactions* would contain more plates than any previous volume, there was 'no want of means for providing illustrations to papers communicated to us for publication'.[76] Again in 1880, Spottiswoode (now president) reassured the fellows that the Society's finances were 'generally good' and 'will suffice for the large claims upon them for printing our publications'.[77] Spottiswoode – and presumably the other presidents – was well aware of the rising costs due to the expansion of the Society's publishing activities, but as long as the Society's finances were strong enough to support those costs, there was no reason for the fellowship to be concerned.[78]

Behind the scenes, at the meetings of the Council and its subcommittees, there was slightly more concern about the financial sustainability of the expanding publication and distribution programme. In 1877, the expense of 'excessive corrections' of proofs by authors was raised, and Council agreed to inform all authors that they might be held liable for the costs involved.[79] The following year, new guidelines for the provision of separate copies tried, once again, to establish a firm limit to what the Society would pay for.[80] This was minor tinkering, however. Far more potentially significant was that, in 1877, the Society decided to issue a tender for its printing services. This appears to have been the first time since 1846 that a formal tendering process had been undertaken, although there had been a renegotiation with Taylor in 1852.[81]

The decision to seek 'tenders for printing the Society's publications' originated with an 1876 request that the Library Committee investigate the expense of printing.[82] The Library Committee oversaw all matters to do with stocking the Society's library and had, thus, taken on responsibility for the gift and exchange list for the *Transactions* and the *Proceedings*. It therefore appears to have been deemed the appropriate body to engage with details of publishing and circulating journals.

Back in 1828, Richard Taylor's expertise in scientific typesetting and his substantial experience in printing for learned societies had made him an obvious choice for the Royal Society. Taylor's natural son, William Francis, kept the firm's own journals flourishing, and formed close friendships with many of the men of science involved in editing them. But by the 1870s, the position of Taylor & Francis as the premier printer for learned societies was being challenged, and it lost several contracts.[83]

Before issuing the tender, Council charged the Finance Committee to take a wider look into the 'receipts and expenditure of the Society for the last few years'. The opening sentence of that report, in March 1877, is striking for placing publications at the centre of the Society's purpose. The committee saw its purpose as establishing the financial basis that was 'necessary to maintain the Society in the same state of efficiency with regard to its *Transactions*, *Proceedings*, and other important work it has in hand'.[84] Grant-making, policy work and cultural diplomacy were still in the future; in the 1870s, the publication of scientific knowledge remained the most important work the Society had 'in hand'.

The Finance Committee identified the main threats to the Society's financial health as the *Catalogue of Scientific Papers* and the consequences of the move from Somerset House to Burlington House in 1867. As well as fitting out the new apartments there were the ongoing costs of heating and lighting the larger premises. The cost of 'maintaining an adequate staff' was also growing. The committee's analysis of the overall financial health of the Society suggested that it was facing a 'prospective deficit of about £200 annually' going forwards, on an estimated annual expenditure of £4,000. The committee had a range of suggestions, including a rise in the fellows' annual membership fees, a reorientation of the investment portfolio, an appeal to fellows for donations to cover the one-off costs of the move, and an appeal to the government for assistance with the *Catalogue of Scientific Papers*.[85] The complete absence of any suggestions for reducing the expenditure of publishing, or for increasing the Society's income by pursuing a more active sales strategy, reiterates our point that the Society did not think of its publication activities in terms of profit or loss, but as a core activity to be sustained however possible. From the archival record, we can estimate that the publications were costing the Society about £500 or £600 more per year in the 1870s than they generated in sales.[86] (There is no evidence that either the Library Committee or the Finance Committee did this calculation, however.)

As the Finance Committee was finalising its report, the Library Committee was moving ahead with the tender, and in May 1877, it received proposals from five firms: Taylor & Francis; William Clowes & Co.; Gilbert & Rivington; Harrison & Sons; and the family firm of the Society's treasurer, Spottiswoode & Co.[87] The proposals were presented to Council in June, at the meeting following the receipt of the Finance Committee report. Nothing survives of the details nor of the discussion about their relative merits. In its previous choices for printers, in the 1790s and 1820s, the Society had preferred a printer with recognised expertise in scientific typesetting and printing, rather than the strictly cheapest option. Given the context of financial concerns in the late 1870s, and the lack of specialist expertise of the successful candidate, it seems likely that cost may have been more important this time round. Council resolved that the printing of the *Transactions* and the *Proceedings* should be transferred to Harrison & Sons.[88]

Harrison & Sons undertook substantial amounts of government work, printing for the Foreign Office and War Office, as well as the official *London Gazette*. In the 1880s, they would win the contract to print the telegraph forms used in Post Offices across the country.[89] They also printed a variety of monthly and weekly periodicals, and had been the Chemical

Society's printer since 1857.[90] Unlike Taylor & Francis, Harrisons was expanding, able to modernise and adapt, and well-equipped with steam and a range of different printing techniques.[91] Printing the Royal Society's *Transactions* and *Proceedings* would be but a small aspect of Harrisons' varied and expansive business ventures. They would go on to print for the Society for even longer than Taylor & Francis had done, serving until 1937.

We do not know what rate per printed sheet Harrisons were charging, nor what Taylor had been charging, but the archival record of bills paid enables us to estimate that the change of printer led to a saving of about a quarter on unit print costs.[92] However, the total cost of printing remained broadly the same as it had been, on average, because of the increase in the number of pages of the *Transactions* and the *Proceedings* each year. The biggest factor in the increasing cost of publications during the Stokes years was not down to the British print trade, but to the Society's editorial policy.

Back in 1852, Sabine had suggested that Council should consider cost in the editorial decision-making process, and cap the number of papers published in any one year.[93] This was essentially what most commercial periodicals did, because a fixed page length per issue was essential for survival with a fixed cover price. But for the Royal Society, limiting the number of papers published per year would have meant either holding some papers over until the following year, or being more selective about which papers to publish. The first option would inevitably lead to a backlog and could be seen as inhibiting the circulation of knowledge. The second potentially meant rejecting papers that were worthy of publication, and might be socially awkward since all papers under consideration were either authored by fellows, or had been communicated by a fellow.

Changing printers was the only development during Stokes's secretaryship that had much chance of making a difference to the cost of the Society's publishing programme. Limiting the number of separate copies, or charging authors for alterations made in proof, was mere tinkering. The Society could have altered its editorial mechanisms by systematically restricting the number of papers published, or their length, or their illustrations. Or it could have stopped giving free copies to its fellows and cut back its philanthropic gifts to research libraries across the British world. But it did none of these things. Instead, the Society's treasurers looked again at their investments and began a regular series of appeals to the government, and to the private philanthropy of wealthy fellows (for example, industrialists) for specific projects. The list of benefactors printed in the *Record of the Royal Society* for the 1880s onwards is testimony to the success of that approach.[94]

In an earlier age, the Society's own resources, from membership fees, property and investments, would have supported its benevolent activities, but the Society's ambitions were now grander and more expensive. It did not help that one of the consequences of the 1847 reforms was a slow shrinking in the number of fellows, coupled with a change in their socio-demographic background. By the end of the century, the typical fellow was no longer a gentleman of comfortably independent means, but a modestly paid university academic. During his presidency in the mid-1870s, the botanist Joseph Hooker had pushed to make the Society more socially inclusive by removing the joining fee paid by new fellows, and reducing the annual subscription paid by all fellows. He recognised the challenge this would pose for Society finances, and thus raised £10,000 for a fund 'in aid of publications and for the promotion of research'. The major donors included industrialists Joseph Whitworth and James Young.[95] Over time, this fund was known both as the 'Publication Fund' and the 'Fee Reduction Fund', because its purpose was to support the publications despite the reduction in fee income.

The Society's own investment portfolio was healthy and growing, but far from infinite.[96] Hooker's fund-raising drive illustrates a new feature of Royal Society finances: in the last three decades of the nineteenth century, it became increasingly adept at securing funding from external sources, such as government departments and private donors. The way in which the Society funded the printing costs of the *Catalogue of Scientific Papers* illustrates this point. In the 1860s and 1870s, the Treasury was persuaded to fund the publication of the first two volumes of the *Catalogue* as a benefit to scholarship. But by 1887, as Stokes told the fellows, the Treasury was less willing to keep funding the apparently never-ending project. It made a grant of £1,000 but left the Society to find the remainder of the funds elsewhere.[97] Help came from the chemical industrialist Ludwig Mond, who made donations of around £18,000 to support the *Catalogue*.[98] (Both the Mond family and the Brunner-Mond company would continue to support the Royal Society and its publications into the 1920s and beyond; see Chapter 12.)

The financial challenges of publishing the growing quantity of research being produced in late nineteenth-century Britain were not unique to the Royal Society. Other learned society publishers also faced challenges, and had to meet them with fewer resources than the 200-year-old Royal Society could muster. The Royal Society's age and status, coupled with its broad generalist remit, put it in a powerful position for persuading government officials or wealthy industrialists. However, the Society's enduring commitment to gentlemanly values from the eighteenth century, in terms of production quality and philanthropic distribution, compounded the challenges it faced.

In his presidential address in November 1887, George Stokes announced a significant 'change in the mode of the publication' of the *Philosophical Transactions*. It would be issued 'from henceforth in two series', because, said Stokes, 'the sciences ... divide themselves very naturally into two groups: mathematics, physics, and chemistry forming one, and the biological sciences the other'.[99] Until this point, the *Transactions* had been the supreme generalist journal. Its definition of 'philosophical' or 'scientific' had narrowed since the days when it carried papers dealing with medicine, philology, antiquities and architecture, but it had nonetheless retained a broad natural-scientific remit. This is all the more remarkable given the increasing tendency to disciplinary specialisation apparent in other contexts in Victorian Britain: the formation of specialist societies for geology, astronomy, chemistry and anthropology; the existence of periodicals of specific interest to chemists, microscopists or telegraph engineers; and the creation of new professorships and degree programmes at the universities. The transformation of 'the *Transactions*' into '*Transactions*, series A' and '*Transactions*, series B' may have been recognition of the distance growing between physical and biological scientists, but it is striking that each series of the *Transactions* remained far broader in disciplinary remit than other research journals. The *Proceedings* did not (yet) split into similar series.

Stokes presented the split of the *Transactions* as an entirely pragmatic decision, pointing out that the annual output 'is a good deal more bulky now than it was at the beginning of the century', so it would be 'desirable' to be able to bind it as two volumes rather than one. This would also allow institutions 'that are concerned with one only of the two groups of subjects' to acquire (and give shelf-space to) only one volume per year. Stokes presented this as a financial saving to institutions 'that are not on our list for free presentation', who might thus be spared 'the expense of procuring the whole "Transactions"'. In reality, given the tiny number of paying customers, the real saving would be to the Society, as it enabled the presentation copies of the *Transactions* to be shared among more institutions (and, potentially, fellows). There was no need to send series A to the libraries of botanic gardens or natural history societies, nor any need to send series B to the libraries of observatories or bureaux of standards. In theory, the print run could be reduced, without significantly reducing the Society's philanthropic reach. (And in 1898, the print run of the *Transactions* was, indeed, reduced to 800.)

The split of the *Transactions* did nothing to change the generous editorial dispensation offered by the Royal Society to those authors whose research met with approval. The thick, white paper, well-spaced type, and bountiful illustrations distinguished both series of the *Transactions* from the typical product of the industrial Victorian print trade. At a time when the pursuit of science was gradually opening up to people from a wider range of social classes, the Society's officers were well aware that few scholars or researchers – let alone amateurs – would be able to afford its price. But the price was in many senses irrelevant, for the *Transactions* was not primarily something to be bought or sold.

During the 1870s and 1880s, the Royal Society expanded its patronage of the publication and circulation of scientific knowledge. The expenditure on publishing was seen as a valuable use of the Society's funds: good scholarship mattered more than cost-cutting. The challenge was to ensure that the Society's finances remained able to support the cost of the publications, especially given the changing demographic of the fellowship. Fortunately, the support of the Banksian *dilettante* gentlemen could be replaced by that of industrialists who recognised the value of science. In the decades that followed, the Society's ability to leverage its historic status into financial support from government or private individuals would become vitally important to the *Transactions* and the *Proceedings*.

Notes

1 T.H. Huxley, 'Anniversary meeting', *Proc* 39 (1885): 295–6.
2 On the work of compiling the catalogue, see Hannah Gay, 'A questionable project: Herbert McLeod and the making of the fourth series of the Royal Society Catalogue of Scientific Papers, 1901–25', *Annals of Science* 70 (2013): 149–74. On the meaning and uses of the catalogue, see Alex Csiszar, 'How lives became lists and scientific papers became data: Cataloguing authorship during the nineteenth century', *British Journal for the History of Science* 50 (2017): 23–60.
3 Stokes, 'Anniversary meeting', *Proc* 43 (1887): 191 [arduous, stray papers]; Stokes, 'Anniversary meeting', *Proc* 46 (1889): 455 [completeness].
4 Stokes, 'Anniversary meeting', *Proc* 46 (1889): 455.
5 These projects were all discussed in Stokes, 'Anniversary meeting', *Proc* 43 (1887): 190.
6 William H. Brock and A.J. Meadows, *The Lamp of Learning: Taylor & Francis and the development of science publishing* (London: Taylor & Francis, revised edn 1998), ch. 5; Melinda Baldwin, *Making 'Nature': The history of a scientific journal* (Chicago, IL: University of Chicago Press, 2015), chs 1–2.
7 List of institutions, printed in the (unnumbered) front matter of *PT* 169.1 (1878).
8 Report of the Committee on the Publication of the Proceedings, in RS CMP/03, 5 November 1863.

9 Taylor & Francis print runs from 1860s are from Brock and Meadows, *Lamp of Learning*, 140.
10 Imogen Clarke and James Mussell, 'Conservative attitudes to old-established organs: Oliver Lodge and Philosophical Magazine', *N&R* 69.3 (2015): 321–36, 324, notes that price was one aspect of the *Philosophical Magazine* that remained constant throughout the nineteenth century.
11 RS CMP/03, 5 November 1863.
12 For the 1870s prices, see Sampson Low, *The English Catalogue of Books*, 9 vols (London: Sampson Low, 1863–1915), vol. 3 (1873–80), 527 and *passim*.
13 RS CMP/09, 5 April 1906, reporting stock of 24,000 copies of the *Transactions* and 198,000 copies of the *Proceedings*.
14 *The Jubilee of the Chemical Society of London. Record of the proceedings together with an account of the history and development of the Society, 1841–1891* (London: Harrison & Sons, 1896), 243.
15 T&F, R&J Taylor Cash Books 1848–57 (re summer and autumn 1857).
16 For instance, T&F, Journal 1873-7, f. 13 lists advertisements among the expenses of *Transactions* 1873 Part 1, and *Proceedings* number 147. The *Times* advertisement was discontinued in 1884, see RS CMP/05, 17 January 1884.
17 Aileen Fyfe, 'The Royal Society and the non-commercial circulation of knowledge', in *Reassembling Scholarly Communications: Histories, infrastructures, and global politics of open access*, ed. Martin Paul Eve and Jonathan Gray (Cambridge, MA: MIT Press, 2020), 147–60.
18 RS CMP/03, 5 November 1863.
19 *PT* 169.1 (1878), front matter.
20 The concept of the 'British academic world' is relevant here, see Tamson Pietsch, *Empire of Scholars: Universities, networks and the British academic world, 1850–1939* (Manchester: Manchester University Press, 2013).
21 Faraday to A. de la Rive, 23 November 1830 (emphasis is ours), Letter 468 in Frank A.J.L. James, ed., *The Correspondence of Michael Faraday*, 6 vols (London: Institution of Electrical Engineers, 1991–2012), vol. 1.
22 RS CMP/03, 30 November 1863.
23 RS CMP/04, 21 May 1874. The Royal Engineers also got an additional six copies of future *Proceedings* for distribution to the foreign libraries of the Corps.
24 Library Committee, 22 January 1874, RS CMB/47/3.
25 Jenny Beckman, 'Editors, librarians, and publication exchange: The Royal Swedish Academy of Sciences, 1813–1903', *Centaurus* 62, no. 1 (2020): 98–110.
26 On taking copies on his travels, see Tyndall to Stokes, 19 June [no year], Stokes Correspondence, CUL MS Add. 7656.
27 For an example of an author being too late asking for copies of an abstract in the *Proceedings*, and being told that the type had already been distributed, see RS CMP/03, 24 October 1861.
28 RS CMP/03, 24 February 1859.
29 A 'printed notice affixed to the proof' was decided upon by Council, RS CMP/04, 18 March 1875.
30 RS CMP/04, 28 October 1875, re E. Ray Lankester, '... the developmental history of the mollusca', *PT* 165 (1875): 1–48.
31 RS CMP/05, 21 March 1878.
32 Tyndall to Lady Mary [Adair?], 8 December 1874, RI MS JT/1/TYP/1/398.
33 RS CMP/05, 21 February 1878 (citing correspondence from 20 December 1877). We have discussed this episode in Aileen Fyfe, Julie McDougall-Waters, and Noah Moxham, 'Credit, copyright, and the circulation of scientific knowledge: The Royal Society in the long nineteenth century', *Victorian Periodicals Review* 51, no. 4 (2018): 597–615. The papers were published between 1859 and 1876, ending with: R. Owen, 'On the fossil mammals of Australia. Part X', *PT* 166 (1876): 197–226.
34 RS CMP/04, 15 December 1870, re a paper by George West Royston-Pigott and George Gabriel Stokes on 'aplanatic images applied to microscopes'; and RS CMP/04, 15 June 1876, re a paper by F.M. Balfour, on 'the spinal nerves in elasmobranch fishes' (presumably linked to his 1878 monograph 'on the development of elasmobranch fishes').
35 RS CMP/05, 21 February 1878 (citing correspondence from 20 December 1877).

36 RS CMP/05, 21 March 1878.
37 Minutes of the Library Committee, 25 February 1875, in RS CMB/47/3: 'The President stated that he had received a letter from Mr Darwin suggesting that the Papers in the Philosophical Transactions should be published and sold separately.' The letter does not appear to survive.
38 Trübner to Royal Society, 11 March 1875, in Library Committee minutes, 1 June 1875, RS CMB/47/3.
39 RS CMP/04, 17 June 1875.
40 Trübner to Royal Society, 11 March 1875, in Library Committee minutes, 1 June 1875, RS CMB/47/3.
41 RS CMP/04, 18 March 1875.
42 RS CMP/05, 24 May 1883.
43 The contract was transferred (for seven years) to Messrs Dulau & Co. in 1894, see RS CMP/07, 25 October and 6 December 1894.
44 'Explanatory notes on the procedure relating to the reading and publication of papers', in RS Year Book (1896/97), 66–9.
45 RS CMP/04, 16 February 1871.
46 RS CMP/04, 17 February 1870.
47 RS CMP/04, 19 April 1877. This edition was printed by Taylor & Francis for the antiquarian bookseller, Bernard Quaritch.
48 RS CMP/04, 16 February 1871.
49 RS CMP/04, 21 March 1872.
50 RS CMP/06, 12 March 1891.
51 For more on reuse of text and images, see Fyfe et al., 'Credit, copyright'.
52 On the copyright debates of the 1890s, see Will Slauter, 'Copyright and the political economy of news in Britain, 1836–1911', *Victorian Periodicals Review* 51, no. 4 (2018): 640-60.
53 Katherine Ford, 'The role of the Royal Society in Victorian literary culture' (PhD, Reading, 2015).
54 Stokes, 'Anniversary meeting', *Proc* 43 (1887): 191.
55 Our analysis of data provided by the Royal Society.
56 Thomas George Bonney, on a paper 'On the origin of the mineral, structural, and chemical characters of ophites …' by W. King and T.H. Rowney, 30 June 1879, RS RR/8/165.
57 John Burdon-Sanderson, on a paper 'Considerations generales sur la digestion stomacale …' by an unknown author, 18 March 1880, RS RR/8/220.
58 Henry J.W.L. Glaisher, on a paper 'On Clairautian functions and equations' by Allan Cunningham, 10 August 1877, RS RR/8/9.
59 On the costs of printing and publishing, see Alexis Weedon, *Victorian Publishing: The economics of book production for a mass market, 1836–1916* (Aldershot: Ashgate, 2003), ch. 3. On new technologies, see Simon Eliot and Jonathan Rose, eds, *Companion to the History of the Book* (Oxford: Blackwell, 2007), ch. 20. It is not clear whether Taylor & Francis used their steam-printing machine to print the Royal Society's periodicals, but it was possible. They did not adopt Monotype machines until the 1930s, long after losing the Royal Society contract. See Brock and Meadows, *Lamp of Learning*, 85–6 (printing machines), 165 (Monotype).
60 Stokes to Huxley, 4 August 1874, in Joseph Larmor, ed., *Memoir and Scientific Correspondence of the Late Sir George Gabriel Stokes*, 2 vols (Cambridge: Cambridge University Press, 1907), vol. 1, 221. He was referring to E. Ray Lankester, 'Contributions to the developmental history of the mollusca', *PT* 165 (1875): 1–48.
61 A. Günther to Stokes, 24 March 1871, RS MM/14/157.
62 RS CMP/02, 8 June 1858 (Owen agreed to contribute).
63 Jonathan R. Topham, 'Redrawing the image of science: Technologies of illustration and the audiences for scientific periodicals in Britain, 1790–1840', in *Science Periodicals in Nineteenth-Century Britain: Constructing scientific communities*, edited by Gowan Dawson, Bernard Lightman, Sally Shuttleworth and Jonathan R. Topham (Chicago, IL: University of Chicago Press, 2020), 65–102. For the later use of photography, see Geoffrey Belknap, *From a Photograph: Authenticity, science and the periodical press, 1870–1890* (London: Bloomsbury, 2016).
64 Richard Goddard, *'Drawing on Copper': The Basire family of copper-plate engravers and their works* (Maastricht: Maastricht University Press, 2016), 287.
65 Production costs from RS Annual Accounts, *passim*.

66 A. Günther to Stokes, 24 March 1871, in RS MM/14/157. This exchange concerned the 13 plates in A. Günther, 'Description of Ceratodus ...', *PT* 161 (1871): 511–71. Ford also illustrated A. Günther, 'Description of the living and extinct races of gigantic land tortoises', *PT* 165 (1875): 251–84.
67 Stokes to Huxley, 8 December 1864, Letter 4709 in *Darwin Correspondence Project*.
68 Huxley to Stokes, 9 December 1864, Letter 4711 in *Darwin Correspondence Project*. The illustrator was J. Erxleben.
69 T.H. Huxley, 'On the osteology of the genus Glyptodon', *PT* 155 (1865): 31–70.
70 W.B. Carpenter to Stokes, 20 June 1872, RS MM/14, f. 163. This presumably refers to Carpenter's 'Report on scientific researchers carried on during the months of August, September, and October 1871, in H.M. Surveyingship *Shearwater*', *Proc* 20 (1871–2): 535–644, received by the Society on 13 June 1872.
71 Sabine (treasurer) to Rosse (president), in RS CMP/02, 19 February 1852.
72 Treasurer's report (by Sabine), in RS CMP/03, 24 February 1859.
73 Treasurer's report (by W.A. Miller), in RS CMP/03, 17 March 1864.
74 RS CMP/03, 5 November 1863.
75 Treasurer's report (by W.A. Miller), in RS CMP/03, 17 March 1864.
76 W. Spottiswoode, 'Anniversary meeting', *Proc* 23 (1874): 51–2.
77 W. Spottiswoode, 'Anniversary meeting', *Proc* 31 (1880): 76–7.
78 See also J. Evans, 'Anniversary meeting', *Proc* 37 (1884): 433. Evans was another treasurer standing in for his president (in this case, Huxley); he described the finances as being in a 'satisfactory condition'.
79 Library Committee, RS CMB/47/4, 16 November 1876.
80 RS CMP/05, 21 March 1878.
81 See RS CMP/02, 11 March 1852. For 1846, see Committee on Expenses of Printing, and Prospective Financial Condition, in RS CMB/86/A.
82 RS CMP/04, 30 November 1876. The Library Committee was asked to investigate printing on 26 October 1876, RS CMP/04.
83 Brock and Meadows, *Lamp of Learning*, 136–7, 140.
84 Finance Committee, 21 March 1877, in RS CMB/2/23/1.
85 Finance Committee, 21 March and 26 April 1877, RS CMB/2/23/1. Only some of these were proposed to Council on 17 May 1877, RS CMP/04.
86 Aileen Fyfe, 'Journals, learned societies and money: *Philosophical Transactions, ca.* 1750–1900', *N&R* 69.3 (2015): 277–99.
87 The receipt of the estimates was recorded in Library Committee, 11 May 1877, CMB/47/4. The minutes note that 'the President stated that the Officers were not prepared to report on the estimates'.
88 RS CMP/04, 7 June 1877.
89 Cecil Reeves Harrison and H.G. Harrison, *House of Harrison: Being an account of the family and firm of Harrison and Sons, printers to the King* (London: Harrison & Sons, 1914), 11–22, 27–8, 47.
90 *Jubilee of the Chemical Society*, 241.
91 On the premises and technologies, see Harrison, *House of Harrison*, 49 ff.
92 Combining the printers' bills with the number of pages per volume and size of print run, the cost per thousand pages printed fell from 0.55 pennies (1865–75) to 0.42 pennies (1880–90).
93 RS CMP/02, 11 March 1852. Sabine succeeded in making a tiny saving by discontinuing the printing of title pages for separate copies.
94 See *The Record of the Royal Society of London* (3rd edn) (London: The Royal Society, 1912), 196–9; and the expanded list in Henry George Lyons, ed., *The Record of the Royal Society of London*, 4th edn (London: The Royal Society, 1940; reprint, 1992).
95 RS CMP/05, 27 June 1878; see also *Record of the Royal Society* (1912), 178 and 196. On Hooker's role, see Ruth Barton, *The X-Club: Power and authority in Victorian science* (Chicago, IL: University of Chicago Press, 2018), 279–80.
96 Fyfe, 'Journals, learned societies and money'; Henry Lyons, 'The Society's finances Part II: 1831–1938', *N&R* 2.1 (1939): 47–67.
97 Stokes, 'Anniversary meeting', *Proc* 46 (1889): 455.
98 Gay, 'A questionable project', 157.
99 Stokes, 'Anniversary meeting', *Proc* 43 (1887): 189.

Part IV
The growth of science, 1890–1950

11
The rise of the *Proceedings*, 1890–1920s

Julie McDougall-Waters and Aileen Fyfe

In 1902, the chemist Henry Armstrong warned that the Royal Society was in 'danger of losing its position as the most important body in this country engaged in the promotion of Natural Knowledge'. He accused it of failing to display leadership, and issuing publications that were 'somewhat trivial in character'.[1] Twenty-five years later, however, the physicist Ernest Rutherford was able to confidently assure the fellows gathered for the Society's anniversary meeting that: 'Our Society is now the most important medium of publication of papers in Experimental and Theoretical Physics and Physical Chemistry in this country.... Anyone who reads our 'Proceedings' cannot fail to be impressed ... by the great variety and importance of the papers appearing in them.'[2]

This chapter investigates the editorial reforms that enabled the journals criticised by Armstrong to become worthy of Rutherford's praise. The next chapter looks at broadly the same period, but will focus on the financial challenges of publishing more and more research. The *Proceedings* is central to both stories: by Rutherford's day, it had replaced the *Transactions* as the Society's main venue for publishing research. For the rest of the twentieth century, the Society's thinking about journal publishing would be focused on the *Proceedings*, and these early decades of the century were crucial to the development of the modern *Proceedings*. The core content of the nineteenth-century *Proceedings* had been summaries of much lengthier research papers, some of which later appeared in full in the *Transactions*, while others did not (because referees judged that their key points had already been adequately conveyed by the so-called 'abstract'). It also carried occasional research papers that were complete in themselves, but were judged either too short for the *Transactions*, or too lacking in the intellectual significance required for a *Transactions* paper. By the 1920s, the *Proceedings* had been freed from both its intellectual inferiority to the *Transactions* and its original mission to carry abstracts, and it had been transformed into a journal focused on the publication of original research papers of up to 24 pages in length. It had also been split into series A and series B, in a partial recognition of the specialisation of scientific research. Henry Armstrong was just one of the voices calling for the reform of the *Proceedings* in the two decades leading up to the First World War. Differing opinions about the best route forward meant that changes took place in a gradual and rather piecemeal fashion, and only fully came to fruition after the war.

The years around 1900 were more broadly a period of organisational and administrative reforms within the Royal Society itself. New officers and staff helped the Society adapt its processes to the new conditions of scholarly life in Britain, including the ongoing growth in scientific research output, the specialisation of research, and the Society's shifting relationship to the wider British academic community.

The late nineteenth-century expansion in the number of professors and researchers employed in universities (and elsewhere) meant that more new research was being generated, and a wider variety of people were becoming involved in scientific research. Recent attempts to quantify the growth of scientific research through analysis of the published output have suggested a growth rate of about 2–3 per cent per annum in the late nineteenth and early twentieth centuries.[3] This fits broadly with the Royal Society's experience, where the number of submissions showed a roughly four-fold increase between 1865 and 1935, equivalent to 2 per cent per annum.[4] This increased the pressure on the Society's hour-long, Thursday evening meetings, on the capacity and expense of its periodicals, and on the editorial labour required to run them.

More and more of the Society's activities came to be divided between physical and biological sciences, as the *Transactions* had been in 1887. In the 1890s, papers for discussion were grouped to create meetings that would appeal either to physical scientists or to those from the biological sciences, rather than the entire fellowship. In 1905, the *Proceedings* was split into two series. The secretaries came to be labelled (informally) as 'physical' and 'biological', rather than 'junior' and 'senior'. The division into 'A side' and 'B side' appeared to be a useful way of structuring the Society's activities, and was a partial recognition of the specialisation of scientific research. But it would also create an ongoing tension as fellows fought to ensure that their 'side' received equitable access to resources, rewards and positions of power within the Society. For most of the twentieth century, the concern would be that the interests and needs of biological scientists were being side-lined.

The demographic composition of the British scientific community was changing more quickly than that of the Royal Society. New universities and university colleges were established in the provinces, colonies and dominions, and an increasing proportion of the research community worked at institutions that were very different from the Society's traditional heartland of Oxford, Cambridge or London. In the 1860s, it might have been possible to claim that the Society's 600 fellows included the majority of British scientists, but by the 1920s, this would have been an implausible claim. The Society's 450 fellows were no longer the socially elite group they had been in Joseph Banks's day, but they were an intellectual or academic elite in a scientific community whose numbers had increased around four-fold over 20 years.[5] There was now a tiny (but increasing) number of women undertaking doctoral and even postdoctoral research in science, but none of them were fellows of the Royal Society.

For the Society's publications, the growing number of research papers being written and submitted meant more editorial work to be done. Most of the growth originated with authors who were not (yet) fellows. The proportion of manuscripts submitted by non-fellows had been increasing through the late nineteenth century; but in the twentieth century, it would massively outstrip those from fellows.[6] And that changed the nature of refereeing from something that (mostly) fellows did for other fellows, into something that Royal Society fellows were doing for the entire British scientific community.

Throughout the nineteenth century, amid the increasing numbers of other types of periodicals and societies, the Royal Society's officers and Council had remained confident that the *Philosophical Transactions* was the most prestigious of British scientific periodicals. But by 1900 that confidence was wavering; or rather, it was being assailed by doubts about whether historic prestige was sufficient. Henry Armstrong was the most visible voice of critique in the archival record of the turn-of-the-century Royal Society. He was an educational reformer and professor of chemistry at the Central Institution at South Kensington.[7] He was closely involved in the Chemical Society, having served on its Council and for a term as president in the 1890s. He also served a term as vice-president of the Royal Society in 1901–2, and his comments on the Society's publications formed part of a substantial critique of the Society's recent organisational reforms, delivered at the end of his term in office.

Armstrong questioned the role and purpose of the Society's journals and put forward a bold vision: use the Society's unusual disciplinary breadth to provide a 'platform for the discussion of borderland problems'.[8] He criticised the Society's 'anachronistic' editorial practices as too slow and conservative to suit the needs of modern researchers. And he reiterated long-standing arguments about the need for an overhaul of the *Proceedings*, in terms of both its remit and its practical publishing arrangements. This chapter begins with the reforms to the editorial system in the 1890s and 1900s, and will then look more closely at each of Armstrong's three areas of concern.

New ways of working

The Royal Society went through a quarter-century of reforms between George Stokes stepping down as president in 1890 and the outbreak of war in 1914. Some were simple changes of procedure and practice within the Society's paid staff and volunteer officers, but others were major organisational changes requiring legal consultation and revisions to the governing statutes. Together, these reforms helped the Society adapt its gentlemanly traditions to the needs of academic scientific research at the turn of the twentieth century. In the 1890s, attention focused upon the organisation of meetings, including the relationship between meetings and the journals. In the first decade of the twentieth century, the focus would shift to making the administrative operations of the Society more efficient. The statutes went through repeated revisions in the 1890s, and were completely overhauled in 1905. Most procedural details were moved from the statutes to 'standing orders', and these were also repeatedly revised, most notably in 1914. There were far fewer reforms after the war, but this is when the earlier changes came to fruition.

From the perspective of the journals, there were some significant changes bundled into the wider reforms of the 1890s. Practices that had been core elements of the Royal Society editorial system since 1752 were changed: the ordinary Thursday meetings of Society fellows ceased to be part of the editorial workflow, and most of the responsibilities of the Committee of Papers were delegated to a series of discipline-based 'sectional committees'. The new editorial process can be seen in Figure 11.1; it remained in place until the late 1960s.

The wave of reforms began with a desire to make the meetings of the Society more interesting. The traditional one-hour meeting had been putting constraints on the reading of papers since Joseph Banks's day and, by the 1880s, it was clearly impossible to 'read' all the papers

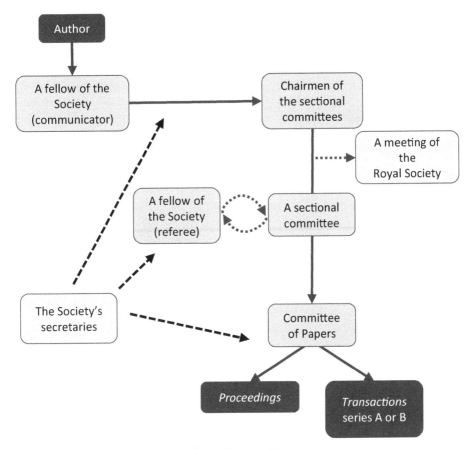

Figure 11.1 The editorial workflow after 1896.

received in any meaningful way. Most were presented only as titles and abstracts, and many fellows found the meetings boring.[9] An 1890 committee proposed altering the arrangements to avoid reading papers which were 'not likely to excite interest'.[10] The consequence of this, in 1892, was a new set of regulations enabling papers to go straight into the editorial process. Only a few, such as those which 'the author is prepared to illustrate by experiments, diagrams &c.' or which 'is likely to give rise to discussion', would now be selected for oral presentation and discussion at meetings.[11] Thus, the 140-year-old role of the weekly meeting as an essential first step in the publication process came to an end.

The legitimacy of this move was challenged in June 1895, when a fellow complained about 'the injustice done him by a foot-note' in a recent paper (by another author) in *Transactions B*. He argued that the Society's president and Council had acted 'in contravention to the statutes' by deciding to publish the paper without first taking it to a meeting of the fellowship. Council sought legal advice from Alfred Kempe, a mathematician, barrister and fellow of the Society who would become its treasurer a few years later.[12] Council argued that not requiring all papers to be read at a meeting was a change 'in the interest of Science', for 'every effort should be made to avoid all unnecessary delay in the publication of papers communicated to the Society' – but Kempe's legal opinion supported the complainant.[13] The episode drew attention to the challenges of adapting to the modern world with statutes that were at least half a century old.

At the time, the Council was in the midst of implementing a report from a 'procedural committee' which proposed an even more significant innovation in editorial practice: the delegation of much of the work traditionally done by the Committee of Papers to a new set of discipline-specific committees. These committees were broadly similar to the 'scientific committees' that had been in operation between 1838 and 1848, but they would now be termed 'sectional committees'.[14] In spring 1895, the Committee of Papers had been paralysed by its inability to resolve the tension between the treasurer's demands for economy, and authors' and referees' claims about the necessity of illustrations (as will be discussed further in Chapter 12). By summer, the plans for reform were well-advanced, but the complaint about the footnote led Council to ask Kempe to look over the proposed regulations. He reported that delegating the powers of the Committee of Papers to other committees 'would be in direct conflict with the statutes, as the latter now stand'. The only form of 'outside assistance' explicitly allowed was that of referees. Kempe's suggestion was to recast the statutes into 'a very general character', and to move all 'details' of procedure into standing orders (which could be amended by Council from time to time).[15] It took until 1905 before that major reform of the statutes occurred, but minor statute changes enabled the new sectional committees to come into operation in late 1896.[16]

Joseph Lister, as president, claimed that the creation of the sectional committees would ensure that editorial decisions were made by committees with discipline-specific expertise, and would thus lead to 'a more secure, and, at the same time, more rapid judgment'.[17] The former claim might possibly be true, but, as Armstrong would later point out, the additional layer of committees meant that 'the machinery of publication has … been complicated rather than simplified'. Armstrong's 1902 memorandum was directed at the sectional committees, which he wanted to remove from the editorial process. (He did not want them to be abolished completely, but to be more effectively deployed 'to relieve the Council of much burdensome work'– particularly the 'discussion of details' relating to the selection of new fellows and medal winners – so that Council would be freer 'to consider questions of broad policy'. This eventually happened in the 1960s, see Chapter 14.[18])

In spring 1914, there would be a thorough revision of all the Society's procedures, resulting in a revised set of standing orders approved by Council.[19] The core elements of the (new) editorial system were left untouched: the sectional committees retained their role, although physics and chemistry became separate committees. The sectional committees could still request reports from referees to aid decision-making. The system was not quite so administratively complex to operate as it seemed on paper, because the sectional committees rarely met. In reality, their significance was in the provision of field-specific advice to the secretaries, and this was usually provided by the chairs of the committees, who acted as editorial advisers or (as the role would later be termed) associate editors.

The revised 1914 standing orders included new procedures to deal with a variety of potentially awkward scenarios. This bureaucratisation sat somewhat at odds with the stated aim of 'simplifying and expediting' procedures, yet it reflected the challenges of scaling up processes that had once operated informally among friends and gentlemen. The new provisions dealt with conflicts of interest on the part of the chairmen of the sectional committees, and gave explicit clarification that referees' reports were to be considered confidential. They also addressed the role remaining to the Society's Council, with so much of the routine work now handled by the sectional committees. Council members no longer needed to meet separately as 'the Committee of Papers', but Council remained the court of appeal for problematic editorial decisions.[20] The types of problems they anticipated included papers with high estimated costs of publication; any instances of contradictory reports from different sectional committees (usually for interdisciplinary papers); and papers that were 'not recommended for publication and have not been withdrawn'. This last reminds us that the Royal Society still tried very hard to avoid rejecting papers: it preferred to suggest politely to the author or communicator that the paper was not well suited to the Society. Most such papers would then be withdrawn, and usually submitted to another journal. But if the author did not take the hint, then Council would have to get involved.[21]

The 1914 revisions also attempted to create a space for strategic thinking about editorial practice and policy. A new standing committee was agreed, to bring together the chairs of the sectional committees on an annual basis. This unnamed committee appears to be the ancestor of the post-1968 Associate Editors Committee (Chapter 14), but it has left little evidence of its actual operation (or existence).

The new 1914 standing orders were the culmination of a second phase of reforms, focused on the Society's internal administrative processes, in the early years of the new century. This phase was driven by the Cambridge physicist Joseph Larmor (secretary 1901–12) and the recently retired director of the Geological Survey, Archibald Geikie (secretary 1903–8, president 1908–13), aided by Robert Harrison, who had become the new assistant secretary in 1896. His predecessor, Herbert Rix, had been described as 'punctual, systematic, upright and conscientious'; but he was also a student of philosophy and theology who longed to retire to a simpler life in the country.[22] Harrison, in contrast, was an experienced scientific administrator. He had studied physics and chemistry at King's College London before becoming a political private secretary, which involved him in the government committee responsible for observations of the 1881 transit of Venus. He then acquired a decade of experience as an administrator, first at the Central Institute for scientific and technical training at South Kensington and then at the Arts Union.[23] Seen through Harrison's eyes, the Royal Society's administrative system was 'inadequate to the status and dignity of the Society'. He set about to build 'a fresh organization' more suited to 'modern requirements'.[24] He hired more staff, acquired typewriters and installed electric lighting.[25]

In 1902, a committee investigating 'the organization of the business of the Society' found that the assistant secretary's lack of autonomous powers severely constrained his ability to deal efficiently with the 'increase in the amount and variety' of work. Harrison worked in the Society's premises each day, but the secretaries whom he was obliged to consult on appropriate courses of action (at that point, Larmor and Michael Foster) had full-time academic jobs in Cambridge. Council agreed to start inviting Harrison to attend its meetings, allowing him to become 'fully acquainted with the policy of the Council and all the details of the Society's business'; and the new statutes of 1905 would give him executive powers.[26] Harrison and his successors thus became able to take action on behalf of the Society – despite their status as staff, not fellows – in a way that none of their predecessors could have done. Over the course of the twentieth century, as busy academics found it increasingly difficult to devote time to the Society, the senior staff would become very influential in guiding its activities and policy, including its publications.

This period also saw a subtle shift in the distribution of work among the honorary secretaries. During the Stokes years, they had divided their responsibilities along the lines of 'internal' and 'external' business. After Stokes's retirement, the physiologist Michael Foster (secretary 1881–1903) continued to focus on the Society's external affairs, so the succession of physicists who served as his junior co-secretaries – including Lord Rayleigh (1885–96) and Joseph Larmor (1901–12) – looked after the meetings and all the publications. After Foster stepped down, this began to change. Arthur Schuster would be known as 'the physical secretary' (1912–19), while John R. Bradford (1908–15) and William B. Hardy (1915–25) served as 'the biological secretary'. However, Bradford and Hardy both acted as military and government advisers during the war, meaning that Schuster still took responsibility for the editorial management of papers in the biological, as well as physical, sciences. In the 1920s, however, we find Henry Dale (1925–35) managing the papers for series B, while James Jeans (1919–29) only managed those for series A. In 1927, the 'Register of Papers' ledger was split into separate ledgers for the A and B sides of the Society's publications, enabling each secretary to see more easily what was going on within their purviews. Both secretaries now performed discipline-defined editorial roles, but they were not 'journal editors' in the modern sense: each had responsibility for both the *Proceedings* and the *Transactions* in their field; and they still had to fit their editorial work around all their other responsibilities as Royal Society secretaries.

Amid all the changes, one reform that did not happen was the admission of women. In 1900, the Royal Society (and several other organisations) received a letter from the campaigning naturalist Marian Farquharson 'requesting that duly qualified women should be eligible for election into the Society'.[27] Farquharson had become the first female fellow of the Royal Microscopical Society in 1885, and would eventually be elected to the Linnean Society in 1908.[28] The Royal Society's Council responded ambivalently to her request, commenting that women's eligibility depended upon 'the interpretation of the statutes'.[29] Eighteen months later, it had to be more decisive when the electrical engineer John Perry submitted a 'certificate of candidature' for Hertha Ayrton. Perry had delivered Ayrton's paper on the electric arc to a meeting of the Society the previous year, and there was no doubt about her scientific credentials.[30]

Council again sought legal advice, explicitly noting that it was willing to alter its statutes if need be. That legal advice, however, was clear: while it remained doubtful whether women in general could be admitted, with or without a statute change, it was certain that married women, having no standing in law, could not be admitted to a royally chartered organisation.[31] Responding to the secretary's letter informing him that Ayrton, as a married woman, could not be considered for election, Perry remarked that 'I feel sure that, as time goes on, there will be more and more pressing need for the invention of a method to enable women to become Members of the Royal Society'.[32] Yet even though the Sex Disqualification Removal Act of 1919 removed the objection to married women, no further women would be proposed for election to the Society until 1944.[33] As a consequence, the Society's editorial decision-making processes remained in the hands of men, despite the small but growing presence of women among its authors.

In July 1912, the Royal Society had celebrated the 250th anniversary of its foundation charters, with representatives of universities and learned societies from all over the world. As president, Archibald Geikie had described the gathering as striking evidence of 'the sympathy which draws together the students of natural knowledge, and unites them into a worldwide brotherhood, inspired by one common spirit of devotion to the study of nature and the search after truth'.[34] Two years later, the outbreak of the First World War would stretch that sympathy to its limits. The Royal Society's authors and readers were primarily from Britain and the wider British academic world, so the breakdown of scholarly networks with Germany and the Austro-Hungarian empire had less impact on the *Proceedings* and the *Transactions* than might have been expected. But the international coalition that had been working to continue, and expand, the *Catalogue of Scientific Papers* disintegrated. The 1914 reforms had been completed before the war was declared, and they marked the end of over two decades of organisational changes.

The war itself appears to have caused fewer upheavals for Royal Society publishing than one might imagine. Most of the fellows and clerical staff were too old to be directly involved in the fighting, though a few of the staff did see active service (and Harrison joined a searchlight crew, serving several nights a week at Hyde Park corner).[35] Many of the fellows were involved in war work of various sorts, and consequently had less time for the Society's work. The Society appears to have been run by Arthur Schuster, with the assistance of Harrison and his remaining staff. Within weeks of the outbreak of war, London newspapers were commenting suspiciously on Schuster's German-Jewish origins (he was a naturalised British citizen and a retired professor of physics at the

University of Manchester), leading the Society to publicly express its 'profound regret at the annoyance to which he has been subjected'.[36]

The challenge of editing, refereeing and managing submissions to the Society's periodicals, despite the constraints on the paid and volunteer labour available during the war, was made easier by the dramatic decline in papers being submitted. In the early 1910s, the Society had been receiving around 200 submissions a year; this had halved by the later years of the war. Military service or scientific war work prevented younger scientists from writing up academic papers. During the Second World War, the Society would help develop processes to enable scientists to claim priority for findings that were too militarily sensitive to be published during war (see Chapter 13); but these issues do not seem to have been a concern during the First World War. The disruption to international communications had only a limited effect on the Royal Society's submissions, because it received relatively few overseas submissions at this time. That said, one Russian foreign member did comment that he planned to 'wait for calmer times' before sending his papers to London.[37]

After the war, Schuster switched roles to become the Society's foreign secretary, and took an active role in rebuilding scientific links between formerly warring countries. Harrison retired, leaving the Society with a key gap in its support staff. After the initial appointee moved on swiftly, the Society appointed Francis Towle, who already had two decades of service to the Society.[38] Whereas Harrison had come in as a new broom, keen to modernise, Towle saw his role as maintaining continuity and traditions. Towle was said to miss the 'formal courtesy' once found in scientific papers, to be 'seriously offended' by those who attended the Society in 'casual dress', and to dislike 'the custom of smoking in the tea-room' before Society meetings. The fact that the serving secretary described him as having a 'conservative tendency', a 'respect for precedent' and a 'suspicion with regard to change' may tell its own story about relationships between the officers and senior staff in the 1920s.[39]

Towle was supported by about half a dozen permanent clerical staff, with temporary staff employed for short-term projects such as cataloguing the library holdings. For Towle's small staff, the publications were just one element within the Society's administrative work, and the Society's increasing involvement in grant-making during the 1920s would lead to an 'increasing volume of work' for them.[40] The Society lacked the core funds that would have been necessary to expand its staff.

Towle had no power to make editorial decisions, but he was integral to the process. He allocated submitted papers to the (now) seven sectional committees, and he liaised with the chairs about the choice of referees, and the recommendations they made. Towle also handled all the correspondence with referees. Once papers had been approved for publication, proofs and corrections appear to have been dealt with by the library assistants. Despite Towle's apparent conservatism, he appears to have supported Jeans and Dale effectively in their editorial roles, and it was during his tenure, rather than that of the reforming Harrison, that *Proceedings A* became the journal Rutherford so praised.

The purpose of the Royal Society journals

In 1902, Henry Armstrong had suggested that the 'proud position' of the Society and its publications was due to 'past history rather than to its present performances', and that its publications could no longer be said to 'represent the high-water mark in all branches of science'. He was concerned that, despite the growth of submissions to the Society, the most interesting or significant new research was being published 'through other channels'.[41] If it was true that scientific authors preferred to publish in the *Philosophical Magazine*, *Nature*, the *Annals of Natural History*, the *Geological Magazine*, the *Proceedings of the Physical Society* or the *Monthly Notices of the Royal Astronomical Society*, then what could the Royal Society do?

One concern was definitely about speed of publication, especially for those in the physical sciences. The expected periodicity for scientific journals had become monthly. The *Transactions* could not compete with this, but the *Proceedings* came close. This is why so many fellows of the Royal Society were frustrated with the *Proceedings*: it appeared roughly monthly, but not necessarily on the same date each month, and, most annoyingly, it ceased issue in the summer (when the Society had no meetings).

But speed was not the only concern. The Royal Society's journals were unusual in their broad disciplinary remit at a time of specialisation. Whether run by learned societies, independent editors or commercial publishers, the vast majority of research journals focused on particular disciplines. Readers were now assumed to have particular interests, and to be relatively uninterested in developments in other fields. *Nature* bucked this trend, but found its readership by offering news, reviews and reports from the world of science, alongside short accounts of new discoveries. The Royal Society's division of the *Transactions* into A and B series had been an acceptance that physical scientists were no longer very likely to want to read detailed, lengthy research papers in the biological sciences; and vice versa. Yet even the A and B series were much more broadly defined than most other journals. Did such journals still have a purpose?

Armstrong suggested that, amid all the specialist societies and journals, the Royal Society and its publications could offer a uniquely 'favourable ... platform for the discussion of borderland problems'. He argued that 'in these days of ultra-specialisation' it is important that problems 'should be fully discussed from all sides', and he bemoaned what he saw as the 'dogmatism' and 'spirit of intolerance' that was arising between researchers in different fields.[42] He disapproved of the division of the *Proceedings*, and remained convinced that researchers should be 'forced together not separated at the present day. The R.S. is the only body which can (but does not) promote union of the sciences'. He thought that the pervasiveness of the A/B split throughout the Royal Society's activities was doing 'serious injury' to science.[43]

Armstrong lost that argument, but he was not the last to argue that the unusual breadth of the Society's journals – even divided – could be presented as their unique selling point. In the early 1920s, for instance, the presidency was held by the physiologist Charles Sherrington (1920–5). He was acutely conscious that research questions in physiology, pathology, bacteriology and pharmacology seemed 'more and more to merge', so that it was often difficult to classify individual papers under particular scientific fields. For more specialised journals or societies, that might create difficulties, but not, he argued, for the 'elastic working' of the Royal Society.[44] Sherrington tried to present *Proceedings B* as a uniquely valuable venue for interdisciplinary research within the biological sciences, but he had as little success as Armstrong. A few years later, Rutherford would attribute the lack of 'so large an expansion' in *Proceedings B* (compared to *Proceedings A*) to the success of the specialist learned societies, arguing that the separation – not merger – of research fields was the dominant trend.[45]

In 1924, an article in *Nature* reported on the ongoing growth of the scientific literature. 'Many new serials have commenced life since the War', 'scientific journals are steadily increasing in number' and there were over 25,000 titles listed in the new *World List of Scientific Periodicals*. (In contrast, the Royal Society's *Catalogue of Scientific Papers* had indexed just 1,400 journals.) The author bemoaned the dispersion of research across so many different journals, and likened it to 'an academic "mixed grill"'. It was difficult for researchers to navigate the literature, and to find the latest research in their field. One problem was that many research fields were now served by multiple journals, rather than having one central venue. But he also pointed to the problem of specialist papers appearing in general journals, where specialist readers might not see them. He suggested that 'the older Academies' might 'set an example' by directing their papers to 'specialised journals' where they might more easily be found.[46] This repeated Armstrong's concern that papers published in the Royal Society's journals were not necessarily being seen by the researchers who would be most interested. A few years later, Rutherford would claim that the Royal Society took a paternalistic pride in 'the healthy and progressive activities of the specialist societies', but the idea that the Society might abandon the field of publishing to those societies does not appear to have been given any serious consideration.[47]

Armstrong's claim that 'many authors prefer to publish through other channels' raises questions about who was actually publishing with the Society in the early twentieth century.[48] As Figure 11.2 shows, there was certainly no shortage of authors who did submit papers to

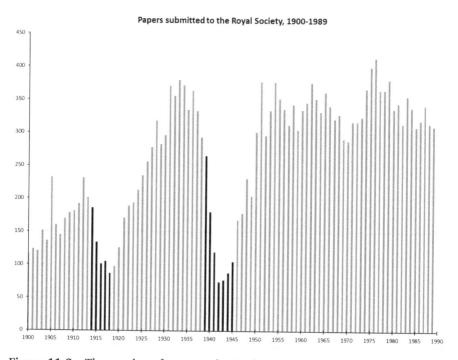

Figure 11.2   The number of papers submitted to the Royal Society, 1900–89 (with the war years in black).

the Royal Society, and there were more, not fewer, in the decades after Armstrong wrote. They were almost entirely British or from the wider British academic world. By the late 1920s, the Society was receiving almost 300 papers a year, numbers that would have shocked the worried treasurers of the 1880s and 1890s.[49] It is difficult to say how the Society's periodicals compared with their competitors, but it is at least clear that there were still plenty of scientific researchers keen to see their papers published by the Royal Society.

One group of researchers who clearly did publish with the Society was those in the physical sciences, and particularly, in physics. About twice as many papers were published every year in *Transactions A* compared to *Transactions B*; and when the *Proceedings* did split in 1905, it would follow the same pattern.[50] The Royal Society had continued to be the main forum for the presentation, discussion and publication of research in physics long after other specialist fields had acquired their own learned societies, and the inherent conservatism of social networks helped it retain that role for decades after the creation of the Physical Society in 1874.[51] Many of the most prominent figures in the Society were from the physical sciences, including William Thomson, Lord Rayleigh and Ernest Rutherford, and they may have attracted or encouraged submissions from their colleagues. The imbalance between the A and B sides of the publications affected the distribution of editorial work: fellows in the physical sciences did more refereeing work.

Another group of authors who still published with the Society was its fellows. Armstrong had claimed that 'the Society is of no proper value to us Fellows as a publishing organisation', but the 'Register of Papers' shows that fellows had certainly not abandoned the Society's publications.[52] The fellowship did not grow, but fellows continued to author (or co-author) between 40 and 60 papers per year throughout the late nineteenth and early twentieth centuries.[53] Armstrong himself published 11 papers in the *Proceedings* between 1900 and 1910 alone. Other fellows who regularly submitted papers in that decade included the physicists Lord Rayleigh and his son R.J. Strutt, astronomer Norman Lockyer, biological statistician Karl Pearson and chemist William Crookes.

Newly elected fellows were conscious of a tacit obligation to consider the Society's periodicals. The electrician Oliver Heaviside submitted his first and only paper for the *Transactions* on 18 June 1891, having been elected a fellow just 12 days earlier.[54] And in 1903, the Belfast chemist John Brown proposed sending his next paper to the *Proceedings*, because of his 'recent election to the Royal [Society]', but admitted to Larmor that he would have preferred to send it to the *Philosophical*

Magazine, because 'I find that papers in Proc. Roy. Soc. are not so well seen as those in Phil. Mag'.[55]

The growth seen in submissions to the Society's journals in the early twentieth century was driven by submissions from non-fellows. By 1930, about three times as many submissions came from outsiders as from fellows themselves.[56] This implies a possible increase in diversity in the origins of papers, but authors of papers were still almost all British, and almost all men. Although still excluded from the fellowship, women could submit papers to its meetings and periodicals so long as they could find a supportive fellow to co-author or communicate their paper. For instance, the astronomer Margaret Huggins co-authored with her husband William throughout this period, and she was joined by a growing number of women who were able to access the Society's networks by virtue of their university training and academic connections, rather than familial links. For instance, the plant geneticist William Bateson was a notable supporter of women scientists, and communicated papers by those who worked with him at Cambridge and, later, at the John Innes Institute, including Florence Durham (his sister-in-law), Dorothea Marryat, Dorothy Cayley and Muriel Wheldale.[57] He also co-authored with some of his female colleagues, including Edith Saunders, director of the Biological Laboratory for Women in Cambridge, and Caroline Pellew, who worked at the John Innes Institute. By the 1920s, women would make up around 7 per cent of all authors submitting to the Society, and – in contrast to the general trend in the Society – they were mostly in the biological sciences.[58]

It is clear that plenty of fellows and non-fellows did still wish to publish their work with the Royal Society, which suggests that there were perceived benefits that Armstrong overlooked. For those in certain fields, the Society's continuing willingness to pay the costs of illustrative plates, and to allow an apparently unlimited number of pages in the *Transactions*, remained a clear benefit that independent journals could not match. Moreover, by focusing on the practical questions of circulation, Armstrong was overlooking other functions that the publications served. If authors had simply wanted publication, there were plenty of other ways to achieve it, but the rising number of submissions from outsiders suggests that many researchers sought the prestige that the Royal Society was perceived to grant. Even John Brown, as a new fellow, saw the possibility of a paper being accepted for the Society's Proceedings as an 'honour',[59] and Armstrong did recognise the social capital that came with publication by the Society, even though he felt that '[i]t would be better if less were made of publication in the Transactions'.[60] The Society's editorial processes might be cumbersome, but successfully navigating these traditional processes, guarded by scientists of acknowledged reputation, could be seen as a badge of achievement. Publishing with the Society could also be a way for junior researchers to attract the attention of the fellows of the Society, which was the first step to election in an organisation that replenished itself through nominations by existing members.

The pros and cons of refereeing

In 1902, Henry Armstrong also proposed that the Society ought to consider 'the appointment of an Editor', as a separate role from that of secretary. The Geological Society and the Chemical Society had appointed editors for their periodicals since the 1840s; in both cases, the appointees were paid members of staff.[61] Armstrong imagined something slightly different, suggesting the editor would be a fellow of the Society, 'conversant with the general trend of scientific enquiry'. For Armstrong, an editor would be able to devote more time and energy to the journals than the secretaries could spare from their multifarious duties. He also thought that an individual editor, with executive powers, would be able to make swifter decisions than were possible in the Royal Society's complex system of committees and referees. Armstrong thought that the new sectional committees had introduced unnecessary complexity; and that 'the old plan of requiring all papers for the Transactions to be submitted to two referees' was 'an anachronism in the majority of cases'.[62] Five years later, in a private letter to Joseph Larmor (then secretary), he referred to the Society's editorial process as its 'prehistoric methods'.[63]

Armstrong's (unsuccessful) call for an editor alerts us to the existence of a discourse that critiqued the use of referees. Outsiders and disgruntled authors had complained of the Royal Society's secret judgements since at least the 1840s[64] and, in the 1920s, an early trade union for scientists would claim that referees were 'anonymous and irresponsible'.[65] Criticisms were also being raised by those within the Society. Armstrong repeated the old concern that secrecy 'too frequently' led to 'ill-feeling', and pointed to the delays caused by refereeing.[66] Others raised concerns about the intrinsic conservatism of the process.

The concern about stifling innovation had been raised publicly by Lord Rayleigh in 1892, when he arranged for the belated publication in *Transactions A* of a paper by John Waterston that had pre-empted James Clerk Maxwell's work on the kinetic theory of gases. Waterston's paper had been declined for publication by Royal Society referees in 1845.

It then languished in the Society's archive until Rayleigh, prompted by reading a summary in an old issue of the *Philosophical Magazine*, rediscovered it. Rayleigh read this episode as an indication that learned societies were not the best channels for announcing 'highly speculative investigations, especially by an unknown author'. The concept of implicit bias was decades in the future, but Rayleigh recognised the conservativism inherent in the refereeing process: while acting as a representative of a prestigious learned society, a referee 'naturally hesitates to admit into its printed records matter of uncertain value'.[67]

The more common criticism of refereeing as a process was the time taken by referees, with authors feeling that referees delayed publication, and referees (according to Armstrong) worrying that too much of their 'valuable time' was being 'practically wasted on such work'.[68] The Society was already encountering problems finding fellows who had time to take on the work of refereeing, particularly in the physical sciences where the ratio of submitted papers to available referees was higher. In 1902, for instance, the chairman of the mathematics sectional committee took it for granted that Joseph Larmor (then secretary) was 'too busy to be asked', but shortly afterwards had to report that, 'I am sorry to say that Lord Rayleigh has also too much in hand'.[69] Only about a fifth of the fellowship did any refereeing in a given year around 1900, and even fewer of them did very much of it. Imogen Clarke has shown how the Society's editorial practices could be heavily influenced by a small network of fellows;[70] but it is equally true that the editorial system relied on that small group of active referees.

Fellows who accepted refereeing duties often struggled to perform them in an appropriate timeframe, and since the single copy of the paper still had to be sent to the referees in sequence, delays could build up. In 1907, a referee returned a paper, noting how 'it has been impossible in the time to do anything in the way of verification or checking, but I have given all the care I could to reading the memoir'.[71] In the 1890s, decisions on papers for the *Proceedings* (mostly made without referees) took 35 to 40 days on average, while those for the *Transactions* (which always involved referees) were taking 80 to 110 days on average. This suggests that the Society's own administrative processes probably took about a month, and that refereeing added seven to 12 weeks to the decision-making process.

Mathematical physicist G.F. Fitzgerald appears to have felt more guilt than some referees, for in 1894 he apologised three times for the time it was taking him to deal with Larmor's discussion of a dynamical theory of the ether.[72] In late February, he told Larmor that he had permission from Rayleigh (then secretary) 'to correspond with you directly'. He went on: 'I am rather busy so that I can only read it by instalments and that does not do it justice, so you must forgive my asking you some things that are idiotic and that I should easily make out if I had time to consider them'.[73] A month later, he told Larmor, 'I have again had time to attack your paper',[74] but it was not until mid-April that he finally submitted his report, apologising that, 'I got off on other things and had almost forgotten all about it'.[75] Even so, Fitzgerald had actually been faster than the first referee, J.J. Thomson, who had kept it right through December and January. Larmor's paper was finally approved for publication in May 1894, almost six months after it was first submitted.[76]

Fellows who did agree to act as referees were becoming fussier about papers that absorbed more time and effort than they deserved. The guidelines for authors that began to appear in the *Year Book of the Royal Society* in 1897 encouraged manuscripts to be submitted 'if possible, type-written, or at least written in a legible hand, and properly prepared as copy for press'.[77] In 1904, a circular was issued with 'a general reminder to authors that papers presented to the Society are understood to be in their final form, carefully revised for press'.[78] Such reminders suggest that the guidance was not proving effective. The Society's stated concern was the cost of author corrections at a later stage in the process, but referees were equally keen to evaluate well-presented papers. One deeply disgruntled referee complained about having 'to wade through the author's rough copy'. He claimed to have agreed to read it only on the understanding that 'it was in a more finished, typewritten state'.[79]

Another reason why refereeing slowed the publication of research was that the practice of passing referees' comments to the authors frequently inspired (or required) them to make revisions to their paper. Recommendations that papers should be corrected for 'type and spelling errors'[80] or 'clerical errors',[81] or that 'the English requires careful revision' were commonplace.[82] Referees also seem to have become more willing to ask for changes that were more complex than the simple deletions of an earlier period. In 1901, for instance, a referee laid out a set of recommendations: four pages should be cut because 'they relate to experiments proposed but not yet carried out'; some tables should be omitted; other tables should be rearranged; a graph should be replotted; details that are 'purely accidental and of no interest to the research' should be cut; and the whole should undergo 'careful verbal revision'. He thought the experimental research was sound and useful, and should be published if the author could 'abridge the paper somewhat' and 'radically revise' its style.[83]

The Society's guidelines continued to state that the original manuscript should be preserved, with additions clearly dated, so that priority claims could be adjudicated if need be. But referees who had to read heavily revised manuscripts did not enjoy the experience. In 1902, one reported that the first 43 pages 'of the original type-written copy have been replaced by 106 pages of manuscript and the remainder of the type-written copy has been altered a good deal'.[84] In 1924, another referee wished that the author had rewritten 'the heavily corrected pages' or had 'the manuscript typewritten'.[85]

Some efforts were made to speed up and standardise the refereeing process. Reducing the time allowed for referees to return their reports from a month to two weeks in the late 1890s had little effect. Around the same time, a printed form was introduced with standardised questions for referees to address, with gaps for the answers to be filled in (see Figures 11.3 and 11.4). At first, referees used to unconstrained, free-form reports were eclectic in the way they filled in the form, but by the 1920s, most were using it (sometimes with a longer report attached as a separate sheet). By organising the referee's response, the printed form made it easier for the editorial team to process the reports.[86] It also had the effect of encouraging brevity from the referees. Most of the questions could be answered with just a couple of words, and the most open question was at the foot of the page, where the available space suggested that responses should be no more than a paragraph. By the 1920s, it was common for positive reports to say simply 'recommended for publication, in full'. Even the negative reports were briefer, concisely noting that papers were 'most unsatisfactory',[87] 'more suited to a technical journal',[88] or 'offers nothing to a well worn subject'.[89]

Despite the delays caused by refereeing, directly and indirectly, and despite worries about irresponsibility or conservatism, the Royal Society did not abandon refereeing. In fact, in the first three decades of the new century, the Society increased its use of referees both absolutely and relatively. The growth in submissions meant that there was more refereeing to be done than ever before. The 'Register of Papers' reveals that the number of reports being written each year rose from about 100 in 1910 to about 400 in 1930. This is a faster rate of increase than can be explained by the number of submissions alone. A higher proportion of papers was now being sent to referees. Before the First World War, 60–70 per cent of papers had been published by the Society without the involvement of referees (and in the *Proceedings*). They had been approved solely on the authority of the chair of the relevant sectional committee and the relevant secretary. There was no formal change of

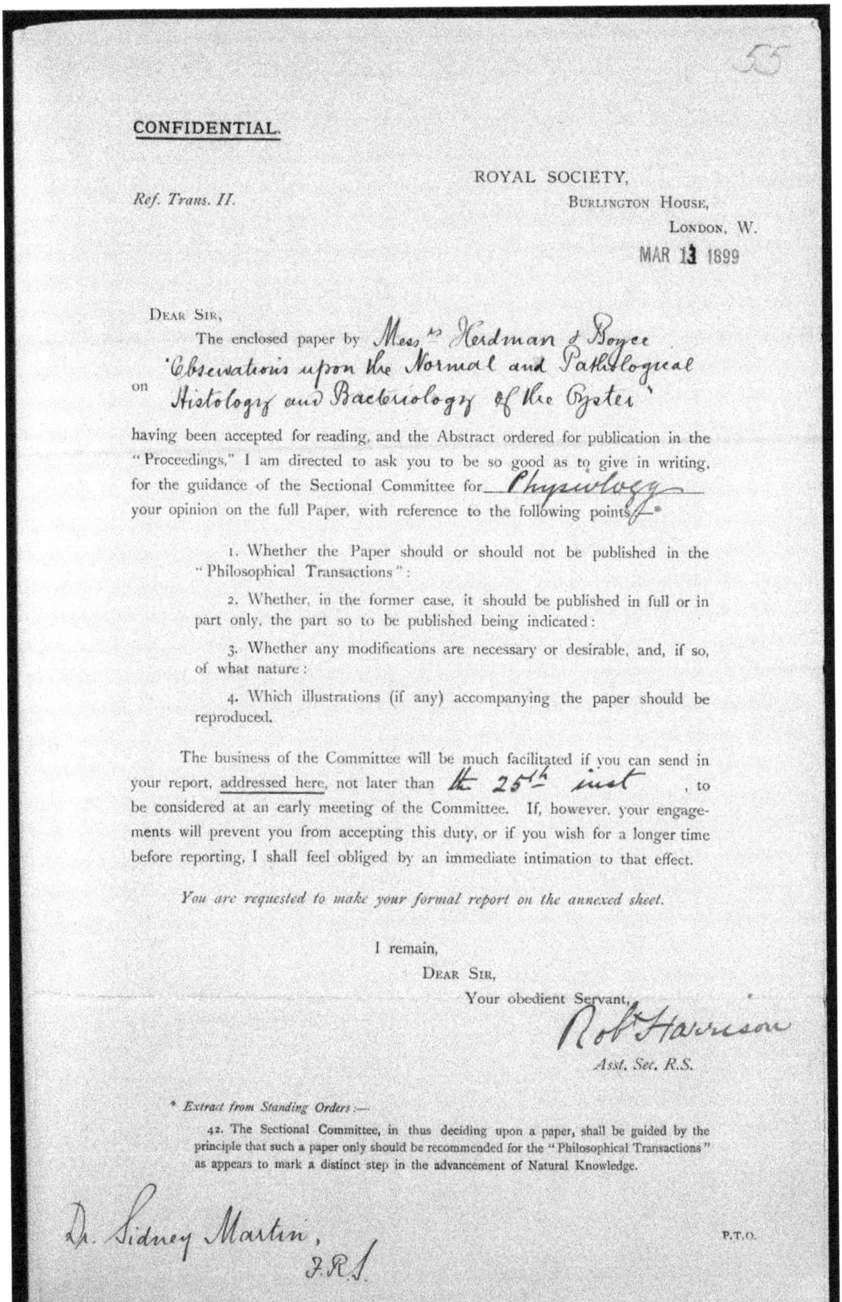

55

CONFIDENTIAL.

ROYAL SOCIETY,

Ref. Trans. II.

BURLINGTON HOUSE,

LONDON, W.

MAR 13 1899

DEAR SIR,

The enclosed paper by Messrs Herdman & Boyce

on 'Observations upon the Normal and Pathological Histology and Bacteriology of the Oyster'

having been accepted for reading, and the Abstract ordered for publication in the "Proceedings," I am directed to ask you to be so good as to give in writing, for the guidance of the Sectional Committee for Physiology your opinion on the full Paper, with reference to the following points:—*

1. Whether the Paper should or should not be published in the "Philosophical Transactions":

2. Whether, in the former case, it should be published in full or in part only, the part so to be published being indicated:

3. Whether any modifications are necessary or desirable, and, if so, of what nature:

4. Which illustrations (if any) accompanying the paper should be reproduced.

The business of the Committee will be much facilitated if you can send in your report, addressed here, not later than the 25th inst, to be considered at an early meeting of the Committee. If, however, your engagements will prevent you from accepting this duty, or if you wish for a longer time before reporting, I shall feel obliged by an immediate intimation to that effect.

You are requested to make your formal report on the annexed sheet.

I remain,

DEAR SIR,

Your obedient Servant,

Robt Harrison

Asst. Sec. R.S.

* *Extract from Standing Orders:—*

42. The Sectional Committee, in thus deciding upon a paper, shall be guided by the principle that such a paper only should be recommended for the "Philosophical Transactions" as appears to mark a distinct step in the advancement of Natural Knowledge.

Dr. Sidney Martin, F.R.S.

P.T.O.

Figure 11.3 Instructions to referees (1899), incorporating questions drafted by treasurer John Evans in 1894 © The Royal Society.

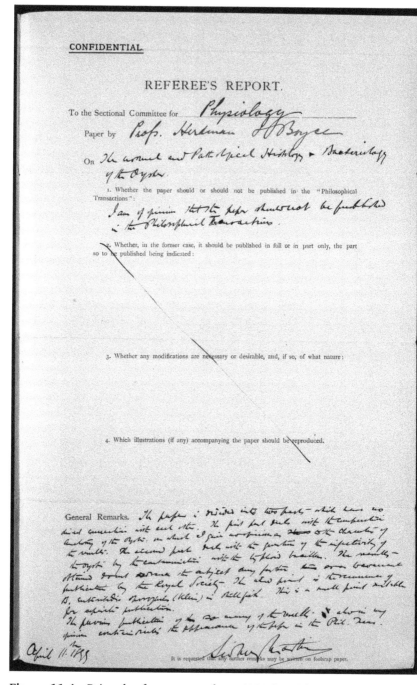

Figure 11.4 Printed referee report form (1899), with spaces for answers © The Royal Society.

policy regarding the use of refereeing, but the 'Register of Papers' reveals that, after the war, the proportion of papers published without referees' reports fell, and kept on falling. By 1930, only around 20 per cent of papers were being published without at least one referee report.[90]

There was nothing in the standing orders to give fellows a 'fast track' to publication, yet some fellows did think that that 'any communication from a Fellow should be published', and this was the case even if its conclusions seemed 'untrustworthy'.[91] The fellows' traditional privilege of communicating papers to the Society had been taken to imply the right to have one's own papers read to the Society (though not necessarily published), but the changes around 1900 forced some readjustment. The new arrangements for meetings meant that only a select few papers could be read and discussed, and the emergence of the *Proceedings* as a fully fledged research journal would prevent its pages being used as an alternative means of acknowledging all communications from fellows. The increasing proportion of papers now being sent to referees implies that 'being a fellow' was not understood to mean being exempt from scrutiny. The privilege that fellows retained was that of judging the work of others.

The papers that were published in the 1920s without the scrutiny of referees had usually been communicated via certain trusted or eminent fellows. As in earlier periods, these were often senior figures in the Society, including the officers: William Hardy (secretary 1915–25), Charles Sherrington (president 1920–5) and Ernest Rutherford (president 1925–30). For instance, in 1921, Rutherford communicated seven papers to the Society, all of which were published in the *Proceedings* and only one was sent to a referee. The six papers published without refereeing were all by authors associated with Rutherford's Cavendish Laboratory, including Ann Davies (later, as Ann Horton, the first tenured woman lecturer at the Cavendish), the Japanese students Takeo Shimizu and Taiji Kikuchi, and the Canadian Etienne Bieler. For papers coming out of the Cavendish, Rutherford's recommendation clearly carried weight. On the other hand, the paper he communicated on behalf of his new son-in-law, mathematician Ralph Fowler, was sent to the mathematician G.H. Hardy for a report. Another trusted communicator was William Henry Bragg, then vice-president of the Royal Society and head of the Royal Institution laboratory. He communicated the first two sole-authored papers by Kathleen Yardley (later Lonsdale), who was then working in his laboratory; they were both published without consulting further referees.[92] In these instances, it was assumed that the fellow communicating the paper was familiar with its contents, and had the appropriate expertise to evaluate it; the communicator was implicitly acting as a referee ahead of time.

In the 1890s, the statute discussing 'communication' stated that 'it shall be the duty of each fellow ... to satisfy himself that any letter, report, or other paper which he may communicate, is suitable to be read before the Society'. In an 1896 circular to all fellows, the secretaries explained that '[s]uitability is here meant to refer in the first instance to the scientific value of the paper, but may perhaps be extended to the style of the paper'.[93] This marks a different vision of the role of communicator from that used in Joseph Banks's day: it was now the 'scientific value of the paper', not the social credentials of its author, that was to be vouched for.

Whether fellows acting as communicators did actually scrutinise the papers as if they were referees is a different question. Both in 1908 and in 1925, there were attempts to require communicators 'to sign an explicit declaration' that they had read and evaluated the paper they were submitting.[94] It was not until the 1930s that questions would be raised about the possible bias of a recommendation from a communicator – who might well be the author's supervisor or laboratory head – compared to that of an independent referee.

Apart from the minority of papers that arrived with the approbation of a particularly trusted (or eminent) fellow, most papers in the 1920s were sent to at least one referee, regardless of whether the authors were fellows or not. Problematic papers, and long papers for the *Transactions*, went to two or three referees. This meant that the Society had to ask more of its fellows: there were more reports to be written every year. The proportion of the fellowship receiving requests to referee did grow slightly, but there were still fewer than a quarter of fellows involved in any given year. The sectional committee chairs appear to have been unable (or unwilling) to expand their pool of active referees to keep pace with the number of reports needing to be written. The relatively small size of the fellowship meant that the pool of potential referees and committee members had not kept pace with the expanding number of submissions to the journals.

Another problem was that fellows might not be willing to referee when asked to do so. In 1923, Charles Sherrington was already noting that 'men – or, for that matter, women' in academic jobs faced competing demands 'upon their strength and time'.[95] This affected the ability of fellows to engage with Society activities, as Rutherford noted when he commented on the 'painful impression' conveyed to authors when their papers were read 'before a very small audience' at Royal Society meetings.

He urged more fellows to see it as their duty to attend meetings, 'even though it may involve some sacrifice of their time and energy, and even of their inclinations'.[96] Fellows willing to undertake refereeing duties and committee work were making similar sacrifices.

The most active referees of the 1920s were not, in contrast to earlier decades, the secretaries (nor the sectional committee chairs). Judging by samples from the 'Register of Papers' for 1921 and 1925, they tended to be recently elected fellows: applied mathematician Geoffrey I. Taylor, physiologist Archibald Vivian 'A.V.' Hill, physicist Frederick Lindemann (later Viscount Cherwell), geophysicist Sydney Chapman and astrophysicist John W. Nicholson. Nicholson was the fellow of longest standing in this group, having been elected in 1917.[97]

Refereeing thus became another aspect of the Society's activities where the early twentieth-century officers faced challenges in getting more of the (busy, distant) fellows involved in any regular capacity. In 1925, for instance, 99 of the 443 fellows did do some refereeing. But 51 of those wrote only one report, whereas 13 fellows wrote five or more reports.[98] The result was that the Society was even more dependent on a core of active and efficient fellows to act as referees: by 1930, almost half of all reports were being written by a 'top 20%' of referees.[99] As president, Rutherford acknowledged that this 'difficult work of adjudication' could be 'a heavy burden', and thanked the fellows for undertaking it 'uncomplainingly'.[100]

The new *Proceedings*

Henry Armstrong's attention in 1902 had been focused upon the *Proceedings*. In common with many fellows over the previous 50 years, he felt that the *Proceedings* had not lived up to its potential to be an effective medium of communication.[101] As long as only *Transactions* papers were required to demonstrate 'a distinct step in the advancement of Natural Knowledge', papers in the *Proceedings* could be seen as less intellectually significant.[102] At the same time, the *Proceedings* was unable to fully capitalise on its faster publication times, and fill a similar niche to that occupied by the *Philosophical Magazine*, because the irregular pattern of its issue and pricing limited its ability to circulate effectively through commercial channels.

The problems caused by the pattern of issue had been recognised since the 1850s, but there appears to have been a total lack of determination to do anything about it. Most recently, the format of the publications had been discussed in the mid-1890s, as part of the series of reforms that led to the creation of the sectional committees. In 1895, Council considered passing a resolution that 'the Proceedings of the Royal Society be published, if possible, more promptly and at more frequent intervals than at present', but this was watered down to being 'issued at short intervals as may be found suitable'.[103]

In 1896, Michael Foster and his co-secretary Lord Rayleigh notified the fellows of changes that they hoped would turn the *Proceedings* into 'a suitable channel for the speedy and brief announcement of new discoveries in all branches of science'. No changes were made to the pattern of issue. Rather, the aim was to prevent the *Proceedings* being perceived as a journal of second-best work. Foster and Rayleigh seem to have believed that the increased willingness to publish independent papers (not just abstracts) in the *Proceedings* in the late nineteenth century had harmed its reputation. They disliked the use of the *Proceedings* as a way for the Society to publish 'papers of a technical character' or those 'which, though embodying accurate and valuable observations, are more or less incomplete'. Such papers had 'been judged unsuitable' for the *Transactions* because referees felt they lacked the 'distinct advance in natural knowledge' that would demonstrate their 'high scientific Standard'; this requirement was printed at the foot of the referee report form (see Figure 11.3).[104] As Foster had told a correspondent the previous year (after resolving a disagreement between referees by suggesting that the contested paper on brain physiology should be published in the *Proceedings*), 'I quite agree as to the undesirability of using the Proc. to publish papers not good enough for the Phil. Trans.', but it was inevitable 'as long as we have Proc. in our present form'. He hinted that, 'I trust we are near the end of a bad system'.[105]

The key innovation was a page limit, to be more strictly applied. The relative merits of six or 12 pages were debated, before a decision for 12 pages was made.[106] This was significantly more than the two pages claimed as a norm in 1892, and was intended to encourage authors to consider writing a short independent paper for the *Proceedings* (as against a lengthy potential *Transactions* paper and an abstract).[107] Foster hoped it would be strictly applied, but the agreed wording still retained the possibility of exceptions. This change may have prevented the publication of (long) rejected *Transactions* papers being published in the *Proceedings*, but it seems unlikely that it met Foster and Rayleigh's other goal of enabling 'the Proceedings to be issued more rapidly'.[108]

The perception of the *Proceedings* as the intellectual inferior of the *Transactions* was partly due to its historical origins as an abstract journal (see Chapter 7), and partly to the absence of any statement equivalent to the requirement for *Transactions* papers to display a 'distinct advance in natural knowledge'. But it was not helped by the somewhat miscellaneous genres of items that appeared in the *Proceedings*. The standing orders developed in 1896 ensured that it continued to be used to record 'the formal business conducted' at each weekly meeting of the Society (including presidents' anniversary addresses, annual reports of Council, and lists of gifts received), as well as 'abstracts of papers', 'short papers', and any other material Council wished.[109] The presence of so much material related to the Society's internal organisation limited the appeal of the *Proceedings* to non-fellows; and the miscellaneity of genres may have undermined the intellectual merit of the 'short papers', by association. That said, there is no doubt that an increasing number of authors were already choosing to publish in the *Proceedings* rather than the *Transactions*.

The miscellaneous nature of the *Proceedings* did give the secretaries flexibility about what could appear in it. It could, for instance, be used to publish 'preliminary notes'. By the late 1890s, as Melinda Baldwin has shown, researchers were finding the letters page of *Nature* to be a useful way of announcing preliminary results, that would later be worked up for fuller publication elsewhere. Physical scientists, the young Ernest Rutherford among them, found the weekly publication schedule useful for establishing priority as they raced to announce the latest discoveries in radioactivity.[110] The practice of preliminary notes had, however, begun much earlier. A handful of physicists (and one or two naturalists) had used the *Proceedings* this way in the 1860s and 1870s, and more did so in the 1880s and 1890s.[111] In contrast to the abstracts that had usually appeared in the *Proceedings*, and summarised completed research, these preliminary notes were a promise of things to come.

Using the *Proceedings* in this way was sufficiently familiar by 1895 that Karl Pearson saw it as a convenient solution, when he realised that his paper on statistical genetics, intended for *Transactions A*, would not be ready for submission until after the Society's summer recess had begun. No publication decisions were made over the summer, so Pearson feared that his paper would be stuck in limbo. He could publish it somewhere else rather than wait, but as he told his friend and mentor, Francis Galton, 'I don't like to publish … elsewhere … I should like to send my paper when completed to the Royal Society'.[112] Pearson was not yet a fellow, but he hoped to be so soon. He compromised by asking Galton to communicate 'this short note', which he hoped 'could be published in this session's *Proceedings* … I mean before the long vacation'.[113] The preliminary note was published; the full paper appeared the following year in *Transactions A* and Pearson was elected to the fellowship.

Some fellows felt that rapid publication of these short notes should become the core purpose of the *Proceedings*: in 1907, the zoologist E. Ray Lankester would propose making it a weekly periodical, with a limit of six pages, and offering guaranteed fast-track publication (without referees, and within a week) for fellows submitting short contributions of a page and a half. Lankester's vision would have kept the *Transactions* as the Society's 'chief publication', able to give 'full and very ample space and illustration to work of high importance'; whereas the *Proceedings* would focus on 'offering the means of *immediate* and *brief* publication to all varieties of scientific work'.[114] In reality, the Society moved in the opposite direction in the decade before 1914, making the *Proceedings* its chief publication and allowing it to carry longer (not shorter) papers.

In 1903, the Society's Council agreed to establish a committee to investigate its arrangements for publishing and circulating scientific research. This was a response both to Armstrong's 1902 memorandum, and to an alternative set of proposals put forward by the physiology sectional committee in 1903.[115] The committee involved no fewer than 24 fellows, including Norman Lockyer (founder of *Nature*), Michael Foster (no longer secretary; also founder of the *Journal of Physiology*) and former treasurer John Evans. They reported in spring 1904.[116] They had a broad remit, but it explicitly included thinking about how to make the *Proceedings* 'a more effective publishing medium', and whether it was 'desirable or feasible to publish parts [of the *Proceedings*] regularly at stated times'?[117]

The report disappointed many reformers by saying nothing about the pattern of issue of the *Proceedings*. It did recommend several significant changes, however. Foster and Rayleigh's page limit was removed, the paper size was increased slightly, and the abstracts of papers communicated to the Society were removed from the *Proceedings*.[118] (The Society continued to print abstracts of papers it received, but they were now issued as separate inserts. The revised standing orders of 1914 would make explicit the requirement for the authors of all papers – not just those potentially for the *Transactions* – to provide a 300-word abstract on submission.[119]) The 1904 report also recommended that 'all papers' published by the Society should be, 'so far as convenient', published in the *Proceedings*.[120] The *Proceedings* was being transformed from a place to publish abstracts, into a journal carrying independent research papers of considerable 'length and complexity' – and this recommendation turned it into the default mode of publication for the Society.[121]

Council did not accept all the recommendations of the 1904 committee. In particular, the report agreed with Armstrong (and not with the physiology sectional committee) by advising that the *Proceedings* should not be split into two series. Yet from 1905, the *Proceedings* was, nonetheless, 'divided into two sections, Physical subjects and Biological subjects, separately paged'.[122] As with the *Transactions*, the decision was probably motivated by the practicalities of distributing copies to fellows and institutions who were interested only in specific fields, and could henceforth be sent the appropriate series of the *Proceedings*, rather than the whole thing.[123] It took little more than a year for the treasurer to comment on the 'increase of matter' being published under the new arrangements.[124]

Henry Armstrong was unsurprisingly disappointed. In 1908, he told the Society's secretaries that there was still much to do to make the *Proceedings* into 'a really useful publication'. He also told them that the money the Society spent on the *Proceedings* each year 'is probably in great measure wasted', and that authors who published in the *Proceedings* did so 'at a considerable sacrifice', because the *Proceedings* 'do not receive the attention they deserve nor do they circulate among scientific workers generally in any regular manner'.[125] Nor was he a lone voice. A year later, Hertha Ayrton remarked that the *Proceedings* 'need a thorough reform. Everyone knows this, and yet, like "everyone's business" at all times it does not get done'. She herself could 'do nothing because I am a mere outsider, being a woman', yet generations of male fellows had also failed to get it done.[126]

The role of the *Proceedings* as the main publishing outlet for the Royal Society was cemented in the revised standing orders issued in 1914. Here, the *Proceedings* was now listed first, ahead of the *Transactions*. These standing orders also granted *Proceedings* papers equal intellectual merit with the *Transactions*: both journals were now described as publishing 'papers of approved merit'. The 1914 rules reimposed a page limit, but it was now set at 24 pages, enabling the *Proceedings* to take over more of the content that would previously have had to appear in the *Transactions*.[127] The changes were reflected in revisions to the report form for referees. The questions were rephrased and reordered, so that the first priority was no longer 'whether the paper should or should not be published in the *Philosophical Transactions*', but 'whether or not the paper should be read before the Society' and 'whether the paper should or should not be published by the Society'. Only then were referees asked to consider whether it should appear in the *Proceedings* or the *Transactions*. They were provided with the text from the new standing orders to guide

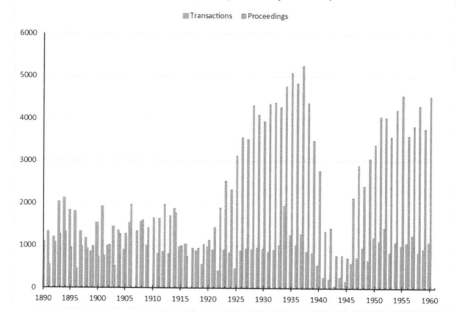

Figure 11.5   Pages printed annually in Royal Society journals, 1890–1960.

that decision, making clear that the length and the presence (or not) of 'numerous or elaborate illustrations' were the only differences between the journals.[128]

The result of these changes was that the *Proceedings* absorbed the growth in submissions seen in Figure 11.2. The majority of papers received by the Society had already been featured in the *Proceedings*, rather than the *Transactions*, long before this, but after the reforms of 1904 and 1914, they did so as mid-sized independent papers written specifically for it, rather than as summaries of a longer paper. Even with the increased page limit, *Proceedings* papers were still markedly shorter than those issued in the *Transactions* series. In the 1920s, the median length of a *Proceedings* paper was about 15 pages, whereas it was between 30 and 40 pages for a *Transactions* paper (and there were occasional monster papers running to 100 pages or more).[129] By the late 1920s, the two series of the *Proceedings* were carrying 10 to 15 times as many papers as those of the *Transactions*, with a total page count triple that of the *Transactions*, as Figure 11.5 shows. This is why almost all discussions about editing and publishing during the twentieth century focused on the *Proceedings*.

Surviving reports from before the reforms to the *Proceedings* (when authors were usually writing at *Transactions*-length) suggest that referees were unimpressed by papers they perceived as long-winded, poorly expressed or badly structured. How much of this was due to a concern for good literary style, or to protect the Society's finances, is unclear. The Society was certainly willing to fund the publication of long papers, but the length had to be justified by intellectual merit. Papers believed to be 'exceedingly diffuse' were unlikely to be published in full, yet even very poorly written papers might have publishable abstracts.[130] One referee complained in 1900 that: 'The numerous experiments are arranged and recorded in such an unsystematic manner and are so mixed up with various unnecessary and confusing details that the paper is unfit for publication in its present state.' The abstract nevertheless appeared in the *Proceedings*.[131]

The willingness of researchers to submit their papers in a form and length that would suit the new *Proceedings* (rather than the *Transactions* or, indeed, a book) suggests that, by the early twentieth century, scientific authors were becoming used to writing in shorter units of prose. Some authors would already have gained these literary skills from writing for other journals. It is clear from Joseph Larmor's correspondence that physical scientists considered the *Proceedings* (and later, *Proceedings A*) in the same category as several journals with restricted page space. In 1902, for instance, Henry Armstrong had asked Larmor's advice on a paper, saying, 'If you think well enough of it pass it on for the R.S. Proceedings', but otherwise, 'I should prefer … to send it either to the Physical [Society] or direct to the Phil[osophical] Mag[azine]'.[132] Another correspondent repeatedly sought Larmor's advice on whether to send his papers to the *Philosophical Magazine* or to *Nature*. When the Society accepted one of his papers, but directed it to the *Transactions*, he appears to have been taken aback, telling Larmor he was 'rather anxious to get the paper published as soon as possible; that was partly why I suggested the Proceedings'.[133]

Authors for whom the Royal Society was but one possible outlet for their papers had to learn to write more concisely than those of an earlier generation who had expected to be published in the *Transactions*. At the same time, referees' own experiences as readers of shorter-format journals may have shifted their feelings about the desirable length of scientific papers. The shift from book-length to *Transactions*-length to *Proceedings*-length may explain the shifts in language noticed by those who have studied the rhetoric and discourse of scientific articles over the nineteenth and early twentieth centuries: the use of complex noun-adjective phrases and abbreviations – later known as technical jargon – in place of extended explanatory sentences was increasing around the turn of the century.[134]

The other significant change in Royal Society publishing in the early twentieth century was that *Proceedings A* came to be seen as more successful than *Proceedings B*. It attracted more submissions, and it probably had more subscribers. The earliest data we have are from 1935 (see Table 13.1), and by that point, *Proceedings A* had about 725 sales, compared to 420 for *Proceedings B*. (These figures are both higher than for the *Transactions*; and they exclude non-commercial circulation.) The Society marketed the two series of the *Proceedings* equally well (or poorly), so these higher sales presumably represent greater interest in *Proceedings A* from readers and librarians in the 1920s and early 1930s. In the first few years after the split of the *Proceedings*, the numbers of articles published each year in the two series had been broadly similar, but from 1912 onwards, there were always more in *Proceedings A*. Its growth was driven by a flood of submissions from physical scientists – especially physicists – during the late 1910s and 1920s. By 1916, there would be twice as many articles published in series A as in series B, and some years in the 1920s would see more than three times as many in series A. When Towle created separate 'Registers of Papers' in 1927, the dominance of the A side was clear: 201 of the 278 submissions to the Society that year were recorded in the A side Register. This numerical dominance of the A side would continue until the late 1980s.

In the 1960s, Henry Dale (biological secretary, 1925–35) was asked for his recollections of Royal Society publishing in the 1920s. He recalled 'an impression' that '[p]hysics, both theoretical and experimental, had indeed come to occupy a place of predominance over the other A side subjects'. He did not think there had been 'any deliberate policy involved', but that it was a consequence of Jeans and Rutherford themselves being so closely involved in the 'tremendous new developments in experimental and mathematical physics' in the 1920s.[135]

Historians of physics have associated the expansion of *Proceedings A* in the 1920s with the rise of the new physics of relativity and quantum mechanics.[136] Some scientific journals accepted papers from both classical and modern perspectives, but *Proceedings A* was seen – especially by those excluded from its pages – to have come down on the modern side. This consolidated its position against the *Philosophical Magazine*, which had for decades been a popular outlet for researchers in the physical sciences, but whose influence was waning. The Royal Society's preference for modern physics reflected the interests of such fellows as James Jeans, Arthur Stanley Eddington and Charles Galton Darwin, as well as Rutherford himself. Plenty of the Society's older fellows stuck to classical physics – including the former secretary Joseph Larmor – but they were no longer intimately involved with the refereeing and editorial work on the publications.

Those who have emphasised Jeans' contribution to the modernisation of *Proceedings A* (including Rutherford) have missed the fact that the divergence between submissions from the physical and biological sciences had already begun in the 1910s, when Arthur Schuster was in charge.[137] It has also over-shadowed the fact that *Proceedings B* was not doing notably badly: it took longer than *Proceedings A* to return to pre-war levels of publishing but, by 1930, it was publishing around the same number of articles per year as it had in 1910.[138] Physical scientists certainly appear prominently on the list of authors who submitted repeatedly to the Society, but so do a handful of biological scientists, including Charles Sherrington, A.V. Hill, botanist Agnes Arber and biochemist Dorothy Moyle Needham. Sherrington's addresses as president in the early 1920s regularly discussed the latest developments in the experimental biological sciences, including Hill's Nobel Prize in physiology, and Ivan Pavlov's research into conditioned reflexes in dogs.[139] However, neither he, nor the biological secretaries William Hardy and Henry Dale, appear to have attracted as many authors to the Society as did Schuster, Jeans and Rutherford. The steady state of *Proceedings B* might have been a relief to worried treasurers and over-worked assistant secretaries, but as success came to be measured through numbers of authors submitting, and circulation, *Proceedings B* would suffer in comparison to *Proceedings A* for most of the twentieth century.

The growth of *Proceedings A* undoubtedly reflects the boom in physics research in the early twentieth century, but it was possible only because the role of the *Proceedings* had changed: it was no longer second-rank to the *Transactions*; it carried full, independent papers, not abstracts; and it carried papers that were long enough to convey detail and argument (without being as long as a *Transactions* paper). The boom in *Proceedings A* also occurred despite the fact that the extension of refereeing to include most papers for the *Proceedings* meant that the average time taken to make an editorial decision about a *Proceedings* paper in the 1920s had grown to about 70 days. It was certainly not as quick as writing to *Nature*, but the flood of submissions suggests that *Proceedings A* offered a format that suited modern physicists.

The growth of submissions to the Royal Society in the early twentieth century reflects the growth of the scientific enterprise, but it also suggests that the Society's journals were seen as desirable places to publish, despite Henry Armstrong's concerns. The period from the early 1890s until 1914 was one of upheaval, in which the 1752 editorial structures were rearranged. It created a more complex editorial system: the sectional committees brought a diversity of expertise and ensured that the system was less reliant on one man than it had been. The piecemeal reforms to the *Proceedings*, consolidated in 1914, enabled it to come of age as an independent research journal, while the *Transactions* slipped into its shadow.

There were far fewer editorial changes after the war. The earlier reforms appear to have created an effective framework for the secretaries in the 1920s to manage the periodicals, even though their core group of active referees and committee members had to tackle an increasingly large editorial load. The unplanned long-term effects of the 1905 split of the *Proceedings* include not only the imbalance between numbers of papers printed in *Proceedings A* and *Proceedings B*, but also the amount of editorial and committee work requested of fellows on the A side of the Society.

The significant invisible change was the increased use of refereeing. We have found no archival evidence calling for referees to be used for *Proceedings* papers as well as *Transactions* papers; nor any rule specifying that papers for the *Proceedings* could be published with just one referee report, while those for the *Transactions* needed two or three reports. But our analysis of the data preserved in the 'Register of Papers' makes clear that, despite the known criticisms of refereeing, the Society did come to use referees more, not less, in the early twentieth century. It is difficult to say whether this expansion of refereeing represents an attempt to constrain the escalating costs of publishing, an attempt to spread the decision-making work and responsibility, or a belief that, as the main public face of the Society, papers in the *Proceedings* now required the sort of close scrutiny that had once been reserved for the *Transactions*. Henry Armstrong's hope for a dramatic streamlining of the 'prehistoric' editorial system did not come to pass. Nor did the Society try to establish a unique role for itself as a venue for discussing 'borderland' or interdisciplinary challenges.

The rise of an independent Proceedings created an identity problem for the *Transactions*. The *Transactions* had once been defined as the venue of publication for uniquely distinguished research (marking 'a distinct step'), but its two series had come to be defined negatively, by their ability to publish papers that did not fit into the *Proceedings*: those that contained 'numerous or elaborate illustrations', or 'which cannot without detriment to their scientific value be condensed' to fit into the *Proceedings*.¹⁴⁰ The absence of a coherent, positive remit for the *Transactions* would be an ongoing issue for decades, but the Society's officers and fellows dealt with it by mostly ignoring it. The *Transactions* was now consistently an afterthought in discussions of the Society's publications. The best Rutherford could find to say about it was that it provided an 'important service', by 'issuing large and elaborately illustrated monographs, the publication of which would be beyond the scope or the resources of the specialist organisations'.¹⁴¹ These lengthy and expensive papers were the ones which were scrutinised by two, or sometimes more, referees, but the relatively small number of such papers submitted raises real questions about whether this 'important service' was in much demand from early twentieth-century scientists.

From the perspective of the Society's secretaries and many of its fellows, the long period of reform and reorganisation that started in the 1890s came to fruition in the 1920s, with *Proceedings A* well-positioned to capitalise on the boom in physics research. The Society's treasurers, however, had a very different experience of scientific journal publishing in these years, as they struggled to find ways to pay for the publication and distribution of more and more scientific papers – and then to do so amid the economic disruption of the First World War and its aftermath. For them, the success of *Proceedings A* in the 1920s did little to help, and may even have made the financial problems worse. It is to their perspective that we now turn.

Notes

1. 'Memorandum by Dr Armstrong', in RS CMP/08, 6 November 1902.
2. [E. Rutherford], 'Address of the president, 1927', *ProcA* 117 (1928): 305.
3. Lutz Bornmann and Rüdiger Mutz, 'Growth rates of modern science: A bibliometric analysis based on the number of publications and cited references', *Journal of the Association for Information Science and Technology* 66, no. 11 (2015): 2215–22.
4. Aileen Fyfe, Flaminio Squazzoni, Didier Torny and Pierpaolo Dondio, 'Managing the growth of peer review at the Royal Society journals, 1865–1965', *Science, Technology and Human Values* 45, no. 3 (2020): 405–29, especially Figure 4A.
5. On this period, including the 'crude estimate' of a four-fold increase in scientific graduates, see David Edgerton and John V. Pickstone, 'United Kingdom', in *The Cambridge History of Science: Volume 8: Modern science in national, transnational, and global context*, ed. David N. Livingstone, Hugh Richard Slotten and Ronald L. Numbers (Cambridge: Cambridge University Press, 2020), 151–91, section on 1890–1914.
6. Fyfe et al., 'Managing the growth of peer review', Figure 4B.
7. William H. Brock, ed., *H. E. Armstrong and the Teaching of Science 1880–1930* (Cambridge: Cambridge University Press, 1973). Armstrong was noted as a 'keen critic', according to his son-in-law Stephen Miall, quoted in E.H. Rodd, 'Armstrong, Henry Edward (1848–1937)', rev. W.H. Brock, in *ODNB*.
8. 'Memorandum by Dr Armstrong', in RS CMP/08, 6 November 1902.
9. Despite the changes, meetings were still being seen as boring in the 1930s; see Charles Galton Darwin, 'The "reading" of papers at meetings of Royal Society', *N&R* 2.1 (1939): 25–7.

10 RS CMP/06, 6 November 1890.
11 RS CMP/06, 18 February 1892. An element of tradition was preserved by insisting that the titles of all papers received were read out at meetings, though this ultimately lapsed. For a more extensive discussion, see Noah Moxham and Aileen Fyfe, 'The Royal Society and the prehistory of peer review, 1665–1965', *Historical Journal* 61, no. 4 (2018): 863–89.
12 The complaint from Dr Frederick Pavy was received on 13 June 1895, RS CMP/07, and concerned a paper by Noel Paton on hepatic glycogenesis in *PTB* 185 (1894): 233.
13 RS CMP/07, 5 July 1895.
14 The committee had reported on 5 July 1894. Implementation was still being discussed on 21 February and 14 March 1895, RS CMP/07.
15 Opinion of A.B. Kempe, in RS CMP/07, 5 March 1896.
16 On statute reforms, see *The Record of the Royal Society of London* (3rd edn) (London: The Royal Society, 1912), 171–2.
17 [J. Lister], 'Address of the president', RS *Year Book* (1896–97), 124.
18 'Memorandum by Dr Armstrong', in RS CMP/08, 6 November 1902.
19 The revised standing orders were presented to Council on 21 May 1914, and approved on 18 June 1914 (when further regulations for the Committee of Papers were presented). See RS CMP/10. The changes were reported in 'Report of the Council 1914', RS *Year Book* (1915), 185.
20 Editorial matters were now to be carried out as part of regular Council meetings, rather than (as had been the case from 1752) in a separate meeting of Council, meeting as the Committee of Papers.
21 RS CMP/10, 18 June 1914.
22 Herbert Rix, *Sermons, Addresses and Essays, with an Appreciation by Philip H. Wicksteed* (London: Williams & Norgate, 1907), xviii.
23 'P.D.R.', 'Robert William Frederick Harrison (1858–1945)', *N&R* 4.1 (1946): 113–20.
24 'P.D.R.', 'Robert Harrison', 116.
25 'P.D.R.', 'Robert Harrison', 117.
26 RS CMP/08, 6 November 1902. The proposal was passed, despite opposition from Henry Armstrong, on 27 November 1902. *Record of the Royal Society* (1912), 172.
27 RS CMP/08, 10 May 1900.
28 Gina Douglas, 'Farquharson, Marian Sarah (1846–1912)', in *ODNB*. See also Peter Ayres, *Women and the Natural Sciences in Edwardian Britain* (London: Palgrave, 2020), ch. 4.
29 RS CMP/08, 31 May 1901.
30 Joan Mason, 'Sarah Phoebe [Hertha] Ayrton [née Marks] (1854–1923)', in *ODNB*; and Joan Mason, 'Hertha Ayrton (1854–1923) and the admission of women to the Royal Society of London', *N&R* 45.2 (1991): 201–20.
31 Legal opinion from W.O. Danckwerts and R.J. Parker (Lincoln's Inn), 5 February 1902, recorded in RS CMP/08, 13 February 1902.
32 Perry to Foster, 20 February 1902, in RS CMP/08, 20 March 1902.
33 See Mason, 'Hertha Ayrton'; 'The admission of the first women to the Royal Society of London', *N&R* 46.2 (1992): 279–300; and 'The women fellows' jubilee', *N&R* 49.1 (1995): 125–40.
34 [A. Geikie] 'Address of the president, 30 Nov. 1912', *ProcA* (1913): 7.
35 'P.D.R.', 'Robert Harrison', 118.
36 RS CMP/10, 29 October 1914. See also R.J. Howarth, 'Schuster, Sir Arthur (1851–1934)', in *ODNB*.
37 Boris Galitzine to Arthur Schuster, 9 January 1915, RS MS/663/1/77. Galitzine was a member of the Petrograd Academy of Science and a foreign member of the Royal Society. He died in 1916, and his paper was never published with the Royal Society.
38 Edwin Deller held the assistant secretary role briefly, in 1920, but left for what became a high-flying administrative career in the University of London. See H.E. Butler (revised by M.C. Curthoys), 'Deller, Sir Edwin (1883–1936)', in *ODNB*.
39 Henry Hallett Dale, 'Francis Alexander Towle, 1874–1932', *Obituary Notices of Fellows of the Royal Society* 1 (1932): 85–87, 86.
40 Dale, 'Francis Towle', 87.
41 'Memorandum by Dr Armstrong', in RS CMP/08, 6 November 1902.
42 'Memorandum by Dr Armstrong', in RS CMP/08, 6 November 1902.
43 Henry Armstrong to the secretaries of the RS, 23 October 1908, RS MS/603/25.

44 [C.S. Sherrington], 'Address of the president, 1925', *ProcA* 110 (1926): 8–9.
45 [E. Rutherford], 'Address of the president, 1927', *ProcA* 117 (1928): 305.
46 F. W. Clifford, 'World list of scientific periodicals', *Nature*, Suppl. to 13 Sept. 1924 (1924): 401–2, 402.
47 [E. Rutherford], 'Address of the president, 1927', *ProcA* 117 (1928): 305.
48 'Memorandum by Dr Armstrong', in RS CMP/08, 6 November 1902.
49 For more discussion of these trends in submission, see Aileen Fyfe, 'Submissions in life sciences vs physical sciences, 1927–1989' (13 February 2018), https://arts.st-andrews.ac.uk/philosophicaltransactions/submissions-in-life-sciences-vs-physical-sciences-1927-1989.
50 Aileen Fyfe, 'The Proceedings in the 20th century' (8 February 2022), https://arts.st-andrews.ac.uk/philosophicaltransactions/proceedings-in-the-20thc.
51 Graeme Gooday, 'Periodical physics in Britain: Institutional and industrial contexts, 1870–1900', in *Science Periodicals in Nineteenth-Century Britain: Constructing scientific communities*, ed. Gowan Dawson, Bernard Lightman, Sally Shuttleworth and Jonathan R. Topham (Chicago, IL: University of Chicago Press, 2020), 238–73; John L. Lewis, *125 Years: The Physical Society and The Institute of Physics* (London: Taylor & Francis, 1999).
52 H.E. Armstrong to Larmor, 15 May 1907, RS MS/603/20. Papers submitted by fellows are clearly visible in the Register, because they have no entry in the column for recording the name of the communicator.
53 Fyfe et al., 'Managing the growth of peer review'.
54 O. Heaviside, 'On the forces, stresses and fluxes of energy in the electromagnetic field', *PTA* 183 (1892): 423–80.
55 Brown to Larmor, 28 February 1903, RS MS/603/171.
56 Fyfe et al., 'Managing the growth of peer review', Figure 4B.
57 On Bateson, see Marsha L. Richmond, 'Women in the early history of genetics: William Bateson and the Newnham College Mendelians, 1900–1910', *Isis* 92, no. 1 (2001): 55–90; 'Women as Mendelians and geneticists', *Science & Education* 24, no. 1 (2015): 125–50.
58 Camilla M. Røstvik and Aileen Fyfe, 'Ladies, gentlemen, and scientific publication at the Royal Society, 1945–1990', *Open Library of Humanities* 4, no. 1 (2018): 1–40, http://doi.org/10.16995/olh.265.
59 Brown to Larmor, 28 February 1903, RS MS/603/171.
60 'Memorandum by Dr Armstrong', in RS CMP/08, 6 November 1902.
61 Gowan Dawson, '"An independent publication for geologists": The Geological Society, commercial journals, and the remaking of nineteenth-century geology', in *Science Periodicals in Nineteenth-Century Britain: Constructing scientific communities*, edited by Gowan Dawson, Bernard Lightman, Sally Shuttleworth and Jonathan R. Topham (Chicago, IL: University of Chicago Press, 2020), 138; *The Jubilee of the Chemical Society of London. Record of the proceedings together with an account of the history and development of the Society, 1841–1891* (London: Harrison & Sons, 1896), 240.
62 'Memorandum by Dr Armstrong', in CMP/08, 6 November 1902.
63 Armstrong to Larmor, 15 May 1907, RS MS/603/20.
64 Alex Csiszar, *The Scientific Journal: Authorship and the politics of knowledge in the nineteenth century* (Chicago, IL: University of Chicago Press, 2018), 152–4.
65 'Philosophical Magazine: report by research committee [of the National Union of Scientific Workers]', *The Scientific Worker* 29–30 (11 March 1922), 29, quoted in Imogen Clarke and James Mussell, 'Conservative attitudes to old-established organs: Oliver Lodge and Philosophical Magazine', *N&R* 69.3 (2015): 321–36, 321.
66 'Memorandum by Dr Armstrong', in CMP/08, 6 November 1902.
67 J.J. Waterston and Lord Rayleigh, 'On the physics of ... free and perfectly elastic molecules ...', *PTA* 183 (1892): 1–79, at 3. On implicit bias, see Carole J. Lee, Cassidy R. Sugimoto, Guo Zhang, and Blaise Cronin, 'Bias in peer review', *Journal of the American Society for Information Science and Technology* 64, no. 1 (2013): 2–17.
68 'Memorandum by Dr Armstrong', in CMP/08, 6 November 1902.
69 E.B. Elliott to Larmor, 10 February 1902 and E.B. Elliott to Larmor, 18 February 1902, both in RS MS/604/416.
70 On the influence of a small group of physicists on *Proceedings A* in the early twentieth century, see Imogen Clarke, 'The gatekeepers of modern physics: Periodicals and peer review in 1920s Britain', *Isis* 106, no. 1 (2015): 70–93.

71 Andrew Gray to Larmor, 18 July 1907, RS MS/604/656.
72 Larmor, 'A dynamical theory of the electric and luminiferous medium', *PTA* 185 (1894): 719–822.
73 G.F. Fitzgerald to Larmor, 27 February 1894, RS MS/604/443.
74 G.F. Fitzgerald to Larmor, 29 March 1894, MS/604/447.
75 G.F. Fitzgerald to Larmor, 17 April 1894, RS MS/604/451.
76 Item 19 (15 November 1893), 'Register of Papers', RS MS/622.
77 'Explanatory notes on the procedure relating to the reading and publication of papers', RS *Year Book* (1897–8), 67–71.
78 RS CMP/09, 19 May 1904.
79 Frederick Orpen Bower, on a paper 'Longitudinal symmetry in Phanerogamia' by Percy Groom, [n.d.] February 1907, RS RR/17/167.
80 John Bretland Farmer, on a paper 'The histology of the cell wall …' by Arthur William Hill and Walter Gardiner, [n.d.] August 1900, RS RR/15/54.
81 Edward Bagnall Poulton, on a paper 'Some aspects of animal colouration …' by J.C. Motham and Frederick William Edridge-Green, [n.d.] June 1917, RS RR/24/68.
82 Hugh Longbourne Callendar, on a paper 'Heat of crystallisation of quartz' by Rames Chandra Ray, [n.d.] May 1922, RS RR/28/2.
83 George James Burch, on a paper 'The application of Maxwell's curves to three colour work …' by Reginald S. Clay, 9 May 1901, RS RR/15/170.
84 Augustus Edward Hough Love, on a paper 'The stability of the spherical nebula' by James Hopwood Jeans, 18 January 1902, RS RR/15/201.
85 Sidney Frederic Harmer, on a paper 'The morphology of a rare oceanic fish, Stylophorus chordates …' by Charles Tate Regan, 10 December 1923, RS RR/31/43.
86 On the power of printed forms, see Lisa Gitelman, *Paper Knowledge: Toward a media history of documents* (Durham, NC: Duke University Press, 2012), ch. 1.
87 William Bate Hardy, on a paper 'Causes of cell-proliferation in healing and in cancer' by H.C. Ross, [n.d.] October 1910, RS RR/18/71.
88 Hugh Longbourne Callendar, on a paper 'On the calorimetry of gases …' by W.E. Fisher, [n.d.] August 1911, RS RR/18/133.
89 Joseph Larmor, on a paper 'An experimental method for the production of vibrations on strings …' by J.A. Fleming, [n.d.] October 1913, RS RR/20/40.
90 Fyfe et al., 'Managing the growth of peer review'.
91 Henry R.A. Mallock, on a paper 'Surface reflexion of earthquake waves' by George W. Walker, 4 August 1917, RS RR/24/84.
92 The editorial progress of Yardley's 1924 and 1925 papers is recorded in 'Register of Papers', RS MS/588/158, ff. 58 and 89. See also Dorothy M.C. Hodgkin, 'Kathleen Lonsdale, 28 January 1903–1 April 1971', *Biographical Memoirs of Fellows of the Royal Society* 21 (1975): 447–84.
93 Circular to fellows, from Michael Foster and Lord Rayleigh, 16 June 1896, copy in Botany Sectional Committee, RS CMB/253.
94 Such a declaration was approved on 20 February 1908, RS CMP/09; but it was once again recommended (by the physiology sectional committee) on 22 January 1925, RS CMP/12.
95 [C.S. Sherrington], 'Address of the president, 1923', *ProcA* 105 (1924): 7.
96 [E. Rutherford], 'Address of the president, 1927', *ProcA* 117 (1928): 306
97 Each of these men referred 12 to 15 papers in the sample years 1921 and 1925, see 'Register of Papers'.
98 From analysis of our *Virtual Register of Papers*.
99 Fyfe et al., 'Managing the growth of peer review'. It had been 40 per cent prior to 1914. Note that this is a top 20 per cent of active referees, and only about a quarter of fellows were referees at all.
100 [E. Rutherford], 'Address of the president [1927]', *ProcA* 117 (1928): 306.
101 Andrew John Harrison, 'Scientific naturalists and the government of the Royal Society 1850–1900' (PhD, Open University, 1989), 7–11.
102 Standing Order 42, in RS *Year Book* (1899), 84.
103 RS CMP/07, 21 February 1895; compare standing order 37, in RS CMP/07, 26 March 1896.
104 Circular to fellows, from Michael Foster and Lord Rayleigh, 16 June 1896, copy in Botany Sectional Committee, RS CMB/253.

105 Michael Foster to unknown correspondent, regarding a paper 'A research into the elasticity of the living brain …' by A.G. Levy [n.d. but probably mid-July] 1895, RS RR/12/175, quoted from Harrison, 'Scientific naturalists', 19.
106 Six pages appears to have been the original plan, see RS CMP/07, 20 June 1895; but 12 pages was given in the Circular to fellows, from Michael Foster and Lord Rayleigh, 16 June 1896, copy in Botany Sectional Committee, RS CMB/253. Twelve pages was the limit stated in the 'Explanatory Notes [for authors]' printed in the *Year Book* from 1897 onwards, e.g. RS *Year Book* (1898), 67.
107 RS CMP/06, 18 February 1892 (for two pages).
108 Circular to fellows, from Michael Foster and Lord Rayleigh, 16 June 1896, copy in the Botany Sectional Committee, RS CMB/253.
109 RS CMP/07, 26 March 1896.
110 Melinda Baldwin, '"Keeping in the race": Physics, publication speed and national publishing strategies in Nature, 1895–1939', *British Journal for the History of Science* 47, no. 2 (2014): 257–79.
111 Five 'preliminary notes' were published in the *Proceedings* in the 1860s, compared to 33 in the 1890s. On preliminary notes in *Nature*, see Baldwin, 'Keeping in the race'.
112 Karl Pearson to Francis Galton, 3 May 1895, RS MM/17/53.
113 K. Pearson, 'Note on regression and inheritance …', *Proc* 58 (1895): 240–2. The longer paper appeared in *Trans A* the following year.
114 'Memorandum by Sir E. Ray Lankester', in Committee of Papers, RS CMB/90/6, 24 October 1907.
115 Report of the Sectional Committee for Physiology in RS CMP/08, 12 February 1903.
116 RS CMP/08, 30 April 1903 and RS CMP/09, 21 January 1904.
117 RS CMP/08, 30 April 1903 (more effective); RS CMP/09, 21 January 1904 (desirable or feasible). It was decided to set up a committee in spring 1903, but nothing seems to have happened until January 1904.
118 See Item 5, Publications Review report, in CMP/09, 17 March 1904.
119 See paragraph 37 of the revised standing orders, in CMP/10, 21 May 1914. For more on abstracts, see Aileen Fyfe, 'Where did the practice of "abstracts" come from?' (8 July 2021), https://arts.st-andrews.ac.uk/philosophicaltransactions/where-did-the-practice-of-abstracts-come-from/.
120 Item 2, Publication Review report, in RS CMP/09, 17 March 1904. The recommendation was adopted RS CMP/09, 28 April 1904.
121 RS CMP/09, 7 December 1905.
122 RS CMP/09, 3 November 1904.
123 RS CMP/09, 30 April 1908.
124 This was being noted within a year, see treasurer's comments in RS CMP/09, 25 October 1906.
125 Henry Armstrong to the secretaries of the RS, 23 October 1908, RS MS/603/25.
126 Hertha Ayrton, 1909, quoted in Harrison, 'Scientific naturalists', 23 from Evelyn Sharp, *Hertha Ayrton: A memoir* (London: Edward Arnold and Co., 1926), 211.
127 Paragraphs 29–35 of the revised standing orders, proposed to Council on 21 May 1914 and approved on 18 June 1914, RS CMP/10.
128 An example of the referee report form used in 1915 (RS RR/1915/22) can be found on the 'Science in the Making' website: https://makingscience.royalsociety.org/s/rs/items/RR_1915_22_.
129 Derived from our analysis of data on pages and articles supplied by the Royal Society publishing team.
130 William Dobinson Halliburton, on a paper 'South-African horse sickness …' by Alexander Edington, 27 September 1900, RS RR/15/35.
131 John Rose Bradford, on a paper 'South-African horse sickness …' by Alexander Edington, [n.d.] May 1900, RS RR/15/36.
132 Armstrong to Larmor, 11 August 1902, RS MS/603/16.
133 James Berkeley to Larmor, 20 August 1908, RS MS/603/85.
134 Alan G. Gross, Joseph E. Harmon, and Michael S. Reidy, *Communicating Science: The scientific article from the seventeenth century to the present* (New York: Oxford University Press, 2002).
135 Henry Dale to H.W. Thompson, 2 April 1964, RS HD/6/10/2/151.

136 Imogen Clarke, 'Negotiating progress: Promoting "modern" physics in Britain, 1900–1940' (PhD, University of Manchester, 2012), ch. 5; 'Gatekeepers of modern physics'.
137 Perhaps unsurprisingly, the 1952 memoir of Jeans celebrated his role in transforming the *Proceedings*, see E.A. Milne, *Sir James Jeans: A biography* (Cambridge: Cambridge University Press, 1952), 34–5.
138 Aileen Fyfe, 'The Proceedings in the 20th century' (8 February 2022), https://arts.st-andrews.ac.uk/philosophicaltransactions/proceedings-in-the-20thc.
139 [C.S. Sherrington], 'Address of the president, 1923', *ProcA* 105 (1924): 12; [Sherrington], 'Address of the president, 1925', *ProcA* 110 (1926): 8.
140 Revised standing orders, 1914, proposed to Council on 21 May 1914 and approved on 18 June 1914, RS CMP/10.
141 [E. Rutherford], 'Address of the president, 1927', *ProcA* 117 (1928): 305.

12
Keeping the publications afloat, 1895–1930

Julie McDougall-Waters and Aileen Fyfe

In 1894, the Royal Society's treasurer, John Evans, warned that its publications were getting so expensive that it would soon be 'impossible to meet such expenditure out of current revenue'.[1] As we saw in Chapter 10, the Society's tradition of distributing most of the copies of the *Transactions* and the *Proceedings* for free to its fellows and the libraries of learned institutions meant that relatively few copies were sold, and the income they generated was far too little to cover all the paper, printing and illustration costs. Over 60 per cent of the publishing costs were covered by the Society's internal funds, such as membership fees and investments.[2] The Society's investments were performing sufficiently strongly that Evans felt able to support the publications at the rate of about £1,800 a year. However, in April 1894, he had discovered that the papers already accepted for publication that year had estimated costs of £2,600, and that there were another 10 papers ('probably costing £300') yet to be considered by the Committee of Papers.[3] Evans arranged to sell £1,000 of the Society's investments to cover the shortfall, but warned that, in the longer term, it was not sustainable to eat into capital regularly.[4]

Part of the problem was that the management of the *Proceedings* and the *Transactions* was divided between several different individuals and committees, who do not appear to have coordinated their activities. The Committee of Papers looked after the content appearing in the journals; the Library Committee made gifts of the printed copies to other learned societies; and the Finance Committee occasionally reviewed the costs. The secretaries were technically in overall charge, but they had many other responsibilities to the Society, and tended to focus on editorial matters. Liaising with printers, lithographers and paper-makers usually fell to the assistant secretary. The treasurer's financial oversight of all Society activities meant that he was perhaps best-equipped to see the bigger picture. Evans was not the first treasurer to warn that the expansion of research was making the Society's philanthropic model of publishing difficult to sustain, but he appears to have been the first to make the Society's Council sit up and take notice.

Evans wanted the Society to reduce its publishing costs, but that proved impossible to achieve. As Figure 11.2 showed, the number of research papers submitted to the Royal Society would keep increasing for the next 40 years (1914–18 excepted), and the Society remained unwilling to put serious constraints on the amount of material it published. These financial difficulties were exacerbated by the material and manpower shortages during the First World War that pushed the costs of paper and printing even higher. By 1920, British consumer prices in general had inflated to about 270 per cent of their 1913 level.[5] The following years brought deflation, but also high unemployment, the General Strike, the Wall Street Crash, and the UK's departure from the Gold Standard in 1931. The same years that saw the flourishing of *Proceedings A*, as celebrated by Ernest Rutherford, were difficult economic times for organisations of all sorts. The Royal Society kept its publications afloat not by cost-cutting but because of its success in securing additional sources of income. By 1930, the Society's expanded journal publishing would be supported by an increased endowment (due to philanthropic gifting), and by both the UK government and the chemical industry.

The difficult economic conditions of the 1920s would force the Society's officers to cut back the global philanthropic distribution of the journals for the first time. At the same time, sales income increased sufficiently that it was able to cover more than half of the costs of publishing. There is little evidence that the Society paid much active attention to the sales and marketing of its journals, but it was during this period that the seeds of its later transition to a commercial model of distribution were (inadvertently) sown. This chapter will investigate the shifting balance between philanthropic and commercial distribution, and income, before and after the First World War.

Before the war

In 1894, John Evans's suggested solution to the spiralling costs of publication was a stricter editorial policy. It was he who proposed the questions on the referee report form (as shown in Figure 11.3), asking whether any section could 'be omitted as being unnecessary' or illustrations 'be reduced in number or extent ... with a view to economy?'[6] Unfortunately, referees of the 1890s and 1900s proved little more willing or able to act as stewards of the Society's finances than their predecessors in the 1850s and 1860s. One of the first to respond to the new questions, in December 1894, was astronomer and mathematician George Darwin. He admitted that some figures in Karl Pearson's paper 'might be largely reduced without much loss', but said they were 'very interesting and should only be cut down if the R.S. really cannot afford to print so many'.[7] The Society's commitment to prioritising the needs of its scholarly authors meant that it had no defence against a referee who insisted that a paper would be 'injured by ... abridgement'.[8] Just a few months later, the Committee of Papers repeatedly found itself unable to reach a decision on a heavily illustrated paper on brain lesions. The committee members were caught between referees arguing for intellectual integrity, and a treasurer demanding more careful financial stewardship. Michael Foster told one of the fellows involved in the affair of the brain lesion paper that 'a very great change' was being planned in the way decisions were made, for the near breakdown of the editorial process demonstrated the urgency of reform.[9] The introduction of the sectional committees (see Chapter 11) did not make it any easier to resolve questions of intellectual versus financial needs, but did divert the problem from the already-packed agenda for Council meetings, and ensured it was addressed by those fellows most likely to understand the real needs of the subject matter in question.

When new standing orders were introduced in 1896, Evans ensured that the Committee of Papers would consider estimated publication costs, as well as sectional committee recommendations, for all *Transactions* papers, these being most likely to be expensive.[10] But there was no guidance on what an appropriate expenditure would be, nor any mechanism for deciding between the papers recommended by the different sectional committees. In June 1901, the new treasurer (and lawyer) Alfred Kempe had to warn the sectional committees that 'in the present condition of the income of the Society and under the existing arrangements as to its expenditure', the Society could only afford to spend about £1,100 per year on the *Transactions*, and the first six months of 1901 had already seen the committees recommend papers that would cost £1,044. He pointed out that the sectional committees' willingness to recommend the publication of more, and more expensive, papers was imposing 'a serious burden ... upon the resources of the Society which it will be difficult to sustain'. Going forward, the sectional committees were tasked 'not to lose sight of' the financial implications of their editorial decisions.[11]

Overall, the treasurers' attempts to include financial considerations in editorial decision-making had little effect. The Society's annual output of print did fall back after the bumper years of 1894 and 1895 that had perturbed Evans, but Kempe's letter shows that the problem soon returned. And when E. Ray Lankester made his 1907 proposal for turning the *Proceedings* into a journal focused on immediate and brief publication (see Chapter 11), he offered it as a possible solution to the fact that 'the amount of material published by the Society is at this moment more than the resources of the Society justify'.[12]

An alternative means of reducing the Society's publishing costs would have been to cut the print run and, specifically, to stop paying for the paper and printing of so many copies of the journals for use as gifts, exchanges and perks. There was no discussion in this period of drastically limiting the Society's generosity, but there were attempts to use its resources more efficiently. The division of the *Transactions*, and then the *Proceedings*, into A and B series made it possible to offer more targeted gifts: rather than giving all institutions and all fellows a complete set of the publications (worth just over £6, in 1909), they could be given a sub-set.[13] This explains why the print runs for both series of the *Transactions* could be reduced from 1,000 to 800 copies in 1898, even though the number of institutional beneficiaries was still rising (and it was specifically the portion of the run issued in bound volumes, mostly for libraries, that was cut).[14] We have no direct evidence of the print run for the *Proceedings* in this period, although there is evidence that it was being discussed around 1908. We might guess that it was around 1,500 or 1,600 copies.[15]

The institutional exchange and gift system was, in fact, at its greatest extent just before the First World War, when over 460 institutions globally benefited from the Society's largesse (see Figure 12.1).[16] Of these, 185 institutions received complete sets of the Society's journals, while the others received just A or B series journals, or just the *Proceedings* (both series). Within London, for instance, the Geological Society, the Chemical Society, the War Office and the King's Library received the full set of publications. However, the Royal Astronomical Society and the London Mathematical Society received only series A, the Linnean Society and Entomological Society received only series B, and the Royal Geographical Society received only the *Proceedings* (but both series). About a quarter of all the free copies went to institutions within Britain; as well as the metropolitan scientific societies, this included virtually all the universities and university colleges, as well as national scientific organisations (the National Physical Laboratory), provincial

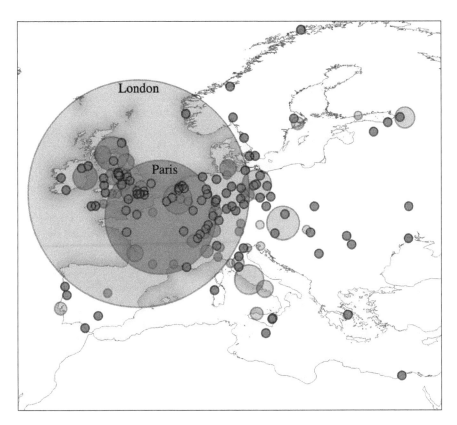
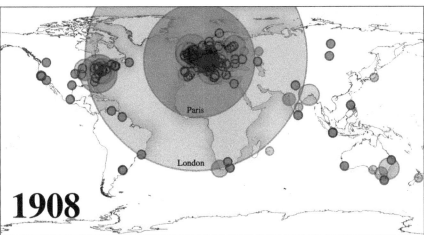

Figure 12.1 Global distribution of free copies of the *Transactions* and/or the *Proceedings* in 1908. The size of the spots is proportional to the number of copies sent to a given city.

societies (the Essex Field Club, Glasgow Natural History Society) and public libraries in Birmingham, Manchester and Cardiff. The Society's willingness to provide copies of its publications to the universities and university colleges indicates its ongoing perception of itself as a patron of scholarship, and not only a participant in an exchange system.

The overseas recipients were predominantly European universities and scientific societies, but copies also went to similar institutions in Canada, Australia, New Zealand, India and South Africa, and to the USA. A handful were sent even further afield, to the observatory at Rio de Janeiro, the university library at Caracas, the imperial university in Tokyo, and the bureau of science in Manila. National academies and universities were usually sent all the Society's journals, whereas observatories, botanic gardens and specialised societies received the A or B journals as appropriate. Around 200 institutions were granted only the *Proceedings*.

A large proportion of the print run of each journal was reserved for free copies for fellows of the Society. Personal copies of journals might not seem as essential for academics with access to a university library as they had been to gentlemen scholars 150 years earlier, but they would remain a great convenience until the days of photocopiers and digital files. The cost of providing this perk had grown significantly since it was introduced in the 1750s, however, for although the actual number of fellows was similar, the amount of printed matter they received was far greater. Moreover, the late nineteenth-century decision to allow fellows to receive their journals by post, rather than having to claim in person, had enabled a higher proportion of fellows to claim their perk (and increased the Society's postage costs). The only attempt to reduce the cost of the fellows' perquisite was to encourage fellows to select only series A or B.[17] And it was not until the late 1940s that the Society calculated the cost of this distribution to fellows (see Chapter 14).[18]

Separate copies, or offprints, were another long-standing instance of the Society's generosity, and the number provided to authors for free or for purchase had been repeatedly discussed and revised in the late nineteenth century. In the early twentieth century, physical secretary Joseph Larmor was particularly concerned about requests for 'large numbers of separate prints' from authors who worked for '[i]nstitutions having publications of their own'.[19] For instance in 1905, the Liverpool Institute of Tropical Research asked for the terms on which it 'could be supplied with from 500 to 2000 extra copies of any communications made to the Royal Society by members of the Institute's Scientific Staff'.[20]

These would then be bound into volumes showcasing the Institute's research, and the scale of the request indicates that these volumes were not made simply for internal use. In 1906, Council fixed the number of 'private' copies permitted to authors at 100 (again); and established a pricing mechanism for 'special cases' in which individuals or their institutions wished to purchase hundreds of copies.[21] This enabled the Society to accede to such requests, without adding to its own expenses. It was not, however, going to make any significant difference to the treasurer's worries.

Printing arrangements

The closest that the Society's officers came to making significant changes to the production and distribution arrangements in the pre-war period was during the 1904 review of the publications (see Chapter 11). As well as discussing the future of the *Proceedings*, the members of that review committee devoted considerable attention to the physical aspects of the Society's publications. They investigated the durability of the paper used, and the 'mode of correction of the proofs of papers'. They recommended improvements to the typesetting of mathematical formulae in the *Transactions*, and a change to larger page size (royal octavo) for the *Proceedings*.[22] The format and layout of both journals had also been discussed in 1895, when 300 fellows had responded to a postal survey on whether the *Transactions* should move to a smaller format (from quarto to octavo), and whether the (octavo) *Proceedings* should use a larger size of paper. At that time, there had been support for enlarging the page size of the *Proceedings*, but mixed feelings about changing the *Transactions*, and nothing actually happened until the 1904 committee once more recommended enlarging the *Proceedings*.[23] It became part of the re-launch of the (divided) *Proceedings* in 1905.

The committee also recommended investigating the possibility of a change of printer, for the first time since 1877. The Society's officers had sometimes felt that Harrison & Sons did not always deal sufficiently promptly with Royal Society business, and fellows and authors had often compared Harrisons' typesetting unfavourably to that of other scientific printers. A particularly scathing critique had come from Oliver Heaviside, whose interpretation of Maxwellian physics was typographically complex. In 1891, he had complained that 'when I received the proof, I was quite shocked! I could not read the formulae without difficulty, and for the first time in my life found my own copy easier to read than the same in print'.[24] It took several months, and several sets of proofs, before Heaviside was happy for printing to go ahead.[25] He remained convinced that Harrisons cared more about 'small things, commas, hyphen, and so on, in the text' than they did about formulae.[26] When the Society's officers tendered for printing services in November and December 1904, the accumulated 'complaints regarding the defective character of the proofs issued of the Mathematical printing' formed the backdrop. The Society's secretary insisted Harrisons address them in their tender document.[27]

We know that seven printers sought to win the Royal Society's printing contract, but only five can be identified from the surviving archival record, and only four (including Harrisons) were seriously considered. In terms of experience with scholarly publishing, the plausible other contenders were Taylor & Francis, Clay & Sons (who printed for Cambridge University Press) and Eyre & Spottiswoode. A bid had also been received from Hazells, who were specialised magazine printers, with no apparent expertise in scholarly publication. Their advertising emphasised their 'highly skilled workmen' and 'most up-to-date machinery', but references to the *Strand Magazine* and the *Daily Mail* suggest a different focus.[28]

Taylor & Francis was still publishing the *Philosophical Magazine*, and still did a lot of work for learned societies, including the new *Science Abstracts* (for the Physical Society). William Francis had died earlier that year, so the business was now being run by his son.[29] Clay & Sons were already known to the Society for their high-quality scholarly work. They printed the later volumes of the *Catalogue of Scientific Papers*, and the 1904 review committee had suggested their typesetting of Lord Rayleigh's collected works as a suitable model for a new 'more compressed' arrangement of the 'mathematical formulae and tabular matter' in the *Transactions*.[30] Eyre & Spottiswoode were the King's Printers, which gave them near-exclusive rights to print the authorised edition of the Bible, and the ability to advertise themselves as 'government, legal & general publishers'.[31] They had printed the first volumes of the *Catalogue of Scientific Papers* for the Society and HMSO, and tendered for the Society's business in both 1876 and 1904.

The financial aspects of the four plausible proposals apparently 'varied within so narrow a range' that the decision had to be determined 'by other considerations'. No record of the discussion survives, so it comes as something of a surprise that the Society decided to reappoint Harrisons. The new agreement included provisos that Royal Society work was to be treated 'as special matter requiring prompt attention, to be entrusted to compositors familiar with such work, with whom it is to take priority of other work', and Harrisons agreed that they would, if requested, engage 'a special reader and special compositors for mathematical work'.[32] The need for these provisos suggests that the Society's officers were not totally reassured by Harrisons but were, perhaps, unwilling to sever the 'long connection'.[33] (As Chapter 13 will show, concerns remained about the quality of service offered by Harrisons, and Cambridge University Press would eventually win the contract in 1936.) The decision not to change printers in 1904 is typical of the Royal Society's approach at the time; there was plenty of discussion, but the Society's officers were not inclined to make drastic changes.

Pricing and sales

Just a month after negotiating the new printing contract (and while treasurer Kempe was applying his legal expertise to the drafting of new statutes for the Society), biological secretary Archibald Geikie turned his attention to the Society's 'arrangements for issuing and advertising' its various publications. He described them as 'inadequate' and 'defective'.[34] Four years earlier, Henry Armstrong had argued that it was 'the irregular manner in which [they] appear' and 'the absence of any fixed price per volume' that prevented the *Proceedings* from circulating effectively in the book trade.[35] Geikie, however, pointed to the piecemeal arrangements for the sale of the Society's publications that had arisen in the decades since it had last formally appointed a 'bookseller' (in the 1820s). He suggested these arrangements be 'consolidated in the hands of a single publishing firm of good standing'.[36]

Geikie's proposal might appear in hindsight to suggest an interest in increasing sales income, but he intended it as another element in the ongoing rationalisation of the Society's administrative arrangements. Customers who wanted to purchase the annual volumes of the *Transactions*, or the issues or volumes of the *Proceedings*, could do so either from the Society's premises at Burlington House or from Harrison & Sons.[37] The sale of separate copies of *Transactions* papers, however, was managed separately; in 1894, this contract had been transferred from Kegan Paul (successor to Messrs Trübner & Co.) to Dulau & Co.[38] Meanwhile, customers 'upon the Continent of Europe' were supplied by Messrs Friedländer and Son, who had been appointed sole European agents for the Society's publications in 1890.[39] Geikie proposed that this fragmented set of distribution channels should be consolidated.

Other learned societies had made such arrangements with a publishing agent. In the 1890s, for instance, Williams & Norgate acted as London agents for several Scottish and Irish learned societies, as well as for the Queckett Microscopical Club. Longman were agents for the Linnean and Zoological societies.[40] Publishers who acted as agents for learned societies (or for overseas publishers) were supposed not merely to stock the publications, but to actively promote and advertise them.

The Royal Society officers initially negotiated with Dulau & Co. They were a long-established firm of 'English and Foreign' publishers and booksellers, who prided themselves on supplying 'Magazines and Periodicals of all Countries'. They already acted as agents for the British Museum (Natural History) and 'several learned societies' and held the Society's contract for sales of separate papers from the *Transactions*.[41] But despite Dulau's undoubted experience in bookselling and wholesaling, and their established distribution networks, the Society decided in January 1906 to cancel its existing agreements with both Dulau and Friedländer, and to authorise Harrison & Sons to sell its publications.

Harrison & Sons had only modest pretensions as a publisher and bookseller (rather than printer), so this appears to be another surprising decision in their favour.[42] The explanation may lie in their willingness to accept a trade discount of just 15 per cent off the retail price rather than the usual 25 per cent. (The difference between the trade price and the public retail price enabled booksellers to make a living. As agent, Harrisons were also paid a commission of 10 per cent of the sales income.[43]) The decision to appoint Harrisons fulfilled Geikie's aim to rationalise the Society's distribution arrangements, but it was unlikely to lead to radical improvements.

The re-launch of the *Proceedings* provided an opportunity to think about its pricing. Retail prices for separate copies and volumes of the *Transactions* depended on the number of pages and illustrations published, which meant that they varied hugely and were impossible to determine in advance. The *Proceedings*, on the other hand, had been offered at a fixed subscription rate since the 1850s; and that rate had been 21*s*. per volume for the last three decades. Setting subscription rates 'per volume' rather than 'per year' was a common technique among publishers of research journals (and would remain so until well after the Second World War) because it protected the publisher's income if more material was published than had been anticipated. In December 1905, Geikie and his co-secretary Larmor proposed a subscription rate that offered a discount of about 25 per cent on the cost of buying the issues separately.[44] Council agreed rates of 15*s*. a volume for *Proceedings A*, and 20*s*. a volume for *Proceedings B*.[45] These discounted subscription rates do seem to have increased the paid-for circulation of *Proceedings A* and *B*, but they would create financial difficulties in the longer term.

As part of this discussion, the basis for calculating the regular prices of the *Transactions* and the *Proceedings* was recorded in the Council minutes. The prices were based on a notional cost per issue, calculated from the cost of printing per sheet, plus a share of the illustration costs. For a *Transactions* paper, the formula for pricing was 4*d*. per sheet (of eight pages) for printing, 1/300 of the cost of illustrations and 3*d*. for covers. For the *Proceedings*, the calculation was 6*d*. per sheet (of 16 pages) for printing, and 1/300 of the cost of illustrations.[46] These calculations imply an underpinning business model that imagined sales to 300 paying customers would cover the entire cost of the illustrations. The notional cost of all the additional copies – about 500 further copies for the *Transactions*, and perhaps 1,200 for the *Proceedings* – was then just print and paper. For a commercial publisher, the additional copies would be priced to bring in a profit, but for the Society, it meant that those copies could be used more freely for gift and exchange. The calculation also shows that the Society did not regard editorial labour as a cost to be recouped, but as a cost to be silently subsidised as part of the Society's commitment to the wide circulation of its research publications. It was also a model that had no profit margin to play with, which meant that the discounted copies offered to the advance subscribers to the *Proceedings* would have to be subsidised by the Society. As Larmor noted, the proposed changes would 'involve a substantial increase of the cost of publications to the Society, unless the sales can be increased'.[47]

Geikie and Larmor also worried that the Society had been spending money printing copies that had never been distributed. Worries about the quantity of surplus stock were not new, but now it was being seen as evidence of poor decision-making about the print run. In 1902, the process of fulfilling a very belated claim from a fellow for a volume of the *Transactions* dating all the way back to 1862 had drawn attention to the warehoused stock.[48] At the very next meeting, the secretaries raised the question of whether the Society was (and long had been) printing too many copies of both the *Transactions* and the *Proceedings*.[49] Holdings of pre-1893 *Proceedings* were immediately thinned.[50] In 1906, Robert Harrison, the assistant secretary, carried out a full investigation and reported that there were 'very large totals' of the *Transactions* (24,000 copies in various forms of binding) and the *Proceedings* (63,000 numbers stitched, and 135,000 copies in loose sheets). They were placing 'an unmanageable demand upon the space for storage in the Society's rooms', as well as costing the Society rent for the stock held at the printers.[51] The quantity of surplus stock could have been used as evidence of the need to reduce the print run, or as support for the recent efforts to sell the journals more effectively. This seems to be the context for some discussion of the appropriate print run for the *Proceedings* in 1908, but – as far as we can tell – once again, nothing was done.[52]

The government grant-in-aid of scientific publications

Geikie and Larmor's investigations in the early years of the twentieth century indicate that the Royal Society's officers began to pay slightly more attention to the commercial book trade than they had done in recent times. Sales income did begin to increase, especially from the *Proceedings*, but it was still far from enough to enable the publications to be self-supporting. It was, therefore, fortunate that the Society had found an alternative solution to the problem of supporting its publications: external funding.

Back in June 1895, the Society's then junior secretary, the physicist Lord Rayleigh, had written to the Treasury, describing 'the financial difficulties attending the adequate publication of scientific papers', and asking for an annual grant from the government 'in aid of scientific publications'. The government already made one-off grants to support the publication of specific research outputs, such as the *Catalogue of Scientific Papers* or the reports from various naval expeditions, but this would be an ongoing commitment to support scientific publishing in general. If the Treasury agreed, Rayleigh promised that the Royal Society would 'take all possible pains to ensure that the money shall be spent in a manner most advantageous to science'. Rayleigh's argument drew upon the fact that the Society already administered £4,000 of government funding every year for scientific research, and that 'assistance in publication' could be seen as an extension of this 'aid to scientific enquiry'.[53]

Rayleigh attributed the 'financial difficulties' to the growth of science. He explained that, although the Royal Society was willing and able to support its publications financially (he mentioned a figure of £1,700 a year), the number of papers that 'after careful consideration, appear worthy of publication' had 'much increased of late'. He explained that the Society was having to realise some of its investments to cover the shortfall, and was considering 'a stricter selection' process. Rayleigh suggested a grant-in-aid of £1,000 or £2,000. Within a fortnight, the Treasury had offered £1,000, and also noted that it had no objection to 'a portion' of the existing research grant being diverted to 'publications showing the results of research'.[54]

The success of Rayleigh's appeal rested on his assertion that the problem affected scientific journals in general, not just the Royal Society. He claimed that 'the scientific journals in this country ... are carried on with great difficulty and in some cases by private enterprise, and at a loss'. In fact, as Rayleigh well knew, there were periodicals run by 'private enterprise', such as Taylor & Francis's *Philosophical Magazine* and Macmillan's *Nature*, which had by this time achieved commercial stability. Rayleigh's concern was with periodicals that published substantial papers of original research, and in the 1890s, that largely meant the transactions, proceedings and memoirs of the learned societies.

According to Rayleigh, such research journals were intrinsically unprofitable because their 'expenses are so great' that the advantage gained by paying authors 'nothing for their contributions' counted for little. Complex typesetting was required for tables and algebra, and 'accuracy and fineness' of illustrations was essential lest 'the scientific value of the papers ... be greatly diminished, or even wholly destroyed'. Rayleigh took it for granted that there were but few readers who were both potentially interested and sufficiently affluent to purchase research periodicals. He said little about institutional purchasers, because so many of the British universities, European specialist learned societies, and national academies all over the world received copies of the *Transactions* or the *Proceedings* either for free, or in return for their own periodicals. With 'so small' a number of potential purchasers, research journals could not generate sufficient income to balance their expenses (and advertisements offered only 'uncertain and insignificant' income). This way of thinking about potential markets underpinned Rayleigh's assumption that substantial research periodicals were necessarily the preserve of learned societies; they needed to be subsidised by someone. But, he said, such publishing programmes now 'exceeded' the 'spending powers' of the learned societies. He pointed in particular to difficulties faced by the Linnean, Zoological, Geological and Entomological societies (all of which dealt with subject areas likely to be costly in illustrations).[55]

For the first three years of the new grant, just over half the government funds were granted to other learned societies, and about a quarter supported the Royal Society's own publications, usually as grants for specific papers in the *Transactions*. For instance, in 1899, C.T.R. Wilson's paper on condensation nuclei formed in gases by X-rays and D.H. Scott's third paper on Palaeozoic fossil plants were among those to benefit. Yet when Kempe took over the treasurership in 1899, he discovered that over £700 of government funding was sitting unused.[56]

His new regulations ensured that any balance not granted to external applicants would automatically benefit the Society's publications.[57]

In the following years, the Royal Society was the biggest beneficiary of the government publications grant. In 1910, for instance, it received £700 of the £1,000 grant.[58] The *Transactions* and the *Proceedings* were not entirely responsible for the Society's need for extra funds for publishing: the *Catalogue of Scientific Papers* had required external funding from government and private philanthropy, and its international successor project was already proving equally expensive. This was the justification when, in late 1914, Council notified the Treasury of its desire both to divert 'a sum not exceeding £500' from the government research grant towards publications, and also to decline external applications for publications support for two years, except those appearing 'especially urgent'.[59]

The fixed value of the government grant, at a time when research activity and publication costs were both growing, meant that it would not prove a long-term solution to the financial problems facing the learned societies. However, the government's willingness to accept that it should provide ongoing funds for the publication of research, as well as the research itself, was a significant marker of state support for science, and one that has resonances in the open access debates of the early twenty-first century. The government's willingness to trust the Royal Society to administer the funds was also an endorsement of the Society's status as the leading voice of British science.

On the other hand, the Society's role as an administrator of government funds could give the impression that it was itself a wealthy organisation. By 1913, this was an impression that the Society actively sought to correct. Then-president Archibald Geikie carefully pointed out that both government grants were for the benefit of the scientific community as a whole, and that the only 'subvention' the Royal Society received 'from the State for its own requirements' was 'a few hundred pounds towards the cost of its publications, together with the use of its rooms in Burlington House, where it sits rent free, but subject to expenditure for internal upkeep and repairs'. Around 60 per cent of the Society's income came from its own investment portfolio, built up over the years from gifts and bequests from fellows and supporters, not from the government. The idea that the Society was wealthy was 'prevailing but mistaken'.[60] That said, the Royal Society undoubtedly had more resources available to it than most of its younger learned society siblings. It had accumulated endowments over its long history, it was well-positioned for access to government funding and it had impressive social, political and scientific networks that could be used to raise support of various kinds. If the Royal Society found it difficult to sustain its publishing activities in the decades around the First World War, it was surely not alone.

The war and afterwards

The First World War and, especially, its aftermath generated a difficult set of circumstances for the Royal Society and for all publishers. It was difficult to think strategically about the journals while coping with shortages, inflation and economic uncertainty. For the Society, the inflating costs of the immediate post-war years would prove to be more of a problem than the shortages and disruption of the war itself. The return to pre-war levels of submissions, and then growth through the 1920s, pushed up the costs of publishing, which were then pushed even higher by inflation.

The wartime problems for the Royal Society's publications programme were much the same as those facing the rest of the print trades.[61] Manpower shortages led to increased printing charges; in November 1915, Harrison & Sons informed the Society of a 10 per cent increase in their rates.[62] Paper rationing was introduced by the government in 1916, limiting publishers to two-thirds of the quantity they had used in 1914. The Royal Society was fortunate that 1914 had been a bumper year for the *Transactions*, so it received a relatively high allowance of paper. At the same time, submissions fell to little more than half their pre-war levels, as Figure 11.2 showed. With less material to publish, the Finance Committee was able to report that it did not see any likelihood of the Society needing more than its official allocation of paper.[63] The Royal Society was thus relatively unsympathetic to a suggestion from the Chemical Society that it should make 'proper representation to the Government' to get scientific societies exempted from the restrictions on paper consumption.[64]

It was the cost, rather than the quantity, of paper that posed problems, and in 1916, the Finance Committee asked the sectional committees to discuss possible austerity measures, yet again recommending 'some limit' on the cost of each research paper approved for publication. But while the zoologists were willing to consider postponing publication of *Transactions* papers until after the war, the joint physics and chemistry committee joined the Chemical Society in urging government intervention to enable societies 'to continue their publications on the same scale as before the war'. With such disagreement within the Society, the final resolution was an unsurprisingly lukewarm expression of the 'desirability of exercising economy'.[65]

The shortage of raw materials for the war encouraged the Society to investigate its stores. Other publishers are known to have sent their old stereotype and electrotype plates to be melted down for the war effort, but it is not clear that the Royal Society's journals had ever been stereotyped.[66] The Society did, however, revisit the 'large surplus stock' of its publications.[67] Despite earlier culls, thousands of copies of papers from the *Transactions* and the *Proceedings* were still lying 'in sheet form in store at Letchworth'.[68] There was also an 'accumulation' of the copper plates engraved over the decades by the Basire family, some of which had been 'in the Society's possession for many years'.[69] Investigation revealed that one and a half tons of old copper plates were apparently 'lying useless on the Society's premises'.[70] The paper and the copper was all recommended for disposal in July 1916.

The outbreak of war disrupted international scientific communication, including the circulation of research journals. The first presidential address of the war had been delivered by William Crookes in November 1914, just a month after 93 German men of science had signed a public manifesto supporting German military action.[71] Crookes insisted that science was 'one of the great bonds of union and peace between Nations', and urged the fellowship, 'even at this desperate moment' to remain 'cosmopolitan in spirit' and maintain a 'dispassionate' attitude towards their colleagues in Germany.[72] Banks and Blagden had worked to keep scholarly lines of communication open during the Napoleonic Wars (see Chapter 7), but during this war, the Society stopped its free circulation to individuals and institutions in enemy countries. The gifts to institutions in Germany, Austria and Hungary would not be restored until 1923 (and then only partially).[73] The feeling appears to have been mutual, and the Society no longer received periodicals from German institutions. It also stopped its paid-for subscriptions to at least a dozen German-language titles, including Leibig's *Annalen der Chemie* and Petermann's *Geographische Mittheilungen*.[74]

It was not just circulation to the German-speaking lands that was affected. By 1916, the Society felt the need to consider what might be done 'to restore the circulation of English scientific journals in allied and neutral countries', though it is not clear what was achieved.[75] And there were other threats to internationalism; there had been 17 gifts to Russian institutions in 1908, but the 1917 Revolution halted them all (and only a handful had been reinstated by 1930).[76] After the war, it took until the late 1920s for international scientific links to be (mostly) restored, with the question of admitting the former 'aggressor' nations to the new International Research Council being particularly hotly contested.[77]

Emergency finances

The end of the war was followed by a period of rampant inflation, with the prices of some goods and services rising to three times their pre-war level. Coupled with a return to pre-war levels of material being published in the journals, this pushed the Society's publication costs higher than ever before. The 1910 publication costs had been around £2,500 but, by 1920, they had risen to over £6,000. Publishing now accounted for 45 per cent, up from 35 per cent, of the Society's total annual expenditure.[78] Over the course of the 1920s, the flood of submissions to *Proceedings A* would push costs even higher. The Society's financial approach in that decade was informed by the report produced by an Emergency Finance Committee in 1920. This was convened after the new post-war treasurer, botanist David Prain, alerted Council to 'the serious situation' arising from the 'increase in the cost of printing' (and also in salaries) 'whilst the income of the Society remained stationary'.[79] The committee confirmed his analysis that '[b]roadly speaking the income of the Society remains as it was in 1914, while the expenditure has practically doubled'.[80]

The Emergency Finance Committee made a series of recommendations to help the Society's finances, most of which involved finding ways to increase income. There were some token efforts at cost-cutting, including a cap on the library budget, and more forcibly restricting fellows to claim only series A or series B. The committee also appealed for 'every possible economy' to be applied to the production of the journals. The question of what seemed 'possible' was, however, limited by a determination to protect the *Transactions* and the *Proceedings* from any 'drastic change' that might 'materially modify' their 'scope and character'.[81] There was no move to cheap paper, fewer illustrations or a more cramped typographical layout. The main outcome seems to have been that, in the years that followed, the size (and costs) of the *Transactions* were indeed kept within tighter limits than before (as Figure 11.5 suggested). But the expansion of the *Proceedings*, and especially *Proceedings A*, meant that the publishing costs kept rising, and the Society seems to have felt powerless to do anything about it.

The only cost-cutting measure of long-term significance was the recommendation that the list of institutions getting free copies of the journals should 'be drastically cut down'.[82] This appears to have been the first time that the Library Committee's frequent ad hoc decisions to add institutions to the free list had come up against the financial limits of the Society as a whole. By 1922, cuts had been made; the number of institutions receiving Society journals had been reduced to just under 350.[83] This may not seem a 'drastic' reduction from the 466 institutions in 1908, but the headline number disguises savings made by restricting some institutions to one series of the *Proceedings* (or the *Transactions*) rather than both. The cuts fell more heavily on younger institutions that were relatively recent additions to the Royal Society's list (many of which received only the *Proceedings*), and on those whose libraries seemed less likely to serve a substantial community of scholars.[84] University libraries generally kept their places on the list, but public libraries did not. The librarian of the Cardiff public library tried to argue that his library served 'practically as a British Museum to the whole of South Wales and Monmouthshire' (Figure 12.2), and had complete sets of both the *Transactions* and the *Proceedings* from 1865 onwards. But his claim that the cessation of the gift would 'be a great loss to scientific readers and students' in Wales was in vain.[85]

This 1921 review appears to have been the first time that the many gifts and exchanges supported by the Royal Society were considered together, as a whole. It encouraged more careful attention to the costs and benefits of this approach, and a consideration of the criteria used to decide on new arrangements, culminating in a 1932 review chaired by former president Charles Sherrington. The Society's library was then struggling with the growth of the scientific literature, which challenged its acquisitions budget and its shelf space, and with the question of its own role in a world where most fellows lived too far from London to consult its shelves regularly and probably had access to a research library in their own university or institution. Sherrington's Library Committee ended up creating a 'general policy with regard to the Library'.[86] This included the non-commercial circulation: it created a policy for dealing with future requests for free copies of the Society's journals, and recommended further cuts to the existing list. This was the first attempt to consider a cost/benefit analysis for the free circulation, though it was not yet expressed in directly financial terms.

Sherrington and his colleagues noted that the current state of the free circulation lists represented 'the accumulated effect of decisions taken over a long term of years'. They presumed that the original aim of gifting the *Transactions* in the eighteenth century had been 'to obtain the publications of other bodies, and at the same time to secure an adequate circulation for those of the Royal Society'. Surveying the current list, they concluded that the desire to ensure that the Society's publications (and, they might have added, its reputation) were known 'in different parts of

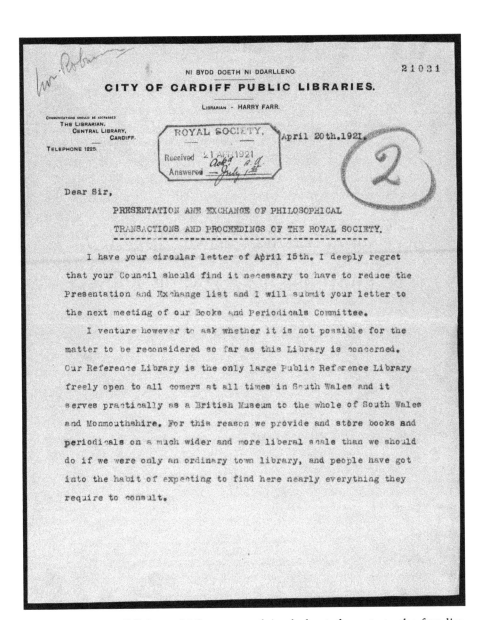

Figure 12.2 Cardiff Central Library complained about the cuts to the free list, 1921 © The Royal Society.

the world' had most often been 'the decisive consideration', rather than the practical finances of stocking the Society's library. Jenny Beckman has shown that the Royal Swedish Academy of Sciences, in contrast, planned their publication exchanges very much around the acquisitions for their own library.[87] Sherrington's committee also recognised that a lack of 'any consistent principle' had led to a situation where it was not obvious why one institution received free copies of the *Transactions* and the *Proceedings* while other similar institutions had to pay 'the ordinary prices'.[88]

A key outcome of Sherrington's 1932 library review was that the 'Exchange List' and the 'Presentation List (or "Free List")' were henceforth to be clearly distinguished. Exchanges would be established only with institutions whose publications were deemed to be of broadly equivalent financial value.[89] Thus, when the new physical secretary attended a Library Committee meeting in 1939, he could report in his diary that 'in the majority of cases', organisations requesting exchanges had 'very little to offer'. His measure of quantity was clearly intellectual as well as financial, for he went on to describe the offered exchanges as 'often only a few pages of badly printed second rate memoirs.... The efforts of a few enthusiasts working under great difficulties perhaps?' Such exchange requests were turned down.[90]

The free list included those institutions that could not reciprocate the Society's gift. Public libraries had already been cut in 1921, so the institutions remaining on the list were mostly universities. Going forward, only universities in the wider British world would still be supplied for free. Foreign universities would not, unless there were 'special reasons'. The universities of Harvard, Caracas, Strasbourg and Peking were among those cut, although Johns Hopkins retained its place. The Society's patron, King George V, and the British Museum were among the few non-university entities that still received free copies. The universities dominated the group of institutions that still received a full set of all four journals. Those 93 fortunate institutions included virtually all the British and Irish universities, as well as those in Adelaide, Brisbane, Melbourne and Sydney, and those in Toronto, Halifax, Edmonton, Fredericton and Montreal.[91]

Figure 12.3 shows the geographical distribution of the non-commercial circulation after the 1932 review. There were now 286 institutions still receiving something from the Royal Society; over 200 of them (70 per cent) were outside the UK. The 1932 review was far from the end of the non-commercial system of journal distribution, but it certainly marks a rationalisation.

External financial support

The cut-backs to the non-commercial circulation would later turn out to be significant, but at the time, the Emergency Finance Committee's

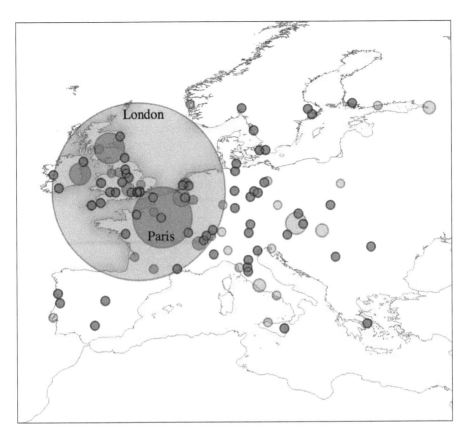
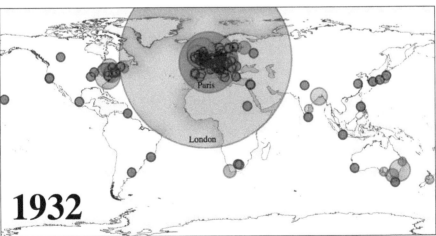

Figure 12.3 Global distribution of free copies of the *Transactions* and/or the *Proceedings* after the 1932 cuts. The size of the spots is proportional to the number of copies sent to a given city.

proposals to raise more income seemed more effective than their attempts to cut costs. Those proposals included a long-overdue increase in the Society's membership fees, and appeals both to the government and 'to private generosity'.[92]

The new appeal to the Treasury began with a request to divert 'a portion' of the main government research grant to support publications. Permission for this had, in principle, been included in the 1895 letter establishing the grant-in-aid of publications, but in 1914 the Society had sought explicit permission for this use of research funds, and it did so again in 1920.[93] Granting such permission, of course, cost the government nothing. However, from the Royal Society's perspective as a leading voice for British science, funding publication at the cost of research could not be a long-term solution.

The creation of the Department of Scientific and Industrial Research (1916), the University Grants Committee (1919) and the Medical Research Council (1920) all suggested that the British government was more willing to support research than it had been before the war.[94] Whether it was willing to support the Royal Society was less clear. In 1913, Geikie had been arguing that the Society deserved better recognition for 'the amount and value of the gratuitous service' it provided to various government departments.[95] This argument was continued by the post-war presidents as they tried to position the Society in the new scientific landscape. Charles Sherrington's 1921 presidential address welcomed the 'long-overdue' establishment of the state-supported organisations, and suggested that they should be seen as complementing – but not supplanting – existing activities. Together with the scientific societies and the universities, they would create a 'triple system' for the 'prosecution of research' in Britain.[96]

The publication of scientific research turned out to be an area where the government proved willing to continue channelling its support via the Royal Society. The £1,000 annual grant continued, but it was utterly inadequate to support the growing needs of all the learned societies that benefited from it. The Royal Society itself only got £60 from the government grant in 1923, despite having a shortfall of £3,700 on its publication costs.[97] The following year, the Society succeeded in getting the publications grant increased to £2,500 annually. Winston Churchill was chancellor of the exchequer in Stanley Baldwin's new Conservative government, and his officials accepted both the need for additional funding and the Royal Society's role as a trusted administrator. The Society was asked to ensure that the quality of publications receiving government funding should not 'depart from the standard which had been imposed before the War'.[98] In later years, the Society's role as a guardian of standards would be explicitly linked to its use of refereeing (see Chapter 14), but at this point, the trust in its ability to evaluate the quality of research was implicit.

The funds of the increased grant-in-aid still had to stretch a long way. In 1925, the Royal Society awarded all £2,500 to other societies and was able to keep nothing for itself. It then drew up revised guidelines warning that funds would only be awarded to cases that demonstrated genuine financial need, as well as intellectual quality.[99] This meant that, in 1927, when the Society received applications from 13 other scientific societies, it felt able to refuse two applications. It awarded £1,700 to the other 11 societies, and kept the remaining £800 for its own publications.[100]

The increased value of the grant-in-aid of scientific publications was undoubtedly a valued sign of governmental support for the Society and for the importance of publishing more generally; but it was far from enough to support the expansion of scientific research publishing taking place in the 1920s. The real reason that the Royal Society's publications were able to stay afloat financially was a wave of philanthropic gifts, partly a result of the Emergency Finance Committee's appeal to 'private generosity', and partly arising from the deaths of wealthy late Victorian industrialists.

Philanthropy had been becoming more important to Royal Society finances since the late nineteenth century. The periodicals were still supported by the income from the Publication Fund (originally the Fee Reduction Fund) established by Joseph Hooker in the 1870s (Chapter 10). This had been topped up by new donations in the 1910s, and was generating a very useful £500 or £600 per year.[101] The Society had also been successful in finding private donors to support specific projects: perhaps most notably, the industrial chemist Ludwig Mond had repeatedly supported the long-running *Catalogue of Scientific Papers* project.

The Royal Society was also coming to be seen as a suitable body to act as trustee for those who wished to leave bequests to science. William Lever (d. 1925) and Henry Wellcome (d. 1936) may have followed the transatlantic example set by Andrew Carnegie and the Rockefeller family by setting up their own charitable trusts, but many others entrusted their money to the Royal Society. They included the industrialists Rudolph Messel and Ludwig Mond, the shipbuilder Alfred Yarrow, and Dr John Foulerton (and his daughter Lucy).[102] A wave of substantial philanthropic gifts meant that, by 1930, the Royal Society held over £600,000 in trust.[103] The administration of these funds added to the burden on assistant secretary Francis Towle and his staff, on top of the work they were already doing managing the Society and its *Proceedings* and *Transactions*, and it was not until the 1930s that the Society learned how to levy an administration charge on these funds, and used it to hire additional support staff (including a separate editorial team for the journals).[104]

Many donors specified how they wished their money to be spent, and it was often to promote research in a particular scientific field. Among those who benefited were the materials scientist Constance Elam, who held a research fellowship from 1924 to 1929, and the cell biologist Honor Fell, whose long career of research support from the Society began with a Messel research fellowship in 1931. There were still no women fellows, but holding grants was a way for these women to become well connected to the Society's networks. By 1936, the Society's committees made grants and awards of around £30,000 a year from these trust funds; but these funds could not be used to support the publications.[105]

Other donors imposed no restrictions, allowing their funds (income and capital) to be used to support the Society's activities in general, and it was this generosity that allowed the treasurers of the 1920s and 1930s to cover the shortfalls in the publication account. The Mond and Messel bequests were particularly valuable in this respect and, from 1923 onwards, transfers of income from these funds routinely appeared in the publication accounts.[106] This enabled the treasurers to avoid eating into capital reserves to fund the publications, as their early 1890s predecessors had been forced to do. The importance of this income was publicly acknowledged in the annual accounts each year; in 1925 (the year in which the Society took no funds from the government grant), the treasurer noted that 'But for these transfers [from Mond and Messel], there would have been a deficit on Publications of £3,716 17s. 7d.'[107]

Ludwig Mond's younger son, Alfred, was also responsible for an early form of industrial sponsorship for scientific publishing. In October 1925, the Society's Council received an offer from the chemical company Brunner, Mond & Co., whose secretary wrote: 'My directors understand the Royal Society has a considerable financial deficit each year on its publications, and that such deficit is met by drawing upon funds which would otherwise be available for scientific research.' The directors offered the Society £500 a year for the next three years to assist in 'the publication of papers in Class A', that is, mathematics and physical sciences (which included chemistry). The Council 'gratefully accepted' this offer.[108] It is not clear whether Alfred Mond consciously made this arrangement in order to release his father's bequest for other purposes, but such was the effect.

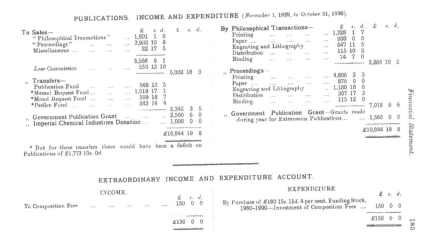

Figure 12.4 The publication finances in 1930 © The Royal Society.

A year later, Brunner Mond became part of the newly formed Imperial Chemical Industries Ltd (ICI). In 1927, Alfred Mond reassured the Royal Society that its funding would continue, and offered to increase ICI's sponsorship of Royal Society publishing to £1,000 a year 'until further notice'.[109] Again, the Council gratefully accepted and, in 1928, Mond was elected a fellow under Statute 12, the category reserved for people who were not themselves researchers but had 'rendered conspicuous service to the cause of Science' and whose election was likely to be 'of signal benefit to the Society'.[110] ICI's sponsorship of the Royal Society journals continued until after the Second World War.

During the 1920s, therefore, the Society had managed to gain substantial support for its publishing programme from both the government and the chemical industry. This double success indicates that publishing the results of scientific research was coming to be seen as a matter of national importance, and an important foundation for industrial development. It was also a reassurance that the Royal Society itself continued to be regarded as a trustworthy facilitator of the publication of research.

The contribution of the various income streams to keeping the Royal Society publications afloat can be seen in the accounts for 1930, as shown in Figure 12.4. That year, funding was needed to cover production and distribution costs of £2,265 for the *Transactions* and £7,019 for the *Proceedings*, or £9,284 overall. Sales income covered almost 55 per cent of those costs, but the remaining £4,281 had to be found elsewhere – and that was a notably larger amount than the treasurers of the 1890s had felt was supportable. In 1930, it was a manageable shortfall, because the Society retained £940 of the government grant funds, received the usual £1,000 from ICI and covered the remaining £2,342 from its own funds, including the Publication Fund and three unrestricted trust funds. External funds were therefore covering 20 per cent of the total publication costs; while the Society's own funds covered the remaining 25 per cent. It is clear that, even with the cut-backs to the non-commercial circulation of the *Proceedings* and the *Transactions*, sales income was nowhere near able to support the publishing costs (even though staff and office costs were silently subsidised by the general operations budget). And, in contrast to the 1890s, it was now the *Proceedings*, not the *Transactions*, whose finances were the greater concern; the annual deficit on the *Proceedings* had grown from around £2,000 in 1920 to over £4,000 by 1929.[111] The intellectual flourishing of *Proceedings A* in the 1920s was not matched by its financial performance, which is why the external financial support was so important.

Sales income

The Society's accounts do reveal that the raw value of sales income generated by the journals was increasing in the early twentieth century. For instance, the annual sales income of the *Proceedings* had grown from around £500 pre-war to almost £4,000 by 1930. Much of that growth is directly linked to the expanding quantity of research papers being printed; the *Proceedings* had expanded from around 1,700 pages to around 4,000 pages a year over the same period (see Figure 11.5), and customers paid more for more content. We get a clearer picture of the sustainability of the publication finances by looking at relative, rather than absolute, income and costs. By 1930, sales income covered almost 55 per cent of the publication costs, whereas sales had only covered around 40 per cent of costs in the years immediately before the First World War (and less than that in the 1890s).[112] This suggests that, while external funding and internal subsidy remained hugely important, sales income was becoming more significant to the Royal Society.

No records of subscriber numbers or individual sales appear to have been kept by the Society in this period, which makes it difficult to tell whether the increased income represents more copies of the *Proceedings* and the *Transactions* being sold, or whether it is the result of the 50 per cent increases in price imposed by the Emergency Finance Committee in 1920.[113] There is, overall, almost no surviving evidence of the Society's engagement with the commercial book trade in the 1920s, nor any strategic attention to advertising or marketing, nor any evaluation of the effectiveness of Harrison & Sons' management of the journal sales.

In 1932, the Library Committee would notice an interesting long-term consequence of the 1921 cuts to the free distribution: 'of institutions from which the free supply was withdrawn ..., a large proportion are known to have become subscribers'.[114] The public librarian in Cardiff had insisted that his library could not possibly afford to pay for the *Proceedings*, but some former gift recipients apparently did become purchasers.[115] We have no list of subscribers against which to check the extent to which this claim was true, but it marks the first time a connection was made between the non-commercial circulation and the potential commercial market for the Society's journals.

If the journals did, in fact, sell more copies in this period, it may have had more to do with the intense interest in the latest physics reported in *Proceedings A*, and with the expansion of higher education institutions in North America, rather than any conscious efforts by the Royal Society.

The Emergency Finance Committee's attention to pricing turned out to be a one-off adjustment. The Society did not re-evaluate the way it set prices, whether for volume subscriptions, individual issues or offprints, and it did not revise its notional cost price again until 1930. It continued to work with a notional cost price per sheet that was purposefully set below the level needed to recoup all costs. All the editorial work done by referees, committee members and staff was still being silently subsidised. So too was some fraction of the *Transactions* illustrations, since the *Transactions* rarely, if ever, achieved the 300 sales that would have paid off the plates under the 1905 calculation. Almost all sales of the *Proceedings* were also subsidised, through the 25 per cent discount on advance subscriptions. This is why net sales income from the *Proceedings* covered just 50 per cent of its costs in 1930, whereas net sales of the *Transactions* covered just over 60 per cent of its costs. The *Proceedings* almost certainly sold more copies (as the 1935 data in Table 13.1 will show), but the *Transactions* copies were less subsidised.

When Ernest Rutherford gave his final anniversary address as president of the Royal Society in November 1930, he noted that the Society's Council had once considered that the best use of its funds was 'to facilitate the publication of the results of research at a cost rendering them accessible to the widest range of scientific workers'. He claimed that this position had not fundamentally changed, and noted that (even with that year's increase in prices) the Society's periodicals were still priced 'low in comparison with that of other scientific publications and in relation to the increased cost of production'.[116]

However, Rutherford followed up by pointing out that if the journals could find a way to generate more sales income, above and beyond the increase caused by the growth of printed matter, this would 'release … a substantial sum' from the Society's other income, which, if not needed to subsidise the publications, could be 'allocated for other purposes'.[117] (He was in the midst of planning for the Mond Laboratory at Cambridge.) Rutherford thus put forward an early version of an argument that became powerful in the 1950s: pursuing sales income from the journals would be a way of releasing the Society's own funds (and those it held in trust) to do important work elsewhere. In the 1970s, this would morph into the argument that if the journals could generate a surplus, the Society would be able to support an even greater range of activities. But in 1930, the idea of the journals making a surplus still seemed remote, and external support from the government and ICI was gratefully accepted.

By the 1890s, it had become apparent that maintaining high production standards for an increasing quantity of output was straining the Royal Society's business model, with its generous discounts and extensive non-commercial distribution. Larmor and Geikie introduced some efficiencies and rationalisations before the war, and the Emergency Finance Committee insisted on price rises and cuts to the philanthropic distribution after the war – but there was no serious re-evaluation of whether the Society's publications strategy was fit for the conditions of twentieth-century science and academia. The *Proceedings* might have been transformed into a useful and effective medium of communication for modern physicists, but the business model underpinning it was still grounded in the gentlemanly, scholarly traditions of the late eighteenth century. Sustaining that model into the twentieth century stretched the capacity of a voluntary organisation of modestly paid professional academics. The Society kept it going by using its socio-political influence to build a more diverse set of income streams to support the publications. Where it had once depended largely on its own membership fees and endowments, the publications were now supported by philanthropic income, government support, industrial sponsorship and a small but apparently growing sales income.

During the 1920s, those managing the Royal Society's publications also had to navigate the changing role of those publications within the Society. In 1913, Archibald Geikie had already noted that meetings and publications no longer comprised 'the whole field of our operations', and Rutherford would see the 1920s as a time in which the Society had 'greatly increased its responsibilities'.[118] The Royal Society was attempting to find new ways to define its position in relationship to the UK government, and to establish a leadership role amid the array of new state-funded organisations supporting research.[119] It was simultaneously reinventing itself as a funding organisation on a larger scale than ever before, thanks to a series of substantial bequests that enabled the Society to support senior research professorships and early career studentships, as well as offering small research grants.

Ernest Rutherford ended his presidency in 1930 believing that the publications would no longer be a source of worry to the Society's secretaries and treasurers. The finances were apparently under control, and he assured the fellows that '[i]t now seems unlikely that there will be any substantial increase in the amount of publication during the next few years'.[120] He was wrong on both points. Nonetheless, it is true that the publications (and meetings) of the Society, which had been its core activities for over 150 years, would no longer dominate its strategy or its officers' attention in the twentieth century, as they had done in earlier times. As we follow the story of the journals further into the twentieth century, we will be dealing with an area of activity that was now largely taken for granted, except at particular moments when it seemed that the health (or not) of the publication finances would affect the Society as a whole.

Notes

1 Evans to Kelvin, 23 April 1894, recorded in RS CMP/07, 26 April 1894.
2 Aileen Fyfe, 'Journals, learned societies and money: *Philosophical Transactions, ca.* 1750 1900', *N&R* 69.3 (2015): 277–99; 'From philanthropy to business: The economics of Royal Society journal publishing in the twentieth century', *N&R* https://doi.org/10.1098/rsnr.2022.0021.
3 Evans to Kelvin, 23 April 1894, recorded in RS CMP/07, 26 April 1894.
4 On the sale of stock, RS CMP/07, 25 October 1894.
5 Inflation figures from 'Measuring Worth' GDP deflator series, see Ryland Thomas and Samuel H. Williamson, "What was the U.K. GDP then?" *MeasuringWorth* (2019) http://www.measuringworth.com/ukgdp.
6 RS CMP/07, 6 December 1894.
7 G.H. Darwin, on a paper 'Contributions to the mathematical theory of evolution' by Karl Pearson, 28 December 1894, RS RR/12/205.
8 Francis Gotch, on a paper 'Degenerations consequent on experimental lesions of the cerebellum' by J.S.R. Russell, 16 January 1895, RS RR/12/223.

9. Michael Foster to Victor Horsley, 12 April 1895, RS RR/12/226 (Horsley was the fellow who communicated Russell's paper). Russell's paper was finally approved, 2 May 1895, RS CMB/90/6.
10. RS CMP/07, 21 May 1896.
11. A.B. Kempe to sectional committee chairmen, June 1901, in Botany Sectional Committee, RS CMB/253.
12. 'Memorandum by Sir E. Ray Lankester', in Committee of Papers, RS CMB/90/6, 24 October 1907.
13. Retail prices are listed in Sampson Low, *The English Catalogue of Books*, 9 vols (London: Sampson Low, 1863–1915), vol. 8 (1906–10), 1472.
14. RS CMP/07, 20 January 1898. The number of bound volumes was reduced from 432 to 250, and the total run was reduced to 800.
15. It had been 1,750 in 1873, see T&F, Journal 1873–77, f. 12. By 1935, that for *Proceedings A* would be 1,400 copies, with 1,060 for *Proceedings B*, see Table 13.1.
16. The map and discussion are based on analysis of the list in RS *Year Book* (1908), 125–42.
17. RS CMP/09, 30 April 1908. This would be reiterated by the Emergency Finance Committee, see RS CMP/11, 15 July 1920.
18. 'Distribution of Royal Society Publications [in 1947]', [undated, but for an early 1948 meeting] in RS OM3/2(48).
19. RS CMP/09, 15 February 1906.
20. RS CMP/09, 26 October 1905.
21. RS CMP/09, 22 February 1906.
22. RS CMP/09, 21 January 1904.
23. For the 1895 survey and outcomes, see complex series of votes and amendments in RS CMP/07, March 14, June 20 and July 5 1895. Specimens for an enlarged (royal octavo) *Proceedings* were approved on 2 July; it is not clear why no further action was taken.
24. Oliver Heaviside to Lord Rayleigh, 23 December 1891, RS MM/17/110. On this episode, see also Graeme Gooday, 'Periodical physics in Britain: Institutional and industrial contexts, 1870–1900', in *Science Periodicals in Nineteenth-Century Britain: Constructing scientific communities*, edited by Gowan Dawson, Bernard Lightman, Sally Shuttleworth and Jonathan R. Topham (Chicago, IL: University of Chicago Press, 2020), 238–73.
25. Unknown correspondent to Harrisons (reporting Heaviside's approval), 27 June 1892, RS NLB/6/682.
26. Oliver Heaviside to H. Rix [asst secretary of RS], 4 June 1892, RS MM/17/111.
27. Geikie to Harrison & Sons, 25 November 1904, RS NLB/29/693.
28. 1910 advertisement for Hazell, Watson & Viney, from *Grace's Guide*, https://www.gracesguide.co.uk/File:Im1910Pa-HWV2.jpg.
29. William H. Brock and A.J. Meadows, *The Lamp of Learning: Taylor & Francis and the development of science publishing* (London: Taylor & Francis, revised edn 1998), 150–2.
30. RS CMP/09, 17 March 1904 (recommendation accepted, 28 April 1904).
31. July 1900 advertisement for Eyre & Spottiswoode, from *Grace's Guide*, https://www.gracesguide.co.uk/File:Im19000714S-Eyre3.jpg.
32. Terms set out in RS CMP/09, 8 December 1904.
33. Such a severance would be 'a shame', according to Geikie to Harrison & Sons, 25 November 1904, RS NLB/29/693.
34. RS CMP/09, 19 January 1905, and (for 'defective') 16 March 1905.
35. Henry Armstrong to the secretaries of the RS, 23 October 1908, RS MS/603/25.
36. RS CMP/09, 19 January 1905.
37. The only publicly advertised option was to purchase from the Society, see Low, *English Catalogue of Books*, vol. 5 (1890–97), 134; but it is clear that Harrisons also sold copies.
38. Low, *English Catalogue of Books*, vol. 5 (1890–97), 103. For the move to Dulau, see RS CMP/07, 25 October and 6 December 1894.
39. RS CMB/47/4, 8 May 1890; and RS CMP/06, 22 May 1890.
40. See their advertisements in Low, *English Catalogue of Books*, vol. 5 (1890–97).
41. On Dulau, see Whittaker's *Who's Who in Business* (1914), from *Grace's Guide*, https://www.gracesguide.co.uk/1914_Who%27s_Who_in_Business:_Company_D.
42. On Harrisons as publishers, see Cecil Reeves Harrison and H.G. Harrison, *House of Harrison: Being an account of the family and firm of Harrison and Sons, printers to the King* (London: Harrison & Sons, 1914), 41–3.

43 RS CMP/09, 18 January 1906.
44 RS CMP/09, 7 December 1905.
45 RS CMP/09, 18 January 1906.
46 RS CMP/09, 7 December 1905.
47 RS CMP/09, 15 March 1906.
48 Claim by Rev. Robert Harley, RS CMP/08, 19 June 1902.
49 RS CMP/08, 11 July 1902. The secretaries also proposed shifting stock more rapidly by dispatching papers to institutions on the presentation list as soon as they were printed (rather than waiting for full volumes to be bound up).
50 Holdings over 10 years old were stripped back to 100 copies; the rest being offered to fellows and institutions, and disposed of. See RS CMP/08, 9 July 1903.
51 RS CMP/09, 5 April 1906.
52 RS CMP/09, 30 April 1908.
53 Rayleigh to the Treasury, in RS CMP/07, 20 June 1895. On the government grant, see Roy M. MacLeod, 'The Royal Society and the government grant: Notes on the administration of scientific research, 1849–1914', *Historical Journal* 14, no. 2 (1971): 323–58.
54 Francis Mowatt (Treasury) to Rayleigh, in RS CMP/07, 5 July 1895.
55 Rayleigh to the Treasury, in RS CMP/07, 20 June 1895.
56 RS CMP/08, 16 February 1899.
57 'Regulations for the Administration of the Government Publication Grant', in RS CMP/08, 15 June 1899.
58 1910 annual accounts, in RS *Year Book* (1911), 188–9.
59 Draft letter from RS to the Treasury, in RS CMP/10, 30 November 1914.
60 [A. Geikie], 'Address of the president, 1 December 1913', *ProcA* 89 (1914): 462–3.
61 On the impact of war, see Sarah Bromage and Helen Williams, 'Materials, technologies and the printing industry', in *The Cambridge History of the Book in Britain: Volume 7: The twentieth century and beyond*, ed. Andrew Nash, Claire Squires and I.R. Willison (Cambridge: Cambridge University Press, 2019), 41–60.
62 RS CMP/11, 11 November 1915. The Society paid Harrisons £676 for printing the *Proceedings* in 1915, and £569 in 1916 (and £572 in 1917).
63 RS CMP/11, 16 March 1916. The committee did suggest reducing the number of free offprints provided to authors to 50.
64 RS CMP/11, 6 April and 18 May 1916.
65 RS CMP/11, 13 July 1916.
66 Jane Potter, 'The book in wartime', in *The Cambridge History of the Book in Britain: Volume 7: The twentieth century and beyond*, ed. Andrew Nash, Claire Squires and I.R. Willison (Cambridge: Cambridge University Press, 2019), 567–79.
67 RS CMP/11, 16 March 1916.
68 RS CMP/11, 13 July 1916.
69 RS CMP/11, 16 March 1916.
70 RS CMP/11, 13 July 1916. Copper plates were sometimes reused decades after their original creation: for an example, see Aileen Fyfe, Julie McDougall-Waters and Noah Moxham, 'Credit, copyright, and the circulation of scientific knowledge: The Royal Society in the long nineteenth century', *Victorian Periodicals Review* 51, no. 4 (2018): 597–615.
71 Roy M. MacLeod, 'The mobilisation of minds and the crisis in international science: The Krieg der Geister and the Manifesto of the 93', *Journal of War & Culture Studies* 11, no. 1 (2018): 58–78.
72 [W. Crookes], 'Address of the president, 1914', *ProcA* 91 (1915): 111–15.
73 Sixteen institutions in former enemy countries were re-admitted to the list in 1923, see RS *Year Book* (1923).
74 'List of incomplete journals in the Society's library, which are now no longer being taken', Library Committee Minutes, RS CMB/47/5, 16 January 1933. This list gives the date on which the RS sets of these journals ceased.
75 RS CMP/11, 7 December 1916.
76 In 1922 and 1923, all Russian gifts were 'in abeyance'. By 1930, five institutions in Moscow, Leningrad and Kazan were once more receiving Royal Society publications, see RS *Year Book* (1931), 165.

77 On Arthur Schuster's involvement, see A.G. Cock, 'Chauvinism and internationalism in science: The International Research Council, 1919–1926', *N&R* 37.2 (1983): 249–88, but see the important correction in Mary Cartwright, 'Note on A. G. Cock's paper "Chauvinism in science"': The International Research Council, 1919–1926', *N&R* 39.1 (1984): 125–8. For wider context, see Frank Greenaway, *Science International: A history of the International Council of Scientific Unions* (Cambridge: Cambridge University Press, 2006) and Robert Fox, *Science Without Frontiers: Cosmopolitanism and national interests in the world of learning, 1870-1940* (Corvallis: Oregon State University Press, 2016).
78 RS *Year Book* (1911), 188–9; RS *Year Book* (1921), 168–9.
79 RS CMP/11, 17 June 1920 (quoted in minutes of the Finance Committee, CMB/86/1/2).
80 Report of the Emergency Finance Committee, in RS CMP/11, 8 July 1920.
81 Report of the Emergency Finance Committee, in RS CMP/11, 8 July 1920.
82 Report of the Emergency Finance Committee, in RS CMP/11, 8 July 1920.
83 Library Committee deferred dealing with the suggestion on RS CMP/11, 4 November 1920. For the 1923 list (printed in late 1922), see RS *Year Book* (1923).
84 The number of institutions receiving both the *Transactions* and the *Proceedings* was cut by 20 per cent, but the number receiving only the *Proceedings* was cut by 30 per cent. In 1908, 189 institutions received both series of the *Proceedings*, but in 1923, only 80 did. (The number of institutions receiving either *Proc A* or *Proc B* had increased, but not to match the shortfall.) See RS *Year Books*, passim.
85 Harry Farr (Librarian, Central Library Cardiff) to Royal Society, 20 April 1921, RS MC/21031.
86 The review started early in 1932, see Library Committee, RS CMB/47/5. By March, the Library Committee had realised the need for a general policy, see RS CMP/13, 3 March 1932. The policy was presented to Council on 21 April 1932, RS CMP/13. See also Marie Boas Hall and Patricia H. Clarke, *The Library and Archives of the Royal Society: 1600–1990* (London: The Royal Society, 1992), 48.
87 Jenny Beckman, 'Editors, librarians, and publication exchange: The Royal Swedish Academy of Sciences, 1813–1903', *Centaurus* 62, no. 1 (2020): 98–110.
88 Report of the Library Committee, RS CMB/47/5; and also in RS CMP/13, 21 April and 19 May 1932.
89 See RS CMB/47/5, 15 March 1932; and the final policy, RS CMP/13, 21 April 1932.
90 Diary of A.C. Egerton, 1938–41, RS AE/2/3, f. 1 (re 15 March).
91 The university exceptions were University College Dublin (formerly the Catholic University), and the relatively new Southampton University College.
92 Report of Emergency Finance Committee, in RS CMP/11, 8 July 1920. The annual subscription was increased from £3 to £5, having last been changed in 1878.
93 Correspondence with Treasury, in RS CMP/10, 30 November 1914; Report of the Emergency Finance Committee, in RS CMP/11, 8 July 1920.
94 Peter Alter, *The Reluctant Patron: Science and the state in Britain, 1850–1920* (New York: Berg, 1987); David Edgerton, *Science, Technology and the British Industrial 'Decline', 1870–1970* (Cambridge: Cambridge University Press, 1996).
95 [A. Geikie], 'Address of the president, 1 December 1913', *ProcA* 89 (1914): 464.
96 [C. Sherrington], 'Address of the president, 1921', *ProcA* 100 (1922): 360, 362.
97 Accounts for 1923, in RS *Year Book* (1924), 178.
98 RS CMP/12, 22 January 1925.
99 'Report of the Council 1926', RS *Year Book* (1927), 173.
100 'Report of the Council 1927', RS *Year Book* (1928), 170.
101 The appeal for new donations is mentioned in *The Record of the Royal Society of London* (3rd edn) (London: The Royal Society, 1912), 179. The income from the fund can be seen in the annual accounts.
102 Peter Collins, *The Royal Society and the Promotion of Science since 1960* (Cambridge: Cambridge University Press, 2015), 57.
103 [E. Rutherford], 'Address of the president, 1 December 1930', *ProcA* 130 (1931): 249.
104 Report of the Finance Committee, in RS CMP/13, 22 May 1930.
105 [W.H. Bragg], 'Address of the president, 30 November 1936', *ProcA* 157 (1936): 712.
106 See, for instance, RS CMP/12, 3 May 1923 and 30 October 1924.
107 Accounts for 1925, in RS *Year Book* (1926), 177.
108 RS CMP/12, 20 October 1925.

109 RS CMP/13, 27 October 1927.
110 RS EC/1928/17.
111 After 1920, the deficit on the *Transactions* remained mostly below £1,000 a year. For more on the state of the deficit, see Fyfe, 'From philanthropy to business'.
112 Fyfe, 'From philanthropy to business', Figure 5.
113 Report of Emergency Finance Committee, in RS CMP/11, 8 July 1920.
114 RS CMP/13, 21 April 1932.
115 Harry Farr (Librarian, Central Library Cardiff) to Royal Society, 20 April 1921, RS MC/21031.
116 [E. Rutherford], 'Address of the president, 1930', *ProcA* 130 (1931): 248–9. On the price increase (of 50 per cent per sheet), see RS CMP/13, 22 May 1930.
117 [E. Rutherford], 'Address of the president, 1930', *ProcA* 130 (1931): 248–9.
118 [A. Geikie], 'Address of the president, 1 December 1913', *ProcA* 89 (1914): 461; [E. Rutherford], 'Address of the president, 1929', *ProcA* 126 (1930): 190.
119 Roy M. MacLeod and E. Kay Andrews, 'The Origins of the D.S.I.R.: Reflections on ideas and men, 1915–1916', *Public Administration* 48 (1970): 23–48; Sabine Clarke, 'Pure science with a practical aim: The meanings of fundamental research in Britain, circa 1916–1950', *Isis* 101, no. 2 (2010): 285–311.
120 [E. Rutherford], 'Address of the president, 1 December 1930', *ProcA* 130 (1931): 249.

13
Why do we publish? 1932–1950
Camilla Mørk Røstvik and Aileen Fyfe

In 1938, William Henry Bragg focused his presidential address on publishing. He described how the volumes of scientific journals had 'swollen till they have become unmanageable', but still 'the papers come pouring in, and the rate of flow even increases'. In other words, Ernest Rutherford had been wrong to think that the growth was over. Like so many presidents and treasurers before him, Bragg noted the 'practical difficulties' of handling this growth. Even with help from the government publication grant, all scientific societies were struggling as their publications 'overstrained' their finances, and they found themselves unable to publish 'worthy material'. But Bragg also raised a broader question: 'Why *do* we publish? Why do we submit papers and why does the Society print them, if they are good enough?' Bragg recognised that there were multiple motivations at work. For the individual researcher trying to develop a career in science (whom Bragg characterised as 'a man'), publishing was a means 'to establish his reputation and position'. Then, there was the desire to show 'what he has done to those who will understand it', and also, though perhaps less consciously, 'the wish that his work may be of service'.[1]

This notion of 'service' was Bragg's nod to the contemporary debate about the 'social responsibility of science'. He had become president in 1935, when, for the first time in living memory, there had been an actual contest for the positions on the Society's Council, partly driven by a desire for the Society to engage more prominently with government. The revolt, led by the Oxford chemist Frederick Soddy, failed, but there is no doubt that Bragg was conscious of a need to unify the fellowship.[2] In this context, Bragg argued that there was a social responsibility to the act of publication. Natural knowledge should be seen as an 'inheritance belonging to mankind', and researchers had a duty to share their discoveries.³

Bragg also emphasised the scientist's responsibility to communicate clearly and effectively with non-specialists. He insisted that scientists were not 'responsible for the uses that are made of our discoveries', but that 'we are, at least, bound to see that our acquired knowledge is rightly stated'. Bragg acknowledged that too many papers in the *Proceedings* and the *Transactions* were 'dull and difficult reading' for all but the tiny number of experts 'working in the same narrow region' as the author (or, in other words, to a Royal Society referee). Bragg wanted the Society's mission to include developing ways to help the results of research 'be sufficiently appreciated, and ... be incorporated with understanding' into the appropriate spheres.⁴ Such a vision would have transformed the Society's publishing activities, and given it a unique place among the many other journals and societies in existence.

Soon after Bragg's speech, ideas about the scientist's role in society would be transformed by the needs of the Second World War. And afterwards, the Society and its publications would have to operate in the changed world of the emerging Cold War, with its emphasis on military and industrial research. The Society's two Empire Scientific Conferences in 1946 and 1948 (the latter specifically focusing on publication) launched a new attention to scientific diplomacy beyond Britain, while the admission of the first women fellows in 1945 was a visible sign that the days of the gentleman-scholar might be ending.

This chapter explores the Society's publications before, during and immediately after the Second World War. It starts by expanding on Bragg's question, by considering not just 'why', but 'what' and 'how' the Society published. The key players include Ronald Winckworth, an Oxford-educated school teacher and naturalist, who became the Society's first 'assistant editor' in 1937,⁵ and the physiologist Archibald Vivian Hill (known as 'A.V.'), a Nobel laureate and Royal Society research professor at University College London, who was the biological secretary of the Society from 1935 to 1945.⁶ For much of Hill's tenure, he served alongside A.C. 'Jack' Egerton, whose diary of his early years as physical secretary gives some sense of the routine. In March 1939, he attended a Thursday evening meeting of the Society, and on Friday he had 'much to see about publications at the RS ...; Hill and I lunched at the Athenaeum together'.⁷ Their lunches continued through the war. Hill had become secretary as the Society was tendering for its printing and publishing needs, and he had been intimately involved with ensuring the success of the bid from

Cambridge University Press. This move would enable the transformation of the *Proceedings* and the *Transactions* into financially viable, internationally oriented scientific journals in the post-war decades.

What do we publish (or not)?

By the 1930s, the journals of the Royal Society found the majority of their authors and readers among members of the newly professionalised world of British academia. Compared to the 1830s, this community was somewhat less socially elite than it had been, and its (limited) international orientation had shifted from Continental Europe toward the wider British world of Canada, Australia, New Zealand and India. (Ties between the British academic world and that of the United States were modest at this time, but on the increase.) Critically, for most of the members of this community, scientific research and publication was now part of their jobs, and something on which career advancement at least partially depended.[8] As mathematician Louis Filon told the Royal Society Council in 1936, 'research qualifications are now more and more insisted upon for appointments to academic and other posts', and in that context, publication in the Society's journals should carry a 'special significance'. Filon noted that journal brands were being used as short-cuts to evaluate the quality of research, in a world where appointment panels might not know candidates personally, and were struggling to distinguish between 'a spate of trivial papers' and 'a few really valuable contributions'.[9] But he feared that the Royal Society's reputation for an 'exceptionally high standard' was fading.

Filon had served a term as vice-president of the Society, and, like Henry Armstrong before him (see Chapter 11), he used the opportunity to set out his concerns in a memorandum. He worried that too much 'routine research' carried out by an ever-increasing group of junior researchers was finding its way into the Society's journals. This, he feared, was not only increasing the 'bulk' (and cost) of the journals, but was threatening their prestige. He was sure that, '[t]wenty or thirty years ago such work would not normally have been either offered or accepted for the Proceedings of the Royal Society'. Filon's vocabulary paints a vivid picture of his idea of what the Society should be publishing: 'valuable' research of 'critical importance' and of 'exceptionally high standard'. Equally, it should *not* be publishing 'routine', 'sound' or 'trivial' research of 'secondary importance', 'the accumulation of data', or 'the elaboration of minor detail'.[10]

Filon's concern was tied to his sense that there was a change in *who* the Society was publishing, which was linked to the growth of research training in the universities.[11] He believed there was a rise in submissions from 'young and comparatively untrained men', whether students undertaking 'research degrees of all kinds' or junior members of staff. The small but steady trickle of submissions from women authors did not merit specific comment. It seems Filon would have preferred to see more papers from established researchers, the sort of men who were, or might soon be, fellows of the Society. Filon appeared to have forgotten that his own first paper in the *Transactions* (with Karl Pearson, in 1898) had appeared just two years after his BA graduation.[12]

The Royal Society's own fellows continued to contribute between 40 to 70 of the papers published each year, most of which now appeared in the *Proceedings* rather than the *Transactions*, but Filon had noticed that the Society was publishing more and more papers by authors who were not fellows. There were around 250 such submissions a year in the 1930s, and the vast majority were published.[13] Filon did not blame the authors for trying, but he questioned whether so many of their papers should have been accepted.

Filon believed that the pre-screening traditionally performed by the requirement that papers be communicated via a fellow was no longer functioning effectively. He claimed that many fellows now experienced a conflict between their academic roles as laboratory heads or supervisors of research students, and their duty to the Society when communicating papers for others. Fellows generally 'find it difficult to refuse the request by a student or a member of his staff to submit a paper', even when the research 'may be comparatively unfamiliar'. He proposed a reminder to all fellows that communicating a paper to one of the Society's journals involved 'a duty as well as a privilege'.[14]

The nature of the communicator's duty was something that had certainly perplexed A.V. Hill, who had personal experience of communicating a paper that he subsequently came to believe was 'fraudulent'. Back in 1922, when Hill was a professor at Manchester, he had agreed to communicate a paper on 'the periodic opacity of certain colloids' by a colleague in pathology named J. Holker.[15] Hill later said that Holker 'used to come and talk to me about it sometimes', but he denied having 'any actual responsibility' for Holker's research. Holker's paper was duly published in the *Proceedings*, and Holker continued his research programme. But Hill found his new claims 'increasingly difficult to swallow'. When the new experiments produced no results when Hill came to observe them, he began 'to seriously doubt' the published findings. He described Holker's work as a 'mixture of great experimental skill and insight with a quite amazing muddle-headedness'. Years later, Hill would reckon the paper was 'as fraudulent as the famous Piltdown skull'.

Looking back, he reckoned there were a few 'obvious defects' in the text, which 'I ought to have made the author correct, but there is nothing obviously wrong'.[16] He was still not sure whether Holker himself was 'deliberately cheating', or whether his assistants were 'fudging the observations'. With hindsight, he now reckoned: 'Obviously I ought to have insisted earlier on witnessing the actual experiments.' But witnessing experiments had never been a normal expectation of fellows communicating papers to the Society, and, as Hill remarked, 'one does not naturally assume that a colleague is a liar or a lunatic'. He wondered whether he should have 'let it be known publicly (in order to save other people's time) that the results were wrong or worse?' But retractions had not yet become a feature of the scientific literature, and Hill said nothing, fearing 'legal action and "damages"'. He was thankful that the experiments had since been 'totally forgotten'.

In 1936, however, Filon's worries about the Society's editorial processes were not about the failure to detect fraud, but its alleged failure to exclude research 'of secondary importance'. He suspected that referees would 'not unnaturally, hesitate' to go against the implied recommendation of the fellow who communicated the paper, particularly if they were 'of some reputation'.[17] For instance, papers communicated by the former president, Ernest Rutherford, still had a smooth passage through the editorial system. In 1935, Rutherford had communicated 15 papers for the *Proceedings* and one long paper for the *Transactions*. Some were from students and staff at the Cavendish or Mond laboratories in Cambridge, while others came from international researchers in Rutherford's network. As in 1921 (see Chapter 12), they were all published. The Royal Society had always had a high acceptance rate, but papers communicated by Rutherford still received less scrutiny; by this point, over 80 per cent of papers for the *Proceedings* were being examined by at least one referee, yet only five of the 15 communicated by Rutherford received that scrutiny.[18]

Filon believed referees should be given clearer criteria, and empowered to recommend rejecting papers. The existing advice said merely that papers could be rejected if, in the opinion of the referee, 'the paper should be more suitably published elsewhere'. Filon wanted the statement to be 'more definite', and suggested that 'some guidance' be provided about 'the type of paper which is desirable or otherwise for publication by the Society'.[19]

Council agreed to remind all fellows about the responsibilities of communicators, and during a wider 1937 revision of the Society's standing orders, they revised the advice to referees and issued a pamphlet of instructions for 'preparing papers for publication'.[20] The new guidance stated that papers published by the Society should contain 'results or methods of critical importance' and be 'of value to others than specialists in the particular subject'. The idea that papers published by the Society should be of special importance might been an attempt to justify the role of the Society's generalist journals in an age of specialisation; some in the 1960s would interpret it that way (see Chapter 14). In the 1930s, however, its main purpose was to avoid unnecessary expense: specialist papers 'could more appropriately be published by some other body'.

The new guidance also repeated Filon's statement that the Royal Society's journals were 'not the proper medium for the mere accumulation of data, or the elaboration of minor details'.[21] The Society subsequently trialled a scheme allowing the deposit in its archives of the vast amounts of data that supported new findings in such fields as X-ray crystallography; even Bragg admitted that most of these data would only be needed by those 'very few readers' who wanted to 'check the detail of the work', and they could consult it in the archives, or request 'photographic copies'.[22]

On the matter of rejecting papers, however, the revised guidelines were less helpful than Filon might have hoped. They did make clear that papers should be recommended for rejection if they featured 'unmistakeable logical fallacy' or evident 'experimental error'. However, they also stressed that rejection was not merited 'merely because a referee disagrees with opinions, theories or conclusion put forward'.[23] This concern with personal biases had not been part of Filon's memorandum, but likely arose from Council's awareness that Frederick Soddy's recent campaign against elitism in the Society had arisen, in part, from his annoyance at the rejection of a paper he had communicated.[24] It was a reminder that rejecting a paper was always going to involve a disagreement between fellows.

Such disagreements were always diplomatically awkward, and could become even more so if one of the parties complained to Council, which still retained an editorial role as a court of appeal. Thus, in 1946, the sectional committee for physics would ask Council to adjudicate on a paper submitted by Chandrasekhara Venkata Raman.[25] The paper had already spent over a year in the editorial process, and had been examined by no fewer than six referees. The sectional committee had eventually concluded that 'the opinions expressed in the paper are unsound from a scientific point of view', but they were worried about the 'possible repercussions' of rejecting Raman's paper. A Nobel Prize-winning fellow, Raman was one of India's most famous scientists, and there might have been political as well as scientific sensitivities in play in late 1946. For all these reasons, it needed the authority of the Council to inform Raman that his paper could not be published 'in its present form'.[26]

Filon's concerns about quality had been put on hold during the war years, but they were resurfacing in 1945 and 1946. There was, once more, a worry about 'undue pressure' arising from 'junior research workers', for whom 'one or more papers in the Proc. Roy. Soc. becomes an almost automatic recommendation in getting posts'.[27] The Society's physical secretary, Alfred 'Jack' Egerton, responded by reporting that 'about fifty percent' of the authors in the *Proceedings* were 'either Fellows or research workers who later became Fellows'.[28] Being or becoming a fellow was assumed to be a strong indicator of the quality of research because the Royal Society's election procedures were predicated on recognising quality. Egerton agreed that many junior researchers submitted 'to get kudos', but believed that the quality remained high, as 'readers in all part of the world know'. This quality, he argued, was assured because 'no Journal in the world has the advantage of so critical and helpful a body of referees' as the fellows of the Royal Society.[29] Such claims would become increasingly common in the post-war years, as learned societies sought to differentiate their research journals from those issued by new commercial players in academic publishing.[30]

In 1945, then-president Henry Dale acknowledged that the pages of *Nature* had replaced the Society's weekly meetings as the favoured venue for announcing new findings, and thus had robbed the meetings 'of much of a factor of interest which they once possessed'. But he feared little could be done. American researchers were 'conspicuously' enthusiastic proponents of 'priority notes' in *Science*, because it was so difficult for them, 'owing to consideration of distance', to gather in person. He thought the trend for British researchers to announce 'real or imaginary discoveries' in *Nature* before submitting the full paper to the Society later was 'deplorable'; but he also thought that it would be 'difficult, if not impossible', to do anything about it. To a suggestion that the Society should refuse to accept submissions that had been previously announced in *Nature*, he was pragmatic; it would create 'an increasing reluctance to bring anything for communication and publication to the Royal Society'.[31]

In 1936, the Society had responded to the perceived need for a publication mechanism faster than the (monthly) *Proceedings* by launching *Abstracts of Papers Communicated to the Royal Society*. It was to publish abstracts rapidly, ahead of review. By announcing all submissions to the Society, it would perform a similar function to that originally undertaken by the *Proceedings* in the 1830s and 1840s, when it contained reports of Society meetings. W.H. Bragg also hoped that making (properly written) abstracts available more widely and rapidly would be a valuable service to the many researchers with a general rather than specialist interest in a field.[32] However, *Abstracts* would fall victim to wartime paper rationing.[33]

Another appeal of *Nature* was that it provided a place for expressing opinions about scientists' role in society and politics. Many scientists in the 1930s, especially those on the Left politically, were arguing that scientists should be more actively involved in politics. This had been an element in Soddy's attempted revolt against the Royal Society Council in 1935, and it was also a theme in the articles written by science journalist James Gerald Crowther for the *Manchester Guardian*, publicly critiquing the Society for being too elitist and unaccountable.[34] Julian Huxley, John D. Bernal and Solly Zuckerman found an alternative venue for discussion by establishing an informal dining club, whose members wrote the anonymous 1940 Penguin paperback *Science in War*.[35]

A.V. Hill was one of those who used the pages of *Nature* to publicly attack the new Nazi regime in Germany, and (with Rutherford and others) he helped establish the Academic Assistance Council, which helped Jewish academics dismissed by the regime; it was given office space in the Royal Society's premises at Burlington House.[36] The Royal Society offered a potential forum for discussing these sorts of issues at the teas that preceded its meetings, but there was no space in its publications for discussions of the difficulties facing Jewish scientists in Germany, or of Goebbels' restrictions on importing foreign publications, or the best ways to mobilise science for the war effort.[37]

It was Hill who, as secretary in 1944, responded to a request from British rabbi Dr Solomon Schonfeld about whether research on 'race hygiene' would be permitted in the Society's journals. Schonfeld helped transport adults and children out of Nazi-controlled areas, and wrote countless letters to British institutions asking for action in their respective fields.[38] He appears to have hoped that the Royal Society might publicly condemn the racial claims made by the Nazi regime. Hill was caught between his personal feelings and the Society's official determination to avoid having 'its scientific prestige … exploited for political or other causes, good, bad or indifferent'. He fell back on the wording of the 190-year-old 'Advertisement' that was still prefaced to the (handful of) bound volumes of the *Transactions* (see Chapter 5), and told Schonfeld that, 'it is an established rule of the Society, to which they will always adhere, never to give their opinion, as a Body, upon any subject, either of Nature or Art, that comes before them'.[39]

Hill explained that the Society's decision to publish an article indicated that its claims had passed a threshold 'of sufficient scientific merit', but that 'even so', the Society itself 'expresses no opinion on the claims made'. Thus:

> The Society publishes scientific papers on pharmacology: but it has never instituted an inquiry into the advertised claims of patent medicine manufacturers. The Society publishes papers on embryology and evolution: but it never, as a Body, expresses an opinion on Fundamentalism.[40]

Thus, he told Schonfeld, the Society could not speak out as a corporate body on racial science, and research papers on 'anthropology, ethnology, or comparative psychology', if submitted via a fellow of the Society, would be considered in the usual way. He offered cold comfort by reassuring Schonfeld that there was little chance of this happening, since 'I am sure that the vast majority of its Fellows would agree with the President that the pseudo-science of "Rassenhygiene" etc. is not a fit subject for its discussions'.[41] For Schonfeld, the Society's insistence on avoiding public pronouncements must have been disappointing.

The Royal Society's officers could see merit in the desire for a publication that enabled better communication with, and among, the fellowship, even if they did not want to provide a forum for controversial debates. The new twice-yearly *Occasional Notices*, in 1937, was to keep fellows 'more fully informed of the activities of the Society'.[42] The new periodical was renamed *Notes and Records of the Royal Society* in 1938. Early issues included suggestions from Charles G. Darwin (director of the National Physical Laboratory) for making the Society's meetings more interesting, and treasurer Henry Lyons shared his archival explorations into the Society's financial history. How much controversy was permissible in its pages was, however, not yet clear. One fellow complained in 1940 about what he saw as 'propaganda'. He acknowledged that *Notes and Records* 'is not a scientific periodical', and 'we must not judge it quite by the standards which we should apply to the *Transactions* or the *Proceedings*', but he feared that the 'dignity' of the Society would be undermined.[43]

The historical material in *Notes and Records* became an important part of the self-presentation and mythologising of the Royal Society itself, as various fellows and historians became increasingly interested in the Society's foundation and early years. Several of the presidential addresses delivered by the biochemist Frederick Gowland Hopkins in the early 1930s touched on historical themes, inspired by the 300th anniversaries of the births of Christopher Wren and Antoni van Leeuwenhoek. An edition of the correspondence of Isaac Newton was begun in 1935, although his tercentenary celebration had to be postponed until after the war. The Society's library was increasingly being used for its historical resources, as well as (or instead of) its current scientific content; and this was assisted by the acquisition of various collections of historic manuscripts, including the papers of Charles Blagden.[44] *Notes and Records* was not considered to be one of the research journals, and its contents did not go through the complex editorial procedures used for the *Proceedings* and the *Transactions*. Like the *Year Book* and the *Obituary Notices*, it was seen as an internal membership publication for circulation 'mainly among Fellows'. These three internal publications had minimal sales, though they did, of course, add to the annual expenditure on publications. Oldenburg's *Transactions* had contained a miscellaneous variety of content types, but by the twentieth century, the single multi-purpose scientific periodical had given rise to a range of publication types. The Royal Society's own range of publications filled several different niches, but it left other niches (including news, controversy and book reviews) to other publishers.

How do we print and publish?

The Royal Society had made no drastic changes to its publications during the difficult 1920s, and this remained superficially true in the 1930s and 1940s. But behind the scenes, there were significant changes in the staffing of the Society's editorial system, and in its practical arrangements for printing and publishing its journals. The 1936 process of tendering for a new printer generated a paper trail that gives us an unusually detailed picture of the state of the Society's journal publishing.

An assistant editor and a publications committee

Since 1925, much of the work of preparing accepted manuscripts for the press had been done by Ronald Winckworth, who was formally a member of the Society's library staff. Winckworth would become the Society's assistant editor in 1937, and the creation of his small team marks the formal beginning of the Society's publishing division. Winckworth was an Oxford-educated schoolteacher whose war service in the Royal Naval Reserve had introduced him to marine zoology. He would remain a keen collector of molluscs and echinoderms, and was active in the

Conchological and Linnean societies. In the Royal Society's library, he became 'masterly' at preparing manuscripts for publication, learned to be a 'painstaking proof reader' and gained 'an expert's knowledge of the niceties of typography'.[45]

In 1932, the unexpected death of Francis Towle created a vacancy, and Winckworth was promoted to assistant secretary (the Society's most senior member of staff). Here, he found himself 'irked by the routine of administration', and soon came to the conclusion that the role's ever-expanding workload – involving the library, the publications and the general administration of the Society – was beyond the capacity of a single person.[46] He had the senior library assistant's role upgraded to 'librarian' in 1935 and, in 1937, he had the publication duties transferred to a new 'assistant editor' role. At that point, he chose to take the editorial role himself, and John Griffith Davies was recruited to the redefined assistant secretary role.

The creation of the assistant editor role meant that, for the first time, the Society's publications had the entire attention and energy of a full-time member of staff. Winckworth also began to create a small editorial team. It originated in 1932, when he became assistant secretary and recruited William Diamond to cover the publications-related work Winckworth had previously been doing. Diamond had served in the Royal Flying Corps, and then studied chemistry, gaining a PhD. He then worked in the imperial civil service before becoming the Society's 'publications clerk'.[47] Diamond would play a key role in the 1936 tender process for the publications, but he subsequently moved on. The role of editorial assistant was retained in the administrative rearrangements, however, and from 1938, Winckworth was assisted by John ('Jock') Courtney Graddon, a BSc graduate from Imperial College who had worked for Cable & Wireless before joining the Society's library in 1932.[48] Neither Winckworth nor his assistants had any (known) prior experience in publishing, though Diamond and Graddon were both university-educated in science. Winckworth's own training presumably came on the job from Towle, and he himself trained Graddon to become his successor. This tradition of internal promotion to the Society's senior editorial post lasted until the 1990s (see Chapter 16). The assistant secretary of the Society still provided some administrative support for refereeing, but the majority of the day-to-day management of the publications (and especially the copy-editing, proofing and printing of papers approved for publication) was now carried out by Winckworth and his team.

Winckworth's job title (after 1937) was 'assistant editor', but there was no 'editor' for him to assist. Throughout this period, ultimate intellectual responsibility for the contents of the *Proceedings* and the *Transactions* continued to lie with Council as the Committee of Papers, but it was delegated on a practical basis to the two secretaries, who each took responsibility for one series of the *Proceedings* and the *Transactions*. They could seek expert input from the chairmen of the discipline-based sectional committees, but this seems to have become more of a choice than the procedural requirement it had seemed in 1896 (see Chapter 11).

In spring 1939, Egerton was learning the responsibilities of his new role as secretary. He may not have had the title of 'editor', but he exercised more executive editorial authority than the distributed structures created in 1896 (see Figure 11.1) suggested.[49] In particular, the sectional committees were peripheral to Egerton's editorial activity. He explained: 'A lot of papers have had to be dealt with; Winckworth brings me them and I have to make up my mind who the right person is to send them to for refereeing.' Another part of his remit was dealing with the reports that 'return from the referees, which often raise conundrums as to what is to be done next'. As an example, he described a case where the negative comments made by Arthur Eddington as referee had been sent to the communicator of the paper, who was 'up in arms at once' about the injustice done to his protégé. Wishing to avoid a 'controversy leading nowhere', Egerton felt that he needed to 'side-track' the paper, which he achieved by gathering additional opinions: from the chair of the sectional committee, and from a referee suggested by him. With no consensus among the referees, in the end the paper was returned to the author with 'comments suggesting amendments or publication elsewhere'. Egerton then added that this was not a typical case; some papers 'of course get passed at once, and I sometimes deal with them on my own'.[50] Egerton's account of his editorial activity suggests a decline in the involvement of the sectional committees and their chairs; and in the late 1960s, the sectional committees formally lost their editorial function (see Chapter 14). Egerton's account also reveals that, despite the increasingly bureaucratised nature of Royal Society editorial practice (with detailed written procedures), the Society's senior officers could still exert substantial individual influence by operating in the spaces outside and between those rules – just as they had done 150 years earlier (see Chapter 6).

The move to Cambridge University Press

A new attention to the physical production of the journals in the 1930s may be due to Henry Lyons (treasurer 1929–39). His election was later described as 'a fresh breeze' that swept through 'the stuffy conservatism' which had become 'the prevailing atmosphere of an ancient Society with a permanent staff devoted to its traditions' (another swipe at the conservatism of Francis Towle). Lyons proved himself willing to spend money improving the 'dingy and depressing' décor, 'archaic' heating and lighting systems and 'mid-Victorian' sanitation of the Society's rooms in Burlington House.[51] And it was in this context that, in 1934, Council appointed a committee 'to enquire into and report on the paper, printing and engraving of the Society's publications'.[52]

Led by the retired zoologist William T. Calman, this ad hoc committee launched an unprecedented investigation into the technical details of the Society's print production processes. Its members interviewed paper merchants, engravers and printers; they sought advice from the Printing Industry Research Association; they discussed the significance of the pH and mineral loading of different papers; they submitted samples of papers to be tested by the printers; and they undertook site visits to grasp the differences between the processes of collotype, photo-litho-offset, deep-etched halftones and photogravure. In July 1934, they submitted a lengthy report to Council, which was adopted in full despite the fact that it implied an increased expenditure of at least £415 a year. The brunt of the report was that different paper was to be used by the printers, and various efforts would be made to improve the quality of images.[53] All in all, the report once more demonstrated the Society's traditional attention to the high physical quality of the publications, rather than any search for cheapness.

Calman's committee focused on paper and illustrations, but its report offers some insights into the Society's deteriorating relationship with its long-standing printers. The committee reported that, 'Messrs Harrisons have frankly admitted that some of the printing has been unsatisfactory', but they hoped that different paper and different ink would enable them to obtain 'better results'.[54] Harrisons offered to reduce their charges, and promised that 'every care will be taken to maintain a high standard of production', but given that Harrisons had just regained the Post Office contract in addition to their work for His Majesty's Stationery Office, it is likely that the Royal Society's work was simply not a priority for the firm.[55] Calman recommended that Harrisons continue as the Society's printers, but insisted that 'a report be made at the end of a year on the quality of their work'.[56]

One legacy of Calman's committee was the 1935 decision to create a standing Publications Committee. This signalled that publications were as important to the Society as its finances and its library. However, editorial policy was notably not part of its remit. The new committee was to meet annually to consider 'any questions relating to the printing, paper, and illustration of the Society's publications, including costs'.[57] Calman became its first chair, and its first major item of business, in December 1935, was to ask William Diamond to organise a tender for the Society's printing services.[58]

For the tender, Diamond prepared a background document which provides the first comprehensive information about the circulation of the Society's periodicals since the nineteenth century (summarised in Table 13.1).[59] It reveals that free distribution remained higher than sales for both the B-side periodicals, but the higher print run and higher sales suggests the emergence of a commercial market for *Proceedings A*. The vast majority of copies of all the journals were now circulating

Table 13.1   Print run, sales and distribution of Royal Society periodicals, c. 1935

	Proc A	*Proc B*	*Trans A*	*Trans B*
Print run				
… in wrappers 'for distribution or sale by parts'	1,325	985	825	675
… in sheets 'for binding into volumes for distribution or sale by volumes'	75	75	75	75
Total print run	1,400	1,060	900	750
Sales				
Sales of parts by subscription	680	380		
… from other sources	20	20	220–240	170
Sales by volume by subscription	c. 25	20		
… from other sources	small	small	50	45
Total sales	725	420	c. 290	215
Non-commercial distribution				
Distribution to FRS	320	270	125	100
Distribution to institutions	200	200–210	120	130
Total non-commercial distribution	520	c. 480	245	230
Remaining on hand	155	160	365	305

as parts or issues, rather than bound annual volumes, implying that receiving publications speedily had become more important than the ease of shipping bound volumes. The relatively large number of copies of the *Transactions* remaining on the warehouse shelves suggests that the Society had not capitalised on the cuts to the non-commercial distribution to institutions in 1932. Fewer copies were being distributed philanthropically, but the Society had not apparently reduced the print run; nor had it found a way to sell those copies to paying customers. Diamond's figures also reveal that barely half of the 459 fellows were claiming one of the *Transactions* series, but more claimed the *Proceedings*; and at least 130 of them must have been claiming both series of the *Proceedings* (despite requests to select one or the other).

Diamond's background document gave no information on how the Society's base retail prices were set, but it did reveal the discounts available for different purchasing options. The most expensive option for acquiring the *Proceedings* was to buy individual issues: this would cost 36s. per volume. A customer willing to wait to purchase the bound volume would pay a shilling less, but customers who subscribed in advance could get the issues for just 30s. It is hardly surprising that the majority of sales of the *Proceedings* were of the discounted advance subscriptions, but it was very problematic for the Society's finances. Diamond's briefing document also failed to mention a downward trend in sales over the previous few volumes.[60]

The list of questions sent to potential printers in early 1936 reveals the particular issues concerning the Publications Committee. Potential printers were asked how they would manage the orders for separate parts and what commission they would take on sales (tasks currently carried out by Harrisons); and also whether and at what cost they would be willing to store the Society's stock of publications or manage the distribution of free copies (tasks done by the Society itself). The potential printers were asked whether there would be any difference in their terms if they were 'entrusted with both printing and publishing', and whether they would be willing to act 'as publishers only'.[61]

By mid-May 1936, Diamond had received responses from eight printers, and the Committee decided to shortlist Harrison & Sons and Cambridge University Press (CUP). During the previous tender process in 1904 (see Chapter 12), Cambridge had lost to Harrisons, despite some admiration for the quality of its scientific typesetting and appreciation of its willingness to publish the later volumes of the (unprofitable) *Catalogue of Scientific Papers*. While Harrison & Sons had developed successful printing operations for labels and postage stamps, the printing house

Figure 13.1 Compositors at the Cambridge University Press Printing House, by J. Palmer Clarke, early twentieth century © Cambridge University Library (UA PRESS/2/4/1).

at Cambridge retained the specialist skills that its scholarly customers were willing to pay for, including hand composition for mathematical typesetting (Figure 13.1).[62] It meant that the contest in 1936 was likely to involve, once more, weighing Cambridge quality against Harrisons' prices.

Both proposals received in early June 1936 were cheaper than the costs from 1934–5 that were being used as a benchmark.[63] However, Calman and the Publications Committee were adamant that they could not commit to a seven-year contract with Harrisons, even though it would be cheaper than what CUP was then offering, and they were no happier with the offer of a five-year contract. However, whether the Society's Council would accept a recommendation for a more expensive option – and one that would necessarily involve some upheaval – was far from certain.

Biological secretary A.V. Hill led the campaign for a move to Cambridge, and actively sought to persuade those 'who need convincing that it is desirable to change'.[64] Those opposed to the move seem to have included 'the office staff', presumably Winckworth and Diamond, who were apparently worried that it might increase their workload,[65] and certain members of Council who, according to Hill, could only see 'the financial side of the proposed change-over', even though the benefits 'cannot be expressed numerically'.[66]

During June 1936, Hill (himself a Cambridge graduate) was in regular, direct contact with the secretary to the Syndics of the Press, Sydney Roberts, and the university printer, Walter Lewis, who is credited with having overseen a 'typographical renaissance' at Cambridge.[67] Hill wrote letters, met them in person, and posed questions by telephone. In other words, Hill devoted a lot of time and energy to helping Roberts and Lewis revise the CUP bid. He explained that persuading Council would be 'perhaps a little easier if we could be sure at least that a change to Cambridge would not cost us more', so that the 'high standard of work' could win the day.[68]

Roberts and Lewis were certainly keen to win the Royal Society contract, and repeatedly revised their proposals. The original proposal would give the Press a 10 per cent commission on sales, and also a 10 per cent fee on the total print costs. In late June, Roberts put forward an alternative proposal that would have involved the Press doing the printing essentially at cost, in return for being allowed a 20 per cent commission on sales.[69] But Hill replied by return of post that this would leave the Society worse off than the original proposal, so Roberts reverted to the original proposal.[70]

Whether by chance or design, the decision on printers was deferred to a meeting in July, giving Hill more time to work out why, despite trying different models, Cambridge's bid kept coming out more expensive than Harrisons. On 30 June, he wrote again to Roberts. The fact that this letter was handwritten on his university notepaper and marked 'personal' (see Figure 13.2) – rather than on his 'secretary of the Royal Society' notepaper – suggests how involved he had become in this process. In this letter, he shared his calculations with Roberts, showing in detail how – for the quantity of different sorts of typesetting involved in a typical annual output of the *Proceedings* or the *Transactions* – CUP worked out £515 more expensive for printing than Harrisons. Hill told Roberts, 'As I said to you before, the Trans. is the deciding factor'.[71]

A revelation came on the morning of 1 July 1936, when a telephone conversation between Hill and the university printer revealed that he had misunderstood how mathematical typesetting was charged. Hill now realised that, whereas Harrisons 'seem to reckon mathematical printing by the inch', Cambridge 'I gather from Lewis, reckon it by the line'. This meant that Hill's calculations for Cambridge now came out £380 less than before. As he commented with relief, 'this is on the right side.... Now

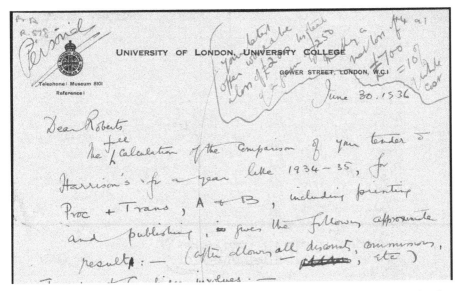

Figure 13.2 Personal letter from A.V. Hill to Sydney Roberts at Cambridge University Press, 30 June 1936 © Cambridge University Library (UA Pr.A.R.578.44).

I think we have come to an end of the comparison of costs and can get down to discussing the other possible advantages'.[72]

Those advantages might well have included the Press's undoubtedly greater experience in distribution, sales and marketing arrangements. Harrisons were primarily printers, though Hill dismissively admitted, 'I think they claim to be booksellers as well as printers and publishers'.[73] Yet the Society's tender document did not ask any specific questions about marketing strategies, subscription management or advertising, and Hill seems not to have tried to make any quantitative comparison. From the Cambridge side, Roberts was certainly interested in selling the journals, since the commission on those sales would be income for the Press. But when he inquired about the potential for expanding the sale, Hill responded, 'I have no idea'.[74] Such a comment from a senior officer of the Royal Society strongly suggests that the Society still did not regard its publications as a (possible) money-making activity.

When the Publications Committee met on 6 July to formulate their final recommendation to Council, Hill was able to present calculations showing Cambridge's printing to be only £160 more expensive than Harrisons' (see Table 13.2).[75] The Committee felt able to describe this as a 'negligible' difference,[76] and noted that more favourable terms for managing the sales would make Cambridge about £100 cheaper overall.

WHY DO WE PUBLISH? 1932–1950   453

Table 13.2 Final estimates of printing charges, presented to Council, 1936

	Harrisons' costs for 1934–35	Harrisons' proposal	CUP's proposal
Transactions	£2,010	£1,860	£2,360
Proceedings	£5,266	£5,060	£4,250
Total	£7,276	£6,920	£6,610
Actual cost of printing once discounts and commissions are included	£7,276	£6,570	£6,730

It assured Council that the potential inconvenience of working with a printer outside London was not 'serious'; that it attached 'very great weight to the experience of the Cambridge University Press in scientific printing'; and was definitely 'unable to recommend the signing of a five years contract with Messrs Harrisons'. Three days later, Council agreed to the move.[77] Hill immediately wrote to Roberts (again, in a personal capacity) to share the good news.[78]

With the decision made, Hill immediately wanted to 'make plans for the transfer, & for getting the most efficient arrangements for the detailed working'.[79] Within a week, the university printer reported that Cambridge physicist Ralph Fowler 'has already been discussing founts and mathematics with our overseer, and is coming again tomorrow'.[80] (Fowler had worked with Hill in operational research during the First World War.) The transfer actually took place in March 1937. It was marked by a meeting between representatives of CUP and the Royal Society to iron out such practical details as the preparation of copy and the cost of reprints, arrangements for insurance and what to do about copies lost in the post. The Society took responsibility for the manufacture of the blocks for illustrations and for standardising and copy-editing the references to cited works; while the Press agreed that its compositors would standardise the use of scientific abbreviations and symbols. A mention of the dispersal of type reveals that the Society's publications were still being printed from loose type, despite the many technical advances in typesetting in the previous half-century.[81] Roberts thanked Hill for his liaison work: 'We are all very grateful for the trouble you have recently taken to facilitate smoother working.'[82]

The staff at the Press paid close attention to arrangements for distribution.[83] They were especially keen to acquire a list of subscribers to the *Proceedings* 'and of purchasers of recent numbers of Transactions', but it proved difficult to extract such a list from Harrisons.[84] A month after the transfer, the Press was 'still waiting', and Roberts urged Hill to 'accelerate its completion', stressing that it was 'a matter of some importance'.[85] By August, however, the Press was able to report that 'the transfer has been effected without any loss to the Society', and, in its first five months, the Press had already marginally beaten the 1935 sales figures for the *Proceedings*. The number of casual sales (that is, of parts or volumes) had decreased slightly, but CUP had already managed to increase the number of subscribers for each series, boding well for future years.[86]

The early days of working with CUP coincided with significant staff changes at the Society, and it is intriguing to wonder whether there was a connection, as queries from the Press may have forced the Society to articulate its habitual processes. The university printer found the Society's lack of a strict publishing timetable frustrating, telling his colleagues that 'some kind of table of dates for receipt of copy, sending out proofs, etc, based on definite publishing days' would be 'a great help'.[87] William Diamond's departure from the Society around this time – and the reorganisation of the senior staff responsibilities – may not have helped.

In summer 1938, just as the relationship with CUP was settling down, the Society was invited to help establish a new publishing company that would have purchased the ownership of the various scientific journals published by Taylor & Francis, and employed Taylor & Francis as printers. This was an attempt to revive the failing fortunes of the publisher of the *Philosophical Magazine*, which was struggling to keep up with technological innovations.[88] However, the Royal Society's Publications Committee found the proposals 'unacceptable', and the president (Bragg) privately remarked that 'the RS does not want to be a publishing business'.[89]

It is difficult to evaluate the impact of the change of printer and publisher on the Society's journals due to the outbreak of the Second World War just two and a half years into the new arrangements. Sales income for both the *Proceedings* and the *Transactions* was slightly up, but the publications account as a whole remained in the red, even allowing for the various sources of external grant income, because the *Year Book*, *Obituary Notices* and *Notes and Records* had virtually no sales; and because the Society had begun to charge the salaries of the editorial staff to the publications account. Printing in Cambridge did, however, turn out to be a fortunate choice when the London print trades took extensive damage during the Blitz. And after the war, it was CUP's attention to international sales and marketing that would help the *Proceedings* (and sometimes the *Transactions*) to break even financially. This would be another key element in transforming the Society's thinking about its journals, from scholarly gifts into products that were bought and sold.

Wartime challenges and post-war needs

The number of papers submitted to the Royal Society's journals fell dramatically during the Second World War, as they had in the previous war. Shortages of printing paper and skilled labour made it difficult to print the journals, and the activities of the Society were disrupted by war service and postal delays. The Society's officers were, again, busy assisting with the national war effort, while its staff tried to keep things going through the Blitz. Concerns about the national security implications of certain areas of scientific work were prominent, and the Society assisted in both censorship and archiving during the war, as well as in scrutinising priority claims afterwards.

Keeping the Society going

The Society had a dozen office staff at the start of the war, over half of whom were female.[90] In autumn 1939, while the Society's historic treasures were sent out of London for safe-keeping, the administrative staff briefly moved to rooms at Hill's *alma mater*, Trinity College, Cambridge; but they returned to London in March 1940. A few months later, the finance team and their 'working files' moved to the home of the Society's new treasurer in Herefordshire, where they would remain until January 1946.[91] Those who remained in London spent the war in the strangely twilight environment of a boarded-up Burlington House. Compulsory fire watching became part of their routine, and Winckworth was noted for his dedication to bomb-watching duties. A building on the Society's land on Basinghall Street was destroyed by fire, but Burlington House was damaged only once, when its windows were shattered in April 1941.[92]

As the male staff (and one of the women) were called to war service, the Society's general office staff became increasingly feminised.[93] Winckworth was too old to serve in this war, but Graddon was called up. A conscientious objector, he initially served as a non-combatant in the Pioneer Corps but transferred to the Royal Army Medical Corps, becoming a Captain. He would be the last of the Society's staff to return, in 1946.[94] Winckworth had technically retired in 1944 but continued to oversee the Society's publications until Graddon was able to take up the assistant editorship.[95]

The Society's ordinary meetings for the reading of papers were suspended during the war.[96] Council was granted emergency powers to hold meetings at any time or place, and to change regulations and normal routines; this was why the 1944 vote on the admission of women could be done by postal ballot.[97] Events involving foreign members became problematic. The 1940 Croonian prize lecturer was to have been the Danish physiologist Schack A.S. Krogh, but he was unable to travel.[98]

The Society's leadership had strong opinions about how science could have been used more effectively in the First World War, and by 1938, Hill and Egerton were already working – with the support of Bragg – to ensure that 'scientific and technical manpower was properly employed in the war that loomed ahead'.[99] Wartime service took officers and fellows away from their usual routines. For instance, in 1940, Hill travelled to Washington as a diplomatic attaché to seek ways for then-neutral United States to provide scientific and technical assistance.[100] Much of that diplomatic work was necessarily secret, but public American aid included a $10,000 gift from the American Philosophical Society to the British scientific community.[101] And in 1940, Julian Huxley liaised with the Rockefeller Foundation about ways to assist the continued publication of scientific research from Europe; the Royal Society subsequently distributed around £3,000 of Rockefeller funding to various British societies and journals.[102] When senior officers were away, other fellows had to be found to cover the duties of absent colleagues, and it fell to the office staff to coordinate everything and keep things going.

Publishing research (or not)

By early 1940, government restrictions on paper threatened to affect scientific publishing, and the Royal Society's Council made plans to reduce the journals to 60 per cent of their previous length.[103] Later that year, they also urged authors to keep their papers brief, and (yet again) limited the provision of offprints.[104] In fact, submissions to the research journals had already fallen dramatically, and it was not because of a lack of paper that the *Proceedings* shrank from its pre-war size of around 4,800 pages a year to under 800 pages a year in 1943, 1944 and 1945, or that only two volumes of the *Transactions* appeared between 1939 and

1945 (see Figures 11.2 and 11.5).[105] Pre-war, Winckworth and his team had usually processed around 100 submissions per year in the biological sciences, and 250 or so in the physical sciences. By 1942 and 1943, that had fallen to fewer than 30 for the biological sciences, and fewer than 50 a year for the physical sciences.[106] It was the Society's other publications that were worst hit by paper restrictions: by 1941, *Abstracts of Papers* had to be discontinued;[107] the *Year Book* was gutted to much slimmer proportions; and no new issues of *Notes and Records* appeared. The move to Cambridge meant that the physical plant involved in printing the journals (see Figure 13.3) was relatively safe from bombing, but many other members of the British book trade had stock, supplies and equipment destroyed.[108] This added to the shortage of manpower and meant that, even in Cambridge, there were 'unusual delays' in publishing due to the 'difficult conditions in the printing industry'.[109]

The Rockefeller funds negotiated by Huxley were to help the many learned societies that had found their income 'inadequate for [the] wartime costs' of publishing.[110] The Royal Society, however, seems to have coped reasonably well, and did not take a share of the Rockefeller money. Imperial Chemical Industries was still making a (reduced) annual donation (£500); the Society continued to take a share of the UK Government's grant-in-aid of scientific publications (usually £700 or £800 a year); and the interest from the Society's own Publications Fund (formerly the Fee Reduction Fund) was around £300 or £400 a year. In ordinary times, these income streams would cover only a fraction of the Society's publication costs, but the reduced scale of wartime publishing meant that the Society actually had sufficient income to support its publications.

During the war, many of the fellows who would usually have acted as referees or committee members were unavailable. In these circumstances, Hill and Egerton as secretaries perhaps relied more on their own judgement, but they still rarely made decisions alone. The chairmen of the sectional committees were valuable advisers, and it is clear that certain authors and communicators were treated as trusted sources. Winckworth (and sometimes Davies) routinely wrote to sectional committee chairs with requests along the lines of: 'Professor Egerton thinks this paper might be accepted without referee. Please let me have your views.' In one such case, the paper had been communicated by Bragg; in another, both authors were said to be 'prominent authorities on the subject'.[111] Egerton clearly appreciated communicators who explicitly confirmed: 'I have been through the paper carefully and consider it suitable for publication.'[112] Egerton himself took on a lot of wartime refereeing, as did Bragg (until

Figure 13.3 Printing machines at the Cambridge University Press Printing House, by D.E. Williams, mid-twentieth century © Cambridge University Library (UA UA Pr.B.42.1.75).

his death in 1942), Charles G. Darwin (who was chair of the Publications Committee) and the London applied mathematician Sydney Chapman.[113]

It was during this period of wartime disruption that we have found the first evidence of women being asked to act as referees. During the pre-war years, the Society had been publishing about a dozen papers a year from women authors and giving grants to women researchers. One of those who benefited was the X-ray crystallographer Kathleen (Yardley) Lonsdale, who had worked in Bragg's laboratory in the mid-1920s. In the early 1930s, Bragg helped her get funding from the Society for equipment and childcare, and he communicated her research to the Society. He reassured her in 1931 that 'your work has been beautifully done, it seems to me and is interesting both as a statement of observations and as a piece of reasoning. Shall I not therefore send it to the Royal Society?'[114] However, since no women had yet been admitted to the fellowship, referees had continued to be all-male. But in August 1939, Agnes Arber refereed a paper on South African fossil plants, and in September 1945, Honor Fell refereed a paper on the effects of irradiating rabbit sperm with X-rays. It was rare but not completely unheard of for non-fellows to be asked to act as referees; they were most commonly men whose election to the Society was already in progress. Both Arber and Fell would ultimately become fellows, but their elections were still distant (1946 and 1952, respectively) at the time they acted as referees.[115]

Arber and Fell were, like Lonsdale, clearly well connected in the Society's networks. They were both Cambridge-based, as were many of the Society's senior leadership. Fell was particularly well known to the Society, not merely as director of the Strangeway Laboratory, but because she was in receipt of long-term salary support (as a Messel and then Foulerton research fellow) from the Society.[116] Indeed, Fell's connections were so good that she was able to act as communicator in all but name. For instance, in 1941, she forwarded a paper by one of her staff to the president Henry Dale, asking him to communicate it for her. Dale made no secret of this when he forwarded it, in turn, to Hill, admitting that he himself was 'not in a position to give any expert endorsement'. Dale had clearly read the paper, for he was frustrated by the *émigré* author's approach to English punctuation ('My pencil began to get active, even on the first page').[117] During the editorial process for this paper, Dale and Fell were both involved in the traditional communicator's role as intermediary between the author and his referee. They even organised a meeting between author and referee to discuss the specimens, and Dale had copies of 'the printed Notes on the Preparation of Papers' sent to Fell, to pass on to the author.[118] Nonetheless, it was Dale's name that appeared on the printed paper as 'communicator'.

The most explicit challenge for the Society's editorial practices arose from worries about the security implications of wartime research, which rapidly undermined Bragg's 1938 vision of scientific research as the shared heritage of all humankind. From 1939, a group of fellows helped the government's Press and Censorship Bureau develop guidelines to deal with research that might have military or security implications.[119] The report forms sent to referees acquired a small red note asking them to comment 'whether or not papers submitted to them contain information which, in their opinion, might be of value to the enemy'. Referees were told that such papers could still be recommended for publication, but would be delayed until 'after the period of hostilities'.[120]

Thus, the Society's archive in Burlington House became potentially relevant to practising scientists once more, as other learned societies (and journals) were felt to lack 'much claim to standing as archivists in such a matter'.[121] In 1937, Bragg had overseen the opening of all the historical sealed notes then held by the Society, including one deposited by Michael Faraday.[122] Unlike those old notes, however, wartime deposits in the archive were not usually sealed, and they were subjected to scrutiny before being accepted. The Society had resolved only to accept papers for deposit (until delayed publication) on the same conditions as for publication; they should be communicated through a fellow, and should be sent to a referee to evaluate 'the merits of the communication'.[123] As with publication procedures, however, there was clearly some flexibility. Several papers on nuclear fission written in 1940 and 1941 by French *émigré* physicists Hans von Halban and Lew Kowarski were deposited in the archives, even though at least one of them was sent in by James Chadwick 'in a sealed envelope'.[124] The Society's record in ensuring these approved-but-archived papers ultimately saw the light of day was mixed. Twenty-five papers were printed during the war for restricted circulation as special Series C and D of the *Transactions* or the *Proceedings*, and all of these were ultimately (though at different times) reprinted for publication after the war.[125] On the other hand, the papers by von Halban and Kowarski were only rediscovered in 2007.[126]

By 1943, the Society started planning for the post-war publication of research, including the adjudication of priority claims for secret war work. It informed the Scientific Advisory Committee to the War Cabinet that the Society's referees would undertake to 'inspect, if necessary, the full text of the original report … on which any claim for priority is based', provided it had been 'properly dated' and 'certified by some person acceptable to the Society'. It offered this service for papers being considered in 'any recognized scientific society or journal'.[127] This commitment was made public in July 1945, and the Society subsequently liaised with a variety of government and military research units, as well as national and international learned societies.[128]

Post-war needs and scientific information

The end of the war proved, once again, to be a time of transition. The admission of Kathleen Lonsdale and Marjory Stephenson to the fellowship in 1945 created the possibility of women as communicators, referees and committee members.[129] The Council elections that year, however, proved almost as contentious as those of 1935, if not as public. The election of the organic chemist Robert Robinson as president was a disappointment to those fellows who had hoped that the Society's wartime work would presage closer links with government in the years to come.[130] Hill finished his secretarial term in 1945, but moved on, as Arthur Schuster had done at the end of the previous war, to the role of foreign secretary. He then found himself acting as physical secretary for six months (despite being a physiologist), when Egerton broke his leg in a skiing accident not long before his own term ended in 1948.[131]

The Society's staff also changed as some returned from war service and others retired. In the publications team, one of Graddon's first acts as the new assistant editor was to hire an editorial assistant. As a physics graduate, William Geraint Evans was well placed to support the A-side journals, and he would work on the Society's publications until 1982.[132] Jean Lamb provided secretarial support successively to Winckworth, Graddon and Evans.[133] This post-war staff team provided stability to the Society's editorial processes for decades. In the wider Society, another big change arose in 1947 when John Griffith Davies was replaced as assistant secretary by David Christie Martin, a young Scottish PhD chemist who had spent the war working for the Ministry of Supply; he would stay at the Society until the 1970s.[134]

Amid the discussions of *The Needs of Research in Fundamental Science after the War*,[135] old questions about the role of the Society's publications, particularly in physics, resurfaced. Egerton took the lead in these discussions. As well as the familiar issues about the prestige of the *Proceedings* and how this was affected by the number of junior scholars publishing in it, there was some consideration of the oft-ignored *Transactions*. Some argued that it 'hardly now occupies a position worthy of its traditions. It would be very desirable that something be done to remedy this'. In 1945, however, Egerton insisted (with little evidence) that the *Transactions* remained 'a most valuable publication for long papers and memoirs of high standard', and he hinted again at its potential value for the publication of 'research work in borderline subjects'.[136] All in all, Egerton urged caution, arguing that it would be 'a mistake to make changes in a hurry now, when there are so many other urgent matters in hand', one of which was financial provisions for the Society in a post-war world.[137]

During 1945, the 'whole question' of government funding for the Society was reviewed. The key change was that the various grants-in-aid awarded to the Society, including that for publications, would be negotiated annually, rather than rolling forward indefinitely until the Society pushed for an increase. The grant for publications had been £2,500 since 1925; it now rose to £10,000.[138] This helped the Royal Society (which took almost £1,400 from the government grant in 1946) as well as the other societies.[139] In 1947, Egerton asked the Treasury to further increase the publications funding to £15,000, explaining that, with the 'volume of scientific literature' returning to normal after the war, 'all the publishing societies are finding themselves in difficulties owing to the rise in costs', which he estimated at around 70–80 per cent. Without government help, he argued that the societies would be 'in great difficulties' and unable to maintain the valuable service they provide to the scientific community.[140] This emphasis on 'service' as the motivation for society publishing activity reiterated the vision set out by Bragg in 1938.

Egerton's 1945 plea to avoid making drastic changes in a rush had pointed out that there would soon be an opportunity 'for wide discussion of many matters', including publications.[141] This was the planned Empire Scientific Conference 1946. Egerton and Hill had been thinking about how to improve scientific collaboration among Commonwealth countries throughout the war. This conference, and the establishment of a British Commonwealth Scientific Office, was the result.[142] In later decades, the Society's international diplomacy would be played out in the context of the Cold War, but at this stage, the Society's focus was on building enduring links between researchers in the wider British world, despite – or because of – the disintegration of the imperial mandate.[143] One outcome of the 1946 conference was a follow-up Scientific Information Conference in 1948.

'Scientific information' was the umbrella term used in the mid-twentieth century for all manner of questions to do with publishing, archiving, searching and accessing the 'mass of periodical scientific information' which was 'so great that no single worker can keep touch with all the literature of even one branch of research'.[144] It had proven impossible to continue the *Catalogue of Scientific Papers* project, and a variety of other 'access fantasies' had proven equally impractical.[145] Without any prospect of a universal solution for searching and accessing all scientific knowledge, the Royal Society had focused on smaller, pragmatic steps. In the 1930s, it had been involved in surveying the holdings and use of the various university and learned society libraries in London, attempting to reduce duplication between libraries; and in discussions about a possible central science library.[146] In 1936, it organised a meeting of scientific societies to discuss the desirability of 'standard abbreviations' and 'a standard system of reference' for citations. It standardised the bibliographic referencing styles used in its own publications.[147] William Bragg's address on publishing in 1938 had been inspired by his participation in a 'Documentation Conference' at Oxford.

The Scientific Information Conference was held at the Royal Society in June 1948, organised by Egerton and the new assistant secretary, David Martin.[148] Almost 150 delegates gathered to discuss their organisational or national needs for scientific information. They represented learned societies, professional library associations and a variety of governmental bodies; there were representatives from Australia, Canada, Ceylon, Eire, India, New Zealand, Pakistan, South Africa, the Caribbean Commission and (by special invitation) the USA. They were joined by a further 250 observers, many representing British scientific societies.[149] The four main topics of discussion were: the publication and distribution of papers reporting original work; abstracting services; indexing and other library services; and reviews and annual reports. The conference spawned a series of subsequent meetings on the topic, including one in Washington in 1958, and it led to the creation of a standing committee on 'scientific information' at the Royal Society.

At the 1948 conference, the Royal Society presented an upbeat account of its own publications. Yet the minutes of the Publications Committee's meeting in March 1948 reveal that the Society's journals were actually facing a familiar litany of problems, particularly printing delays and high costs. The staff developed new paper technologies for tracking progress, and noting 'unusual delays'. As Figure 13.4 shows, 'time taken from receipt to publication' had joined 'number of papers published' as a performance indicator for *Proceedings A*. Authors were blamed for causing delays, with the staff feeling that too few scientists could prepare a manuscript properly for the press. The Society's editorial team produced a leaflet *Notes on the Preparation of Papers* offering guidance to authors on abbreviations and references, and warning that they alone were responsible for 'grammatical correctness'. Even so, too many papers required heavy copy-editing from Evans, and some had to be returned for retyping when they were 'presented in a state which would make it difficult for the compositor to set'.[150]

CUP insisted that high charges were 'necessary' to maintain the 'high standard' of printing. The prices of the Society's periodicals were (again) raised by 50 per cent, resulting in an almost doubling of sales income between 1947 and 1948.[151] But the following year, sales income grew even further, probably as a result of the opening of the Press's branch in

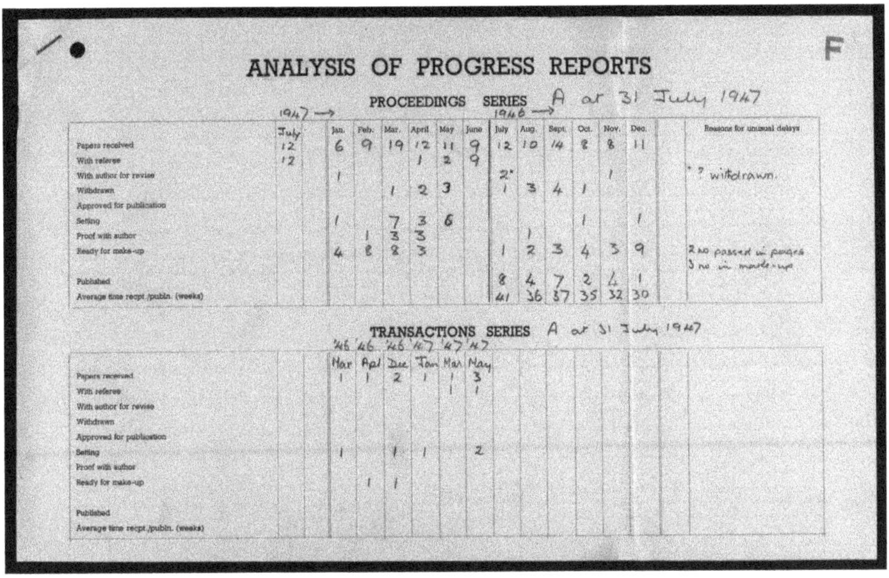

F

ANALYSIS OF PROGRESS REPORTS

PROCEEDINGS SERIES A at 31 July 1947

1947 → 1946 →

	July		Jan.	Feb.	Mar.	April	May	June	July	Aug.	Sept.	Oct.	Nov.	Dec.		Reasons for unusual delays
Papers received	12		6	9	19	12	11	9	12	10	14	8	8	11		
With referee	12					1	2	9								
With author for revise			1						2*				1			* ? withdrawn.
Withdrawn					1	2	3		1	3	4	1				
Approved for publication																
Setting			1		7	3	6					1		1		
Proof with author				1	3	3				1						
Ready for make-up			4	8	8	3			1	2	3	4	3	9		2 no passed in pages 3 no in make-up
Published									8	4	7	2	4	1		
Average time recpt./publn. (weeks)									41	36	37	35	32	30		

TRANSACTIONS SERIES A at 31 July 1947

			'46 Mar	'46 Apl	'46 Dec	'47 Jan	'47 Mar	'47 May								
Papers received			1	1	2	1	1	3								
With referee							1	1								
With author for revise																
Withdrawn																
Approved for publication																
Setting			1		1	1		2								
Proof with author																
Ready for make-up				1	1											
Published																
Average time recpt./publn. (weeks)																

Figure 13.4 Analysis of editorial progress, prepared for the Royal Society's Council, in August 1947. On the A side, the *Proceedings* was receiving more submissions than the *Transactions* (but the opposite was true for the B side in 1947) © The Royal Society.

New York.[152] In 1949, the Society's treasurer was able to report that the publications account was in surplus by almost £4,000, something that had been unheard of before the war. This reported surplus was artificially inflated by the receipt of almost £3,500 from the government grant and the restoration of the ICI grant to its pre-war level of £1,000, yet it was indeed the case that both the *Proceedings* and the *Transactions* had not merely broken even themselves, but generated sufficient surplus to cover the costs of the internal publications, office expenses and salaries.[153] Over the next few years, the overall account was more usually in deficit, but the *Proceedings* itself managed (usually) to stay in surplus. This was a huge contrast to the financial situation facing the *Proceedings* in the 1920s and early 1930s.

Through the tumultuous years of the 1930s and 1940s, the Royal Society's editorial processes remained formally the same as they had been since the start of the century. Council, as the Committee of Papers, was only occasionally involved in the routine affairs of the journals, called upon to resolve unusually difficult or awkward cases. The sectional committees, and their chairs, still assisted with decisions, but there are hints that the secretaries, especially during the war, were more often taking executive editorial actions; when they consulted the sectional committee chairs, it appears to have been a choice, not a requirement. The role of communicators and referees did not change. The questions raised by Louis Filon and others, about the purpose, mission and prestige of the Society's journals – especially the *Transactions* – were not really addressed. Despite the officers' awareness that the demographic of contributors to the Society's journals was changing, they remained publicly confident that the Society's editorial system allowed it to identify high-quality research.

In 1936, after decades of partial dissatisfaction, the Society finally decided to move its printing away from Harrison & Sons. It is easy to see that there were shared scholarly values between the Society and a university press, but it is equally important to note that appointing CUP as publisher as well as printer might finally enable the Society to see the benefits of having an agent actively promoting its journals, as Archibald Geikie had wanted back in 1905 (see Chapter 12).

The war, however, had disrupted all immediate plans. The Society emerged from the war years with a small permanent staff team to support its publications, and with an apparently strong relationship with CUP, whose compositors and printers could deliver the high-quality product the Society wanted. Both the Society and the Press, however, would have to adjust to a scientific and academic world in which North American institutions were more important than ever before, and in which the output of scientific research grew even more rapidly than it had done in the 1920s and early 1930s.

In 1947, the Royal Society analysed the free and paid-for circulation of its periodicals. The general trends from the 1935 analysis (see Table 13.1) remained true: the physical sciences were still booming, and for both periodicals, series A had a higher circulation than series B. However, although *Proceedings A* still had more subscribers than its sibling, the gap was closing. The Press had grown the sales of *Proceedings B* (almost 560 subscribers, up 32 per cent) more than *Proceedings A* (just over 780 subscribers, up 8 per cent).[154] Two years later, the Press reported on its progress in the emerging North American market; they had only managed to find 30 or so subscribers to the *Transactions*, but the new office already had 193 subscribers to *Proceedings A* and 134 to *Proceedings B*.[155] One of the advantages of the United States as a market for the Royal Society was that few of its universities and research institutions were accustomed to receiving free copies of the Society's publications, and they could be treated as customers from the start.

Sales and sales income were undoubtedly becoming more important to the *Proceedings*, but the non-commercial circulation had not disappeared. For *Proceedings B*, as well as *Proceedings A*, the free circulation to fellows and institutions was now numerically less than the paid-for circulation, but the majority of copies of the *Transactions* still circulated outside commercial channels. In fact, despite the earlier cuts to institutional gifts, the free circulation of the *Transactions* had actually increased, because more fellows were claiming their members' copies. Nonetheless, the efforts of the CUP's marketing team were opening the eyes of the Society's staff and officers to the potential opportunities for expanding sales. By the early 1950s, it would begin to seem possible that the days of the *Proceedings* and the *Transactions* being a drain on the Society's finances might be over.

Notes

1 [Bragg] 'Address of the president, 1938', *ProcA* 169 (1938): 13–14.
2 Jeff Hughes, '"Divine right" or democracy?' The Royal Society "revolt" of 1935', *N&R* 64, Suppl. 1 (2010): S101–S117. For the social and political context of the Royal Society 1939–49, see Jennifer Rose Goodare, 'Representing science in a divided world: The Royal Society and Cold War Britain' (PhD, University of Manchester, 2013), ch. 1. More generally, see Gary Werskey, *The Visible College: A collective biography of British scientists and socialists of the 1930s* (London: Allen Lane, 1978).
3 [Bragg] 'Address of the president, 1938', *ProcA* 169 (1938): 15.
4 [Bragg] 'Address of the president, 1938', *ProcA* 169 (1938): 14–16. On Bragg, see E.N. da Costa Andrade and Kathleen Londsdale, 'William Henry Bragg, 1862–1942', *Biographical Memoirs* 4, no. 12 (1943): 276–300.
5 John D. Griffith Davies, 'Ronald Winckworth 1884–1950', *N&R* 8.2 (1951): 293–6.
6 Bernard Katz, 'Archibald Vivian Hill, 26 September 1886–3 June 1977', *Biographical Memoirs* 24 (1978): 71–149; Archibald Vivian Hill, 'Memories and reflections' (Churchill College, Cambridge, 1974), especially ch. 1 section 12 ('Jack Egerton as Secretary of the RS').
7 Diary of A.C. Egerton, 1938–41, RS AE/2/3, f. 1 (re 15–16 March 1939).
8 On the British academic world, see Tamson Pietsch, *Empire of Scholars: Universities, networks and the British academic world, 1850–1939* (Manchester: Manchester University Press, 2013). On career progression, see Tamson Pietsch, 'Geographies of selection: Academic appointments in the British academic world, 1850–1939', in *Mobilities of Knowledge*, ed. Heike Jöns, Peter Meusburger, and Michael Heffernan (Cham: Springer International Publishing, 2017), 157–83; and on publications as a form of evaluation, see Alex Csiszar, 'How lives became lists and scientific papers became data: Cataloguing authorship during the nineteenth century', *British Journal for the History of Science* 50 (2017): 23–60; and *The Scientific Journal: Authorship and the politics of knowledge in the nineteenth century* (Chicago, IL: University of Chicago Press, 2018), chs 4–5. On research within academic careers, see also Bruce Macfarlane, 'The spirit of research', *Oxford Review of Education* 47, no. 6 (2021): 737–51.
9 'Memorandum by L.N.G. Filon', in RS CMP/14, 9 July 1936. On Filon, see G.B. Jeffery, 'Louis Napoleon George Filon, 1875–1937', *Obituary Notices of Fellows of the Royal Society* 2, no. 7 (1939): 501–9.
10 'Memorandum by L.N.G. Filon', in RS CMP/14, 9 July 1936.

11 Ku-ming Kevin Chang and Alan Rocke, *A Global History of Research Education: Disciplines, institutions, and nations, 1840–1950*, History of Universities Series (Oxford: Oxford University Press, 2021).
12 Karl Pearson and Louis Napoleon George Filon, 'Mathematical contributions to the theory of evolution. IV', *Proc* 62 (1898): 173–6. Filon had a further three papers in the *Transactions* by 1903.
13 Aileen Fyfe, Flaminio Squazzoni, Didier Torny and Pierpaolo Dondio, 'Managing the growth of peer review at the Royal Society journals, 1865–1965', *Science, Technology and Human Values* 45, no. 3 (2020): 405–29.
14 'Memorandum by L.N.G. Filon', in RS CMP/14, 9 July 1936.
15 All quotations in the next two paragraphs are from Hill's retrospective (c. 1971–4) account of the affair, which appears in his unpublished memoir in section 87, 'Spooks, frauds, ectoplasm and self-delusion', in Hill, 'Memories and reflections'.
16 The paper was also examined by three referees, as recorded in the 'Register of Papers', RS MS/588, ff. 28–9.
17 'Memorandum by L.N.G. Filon', in RS CMP/14, 9 July 1936.
18 In 1935, 94 per cent of submissions were published (some of the others may have been voluntarily withdrawn, rather than rejected). For the prevalence of refereeing, see Fyfe et al., 'Managing the growth of peer review', Figure 3.
19 'Memorandum by L.N.G. Filon', in RS CMP/14, 9 July 1936.
20 'Notes on the Preparation of Papers to the Royal Society' appears to have been first issued around this time. In 1938, it was noted as being revised (RS CMP/15, 7 July 1938) and it would be revised repeatedly after the war.
21 Revised standing orders, RS CMP/15, 4 November 1937.
22 RS *Year Book* (1940), 214; RS CMP/15, 30 March 1939.
23 Revised standing orders, RS CMP/15, 4 November 1937.
24 Hughes, '"Divine right" or democracy?'.
25 S. Bhagavantam, 'Chandrasekhara Venkata Raman, 1888–1970', *Biographical Memoirs* 17 (1971): 564–92.
26 RS CMP/17, 30 November 1946. Raman's paper was received on 7 August 1945. It was refereed by Owen Richardson, W. Wilson, Charles G. Darwin, A.H. Wilson, W.T. Astbury and E.N. da Costa Andrade. The final decision is noted in the 'Register of Papers' as 'withdrawal requested', RS MS/611, f. 42. It was not refereed by Kathleen Lonsdale or Max Born, with whom he was then engaged in controversy, see Rajinder Singh, 'Sir C.V. Raman, Dame Kathleen Lonsdale and their scientific controversy due to the diffuse spots in X-rays photographs', *Indian Journal of History of Science* 37, no. 3 (2002): 267–90.
27 P.M.S. Blackett, 'Note on request to Council' (c. 1945), in RS AE/1/5/1. Blackett did not question the quality of the *Proceedings*, but he queried whether its undoubted prestige was skewing the field in an unhealthy manner.
28 A.C. Egerton, 'Note on request to Council' (c. 1945), in RS AE/1/5/1. His figures showed that 24 per cent of A-side authors in 1938, and 35 per cent in 1941–5, had been fellows. Given the short window of opportunity, his estimate of 50 per cent fellows or fellows-to-be seems generous.
29 A.C. Egerton, 'Note on request to Council' (c. 1945), in RS AE/1/5/1.
30 [Bragg], 'Address of the president, 1938', *ProcA* 169 (1938): 13. See also Noah Moxham and Aileen Fyfe, 'Royal Society and the prehistory of peer review, 1665–1965', *Historical Journal* 61, no. 4 (2018): 863–89.
31 Dale to Morris Travers, 24 July 1945, RS HD/6/1/3/143.
32 RS CMP/14, 18 June 1936. See also [W. Bragg], 'Address of the president, 1938', *ProcA* 169 (1938).
33 They were discontinued in 1941, see RS *Year Book* (1942), 90.
34 Allan Jones, 'J.G. Crowther's war: Institutional strife at the BBC and British Council', *British Journal for the History of Science* 49, no. 2 (2016): 259–78; Hughes, '"Divine right" or democracy?'.
35 Werskey, *The Visible College*, 263–4; Jon Agar, *Science in the Twentieth Century and Beyond* (Cambridge: Polity Press, 2013), 273.
36 On *Nature*'s public condemnations of the Nazi regime, see Melinda Baldwin, *Making 'Nature': The history of a scientific journal* (Chicago, IL: University of Chicago Press, 2015), 139; and see, for instance, A.V. Hill, 'International status and obligation of science

[Thomas Huxley memorial lecture]', *Nature* 132, no. 3347 (1 December 1933): 952–4. On the AAC, see Jack A. Rall, 'Nobel Laureate A. V. Hill and the refugee scholars, 1933–1945', *Advances in Physiology Education* 41, no. 2 (2017): 248–59.
37 On the circulation of journals during the war, see Pamela Spence Richards, '"Aryan librarianship": Academic and research libraries under Hitler', *Journal of Library History* 19 (1984): 231–58; 'The movement of scientific knowledge from and to Germany under National Socialism', *Minerva* 28, no. 4 (1990): 401–25; and Michael Knoche, 'Scientific journals under National Socialism', *Libraries & Culture* 26 (1991): 415–26.
38 On Schonfeld, see David H. Kranzler, 'Schonfeld, Solomon', in *ODNB*, and David Kranzler, *Holocaust Hero: Solomon Schonfeld* (Jersey City, NJ: Ktav Publishing House, 2004).
39 Hill to Schonfeld, 19 October 1944, RS HD/6/1/2/126. On the RS officers and scientific neutrality (including this 'advertisement') in the 1940s, see Goodare, 'Representing science', 74–6.
40 Hill to Schonfeld, 19 October 1944, RS HD/6/1/2/126.
41 Hill to Schonfeld, 19 October 1944, RS HD/6/1/2/126.
42 RS *Year Book* (1938), 211.
43 G.H. Hardy, 'Letter to the editor', *N&R* 3.1 (1940): 96.
44 Marie Boas Hall and Patricia H. Clarke, *The Library and Archives of the Royal Society: 1600–1990* (London: The Royal Society, 1992), 47–9.
45 Davies, 'Ronald Winckworth', 295.
46 Davies, 'Ronald Winckworth', 295.
47 For assistance tracking down William E. de B. Diamond (1898–1993), we thank Morag and Alastair Fyfe. Diamond went on to have a career as an administrator for professional technical organisations. It remains unknown when or where he gained his MA and PhD degrees.
48 On Graddon (1907–82), see David Martin, 'The retirement of Mr J. C. Graddon, assistant editor to the Society', *N&R* 27.2 (1973): 325–6; and 'Comrades of old: Royal Society staff pensioners, biographical sketches' (privately circulated for The Royal Society Former Staff Association, 2015), 1–2. Peter Cooper and the Former Staff Association of the Royal Society have been enormously helpful in helping us find out about former members of the publications staff. They kindly shared with us their newsletter, as well as this collection of biographical sketches.
49 On 'distributed editorship', see Aileen Fyfe, 'Editors, referees and committees: Distributing editorial work at the Royal Society journals in the late nineteenth and twentieth centuries', *Centaurus* 62, no. 1 (2020): 125–40.
50 Diary of A.C. Egerton, RS AE/2/3, late April 1939 (between passages dated 17 April and 4 May). The paper had been communicated by Edward A. Milne, whose cosmological work was critical of Eddington.
51 Henry Hallett Dale, 'Henry George Lyons, 1864–1944', *Obituary Notices of Fellows of the Royal Society* 4, no. 13 (1944): 795–809, 804. For Lyons' own account of his treasurership, see Henry Lyons, 'On the finances of the Royal Society for the ten years 30th November 1929 to 30th November 1939', in RS *Year Book* (1940), 222–7.
52 RS CMP/14, 1 March 1934.
53 Report of Publications Committee, in RS CMP/14, 5 July 1934.
54 Report of Publications Committee, in RS CMP/14, 5 July 1934.
55 For the history of Harrison & Sons at this point, see section 2 of the document prepared for the firm's listing as a public company, c. 1947, item 1272/12 in Harrison archive, City of Westminster Archives Centre.
56 Report of Publications Committee, in RS CMP/14, 5 July 1934.
57 RS CMP/14, 9 May 1935.
58 RS CMB/329, 17 December 1935.
59 A copy of 'Sales and Distribution of Royal Society Periodicals' survives in CUP PrA/578/42.
60 Lyons (treasurer) later sent the figures for the last three volumes to CUP, who compared their own sales. See Kingsford (CUP) to Lyons, 19 August 1937, CUP PrC/R229.
61 A copy of the questions survives in CUP PrA/R578/42.
62 For CUP and Roberts, see David McKitterick, *A History of Cambridge University Press: Volume 3, new worlds for learning, 1873–1972* (Cambridge: Cambridge University Press, 2004), ch. 12.
63 RS CMB/329, 3 June 1936.

64 Hill to Cameron (Master of Gonville & Caius, syndic), 17 June 1936, CUP PrA/R578/39.
65 Hill to Roberts, 9 July 1936, CUP PrA/R578/46.
66 Hill to Roberts, 25 June 1936, CUP PrA/R578/41.
67 McKitterick, *History of Cambridge University Press 3*, ch. 11.
68 Hill to Cameron, 17 June 1936, CUP PrA/R578/39.
69 [Roberts?] to Hill, 23 June 1936, CUP PrC/R229.
70 Hill to Roberts, 25 June 1936, CUP PrA/R578/41; Hill to Roberts, 1 July 1936, CUP PrA/R578/45.
71 Hill to Roberts, 30 June 1936, CUP PrA/R578/44.
72 Hill to Roberts, 1 July 1936, CUP PrA/R578/45.
73 Hill to Roberts, 19 June 1936, CUP PrA/R578/43.
74 Hill to Roberts, 29 June 1936, CUP PrA/R578/43.
75 These are the firms' final offers, as reported by the Publications Committee to Council, RS CMB/329, 6 July 1936.
76 RS CMB/329, 6 July 1936.
77 RS CMB/329, 6 July 1936 and CMP/14, 9 July 1936.
78 Hill to Roberts, 9 July 1936, CUP PrA/R578/46.
79 Hill to Roberts, 9 July 1936, CUP PrA/R578/46.
80 [Lewis?] University printer to assistant secretary of the RS, 16 July 1936, CUP PrC/R229. The typefaces were agreed RS CMP/14, 5 November 1936; see also [Roberts?] to Kingsford, 2 November 1936, CUP PrC/R229.
81 Hill to CUP, 12 March 1937, CUP PrA/R578/54.
82 CUP [Roberts?] to Hill, 7 April 1937, in CUP PrC/R229.
83 On the cost of postage for the *Proceedings*: CUP to Lyons, 23 February 1937, CUP PrC/R229; and Hill to CUP, 12 March 1937, CUP PrA/R578/54. On the cost of packing: Kingsford to Roberts, 15 March 1937, CUP PrA/R578/55.
84 Hill to CUP, 12 March 1937, CUP PrA/R578/54. The Press had started asking about this list as early as November, see Unknown to JL Kingsford, 2 November 1936, CUP PrC/R229.
85 CUP [Roberts?] to Hill, 12 April 1937, CUP PrC/R229.
86 Kingsford to Lyons, 19 August 1937, CUP PrC/R229.
87 'Memo by W.L.' [Walter Lewis?], [April 1937], CUP PrA/R578/57.
88 William H. Brock and A.J. Meadows, *The Lamp of Learning: Taylor & Francis and the development of science publishing* (London: Taylor & Francis, revised edn 1998), 171 ff.; Imogen Clarke, 'Negotiating progress: Promoting "modern" physics in Britain, 1900–1940' (PhD, University of Manchester, 2012), ch. 5.
89 RS CMP/15, 7 July 1938; William Bragg to Frank Smith, 11 May 1938, RS MDA/B/3/1. See also Brock and Meadows, *Lamp of Learning*, 174.
90 The office staff were listed near the start of each *Year Book*. In 1939, there were five men and seven women.
91 RS *Year Book* (1940), 220–1; RS *Year Book* (1941), 93; RS *Year Book* (1945), 128–9.
92 RS *Year Book* (1942), 94; Davies, 'Ronald Winckworth'; J.S. Rowlinson and Norman H. Robinson, *The Record of the Royal Society: Supplement to the fourth edition, for the years 1940–1989* (London: The Royal Society, 1992), 4.
93 The woman called to war service was Violet Carruthers (née Markham), who had a long career of public service, see Helen Jones, *Duty and Citizenship: The correspondence and papers of Violet Markham, 1896–1953* (London: Historian's Press, 1994).
94 'Comrades of old', 1.
95 Davies, 'Ronald Winckworth', 295.
96 [Bragg], 'Address of the president ,1939', *ProcA* 173 (1939): 298.
97 RS *Year Book* (1942), 88–9.
98 RS *Year Book* (1941), 93.
99 Hill, 'Memories and reflections', in ch. 1 section 12 ('Jack Egerton as Secretary of the RS'). See also William McGucken, 'The Royal Society and the genesis of the Scientific Advisory Committee to Britain's War Cabinet, 1939–1940', *N&R* 33.1 (1978): 87–115; David Edgerton, *Warfare State: Britain, 1920–1970* (Cambridge: Cambridge University Press, 2005), ch. 4.
100 Hill, 'Memories and reflections', in ch. 1 section 12. For a similar trip by Darwin, see RS *Year Book* (1941), 86.
101 Reported in *N&R* 3 (1940): 105–6.

102 Early negotiations can be seen in a letter from Julian Huxley to Egerton, 1 March 1940, RS AE/1/11/3. Distribution of the Rockefeller grant was reported in RS *Year Book* (1943), 90.
103 RS CMP/15, 11 January 1940.
104 RS *Year Book* (1941), 90 [the *Year Book* for 1941 was published in late 1940].
105 Aileen Fyfe, 'The Proceedings in the 20th century' (8 February 2022), https://arts.st-andrews.ac.uk/philosophicaltransactions/proceedings-in-the-20thc.
106 Aileen Fyfe, 'Submissions in life sciences vs physical sciences, 1927–1989' (13 February 2018), https://arts.st-andrews.ac.uk/philosophicaltransactions/submissions-in-life-sciences-vs-physical-sciences-1927-1989.
107 RS *Year Book* (1942), 90.
108 Andrew Nash, Claire Squires, and Ian Willison, eds, *The Cambridge History of the Book in Britain, Volume 7: The twentieth century and beyond* (Cambridge: Cambridge University Press, 2019), ch. 23; Valerie Holman, *Print for Victory: Book publishing in Britain 1939–1945* (London: British Library, 2008).
109 RS *Year Book* (1944), 91.
110 Julian Huxley to Egerton, 1 March 1940, RS AE/1/11/3.
111 Memoranda, on a paper 'The torsional flexibility of aliphatic chain molecules' by Alex Muller, 1 November 1939, RS RR/66/202; and Memoranda, on a paper 'Statistical thermodynamics of super-lattices' by Ralph Howard Fowler and Edward Armand Guggenheim, 1 November 1939, RS RR/66/82.
112 Memoranda, on a paper 'The neutrons from the disintegration of fluorine by deuterons' by T.W. Bonner, 1 November 1939, RS RR/66/29. The conscientious communicator was J.D. Cockroft.
113 Our analysis of the *Virtual Register of Papers*.
114 Bragg to Lonsdale, quoted in Jennifer Wilson, 'Dame Kathleen Lonsdale (1903–1971): Her early career in X-ray crystallography', *Interdisciplinary Science Reviews* 40, no. 3 (2015): 265–78, at 273.
115 Joan Mason, 'The admission of the first women to the Royal Society of London', *N&R* 46.2 (1992): 279–300 and 'The women fellows' jubilee', *N&R* 49.1 (1995): 125–40; Georgina Ferry, 'The exception and the rule: Women and the Royal Society 1945–2010', *N&R* 64, suppl. 1 (2010): S163–S172.
116 Audrey Glauert, 'Fell, Dame Honor Bridget (1900–1986), cell biologist', in *ODNB*.
117 Dale to Hill, 24 June 1941, RS HD/6/1/3/22. The author was the German bacteriologist Carl Franz Robinow.
118 Dale to Hill, 7 August 1941, RS HD/6/1/3/27.
119 RS CMP/15, 11 January 1940.
120 See Referee Reports 1939–1940 in RS RR/1939, RR/1940.
121 F.G. Young to Dale, 28 January 1944, RS HD/6/1/2/47.
122 [Bragg], 'Address of the president, 1937', *ProcB* 124 (1937): 385. See also Alex Csiszar, *The Scientific Journal: Authorship and the politics of knowledge in the nineteenth century* (Chicago, IL: University of Chicago Press, 2018), 167.
123 RS CMP/15, 14 December 1939.
124 The papers are in RS Manuscripts General, MS/850, along with correspondence with their communicators Blackett, Pye and Chadwick.
125 Rowlinson and Robinson, *Record of the Royal Society* (1992), 105.
126 On the rediscovery, 'Nuclear reactor secrets revealed', BBC News blog (1 June 2007): http://news.bbc.co.uk/1/hi/sci/tech/6709855.stm.
127 RS CMP/16, 30 November 1943.
128 RS CMP/17, 12 July 1946, appendix D; 17 January 1946, appendix A.
129 Camilla M. Røstvik and Aileen Fyfe, 'Ladies, gentlemen, and scientific publication at the Royal Society, 1945–1990', *Open Library of Humanities* 4, no. 1 (2018): 1–40, http://doi.org/10.16995/olh.265
130 Peter Collins, *The Royal Society and the Promotion of Science since 1960* (Cambridge: Cambridge University Press, 2015), ch. 1.
131 Hill, 'Memories and reflections'.
132 'Comrades of old', 20–1.
133 Mentioned in passing in 'Comrades of old', 1 and 3.
134 Davies re-married and decided to change career, see Archibald Vivian Hill, 'J. D. Griffith Davies 1899–1953 (Assistant Secretary of the Royal Society, 1937–1946)', *N&R* 11.2 (1955):

129–33. On Martin, see Harrie Massey and Harold Thompson, 'David Christie Martin, 1914–1976', *Biographical Memoirs* 24 (1978): 390–407.
135 Egerton, 'The needs of research in fundamental science after the war', printed by the Royal Society (January 1945); Collins, *The Royal Society and the Promotion of Science*, 2–3.
136 Egerton, 'Note on request to Council' (c. 1945), RS AE/1/5/1.
137 Egerton, 'Note on request to Council' (c. 1945), RS AE/1/5/1.
138 RS *Year Book* (1946), 127.
139 Annual accounts for 1946, in RS *Year Book* (1947), 144.
140 Egerton to Symons (Government Treasurer), [undated draft 1947], in RS AE/1/3/3.
141 Egerton, 'Note on request to Council …' (c. 1945), RS AE/1/5/1.
142 Hill, 'Memories and reflections'.
143 Collins, *The Royal Society and the Promotion of Science*, ch. 7. See Pietsch, *Empire of Scholars*.
144 'Statement on existing Collation and Abstraction of Current Scientific and Technical Literature', in RS CMP/14, 15 June 1933.
145 Csiszar, *The Scientific Journal*, ch. 6; Robert Fox, *Science Without Frontiers: Cosmopolitanism and national interests in the world of learning, 1870–1940* (Corvallis, OR: Oregon State University Press, 2016).
146 On the idea of a central science library (in a response to the Economic Advisory Council), see 'Statement on existing Collation and Abstracting of Current Scientific and Technical Literature', in RS CMP/14, 15 June 1933.
147 Plans for the meeting are in RS CMP/14, 4 April 1936.
148 Brian Vickery, 'The Royal Society scientific information conference of 1948', *Journal of Documentation* 54, no. 3 (1998): 281–3; Dave Muddiman, 'Red information scientist: The information career of J.D. Bernal', *Journal of Documentation* 59, no. 4 (2003): 387–409; Alistair Black, Dave Muddiman and Helen Plant, *The Early Information Society: Information management in Britain before the computer* (Aldershot: Ashgate, 2007), Part II; and Goodare, 'Representing science', 86–91.
149 Numbers at 31 May were reported as 123 UK delegates, 22 overseas delegates, and 247 other attendees, see Scientific Information Conference Committee, RS CMB/56/5, 31 May 1948.
150 Report of the Publications Committee, 24 March 1948, in RS CMP/16, 8 April 1948.
151 RS *Year Book* (1948), 159; RS *Year Book* (1949), 196 (and see also p. 174); and RS *Year Book* (1950), 201.
152 On CUP's new American branch, see McKitterick, *History of Cambridge University Press 3*, ch. 15.
153 Annual accounts for 1949, in RS *Year Book* (1950), 201.
154 'Distribution of Royal Society Publications 1947', RS OM/2 (48).
155 North American subscriptions [only], 13 December 1949, RS OM/68 (49)

Part V
The business of publishing, 1950–2015

14
Selling the journals in the 1950s and 1960s

Camilla Mørk Røstvik and Aileen Fyfe

The Scientific Information Conference of 1948 established the Royal Society as a leader in contemporary debates about the publication, circulation and access to scientific research in the British academic world. Whether the Society could maintain that position remained to be seen. It published just four scientific journals, while, by the late 1960s, Elsevier had 27 journals and Pergamon Press had over 70.[1] The Society's international perspective had been focused on Britain and, more recently, the Commonwealth, rather than on the scientific and economic power of the United States of America. And its journal-publishing model was based on subsidy and external grants, whereas the new journal publishers embraced a commercial model. The Society's role in scientific and cultural diplomacy during the Cold War shifted its international perspective.[2] Its approach to publishing would change, too.

The flood of government funding for civilian and military research in the USA, Europe and the USSR benefited both laboratories and libraries in the 1950s and 1960s.[3] It drove the production of even more research to be published, and enabled librarians to purchase the (many) journals to which their researchers wanted access. It was in this context that scientific journal publishing became big business. Entrepreneurial new entrants to journal publishing – including Robert Maxwell's Oxford-based Pergamon Press and the Dutch publishers Elsevier – sought an international market for English-language journals catering to emerging research specialisms. Their success disrupted the context in which learned society publishers, including the Royal Society, operated.

Two things distinguished contemporary debate about the growth of scientific publishing in the Cold War from earlier worries about growth. First, the pace of change was faster. The term 'big science' was coined in 1961 to describe the expensive, long-term, team-based international projects that appeared to typify post-war research.[4] Just the year before, *Nature* had reported, under the heading 'How many more new journals?', that the growth of science was so great that the Science Museum Library was having to subscribe to 700 new journals every year.[5] There were now estimated to be 30,000 journals publishing scientific research and, in 1963, Derek de Solla Price estimated that those journals were publishing about six million articles a year, increasing by half a million each year. Price's quantitative work on the expansion of 'big science' suggested exponential growth.[6]

Second, much of the expansion was due to commercial firms. The publication of original research in Britain had previously been dominated by learned societies and research organisations, whose scholarly interests predisposed them to treat learned journals with goodwill.[7] The handful of commercial firms involved before the war had tended either to mix short research papers with more marketable news and views (as did *Nature*),[8] or treated their journals as loss-leaders that might bring contacts 'with the prospective authors of profitable books'.[9] The new players, on the other hand, were aiming to profit from the publication of research. They inspired other firms to move into journals. As well as the newer firms of Pergamon, Butterworth Scientific and Academic Press, older firms, such as Cambridge University Press and Taylor & Francis, became more involved in journal publishing (not just printing).[10] By 1960, it was said to be easy to set up a new journal, as it required only a 'relatively small' capital outlay, and 'there is often a quick return'.[11] It was against this background of new publishers, new journals and a new-found ability to make journals profitable, that learned society publishers struggled both to defend their role and to make ends meet.

In 1957, the Royal Society's assistant secretary, David Christie Martin, told a London audience that, although 'several commercial publishing houses' were discovering how to make 'quite a bit of money' from scientific journals, it was crucial that learned societies should 'continue to predominate in scientific journal publication'. According to Martin, 'the moment [that] commercial gain began to dominate ... the welfare of the scientific community would suffer'.[12] Martin's robust defence cast learned societies as 'guardians' that could safeguard the quality of scientific publications, and protect the interests of the scientific community. Commercial firms had different motives, and could not be trusted.[13]

David Martin was the central figure in the management and modernisation of the Royal Society in this period. Despite not being a fellow, Martin wrote and spoke as a representative of the Society. As early as 1947, he was noting that, '[w]e must safeguard our rights to defy the bureaucrats when we feel it is in the interest of science'.[14] His diary reveals a man as likely to be found in meetings in Whitehall as examining the Society's experimental printing unit, or chasing the president for a signature along the corridors of Burlington House. John C. Graddon was assistant editor throughout this period, but is less visible in the archival record. He is said to have displayed a 'careful attention to detail' in dealing with authors and their manuscripts, and a 'steadfast maintenance of all that is best in the writing of scientific papers'.[15] Graddon and Martin remained in their roles until the 1970s.

David Martin served under seven presidents, but it was, as usual, the secretaries who were more intimately involved with the publications. The Society now had what were essentially four separate research journals (in addition to its more internally focused publications), but they were still being managed by the two secretaries as part of their responsibility for all things to do with either the physical or biological sciences. In contemporary discussions about the journals, the secretaries were increasingly referred to (informally) as editors; but the lack of an editor per journal would later be seen as having prevented the journals – and especially *Transactions A* and *Transactions B* – developing distinct identities (see Chapter 15). The key figures were the botanist Edward Salisbury (biological secretary 1945–55) and the physiologist G. Lindor Brown (biological secretary 1955–63), both of whom were active in the Society's international efforts to improve the circulation of scientific information; and the mathematician James Lighthill (physical secretary 1965–9), who drove the reforms to the Society's editorial procedures in 1968. A notable innovation in this period was the involvement of some of the new women fellows in committee work: Mary Cartwright and Kathleen Lonsdale both served short terms on Council, and Cartwright was on the Publications Committee from 1959 to 1962.[16]

In the late 1950s and early 1960s, David Martin encouraged the Royal Society to promote the idea of 'self-help' or 'financial independence' for learned society publishing. This chapter examines how the Society managed to end the 1950s with modest publications surpluses, despite having started the decade fearing that its journals were in imminent financial crisis. This financial transformation gave the Society a position of authority from which Martin could argue that learned societies should be able to survive both without the crutch of government subsidy, and without seeking assistance from commercial firms. The Royal Society changed its approach to the business side of journal publishing during the post-war years, but, as we will see, its experience on the editorial side was somewhat different: the steady stream of submissions to its journals did not reflect the rapid growth in scientific research. This provided some breathing space for reflection on the editorial processes, though it would become a new worry by the 1970s.

Becoming financially independent

David Martin's later confidence that '[i]f one took a firm grip, one need not run to commercial organizations for help in publishing', was grounded in the Royal Society's experiences as it coped with the difficult years of post-war recovery.[17] In 1947, the Society had told the government that the publication of British scientific research was being hindered by shortages of manpower (especially for technical typesetting), paper and fuel.[18] These problems took years to resolve: in summer 1951, for instance, Martin reported that the price of paper had gone up by 90 per cent and printing charges were about to go up by 15 per cent.[19] Rationing did not end until June 1954. During these years, both the Royal Society and Cambridge University Press were independently seeking ways to reconfigure the finances of their journals. In November 1953, their concerns collided when the CUP asked to renegotiate the terms of their agreement. A lack of viable alternatives enabled Martin to convince Council to take a dramatic step: the Society took control of its own sales and marketing from 1954.

Financial worries

In the late 1940s and early 1950s, CUP was increasing the sales, and sales income, from the *Proceedings* and the *Transactions*, helped by its new branch in New York. Expanding the international marketing and distribution of the journals was in the financial interests of both parties, since the Press claimed a 10 per cent commission on all sales. In 1950, they were discussing possible sales for the *Proceedings* in Japan.[20]

But, between 1949 and 1952, the production costs for the Society's publications increased from around £16,000 a year to over £35,000 a year. Part of the reason was that the number of papers printed each year was creeping back up, and that the print runs of the journals were increased in response to CUP's success at improving sales. The print run for *Proceedings A* had been 1,450 in 1947, but reached 2,275 in 1953. Acquiring paper was still 'extremely difficult', so changes to typefaces or page layouts were repeatedly suggested during 1952 and 1953.[21] But the university printer described his own 1953 specimen layout as 'very grim indeed', and lacking the 'impressiveness' due to the *Transactions*.[22] And Graddon believed it to be his duty 'to preserve to the best of my ability' the old ways. He doubted that any savings would be sufficient to justify 'the resulting deterioration in the traditionally high standard of style and legibility of the Society's publications'.[23]

Printing (including typesetting) was the largest element of the Society's production costs. Martin helped to create a Consultative Committee for Co-operation with Printing Organizations, on behalf of learned society publishers, which estimated that at least 350 more compositors were needed to meet the expanding requirements of scientific publishing.[24] These conversations drew Martin's attention to the new possibility of printing text using photolithography, rather than letterpress composition.[25] Lithography had initially been used for illustrations, but it was now possible to create lithographic plates from anything that could be photographed, including typewritten text. In 1951 Martin convinced the Royal Society's Council to fund the equipment and staff for an 'Experimental Pro-Printing Unit' to test these techniques.[26] By 1955, the unit had grown to three members of staff, and in 1957, they acquired a small printing press.[27] Thus, for the first time in its history, the Royal Society became a printer. The in-house printing unit became a core part of the Society's support services, extensively used on internal administrative paperwork. The *Proceedings* and the *Transactions*, however, continued to be printed at Cambridge, where skilled compositors used a combination of machine-set and hand-set letterpress to produce the high-quality appearance desired by the Society.

CUP's sales efforts were doing well. In 1951, and again in 1952, the Society received more income from sales of its journals than ever before. From 1949 onwards, the *Proceedings* had been regularly breaking even, and the *Transactions* sometimes did. But the Society's publication account included more than the sales income and production costs for the *Proceedings* and the *Transactions*, as Table 14.1 makes clear.[28]

The Society now charged some of the indirect costs (notably, the salaries of Graddon and his team, and some of their office expenses) to the publication account. And the internal publications (that is, the *Year Book*, *Notes and Records* and *Obituary Notices*) were also counted as publication expenditure, and were equally affected by the high costs of print and paper. These other costs ensured that the total publications expenditure continued to outstrip the sales income from the research journals, despite their new-found success. Grant income – from ICI, from the UK government and from the Society's publication fund (see Chapter 12) – remained essential to the overall sustainability of the publications account. As Table 14.1 shows, in 1951, the publications account showed a modest

Table 14.1 Publication finances in 1951

	Income	Expenditure	Surplus
Transactions	£5,904	£5,661	£243
Proceedings	£17,015	£14,286	£2,729
Grant income	£2,335		£2,335
Other publications	£360	£1,944	−£1,585
Salaries and office expenses		£2,857	−£2,857
Total	£25,613	£24,748	£865

surplus of £865, but that would have been a deficit of £1,470 without the grants income. ICI had been willing to restore its contribution to £1,000 after a pre-war cutback, but the long-term availability of government support was in doubt. The immediate post-war negotiations had vastly increased the grant-in-aid of scientific publishing (Chapter 13), but by the early 1950s, the annual negotiations with the Treasury were being conducted against the backdrop of the austerity measures implemented by the Chancellor of the Exchequer, R.A. Butler. Treasury officials told Martin that they appreciated 'your difficulties with the journals of the learned societies' but were 'under instructions … to make every possible economy of civil expenditure'. In both 1951 and 1952, they had cut support for scientific publications.[29]

It was against this background that Martin warned the officers in early November 1953 that there was likely to be a deficit of £5,000 on the publications account in 1954.[30] To cover that, the Society would have to draw unusually heavily upon the (potentially limited) government grant. Martin offered some suggestions for saving money, such as reducing the number of free offprints given to authors, or cancelling *Notes and Records*.[31] But before the officers got around to discussing these suggestions, a letter arrived from CUP that 'put a completely new complexion on the problem'.[32]

Like the rest of the British print trades, CUP had suffered from post-war labour shortages, and had to find capital to install new equipment. There were also tensions between the needs of its printing house and the publishing division, as its publishing team began to experiment with paperbacks and to expand their line of scientific journals.[33] The Syndics of the Press had traditionally been content that 'subsidising a number of learned Societies' was a valid part of its scholarly mission. In the post-war world, however, the Press 'reluctantly' decided it could not continue offering generous arrangements for society journals that were, in fact, 'insufficient to cover the publishing expenses'.[34]

Thus, in late November 1953, the Press informed the Royal Society that it wished to revise the terms for printing and publishing its journals. It intended to apply a standard set of charges to all 34 of the learned society journals it printed: a 12.5 per cent commission on production costs, and a 15 per cent commission on sales. The Press admitted this would make the Royal Society worse off by about £700 annually, but pointed to the value of its New York office in improving sales. It also argued that the arrangements would remove the current inequity in which 'the youngest Societies' with the most recent contracts were 'helping to pay for their older, and perhaps more wealthy, brethren'.[35] For once, an appeal to its seniority did not find a welcome reception at the Royal Society.

Transforming the publication finances

There was little chance of negotiating over CUP's proposal for an increased commission on the printing costs: everyone acknowledged that these were difficult economic times, and there was no guarantee that a tender process would generate any better offers. Moreover, the Society remained broadly happy with the scholarly standards at which the university printer excelled. There would be occasional niggles in the relationship between Press and Society over the following decades – often around delays – but there was no real desire to move the printing of the journals away from Cambridge. But the letter from the Press did inspire David Martin to make enquiries 'about the possibility of having an alternative publisher'.[36]

Since taking over from Harrisons in 1936, CUP had increased the sales figures for the Society's journals both domestically and internationally. The sales had, however, slumped somewhat in 1953, and the Society was not entirely convinced that the Press could not do better, even though the Press insisted that 'our unspectacular methods have produced good results'.[37] Martin's enquiries evolved into the idea of the Society 'selling its own publications'.[38] This did not mean a return to the days when the Society's clerk sold copies to individual callers at the Society's premises, but the creation of a dedicated publications sales team.

In March 1954, Martin reported that he believed that 'it would be possible for the Society to conduct its own sales, provided it could have access to expert knowledge of sales promotion'. He estimated a capital investment of perhaps £1,500, and an ongoing salary charge of about £1,000 a year.[39] Given the long-running and inconclusive debates about whether to change the typeface and margins of the *Transactions*, it is somewhat astonishing that the Society's officers took barely six months to act upon this far more radical suggestion. Martin was clearly persuasive, and Council confidently assured the fellowship that setting up a sales team would be more economical than employing a publisher.[40]

At the same time, Martin was overseeing another review of the Society's non-commercial distribution. Authors' provision of free offprints was halved,[41] and there was a thorough overhaul of the exchange and gift list; the first since 1932 (Chapter 12). Martin sought, for the first time, to quantify the cost and benefit of the list of 'presents'. He calculated that the Society was spending £3,286 a year in sending its own journals to other societies, universities and scientific institutions; but the journals it received in return, for its library, were worth a mere £931.[42] He recommended greater scrutiny of the return-value of any continuing exchanges, and proposed to save £1,400 by stopping almost all the gifts: the Queen would still receive complimentary copies, but the British and Commonwealth universities would now have to pay for their copies.[43] Council agreed, though 'with great reluctance'.[44]

The 1954 review marked the end of the Society's extensive out-of-commerce distribution to institutions across the world. The free list was gone, and by the early 1970s, only about 40 institutions remained part of an exchange network.[45] Martin had presented the cuts as an effort to reduce the amount the Society paid on printing and shipping, but the result was that university libraries became customers. Sales now became the core form of circulation for the Society's journals.

The only significant non-commercial distribution to survive after 1954 was the free provision to fellows. At £4,580 a year, in 1948, this cost the Society more than the gifts and exchanges, but removing a membership perk dating back to 1752 would have been seen to undermine 'the rights of Fellows'. To economise, fellows were invited to apply for copies of particular papers of interest, rather than automatically receiving the whole series of the *Proceedings* or the *Transactions*, and they were invited to send back any unwanted copies so that they could be sold to libraries filling gaps in their back-runs.[46]

The Society's new publications sales team began work in October 1954. There is no evidence that Martin followed a suggestion from Council to approach the scientific publisher Butterworths for advice on marketing scientific journals.[47] The new sales team was led by John Boreham, who had already been working at the Society since 1946. He was assisted by Gladys Dance (née Glover).[48] Like their editorial colleagues, Boreham and Dance both remained at the Society until the 1970s. Neither was scientifically educated, but both had retail experience, and they appear to have learned the academic publishing business on the job.

Colleagues later remembered the sales team as having 'an unprecedented' success. In the short term, income from sales grew from £27,500 in 1953 (with CUP) to £58,500 in 1955, but little of this can be directly attributed to Boreham and Dance.[49] The Society no longer lost 10 per cent of its sales income to CUP; that commission had amounted to £2,700 in 1953, which was more than Martin had intended to pay the sales team. The Society's treasurer also reversed the 1930s decision to charge salaries and office costs to the publication account, and once more allowed the silent subsidy of these overheads from the Society's operations budget. And the Society followed CUP's advice, and increased the prices of all its journals.

In the longer term, however, Boreham and Dance did make a difference, both to sales income and to circulation. By 1967, the Society's sales income had reached £133,000 a year.[50] Some of this increase arose from a new habit of regularly reviewing the Society's notional 'cost per sheet' (used to calculate retail prices) to compensate for inflation. Before the Second World War, price increases had been once-a-decade occurrences, if that. After the war, inflation came to seem normal; the retail price index doubled from 1947 to 1970. Price increases became a regular feature of journal publishing.[51] Table 14.2 shows how the volumes of the Society's journals became more expensive; and it should be remembered that subscribers would be buying several volumes per year.[52] By the late 1960s, the secretary would comment that *Proceedings A* had become a journal that was read only in university libraries, because 'no-one (except fellows and Foreign Members) can afford personal copies'.[53]

Table 14.2 Retail prices per volume

Year	*Proceedings A*	*Transactions A*
1947	£2 5s. 0d.	£2 11s. 6d.
1954		£4 7s. 6d.
1957	£3 5s. 0d.	
1962	£4 0s. 0d.	£11 17s. 6d.
1970	£6 5s. 0d.	£14 5s. 0d.

Note: the number of volumes issued each year was variable, but the volumes were roughly the same length.

This was a distinct change from the days when the Society had priced its journals and separate copies low to reach individual junior scientists.

There is no doubt that Boreham and Dance increased the number of copies sold, not just the income received. They reported subscriber numbers to the officers quarterly, and the figures were printed annually in the *Year Book* with pride. The trend was consistently upwards through the 1950s and 1960s, as is clear from Figure 14.1 (which was produced for the officers in the early 1970s, when there were hints that growth was faltering).[54] From the start, Boreham and Dance actively chased lapsed subscribers, particularly those who had been dealing with CUP's American office.[55] They compiled specialised mailing lists of individuals, institutions and organisations who could be sent promotional circulars and postcards advertising the contents of the latest issues.[56] They drew upon the expert knowledge of the Society's fellows to target their new promotional materials, and they asked fellows to encourage 'more libraries and other institutions' to subscribe to the journals.[57] Boreham was able to report in February 1955 that at least 55 of the institutions that previously received the journals as gifts had become purchasers.[58] By late 1956, the Society believed that 'most' of the nearly 180 institutions affected by the cut of the free list 'now purchased the publications'.[59]

Another sign of the new approach to marketing was the introduction of an advance subscription rate for the *Transactions* in 1956, which encouraged regular orders and simplified Dance's paperwork.[60] As with the *Proceedings*, purchasers could now subscribe in advance, and the finances suggest that the discount for annual subscribers was no longer set below cost price. The subscription was 'per volume' not 'per year', which left subscribers in some uncertainty about how many volumes would be issued per year. Librarians were said to be 'resigned to the inevitability' of this system, because journal publishers – from Taylor & Francis to Pergamon Press – saw this 'open-ended' system as the only way to offer some sort of advance fixed price without also imposing a hard cap on the amount of material published each year.[61] The Royal Society had long opposed a cap because its fellows and officers did not want to be unable to publish good research papers; Pergamon, on the other hand, sought to take advantage of the rapidly growing output of science by publishing as many papers as it could.

As well as managing sales of the *Proceedings* and the *Transactions*, Boreham and Dance began to look for other possible methods of income generation. In contrast to the 1920s, this did not mean targeting potential donors or sponsors, but identifying other Society assets that could be monetised. The *Year Book*, *Notes and Records* and *Obituary Notices* were

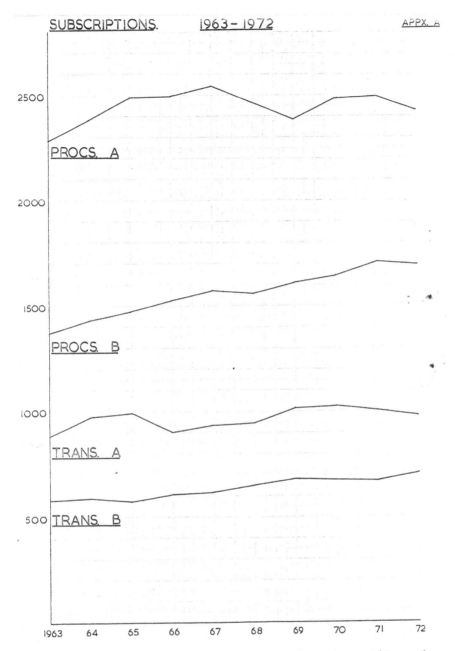

Figure 14.1 Subscriptions to Royal Society journals, 1963–72. This graph was produced by the sales team © The Royal Society

now advertised for public sale (with the latter renamed *Biographical Memoirs of the Royal Society*), and the regular orders for them increased five-fold within 10 years.[62] The Society's back-run also had potential, particularly now that photographic techniques could be used either to generate lithographic plates (enabling reprints without resetting the type) or to create microfilm or microfiche editions. For instance, in 1954, the Society signed an agreement allowing Butterworths to reprint the first eight volumes of the *Catalogue of Scientific Papers* (1800 to 1873), for a royalty on sales.[63] Four years later, the Society began reprinting *Proceedings A* (1939 to 1956) so that it would be 'available throughout the world to all who wish to purchase it', or, more specifically, so that it could be sold to the many new institutional subscribers who did not have the recent back-run.[64] The Society also signed a deal giving Johnson Reprints Co. Ltd and the Kraus Reprint Corporation exclusive rights to produce a new print edition of the *Philosophical Transactions* (1665–1750), in return for a 15 per cent royalty on sales.[65] From the mid-1950s, projects like these generated modest additional income for the Society's publications account.

In Britain, the Society's customary claim to own the rights of the historical *Transactions* and *Proceedings* was still widely recognised, despite the ambiguity about the legal ownership of the pre-1752 rights (Chapter 5) and even though the early volumes would be long out of copyright if copyright had applied.[66] In the United States, however, the Society's historic prestige had less influence, and in 1954, it learned that an American microfiche company was issuing the pre-1905 *Proceedings* without permission. The officers admitted that they had 'no legal redress' against the Microcard Foundation, but insisted that 'an objection should be lodged'.[67] A similar issue arose in 1963, when a Dutch firm, B. De Graaf, began to reprint the same early volumes of the *Transactions* that were covered by the Johnson/Kraus deal, leading the authorised reprinters to inform the Society that they would have to reduce their royalty payments, in view of the competition.[68] The fact that Johnson/Kraus had originally agreed to pay any royalties at all on such long out-of-copyright material indicates an ongoing respect for the Society's moral ownership. The case also suggests the Society's naivety in signing an agreement for exclusive rights that it had no legal power to grant.

Together, all these changes – the cuts to the free list, the increased prices, the targeted efforts at international marketing, and the new income streams from publications – meant that the Royal Society's publication finances looked healthier at the end of the 1950s than they had done at the start. Publication sales consistently generated around

£20,000 a year beyond the production costs in the late 1950s, and this had risen to around £50,000 a year in the late 1960s.[69] The Royal Society ceased to take a share of the government grant-in-aid for scientific publications after 1955 (though it continued to disburse the funds to other societies);[70] and the annual donation from ICI came to an end in 1957. By 1960, the Society's journals could be regarded as financially self-supporting for the first time since Oldenburg's death.

Scientific information, self-help and commercial interests

While the Society's staff spent the 1950s transforming the financial basis of its journals, its fellows and officers were working to promote the effective circulation of 'scientific information', and to protect the role of society publishers in the changing – and increasingly commercialised – landscape of scientific journal publishing.

The Royal Society had an active Scientific Information Committee through the 1950s and 1960s. Its members undertook various activities internally, nationally and internationally to promote the effective collating and sharing of information, and encouraging collective action aimed at improving access to scientific publications in a complex world of international but specialised research. It worked on the standardisation of symbols, citations and abstracts; it represented the Society at later scientific information conferences (such as the one in Washington DC in 1958); and it brought together representatives of British learned societies and journals to discuss shared interests (particularly in a series of 'conferences of editors' in the 1960s and early 1970s).

One of the Society's most influential achievements was the 1950 Fair Copying Declaration.[71] This was a response to concerns that strict interpretation of copyright legislation might prevent librarians and scientists using the new technology of photocopying to copy journal articles. Rather than seeing photocopying as a threat to journal sales, the Royal Society regarded it as an extension of the long-standing circulation of 'separate copies' through personal scholarly networks. The Society brokered a voluntary code that allowed photocopying of scientific articles from signatory journals, for purposes of research and private study. The original subscribers included 100 learned societies and eight other publishers (including CUP).[72] The Society gave evidence about the 'copyright problems which adversely affect scientific work' to a government enquiry, and the UK Copyright Act 1956 included a section on fair copying for educational purposes that was closely based on the Fair Copying Declaration.[73]

In the early 1950s, the Society also issued lists of research periodicals and abstracting periodicals for librarians. It updated its *Notes on the Preparation of Papers* for scientific authors, which had sold 4,000 copies by 1951. It also coordinated a multi-society effort to develop a standard set of 'symbols, signs and abbreviations' (and citation styles) for scientific publications.[74] Its biggest effort to display leadership by coordinating other learned societies occurred from 1955, when the Society worked closely with the Nuffield Foundation on a project 'for the support of learned journals'.[75] This culminated with the publication in 1963 of a booklet entitled *Self-Help for Learned Journals*. In June 1963, the Society convened a meeting of British learned societies to discuss the booklet, and to encourage greater cooperation around 'scientific publications'.[76]

The origins of the learned journals project lay in the difficulties facing learned society publishers in the early 1950s, when it seemed 'impossible to continue to issue learned journals on the old terms'. Costs were still rising post-war, and, as the Royal Society had discovered, those who printed and published for the societies could no longer offer their 'old-time facilities' and goodwill. Some feared that learned society journals 'were on the way to extinction'.[77] In addition to the economic threat, learned societies also had to defend their own role in journal publishing as commercial firms were becoming newly active and successful. In December 1953, when the Royal Society was worrying about the ultimatum from CUP, David Martin invited representatives from several other societies to discuss the problems facing learned journals. They had agreed that 'there was a need for someone, knowledgeable both of the publishing trade and scientific journals' to take a close look at the situation, and to advise learned societies.[78]

By spring 1955, the Royal Society was well on its way to solving its own problems, but it remained concerned with the wider situation. A document produced by David Martin listed 25 learned societies involved in publishing, and classified them from 'a financial point of view' as 'successful', 'borderline' or 'in difficulties'. The Royal Society was one of just seven societies Martin labelled 'successful'. He believed that over half those he listed were 'in difficulties', including the Linnean and Zoological societies, the Royal Astronomical Society and the Physical Society.[79] This apparent crisis was why the Royal Society persuaded the Nuffield Foundation to provide £20,000 over five years to assist learned journals in the sciences.[80] The Nuffield Foundation had been created in 1943 by car manufacturer William Morris, who had been elected a fellow of the Society in 1939, under the Statute 12 provision for those who had performed conspicuous public service to science.[81] His Foundation supported a variety of Royal Society projects in the 1950s, including Commonwealth scholarships and individual research projects. The support of learned journals was another aspect of its work to support capacity-building in science.[82]

A 'Scientific Advisory Committee' was set up to guide the use of the Nuffield funding, and it was packed with Royal Society representatives. The most active individuals were David Martin and the biological secretaries, first Edward Salisbury and then Lindor Brown. In contrast to the government grant, the Nuffield Foundation allowed flexibility in how its funds could be spent. The Committee did make some direct grants to societies in need, but most of its funding was spent on surveying the landscape and providing expert advice. One of its first actions in 1955 was to ask 'someone knowledgeable' to look closely at the situation facing learned journals.

Robert Lusty, deputy director of the publishers Michael Joseph, examined the financial situation of eight learned societies (including the Royal Society), selected to represent a range of experiences. His privately circulated report concluded that 'the only form of self-help' possible for struggling journals would be to focus on distribution, sales and publicity. He knew that some people, including Martin, held out hopes of a technical revolution in printing that would significantly reduce costs, but he felt it was unlikely in the foreseeable future. Examining the editorial and administrative processes at the societies, Lusty was astonished to see so much 'slogging clerical work' being done by 'professors and scientists'. He concluded that there would be no easy savings in editorial or administration, since 'material for the journals was unpaid for', 'editorial work was unpaid for', and few societies even attempted to make any allowance for 'overheads' in their accounts. Thus, struggling learned journals had no choice but to focus on increasing their revenue. For Lusty, that meant increasing sales, raising prices, or both.[83]

Lusty's thoughts on 'publications and sales promotion' were discussed at a meeting of the Scientific Advisory Committee in 1956, and as a result, the Committee made the significant decision to hire a publishing consultant (or, 'liaison officer') to offer tailored advice to help individual societies 'increase efficiency and revenue'.[84] Grants would also be offered to help societies act upon the advice. Charles Hutt, formerly of Butterworths, volunteered his services and began work with four societies in 1956. However, the Committee became uneasy about his commercial aspirations (he would soon begin working for Pergamon), and were glad to be able to replace him with Frank V. Morley. An American in Britain,

Morley was extremely well connected in literary publishing circles, having spent 30 years in publishing on both sides of the Atlantic. But his other credentials were equally important: he was the son of a university mathematics professor (and journal editor), and was himself a Rhodes scholar with a DPhil in mathematics. Morley was thus well equipped to be a plausible 'liaison' between the worlds of publishing and science. In spring 1957, he was informally interviewed by David Martin and Edward Salisbury, who reported that Morley 'had the right approach'.[85] Morley would work for the Nuffield Foundation until at least 1961. During that time, he engaged with at least 28 society publishers.

Morley's final report to the Nuffield Foundation was the basis for the booklet *Self-Help for Learned Journals* discussed at the meeting convened by the Royal Society in 1963. In it, he presented (anonymised) examples of pricing arrangements, publicity campaigns, and agreements with printers. He argued for more conversation between the parties involved, so that both editorial staff and authors really understood what printers needed, and when. He offered suggestions for streamlining procedures; recommended banning authors from making proof corrections; discussed the problems of dealing with an unpredictable annual output; and had lengthy sections on mailing lists, promotional material and setting non-member subscription rates.[86] The Royal Society was not one of the anonymised societies featured in *Self-Help*, but it appeared repeatedly as a source of advice and information.

Like Lusty, Morley believed that the only real solution to the financial problems facing learned society publishers was to acquire additional revenue, and he did not mean appeals to government or philanthropy. Morley discussed pricing, sales and subscribers, and the possibility of monetising a society's 'hidden assets', such as its back-run. He included a brief discussion of the (limited) potential for advertising income, but there was no hint of the idea of levying a page-charge on accepted authors, as the American Chemical Society had just begun to do, nor of charging a processing fee on all submissions, as the US National Academy of Sciences would later do.[87]

Morley's vision closely matched that adopted by the Royal Society's new sales team. The 1920s' focus on philanthropic donations and government support had been replaced by the notion of 'self-help'. Even as the Royal Society continued to disburse government funding to other societies, it was urging those societies to learn 'to stand on their own financial feet'. In 1957, the application process had been revised to reward societies that were deemed to be 'helping themselves by bringing up subscriptions and selling prices to present-day levels', rather than those who 'never appeared to be doing anything to help themselves'.[88] That same year, Martin warned those who relied on annual subventions from government that they were 'living in a fool's paradise'.[89]

The response to *Self-Help*, and to the Royal Society's discussion meeting, was mixed. The meeting was chaired by Howard Florey, then-president of the Society, but he personally found the booklet 'incredibly tedious', and its author 'pompous'. He saw the main purpose of the meeting as dispelling 'the idea that the Royal Society would not do anything to help other Societies when asked'.[90] At least one of the fellows who attended the meeting objected to the 'paternalistic' tone of *Self-Help*, and detested its emphasis on harsh economic realities. He did not wish to see scientific journals published 'on the cheap', and seemed unable to accept that it might no longer be possible to maintain the tradition of producing 'the best possible publications for the science that we regard as valuable'.[91] But other commentators welcomed the effort to share best practice, with one reviewer appreciating that the booklet's aim was 'to make us help ourselves and not to encourage us to seek assistance from outside'.[92]

The society representatives at the 1963 meeting also discussed a proposed 'Code for the Publication of New Scientific Journals' that had been prepared by the Royal Society's Scientific Information Committee, led by Lindor Brown (Figure 14.2).[93] The draft Code accepted that the development of 'new fields and disciplines' meant that 'new journals are necessary', but sought to lay down guidelines for who should create and run them. The Committee was worried by 'the present tendency for commercial publishers to initiate new scientific journals in great numbers' and unsurprisingly, thought that a learned society was 'ideally, the best body' to 'start and run a journal'.[94]

David Martin's rhetoric in the late 1950s contained a sense that scientific or scholarly interests were (or would be) threatened by the rise of commercial interests. The exact nature of the threat was, however, only partially articulated, as were the reasons why it would be preferable to have 'primary journals of original work' run by learned societies. In a 1957 speech, Martin offered two reasons in favour of society control: one was about access and pricing, the other about quality. Martin took it for granted that societies would aim to circulate research 'as widely and cheaply as possible'. He was worried that commercial firms, who 'had another aim in life', would try to 'take over the whole job and charge higher prices for the results', and he wanted 'safeguards' to stop the search for profits coming into conflict with the scholarly mission.[95] As for the 'quality of scientific content', Martin believed that this depended upon 'high-class refereeing', and while a society had its members to draw

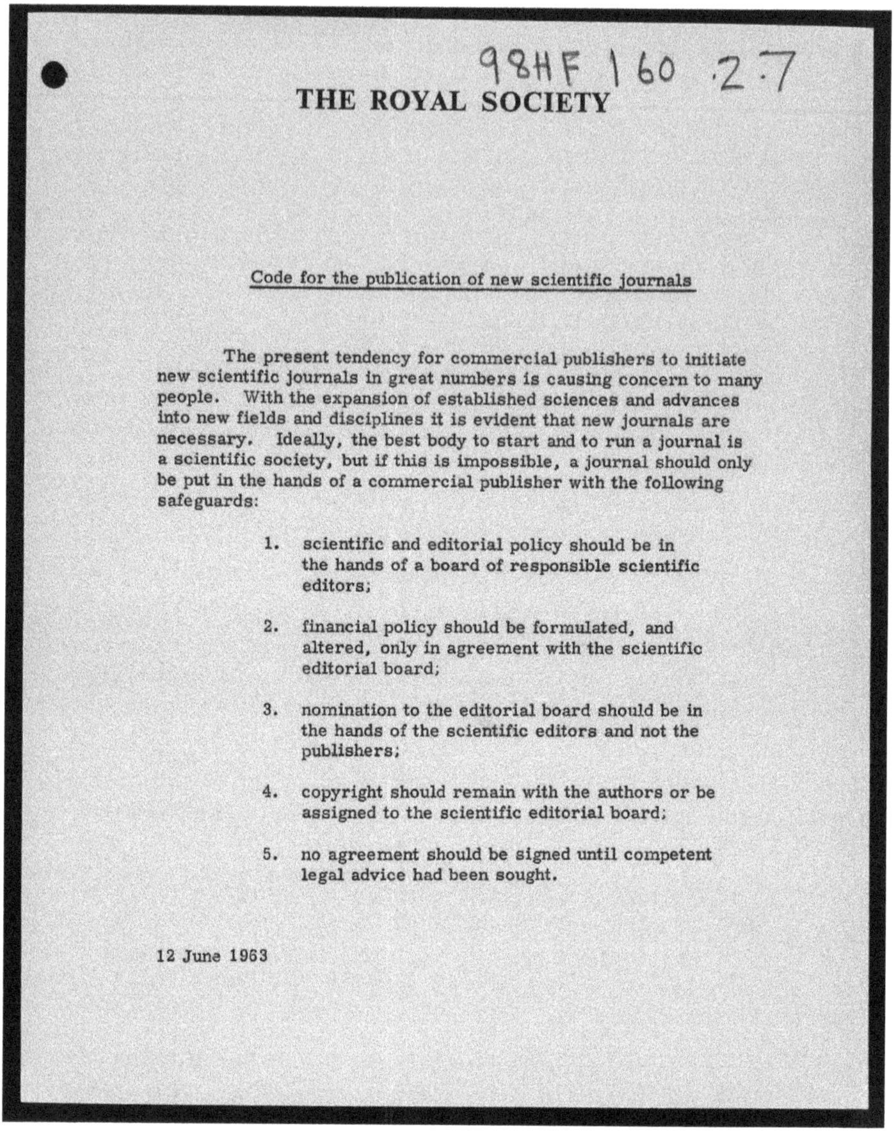

> **THE ROYAL SOCIETY**
>
> Code for the publication of new scientific journals
>
> The present tendency for commercial publishers to initiate new scientific journals in great numbers is causing concern to many people. With the expansion of established sciences and advances into new fields and disciplines it is evident that new journals are necessary. Ideally, the best body to start and to run a journal is a scientific society, but if this is impossible, a journal should only be put in the hands of a commercial publisher with the following safeguards:
>
> 1. scientific and editorial policy should be in the hands of a board of responsible scientific editors;
> 2. financial policy should be formulated, and altered, only in agreement with the scientific editorial board;
> 3. nomination to the editorial board should be in the hands of the scientific editors and not the publishers;
> 4. copyright should remain with the authors or be assigned to the scientific editorial board;
> 5. no agreement should be signed until competent legal advice had been sought.
>
> 12 June 1963

Figure 14.2 Draft code for new journals, 1963 © The Royal Society.

upon, Martin assumed (wrongly, as it turned out) that commercial firms had no such access to expert advice.[96] This feeling was shared beyond learned society circles; at a meeting in 1960, a representative from the Publishers' Association was willing to admit that publishers 'may not really be capable of judging quality'.[97] Others assumed that journals run by commercial firms might find it 'hard to get sufficient good manuscripts to keep them going', because they were not firmly grounded in an academic community.[98]

Frank Morley offered a third reason why research journals should be managed by societies: that society ownership offered the best long-term security. A publishing firm might 'grow weary' of a particular journal, and decide to abandon it or to transform it in some way. A society, on the other hand, was 'a corporation which believes in what it is doing and is determined never to die'. Thus, it would never let its publications die, or change beyond recognition.[99]

The draft Code of 1963 was an attempt to lay down the 'safeguards' that Martin had asked for. It offered five principles that would keep control of scientific journals firmly in the hands of the academic community, of which the most fundamental was a strong editorial board with academic members. The board was explicitly given control of 'financial policy' as well as 'scientific and editorial policy', but no preference was stated for any particular business model, and there was no explicit protection for the mission of wide circulation and cheap access. The fourth point would have kept the enforcement (or waiver) of copyright in academic hands. As it had with the Fair Copying Declaration, the Society aimed to ensure that strict enforcement of copyright did not hinder the circulation of research. It is, however, difficult to imagine how this Code could have been enforced. A society transferring ownership of its journal to a commercial publisher might possibly be in a position to craft a binding legal agreement for the journal's future operation based upon these principles, but there was little the Royal Society, or anybody else, could do to stop individual scientists agreeing to serve on editorial boards for commercial publishers, nor to make them refuse to referee papers for commercial journals, nor, indeed, to stop them sending their work there for publication.

Editorial matters

Authors and submissions

There was a new attention to the number of papers *submitted* to the Royal Society during the 1950s and 1960s, in addition to the long tradition of reporting the number of papers *published* each year. For Martin and the new sales team, the health of the *Proceedings* and the *Transactions* was now clearly measured in sales figures, but the fellows of the Society who sat on publishing-related committees usually regarded the quantity and quality of the published papers as the important measure. Graddon now made quarterly reports on submission numbers, and these came to be used as a proxy for the journals' popularity with scientific authors, just as 'number of subscriptions' became a proxy for 'readership'.

By the mid-1950s, the overall number of submissions received by the Society had returned to pre-war levels, but there were two important differences.[100] First, the biological sciences recovered more slowly than the physical sciences and did not in fact return to 1930s' levels until the 1970s. Thus, the existing imbalance between the A and B series of the *Proceedings* and the *Transactions* was not merely continued but worsened after the war. Second, the overall numbers were stable, not rising, with around 330 papers being submitted each year in the late 1950s and early 1960s. This lack of growth – amid clear evidence for near-exponential growth of scientific research and the widespread creation of new journals – led to a dawning awareness within the Society that its journals might have to compete for the attention of authors.

The concerns were strongest for *Proceedings B* and *Transactions B*. In 1954, a memorandum from the Cambridge professor of experimental medicine, Robert McCance considered why *Proceedings B* was 'no longer the pre-eminent journal we should all like it to be'. McCance wondered whether the broad range of fields covered by the Society's journals could mean that they had 'no strong appeal' to anyone. He suggested that the insistence that papers be communicated via a fellow might be deterring certain authors. McCance even wondered if the Society's biological referees were harsher than those in the physical sciences, but found it hard to see why this should be so.[101]

A new issue raised by McCance was the link between circulation, reputation and submissions. He believed many authors, 'especially those from the Commonwealth and abroad', were 'reluctant' to send their papers to a journal 'which is so little known – owing to its small circulation'. The Society had tended to think of its international distribution efforts as a means of spreading the reputation of the Royal Society, but McCance was pointing out that a wider circulation would also raise awareness among a wider pool of potential authors. He was right. In 1950, about 90 per cent of authors of papers in Royal Society journals had been based at UK institutions, with a further 6 per cent in the Commonwealth. By 1970, the UK share had declined to 71 per cent, due to substantial growth in submissions (from a very low starting point) from the USA and the rest of the world.[102] This was the beginning of the internationalisation of the Society's journals.

Most of McCance's suggestions involved raising awareness of *Proceedings B* among potential authors, and he thought fellows should be reminded that 'they have certain responsibilities in the success and prosperity of their own journal'. There was in fact a higher engagement rate of fellows acting as communicators than there had been before the war – over 350 of the 560 or so fellows acted as communicator at least once in the 1950s – but the distribution was very uneven. Every single one of the 25 most frequent communicators worked in the physical sciences. Some of them, notably Nevill Mott, Harrie Massey and Rudolf Peierls, each communicated over 50 papers over the decade. McCance wanted more of the biological fellows to take this sort of active interest in 'the welfare of the journal'. He also suggested nominating 10 fellows to act as a 'Board of Editors'; their names would appear on the journal, and they would assist in the selection of referees. McCance noted that 'many successful and progressive scientific journals' had, over the last two decades or so, come to be 'managed in this way'.[103] The chairs of the Society's sectional committees arguably already performed the roles that McCance outlined, but they did not do so visibly, nor was it the sole focus of their role.

Something that McCance did not mention on his list of problems with *Proceedings B* was the time it took to get scientific discoveries into print. It was, however, clearly a concern, for 'time to publication' was regularly recorded and published alongside other key indicators in the *Year Book*. There had been 'serious and discouraging' delays in the immediate post-war period, when the Society had told the government that there was an average delay of eight to 10 months, and 'a large number of scientists cannot be certain of having their work published within two years'.[104] David Martin later claimed that the Society had taken 'energetic steps' to improve things, but it did not last. The average time taken from receipt to publication of papers in *Proceedings A* had been cut to under 22 weeks in 1954, but by 1964 it was almost 35 weeks and rising.[105] The *Year Book* now began to carry the *range* of times (see Figure 14.3), perhaps hoping that readers would focus on the possibility of appearing in *Proceedings A* in just 23 weeks, rather than noting the average of 39.5 weeks, let alone the 88 weeks taken by the slowest paper that year.[106] In 1963, Frank Morley claimed that British journals were faster 'than can be guaranteed in most countries', but it is unclear what evidence underpinned that claim.[107]

CUP was blamed for some of the delays in the printing of the Royal Society journals. In 1957, for instance, Martin accused the Press of making 'all the usual excuses', from the difficulty of mathematical typesetting to 'the holiday period, examination papers, illness, etc'. The Press claimed Martin had presented an exaggerated and 'very black picture'.[108] No sooner had things been smoothed over than an eight-week

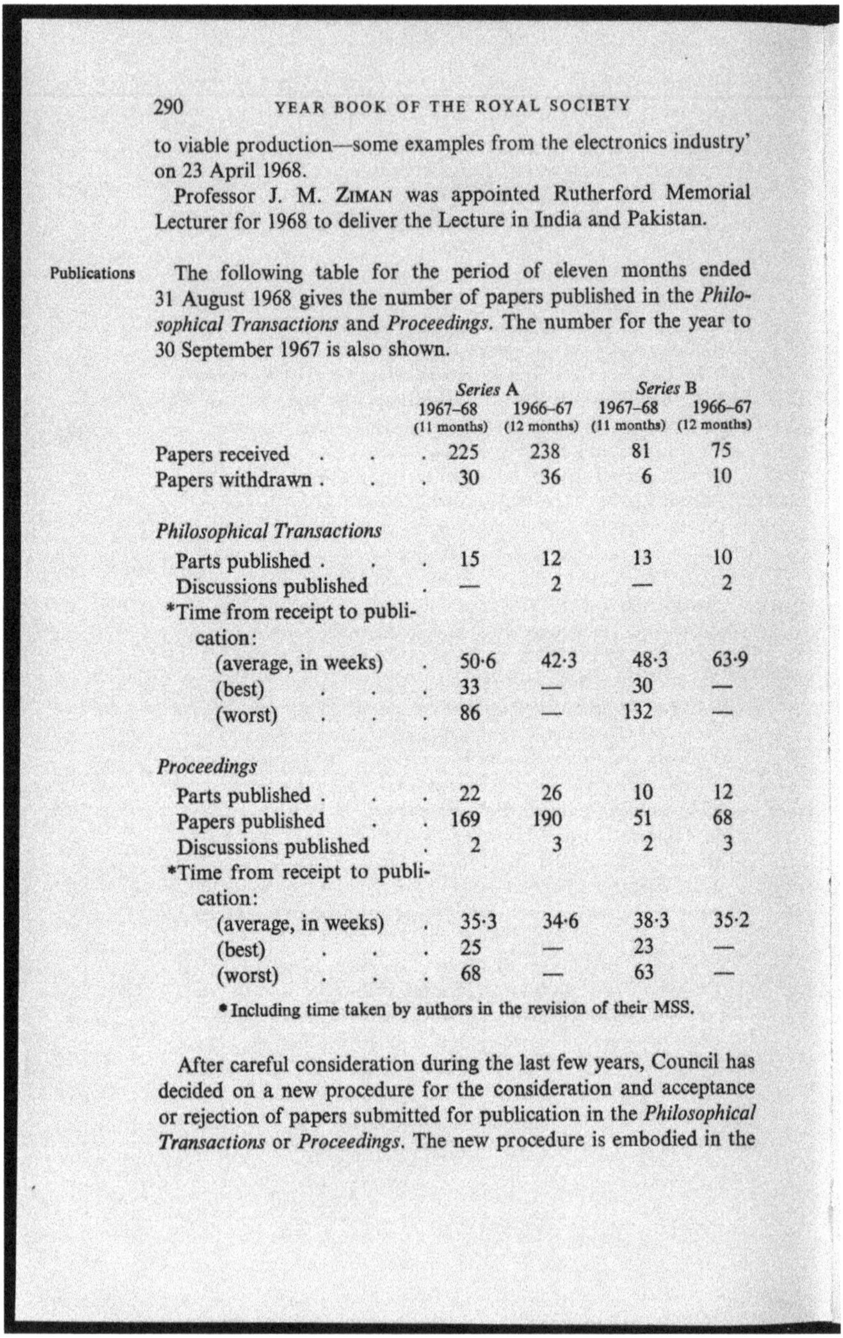

290   YEAR BOOK OF THE ROYAL SOCIETY

to viable production—some examples from the electronics industry' on 23 April 1968.

Professor J. M. ZIMAN was appointed Rutherford Memorial Lecturer for 1968 to deliver the Lecture in India and Pakistan.

Publications The following table for the period of eleven months ended 31 August 1968 gives the number of papers published in the *Philosophical Transactions* and *Proceedings*. The number for the year to 30 September 1967 is also shown.

	Series A		Series B	
	1967–68 (11 months)	1966–67 (12 months)	1967–68 (11 months)	1966–67 (12 months)
Papers received	225	238	81	75
Papers withdrawn	30	36	6	10
Philosophical Transactions				
Parts published	15	12	13	10
Discussions published	—	2	—	2
*Time from receipt to publication:				
(average, in weeks)	50·6	42·3	48·3	63·9
(best)	33	—	30	—
(worst)	86	—	132	—
Proceedings				
Parts published	22	26	10	12
Papers published	169	190	51	68
Discussions published	2	3	2	3
*Time from receipt to publication:				
(average, in weeks)	35·3	34·6	38·3	35·2
(best)	25	—	23	—
(worst)	68	—	63	—

* Including time taken by authors in the revision of their MSS.

After careful consideration during the last few years, Council has decided on a new procedure for the consideration and acceptance or rejection of papers submitted for publication in the *Philosophical Transactions* or *Proceedings*. The new procedure is embodied in the

Figure 14.3 Key performance indicators for the Royal Society journals, *Year Book*, 1969, p. 290 © The Royal Society.

dispute in the printing industry in 1959 brought new problems.[109] The overall impression is that, despite Martin's claims of active management, the Royal Society appeared to be powerless to improve its publication times.

Morley would suggest 'a maximum of six months' from submission to publication as reasonable for a primary research paper, but this was rather longer than some scientific researchers in fast-moving fields would want.[110] For instance, in 1951, one of the physics fellows once again proposed that the Society should facilitate the rapid publication of preliminary notes. Rather than transform the *Proceedings*, he proposed the launch of a new fortnightly journal, to be called *Letters to the Royal Society*, that would carry 'short statements similar to letters to the editor in *Nature*'.[111] Other learned organisations were having similar debates. In 1960, the National Academy of Sciences in the USA reformed its own (monthly) *Proceedings* (*PNAS*): it would focus on the 'prompt publication' of what were initially termed 'brief accounts of important current researches', but soon became 'brief first announcements of the results of original research'.[112] Like the Royal Society journals, *PNAS* only accepted contributions submitted from or via a member, but its more rapid publication times meant that it became the model for many fellows of the Royal Society in the 1960s and 1970s. But, as the Society's secretary pointed out in 1967, weekly *Nature* and monthly *PNAS* achieved their rapid publication times because they were publishing 'shorter, unedited, preliminary communications', rather than lengthy, refereed research papers.[113] *PNAS* limited its authors to eight pages, whereas the *Proceedings* of the Royal Society still allowed 24 pages; *PNAS* also saved time by expecting the member communicating the paper to act as a referee ahead of submission; and by not letting authors check page proofs.[114] The Royal Society officers feared that a speeded-up publication process would risk the 'degeneration' of its high standards.[115]

Referees

The Society's ongoing commitment to the use of referees certainly slowed its publication times, in addition to the time taken for careful copy-editing and proofing. As had been true for over a century, some referees were slow to report; sometimes it was difficult to find fellows willing to referee certain papers; and sometimes authors were slow to make the changes recommended by the referees. But the Society's fellows and staff believed that refereeing was even more important in the 1950s than before, because it distinguished learned society publishing from the new journals of 'unscrupulous' commercial publishers.

Fellows who acted as referees accepted and understood this special role: one physiologist acknowledged that: 'A scientific article in the *Philosophical Transactions* carries authority far greater than that of the author himself. Thus any referee and, of course, The Royal Society itself, have great responsibility.'[116] Yet, despite the pre-war calls for greater clarity around the criteria for acceptance, many post-war referees were still struggling with ambiguity. When asked to read and report on a 100-page manuscript in less than two weeks, a physiologist asked the secretary: 'Help me! What is the standard of the Phil Trans? … It seems to me hardly believable that the paper could be accepted: I just can't understand how it came to be presented! But what is the "atmosphere" about such things? … I am, sorely, puzzled.'[117]

It was not uncommon for referees to differ in their interpretations of significance or appropriateness for the Society. Compared to the 1850s, the fellowship in the 1950s was more fragmented. Fellows associated with their specialist disciplinary communities, and they were employed by a wide array of universities and research institutes around Britain. Establishing shared norms for the acceptance of papers was consequently more difficult than it had been when most of the active referees could expect to socialise regularly at Burlington House on Thursday evenings.

In November 1951, for instance, the Society received a long paper by a newly elected fellow, Alan Turing, proposing a mathematical model for certain biological processes. The physiology committee initially had trouble finding a referee for this cross-disciplinary paper; A.V. Hill and Laurence Hogben both declined. Mathematical geneticist J.B.S. Haldane reported that the 'central idea' was 'sufficiently important', but he was 'equally clear' that it should not be published as it stood. He felt that the sections on basic biology should be substantially trimmed back, as many of the facts were those of 'elementary textbooks'. Some of the mathematics was definitely 'important', but needed to be explained in words for 'non-mathematical readers'. And then there were discussions of biochemistry where Haldane felt that 'it may well be that I have missed the main points … but if so, I think other readers would do so'. The second referee, Charles G. Darwin, director of the National Physical Laboratory, found the mathematics 'not very deep' and lacking 'any new principles', and he too found certain passages difficult to follow. But, unlike Haldane, he advised that it was 'well worth printing, because it will convey to the biologist the possibility of mathematical methodology more definitely than has often been done hitherto'.[118] The paper was passed for publication and appeared in *Transactions B* in August 1952. In the abstract to the published paper, Turing responded to Haldane's remarks by defending his inclusion of 'elementary facts ... which can be found in text-books' as necessary for those readers who did not have the 'good knowledge of mathematics, some biology, and some elementary chemistry' needed for a 'full understanding' of his argument. Turing's paper is now regarded as a landmark application of mathematical modelling to biological processes, but at the time, the referees found it difficult to evaluate.[119]

The practice of refereeing papers continued to work much as it had always done, but there were two significant developments in the 1950s and 1960s. The first was a sense that refereeing had become an unwelcome burden on busy researchers; and the second was the involvement of the new women fellows in what had hitherto been the entirely masculine editorial processes of the Society.

In July 1950, the biochemist Neil Adam sent in a damning report on a 'really shocking' paper, whose only 'reasonable' section had already been published in another journal. He complained: 'During the last five very busy weeks I have had five papers [to referee], not one of which was fit for publication in a first-class scientific journal.' As Louis Filon had done in 1936 (Chapter 13), Adam blamed the submission of these weak papers on the fellows who communicated them, whom he characterised as 'pot-hunting' PhD supervisors so eager to help their students gain that 'little extra prestige' that they ignored 'the standard our Society requires'. Adam thought the supervisors ought to 'take more trouble' to 'exercise their undoubted critical powers and have the papers put into proper shape, or in some cases stopped, before sending them in'.[120]

But where Filon had been worried that the submission and publication of weak papers would hurt the reputation of the Society's journals, Adam was equally concerned about the amount of his own time that was being wasted reading and commenting on papers that would not be published. He pointed out that 'the job of examining unsatisfactory communications is very exacting and time-consuming':

> If I get much more heavy refereeing like this, it is goodbye to any chance of doing real scientific work myself (and please remember that many other societies try and make me referee their papers also!). If I could only get some uninterrupted time, I could do real work of ten times the value of the sort of rubbish I have been required to report on lately.[121]

Thus, Adam told Martin, 'I warn you, you can expect a strike of referees or at least one of them' and added, 'For mercy's sake, don't send me any more papers to referee for a long time!' His appeal worked in the short term, but by 1953, half a dozen or so papers were once more being sent to his Southampton office each year.

By the 1960s, the secretaries were becoming worried about the number of fellows who were declining requests to act as referees. A 1967 report revealed that about a quarter of those asked declined ('but this includes many who did so because of pressure of work or absence abroad or on holiday').[122] Adam's willingness to accept repeated requests, in the face of his own limited time and his despair at the standard of some of the papers, suggests a strong commitment to the Society and its journals, and for the Society's secretaries and sectional committee chairs, such fellows were valuable resources. But with more fellows declining to act, the committed core group ended up doing a higher proportion of the refereeing work.[123] Seventeen years after Adam's threat to strike, the secretary admitted that 'there is a tendency to overburden Fellows who prove to be exceptionally good and conscientious referees'. This comment accompanied the Society's first attempt to analyse the distribution of refereeing work, which revealed that, in 1965, almost a quarter of the papers had been dealt with by just 10 per cent of the referees, whereas 79 referees had reported on only a single paper.[124]

One of those 'good and conscientious' referees was the X-ray crystallographer Kathleen Lonsdale. By 1965, the Society still had only 18 women fellows and, as with their male colleagues, the majority did very little refereeing.[125] The chances of a woman author being evaluated by a woman referee were slim (though see Figure 14.4 for an exception). The 'Register of Papers' shows that Honor Fell, mathematician Mary Cartwright, X-ray crystallographer Dorothy Crowfoot Hodgkin, botanist Helen Porter and biochemist Rosalind Pitt-Rivers all did some refereeing in this period, but Lonsdale was by far the most active. In 1955, for instance, she refereed eight papers (including one by Hodgkin), while Cartwright and Hodgkin each did just one.

The pronouns used in Society paperwork for referees, communicators and committee members were consistently male, and when they were performing the privileged roles of fellows, women were treated as 'honorary gentlemen'; their gender was ignored. It took until the mid-60s before the standard printed report form sent to referees was amended from 'Dear Sir' to 'Dear Sir or Madam'. Even when the old forms were sent to Lonsdale, they were not corrected by hand. This casual neglect of the gender of women fellows was at odds with the Society's determination to acknowledge gender when women were authors. In the 'Register of Papers' and on the forms sent to referees, women authors (including Lonsdale and

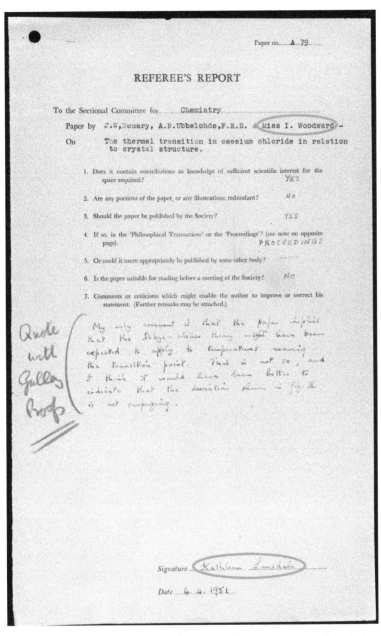

Figure 14.4 A rare example of a paper with a woman author being evaluated by a woman: report by Kathleen Lonsdale on a paper whose co-authors included Miss I. Woodward, 1951. The annotation in red makes clear that Lonsdale's comments were to be conveyed to the authors (with the galley proofs) © The Royal Society.

other fellows) were given their full first names, and sometimes 'Miss' or 'Mrs' (as in Figure 14.4), while men were reduced to their initials. By the 1970s, the Society received complaints from female authors, but its staff believed this to be the appropriately 'gentlemanly' attitude.[126] The practice of marking out women authors in this way continued until 1990.[127]

Reforming a 'cumbersome' system

In June 1967, Council considered proposals to revise the editorial 'policy and procedures' for its journals, to make them better adapted to the needs of post-war scientists. The proposals were presented by the then-secretaries, the applied mathematician James Lighthill and the experimental pathologist Ashley Miles. Despite strenuous efforts by Lighthill, the purpose and remit of the journals remained unchanged; but significant changes to the procedures were agreed by December 1968.

During 1963 and 1964, the unresolved concerns about the B-side journals had led the biological sectional committees to make some modest suggestions for reform. One outcome was that a letter was sent 'to heads of appropriate university departments' letting them know that authors who wished to submit to the Society's journals 'might approach a Fellow, not personally known to them' to act as communicator.[128] This was an attempt to reach a wider pool of potential authors, beyond the Society's networks. Ashley Miles later reported that this 'exhortation' had been 'moderately successful but could be more so'.[129] Discussions about reform continued somewhat aimlessly, because the underlying concerns were not shared by the physical scientists. Things changed with the appointment of James Lighthill as physical secretary in 1965.[130] By June 1967, he had amassed 10 pages of ideas for reform.

Lighthill had two sets of concerns. First, he was frustrated with the day-to-day operation of the editorial processes, which were still operating under procedures drawn up in the 1890s (Chapter 11). On this matter, Miles agreed, and the two secretaries pushed through significant reforms in editorial procedure, which took effect at the end of 1968. Lighthill's second ambition was to clarify, and perhaps reorient, the purpose and remit of the Society's journals. Despite occasional proposals, such as those from McCance in 1954, this does not seem to have received any serious attention at the Society since the 1905 and 1914 reforms that created the modern *Proceedings* (Chapter 11). But Miles did not share this vision, and the opportunity for change was missed.

The secretaries were in broad agreement that the current procedures for choosing referees and evaluating their reports were 'cumbersome and, to some extent, outmoded'.[131] The Society's two-century commitment to distributed and collective – not individual – decision-making was well known to slow things down, especially when individuals were unavailable 'during the summer holidays'.[132] Some improvements were relatively easy, such as corresponding directly with the author rather than routing all correspondence through the fellow acting as communicator. The more difficult questions revolved around the roles of the sectional committees and Council. In theory, since 1896, all publication decisions had to be approved by the relevant committee and also by Council. In reality, for straightforward decisions to publish, those ratifications had become automatic and after-the-fact. But decisions regarding problematic papers usually did go to the committees, and could take weeks or months to finalise.

The underlying issue was that the Society still tried very hard to avoid formally rejecting papers that had been authored or communicated by one of its fellows; and its processes were set up to ensure that it could only be done via a collective decision, not by an individual (whether secretary or sectional committee chair). Papers that referees found to be weak or problematic were traditionally 'withdrawn' by the fellows who communicated them, but, like Egerton before him, Lighthill discovered that fellows were willing to 'fight back'. Without a compliant communicator, the secretary's options were either to seek reports from more referees until the communicator gave way under the force of opinion, or to request Council to make a final adjudication. Lighthill felt that Council meetings were far too busy to consider 'detailed transcriptions of the evidence leading to rejection', but found that persuading communicators to withdraw took an 'excessive amount' of time from both 'referees and editors'.[133] He and Miles both wanted 'increased powers' for the secretaries, and this principally meant the power to reject papers on their own authority.

The secretaries also raised the problems of identifying suitable referees. Lighthill wrote that the current system 'seems to assume that the Physical Secretary is a sort of scientific Pooh Bah, knowing all the "A" side sciences and everyone in them; and that the five Sectional Committee Chairmen can field all the mistakes he may make. Neither assumption is justified.' He believed that six people could not possibly ensure 'personal, and preferably up-to-date, knowledge of [every] field and those working in it'.[134] He suggested that each secretary needed about 12 'helpers'. In similar vein, Miles suggested creating an editorial board for each secretary.

The secretaries convinced the Council that these changes would make it easier to 'deal expeditiously' with papers submitted to the

Society.¹³⁵ From December 1968, the editorial procedures were revised, as can be seen in Figure 14.5. The sectional committees ceased to be involved with publication decisions. The secretaries were granted the power to reject papers. They each received an editorial board of 12 fellows to assist with 'referee selection and evaluation' and generally support the interests of the journals.¹³⁶ These fellows were labelled 'associate editors' (not to be confused with Graddon's job as 'assistant editor'). The associate editors had the personal power to accept papers on the basis of referees' reports, under the supervision of the relevant secretary, but not to reject. The task of dealing with 'potent and vocal' communicators was still recognised as requiring a higher level of authority.¹³⁷ Almost incidentally, the new procedures also disbanded the Publications Committee (from the 1930s), and allowed the secretaries and associate editors to send papers to referees who were not fellows and/or not resident in Britain. It remained rare for papers to be sent overseas for refereeing until the age of email, but expanding the pool of potential referees beyond the 700 fellows of the Society was a practical solution to the burden placed on those fellows by the number of submissions.

The new associate editors replaced the committee-based process that had originally been created in 1752 and augmented in 1896. All the surviving evidence suggests that the Committee of Papers had ceased to have much meaningful role in decision-making (except rejections) long before 1968, with secretaries from George Stokes to James Jeans and James Lighthill managing most of the editorial decisions. Despite the decline of committee decision-making, however, responsibility for the Society's editorial decisions did remain 'distributed'. In that sense, the 1752 determination that one single individual should not be entrusted with the reputation of the Society and its *Transactions* had not disappeared, but had come to be enacted through the involvement of multiple individuals: as communicators, as referees and as associate editors and secretaries. In broad terms, this was a model that was gaining wider currency in the world of journal publishing. For instance, when Robert Maxwell persuaded scientists to edit journals for Pergamon Press, their first step was usually to create an editorial board, thus generating the appearance of a community of scholars behind the journal (and also, by strategic inclusion of scientists from the UK, the US and Europe, forging pathways towards an international community of authors and readers).¹³⁸ The use of referees (or reviewers) was also making an appearance at other journals, but practices varied widely in the 1960s.¹³⁹

At the Royal Society, the 1968 reforms marked the end of the tradition that only fellows could be involved in the evaluation of research

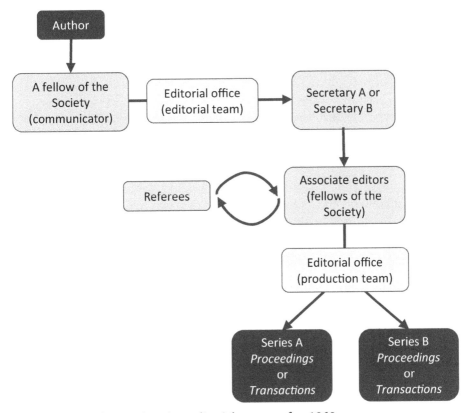

Figure 14.5 The Royal Society editorial process after 1968.

submitted to the Society. David Martin's claim that the refereeing undertaken by learned societies was a guarantee of quality now rested on the Society's ability to choose appropriate referees, rather than on the status of those referees as fellows of the Society. Nonetheless, the choice of referees remained highly dependent upon the personal connections and networks of the secretaries and associate editors. This may be one reason why there was no improvement in the number of women scientists asked to referee after 1968, even though it was now theoretically possible to ask women scientists other than the 20 or so among the fellows.[140] Personal connections were important not just for choosing appropriate referees but for interpreting their reports. As Lighthill had explained, referees' recommendations were 'seldom ... clear-cut', and it was important to check closely for 'some unintentional clue' that might reveal a bias that would lead a recommendation to be 'treated with caution'. For this, 'knowledge of the field and the personalities involved helps enormously'.[141]

Lighthill's desire to reform the Society's journals had not arisen solely from his desire to make the editorial process smoother. He was also concerned that a review of the purpose and mission of the journals was long overdue. He asked:

> It is essential to ask oneself: at a time when the number of specialized journals has increased enormously, and left practically no sub-specialism without its excellent opportunities for publication of good papers, what publication role remains for the Royal Society?[142]

He seems to have particularly feared that the move from personal to institutional copies of journals would reduce the readership for papers in generalist journals because 'library browsers' were more likely to focus on journals in their specialist fields.[143]

In trying to discern a remit for the Society's journals, Lighthill turned to the questions asked of referees, originally devised in the 1890s (see Figure 11.3). He noted the emphasis on publishing work of 'critical significance', and not the mere 'accumulation of data'. He was particularly interested in the question about whether certain papers might be better suited to a specialist society. He seems to have been unaware that this had its roots in the treasurer's financial worries (see Chapter 12), and interpreted it to mean that papers published by the Royal Society (rather than a specialist society) should appeal to readers working in a variety of fields. He shared Henry Armstrong's 1902 idea that a distinctive mission for the Society's journals might be 'to counteract the trends to over-specialization'. But, rather than suggesting a focus on interdisciplinary research, he argued instead that there was a gap in the market for a service along the same lines as that offered to the general public by 'the high-class popular-science journals', but aimed at 'professional scientists who are experienced, active, able to read original papers and keen to get an idea of what's new in science from the most up-to-date and authentic available sources'.[144] Lighthill seems to have been taking his inspiration from *New Scientist* (f. 1956) rather than *Nature* or *PNAS*.

However, Ashley Miles did not agree with the idea that papers published by the Royal Society needed to be 'of immediate interest to a wide circle of scientific readers'. He believed that research papers would inevitably be specialised, and the key thing was to insist that they 'report first class substantial pieces of work'.[145] Nor did Lighthill's vision find support elsewhere in the Society (though it would resurface in the 1990s, with the creation of *Science and Public Affairs*). Thus, the *Proceedings* and the *Transactions* moved into the new decade with the same general remit as they had had since the 1910s. By the 1980s, many fellows would feel that a serious review was long overdue.

In less than 15 years, the new sales-focused approach to publishing had become utterly engrained in the thinking of the Society's officers and fellows. In the midst of the 1967 discussions of editorial changes, one of the secretaries referred to sales as the 'means of disseminating the knowledge embodied in its journals as widely as possible'.[146] The Society's long history of disseminating knowledge outside the commercial marketplace was forgotten. In this new world of publishing, an increased flow of submissions would be seen not as a problem, but as an opportunity to sell more content and generate 'welcome financial gains' to the Society.[147] With its funds freed from the burden of supporting the journals, the Society would be able to undertake a far wider range of other activities. From the 1960s onwards, its fellows and staff were active and innovative in finding ways for supporting scientific research, shaping policy and facilitating international collaboration and communication.[148]

The prominent role played by David Martin, both in the (internal) transformation of the Society's publishing finances and in the (public) collaboration with the Nuffield Foundation, demonstrates the increased importance of senior staff in the activities and policy of the Royal Society in the twentieth century. In 1976, the then-president would write that, 'the great development the Society has undergone since the last war is due more to David Martin than to any other person'.[149] He drove the Society's transition to a commercial model of publishing, and by creating the publications sales team, enabled the Society to become the publisher of its own journals. In contrast, many other learned societies chose to work with a publishing partner, and, by the 1970s, over 60 per cent of UK learned society journals would be published in collaboration with a commercial publisher.[150]

In editorial matters, in contrast, the drive for reform came from the fellows and officers of the Society. The secretaries had personally experienced the frustrations and delays of working within a system that had been devised in an earlier age, where it seemed that everyone had known everyone else, and speed was far less important. The 1968 reforms simplified the workflow and, in the short term, improved the publication times. (In the medium term, the labour shortages, postal delays and strikes of the 1970s would slow things down again.) However, the failure to engage with Lighthill's desire to rethink the purpose of the Royal Society journals was a missed opportunity, even if his specific proposal did not garner support. There continued to be a significant uncertainty of purpose for generalist journals in an age of specialisation, and for lengthy, carefully scrutinised research papers in an age of rapid priority-seeking. In particular, did the *Transactions* have a role other than a place to put things that did not quite fit in the *Proceedings*?

In November 1965, the Society had celebrated the 300th anniversary of the *Transactions* with an exhibition and soirée attended by Queen Elizabeth the Queen Mother and King Gustav Adolf of Sweden.[151] The celebrations were notably backward-looking: the Society was justifiably proud of the seventeenth-century origins of the *Transactions*. Both the exhibition and the anniversary article in *Notes and Records* by Edward Andrade focused on the pioneering achievements of Oldenburg and the activities of the early fellows of the Society, as reported in the early *Transactions*. Amid all the enthusiasm for the late seventeenth century, it perhaps went unnoticed that little was said about the relevance of the *Transactions* to twentieth-century scientists. Other than pointing out that Thomas Henry Huxley, a century earlier, had valued the *Transactions*, Andrade seemed unable to offer even the most anodyne comment about its contemporary role.[152] This nicely encapsulates the Society's uncertainty about what to do with the *Transactions*, which would not be resolved until the 1990s. Fortunately, the new-found financial stability of the journals made it easier to ignore any ongoing worries about their relevance or usefulness.

Notes

1 Elsevier figures from 1969, see Einar H. Fredriksson, 'The Dutch publishing scene: Elsevier and North-Holland', in *A Century of Science Publishing: A collection of essays*, ed. E.H. Fredriksson (Amsterdam: IOS Press, 2001), 68; Pergamon figures for 1964 cited in Martin Richardson, 'Journals', in *The History of Oxford University Press, Volume IV: 1970 to 2004*, ed. Keith Robbins (Oxford: Oxford University Press, 2017), 429.

2 Peter Collins, *The Royal Society and the Promotion of Science since 1960* (Cambridge: Cambridge University Press, 2015), esp. chs. 1–2, 7 and 10. See also Jennifer Rose Goodare, 'Representing science in a divided world: The Royal Society and Cold War Britain' (PhD, University of Manchester, 2013).

3 Roger L. Geiger, 'Science, universities, and national defense, 1945–1970', *Osiris* 7 (1992): 26–48; Loren R. Graham, 'Big Science in the last years of the Big Soviet Union', *Osiris* 7 (1992): 49–71; David Edgerton, *Warfare State: Britain, 1920–1970* (Cambridge: Cambridge University Press, 2005), ch. 6; Jon Agar, *Science in the Twentieth Century and Beyond* (London: Polity Press, 2013), Part 3.

4 Alvin M. Weinberg, 'Impact of large-scale science on the United States', *Science* 134, no. 3473 (1961): 161–4; James H. Capshew and Karen A. Rader, 'Big Science: Price to the present', *Osiris* 7 (1992): 2–25; Peter Galison and B.W. Hevly, *Big Science: The growth of large-scale research* (Stanford, CA: Stanford University Press, 1992).

5 Derek Richter, 'How many more new journals?' *Nature* 186 (1960), 18. This was a meeting of the Scientific Publications Council, sponsored by the Ciba Foundation.

6 Derek J. De Solla Price, *Science since Babylon* (New Haven, CT: Yale University Press, revised edn 1975); and *Little Science, Big Science* (New York: Columbia University Press,

1963). For a modern reassessment, see Lutz Bornmann and Rüdiger Mutz, 'Growth rates of modern science: A bibliometric analysis based on the number of publications and cited references', *Journal of the Association for Information Science and Technology* 66, no. 11 (2015): 2215–22.

7 Andrew Nash, Claire Squires and Ian Willison, eds, *The Cambridge History of the Book in Britain. Volume 7: The twentieth century and beyond* (Cambridge: Cambridge University Press, 2019), chs 17–18.

8 Melinda Baldwin, *Making 'Nature': The history of a scientific journal* (Chicago, IL: University of Chicago Press, 2015).

9 Tom Clare, of Edward Arnold Ltd, quoted in Richter, 'How many more new journals?', 18.

10 H. Kay Jones, *Butterworths: History of a publishing house* (London: Butterworth & Co., 1980); David McKitterick, *History of Cambridge University Press. Volume 3: New worlds for learning, 1873–1972* (Cambridge: Cambridge University Press, 2004), chs 17–18. William H. Brock and A.J. Meadows, *The Lamp of Learning: Taylor & Francis and the development of science publishing* (London: Taylor & Francis, revised edn 1998), ch. 7. Oxford University Press did not seriously expand its journal publishing until the 1980s, see Keith Robbins, ed., *The History of Oxford University Press. Volume IV: 1970 to 2004* (Oxford: Oxford University Press, 2017), ch. 15.

11 Tom Clare of Edward Arnold Ltd, reported in Richter, 'How many more new journals?', 18.

12 D.C. Martin, 'The Royal Society's interest in scientific publications and the dissemination of information', *Aslib Proceedings* 9, no. 5 (1957): 127–41, 135.

13 On the contemporary sense of ambivalence towards commercial journals, see Brock and Meadows, *Lamp of Learning*, 193.

14 Martin's diary 1947, RS MS/869, 29.

15 David C. Martin, 'The retirement of Mr J. C. Graddon, assistant editor to the Society', *N&R* 27.2 (1973): 325–6, 325.

16 Cartwright was on Council 1956–7; as was Lonsdale in 1961–2, during which time she was appointed one of the vice-presidents.

17 Martin, 'The Royal Society's interest', 139–40.

18 'The publication of results of scientific research in the United Kingdom' [1947], in RS CAP/1947/7/6, 5.

19 RS OM2/47(51), 12 July 1951.

20 For North America, see 'Note of decisions taken at a discussion with representatives of the CUP on 13 December 1949', in RS OM/68(49). For Japan, see 'Notes on a discussion with representatives of the CUP on 9 October 1950', in RS OM/47(50). On CUP's new American branch, see McKitterick, *History of Cambridge University Press 3*, ch. 15.

21 Monthly report on papers received for publication, 11 January 1951, in RS OM/6(51).

22 Crutchley to RS, 4 February 1953, in RS OM/8(53).

23 RS OM/17(52), 25 March 1952.

24 Memorandum regarding scientific publication from Assistant Secretary, 12 July 1950, in RS OM/38(50); also RS *Year Book* (1952), 196.

25 On lithography in this period, see Sarah Bromage and Helen Williams, 'Materials, technologies and the printing industry', in *The Cambridge History of the Book in Britain. Volume 7: The twentieth century and beyond*, ed. Andrew Nash, Claire Squires and I.R. Willison (Cambridge: Cambridge University Press, 2019), 41–60.

26 Memorandum regarding Experimental Pro-Printing Plant, [undated, perhaps December] 1951, in RS OM/66(51).

27 RS *Year Book* (1957), 202. The staff were Colin Ronan, Derek Machin and Maurice Oak, all of whom are mentioned in 'Comrades of old: Royal Society staff pensioners, biographical sketches' (London: privately circulated for The Royal Society Former Staff Association, 2015).

28 The table is recalculated from figures in the 1951 accounts in RS *Year Book* (1952), 221. See also Aileen Fyfe, 'From philanthropy to business: The economics of Royal Society journal publishing in the twentieth century', *N&R* https://doi.org/10.1098/rsnr.2022.0021, Figure 2A.

29 Treasury to RS, 8 January 1951, in RS OM/4(51); and Treasury to RS, 27 November 1951, in RS OM/61(51).

30 'Consideration of estimated deficit on RS publications account', 4 November 1953, uncatalogued but in RS OM3.

31 'Consideration of estimated deficit …', 4 November 1953, uncatalogued but in RS OM3.

32 RS OM2/62(53), 30 November 1953.
33 McKitterick, *History of Cambridge University Press 3*, chs. 14 and 16.
34 Kingsford to RS, 26 November 1953, in RS OM3/61(53).
35 Kingsford to RS, 26 November 1953, in RS OM3/61(53).
36 RS OM2/62(53), 30 November 1953.
37 CUP to RS, 16 December 1953, in RS OM3/2(54).
38 RS OM2/15(54), 4 March 1954.
39 RS OM2/15(54), 4 March 1954.
40 'Report of Council, 1954', in RS *Year Book* (1955), 188.
41 Draft letter from RS Council to fellows, [undated 1954, uncatalogued] in RS OM3; also RS CMP/19, 1 April 1954; and RS *Year Book* (1955), 188.
42 'Revision of the lists of exchanges and gifts of the Royal Society's publications', 2 March 1954, in RS OM3/14(54).
43 'Recommended reductions in exchanges and gifts of the Royal Society's publications' [undated, spring 1954], in RS OM3/16(54).
44 RS *Year Book* (1955), 188.
45 [Theobald], 'A report on the sales and distribution of publications', 9 May 1973, RS RMA/199.
46 Draft letter from RS Council to fellows, [undated 1954, uncatalogued] in RS OM3; Council minutes, 1 April 1954, RS CMP/19; and *Year Book* (1955), 188. 'Distribution of Royal Society publications [in 1947]', [undated, but for an early 1948 meeting] in RS OM3/2(48).
47 RS OM2/15(54), 4 March 1954.
48 'Comrades of old', 11 (for Boreham, 1914–98); and 33–4 (for Dance, 1914–2013).
49 See Fyfe, 'From philanthropy to business', Figure 2A.
50 RS AB/3/1/33.
51 On Pergamon price rises in the 1960s, of 5–10 per cent annually, see Brian Cox, 'The Pergamon phenomenon 1951–1991: Robert Maxwell and scientific publishing', *Learned Publishing* 15 (2002): 273–8, 275.
52 We have compiled this subscription price data from scattered sources, mostly minutes of Officers' Meetings and the Library Committee.
53 [James Lighthill], 'Possible changes in arrangements relating to *Proceedings* and *Transactions*', 7 June 1967, p. 7, in RS C/84(67).
54 For our *longue durée* picture of circulation trends, see Figure 3 in Fyfe, 'From philanthropy to business'.
55 Publication Sales Report, 1 February 1955, in RS OM3/6(55).
56 Publication Sales Reports: 1 February 1955, in RS OM3/6(55); and 12 April 1955, in RS OM3/17(55).
57 Draft of letter from Council to fellows, (undated 1954, uncatalogued) in RS OM3.
58 Publication Sales Report, 1 February 1955, in RS OM3/6(55).
59 Learned Journals Committee, RS CMB/96, 7 December 1956.
60 Martin, 'Sales of Philosophical Transactions', 7 June 1955 in RS OM3/26(55); 'Report of Council 1955', in RS *Year Book* (1956), 200.
61 Frank Morley, *Self-Help for Learned Journals: Notes compiled for the Nuffield Foundation* (London: The Nuffield Foundation, 1963), 6; Robert N Miranda, 'Robert Maxwell: Forty-four years as publisher', in *A Century of Science Publishing: A collection of essays*, ed. E.H. Fredriksson (Amsterdam: IOS Press, 2001), 77–89. Brock and Meadows suggest it was invented by the commercial publishers, but the Royal Society had been doing this for decades, see Brock and Meadows, *Lamp of Learning*, ch. 7.
62 RS *Year Book* (1956), 200; and subsequent *Year Books*.
63 'Report of Council, 1955', in RS *Year Book* (1956), 199.
64 'Report of Council, 1958', in RS *Year Book* (1959), 208.
65 This already-existing deal is referred to in RS CMP/21, 18 July 1963. See Charles Chadwyck-Healey, *Publishing for Libraries: At the dawn of the digital age* (London: Bloomsbury, 2020), ch. 2.
66 On the Society's ownership claim, see Aileen Fyfe, Julie McDougall-Waters, and Noah Moxham, 'Credit, copyright, and the circulation of scientific knowledge: The Royal Society in the long nineteenth century', *Victorian Periodicals Review* 51, no. 4 (2018): 597–615 and Aileen Fyfe, 'The production, circulation, consumption and ownership of scientific

67 RS OM2/15(54), 4 March 1954.
68 RS CMP/21, 18 July 1963.
69 Fyfe, 'From philanthropy to business', Figure 2A.
70 RS CMP/28, 3 November 1955.
71 On the Society's historical lack of engagement with copyright, see Fyfe et al., 'Credit, copyright'.
72 RS *Year Book* (1952), 195.
73 RS *Year Book* (1957), 213 and RS *Year Book* (1958), 206. More generally, see Brad Sherman and Leanne Wiseman, 'Fair copy: Protecting access to scientific information in post-war Britain', *The Modern Law Review* 73, no. 2 (2010): 240–61; Leanne Wiseman and Brad Sherman, 'Facilitating access to information: Understanding the role of technology in copyright law', in *The Evolution and Equilibrium of Copyright in the Digital Age*, ed. Susy Frankel and Daniel Gervais (Cambridge: Cambridge University Press, 2014), 221–38. The 1911 Act had been written with hand-copying in mind.
74 RS *Year Book* (1951), 178 and RS *Year Book* (1952), 196–7.
75 Learned Journals Scientific Advisory Committee, Nuffield Foundation archives, L.J.1, 15 April 1955.
76 Documentation for this meeting survives in the Florey papers, RS HF/1/17/3/2; including the invitation from the RS secretaries to the secretaries of other societies, 10 May 1963, RS 98HF.160.2.2.
77 Morley, *Self-Help for Learned Journals*, 1.
78 Unconfirmed notes of meeting on 8 December 1953, quoted in 'Notes on an informal meeting to discuss publication and sales promotion of scientific journals held at the Royal Society', 19 July 1956, copy from the archives of the Royal Anthropological Institute, 40/2/13.
79 See Aileen Fyfe, 'Self-help for learned journals: Scientific societies and the commerce of publishing in the 1950s', *History of Science* 60, no. 2 (2022), 255–79, Table 2 (from a list originally presented to Nuffield L.J.1, 15 April 1955).
80 For a fuller discussion of this project, see Fyfe, 'Self-help for learned journals'.
81 RS EC/1939/24.
82 On the Commonwealth scholarships, see Collins, *The Royal Society and the Promotion of Science*, 236. On the activities of the Foundation in the early 1950s, see Ronald W. Clark, *A Biography of the Nuffield Foundation* (London: Longman, 1972), ch. 4.
83 Lusty Report, quoted in Morley, *Self-Help for Learned Journals*, 1–2.
84 Nuffield L.J.2, 16 March 1956.
85 Nuffield L.J.3, 15 March 1957.
86 Morley, *Self-Help for Learned Journals*, contents page.
87 The American Institute of Physics had been using page charges from the early 1930s; the *Journal of the American Chemical Society* began using them in January 1963. See Tom Scheiding, 'Paying for knowledge one page at a time: The author fee in physics in twentieth-century America', *Historical Studies in the Natural Sciences* 39, no. 2 (2009): 219–47; Marianne Noel, 'Building the economic value of a journal in chemistry: The case of the Journal of the American Chemical Society (1879–2010)', *Revue française des sciences de l'information et de la communication*, no. 11 (2017).
88 For the new RS attitude to the grant process, see Ad hoc Committee on the Economics of Learned Journals, 20 November 1957, RS CMB/148b.
89 Martin, 'The Royal Society's interest', 136.
90 For 'pompous' and leadership, see [Florey] to J.Z. Young, 14 June 1963, RS HF/1/17/3/2/10. For 'tedious', see annotation on 'Memorandum for PRS', 12 June 1963, RS HF/1/17/3/2/5.
91 Young to Florey, 13 June 1963, in RS HF/1/17/3/2/9.
92 J.C. Gardiner, 'Review of self-help for learned journals', *Proceedings of the Botanical Society of the British Isles* 5 (1963/4).
93 'Code for the publication of new scientific journals', 12 June 1963, RS HF/1/17/3/2/7.
94 'Code for the publication of new scientific journals', 12 June 1963, RS HF/1/17/3/2/7.
95 Martin, 'The Royal Society's interest', 140, 134–5.
96 'The Royal Society's interest', 134–5.
97 Richter, 'How many more new journals?'

98 D. Richter, reported in 'How many more new journals?', 18.
99 Morley, *Self-Help for Learned Journals*, 17–18.
100 Numbers of submissions, and of papers published, were printed in the *Year Book* from 1957. See Aileen Fyfe, 'Submissions in life sciences vs physical sciences, 1927–1989' (13 February 2018), https://arts.st-andrews.ac.uk/philosophicaltransactions/submissions-in-life-sciences-vs-physical-sciences-1927-1989.
101 R.A. McCance, 'Memorandum on the Proceedings of the Royal Society, Series B', February 1954, in RS C/20(54).
102 Table 10.2 in Collins, *The Royal Society and the Promotion of Science*, 254.
103 McCance, 'Memorandum on the Proceedings … Series B', February 1954, in RS C/20(54).
104 Martin, 'The Royal Society's interest', 134 ('serious and discouraging'); 'The publication of results of scientific research in the United Kingdom' [1947], RS CAP/1947/7/6.
105 Aileen Fyfe, 'Time taken to publish' (8 February 2022), https://arts.st-andrews.ac.uk/philosophicaltransactions/time-taken-to-publish.
106 RS *Year Book* (1970), 301.
107 Morley, *Self-Help for Learned Journals*, 4.
108 Martin to Crutchley, 22 October 1957, in RS RMA/1015.
109 Regarding the speed of publication 1947–60 in ED7, RS RMA/1015; for the strike, see T.E. Allibone to Martin, 2 March 1960, RS P/01/JCG/AE.
110 Morley, *Self-Help for Learned Journals*, 4.
111 The suggestion came from Oxford-based émigré physicist Franz Simon. The officers initially decided to take no action, RS OM2/31(51), 3 May 1951. See also RS CMP/18, 8 November 1951.
112 'Information to contributors', *PNAS* 46 (January 1960), back matter; and 'Information to contributors', *PNAS* 55 (March 1966), back matter.
113 [Ashley Miles], 'Notes on *Proceedings* and *Philosophical Transactions*' [June 1967], p. 4 in RS C/84(67).
114 'Information to contributors', *PNAS* 46 (January 1960), back matter.
115 RS *Year Book* (1968), 277.
116 R.C. Garry to G.L. Brown, 29 November 1960, RS GLB/65/41/6.
117 R.C. Garry to G.L. Brown, 20 November 1960, RS GLB/65/41/5.
118 The initial approaches to referees are recorded in Register of Papers B (1936–60), RS MS/613, f. 77. Referee reports on paper B62, in RS RR/1950-51.
119 Alan M. Turing, 'The chemical basis of morphogenesis', *PTB* 237 (1952): 37–72. See also Philip Ball, 'Forging patterns and making waves from biology to geology: A commentary on Turing (1952) "The chemical basis of morphogenesis"', *PTB* 370 (2015).
120 N.K. Adam to D.C. Martin, 15 July 1950, with report on paper A128 in RS RR(A)/1950 (withdrawn). On this letter, see Camilla Mørk Røstvik, '"I am seriously tempted to burn some of the papers which reach me for an opinion"', *Times Higher Education* (2016), https://www.timeshighereducation.com/features/workload-survival-guide-for-academics.
121 N.K. Adam to D.C. Martin, 15 July 1950, with report on paper A128 in RS RR(A)/1950 (withdrawn).
122 [Miles], 'Notes on *Proceedings* …' [June 1967], p. 2 in RS C/84(67).
123 Aileen Fyfe, Flaminio Squazzoni, Didier Torny and Pierpaolo Dondio, 'Managing the growth of peer review at the Royal Society Journals, 1865–1965' *Science, Technology and Human Values* 45, no. 3 (2020): 405–29.
124 [Miles], 'Notes on *Proceedings*…' [June 1967], p. 2 and Table 2 in RS C/84(67).
125 Camilla M. Røstvik and Aileen Fyfe, 'Ladies, gentlemen, and scientific publication at the Royal Society, 1945–1990', *Open Library of Humanities* 4, no. 1 (2018): 1–40, http://doi.org/10.16995/olh.265.
126 Our interview with Bruce Goatly, 20 January 2016.
127 See Røstvik and Fyfe, 'Ladies, gentlemen, and scientific publication'.
128 'Report of Council 1965', in RS *Year Book* (1966), 260–1.
129 [Miles], 'Notes on *Proceedings* …' [June 1967], in RS C/84(67).
130 On Lighthill's later work, see Jon Agar, 'What is science for? The Lighthill report on artificial intelligence reinterpreted', *British Journal for the History of Science* 53, no. 3 (2020): 289–310.
131 'Proposed future policy and procedures for Royal Society *Philosophical Transactions* and *Proceedings*' [for Council, 15 June 1967], in RS C/84(67).

132 [James Lighthill], 'Possible changes in arrangements relating to *Proceedings* and *Transactions*', 7 June 1967, p. 9 in RS C/84(67). On distributed editorship, see Aileen Fyfe, 'Editors, referees, and committees: Distributing editorial work at the Royal Society journals in the late nineteenth and twentieth centuries', *Centaurus* 62, no. 1 (2020): 125–40.
133 [Lighthill], 'Possible changes ...', 7 June 1967, pp. 4 and 9 in RS C/84(67).
134 [Lighthill], 'Possible changes ...', 7 June 1967, p. 9 in RS C/84(67).
135 'Proposed future policy ...' [for Council, 15 June 1967], in RS C/84(67).
136 RS CMP/22, 15 June 1967. See also RS CMP/23, 9 May 1968; new standing orders 1969, in RS CMP/23, 4 April 1969; and 'Note for the guidance of associate editors', [early 1969?], RS RMA uncatalogued.
137 [Lighthill], 'Possible changes ...', 7 June 1967, p. 9 in RS C/84(67).
138 Brock and Meadows, *Lamp of Learning*, 205. And see, for instance, the large and very international editorial board listed on the title page of the first volume of *Journal of Inorganic & Nuclear Chemistry* 1 (1955). See also Joe Haines, *Maxwell* (London: McDonald & Co., 1988), ch. 5.
139 The front matter of journals from the 1960s and 1970s sometimes indicates that referees were being used, but it was still far from a standard practice. See Melinda Baldwin, 'Credibility, peer review, and Nature, 1945–1990', *N&R* 69.3 (2015): 337–52; David Pontille and Didier Torny, 'From manuscript evaluation to article valuation: The changing technologies of journal peer review', *Human Studies* 38, no. 1 (2015): 57–79.
140 There were 18 women fellows in 1965 (2.75 per cent), and 24 in 1975 (3.04 per cent); on women as referees, see Røstvik and Fyfe, 'Ladies, gentlemen, and scientific publication'; and Aileen Fyfe, 'What history tells us about diversity in the peer review process' (12 September 2018), https://arts.st-andrews.ac.uk/philosophicaltransactions/diversity-in-the-peer-review-process.
141 [Lighthill], 'Possible changes ...', 7 June 1967, p. 9 in RS C/84(67).
142 [Lighthill], 'Possible changes ...', 7 June 1967, p. 8 in RS C/84(67).
143 [Lighthill], 'Possible changes ...', 7 June 1967, p. 7 in RS C/84(67).
144 [Lighthill], 'Possible changes ...', 7 June 1967, p. 8 in RS C/84(67).
145 [Miles], 'Notes on *Proceedings* ...' [June 1967], p. 1 in RS C/84(67).
146 [Miles], 'Notes on *Proceedings* ...' [June 1967], p. 1 in RS C/84(67).
147 [Lighthill], 'Possible changes ...', 7 June 1967, p. 3 in RS C/84(67).
148 As described in Collins, *The Royal Society and the Promotion of Science*.
149 Lord Todd, 'Address at the memorial service for Sir David Christie Martin ...', *N&R* 32.1 (1977): 1–3.
150 Alan Singleton, 'Learned societies and journal publishing', *Journal of Information Science* 3, no. 5 (1981): 211–26, cited in Richardson, 'Journals', 429. (Eighty-eight of the 143 journals published by 121 learned societies were published with a commercial partner.)
151 'Conversazione to mark the 300th anniversary of the publication of the Philosophical Transactions of the Royal Society, Hooke's Micrographia, and Evelyn's Sylva', *N&R* 21.1 (1966): 12–19.
152 Edward Neville Da Costa Andrade, 'The birth and early days of the *Philosophical Transactions*', *N&R* 20 (1965): 9–27.

15

Survival in a shrinking, competitive market, c. 1970–1990

Camilla Mørk Røstvik and Aileen Fyfe

In 1987, the Royal Society's publishing staff believed that academic publishing was 'in general, in a healthy state'.[1] The previous 25 years had seen the 'discovery' and then 'exploitation' of the new, well-funded market for scientific journals.[2] Commercial publishers had been prominent – Elsevier now published almost 600 journals, and Pergamon had around 400 journals – but the Society's staff was confident that society publishers were 'maintaining their positions'.[3] The Royal Society might only have four journals (plus its various internally oriented periodicals), but its detailed study of the 'scientific information system' had reported that 65 per cent of all primary research journals in the UK were still issued by learned societies.[4]

Since 1972, society publishers had been collaborating and sharing best practice – as Robert Lusty had recommended in 1955 (Chapter 14) – through the Association of Learned and Professional Society Publishers (ALPSP).[5] The fact that the Royal Society was involved, but had not been a founding member, is an indicator of the way that leadership among society publishers was shifting towards those with bigger publishing programmes, notably the Institute of Physics and the Royal Society of Chemistry. These organisations had been formed – in 1960 and 1980, respectively – from mergers of older institutions (including the Physical Society, and the Chemical Society). Their advantages of scale helped both organisations to pursue growth strategies for their publishing divisions.[6] The Royal Society's journals retained whatever prestige accrued from their history, but it would be difficult to present the Society as leading the way for society publishers in the 1970s or 1980s.

The creation of the ALPSP distinguished the mission-driven society approach to journal publishing from the avowedly commercial approach of the members of the STM Group of publishers (1969; later, the International Association of Scientific, Technical, and Medical Publishers).[7] Yet the separation between society and commercial publishers was less than it had been when David Martin worried about commercial influences in the 1950s (see Chapter 14). First, while the Royal Society had become self-supporting, and generated modest surpluses, some societies had learned the benefit of taking an 'increasingly commercial' approach to their publishing, and were generating enough income from their publications 'to make regular and substantial contribution to their other activities'.[8] Society publishers might do different things with their 'profits', but on a day-to-day level, their practices might not be so very different from the commercial firms.

Second, in the late 1970s and 1980s, academic publishers of all stripes had to find ways to cope with the effects of the oil crisis, high inflation and exchange rate fluctuations, and with government cuts to research budgets on both sides of the Atlantic that made it more difficult to sell journals to librarians. Publishers responded by increasing prices and, by 1990, librarians would be referring to the unaffordability of journals as a 'serials crisis'.[9]

At the Royal Society, the enthusiasm generated by the buoyant sales figures of the early 1960s had, by the 1980s, been replaced by worries about subscription numbers.[10] An apparently irreversible decline in subscriptions threatened the Society's relatively new business model, but the accounts did remain in surplus throughout this period. The continuing viability of the sales-based business model, in the absence of the booming subscriber numbers that had made it possible in the first place, was largely due to price rises and, in the 1980s, falling production costs caused by changes in the way the Society's typesetting was done.

The Society's staff and officers worried about the place and prestige of its journals in the wider journal landscape, which affected both subscriptions from libraries and submissions from authors. In 1987, they noted the general problem of the increase in number of journals, and the more specific problem facing the Royal Society's journals. Most learned societies issued specialist journals in their field; in contrast, the Society's 'multi-disciplinary journals' seemed 'especially vulnerable'.[11] The Royal Society had not launched any new scientific journals since it split the *Proceedings* into A and B series in 1905. Other publishers in this period were creating new journals by splitting existing ones or launching totally new ones. The Chemical Society, for instance, split its journal into three series in 1966 (and added a fourth in 1969), and Pergamon Press grew its journals list from 59 in 1960 to 418 in 1991.[12] The Royal Society's worry was that, in this packed landscape of specialised journals, neither libraries nor authors were interested in generalist journals.

This fear was heightened by personal experiences such as that of the associate editor who had been visiting France and Italy, and reported that he 'had been unable to find a library that subscribed' to the *Proceedings*, 'or anyone who wished to have a paper published there'.[13] Nor did the Society's journals score well on the new metrics developed by Eugene Garfield's Institute of Scientific Information in Philadelphia. It had begun offering a 'Journal Citation Report' as part of its 'Science Citation Index' in 1975, and in 1987, its report on 'journal impact factors' ranked the *Proceedings of the National Academy of Sciences* (*PNAS*) at 12th, while the *Proceedings of the Royal Society* was only 73rd, and the *Philosophical Transactions* was not in the top 150.[14] By 1988, a committee would admit that the journals had 'not developed in the way that other, more successful, scientific journals have done so over the past two decades', and that 'a review and revision of the journals is well overdue'.[15]

There had been a Publications Policy Review in 1973, but it was the Publications Policy Review of 1987–8 that culminated in a major relaunch of the Society's journals in 1990. Although it was fellows of the Society who sat on the review committee, the advice and expertise of senior staff were key to the reforms. By the mid-1980s, there had been a generational changeover in the staff. After the retirement of Gerry Evans in 1982, the editorial department was being run by Bruce Goatly. At the more senior level, David Martin had died in 1976; his deputy replaced him, until retiring himself in 1985. The new executive secretary, Peter Warren, launched a series of administrative reviews and efficiency-seeking exercises, and the review of the journals occurred in that context.[16] Both Goatly and Warren were internally promoted, but, compared to their predecessors, they were relative newcomers, having only joined the Society in the late 1970s.

This chapter begins by examining how the Society's publishing finances survived despite the shrinking subscriptions market. It then turns to the concerns about the flow of submissions, and what could be done about it, before turning to the reforms of the late 1980s. Those changes in editorial practice and in journal branding laid the groundwork for the early twenty-first-century internationalisation of Royal Society publishing.

Selling journals in a shrinking market

The first Publications Policy Review was instigated by the electron microscopist James Menter, who became treasurer of the Royal Society in November 1972. At this point, it was not yet apparent that journal subscription numbers were settling into a long-term decline, and Menter was optimistic about the 'possibilities' of publishing becoming 'a major source of income' for the Society.[17] He was worried about the 'sequence of deficits' in the Society's annual accounts, and had undertaken 'a thoroughgoing examination of the Society's costs and income'.[18] He estimated that, with rising costs and the Society's expanding array of activities, he would need to find 'an additional £20,000' each year to keep the Society 'financially stable'.[19] And he thought that the publishing team might be able to generate some of that income, while 'always having proper regard to the constraints imposed by the Society's position as a learned body'.[20] Menter's desire for the journals to generate income was a change of tone from David Martin's emphasis on 'self-help' and sustainability in the years around 1960, but it reflected the experience from such organisations as the Chemical Society, where publications 'contributed significantly to the financial stability of the Society' and were 'its largest revenue-gathering activity'.[21]

The focus of the Publications Policy Review's discussions, in winter 1973–4, was, therefore, financial not editorial. They took place against a backdrop of recession, the oil crisis and escalating costs for paper, printing and postage.[22] As was usual for such committees, there were some trivial attempts at cost-cutting (including more limits on the number of free publications that could be taken by fellows), but the overall strategy depended on generating more income.[23] The key difference from past review committees, such as the 1920 Emergency Finance Committee (Chapter 12), was that the extra income was not necessary to break even, but to make a surplus.

The main recommendation was, again, to increase the selling prices of the journals. As we saw in Chapter 14, the Society's sales team had already been raising prices through the 1950s and 1960s. With the high inflation of the 1970s, it was even more crucial to ensure the prices were adjusted to cover the inflating costs. Menter's committee recommended the institution of annual (rather than sporadic) reviews of pricing. The price of a volume of the *Transactions* had been £19 in 1973, but rose to £70 in 1983, and a volume of the *Proceedings* rose from £9.50 to £34 over the same period.[24] These rises were broadly in line with inflation, and they kept the publication finances stable in real terms.

By the mid-1970s, it was becoming clear that the post-war boom in sales of the *Proceedings* and the *Transactions* had ended. The work of the sales team changed from persuading librarians to take out new subscriptions to journals, to persuading them to renew existing subscriptions from their over-stretched budgets. Menter's Publication

Policy Committee did realise that the desire to raise prices carried the 'real danger of losing subscriptions'.[25] *Proceedings A* lost almost a fifth of its subscribers between 1972 and 1979; the loss across all the journals averaged 11.5 per cent.[26] The Society continued to publish subscription numbers in its *Year Book* until the late 1980s and so the steady decline was publicly visible. It sought to reassure fellows by pointing out that the problem was not unique to the Royal Society, and 'many other scientific journals' were affected.[27] The Chemical Society, for instance, was also experiencing a 'downward trend in circulation'.[28]

The treasurer urged that 'renewed efforts should be made to promote sales', but the sales team struggled.[29] It was now led by Richard Theobald, who argued that his team were hampered by some of the Society's long-standing practices. Despite decades of pleas for a regular monthly periodicity for the *Proceedings*, all the Society publications continued to be 'issued at irregular intervals'. And the Society had reached an awkward compromise between librarians' desire for an annual subscription, and its own long-standing unwillingness to cap the total amount to be printed per year: the cost of a subscription was set for the year on the basis of 'the number of volumes *estimated* to be published'. These were hardly new problems, but they seemed worse now because librarians had so 'little room for manoeuvring' in their annual budgets; and because they had come to be out of step with the practices of other publishers. Theobald claimed that 'very few other societies of any consequence' still operated as the Royal Society did.[30] (The Royal Society finally moved to annual subscriptions for its journals in 1986.)

The Society's editorial team had to provide the sales team with estimates of output 18 months in advance, to enable renewal notices to be sent out to subscribers in August each year. The problem with this system was made glaringly apparent in 1977, when it was 'completely upset' by *Transactions B*.[31] Subscribers to *Transactions B* had been asked, in August 1976, to pay £130.40 for the four volumes estimated for 1977. They had only paid for three volumes of content in 1976, so this was already an increase.[32] But by late May 1977, the editorial team realised that they had already published enough material to fill the planned volumes, and had more material in press. Theobald reported that publication 'will continue' for the rest of the year, 'but in effect we have no subscribers' to pay for the additional material. His suggested plan of action focused on giving 'the best service and minimum inconvenience to subscribers', and trying to 'alienate them as little as possible'. For the longer term, his key desire was that the editorial team exercise 'a simple control of the maximum amount published in a period'.[33] Once more, the Society's long-standing ambition to publish all the good papers that came in was creating problems, but whereas, in the 1890s, the treasurer had wanted controls to limit the cost to the Society, Theobald wanted limits to make it easier to sell the publications.

For Bruce Goatly, 'the year we over-ran on *Trans B*' stuck in his memory as 'a bad year'. Goatly had been a postdoctoral researcher in biochemistry at Hull University before joining the Society as a copy-editor in 1977. He worked with the assistant editor, Gerry Evans, and, six years later, would take over the senior role when Evans retired. Goatly recalled that '[f]inances … were never really the worry … As far as I remember, we always ran at a surplus'. He was aware that there were 'relatively few subscribers', but, as a member of the editorial team, sales was someone else's problem.[34] For Theobald, in contrast, 'the commercial aspects of publishing are inescapable'.[35]

In their efforts to understand the falling circulation, Theobald's team analysed subscription data. They noticed that the *Proceedings* journals were more affected than the *Transactions* journals, and that the A-side (physical science) journals were more affected than the B-side (biological science) journals.[36] The decline in subscriptions in the 1970s mirrored their rise in the 10 years from 1955: *Proceedings A* had risen fastest, and fell most dramatically; but the decline in *Transactions A* and *B* was just a few percentage points, as their rise had been.[37]

The earlier growth had been due to international institutions, particularly in North America, and by the 1970s, almost 90 per cent of subscriptions originated outside the UK (which is why exchange rate fluctuations became a real worry).[38] The United States was the biggest market, accounting for 38 per cent of subscriptions in 1973, followed by the UK with 11 per cent, and Japan with 9 per cent of subscriptions. When the sales team examined the trends in 1980, they noticed that 'the greatest falling off in subscriptions' was in the US, where subscriber numbers (across all four journals) had fallen from 1,932 to 1,780 from 1973 to 1979.[39] By 1986, the number of US subscribers had fallen to 988; it remained the Society's largest market, but now accounted for only 28 per cent of all subscribers, only slightly ahead of the 24 per cent from the UK. Japan was still in third place, with 12 per cent of subscribers, and the Society could also boast 23 subscribers in the USSR, 10 in Brazil, five in Kenya and three in Iraq.[40]

The 1986 circulation figures are shown in Table 15.1.[41] The subscription figures for individual journals had fallen back to roughly where they had been around 1960. It is also notable that the Society's once-extensive non-commercial circulation of knowledge had shrunk to a

Table 15.1 Circulation of the Royal Society research journals in 1986 (number of copies)

	Subscriptions	Free issue to fellows	Gifts and exchanges
Transactions A	768	109	42
Transactions B	615	92	43
Proceedings A	1,476	279	42
Proceedings B	1,223	296	44

few dozen copies, mostly exchanged with other national academies.[42] The sales team reported that there were no longer any individual subscribers to the research journals. All subscriptions now came from an institution of some sort. Most were universities. Theobald's 1980 analysis had noticed that much of the loss in the US had come from 'industrial and governmental institutions', rather than from universities.[43] By 1987, there were still a few subscriptions from American governmental bodies and industrial organisations, and from UK learned societies, but the vast majority of subscribers were universities (202 in the UK, and 628 in the USA).[44] The only remaining readers who had personal copies of the Society's journals were the fellows, 776 of whom took up their rights to receive one of the research journals, in addition to *Notes and Records, Biographical Memoirs* and the *Year Book*. As Table 15.1 shows, *Proceedings A* and *B* were the most popular choices.

    The Society's officers and staff understood that all journals publishers were seeing declines in subscriptions in this period. A former staff member of Pergamon Press, for instance, recollected that 'European journal publishers fell abruptly in the esteem of the US library market' after the ending of fixed exchange rates (with the collapse of the Bretton-Woods agreement in 1971), and that this was particularly difficult at a time of falling research budgets.[45] At the Royal Society, there was the additional issue of the long-standing concern about the performance of multi-disciplinary journals.

    The Society's sales team experimented with ways to re-package its content for different markets. The Publications Policy Review had suggested that papers presented at the thematic 'discussion meetings' that the Society had begun holding in the 1960s could, in addition to being published in the appropriate series of the *Transactions*, be produced as books 'normally cloth-bound, and marketed widely'. By the 1980s, about a dozen of these books were issued each year. Sales were usually around

100 to 300 copies, but production costs were 'quite low' (akin to printing extra copies of the issue of the *Transactions*, with special binding), and it was seen as 'a successful venture'.[46] The sales team also asked librarians at universities and research institutions in the United States whether they would be interested in regular issues of curated collections, in such areas as mathematics, experimental physics, chemistry, engineering, or meteorology and earth sciences. The response was disappointing: of the 4,721 circulars sent out, a mere 71 replies were received, and of those, only 50 expressed any interest in the scheme. And three of them were existing subscribers who planned to save money by switching to the cheaper, more specialised package.[47]

By 1984, the fall in subscriptions had become sufficiently concerning that the secretaries and officers had begun to 'consider the desirability of instituting a more commercial approach to the journals'. They wondered about 'pursuing a more vigorous advertising policy to increase sales'.[48] They even hired an external consultant to examine whether it still made sense to manage sales (including subscription fulfilment and distribution) in-house. The alternatives included transferring sales and marketing back to Cambridge University Press (who still printed the journals), or outsourcing to an organisation such as Turpin Transactions Limited, which offered storage, marketing and distribution services to other learned societies from its Letchworth distribution centre.[49] Yet, all that seems to have happened was a recommendation for better marketing efforts, and 'a more rational pricing system, based on real costs and quantities' for the *Proceedings* and the *Transactions*.[50]

The lack of urgency to improve the subscription numbers can be explained by the fact that treasurers could still regularly report that the sales income 'has amply covered production and staff costs'.[51] The continuing decline in subscriptions in the 1970s meant that the heady days of big 1960s-style surpluses were at an end, but there was no fear that Royal Society publishing might fail to cover its costs. The treasurer was sufficiently sanguine to change the accounting policy, so that, from 1981, salaries, office expenses and overheads were charged to publications once more (as they had been from 1936 to 1955) rather than being silently subsidised by the Society's general operations budget. By 1986, 30 per cent of the publication expenses would be due to staff costs and overheads. Yet, even with these additional costs, the sales income generated in the 1980s was still enough to recover at least 120 per cent of costs.[52]

The ability of the publications team to keep delivering surpluses, despite the falling circulation and the requirement to cover staff costs and overheads, can be traced to falling costs. Rising costs, due to factors

Table 15.2 Production costs, 1936 and 1986

	1936	1986
Paper	14%	5%
Typesetting, printing and binding	60%	55%
Art work	15%	1%
Staff costs	6%	23%
Post and packing	5%	9%
Total expenditure	£10,926	£728,894

(mostly) outside the Society's control, had been the dominant theme of the past 100 years. Our analysis of the Society's expenditure suggests that the production costs of its journals began to fall from the mid-1970s. This was due partly to the emergence of new technologies, and partly to choices made by the Society.

Between 1936 and 1986, the Society's total expenditure on its publications grew enormously, due to the increased output of printed matter and the effects of inflation. But the importance of different sorts of costs also changed, as Table 15.2 shows. Over half the expenditure was due to printing, typesetting and binding – that is, all the tasks carried out for the Society by CUP. That remained true in 1986. Staff costs became much more significant, due to the increased number of personnel working in the editorial, sales and distribution departments, and their expenses and overheads. On the other hand, paper and artwork costs fell.

The editorial team managed all aspects of the production of the journals; from 1982, Bruce Goatly was the assistant editor. He later claimed to have little awareness of the finances, but he did admit to a sense that 'we needed to be … responsible', which he understood as pressing both CUP and the Society's paper merchants hard on their prices every year, so they could 'try to maintain quality and at the same time control costs'.[53] There were many technological changes that helped reduce the costs of publishing in the late twentieth century but, disappointingly, there is no surviving evidence about the extent to which the Society's staff discussed the new options with the Press (or, indeed, about the Society's relationship with the Press in this period more generally).[54]

The Society could have reduced its expenditure on paper decades earlier; wood-pulp paper had been available since the late nineteenth century, and was cheaper than paper made from cotton and linen rags. However, the Society's commitment to eighteenth-century production values had seen it continuing to buy 100 per cent rag paper until the 1960s. It then began buying paper with gradually increasing proportions of wood-pulp, and in 1976 moved to a 100 per cent wood-pulp-based archive paper.[55]

The most dramatic changes were due to the widespread adoption of offset lithographic printing (as against the relief printing traditionally used for letterpress), coupled with the use of photography to create the printing plates. The ability to photograph an artwork for printing made it much quicker and cheaper to print illustrations than it had been when each figure had to be copied by hand onto a lithographic stone (or engraved into a block of wood or a copper plate). It appears to have been only in 1976 that CUP began to use this process for the Society's journals, but the results were described as 'very satisfactory'.[56]

David Martin had known, back in the 1950s, that this process could also, in principle, be used to print text (Chapter 13), and it had been used on newspapers and magazines in the 1960s. The key moment was the 1970s development of computer systems that could generate text directly on photographic film (rather than having to photograph text produced by an adapted typewriter). Goatly remembers visiting the printing house in Cambridge with Gerry Evans, in the late 1970s. At that point, most of the text of scientific papers was being set using hot-metal Monotype machines, but Goatly was 'impressed to see the mathematical typesetting', which was still being done by hand. This ability to combine the typesetting machinery of the nineteenth century with hand-composition methods that had changed little since the fifteenth century was what helped Cambridge retain its 'very high reputation' for typesetting of a 'very high quality'.[57]

But things were about to change. The adoption of computer-aided photocomposition separated the previously linked skills of typesetting and printing, and this eventually meant that the Royal Society's typesetting would no longer need to be done by CUP (as we will see in Chapter 16). During the 1980s, however, Goatly and his team were involved in a lot of experimentation with how best, and where, to use the new techniques. Goatly later used his experience in what became known as 'desktop publishing' to write a series of articles for *BBC Micro User* magazine, and a book.[58]

The Royal Society acquired its first computer system in 1981: a five-terminal Jacquard J100 system with 128k of RAM and two 8-inch floppy disk drives. It was used initially for word-processing, mailing list management and payroll, and by 1984 would be maintaining the subscription list for the journals. The Society's staff experimented with using the word-processing software to prepare camera-ready copy for in-house reports, and in 1982, they proudly reported that the annual report had been 'printed direct from text held on the computer; the disk with the text file on it was sent to the printers'.[59]

For the *Proceedings* and the *Transactions*, authors submitted their papers as typed copies; and Goatly told the Royal Society's officers in 1987 that 'there has so far been very little indication from authors of a wish or willingness to provide their typescripts' in electronic form.[60] This meant that, when *Proceedings B* became the first of the Royal Society's journals to be typeset by computer in 1981, operators at CUP had to re-key the text into the computerised system that printed the film that was used to etch the lithographic plates for printing. By 1984, all the research journals were being set this way.[61]

Goatly predicted that authors in the biological sciences were likely to be the first to move to electronic submission, because word-processors of the 1980s could not generate the symbols needed for the 'often highly mathematical or technical nature' of papers in the physical sciences. (Typewriters were somewhat better, as some specialised models could produce mathematical symbols; but authors wanting to use chemical equations had to hand-write them into gaps left by their typists.) Goatly presumed that the 'more straightforward subject matter' of the biological sciences would be easier to generate on the 'commonly-available word-processing systems' of the mid-1980s. Specialised typesetting software, such as TeX (released in 1978), could cope with the specialist symbols, but Goatly felt that users required 'an advanced knowledge of typesetting', and it was unlikely 'authors' secretaries (or authors themselves)' would be able to use it.[62] The assumption that many scientists had their papers typed up by a secretary reminds us that the advent of personal computers would redistribute labour within the university, as well as in the world of research journal publishing.

These changes in typesetting and art-reproduction technologies, coupled with the Society's long-overdue decision to move to wood-pulp paper, meant that the inflation-adjusted unit costs of producing the Society's journals were falling well before the digital revolution of the 1990s.[63] This was why the treasurer was able to report that the sales income 'has amply covered production and staff costs' despite the continuing decline in the number of subscriptions.[64]

Attracting authors in a competitive world

Richard Theobald and his sales team worried about subscriptions, but Bruce Goatly and his editorial team spent the 1980s wondering how to increase submissions of high-quality papers. This was the measure that particularly bothered the fellows who acted as associate editors, because it was seen as a proxy for the scientific success and reputation of the journals.[65] In the 1950s and 1960s, the lack of growth in submissions had been overshadowed by the discovery that the journals could be financially self-supporting. By the 1980s, however, the actual decline in submissions was recognised to pose a reputational problem (see Figure 11.2).

With hindsight, there were two shifts in the patterns of submissions in the 1970s and 1980s that would be enormously significant for the post-1990 Royal Society: submission rates in the biological sciences started to grow after decades of lagging behind the physical sciences; and so did submissions from international authors. These trends would ultimately transform the focus of the Royal Society's publishing activities from mostly British, and best known for the physical sciences, to international, with strengths in biological and interdisciplinary research. But, at the time, these changes do not seem to have been apparent to participants. For instance, a 1987 report compared data on submissions by subject area since the late 1970s, but the rises in botany and zoology, and the decline in physics, did not raise comment in the accompanying narrative, even though this was something that the Society had been actively trying to improve since the 1950s (Chapter 14).[66]

The same report examined the geographical origins of authors and suggested that efforts since the early 1970s to 'encourage and facilitate' submissions from overseas authors had been moderately successful.[67] Between 1980 and 1986, the proportion of UK-based authors had declined from 67 per cent to 58 per cent, and there had been a concomitant increase in submissions from the US (from 14 per cent to 19 per cent) and Australia (from 4.5 per cent to 7.4 per cent). Japan may have been the third-largest library market for the Society's journals, but it only generated one or two submissions a year. The commentary in the 1987 report tentatively attributed the growth in overseas submissions from English-speaking countries to the 'brain drain', assuming that it was expatriate British scientists who were continuing to send their papers to the Royal Society even though their search for academic jobs had taken them abroad. It also noted the possibility that 'the increasing imposition of page charges in the USA' might make British learned society journals seem attractive to US-based authors.[68] It is also possible that two decades of strong sales to the US had helped raise awareness of the Society's journals among American researchers.

Both the associate editors and the staff seem to have been more interested in what the submission data might reveal about the viability of 'general multi-disciplinary journals'.[69] For instance, the data made clear that the Society's journals were not, in practice, as broad in coverage as they were supposed to be: in 1986, mathematics and zoology were the most common topics, and there were very few papers in earth sciences, genetics or biochemistry. Goatly also reported a perception that *Proceedings A* focused on fluid mechanics; *Proceedings B* was full of physiology; and *Transactions B* was the place for long papers in palaeontology and anatomy.[70] The uneven subject coverage of the *Proceedings* and the *Transactions* (both real and perceived) affected the willingness of researchers to read and submit to the journals, and of librarians to purchase them. For instance, one associate editor reported that 'hardly any Pure Mathematician would include [the Society's journals] in his normal browsing through the current literature, and very few would think of submitting their own papers there'.[71]

One option was for the Society to accept that its journals were not in fact generalist, and market them appropriately. Goatly pointed out that the Royal Society of Edinburgh had taken this route, turning its *Transactions* into a journal focused on the earth sciences while its *Proceedings* was devoted to mathematical sciences. Another option would be to develop a distinctive remit, as both Henry Armstrong and James Lighthill had argued. The National Academy of Sciences in the US had done this for its *Proceedings* (*PNAS*), with its focus on rapid publication of very short articles. *PNAS* had so often been held up as a possible model for the Royal Society that, in 1980, Goatly had been sent to Washington on a week-long fact-finding mission to 'look into' how the National Academy's publishing operated.[72] That visit informed his 1987 discussion of the difficulties that *PNAS* had faced: in the late 1960s, it had struggled to maintain the rapid publication that its identity depended on, and went through a reorganisation (of format and staffing) in 1970. It had briefly experimented with splitting *PNAS* into two series (similar to *Proceedings A* and *B*) in the early 1980s, but returned to an officially broad-spectrum approach even though *PNAS* was widely seen as focused on the biological sciences.[73]

A different option for the Royal Society was to try to increase submissions in areas that were currently under-represented, so that the journals became genuinely as broad-ranging as they were supposed to be. Several associate editors identified their own fields as being under-represented. One claimed that 'hardly any papers on Solid State Physics are published'; another said that 'the journals publish so few papers in Genetics'; and a pharmacologist described 'a steady trickle of papers in my field'. He added that one of his colleagues had described *Proceedings B* as 'where he sent things that he thought might not get into the *Journal of Physiology*, and I suspect that he is not the only one'.[74] Being seen as second-choice to the specialist journals was not how the Royal Society liked to think of its publications.

Providing better coverage of all areas of research could only be done if the Society received more papers in general, and especially in particular fields. Taking active steps to increase submissions was novel for the Royal Society, and it marked a significant shift from the tone of the discussions from the 1920s and 1930s, when the concern had been 'too many papers' and communicators had been urged to be more active as gatekeepers. To increase submissions, the Society's staff tried two main approaches: to persuade the fellowship to be more involved in the journals, either as authors themselves or by encouraging (or communicating) papers by authors among their acquaintance; and to encourage authors to submit, by making it easier to do so, and by advertising attractive features of publishing with the Society.

The days when the majority of papers published by the Society were authored by its fellows were long gone. The proportion of papers authored by a fellow was around 20–25 per cent, in contrast to more than 60 per cent in the nineteenth century. That proportion was, however, a slight increase since the Second World War, though the 1973 Publications Policy Committee would have liked to see it grow much more.[75] The slight growth appears to have been due to the growth of co-authorship. In the late nineteenth century, over 80 per cent of papers submitted to the Royal Society had just one author. By the 1920s and 1930s, that had declined to around 60 per cent, and by the 1980s, it had fallen to just 40 per cent. The majority of papers were now co-authored by small groups of researchers (between two and five authors). In that context, a fellow who might, in an earlier era, have communicated a sole-authored paper by a protégé, seems more likely to have become a co-author.[76] Goatly reported that only 9 per cent of the fellowship had communicated one or more papers in 1982.[77] The involvement of fellows in facilitating a flow of submissions to the Royal Society journals seems, therefore, not to have increased, but to have changed in nature.

The staff and associate editors tried to encourage fellows to communicate more papers, asking them to actively solicit papers rather than merely responding to requests from junior colleagues and students. They tried to make the act of communication more attractive by explaining that it could involve – if desired – much more than just acting as 'a mailbox'. Fellows were told that, as communicators, they could submit their own 'referee report' alongside a paper, or recommend the referees to whom it should be sent; and that they could discuss papers with the referees (as long as they preserved the referees' anonymity in discussions with authors).[78] The fact that this would have involved already-busy fellows in more work seems to have gone unremarked. A 1984 attempt to get more submissions from the physical sciences by writing personally to 70 fellows in appropriate fields had 'a generally disappointing response'. It was not reassuring that some of these fellows reported that they felt 'other journals' offered 'a more appropriate readership' for their students and colleagues.[79]

The associate editors (and the secretaries) were the fellows most closely involved with publications. When the role was established in 1968, the guidance for the new associate editors had been entirely focused on their role in selecting referees, accepting papers for publication, and ensuring that adequate records were kept so that the secretaries and staff could keep track of the progress of each paper.[80] By the 1980s, however, the role was evolving. The associate editors met annually as editorial boards for the A-side and B-side journals, but they were also asked for advice on promoting the journals. In 1984, it was suggested that including some of the Society's foreign members, even if they could not attend the board meeting, might help to draw in more papers from overseas. It was even suggested that non-fellows might be appointed to acquire coverage in certain subjects.[81] No action was taken, but the suggestions reveal that the ideal associate editor was expected to take an active role in soliciting papers for the journals, not merely in arranging their evaluation.

As well as persuading fellows to be more involved, the Society's staff sought to encourage authors to submit by promoting the 'advantages and attractions' of publishing with the Society. The 1973 Publications Policy Committee suggested emphasising its worldwide readership, the absence of page charges on authors, and rapid publication.[82] The first of these was hardly unique in the world of 1970s academic journal publishing, where Pergamon and Elsevier had been joined by Wiley, Blackwell, and even the hitherto German-language firm of Springer, in producing English-language journals for an international market of academic libraries stretching from North America to Japan.[83] In such a context, potential authors might appreciate reassurance that the Royal Society's journals were not parochially British. The absence of page charges was a genuine difference between the Royal Society and many of the American learned society publishers, and might well be important to authors with limited access to funds from their employers or funders.[84] Rapid publication was claimed by many journals, though definitions of 'rapid' differed. Although the 1973 committee had thought the Society's publication speeds were worth advertising, by the 1980s, the staff were less sure. A report on 'the scientific information system' commissioned by the Society in 1981 had revealed that, when choosing a journal in which to publish, scientists cared most about its 'scientific standard and reputation' and subject specialisation, and then about its circulation, speed of publication, and its physical appearance and quality. Reflecting on that report in 1987, Goatly felt able to claim that the Royal Society 'scores well' only on scientific reputation and physical quality.[85]

With no appetite to make the Society's journals more specialised, and no ability to do much about circulation, the editorial staff focused on speed of publication. The average time taken to publication for the *Proceedings* and the *Transactions* had been at a 10-year low at the time of the 1973 publications review: it got worse through the 1970s. By 1979, *Transactions* articles were, on average, taking somewhat over a year; while articles for *Proceedings A* averaged 43 weeks and those for *Proceedings B* averaged 37 weeks. This was the worst time for *Proceedings A* since the Society started keeping records in the mid-1950s.[86] This was why Goatly was sent to Washington in 1980 to find out how *PNAS* managed to publish so rapidly.

One difference was *PNAS*'s approach to copy-editing and proofing. At the Royal Society, 'first proofs are checked by the author; revised page proofs are checked in the Society's editorial office ... [and] as a final precaution an advance copy of each printed issue is checked in the editorial office before approval is given for the copies to be despatched'.[87] In Washington, the pressure on 'getting stuff done' meant that these processes were run, in Goatly's opinion, 'too fast'. He felt proud to work for an organisation that would 'take the time necessary to achieve top quality'.[88]

Another difference was the way that communicators and referees fitted into the National Academy's editorial system. As with the Royal Society, papers could only be submitted to *PNAS* via a member of the Academy. The difference was that that member had to submit two referee reports alongside the paper, thus shifting refereeing into the pre-submission stage and dramatically improving the statistics on time from submission to publication.[89] The role played by the member of the National Academy thus combined that of the Royal Society 'communicator' with some elements of the role of an 'associate editor'. Given how few fellows of the Royal Society were currently acting as communicators, it is perhaps not surprising that the Society did not try to make the role more burdensome.

There were no suggestions that the Society should relax its refereeing criteria to make it either easier or faster to publish in its journals. It remained ideologically committed to the importance of high-quality refereeing; and furthermore, the use of referees – which had been for decades associated with learned society publishing – was now coming to be more widely used in the English-speaking academic world. Many journals established in the 1960s and 1970s had mimicked the practices of learned society journals by asking academics to act as referees for their submissions; and by 1973, even the (new) editor of *Nature* felt that he ought to take advice from referees, rather than relying on his own opinions as his predecessors had done.[90] In the US, the use of referees was coming to be known as 'peer review'; but it was not a term used in Royal Society discussions before 1990.[91]

The associate editors were responsible for selecting referees, and the volumes of the 'Register of Papers' – still being maintained, more than a century after it was started – reveal that, in the 1980s, the Society's staff usually sent a paper to an associate editor within a few days of receiving it, and the associate editor usually sent it to a referee a week or so after that. As ever, the real issue was how quickly the referee would report, which ranged from a few days to a few months. In contrast to earlier periods, one referee was more common than two, for the *Transactions* as well as the *Proceedings*. The pattern of dates in the 'Register of Papers' shows that, when multiple referees were used, they were still consulted sequentially rather than simultaneously: authors were now asked to submit two copies of their papers (typewritten, double-spaced, with the sheets 'serially numbered and securely clipped together'), but only one of those copies was trusted to the postal system.[92] The 1969 reforms meant that 'a significant number of referees' now came from 'outside the Fellowship', but they were still mostly UK-based.[93] This was partly because the choice of referees remained highly dependent on the personal knowledge and networks of the secretaries and associate editors, but there was also an intentional effort 'to keep refereeing to this country', because, in the days before electronic mail, international post would have added delays.[94]

Despite updates to the referee report form, the questions asked of referees remained broadly similar. The legacy of John Evans's concerns about cost (see Chapter 11) could still be seen in the request that referees should point out any 'desirable condensation of unnecessarily long texts or too numerous illustrations'. However, referees were now asked to recommend one of four options, rather than the binary 'publish' or 'not'. Acceptance could be 'subject to the author's considering my suggestions for improvement'; or the referee could defer judgement, asking authors to 'revise the paper and return it for further consideration'. Rejected papers could be rejected outright, or be invited for resubmission 'after revision to meet the criticisms'.[95] These options demonstrate that making revisions had become far more usual and expected than it had been in the mid-nineteenth century.

In 1982, the Society's annual report told the fellows that improving publication times was 'the most useful and urgent action' that could be taken 'to safeguard the competitive position' of the journals. Throughout the 1980s, the time taken for the different stages of production was published in the annual reports.[96] These statistics revealed that it was actually printing and publishing, not editorial decision-making or authors' revisions, that was largely responsible for the slow publication times of the early 1980s: the editorial process was usually over within two to three months, but the printing could take another five months for the *Proceedings* journals. It often took six months for *Transactions* papers.[97]

During discussions with CUP, the Royal Society's secretaries discovered that scheduling was the key issue. The Society's journals were the only ones printed by the Press that were 'not governed by scheduled timing of various stages, linked to regular output'. This meant that other journals had 'the necessary printing capacity reserved for them', but the *Proceedings* had 'to be fitted in as opportunity offered'. More urgent jobs were often given precedence.[98] It was the shock of these slow times that, after 130 years of requests from the fellowship, finally moved *Proceedings A* and *Proceedings B* to a regular monthly publication schedule. The move to a regular, monthly publication schedule in January 1982 allowed the Press to fit *Proceedings A* and *B* into its work schedule; and this immediately cut several weeks from the time taken to publish papers in the *Proceedings*. In 1983, the receipt to publication time was 30 weeks for *Proceedings A*, and 27 weeks for *Proceedings B*.[99] Maintaining these times was, however, a constant struggle.

The Society considered one other method of making its journals more attractive to authors. Its officers and staff were aware that insisting that papers be communicated via a fellow was 'probably a disincentive' for authors, including the many scientists outside the British Isles.[100]

From November 1973 until March 1977, the Society experimented with allowing authors to submit their papers direct to the editorial office.[101] It was not, however, deemed a success. When Goatly reflected on the results 10 years later, he reported that the desired flood of high-quality papers had not arrived. Only one paper that arrived via the direct submission route was 'eventually' published. And the disadvantage was the 'large increase in "crank" and marginal papers, which added considerably to the administrative burden'.[102] For the time being, the administrative cost of removing pre-submission gatekeeping outweighed the desire to encourage more authors to submit to the Society.

None of these efforts to increase submissions – by appealing to the fellowship or by attracting authors – was successful in the long run. Between 1950 and 1990, the number of papers submitted to the Royal Society remained broadly steady, with a low of 291 papers in 1969 and a high of 414 papers in 1976. Mostly, the Society received somewhere between 320 and 370 papers a year. It was only when the communicator requirement was finally removed in 1990 that the Society would see any significant increase in submissions. The broadly steady number of submissions, however, disguises the fact that the balance between the physical and biological sciences was shifting; the numbers received in the physical sciences had been falling from a peak in the early 1950s, whereas biological submissions had been slowly growing over the same period. In the 1980s, there were still more submissions to the A-series journals than to the B series, but only by about 1.5 times; there had often been three or four times as many physical science submissions in the 1950s.[103] There is little or no evidence that this was due to any positive strategy by the Royal Society to attract authors in the biological sciences. It is more likely due to the post-war explosion of research in molecular biology and genetics.

Revamping the journals

By the mid-1980s, there was a new determination at the Royal Society to do something about the journals.[104] In May 1987, Council was presented with the most thorough overview of current publications practice and policy that it had ever received, the heart of which was a 20-page document compiled by Goatly with input from Theobald in sales. It described the current procedures for obtaining, refereeing and publishing papers, and provided the data and commentary on submissions, time taken, costs and subscriptions that we repeatedly quoted in the previous section. It also included comments from the associate editors, reflections on the general state of academic publishing, and an outline of past efforts to establish a publications policy for the Royal Society going back to 1935.[105] It was followed by a new Publications Policy Review. As well as subscriptions and sales, it considered the mission, purpose and identity of the journals, much as James Lighthill had wished to do two decades earlier (see Chapter 14). The *Proceedings* and the *Transactions* were still superficially healthy, both financially and scientifically, but they were not seen as world-leading, whether measured by submissions, subscriptions, the impact factor or the anecdotal remarks of fellows and their colleagues.

The first hint of changes came shortly after Gerry Evans's retirement, when the associate editors proposed an 'experiment' with 'the formal delegation of editorial responsibilities ... to two Editors'.[106] With the exception of Edmond Halley's first editorship in the late 1680s (see Chapter 2), this would mark the first time that editorial responsibility for the *Transactions* and the *Proceedings* had rested with someone other than the secretaries of the Royal Society. The motivation was the desire to give the journals more attention than the busy secretaries, with their expanding workloads, could provide. Giving associate editors the power to choose referees and accept papers had done little to lessen the day-to-day paperwork. Goatly recalled that, in the days before email, the secretaries needed to be at their desks at Carlton House Terrace every day to deal with papers.[107]

In early autumn 1983, two of the ordinary fellows, the quantum physicist Paul Matthews and the zoologist Brian Boycott, took over editorial duties for the series A and series B journals, respectively. Their division of labour shadowed that of the secretaries, with each 'acting editor' taking responsibility for both the *Proceedings* and the *Transactions* in their area. Nine months later, as they reflected on their experiences, Boycott was full of opinions and keen for the journals 'to go more entrepreneurial'. His 'ideal would be to increase sales, publish more and better papers and increase the rejection rate', though he acknowledged it would take 'drastic change' to counter the image of the *Proceedings* as a place 'where good but original and currently unfashionable papers could be published'.[108]

Boycott's enthusiasm was itself a demonstration of the benefits that fresh eyes, focused on the journals, might bring. The introduction of 'acting editors' had apparently 'improved the steady flow of papers through the processes up to acceptance', even though Boycott estimated he had only spent two to three hours work per fortnight on editorial work.[109] The 'experiment' was thus declared 'most successful', and Council decided 'to put this arrangement on to a permanent basis', with five-year editorial terms.[110] The discussion of practicalities ranged from whether the editors should be chosen from Council members (or, if not, then allowed to attend Council meetings), to pragmatic limits on the geographical location of the editors, given the need to travel to the Society to do the paperwork.[111]

Boycott felt that the 1983 experiment had 'represented little more than the delegation by Council of a signature', and he imagined the possibility of 'a more active editorial post and a restructured editorial board'. He suggested editorial boards might 'go international' by drawing on fellows and foreign members resident outside the UK, and might even 'draw upon experience outside the Fellowship'. Boycott wanted the editors to be 'named on the journal covers', so that they would 'take responsibility for making decisions'. He also wanted them to 'be able to operate without much constraint from Council', although it is not clear what constraint he had experienced, given that Council 'seldom intervened' despite still technically being the Committee of Papers. Boycott's thoughts on responsibility were somewhat tangled: he wanted named editors to be responsible for decisions and he wanted to remove the last vestige of Council control; but he also felt that 'it would be important for each Editor to have the backing of an authoritative editorial board'.[112] Much as Henry Armstrong had suggested in 1902 (see Chapter 11), Boycott believed named editors would bring advantages in terms of workflow, enthusiasm and commitment, but still had to be combined with the impression of a community of scholars having overall responsibility. His most radical suggestion was arguably the idea of inviting non-fellows to join the editorial boards, for, combined with the growing use of non-fellows as referees, it would further reduce the sense in which the *Proceedings* and the *Transactions* were distinctively 'Royal Society' journals, rather than anybody's journals.

The Royal Society in 1985 was not, it seems, ready for all of Boycott's ideas. Matthews continued as acting editor until his death in 1987; but Boycott did not. The B-side editing returned to the biological secretary; whereas new 'acting editors' were found to replace Matthews on the A side.[113]

The discussions about editors and editorial boards were interrupted by a substantial reorganisation of the Society's administrative structures, implemented by the new executive secretary in 1985–6. Dr Peter Warren had been the deputy executive secretary since 1977, but was appointed to the senior role in 1985. He had previously worked in the scientific civil service, and would prove an effective administrator.[114] He reorganised the 95 Royal Society staff into three 'divisions', each under the management of its own 'assistant secretary'; he encouraged the review of long-standing procedures; and, with incoming president George Porter, he drove the development of the Society's first corporate plan in 1986. According to that plan, the Society's first aim (of eight) was 'to promote the exchange and development of scientific ideas and knowledge', through discussion meetings, lectures and the publication of learned journals.[115] In the new organisational structure, the publications team was allocated to the 'National Affairs' division, despite the increasingly international aspects of submissions and subscriptions.

By late 1986, Council was able to return its attention to the journals, and asked the secretaries to review the entire 'health of the Society's publications', editorially and financially. This was the stimulus for Goatly and Theobald to compile their detailed report, presented to Council the following year. Council members were told that academic publishing was 'in general, in a healthy state', but there had been significant changes over the previous 25 years, and the Royal Society's journals were not coping well.[116] The report convinced Council to create another ad hoc Publications Policy Committee.

At the first meeting of that committee, in July 1987, the physical secretary Roger Elliott told the group that they had two aims: to consider both the ways 'in which the journals could best serve the needs of the scientific community', and 'their important financial contribution to the Society'.[117] Financially, the worry was that the current publications surplus might 'be unwittingly jeopardized by changes in editorial policy which adversely affected subscriptions or sales, either in the short or long term'.[118] The review group noted that the fellows who served as editors and associate editors were 'not involved in matters relating to the journals' finance, subscribers or production';[119] and within the staff, there was little coordination between those dealing with editorial and production issues, and those responsible for sales and marketing. The review recommended the creation of a new Publications Management Committee to involve fellows in the discussions about strategic direction. At the staff level, Bruce Goatly was appointed Head of Publishing in 1989, with responsibility for both the production team (managed by Chris Purdon), and the sales team (now managed by Vivien Clarke).

The 1987 Publications Policy Committee spent the majority of its time, however, on the problem of why researchers did not apparently send their best work to the Royal Society. They saw two main issues: the requirement to submit via a fellow; and the structure and identity (or lack thereof) of the journals. They were lukewarm about changing the communication system, worrying that allowing direct submissions

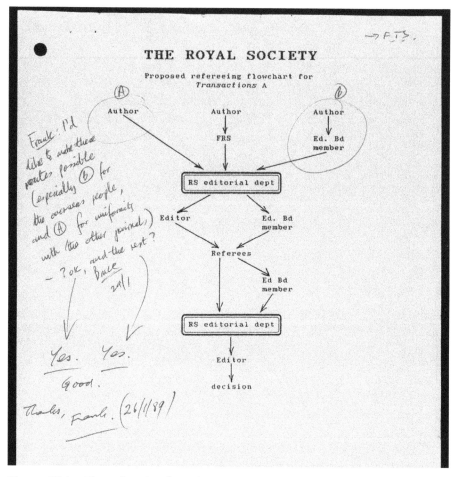

Figure 15.1 The refereeing flow chart proposed for *Transactions A* in 1989 (with annotations by Bruce Goatly and editor Frank Smith) © The Royal Society.

would simply increase the number of unsuitable papers,[120] even as they admitted that the need to know a fellow or foreign member 'was (or was perceived to be) an obstacle' to many potential authors.[121] They eventually recommended that direct submissions to the editorial office be allowed, but imagined this would be a 'secondary route' and 'authors should be advised that if papers were submitted through a Fellow the process would be much quicker'.[122] Figure 15.1 shows how multiple submission routes (including submission via an editorial board member) were being imagined in 1989.

When the new-look journals launched in 1990, the various different submission routes were listed inside their covers, but the promotional

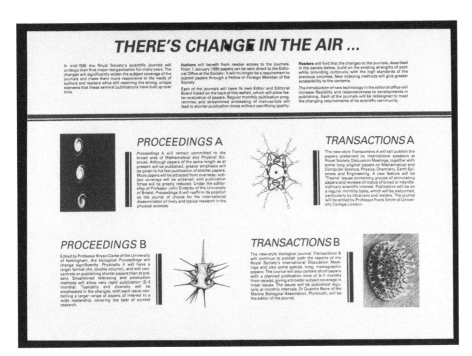

Figure 15.2 'There's change in the air . . .': promotional material for the 1990 relaunch © The Royal Society.

material (Figure 15.2) emphasised that authors would henceforth be able to send their submissions 'direct to the Editorial Office.... It will no longer be a requirement to submit papers through a Fellow'.[123] Direct submission very quickly became the norm, and thus, a significant aspect of the very identity of the Royal Society's publications slipped silently away.

The Publications Policy Committee was far more confident about its plans to deal with the identity of the journals. As one staff member put it later, 'there had been fuzzy lines between journals, and no one understood the differences. Even the fellowship.'[124] The difference between series A and series B was clear enough, but was there anything other than length of papers to distinguish the *Proceedings* from the *Transactions*?

The committee's key recommendation was that the Society's research journals would henceforth be edited, managed and marketed as four separate titles. They would therefore be able to develop distinct identities. At an organisational level, this involved appointing four 'committed and enthusiastic' fellows as editors, each with an 'effective' editorial board. Editors would have day-to-day control and substantial independence, and would coordinate where necessary via the new

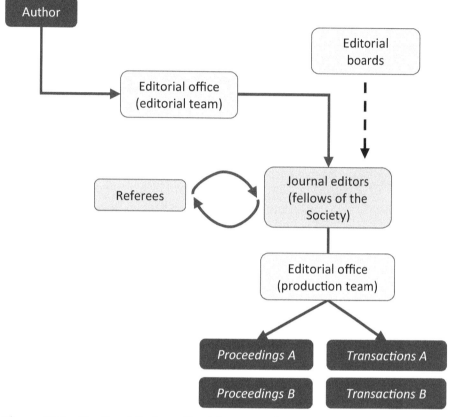

Figure 15.3 The Royal Society editorial process after 1990.

Publications Management Committee. They were to be, as Boycott had hoped, far more than 'delegated signatures'; and in recognition of their 'time and energy' they would be paid a modest honorarium and 'provided with assistance in the form of a contribution towards secretarial help, loan of word processors etc'.[125] The structure of the new editorial system can be seen in Figure 15.3.

Once the outline of the new model had been approved by Council, the four journal editors were appointed: John Enderby (*Proceedings A*), Bryan Clarke (*Proceedings B*), Frank Smith (*Transactions A*) and Quentin Bone (*Transactions B*). The first three were university professors, whereas Bone worked at the laboratory of the Marine Biology Association. The editors were all in their late fifties, and as relatively new fellows were, presumably, enthusiastic. Despite the earlier worries about location, only Smith was based in London; the others worked in Bristol, Nottingham and Plymouth. By autumn 1988, these men were choosing their editorial boards, and took charge of the detailed planning for a June 1990 relaunch.[126]

Each journal was to have 'a more distinct and clearly defined character', and this included its appearance as well as its contents. The 'tradition that the four journals should all be of similar design' would be broken.[127] There were to be new cover designs (see Figure 15.4), and decisions had to be made about typography and page layouts. Some journals changed page size: *Proceedings B* moved from B5 to the larger A4 (with double columns), while *Transactions A* moved from A4 to B5 (single column).[128] Chris Purdon, as the newly appointed production manager, recalls a 'very hard' 18 months in which staff had to plan for the relaunch at the same time as settling into their new organisational structure, as well as keeping the existing journals running. His own role felt like 'two jobs in one', and the 'enormous change' was almost 'overwhelming'.[129]

The relaunch was presented to fellows and to the public as a way to make the journals 'more responsive to the needs of readers and authors', but the announcement of 'substantial changes' was balanced by a careful emphasis on retaining 'the strong, unique elements that the journals have built up over time'.[130] The promotional material (such as, for example, Figure 15.2) made generous use of such phrases as 'will remain committed to' and 'will continue to', and in the descriptions of the four journals, only *Proceedings B* was presented as changing 'significantly'. After decades of watching *Nature* and *PNAS* with something like envy, Bryan Clarke and the biological scientists in the Royal Society had decided that rapid publication of short papers had to be the way forward for *Proceedings B*: it would henceforth focus on papers of under 10 pages and aim for publication in two to three months. They had also intended to mimic *PNAS* by encouraging authors to submit via fellows and editorial board members, and for them to arrange refereeing before formal submission; but, as with the other journals, direct submission became the main route.[131]

Proceedings A, on the other hand, changed relatively little. John Enderby and his board remained confident in its 80-year history as 'the journal of choice for the international dissemination of lively and topical research in the physical sciences'. The new *Proceedings A* would aim for wider subject coverage, more overseas papers, and faster publication times, especially for shorter papers, but would otherwise remain largely the same.[132]

For *Transactions A* and *B*, the most dramatic change was the decision to copy *Proceedings A* and *B* by moving to a regular monthly publication

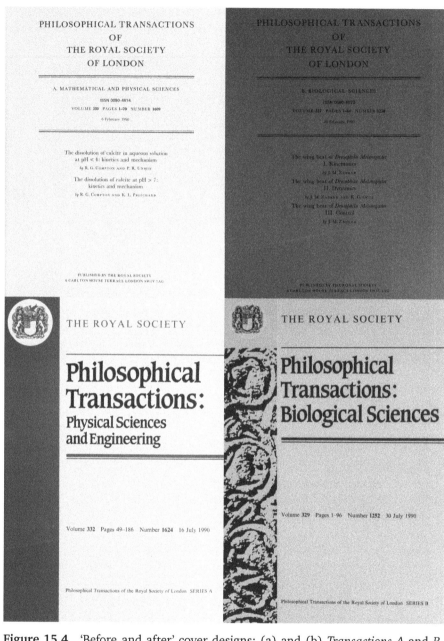

Figure 15.4 'Before and after' cover designs: (a) and (b) *Transactions A* and *B*, February 1990; (c) and (d) *Transactions A* and *B*, July 1990 © The Royal Society.

schedule. (The *Transactions* had not been monthly since the 1670s.) The promotional material hoped this would 'be welcomed, particularly by librarians and readers'. Both *Transactions A* and *B* would continue to publish a mix of types of material, most notably, the collections of papers originating from discussion meetings. Frank Smith planned to add a 'new feature' to *Transactions A*, with thematic issues 'on topics of broad or interdisciplinary scientific interest', similar to those from the discussion meetings but without an actual meeting. The emphasis on 'broad or interdisciplinary' interest reflected the old argument that one way of distinguishing Royal Society publishing from that of other, more specialised societies might be to focus on 'areas which are interdisciplinary such as in the frontier between mathematics and physics'.[133] It also created a meaningful distinction between *Transactions A* and the 'lively and topical research' in *Proceedings A*. *Transactions B* did not initially plan to publish thematic issues, but a mixture of 'short' papers and 'some special, long, monographic papers'.[134]

The public emphasis during the relaunch was about benefits to readers and authors, increased 'flexibility and responsiveness' and clearer mission. Behind the scenes, the Publications Policy Committee had not forgotten that it also had a remit to consider the financial situation of the publications. It was certainly hoped that the relaunch would stimulate sales as well as submissions, and the committee also aimed to reduce costs by about £15,000–£20,000 a year. The new design for the journals certainly modernised their appearance, but using different paper sizes, smaller type, less expensive binding and the double-column layout for *Proceedings B*, would all help to reduce costs. The committee also had high hopes for new technologies, suggesting that there might be savings from accepting papers from authors 'in magnetic form and transfer to printers by disk or tape' or 'use of electronic mark-up by copy-editors' or 'use of optical character readers'.[135]

As all participants acknowledged, the transformation of the Royal Society's journals in 1990 was long overdue. The appearance of the journals was dated, but the real problem was that the Society had for so long avoided coming up with a strategy for publishing generalist journals in an age of specialisation; for improving the uneven coverage in its supposedly multi-disciplinary journals; or reversing the declining enthusiasm for Royal Society journals among both authors and librarians. Most of the issues discussed in the 1980s had been raised many times since the 1960s (and in some cases, since the start of the century), but the difference was that the Society's leadership finally recognised the need for change, and the Society's senior staff were determined to make it happen. As one would later recollect, 'Looking at the report now you may think "Is that all they did?", but to us it felt radical'.[136]

The 1987 Publications Policy Committee also laid out a three-part mission for Royal Society publishing. The incoming editors were tasked with ensuring 'that the publications continue to be worthy of the Society'; 'that they continue to be financially successful'; and 'that they provide a valuable service to the scientific community'.[137] It was not enough for the journals to provide a service to the scientific community by disseminating research, as thousands of other journals also did. The Royal Society hoped that its journals would perform this role well enough to be a credit to the organisation, its broad remit and its long history. It remained important, as it had in 1752, that the *Proceedings* and the *Transactions* were journals by and of the Royal Society. As the traditional linkages between the fellowship and the editorial work were gradually removed – with the relaxation of rules limiting the selection of referees and editorial board members to the fellowship, and the removal of the communicator requirement – it became more important than ever to decide what it meant for the journals to be 'Royal Society journals'.

The difficult financial climate facing the Royal Society and other academic institutions in the 1970s and 1980s underpinned the explicit desire for the journals to be 'financially successful'. In the days of David Martin, this would have meant that the journals covered their costs and were self-supporting, but by the 1980s, it came to mean generating enough surplus to make a 'financial contribution' to the Society's finances. The then-president Andrew Huxley reminded the fellows that although the Thatcher government's 'very substantial reduction of funds for universities' might make the 1970s seem to have been 'a period of generous funding', in reality, 'the cuts of 1981 came on top of a long period of deepening austerity'.[138] The hope that journal publishing might enable learned societies to survive these wider economic challenges facing scientific research would shape the Royal Society's response to the digital revolution in publishing in the 1990s and beyond.

That said, the 1987 review could be seen as toning down the 1970s emphasis on money because it explicitly acknowledged the scientific, or intellectual, value of publishing. As Peter Cooper put it, it was accepted that 'the publishing surplus was not the end all and be all'.[139] Thus, while the new Publications Management Committee did regularly scrutinise the publication surpluses in the early 1990s, it also accepted that 'the success of the journals should be measured in terms of scientific esteem as well as income'.[140]

This was just as well, because the relaunch was an expensive undertaking. It was predicted to involve a one-off expenditure of around £40,000, almost half of which would be on publicity (with the remainder on new equipment and the redesign of the journals). The effects can be seen in the reduced publishing surpluses in 1990 and 1991. The new-style journals also entailed new recurrent costs, including travel costs for editorial board meetings, and the honoraria and secretarial support for the four editors.[141] By the mid-1990s, money would again come to dominate the aims of the publishing programme.

As would become clear in the years that followed, the revamp had been very successful in attracting more submissions of papers, particularly in the biological sciences.[142] Unfortunately, it had no effect on 'the continuing decline in subscriber numbers', though the 1991 annual report claimed that 'an immediate turnaround in numbers' had not been expected. The treasurer was pleased that the surplus income from publishing was 'much improved', but it was due to 'great efforts' to reduce costs rather than increased subscribers. His comment that the future was likely to involve being 'especially vigilant in controlling costs' and 'further improving efficiency' was prescient.[143] With hindsight, it might also be noted that 1990 was an unfortunate time to invest substantial sums in a revamp that would almost immediately face the challenges of new digital communication and publishing technologies.

Notes

1 'The Society's publications', 11 June 1987, p. 12 in RS C/87(87).
2 Robert Campbell (formerly of Blackwell) describes a 'discovery' phase c. 1946 to the 1960s, followed by 'exploitation' that lasted until the serials crisis hit in the late 1980s, see Robert Campbell, Ed Pentz and Ian Borthwick, eds, *Academic and Professional Publishing* (Oxford: Elsevier, 2012), 3–4. See also, Einar H. Fredriksson, ed., *A Century of Science Publishing: A collection of essays* (Amsterdam: IOS Press, 2001); and Michael Mabe and Anthony Watkinson, 'Journals (STM and humanities)', in *The Cambridge History of the Book in Britain. Volume 7: The twentieth century and beyond*, ed. Andrew Nash, Claire Squires and Ian Willison (Cambridge: Cambridge University Press, 2019), 484–98.
3 'The Society's publications', 11 June 1987, p. 12 in RS C/87(87). For Elsevier, see Einar H. Fredriksson, 'The Dutch publishing scene: Elsevier and North-Holland', in *A Century of Science Publishing: A collection of essays*, ed. E.H. Fredriksson (Amsterdam: IOS Press, 2001), 61–76, 74; for Pergamon, see Robert N. Miranda, 'Robert Maxwell: Forty-four years as publisher', in E.H. Fredriksson, *A Century of Science Publishing* (Amsterdam: IOS Press, 2001), 77–89, 78.
4 RS Annual Report (1981), 8 (reporting results of a study undertaken by the RS from 1978 to 1981). See also, J.F.B. Rowland, 'Economic position of some British primary scientific journals', *Journal of Documentation* 38, no. 2 (1982): 94–106.

5 On the early history and achievements of ALPSP, see Peter Williams, 'Twenty-five years – and still centre stage: A brief review of the history of ALPSP with a clarification of the aims and objectives of the Association', *Learned Publishing* 10, no. 1 (1997): 3–6; and Kurt Paulus, 'ALPSP: Bring on the half century!', *Learned Publishing* 25, no. 3 (2012): 167–74. For the Royal Society of Chemistry perspective, see David Hardy Whiffen and Donald H. Hey, *The Royal Society of Chemistry: The first 150 years* (London: The Royal Society of Chemistry, 1991), 156.

6 John L. Lewis, *125 Years: The Physical Society and the Institute of Physics* (London: Taylor & Francis, 1999), ch. 8; Whiffen and Hey, *Royal Society of Chemistry*, chs 4 and 6.

7 Paul Nijhoff Asser, 'A new society of publishers', *Scholarly Publishing* 2, no. 3 (1971): 301–5; Lex Lefebvre, 'The story of STM', *Serials* 7, no. 1 (1994): 53–5.

8 'The Society's publications', 11 June 1987, p. 12 in RS C/87(87).

9 Kimberley Douglas, 'The serials crisis', *The Serials Librarian* 18, no. 1–2 (1990): 111–21.

10 See Aileen Fyfe, 'From philanthropy to business: The economics of Royal Society journal publishing in the twentieth century', *N&R* https://doi.org/10.1098/rsnr.2022.0021, Figure 3.

11 'The Society's publications', 11 June 1987, p. 11 in RS C/87(87).

12 Brian Cox, 'The Pergamon phenomenon 1951–1991: Robert Maxwell and scientific publishing', *Learned Publishing* 15 (2002): 273–8, 276 and 278.

13 Associate editors meeting 1985, AE(B)7(85) in RS RMA/729 (Deep Store), box of Associate Editors meetings 1970s–1980s.

14 'The Society's publications', 11 June 1987, p. 12 in RS C/87(87). On Garfield and ISI, see Blaise Cronin and Helen Barsky Atkins, *The Web of Knowledge: A Festschrift in honor of Eugene Garfield*, ASIS Monographs (Medford, NJ: Information Today Inc., 2000); on impact factor, see Éric Archambault and Vincent Larivière, 'History of the journal impact factor: Contingencies and consequences', *Scientometrics* 79, no. 3 (2009): 635–49; and Vincent Larivière and Cassidy R. Sugimoto, 'The journal impact factor: A brief history, critique, and discussion of adverse effects', in *Springer Handbook of Science and Technology Indicators*, ed. Wolfgang Glänzel, et al. (Cham: Springer International Publishing, 2019), 3–24.

15 'Proceedings and Transactions: Proposals and recommendations' [late 1988] in RS CMB/328b.

16 On the Society in the 1980s, see J.S. Rowlinson and Norman H. Robinson, *Record of the Royal Society: Supplement to the fourth edition, for the years 1940–1989* (London: The Royal Society, 1992), 30–8; Peter Collins, *The Royal Society and the Promotion of Science since 1960* (Cambridge: Cambridge University Press, 2015).

17 'Review of the Society's finances', in RS OM2/16(73), 26 January 1973; and Publications Policy Committee (hereafter PPC), 21 June 1973, RS CMB/328.

18 RS CMP/24, 22 March 1973.

19 PPC minutes, 21 June 1973, RS CMB/328.

20 'Review of the Society's finances', in RS OM2/16(73), 26 January 1973; and RS CMP/24, 22 March 1973. Menter also suggested seeking an increase in the annual parliamentary grant and increasing fellows' annual subscriptions. See also Collins, *The Royal Society and the Promotion of Science*, 63–4.

21 Chemical Society, Annual Report (1970–71), 5, RSC AR1147.

22 A sense of the urgency and 'increasingly acute' problems can be seen in Chemical Society, Annual Report (1973–74), 14, RSC AR2123.

23 PPC minutes, 30 January 1974, RS CMB/328.

24 'Trends in subscriptions and prices of journals [1973–1986]', Annex C to 'The Society's publications', 11 June 1987, in RS C/87(87).

25 PPC minutes, 30 January 1974, RS CMB/328.

26 'Trends in subscriptions for Proceedings and Philosophical Transactions', RS OM3/88(80).

27 RS Annual Report (1984), 10.

28 The Chemical Society, Annual Report (1978), 11, RSC AR2127.

29 Finance Committee report, in RS CMP/25, 10 July 1975.

30 [R. Theobald], 'Subscriptions to Philosophical Transactions and Proceedings', 28 May 1977, RS OM/79(77) [copy in AB/2/20/3].

31 [Theobald], 'Subscriptions …', 28 May 1977, RS OM/79(77).

32 'Subscription renewals to journals of the Royal Society', August 1976, included as Appendix A to [Theobald], 'Subscriptions …', 28 May 1977, RS OM/79(77).

33 [Theobald], 'Subscriptions …', 28 May 1977, RS OM/79(77).
34 Our interview with Bruce Goatly, 20 January 2016.
35 [Theobald], 'Subscriptions …', 28 May 1977, RS OM/79(77).
36 'Trends in subscriptions for Proceedings and Philosophical Transactions', RS OM3/88(80).
37 See Figure 3, in Fyfe, 'From philanthropy to business'.
38 In 1973, just 11 per cent of subscriptions were to the UK, see 'Trends in subscriptions …', RS OM3/88(80).
39 'Trends in subscriptions …', RS OM3/88(80).
40 'The Society's publications', 11 June 1987, p. 9 in RS C/87(87).
41 'The Society's publications', 11 June 1987, p. 2 in RS C/87(87).
42 'The Society's publications', 11 June 1987, p. 2 in RS C/87(87).
43 'Trends in subscriptions …', RS OM3/88(80).
44 Annex D to 'The Society's publications', 11 June 1987 in RS C/87(87).
45 Cox, 'The Pergamon phenomenon', 276.
46 'The Society's publications', 11 June 1987, p. 10 in RS C/87(87). On the new style of meetings, see Collins, *The Royal Society and the Promotion of Science*, 254–6.
47 'Inquiry into demand for wider distribution' [circular letter, undated, signed by D.C. Martin, i.e. pre-1976] and 'Pilot scheme to test feasibility of accepting Standing Orders for specific categories of Philosophical Transactions' [survey results], in RS RMA (offsite), uncatalogued.
48 'Ad hoc meeting on RS publications policy, 25 May 1984', p. 4 in RS C/155(84).
49 On Turpin (and its connections to the Royal Society of Chemistry), see Whiffen and Hey, *Royal Society of Chemistry*, 153–9.
50 For the external consultant, see 'The Society's publications', 11 June 1987, p. 3 in RS C/87(87).
51 RS Annual Report (1984), 10.
52 For the trends in expense recovery rate, see Figure 5 in Fyfe, 'From philanthropy to business'. On staff costs, see Figure 4B.
53 Our interview with Bruce Goatly, 20 January 2016.
54 For an overview of the technological changes, see Sarah Bromage and Helen Williams, 'Materials, technologies and the printing industry', in *The Cambridge History of the Book in Britain. Volume 7: The twentieth century and beyond*, ed. Andrew Nash, Claire Squires and I.R. Willison (Cambridge: Cambridge University Press, 2019), 41–60; and Paul Luna, 'Technology', in *History of Oxford University Press: Volume IV*, ed. Keith Robbins (Oxford: Oxford University Press, 2017), 179–204.
55 'The Society's publications', 11 June 1987, p. 8 in RS C/87(87); and our interview with Goatly, 20 January 2016; Rowlinson and Robinson, *Record of the Royal Society* (1992), 108.
56 RS Annual Report (1981), 18; Rowlinson and Robinson, *Record of the Royal Society* (1992), 108.
57 Our interview with Goatly, 20 January 2016.
58 For *Desktop Publishing for the Archimedes* (1991), see 'Bruce's book page', http://www.goatly.com/DTPbook/index.html.
59 See 'Office services' in RS Annual Report (1981), 50; RS Annual Report (1983), 51–2 (re 1982 report being printed from computer); and RS Annual Report (1984), 52–3.
60 'The Society's publications', 11 June 1987, p. 9 in RS C/87(87).
61 RS Annual Report (1981), 18; and 'The Society's publications', 11 June 1987, p. 8 in RS C/87(87).
62 'The Society's publications', 11 June 1987, p. 9 in RS C/87(87). Barbara Beeton, 'Developments in technical typesetting: TeX at the end of the 20th century', in *A Century of Science Publishing: A collection of essays*, ed. Einar H. Fredriksson (Amsterdam: IOS Press, 2001), 191–202.
63 Fyfe, 'From philanthropy to business', Figure 2C.
64 RS Annual Report (1984), 10.
65 The claim that success can be 'measured by submission rates' was made explicit in RS Annual Report (1994), 21.
66 'The Society's publications', 11 June 1987, p. 6 in RS C/87(87).
67 'The Society's publications', 11 June 1987, p. 3 in RS C/87(87), describing outcomes of the 1973 policy review.

68 'The Society's publications', 11 June 1987, pp. 5–6 in RS C/87(87).
69 'The Society's publications', 11 June 1987, p. 3 in RS C/87(87).
70 'The Society's publications', 11 June 1987, pp. 4 and 6 in RS C/87(87).
71 Frank Bonsall, 'Comments received from Associate Editors', Annex A to 'The Society's publications', 11 June 1987, p. 14 in RS C/87(87).
72 Our interview with Goatly, 20 January 2016.
73 'The Society's publications', 11 June 1987, p. 12 in RS C/87(87).
74 'Comments received from associate editors' (Michael Pepper, Bryan Clarke and David Colquhoun), Annex A to 'The Society's publications', 11 June 1987, pp. 14–16 in RS C/87(87).
75 Publications policy report, in RS CMP/24, 12 July 1973.
76 Statistics are from our analysis of sample years (1970, 1975, 1980, 1985) from the 'Register of Papers'. Co-authorship by fellows is suggested by the fact that, although over 20 per cent of papers had a fellow as an author, only about 12 per cent of all authors were fellows.
77 'The Society's publications', 11 June 1987, p. 6 in RS C/87(87).
78 RS CMP/26, 3 March 1977.
79 'Ad hoc meeting on RS publications policy, 25 May 1984', in RS C/155(84).
80 See new standing orders, in CMP/23, 4 April 1969; and 'Note for the guidance of associate editors', [?early 1969], RS RMA (offsite) uncatalogued.
81 'Ad hoc meeting on RS publications policy, 25 May 1984', in RS C/155(84).
82 Publications policy report, in RS CMP/24, 12 July 1973.
83 Mabe and Watkinson, 'Journals'. On Springer's move to English-language publishing, in the late 1960s (including the Japanese demand), see Heinz Goetze, *Springer Verlag: History of a scientific publishing house,* Part II *1945–1992* (Berlin; Heidelberg: Springer, 1996), ch. 3.
84 On American learned societies at the time, see Robert H. Marks, 'Learned societies adapt to new publishing realities: A review of the role played by U.S. societies', in *A Century of Science Publishing: A collection of essays,* ed. E.H. Fredriksson (Amsterdam: IOS Press, 2001), 91–5; Marianne Noel, 'Building the economic value of a journal: The case of the Journal of the American Chemical Society (1879–2010)' [in French], *Revue française des sciences de l'information et de la communication,* no. 11 (2017).
85 'The Society's publications', 11 June 1987, p. 6 in RS C/87(87), including paraphrase of the 1981 report. The findings were reported in J.F.B. Rowland, 'The scientist's view of his information system', *Journal of Documentation* 38, no. 1 (1982): 38–42.
86 RS Annual Report (1979), 10–11. For the longer trend, see Fyfe, 'Time taken to publish' (8 February 2022), https://arts.st-andrews.ac.uk/philosophicaltransactions/time-taken-to-publish.
87 'The Society's publications', 11 June 1987, p. 8 in RS C/87(87).
88 Our interview with Goatly, 20 January 2016.
89 'The Society's publications', 11 June 1987, p. 12 in RS C/87(87).
90 David Pontille and Didier Torny, 'From manuscript evaluation to article valuation: The changing technologies of journal peer review', *Human Studies* 38, no. 1 (2015): 57–79; Melinda Baldwin, 'Credibility, peer review, and Nature, 1945–1990', *N&R* 69.3 (2015): 337–52.
91 Melinda Baldwin, 'Scientific autonomy, public accountability, and the rise of "peer review" in the Cold War United States', *Isis* 109, no. 3 (2018): 538–58.
92 'Instructions to authors' (approved 13 May 1971; last amended 13 October 1983), from the backmatter of *ProcB* 225 (1985): 503–6.
93 'Ad hoc meeting on RS publications policy, 25 May 1984', in RS C/155(84).
94 Our interview with Goatly, 20 January 2016. On the gendered implications of relying on personal networks, see Camilla M. Røstvik and Aileen Fyfe, 'Ladies, gentlemen, and scientific publication at the Royal Society, 1945–1990', *Open Library of Humanities* 4, no. 1 (2018): 1–40, http://doi.org/10.16995/olh.265.
95 See, for instance, referee report forms from 1980 in RS RR/1980A.
96 RS Annual Report (1982), 15.
97 'Times taken (median values) for stages in the course of publication [1982–86]', Annex C in 'The Society's publications', 11 June 1987, p. 18 in RS C/87(87).
98 RS Annual Report (1982), 15.

99 RS Annual Report (1983), 16.
100 'The Society's publications', 11 June 1987, p. 6 in RS C/87(87).
101 RS CMP/24, 12 July, 11 October and 8 November 1973; and RS CMP/26, 3 March 1977.
102 'The Society's publications', 11 June 1987, p. 6 in RS C/87(87). For the contemporary evaluation (which also mentioned increased office work, and 'exceptionally troublesome' papers), see 'Joint meeting for the associate editors for the biological and mathematical & physical sciences', 28 January 1977, RS AE (AB)3(77).
103 See Aileen Fyfe, 'Submissions in life sciences vs physical sciences, 1927–1989' (13 February 2018), https://arts.st-andrews.ac.uk/philosophicaltransactions/submissions-in-life-sciences-vs-physical-sciences-1927-1989.
104 This is briefly discussed in Collins, *The Royal Society and the Promotion of Science*, 252–3.
105 'The Society's publications', 11 June 1987, in RS C/87(87).
106 RS Annual Report (1983), 10; Rowlinson and Robinson, *Record of the Royal Society* (1992), 108.
107 Our interview with Goatly, 20 January 2016.
108 'Ad hoc meeting on RS publications policy, 25 May 1984', report in RS C/155(84), 2 July 1984.
109 'Ad hoc meeting ... 25 May 1984', report in RS C/155(84), 2 July 1984.
110 RS Annual Report (1984), 10, 16–17.
111 'Ad hoc meeting ... 25 May 1984', report in RS C/155(84), 2 July 1984.
112 'Ad hoc meeting ... 25 May 1984', report in RS C/155(84), 2 July 1984.
113 See the pages 'Editors and Associate Editors' in the RS *Year Books* for 1986-9. J.T. Stuart covered the A-side editorial work in 1987, and Frank Smith took over in 1988.
114 'Comrades of old: Royal Society staff pensioners, biographical sketches' (privately circulated for The Royal Society Former Staff Association, 2015), 9–10; Rowlinson and Robinson, *Record of the Royal Society* (1992), 27 and 37. For Warren's own memories of his appointment, see edited transcript of a conversation between Peter Collins and Peter Warren on 6 January and 25 March 2009, in RS R82576, vol. 4, 862–3. The job had been openly advertised in *New Scientist* (7 June 1984).
115 Rowlinson and Robinson, *Record of the Royal Society* (1992), 37.
116 'The Society's publications', 11 June 1987, p. 12 in RS C/87(87).
117 Publication Policy Committee (PPC2), 13 July 1987, RS CMB/328b.
118 'Publications policy/a note by the secretaries', 6 May 1987, in RS C/87(87).
119 PPC2, 13 July 1987, RS CMB/328b.
120 PPC2, 2 March 1988, in RS CMB/328b.
121 PPC2, 13 July 1987, in RS CMB/328b.
122 PPC2, 21 December 1987 and 2 March 1988, in RS CMB/328b.
123 'There's change in the air ...', Royal Society marketing material [c. 1990] [RS archive uncatalogued, donated by Chris Purdon].
124 Our interview with Peter Cooper, 3 June 2016.
125 PPC2, 21 December 1987, in RS CMB/328b; 'Publications review, 1988/Summary of conclusions', reproduced in file at RS C/31(95).
126 PPC2, 2 March 1988, in RS CMB/328b; 'Review of the Society's publications, by the secretaries', *Royal Society News* (January 1989), 1.
127 'The Society's publications', 11 June 1987, pp. 8–9 in RS C/87(87); 'Publications review, 1988/Summary of conclusions', reproduced in file at RS C/31(95).
128 Since 1967, the *Transactions* had been printed on A4, while the *Proceedings* had used the smaller B5 size, Rowlinson and Robinson, *Record of the Royal Society* (1992), 108.
129 Our interview with Chris Purdon, 18 March 2016.
130 'Review of the Society's publications, by the secretaries', *Royal Society News* (January 1989), 1. The same text was used in the advertising leaflet prepared for the relaunch.
131 'Notice to authors: submission', inside front cover of *ProcB* 242 (1990) [22 December].
132 'Announcement' (December 1989), inside front cover of *ProcA* 427 (1990) [8 January].
133 Comment by Associate Editor Michael Atiyah, in Annex A, 'The Society's publications', 11 June 1987, p. 14 in RS C/87(87).
134 'There's change in the air ...', Royal Society marketing material [c. 1990] [RS archive uncatalogued, donated by Chris Purdon].
135 'Proceedings and Transactions: proposals and recommendations' [late 1988], in RS CMB/328b.
136 Our interview with Peter Cooper, 3 June 2016.

137 'Publications review, 1988/Summary of conclusions', reproduced in file at RS C/31(95).
138 [A. Huxley], 'Address of the president, 30 November 1984', *ProcB* 223 (1985): 407.
139 Our interview with Peter Cooper, 3 June 2016.
140 Publications Management Committee, 10 December 1991, in RS PMC/45(91).
141 'Proceedings and Transactions: Proposals and recommendations' [late 1988], in RS CMB/328b.
142 On the step-change in submissions, see Aileen Fyfe, 'More submissions, more rejections: The Royal Society journals since the 1950s' (18 June 2020), https://arts.st-andrews.ac.uk/philosophicaltransactions/more-submissions-more-rejections-the-royal-society-journals-since-the-1950s.
143 RS Annual Report (1991), 17 and 19.

16
Money and mission in the digital age, 1990–2015

Camilla Mørk Røstvik and Aileen Fyfe

The staff and officers of the Royal Society interpreted the dramatic rise in submissions to the journals, after the 1990 relaunch, as an indicator of the success of 'the restyling, the redefinition of content and the revised and faster publishing schedules'. Just a year after the relaunch, submissions were up by 67 per cent; the target publication times were being met; papers were being submitted from a wider mix of subjects, and 'from all parts of the world'; and the Society had published 'several papers of outstanding international importance'.[1] However, it was not an unmixed blessing, and the relaunch did nothing to reverse the decline in subscriber numbers nor, despite the treasurer's initial cautious optimism, to significantly improve the financial situation. Moreover, the new-look journals, with their new editors, were barely bedded in when the Royal Society – and all other academic journal publishers – had to work out how to respond to the digital revolution.

The Royal Society's Scientific Information Committee had been eagerly anticipating 'revolutionary methods of reproducing scientific communication' since the 1960s.[2] As things turned out, microfilm and photocopying had only modest impacts on a world that remained dominated by the print-on-paper journal; but by the 1990s, it seemed that computers were finally going to bring about that revolution. Most of the Society's fellows, authors and reviewers worked in research institutions, and were early adopters of computing technologies. In 1995, 38 per cent of fellows reported that they had access to CD-ROMs; 46 per cent had electronic mail; and 65 per cent had access either to a personal computer or to the terminal of a mainframe computer.[3]

To investigate the impacts of these changes on the 'information needs, and current and future information retrieval methods', the

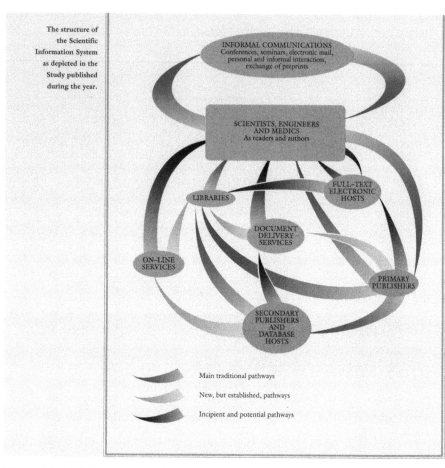

Figure 16.1 'The structure of the Scientific Information System in the UK', from RS Annual Report, 1993 © The Royal Society.

Scientific Information Committee commissioned a study of the changes in 'the scientific information system' (see Figure 16.1), including a survey of 4,500 scientists and 600 librarians. The 1993 report predicted that some of the new technologies could be seen adapting or improving existing practices. For instance, electronic mail and the electronic exchange of preprints were an extension of long-standing practices of researchers corresponding with each other by post, and enclosing printed 'separates' or 'offprints' of their latest papers. Similarly, electronic authoring, editing and typesetting software had already proved able to fit into the production of traditional print journals. There were, however, other technological developments that appeared potentially more disruptive to the established systems of libraries and publishers: these included databases, 'on-line services' and 'full-text electronic hosts'.[4]

The technological changes now referred to as the 'digital revolution' involved an array of different technologies, used by different groups of people. The Society's authors began to submit their papers as electronic files, first by disk and later by email; and they would eventually come to receive their copy-editing queries and proofs by email too. But the staff were most closely affected. Computerised typesetting and art generation changed the production workflows for the printed journals, and tasks that had previously been bundled into the remit of traditional printing houses could be unbundled and redistributed. This changed the nature of the work done by the Royal Society's editorial and production teams, and led to the end of the printing contract with CUP, dating from the 1930s (Chapter 13). Creating electronic editions of the journals – and, ultimately, launching electronic-only journals – required the Society to develop new relations with an array of contractors and digital service providers. Meanwhile, the in-house staff worked with new editorial management software that helped track and optimise editorial workflows. This helped the staff to process the exponential growth in submissions, but also enabled closer attention to staff efficiency.

In 1995, a publishing review would recast the three-part mission for Royal Society publishing, established in 1987 (Chapter 15) into a set of twin objectives. This had the effect of making income-generation a more prominent objective once more, and that had two major consequences for Royal Society publishing in the 1990s and beyond. First, it drove yet another round of organisational cultural changes in the publications department. The years around 1995 saw substantial staff turnover as the Society brought in a management team with external experience to create a new professional culture of publishing. Second, it informed the Society's engagement with the new digital technologies. Some individual fellows hoped that electronic journals would enable the wider, non-commercial circulation of research, but the Society's corporate approach in the late 1990s and early 2000s was based upon selling scientific content.

The Society spent the 1990s trying to work out how to adapt its existing journals to the uncertain promise of electronic publishing. By the early twenty-first century, it had begun to look beyond the *Transactions* and the *Proceedings*. The Society had avoided the enthusiasm for starting new journals, or splitting (or 'twigging') existing ones, in the 1970s, but it would launch five new journals between 2003 and 2014. The success of these new ventures meant that the biological sciences and interdisciplinary research became the main areas of Royal Society publishing, in total contrast to the dominance of the physical sciences in the days of George Stokes (Chapter 9) or Ernest Rutherford (Chapter 11).

These changes occurred in the context of highly public debates about peer review, and the rising prominence of the open access movement, in which the tensions between the 1995 twin objectives – of money and mission – were revealed.

From 1990 relaunch to 1995 review

After the relaunch, the number of papers submitted doubled immediately, and, as Table 16.1 shows, it continued to increase.[5] Royal Society publishing was finally experiencing the growth in the global output of research from which it had been strangely isolated since 1945.[6] The proportion of authors who were not based in the UK had already risen to 51 per cent in 1990, and it kept rising, reaching 76 per cent by 2010.[7] The Society attributed the increased submissions to the appeal of the new identities of the journals, though the substantial expenditure on publicity surely helped.[8]

The decisive factor for increasing the number and diversity of submissions was the decision to accept them directly from authors. The subsequent decision to remove any mention of submitting via a fellow from the guidance to authors also helped to convey the message that the Society's journals were possible publications venues for any scientist, not just those who were part of the personal networks of fellows of the Society.[9] A fast-track for papers by fellows themselves was still proposed from time to time, but by the early twenty-first century, it was perceived as out of step with efforts to increase diversity in science. For instance, when in 2014, the US National Academy of Sciences began to allow

Table 16.1 Submissions to Royal Society journals, 1988–2013 (scientific research journals only)

Year	A side	B side	Interdisciplinary	Total
1988	199	115		314
1993	405	446		851
1998	317	1,035		1,352
2003	389	1,667		2,056
2008	510	2,650	490	3,650
2013	829	4,552	1,211	6,592

direct submission to *PNAS*, in addition to the traditional route, the then-president of the Royal Society privately commented that he thought the Royal Society was wise not to have retained an inside-track for fellows.[10] The Society's publishing director agreed: 'The last thing we should do is give ammunition to those ill-informed critics who already think the RS is an old boys club.'[11] The Royal Society's authors were still predominantly men, but the proportion of women authors had risen from less than 4 per cent in the mid-1980s to around a third of authors in the 2010s.[12]

The flood of submissions was accompanied by an increase in the rejection rate: it had historically been around 10 per cent, but in 1992, it was reported as 'around 30–40% for all journals', and the following year, it was 'nearing 50%'.[13] The Society did publish more papers, but it also rejected many more.[14] This was partly because of the limits of space available in the printed journals, but it also reflected the greater variation in the papers received now that communicators no longer filtered out the '"crank" and marginal papers'.[15] It was around this time that the Royal Society began to present the rejection rate as a proxy for the quality of papers. In the 1950s, David Martin had been able to claim that refereeing was in itself a guarantee of high standards (Chapter 14); but in a world in which virtually all Anglophone scientific journals had adopted peer review, the simple fact of using referees was no longer a distinguishing factor. And the 1968 reforms (Chapter 15) meant that the Society could no longer claim to have a particularly distinguished or superior pool of referees: fellows still did some refereeing, but most of the referees were now ordinary scientists, who were probably also refereeing for other journals. In this context, the rejection rate appeared to be a metric that could distinguish between journals that all used referees. The Royal Society claimed that the increased rejection rate in the 1990s demonstrated that its journals were 'back to their leading position'.[16]

At the same time, the Society's staff made plans to capitalise on the increased submissions by publishing (and selling) more content. As Table 16.1 makes clear, the biggest growth area was the biological sciences, driven by the rapid advances in genetics and biomedical sciences. From 1995 onwards, submissions in the biological sciences vastly outstripped those in the physical sciences. In 1998, *Proceedings B* moved from monthly to fortnightly issues 'to meet the excessive demand'.[17] This provided the capacity to double its output. (*Proceedings A* remained monthly.)

The creation of distinct identities for each research journal had been only partially successful, as both *Transactions A* and *Transactions B* continued to carry a mix of content, some of which might be thought to overlap with the types of articles that could appear in the *Proceedings* journals. As well as separate research papers, both *Transactions* journals also carried thematic content: this originated with reports of discussion meetings held at the Society, but at the relaunch, *Transactions A* also invited 'theme' issues, containing 'groups of stimulating papers and reviews on topics of broad or interdisciplinary scientific interest'.[18] This reduced the space available for submitted papers: referees for *Transactions B* were warned that 'there is space to publish only two of these [25-page research papers] each year' and that there was 'competition between such papers for publication'.[19] In view of this, and the overlap in remit with the *Proceedings*, in 1994, the board of *Transactions A* decided to stop accepting research papers, and to focus instead on thematic review issues; and in 1995, the board of *Transactions B* agreed to follow suit.[20] *Transactions A* and *Transactions B* would now have a new purpose surveying the research in a topic or field, and it would be clearly distinct from the role of *Proceedings A* and *Proceedings B* in publishing cutting-edge research papers. From this point on, the Royal Society no longer counted *Transactions A* and *Transactions B* as 'research journals'. The board of *Transactions A* hoped that it would now become 'an "archival" journal in the best sense'.[21]

The opportunity to act as guest editors, to curate a thematic review issue, proved popular with scientists, and there was sufficient demand that both series moved to a fortnightly periodicity in 2008. It also proved successful with librarians, and was marketed both as a periodical and as separate books. As the Society's sales manager commented, the books were 'easy to sell – thematic, topical – and, being commissioned, you can control the content and quality'. The disadvantage was that 'there were more production issues ... than with other journals', because 'if one paper was late, the whole thing was late'.[22]

For *Proceedings A* and *B*, the flood of submissions – of more varied quality – created challenges for the editorial team and the referees. The staff developed software to improve their ability to track the editorial progress of all the submissions, and the paper volumes of the 'Register of Papers', that had been maintained since the 1850s (Chapter 9), were discontinued. By 1990, referees for *Proceedings B* were already being asked to discriminate between five shades of publishable papers: 'outstanding ... of the highest international importance ... must be published'; 'excellent ... important ... should be published'; 'very good ... well worth publishing'; 'good ... worth publishing'; and 'acceptable'.[23] By 1994, this had been turned into a numerical grading system for rapid, and potentially automated, processing in the office.[24] These finely graded distinctions indicated an intention (or a need) to be more selective than in the days of distinguishing between papers that were 'significant' or merely 'worthwhile' (Chapter 13).

In the early years of the new system, a handful of papers absorbed a vast amount of editorial time and energy. An immediate test of the new arrangements came from submissions on the relationship between HIV and AIDS, an issue whose medical and activist dimensions were then attracting significant media attention. Both *Proceedings B* and *Transactions B* published articles dealing with the transmission dynamics and mathematical modelling of HIV, but a 1990 submission on 'AIDS in the Bronx' by a known activist was swiftly rejected by all its referees.[25] But scholarly etiquette would not allow authors with strong academic credentials to be summarily dismissed, and this gave the referees some awkward cases. For instance, the epidemiologist Gordon Stewart (1919–2016) was an emeritus professor at the University of Glasgow and had advised the World Health Organization on AIDS.[26] In the 1990s, Stewart publicly criticised the consensus view, arguing that 'there is more to AIDS than just HIV' and claiming that the medical establishment had been guilty of feeding 'undue alarm and anxiety' with 'panic statements' about the spread of AIDS. He complained that specialist scholarly journals were refusing to publish 'even verifiable data that puts the conventional view into question'.[27]

One of the journals that declined to publish Stewart's work was *Transactions B*: a paper submitted in November 1990 was sent to two referees, both of whom firmly rejected it. One of the referees described it as 'clothed in the trappings of conventional scholarship' but 'essentially crackpot', and added: 'My life is too short for this.' They refused to engage with the details of the paper, claiming it was 'a vulgar view of science' to hold that 'any idea, no matter how wild, deserves a measured consideration and response'. For this referee, 'crackpot' science was apparently instantly recognisable, and did not deserve the usual scholarly etiquette of 'measured and careful rebuttal'. Stewart protested against his rejection, both by letter and by phone, and the paper continued on what one referee called 'the merry-go-round of more refereeing/revising/rejection' until the end of 1992, when it finally disappeared (unpublished) from the *Transactions B* files.[28]

The episode makes clear that, even without communicators pre-screening the papers, the Society's referees, editorial board members and editors still guarded access to the pages of the journals. Whether those processes were perceived as protecting the quality of published scientific research, or as blocking the publication of alternative (or, potentially, innovative and exciting) ideas, depended on one's perspective.

The financial consequences of the 1990 relaunch were less clear-cut than its effects on submissions. It did not reverse, nor even halt, the decline in subscriber numbers. By 1994, the circulations of the journals had fallen to levels last seen in the 1950s: *Transactions A* had 622 subscribers; *Transactions B* had 496; *Proceedings A* remained the most popular journal, with 1,125 subscribers; and *Proceedings B* had 977.[29] Once the one-off costs of the relaunch had been absorbed, the Society's annual accounts did show the publication surplus as increasing each year, but this disguised very different performances from the individual journals. For instance, *Proceedings A* made an average surplus of £174,000 in the three years 1992–94, whereas *Proceedings B* averaged barely £12,000 over the period (and had a deficit in 1994).[30] *Proceedings B* was experiencing the problems of its new popularity with authors: there was more editorial work to be done, and the higher rejection rates meant, on average, higher editorial costs for each paper published. Fortunately, technological changes were finally reducing production costs.

The typesetting for the Society's journals was part of the service supplied by its printer; and since 1937, this had been done by the highly skilled (but expensive) compositors at CUP. They began using photocomposition in the late 1970s (Chapter 15), and soon moved to computerised photocomposition. But once typesetting could be done by a computer operator, rather than a skilled artisan, it became possible for it to be done elsewhere. The work of typesetting shifted from the Press to the Society's staff and authors, and later to outsourced services. In 1992, the Royal Society began accepting papers from authors as word-processed files on floppy disks. These were then processed into the TeX typesetting language by Royal Society staff (see Figure 16.2). The following year, the Society began accepting disks in TeX format, usually from authors in the physical sciences who were willing to learn TeX to be sure that the technical symbols and characters in their papers were correct. By 1994, over 80 per cent of papers were being typeset in TeX by either the Society's authors or staff.[31] The editorial team then used an Imagesetter to turn the TeX files into photo-film that would be sent direct to the printers to create plates for offset lithographic printing.[32] These changes reduced the Society's bill from the Press, but put new expectations on its editorial staff. By 1996, the head of publishing was struggling 'to recruit and retain suitably qualified staff'.[33]

With the Press now only printing – not typesetting – the key justification for working specifically with the CUP, rather than any other

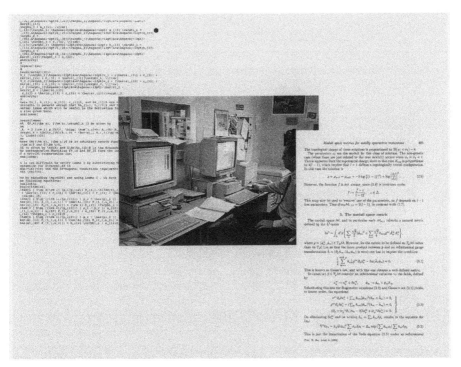

Figure 16.2 'New typesetting equipment enables papers for the journals to be processed entirely in-house using the TeX mathematical typesetting system', from RS Annual Report, 1994 © The Royal Society.

printer, disappeared. In the early 1990s, the Society began to issue regular competitive tenders for the printing of each journal, breaking its three-century tradition of appointing a single printer, long-term, for all its printing. The first tender for *Transactions B* was won by Alden Press, a family-owned firm in Oxford, whose quotation was apparently £20,000 a year cheaper than CUP's, even after CUP had been given a chance for 'reconsideration and reduction in the light of other tenders'.[34] The changing technologies were proving difficult for many long-established printers in this period: Oxford University Press closed its printing house in 1989; Alden went out of business in 2008; and CUP would close its printing house in 2012. Cambridge had retained the contracts for some of the Royal Society's journals, but they had become one of a pool of printers used by the Society, alongside firms such as Cambrian Printers in Aberystwyth and Latimer Trend Ltd in Plymouth.[35]

Cutting production costs while selling more content enabled Bruce Goatly and the Society's publishing team, in the early 1990s, to deliver publication surpluses for which they were routinely thanked by the Society's treasurers. However, since the Society's accounts were not normalised for inflation, the year-on-year increases in the publication surpluses represented no increase in actual spending power for the Society.[36] These relatively unimpressive financial results mattered in the wider context in which the Society was operating in the early 1990s. The annual addresses of then-president Michael Atiyah were full of concerns about the Conservative government's policies of privatisation and deregulation. Universities were increasingly subjected to 'financial stringency' and government evaluation, and Atiyah insisted that the Royal Society 'must stoutly resist' any such 'improper interference' in its affairs. One of the ways in which he and his fellow officers sought to strengthen the Society's independence from government was by 'seeking to broaden its financial base', and in particular, to reduce the Society's reliance on its parliamentary grant-in-aid (which now accounted for over 80 per cent of the Society's income). A major fundraising campaign, *Project Science*, generated £23 million from philanthropic donors.[37] The Society also began to develop its 'trading activities': conferences, catering and publications.[38]

The Society's desire for income would be a key influence on the 1995 Publications Review, but there were two others. One was the external review of the Society's administration, carried out in 1993, which recommended more professional, managerial structures.[39] This led to the dismantling of the organisational structure created by Peter Warren in 1986 (Chapter 15), and the creation of a new structure involving seven 'sections' and a range of support units. The publishing team had been excluded from this review, because its 'objectives, financial culture and external environment' were seen as 'substantially different' from the rest of the Society.[40] Thus, publishing needed to have its own review.

The other factor underlying the 1995 review was that, just as it began its deliberations in October 1994, the head of publishing, Bruce Goatly, left the Society.[41] His role was covered by the production editor, Chris Purdon, but he himself was exploring other opportunities, and would accept a publications manager position at another learned organisation shortly after the review group reported.[42] Deciding how to fill the senior management position in publishing was another element of the review's remit.

The Publications Review was motivated by the problems of 'declining income from the journals while submission rates soar'; and a 'general recognition' that publishing was changing 'at a considerable pace' and the Society would need to 'respond appropriately and quickly'.

A review group of officers and staff was chaired by the chemist John Rowlinson, then serving as physical secretary. Rowlinson's committee also included the author of the 1993 study of the 'scientific information system', and a representative from the Institute of Physics publishing division.[43] The Society's leaders in the 1990s seem to have been more willing than their predecessors to accept that their own undoubted scientific expertise was not necessarily sufficient for running an organisation with over a hundred staff and a turnover of several million pounds; or for running a publishing business with 19 staff.[44]

The review group met with the Society's four editors, visited two other publishing organisations, and surveyed fellows' attitudes to the publications (which revealed that only 28 per cent of respondents had ever submitted to one of the Society's journals, and that fellows were far more likely to read *Notes and Records* and *Biographical Memoirs*, than either the *Proceedings* or the *Transactions*).[45] In March 1995, the group made 23 recommendations to Council. They acknowledged the 'considerable success in attracting more papers', but called for clearer strategies to capitalise on it. They recommended that the Society should actively investigate and explore the new electronic technologies for dissemination, either 'alone, in consortia or in collaboration with outside organizations'; and that each year, about £50,000 of the publication surplus should be set aside (rather than added to the Society's general funds) to create a 'development budget' for the publishing department. Without reinvestment in the business, the publishing team was struggling to 'maintain its effectiveness and competitiveness'.[46] The review group did not recommend transferring any editorial, production or marketing processes 'to outside agents', despite suggestions that marketing and distribution could be outsourced.

The review group's report noted that the publications were a means for the Society 'to further its objective of disseminating knowledge and of bringing financial return to the Society'.[47] By October 1996, the newly created Publishing Board would refer to the aims 'to disseminate science effectively and to contribute financially to the Society' as its 'twin objectives'.[48] The 1987 objective that the publications be 'worthy of the Society' became implicit, and generating income for the Society thus became one of two (rather than three) goals for the publishing team. The Society's response to new technologies would consistently be driven by finances, rather than the possibility of disseminating science more effectively.

In order to achieve the financial goal, the review group insisted that a new culture of professional management was needed for the publications. Back in 1989, Bruce Goatly, then recently appointed as head of publishing, had told one of the new journal editors: 'We have to be businesslike now: it may go against the grain to be so ungentlemanly and commercial, but publishing is a hard commercial world!'[49] But there was a world of difference between fellows and staff trying to be more 'businesslike', and the desire of the 1995 review group to enable a 'more flexible and pro-active approach' to new opportunities, provide clearer strategic management and 'create more corporate unity'.[50] The review group saw two problems with the way things were being run: the unclear division of responsibilities between staff, editorial boards and the Publications Management Committee (which had enabled some boards to take courses of action 'without being fully aware of the financial consequences'); and the 'lack of integration' arising from the absence of a 'single head covering all aspects from paper acceptance to sales and marketing'.[51]

The review group dealt with the first concern by replacing the Publications Management Committee with 'a substantially more independent Publishing Board' that included external members with 'practical and commercial publishing skills'.[52] The Publishing Board was to focus on strategy, while day-to-day management would fall to a new executive group formed of senior staff and fellows meeting on a frequent basis.[53] The review group also insisted that a director of publishing, 'with experience of managing an entire STM publishing operation in a commercial or semi-commercial environment', should be appointed 'as soon as possible'.[54]

The desire for someone with significant experience in the publishing industry broke the Society's decades-long tradition of hiring science graduates and training them in editing and publishing on the job. (This was how Graddon, Evans and Goatly had all learned their roles.) The Society still wanted someone with 'a sound academic background', but they should also have 'first-class training and experience within academic publishing or a learned society'.[55] Subsequent advertisements for managers to support the new director would specify at least three years' experience in publishing.[56]

The new director of publishing was Dr Ruth Glynn, formerly of Oxford University Press.[57] Glynn's academic background was in classics, but she had over 15 years' experience in scholarly publishing. Particularly notable was her expertise in electronic publishing, an area that the Royal Society had specifically mentioned in its job description.[58] Glynn seems to have begun her publishing career with the Oxford University Computing Service, which pioneered the application of computing to humanities research; used the new photo-typesetting equipment (such as the Monotype Lasercomp) to print editions of classical Greek works more cheaply and quickly than had been possible with traditional typesetting; and explored the use of CD-ROMs to distribute texts electronically. By the late 1980s, these efforts had inspired the creation of an electronic texts division at Oxford University Press, where Glynn latterly worked.[59]

The publishing team that Glynn was to lead included plenty of women, but she was the first woman to hold such a senior position. In 1991, all five sales and marketing staff had been women, as were both staff in the publishing finance and accounts department, and five of the nine members of the editorial and production team. Only the stock and despatch team was entirely male.[60] One female member of staff remembers the office as being 'dominated by women', where they joked that it would have been nice to have a man around once in a while.[61] This contrasted with the fellowship of the Society, which was still overwhelmingly male. In 1991, developmental biologist Anne McLaren had been elected as foreign secretary of the Society, becoming the first woman in its history to hold one of the elected officer roles.[62]

Glynn's tenure as director of publishing was short. Her main legacy was the recruitment of the three managers who would lead publishing into the electronic age. Phil Hurst, Charles Lusty and John Taylor all joined the Society in January 1996. Hurst was a chemistry graduate who had already held a senior editorial position in pharmaceutical publishing, whereas Taylor, who took on marketing, had previously worked in Fleet Street. When Glynn left, the Society did not re-advertise her post but asked Taylor to be the senior manager. (John Taylor is not to be confused with Stuart Taylor, who would become head, and later director, of publishing in 2006.[63])

Hurst and Lusty understood their appointments as part of an intentional culture shift in Royal Society publishing. Hurst felt that, even after the relaunch of the journals, the Society had still been taking 'an amateur's approach to publishing'. In an organisation whose fellows were themselves experienced authors, reviewers and editors, it seemed that 'everyone's an expert on publishing', and there was an 'old boy's network club-y approach'.[64] Hurst later reflected that, 'the past twenty years have been very commercial.... It sounds like we are bragging, but there wasn't business-discipline [before]'.[65] 'Things changed', recalls Lusty, 'after bringing in three guys with ideas.'[66] Lusty was excited by his new job, and confident that he could make the production processes more efficient: 'I could see straight away that I could make an impact on the cost base.'[67] One of the first things he did was to negotiate 'for a big reduction of 50%' in CUP's printing charges, noting that '[t]hey didn't want to lose their prestigious customer'.[68] The new management team in publishing understood that their remit was to make publishing financially remunerative for the Society.

The changes were not comfortable for everyone. For instance, Taylor was tasked with shutting down the storeroom from which journals were distributed; this led to redundancies. Meanwhile, a salary review led to the downgrading of some of the existing staff. 'We came in and saw lots of fat', noted Hurst. 'It was easy, to increase profit, to cut waste. It happens when someone new comes in.'[69] Thus, for many of the existing staff, it was a traumatic time. As one member of senior management recollected, 'Everybody hated management'.[70]

The changes of the mid-1990s included a conscious shift away from the previous under-investment, when the publishing team had explored the new authoring and typesetting technologies on a shoe-string budget. The 'policy of optimising in-house skills' pursued by Goatly and his team had initially been necessary because the technologies were so new that there were few third-party services or off-the-shelf products available. By the late 1990s, the increased availability of digital service providers, commercially produced software packages and specialist typesetting companies meant this was no longer necessary. The 1995 Publishing Review encouraged a change of approach, while diplomatically admitting that the earlier approach had avoided 'reliance on outside suppliers for day-to-day management', and avoided the risk of investing in 'high capital cost technologies' that might soon be outdated.[71] Going forward, the publishing team would have a budget for development, and were encouraged to explore options for buying in services or products.

For instance, the Society's editorial staff had initially developed their own electronic replacement for the 'Register of Papers', to run on the Society's mainframe computer. Then, in the late 1990s, they collaborated with other learned societies to develop an Electronic Submission and Peer Review system (ESPERE).[72] By the 2000s, however, once Windows-based PCs were in widespread use, firms like Thomson Reuters and Aries Systems were able to find a market niche in providing standardised software and services to hundreds of learned societies, university presses and publishers. The product development budgets of these firms enabled them to create far more sophisticated systems than the learned societies could do themselves. In 2006, the Society moved its online manuscript submissions and peer review process onto Thomson Reuters' *Manuscript Central* system.

Computerised typesetting was another process that had been done in-house in the late 1980s and early 1990s. Staff shortages then forced the Society to outsource some of the typesetting work in 1995. In the short term, this seemed expensive, but it solved the problem of having to recruit copy-editors able to programme mathematical formulae and symbols in TeX. In the longer term, it opened up the possibility of using competitive tenders to reduce costs. In 1997, Dobbie Typesetting of Tavistock was appointed to set the B-side journals (where authors were less likely to submit in TeX format), and this was expected to save at least 25 per cent.[73] And a few years later, the ease of transferring digital files over the internet made it possible to outsource the typesetting to TechSet in India, and take advantage of cheaper labour in the Global South.[74]

For centuries, the relationship with its printer had been the Society's most significant external partnership. The arrival of computerised typesetting disrupted that relationship, and the effect was exacerbated by the move to electronic publishing, which involved developing relationships with dozens of new and unfamiliar contractors and service providers, of which internet service providers and journal hosting platforms were only the most obvious.

First steps in electronic publishing

The 1995 review group had argued that the Royal Society ought, 'as the academy of science', to participate actively in the transition to 'electronic publishing'.[75] The advertisement for Phil Hurst and Charles Lusty's positions informed potential applicants that the Society intended to embark on an 'innovative and ambitious publishing programme', with particular emphasis on 'computer technology' and a 'rapid movement into electronic publishing media'.[76] When Lusty started as production manager a few months later, he found the Society a 'very exciting' place to work, because it was the 'dawn of a completely new age'.[77]

But it was still very early days. Lusty and Hurst arrived before the Society's treasurer had embarked on the 'full computerization' that would introduce Windows-based personal computers throughout the Society's administrative units, and create an internal network by 1997.[78] Prior to that, the publishing staff were working on three different computer systems. Copy-editing and proofing work was done on PCs (in WordPerfect), as was the processing into TeX. The artwork and desk-top publishing was done on an Apple Mac. The third system was the Society-wide NCR mainframe that had replaced the original Jacquard system:

the sales and marketing staff used this to manage subscriptions, and the editorial staff had an 'internally-developed administrative system' for monitoring the refereeing and production workflows.[79] Despite the computers, paper continued to be everywhere. Even in 2002, the first impressions of a new editorial assistant were of '[w]ood panelling, filing cabinets, [and] orange files'.[80]

The review group had admitted that it was impossible to be 'certain how, or how quickly' the transition to electronic publishing would happen, and this uncertainty made it difficult for the Society to establish a coherent strategy.[81] Phil Hurst felt that the fellows involved in publishing seemed to be preoccupied with the potential of new electronic formats, and, rather than developing a coherent strategy, had a 'knee-jerk reaction' to each new possibility. He felt that the Society 'lost its way' in the mid-1990s, as 'editors did their own things, like the CD-ROM'.[82]

The first personal computers featuring CD-ROM drives had begun appearing in the early 1990s. In 1995, the neuroscientist Patrick D. Wall took over as editor of *Transactions B*, and wondered whether CD-ROM disks might be the solution to the long-standing problem of printing vast amounts of detailed data that few people would consult. Like William Bragg half a century earlier (Chapter 13), he proposed a 'two-tier publishing' system, but with the datasets being issued on CD-ROMs rather than deposited in the Society's archive. For Wall, the idea that the data could also 'be made available on the Internet' seemed a secondary consideration.[83] The physicality of the disks meant they could be sold and distributed in a manner analogous to the printed journals. However, as Oxford University Press was discovering, the sales of CD-ROMs did not live up to early ambitions.[84]

The new publishing team would take 'the view that the web was the way forward'.[85] The challenge was that it was more difficult to imagine a business model for electronic journals hosted on computer networks. As a Royal Society committee had noted in 1993, 'We know how to give electronic journals away, [but] we have no idea how to sell them.'[86] Selling them was, however, seen as essential. Subscriber numbers were still falling, and a 1997 report noted that 'in common with most academic publishers, the RS is currently taking an increasing amount of cash income from a shrinking number of customers'. The 'whole industry' was 'gearing up' for online delivery of journals 'in the hope/expectation that these new services will become a revenue stream'.[87]

The *Transactions A* editorial board wondered whether printed journals would 'become unnecessary, with scientists publishing openly over the Internet'.[88] The board members would have been aware that researchers in high-energy physics were already using computer networks to share their research. Since late 1991, the Los Alamos laboratory had run an automated email service that allowed researchers to deposit or access preprints (as TeX files) on an FTP server. The service acquired a web interface in 1993, and became known as arXiv.org.[89] This differed from the model the Royal Society was seeking because it was not about sales, and it focused on the circulation of preprints. The Society had been enabling the informal circulation of separate copies, or offprints, of papers from its journals since the late eighteenth century (see Chapter 7), but these were the final, peer-reviewed and published versions. From the 1960s, if not earlier, various communities of scientific researchers had also begun sharing early versions of their research papers through correspondence networks, as typescripts, mimeographs or photocopies.[90] This was what arXiv.org facilitated.

The *Transactions A* board decided that the absence of any 'refereeing and quality control system' meant that papers deposited in arXiv.org did not count as published, and that electronic preprints were not in fact competing with journals (electronic or not). They could then decide that, although prior publication had, since the days of Joseph Banks (Chapter 6), precluded papers being considered for publication in the *Transactions*, this did not apply to papers 'pre-published on electronic servers, such as that at Los Alamos'[91] In the 1990s, preprints, even when electronically circulated, were still seen as part of long-standing, informal practices of scholarly communication.[92]

The best way for learned society publishers to adapt to the new possibilities was much debated. In 1992, Cambridge mathematician Peter Swinnerton-Dyer, and former chair of the now-abolished University Grants Committee, outlined a potential future publishing system in which a government-funded data network hosted electronic journals edited by the various learned societies.[93] The Royal Society was sufficiently interested in his ideas to invite him to join the Publications Management Committee. His 1993 advice to the Society was that it would be easier, in the short term, to set up a new electronic journal than to convert the existing journals.[94] Despite this, the Society's first steps into electronic publishing would focus on creating electronic editions of its *Proceedings* and *Transactions* journals. These were perceived as add-ons to the print editions, whose additional costs could be passed on to institutional subscribers who wanted electronic access. There was no attempt, at this point in the Society's history, to reimagine what a journal might be, or how editorial practices might operate differently.

The 1995 review advised that the Society 'with all urgency' establish an internet server.[95] This was initially imagined as a way of enhancing the global presence of the journals, providing information for potential authors as well as for librarians seeking subscription information, but there were hopes that it would become possible to provide access to the full-text contents of the *Proceedings* and the *Transactions*. The Royal Society's first web pages appeared in 1996, hosted on a server run by the British Academy, its counterpart national academy for the humanities and social sciences.[96] By summer 1996, the publications team was working on a set of '40 to 50' web pages that would be independently hosted by UUNet Pipex, the UK's first commercial internet service provider. The team was excited that, 'with 2Mb of memory available', there would be sufficient room to expand 'towards full electronic publishing'.[97] Electronic access to the full text of the current issues of *Proceedings A* and *B* and *Transactions A* and *B* began in 1997.

The move to electronic authorship and typesetting meant that the Royal Society already had electronic versions of the articles it published; but, although physicists and mathematicians might be happy to use TeX files, most researchers wanted papers in a more easily readable format. Plain text was possible; and it was also possible to convert TeX files into an electronic format that replicated the appearance of the printed page. Adobe's PDF format would become the preferred way of doing this, once it ceased to be a proprietary format. The publishing team was most interested in new 'mark-up' languages, such as HTML. These formats supported features, such as hypertext links, that were not possible on the printed page and, in the late 1990s, it was hoped that these 'added value features' would persuade subscribers of the value of paying extra for electronic access.[98] By the 2010s, enhanced HTML would allow articles to incorporate non-textual elements such as animations of molecular structure in three dimensions or audio and video clips, but the then-publishing director noted that researchers still preferred 'simply to read the PDF version'. Publishing professionals might be disappointed that researchers were so 'conservative in their information consumption', but it demonstrates that printed journals had an enduring legacy.[99]

Creating electronic versions of the articles was only part of the challenge facing Royal Society publishing in the late 1990s. Even once the Society had a web server to host them, how would subscriber access be managed? The problem was not unique to the Royal Society and, by 1996, a number of systems were under development that could aggregate content from multiple publishers and provide readers with a single interface that promised to be more convenient than having to cope with multiple passwords for each publisher's website. *JournalsOnline*, a government-funded system for UK higher education, was in competition with several commercial services, such as Blackwell's *Electronic Journal Navigator*, SwetsNet, and OCLC FirstSearch *Electronic Collections Online*. Unlike *JournalsOnline*, these systems charged university libraries a fee, and they were more successful in persuading journal publishers to sign up.[100] The Royal Society's new attention to wider trends in the academic publishing industry – including the presence of two directors of Blackwell on its Publishing Board – meant it was aware of at least some of these options and, in early 1997, it decided that Blackwell's *Navigator* would be a relatively easy and low-cost entry into electronic publishing.[101] The decision to use systems that had been developed by and for commercial publishers marks a shift in the approach of Royal Society publishing: it was an admission of shared aims between society publishers and commercial publishers that would have been unimaginable in the 1950s.

It turned out that requiring subscribers to access the Society's journals through 'aggregation services' (that is, *Navigator*) 'attracted unfavourable comment', and so in 1998, the Society began plans to offer direct access to its journals.[102] Just as printing had traditionally involved working with an external partner with the appropriate expertise, so did journal hosting. The Society initially chose to work with Turpin Online Publishing Services, a wholly owned subsidiary of the Royal Society of Chemistry and an offshoot of its warehouse and distribution service.[103] When Turpin withdrew from providing online hosting services soon after, the Royal Society had a series of short-lived hosting partners until settling from 2004 with MetaPress, and then, from 2009, with HighWire Press, a spin-off from Stanford University library.[104]

For access to the historic back-run of the *Transactions* and the *Proceedings*, the Society pursued a different route. Its collaboration with JSTOR mirrored earlier arrangements with publishers of facsimile reprints and microfilm editions (see Chapter 14). JSTOR (for 'Journal Storage') was a non-profit organisation that aimed to create a digital archive of journal back-runs. It had been developed by US university libraries, with funding from the Andrew W. Mellon Foundation. In March 1998, JSTOR representatives went to London to persuade Royal Society officers and staff that the uniquely long run of the *Transactions* (and the *Proceedings*) should be the core of its 'General Science Collection'. John Taylor later reported that the project involved photographing 700,000 journal pages, running them through OCR software (and re-keying the passages with unusual typographic characters or foreign languages), and delivering the data on 300 CD-ROMs. The effort was, he said, 'mind-boggling'. Whereas eighteenth-century scholars had relied heavily on abridgements to access old papers in the *Transactions* (Chapter 4), from February 2000, their modern counterparts could 'call up the full text of papers to their desktops, 24 hours a day, seven days a week, from the comfort of their university libraries, offices and homes' (if their library had a JSTOR subscription). The Society agreed to add further papers to JSTOR each year, three years behind current publishing.[105] The collaboration with JSTOR enabled the Society to have its massive back-run digitised at no direct cost to itself, and it received a usage-based royalty derived from the subscriptions paid by university libraries to JSTOR.

In the post-1995 climate at the Royal Society, it was taken for granted that the Society would not be providing electronic editions of its journals free to all internet users. The possibility of seeking additional streams of income – from government, industry or philanthropy – to fund the costs of publishing (as had been done in the 1920s, see Chapter 12) does not appear to have been discussed, even though (or perhaps because) the Society was already fundraising for a range of other projects. The era of subsidising journal publishing and distribution had passed from the Society's institutional memory with the wave of staff retirements in the 1970s and early 1980s. The clear expectation was that journals – or access to them – were commodities to be sold.

Embracing digital publishing

By the year 2000, the Royal Society could, in fairness, claim to have made the transition to electronic publishing. Current issues of the research journals, *Proceedings A* and *B*, and the thematic review journals, *Transactions A* and *B*, were available electronically over the internet as well as in print, and the accessibility of the historic back-run had been greatly enhanced by its inclusion within JSTOR. However, in many ways, Royal Society publishing had not changed all that much. The journals were still edited as printed products, with groups of articles gathered into issues of a standard length and published at regular intervals. And they were still produced as printed products, with workflows designed around the needs of typesetters and printers, with the electronic editions bolted on afterwards.

It would not be until the early twenty-first century that the Society became a digital-led publisher. Charles Lusty led the conversion of the production processes to an XML-first workflow, in which the creation of an electronic edition was planned from the start. Meanwhile, Phil Hurst and his editorial team were investigating opportunities for product development. In 2003, *Biology Letters* became the first of a wave of new Royal Society research journals that were designed and created for the digital age.

This second phase of engagement with the digital world was aided by new leadership, after John Taylor's retirement in 2004. In seeking a new head of publishing, the senior management were, as Phil Hurst put it, keen to recruit 'new blood' from 'a commercial society publisher', such as the Institute of Physics or the Royal Society of Chemistry.[106] The Royal Society's willingness to learn from other societies, rather than expecting to lead the way, reflects the shift in its position in the landscape of society publishing over the second half of the twentieth century.

The new appointee was Ian Russell, who had spent 11 years in the editorial team at the Institute of Physics, most recently as assistant director, where he had been at the forefront of 'e-content delivery' for 40 journals. From his perspective, the Royal Society's four journals seemed to have 'been left behind', despite having been 'exceptionally strong in the 1960s and 70s'. He saw the Society's journals as a 'fantastic brand' that could be developed, and he appreciated the support of the senior management team, who gave Russell the opportunity to 'really invest' and to hire 'some new people to do the editorial development activities'.[107] This meant that Hurst – who had previously been doing 'all the journal development across the portfolio' – was promoted to manage four staff, each of whom combined the day-to-day management of submissions and peer review with strategic thinking about journal development. Hurst later remembered Russell as having been 'very proactive and energetic', but Russell only stayed for two years before moving on to a senior role with ALPSP.[108]

His successor was Dr Stuart Taylor, who had an academic background in pharmacology and 18 years' experience in scientific and medical publishing at Blackwell. Taylor continued Russell's commitment to product development and author-centred service and, by creating new journals and increasing the periodicity of existing ones, he found ways to publish more of the continuing flood of submissions. By selling more content and focusing on production costs and on staff efficiency, he was able to deliver the commercial success that the Society's leadership sought.

These successes were, however, delivered against a political context that had changed from the mid-1990s. The threat of direct government intervention in Royal Society affairs had been headed off, but government regulation and controls on scientific research increased, from legislation controlling biomedical research to parliamentary investigations of the public understanding of science and the role of peer review.

One of the most significant interventions was the growing political support for the open access movement, whose vision of free digital access to the latest research challenged the sales-focused business model that had underpinned academic journal publishing since the 1950s and 1960s. Support for open access from the research community was often associated with protests against the alleged exploitation of academics by commercial companies, regularly exemplified by the high profit margins of the Elsevier corporation. Having bought Pergamon Press in 1991, Elsevier had become the biggest publisher of academic journals in the world. In 2013, it published 24 per cent of the global output of scientific research, and its scientific, technical and medical division returned a profit margin of almost 40 per cent.[109] The Royal Society's publishing business was much smaller in absolute terms, yet in the 2010s, it too reported annual surpluses around 40 per cent of turnover (see Figure 16.3).

The undoubted commercial success of the Society's publications in the twenty-first century was delivered by a publishing director who was himself personally sympathetic to the open access movement. Stuart Taylor joined the Royal Society just as open access was gaining political traction in the UK. In his 2006 job interview, he recollected starting to answer a question about open access with enthusiasm, 'assuming that as an academic body this would be a very pro-OA organisation'. But then, 'it was clear from the expressions of the people interviewing me that they weren't necessarily of the same view, and they thought OA was primarily a threat. So I had to change my tack a bit.'[110] Over the following decade or so, Taylor was able to push Royal Society publishing further towards open access, but his task was complicated by the absence of guidance on how to balance the twin objectives of disseminating science and making money.

Product development

Even before the arrival of Russell and Taylor, the Royal Society publishing team had begun to look for new opportunities. Phil Hurst remembers that, once the existing journals had moved online, '[a] lot of us got into journal development'.[111] That could mean developing the existing journals to ensure they remained attractive to authors and to library purchasers, but it might also mean the creation of new journals. Launching new journals under an existing 'brand' was something that other publishers – including the Institute of Physics and the Royal Society of Chemistry – were doing successfully, and it was an attractive commercial option, since it created an additional product for which libraries could be persuaded to pay.

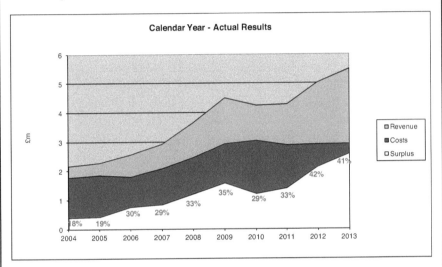

Figure 16.3 Publishing surplus, from Publishing Director's Report to the officers for year-end 2013 © The Royal Society.

During the twentieth century, the Royal Society had not taken this route. Its only new periodicals since 1905 had been aimed at its internal audience of fellows: *Notes and Records*, *Biographical Memoirs* and *Royal Society News*. Its greatest innovation was the magazine *Science & Public Affairs* (*SPA*), which began in May 1986 as an annual publication 'dealing with the wider implications of science and technology in Government, industry, commerce and daily life'.[112] It reflected the contemporary interests in the 'public understanding of science', and proved popular with the fellowship: by 1989, almost 40 per cent of fellows were choosing to receive free copies of the now-quarterly *SPA* rather than either the *Proceedings* or the *Transactions*.[113] It was not intended to be fellows-only, and its initial announcement suggested a potential audience ranging from those working in 'industry, commerce or Government' to teachers, academics and 'the intelligent and knowledgeable "person in the street"'.[114] But targeting such a variety of readers was not straightforward: in 1989, its editorial board noted that *SPA* was not competing with *New Scientist*, but still hoped it might 'possibly even [be] sold through newsagents'.[115] *SPA* was not a financial success, nor a good fit with the other publications of the Society. In the early 1990s, it was transferred to the British Association for the Advancement of Science, which continued to run it with financial support from the Committee on Public Understanding of Science.[116]

The interest in product development in the early 2000s led to the launch of two new academic journals: *Biology Letters* (2003) and *Interface* (2004). With *Biology Letters*, the Society finally responded to the oft-spoken desire among its biological fellows for the Society to publish something akin to *Nature* or *PNAS*. By publishing short papers online, it would help scientists 'get validated research disseminated as quickly as possible'. It was initially presented as a digital extension of *Proceedings B*: it shared the same editor, and abstracts of its letters were printed in *Proceedings B*. This allowed the Society's marketing team to present *Biology Letters* as an extra service for existing *Proceedings B* subscribers, rather than trying to find subscribers for a completely new product.[117] The plan included printed abstracts and a 'printed archive' four times a year because market research had suggested that 'whilst biologists value electronic publishing as a quick way to publish findings, they still want a printed publication because this is a key requirement for accreditation by funding bodies'.[118] *Biology Letters* established itself quickly, and by 2005 had become an independent journal.

The publicity material for *Biology Letters* hinted at some ambivalence about purely electronic journals: potential authors were reassured that the electronic papers would be 'fully citable', and would undergo 'rigorous peer-review'.[119] Peer review was a particularly sensitive topic in the years around 2000, because a series of highly public debates around BSE and genetically modified foods had focused attention on the nature of scientific knowledge. There was a parliamentary inquiry into the supposed 'crisis of confidence' in science and scientists in 1999–2000.[120] Peer review emerged from these debates as the 'cornerstone of the quality assurance process in science', with the parliamentary inquiry endorsing the Royal Society's advice to journalists that they should 'treat with healthy scepticism work that has not been approved through peer review, including information that can be accessed through the internet'.[121] The fact that the Royal Society had felt it necessary to create guidelines for journalists, to create its own Press Office and, subsequently, to support the establishment of the Science Media Centre (2002) and the Sense about Science campaign (2002), indicates a perceived need to explain the processes of constructing scientific knowledge.

It was also around this time that the Society started emphasising its own 'unparalleled record' in peer review.[122] Refereeing had long ceased to be unique to learned society journals, but the Royal Society now claimed the distinction of being responsible for 'the invention of the peer-reviewed scientific journal'.[123] In 2000, the then-president emphasised the role of the Society's referees in 'sifting' submissions and distinguishing 'worthwhile science' from other sorts of material, which he claimed would 'confer … authenticity' on research findings.[124] The standard adjectives used in the marketing of the Society's publications were now 'peer-reviewed', 'excellent' and 'authoritative'.[125]

It was particularly important to emphasise that *Biology Letters* would be rigorously peer-reviewed, because in 2003, the Society was being accused of being a 'corporate rent boy' for promoting the interests of international bio-technology firms.[126] The acrimonious and very public controversy was about the Society's role in evaluating and publishing results from the first large-scale UK field trials of genetically modified crops, intended to help the government decide whether to lift its 2000 moratorium on commercial planting of such crops.[127] Given the public interest in the field trials, and after accusations were made that the Society's peer review process was not as stringent as it should be, the president of the Society, Robert May, took to the pages of the *Guardian* newspaper to walk readers through the 'normal practice' for reviewing scientific papers for potential publication. He explained how the journal editor, 'acting independently from the society's governing body', would select 'at least two referees', who would be 'unpaid' but would have relevant expertise, and who would prepare a report about 'whether the appropriate methods were used (and are written up in a way that they could be replicated) and whether the results are accurate'. May carefully did not claim that papers which passed peer review were unchallengeable, but he insisted peer review was 'the primary quality control mechanism applied to the results of new scientific research'.[128]

The other area identified by the Society's market research was 'multi-disciplinary and cross-disciplinary work spanning the physical and life sciences'.[129] Like rapid publication in the bio-sciences, this was hardly a new suggestion for the Royal Society (see Chapters 11 and 14), but it was finally taken seriously with the launch of *Interface*. For Charles Lusty, interdisciplinary research seemed 'a natural place for us to move into'.[130] Stuart Taylor would come to think of the new journal as a '*Proceedings C*, where C is your cross-disciplinary'.[131] *Interface* was launched in 2004, promising to 'draw together a broad spectrum of scientists from a variety of backgrounds' working at 'the interface between the physical sciences and life sciences'.[132] It proved so attractive to authors that its initial bi-monthly periodicity increased to monthly in 2008. Its success inspired the creation of *Interface Focus* in 2011, as a *Transactions*-style series of thematic issues.

Biology Letters might be seen as reactive to trends elsewhere in life sciences publishing, but with the launch of *Interface*, Royal Society publishing was actively leading in a new area. As Table 16.1 showed, the biological sciences were the source of almost 70 per cent of the papers submitted to the Society in 2013. And the new interdisciplinary journals were already attracting more submissions than the established journals in the physical sciences.[133] Most of the income, however, continued to be generated by 'the big four, the two transactions and the two proceedings'.[134]

The willingness of the early twenty-first-century publishing team to launch new journals, and to expand existing ones (*Transactions A* and *B* became fortnightly in 2008) stood in stark contrast to the financial fears of their predecessors a century earlier. The difference was that the publishing finances around 2010 looked exceedingly healthy. The reasons for this were, as ever, two-fold: costs and income. As well as keeping up the pressure on production costs, Stuart Taylor paid close attention to editorial costs, particularly as the staff cohort grew to more than 20 people.

In its 2004 submission to government on the peer review process, the Society had noted that 'even though the referees are volunteers, managing the peer review process to ensure timeliness and quality is administratively expensive'. The peer review process was estimated to account for '42% of our publishing staff costs'. This was not simply about tracking the progress of papers through the refereeing, revision and production processes. The Society's sub-editors checked that 'language is unambiguous, correct style and nomenclature has been applied and illustrative material is of the required standard', and for all those for whom 'English is not their first language', they ensured 'that the finished paper is understandable to the global scientific community'. Thus, the Society concluded, the publication of scientific research papers could not be done for free.[135] But, as Stuart Taylor remarked, 'the executive team didn't really know if we were doing it effectively or not'.[136] He added a new metric to the performance indicators that had been reported to Council since the 1950s: staff efficiency. After the move to the new (commercially produced) 'fully electronic peer review system', the number of articles that could be processed by each editorial assistant more than doubled. By 2014, each editorial assistant was, on average, dealing with almost 1,200 submitted articles a year.[137]

This focus on efficiency also improved the publication times, which began to be measured in days rather than weeks. Referees were chased for faster reports, and authors were given tighter deadlines for proofs. In 2001, the mean publication time had been over 300 days from receipt to publication; at almost 43 weeks, this compared poorly with the average of 35 weeks for *Proceedings A* and *B* in 1967 (compare Figure 14.3). By 2013, improvements in both the editorial and the production processes had dramatically reduced the time taken to publication: it had fallen below 100 days, or 14 weeks (Figure 16.4). Taylor proudly reported that 'we are considerably faster than all our major competitors (most notably *Nature*, *Science* and *PNAS*)', though he also warned that 'we are probably close to the limit of what is possible'.[138]

Taylor also introduced metrics for the costs per article published, and per article downloaded. In 2014, he was able to report that the cost per article published had fallen from over £500 in 2009 to about £350, which in turn had kept the total publishing expenditure 'flat for five years, despite increasing output'.[139] In a total transformation from earlier decades, paper and printing costs were now relatively insignificant. Salaries were now the largest cost item, followed by typesetting, XML, and the hosting costs for the digital platform.[140] Print costs had by this time become relatively insignificant.

By the 2010s, the relationship between print and online editions of the journals had flipped: the print edition had come to be seen as the 'extra' to an online subscription. Taylor saw it as 'a bit of an anachronism in science publishing', and was keen to discontinue print entirely. He purposefully increased the surcharge for the print edition: 'It's put off about 80% of [the subscribers], but there are some who still insist on having print'.[141] The print runs of the *Transactions*, the *Proceedings* and the other journals were so small that the Society's various printers could use digital printing, which was 'more economical than traditional printing for low print runs (particularly where colour is concerned)'.[142]

Publication times

Speed of publication is typically ranked first or second in terms of importance by authors choosing a journal for submission. We have therefore made considerable efforts in the last ten years to reduce publication time as much as possible, partly by the introduction of a fully online submission system, but also by working on each stage of the process and optimising efficiencies as much as possible. We are considerably faster than all our major competitors (most notably *Nature, Science* and *PNAS*) and we are probably close to the limit of what is possible without either incurring extra cost or taking shortcuts. This is a key factor in attracting authors to our journals. There is a small increase (two days) from 2012 to 2013 which is the result of a change in the way we record resubmitted articles to provide greater transparency, rather than a slowing of our processes.

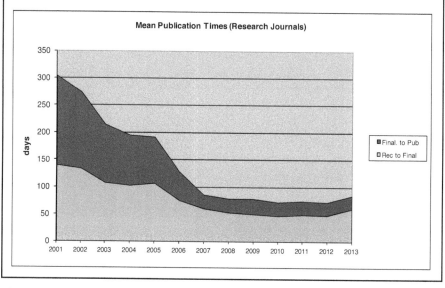

Figure 16.4 Publication times, from Publishing Director's Report to the officers for year-end 2013 © The Royal Society.

Throughout this period of expanding output, the numbers of institutions subscribing directly to one or more of the Society's journals (in any format) kept on falling, though price increases meant that the total subscriptions income still grew. Old patterns of geographical distribution remained broadly true: 39 per cent of subscribers were in the USA; 22 per cent in Continental Europe; 15 per cent in Asia (mostly in Japan); and 11 per cent in the UK.[143] About 80 per cent of publishing income in 2014 was generated by subscriptions, but the Society's publishing income was also boosted by deals with consortia of institutions.[144] Negotiating collectively allowed many institutions in India, Brazil and China access to British (and American) journals for the first time. The Society also received a small income from JSTOR royalties for access to the historic back-run. Altogether, in 2014, the Society received an average revenue per article of £2,250. That was apparently lower than an 'industry average of £3,000' per article[145] but, with an average cost of just £350 per article, it is clear why Royal Society publishing was generating a substantial surplus in the 2010s.

Since the 1950s, the effectiveness of the Society's desire to disseminate research had been measured by the number of subscriptions. The new ways of accessing journals made this a relatively meaningless measure, and forced the development of new indicators. For instance, Taylor reported to Council in 2014 that 'we comfortably exceeded our target ... with a total of 16.3m article downloads by the end of the year'.[146] The Society also began to publicise the list of countries which had access to its journals. This was an extensive global list, because it participated in a number of schemes to make scientific journals more easily available to institutions in the developing world. The ability to grant access to research digitally, without paying for extra printing and shipping costs, enabled the Royal Society to rediscover some of its historic philanthropy (see Chapter 10). In 2003, for instance, it had joined the UN Programme for the Enhancement of Research Information.[147] By 2015, it was participating in five schemes making journals available 'immediately and free of charge to the world's poorest nations'. Institutions in over 100 countries were eligible for free access to the Society's journals, while others were eligible for highly discounted access.[148] Free online access for the developing world was possible because the income from paying customers, mostly located in the Global North, more than covered the production and editorial costs of the journals.

Open access

The possibility of global circulation of knowledge was a clear benefit of the arrival of digital publishing. It had been foreseen by the signatories to the Budapest Open Access Initiative (2002), who argued that new technologies would make 'the world-wide electronic distribution of the peer-reviewed journal literature' financially practical, so that 'all scientists, scholars, teachers, students, and other curious minds' could have 'completely free and unrestricted access' to scholarly research.[149] Over the next 15 years, the open access movement gained increasing political traction with funding agencies in the UK and Europe.[150] Learned society publishers were largely protected from the criticisms of profiteering because they used the surplus from their publications to support the research community, but if the model of academic publishing were to change significantly, the financial basis of learned society publishing would be forced, once more, to change.

The Royal Society took its first steps into open access publishing in 2006, and launched its first wholly open access journal in 2011. Those first steps were, however, taken under duress, for the objective of making money from publishing was beginning to work rather successfully: publishing generated an 18 per cent surplus in 2004, rising to 30 per cent by 2006 (see Figure 16.3). A contemporary commentator noted that Ian Russell had to navigate 'a very difficult path between what the membership think and what the publishing division think', and praised him for his 'extremely diplomatic' negotiation that enabled the Society to 'experiment' with open access.[151]

Those involved in the administration of the Royal Society initially saw the open access movement as a financial threat, and their public statements indicated how thoroughly the sales-driven model of circulating knowledge had become normalised at the Society. With a dubious grasp of their own history, Royal Society leaders claimed that open access would be 'the biggest change in the way that knowledge is exchanged since the invention of the peer-reviewed scientific journal 340 years ago'.[152] In 2004, just as Russell was taking over, there was a UK parliamentary inquiry into 'the provision of scientific journals to the academic community and wider public'. The Royal Society's submission pointed out that the Society was already doing a great deal to make the results of scientific research widely available: the full contents of its journals were 'made available free of charge online 12 months after publication'; its journals were freely available online to institutions in 15 developing countries; and its 'liberal copyright policies' allowed authors to self-archive their papers on their university websites.[153] The Society's leadership might be supportive of Russell's desire to develop the journals, but they were extremely lukewarm about any more radical changes to the model of academic publishing that might diminish the publishing surpluses that learned societies used to support their conferences, grants and public engagement activities.[154]

Following the parliamentary inquiry, the UK research councils issued a supportive position statement on open access in 2005, requiring researchers funded by them to deposit copies of their published articles into open access repositories, and allowing the costs of publishing in open access journals to be included in grant applications.[155] In response, the Royal Society issued its own position statement, again stressing the risks of dramatic changes with unclear consequences, while simultaneously insisting that it was 'as committed now as it was when it was founded to promoting the exchange of knowledge, not just between scholars, but with wider society'. It worried about the potential loss of high-quality peer review; about the potential inequity of introducing a system that would require authors to pay to be published; and it was very worried about the loss of publication income for learned society publishers, which 'would lead to a reduction in, or cessation of' the current activities of such societies, many of which (including conferences and public lectures) were themselves a means of exchanging knowledge.[156] The Society's position was paraphrased in a headline in the *Guardian* newspaper: 'Keep science off the web, says Royal Society.'[157]

The position statement generated a backlash. Within a fortnight, 46 fellows published an open letter to the Society, accusing it of 'putting the concerns of existing publishers (including the Society itself) ahead of the needs of science'. The fellows believed the Society was mistaken to be 'warning ominously of "disastrous" consequences for science publishing' and ought instead to support the research councils' position.[158] In response, the Society's new president, astronomer Martin Rees, insisted that the Society was not being 'negative' towards open access, but was legitimately worried that the new alternatives to 'the established subscription model of publishing' had not yet been 'fully explored' or 'shown to be viable in the long run'.[159] The spat drew attention in the scientific press, and among open access campaigners. The Society was generally portrayed as unenlightened; insiders claimed it had been 'misrepresented'.[160]

It was in the aftermath of that debate that the Society began offering immediate open access to articles in *Transactions A* and *B*, *Proceedings A* and *B*, *Biology Letters* and *Interface*, if the author was willing to pay a fee. The first open access paper, modelling the way the brain deals with novel stimuli from auditory and visual sources, appeared in *Proceedings B* in June 2006.[161] Phil Hurst recollects the Society 'responding proactively to the external situation rather than driving it'.[162] The real driver was the Wellcome Trust's announcement that it would require all publications arising from its funded research to be open access from October 2006, and that first paper was funded by the Wellcome.

Despite the *Financial Times* enthusiastically claiming (again, with a poor grasp of history) that the Society's experiment would 'tear up its 340-year-old business model',[163] the Society's official attitude to open access remained lukewarm. A vice-president was quoted as saying that there was 'still a lack of evidence' about the long-term sustainability of such a system, and the press release stressed that the model was 'being tested ... to see if it provides a viable way of sustaining the costs of peer review and other aspects of journal production'.[164] *The Scientist* reported the Society's press officer speaking of dipping 'a cautious toe into the waters of open access publishing'.[165]

By turning its journals into 'hybrid journals', in which author fees would enable some articles to be immediately available free to all readers while other articles would remain accessible only to readers with a subscription, the Royal Society was following the lead set by commercial publishers. The specific details of its model led to some critique, however. Open access campaigner Peter Suber queried the absence of a waiver scheme for researchers without funds, and noted that the fees were relatively high. Whereas Springer and Elsevier were both then charging $3,000 per paper, the Royal Society fees were initially charged per page: £300 per page for articles in the biological journals and *Interface*, and £200 per page for the other journals. It did not take commentators long to work out that the Society's first open access paper would have been charged £3,000, then equivalent to $5,500.[166]

Stuart Taylor was appointed as head of publishing later that year. As he settled into his role, he sought to persuade Council that the Society ought to be taking a lead on matters of open access, as part of its historic duty to circulate scientific knowledge. He had some support: the chair of the Publishing Board in the 2010s, engineer and entrepreneur Michael Brady, said: 'I mean, a Society that proclaims itself to be the leader of science in the country, really ought to be leading intellectual discussions about Internet and OA and etc. ... rather than just being dragged along reluctantly after everyone else.'[167] The Society's shifting attitude was publicly signalled when it joined the Open Access Scholarly Publishers Association in 2011, and launched its first wholly open access journals.

Support for open access was strongest in the biomedical sciences, where funding bodies, including the Wellcome Trust, proved willing to pay the publication charges. The Royal Society's first open access journal, therefore, was *Open Biology*. It launched in September 2011, and was also the Society's first online-only journal. It particularly sought to attract researchers working in cell biology, molecular biology, microbiology or genetics, since these were fields that did not yet submit many papers to the Royal Society.[168] *Open Biology* charged a per-article fee to authors of accepted manuscripts; this Article Processing Charge (APC) was originally set at £1,200 ($2,160) per article, but was entirely waived until 2014.[169] In response to the long-standing worry about author fees being a barrier to authors who could not pay, the Society offered fee waivers to corresponding authors in the 69 countries on the Research4Life list of the world's poorest nations: this list included almost all of Africa (except South Africa) as well as parts of southern and central America, and south-east Asia.[170] The APC on *Open Biology* could be lower than for the Society's hybrid journals due to the cost-benefits of creating a new, born-digital journal rather than adding an option to an existing print-and-online journal.

Open Biology's founding editor, David Glover, drew attention to two other original features. As an online-only journal, *Open Biology* did not need to hold on to accepted articles until sufficient pages had accumulated for a printed issue. It could instead adopt a 'continuous publication model' and publish articles as soon as possible. Nonetheless, the legacy of the traditional 'periodical' survived in the organisation of the articles on the digital platform, where they were presented as if they were in monthly issues and annual volumes. *Open Biology* also published its papers under a Creative Commons licence (CC-BY), 'leaving copyright with the authors, but allowing anyone to download, reuse, reprint, modify, distribute, and/or copy articles provided the original authors and source are cited'.[171] In 1990, the Society had started insisting on authors transferring their copyright to the Society; this had been seen as necessary in order to protect the Society's commercial interests in income from photocopying and electronic dissemination, and followed industry practice. But transfer of copyright would come to be seen as symptomatic of the unfair commercial exploitation of academic authors and, from 2006, the Society no longer insisted. The use of Creative Commons licences for *Open Biology* made this shift more visible.

In October 2014, *Royal Society Open Science* became the Society's newest journal. It was marketed as a 'fast, open access journal publishing high quality research' from any area of science, engineering or mathematics (not necessarily interdisciplinary).[172] With its broad remit, *Royal Society Open Science* mimicked the new 'mega-journals' that took advantage of the capacity of digital platforms to publish far more content than print-based journals. Like *Open Biology*, it was an entirely online publication, with a continuous publication schedule, and its articles were free to read immediately on publication.

The most influential of mega-journals was *PLOS ONE*, created in 2006 by the US non-profit organisation Public Library of Science (PLOS). It covered a vast disciplinary remit, and had proved hugely popular: by 2013, it was publishing over 30,000 articles a year.[173] That year, the Royal Society's journals published just 2,100 articles.[174] PLOS also pioneered a new approach to peer review. Its founders argued that the space limitations of printed journals had made editors and referees overly selective, with the result that many papers presenting solid research were unnecessarily rejected or delayed as their authors tried one journal after another. *PLOS ONE* announced that it would publish all papers that were judged to be sound and valid contributions to research. Its referees were asked to evaluate only the rigour and technical soundness of a paper, and not to comment on its perceived significance. This approach became known as 'technical' or 'objective' peer review.

At the Royal Society, Stuart Taylor had tried to introduce objective peer review for *Open Biology*, but had been unable to persuade the Society's leadership that it would not result in 'a journal of rejects'.[175] The officers shared Louis Filon's desire that the Society should not be perceived as a publisher of second-rate articles (see Chapter 13). In 2014, however, Taylor won the argument for *Royal Society Open Science*. The Society described the new peer review system as 'publishing all articles which are scientifically sound, leaving any judgement of importance of potential impact to the reader'.[176] The use of two different peer review regimes appears to create a hierarchy of prestige among the Royal Society journals in the early twenty-first century, similar to that between the *Transactions* and the *Proceedings* before 1914 (see Chapter 11). Whether *Royal Society Open Science* will evolve into the Society's main research journal, as the *Proceedings* did in the twentieth century, remains to be seen.

By 2015, over 30 per cent of the Society's published output was immediately open access on its journal platform: all the content of *Open Biology* and *Royal Society Open Science*, and about 18 per cent of the content in the older hybrid journals.[177] Taylor was keen to 'flip' the older journals from subscription to open access one by one, but, with APCs proving expensive for funders and arguably inequitable for researchers, the long-term financial sustainability of open access publishing remained uncertain. *Royal Society Open Science* was initially able to waive APCs for all authors thanks to cross-subsidy from the other Royal Society journals; but, like the free access offered to the Global South, which depended upon the other journals generating income, it was not a route towards flipping all the journals.[178] Charles Lusty recognised the need to think creatively about 'another business model' for open access, including exploring other sources of income.[179] Covering costs might be possible, but whether it was also possible to generate a surplus was not yet clear. Back in the 1950s, David Martin had found it remarkably easy to convince the Royal Society's Council to shift to a commercial model of publishing, but in the 2010s, Stuart Taylor found it far more difficult to reverse the process, admitting that 'there would be quite a big impact financially … [and] ultimately I have to do what Council wants me to do'.[180]

The 1995 Publishing Review had cemented the Royal Society's growing interest in publishing as income-generation into one of the 'twin objectives'. The desire not merely to cover costs but to generate a surplus determined the Society's early engagement with electronic journals, the internet and open access. While there were individuals – both staff and fellows – who urged alternative approaches that would focus upon the goal of circulating research, they found it difficult to gain traction against the commercial focus established in 1995. That commercial focus formed the framework against which all those involved with the publications have had to respond to huge changes in academic publishing. As the chairman of the Publishing Board put it, Royal Society publishing had become 'at the heart ... a small business'.[181]

The determination to run Royal Society publishing in a more businesslike manner had some false starts, but it ultimately proved highly successful. The publishing operation had been generating a surplus of around 20 per cent of its income in the mid-2000s; by the mid-2010s, that had climbed to around 50 per cent, representing over £3 million of unrestricted income for the Society.[182] The journals, both old and new, attracted increasing numbers of authors internationally. Their papers were processed and published more quickly than before, and were rapidly available to readers all over the world.

One thing that changed very little, however, was the academic side of the editorial process. Even with the introduction of objective peer review on its newest journal, decision-making still depended on the volunteer labour of thousands of researchers each year. The Society was only able to cope with the flood of submissions because it was willing to look far beyond its own fellowship to find referees and editorial board members. This meant that the diversity of people involved in publishing was somewhat better than that of the Society's own fellowship. The adoption of electronic submission and the widespread availability of email meant that referees could now be based almost anywhere in the world; and the same came to be true of editors: in 2015, ecologist Spencer Barrett felt able to accept the editorship of *Proceedings B*, despite being based in Canada, thousands of miles away from Carlton House Terrace.[183]

The gender imbalance also began to shift in these decades. The Society appointed its first female journal editor in 2008, when ecologist Georgina Mace became editor of *Transactions B*. She was succeeded by geneticist Linda Partridge, but none of the other journals had had a female editor by 2015. Journal editors continued to be appointed from the fellowship – unlike referees or editorial board members – and were almost the only remaining link between the journals and the fellowship. But with a fellowship that was still 93 per cent male in 2015, the chances of more female editors were slim. The publishing staff, on the other hand, were 63 per cent female, and in 2011 the Society appointed Dr Julie Maxton as its first female Executive Director, the senior staff position once held by David Martin.[184]

During her editorship of *Transactions B*, Mace consciously sought to expand the editorial board to make it more international and diverse.[185] Since board members helped to choose referees, this also helped with the diversity of the pool of referees, in terms of institutional location, career stage, gender and ethnicity. It was only in the 2010s, that the Royal Society in general started paying serious attention to its lack of diversity (in various senses), appointing a diversity officer and reporting its gender statistics annually.[186] Thus, when Barrett inherited an editorial board with only 24 per cent women, he immediately undertook 'a concerted effort' to target more 'excellent female' researchers.[187] This marks a significant shift from the days when refereeing was always done by the fellows.

During the early twenty-first century, the Royal Society became more adept at expressing its strategic priorities. The role of its journals and its publishing team, however, seemed uncertain. Prior to 2006, the journals had been presented as part of a mission 'to support science communication and education, and to communicate and encourage dialogue with the public'.[188] The new strategic plan for 2006–11 had no equivalent goal. The ambition to 'increase access to the best science internationally' was concerned with networking and capacity building in the Global South, rather than with open access; and publishing appeared in the annual reports somewhat awkwardly under 'invest in future scientific leaders and in innovation'. It is an indication of the different priorities of a Society that had become increasingly outward looking, and active in both international scientific diplomacy and national policy-making.[189]

The reports of those years described the aims of Royal Society publishing as 'to publish high quality science, provide first class service to its authors and deliver a financial contribution to the Royal Society'.[190] This formula notably retained the 1995 financial objective, but replaced the ambition 'to disseminate science effectively' with a focus on the needs of authors rather than readers. This interesting choice of words appeared just as the Society's publishing team prepared to engage with open access by launching *Open Biology*. It signalled the ongoing tension between those who hoped the Society would lead on open access, and those who focused on the risks and uncertainties of a new publishing model.[191] Stuart Taylor noted that Council had supported his plans for the new open access journals because 'so far I've been able to do things while still getting extra income, so they are quite happy, because they have everything, both OA and increased profit'.[192] He believed it would not last. In future, the Society would surely 'have to accept lower profits, considerably lower profits'. 'There are', he said, 'a number of choices to make, a choice between what's good for science and what's good for generating surplus. They are not always the same thing. And without a clear steer from Council it can be quite hard to balance those two things.'[193] In 2015, that discussion about the balance between commercial imperatives and the mission 'to serve science well and to disseminate results' had not been resolved.

Notes

1 RS Annual Report (1991), 17 and 19.
2 Report of Conference of Editors (held on 2 November 1964), in RS CMB/150.
3 'Analysis of replies to questionnaires', in Annex C of 'Publications Review Group: Final Report', 29 March 1995, RS C/31(95).
4 A pilot study by A.J. Meadows was commissioned in 1991, see RS Annual Report (1991), 16. It was followed by Bryan R. Coles, 'The scientific, technical and medical information system in the UK: A study on behalf of The Royal Society, The British Library and The Association of Learned and Professional Society Publishers' (London: British Library, 1993), reported in RS Annual Report (1993), 18–19.
5 The Society changed the way it thought about the *Transactions*: since 1994, it has not been counted as one of the 'research journals'. The figures in Table 16.1 include the *Transactions* before 1994, but not afterwards.
6 Lutz Bornmann and Rüdiger Mutz, 'Growth rates of modern science: A bibliometric analysis based on the number of publications and cited references', *Journal of the Association for Information Science and Technology* 66, no. 11 (2015): 2215–22.
7 Table 10.2 in Peter Collins, *The Royal Society and the Promotion of Science since 1960* (Cambridge: Cambridge University Press, 2015).
8 Relaunch expenses of £35,726, see publication accounts, RS *Year Book* (1991), 40.
9 In March 1991, the Publications Management Committee [hereafter PMC] 'agreed that submission via a Fellow or Foreign Member of the Society would no longer be an explicit route for any of the four primary journals'. Cited in *Transactions A* editorial board minutes, 14 June 1991, RS CMB/365/TA/10(91). However, three of the four journals offered the option of submitting via an editorial board member (who was often but not always a fellow), see 'Procedures for submission of papers' in *Year Book* (1997), E59.
10 Email from M.J. Rees to S. Taylor, 18 February 2014 [shared by Taylor].
11 Email from S. Taylor to M.J. Rees, 18 February 2014 [shared by Taylor].
12 Other forms of diversity were not recorded until recently: in 2017, 69 per cent of authors responding to a questionnaire identified as white, rather than black, minority or ethnic, see Royal Society Diversity Report 2017, 27, available at https://royalsociety.org/diversity. For historic gender figures, see Camilla M. Røstvik and Aileen Fyfe, 'Ladies, gentlemen, and scientific publication at the Royal Society, 1945–1990', *Open Library of Humanities* 4, no. 1 (2018): 1–40, http://doi.org/10.16995/olh.265.
13 RS Annual Report (1992), 16; Annual Report (1993), 19.
14 See graphs in Aileen Fyfe, 'More submissions, more rejections: The Royal Society Journals since the 1950s' (18 June 2020), https://arts.st-andrews.ac.uk/philosophicaltransactions/more-submissions-more-rejections-the-royal-society-journals-since-the-1950s.
15 'The Society's Publications', 11 June 1987, p. 6 in RS C/87(87), with reference to the 1970s experiment with direct submissions (see Chapter 15).
16 RS Annual Report (1993), 19.
17 RS Annual Report (1994), 22.
18 'There's change in the air' promotional material (c. 1990), see Figure 15.2.

19 Referee report form for *Transactions B* (c. 1990), examples in RS RR/90/TB [in box ARB848, Deep Store].
20 *Transactions B* editorial board, 21 November 1991 and 10 April 1995, in RS CMB/365. [Its first editor, Quentin Bone, had retired in 1994.]
21 *Transactions A* editorial board, 11 May 1994, TA/12(94) in RS CMB/365.
22 Our interview with Charles Lusty, 25 February 2016.
23 RS *Year Book* (1990), 272.
24 RS Annual Report (1994), 22.
25 RS RR/90/TB/4 [in box ARB848, Deep Store].
26 A. Shaw, 'Obituary: Gordon Stewart', *Herald Scotland* (15 November 2016), https://www.heraldscotland.com/opinion/obituaries/14904952.Obituary___Gordon_Stewart__expert_in_public_health_and_colleague_of_Fleming_who_helped_pioneer_the_use_of_penicillin.
27 G. Stewart, 'Conspiracy of humbug hides the truth on AIDS', *Sunday Times* (7 June 1992). This and other writings by Stewart can be found on 'Virus Myths', https://www.virusmyth.com/aids/hiv/gshumbug.htm.
28 Referee reports and notes, RS RR/90/TB/80 (November 1990) [in box ARB848, Deep Store].
29 Aileen Fyfe, 'From philanthropy to business: The economics of Royal Society journal publishing in the twentieth century', *N&R* https://doi.org/10.1098/rsnr.2022.0021, Figure 3.
30 'Publications Review Group: Final Report', 29 March 1995, p. 14 in RS C/31(95).
31 RS Annual Report (1992), 17 and RS Annual Report (1993), 19; Publication Management Committee meeting (1 December 1994), RS CMB/367.
32 RS Annual Report (1993), 19. For the technologies, see Sarah Bromage and Helen Williams, 'Materials, technologies and the printing industry', in *The Cambridge History of the Book in Britain. Volume 7: The twentieth century and beyond*, ed. Andrew Nash, Claire Squires and I.R. Willison (Cambridge: Cambridge University Press, 2019), 41–60.
33 Publications Executive Committee [hereafter PEC], 9 December 1996, RS PEC/25(96).
34 RS PMC/32(91), 20 June 1991, and RS PMC/45(91), 10 December 1991. On the history of Alden Press, see Chris Koenig, 'Printing firm hits trouble', *Oxford Times*, 1 December 2008.
35 Our (second) interview with Charles Lusty, 20 July 2016.
36 Fyfe, 'From philanthropy to business', Figure 5. From 1992 on, inflation was below 4 per cent, see Samuel H. Williamson (2019), 'Annual inflation rates in the United States, 1775–2018, and United Kingdom, 1265–2018', *MeasuringWorth*, www.measuringworth.com/uscompare/.
37 Michael Francis Atiyah, 'Address of the president, 1991', *N&R* 46.1 (1992): 155–69, at 156–8; and 'Address of the president, 1995', *N&R* 50.1 (1996): 101–13, at 107. See also Collins, *The Royal Society and the Promotion of Science*, 285–6.
38 See Fyfe, 'From philanthropy to business', Figure 1.
39 The review was led by Austin Bide, the former chief executive of Glaxo. See Collins, *The Royal Society and the Promotion of Science*, 277.
40 'Publications Review Group: Final Report', 29 March 1995, p. 3, RS C/31(95).
41 Our interview with Bruce Goatly, 20 January 2016.
42 C. Purdon to P. Warren, letter of resignation, 5 April 1995 [Purdon's private collection]. Also, our interview with Chris Purdon, 18 March 2016.
43 The list of committee members is annexed to 'Publications Review Group: Final Report', 29 March 1995, RS C/31(95).
44 Royal Society turnover in 1990 was £2.3m. The publishing staff were listed in RS *Year Book* (1991), 367.
45 'Analysis of replies to questionnaire', Annex C to 'Publications Review Group: Final Report', 29 March 1995, RS C/31(95).
46 'Publications Review …', 29 March 1995, pp. 1–2, RS C/31(95).
47 'Publications Review …', 29 March 1995, p. 7, RS C/31(95).
48 'The future of publishing', Publication Board minutes, 14 October 1996, RS CMB/417.
49 Goatly to Frank Smith (editor of *Transactions A*), 17 January 1989, in Smith's papers, RS CAX/other/06.
50 'Publications Review …', 29 March 1995, p. 2, RS C/31(95).
51 Publications Review …', 29 March 1995, p. 14, RS C/31(95).
52 'Publications Review …', 29 March 1995, p. 2, RS C/31(95).

53 'Publications Review ...', 29 March 1995, pp. 13–14, RS C/31(95).
54 'Publications Review ...', 29 March 1995, p. 2, RS C/31(95).
55 Job description for Director of Publishing, 13 June 1995 [Purdon collection].
56 Royal Society advertisement, *The Guardian*, 14 October 1995 [Purdon collection].
57 Glynn is unfortunately a shadowy figure in this story. She did not stay long at the Society, so there are few records about her work there; and we have not been able to speak to her. These brief biographical details are derived from her academic publications (a 1981 article and a 1986 review in the *American Archaeological Journal*), and the many acknowledgements to her in books (mostly classical or literary) published by OUP in the 1980s.
58 Job description for Director of Publishing, 13 June 1995 [Purdon collection].
59 On OUP in this period, see Keith Robbins, ed., *History of Oxford University Press. Volume IV: 1970 to 2004* (Oxford: Oxford University Press, 2017), 294–6; and Paul Luna's chapter on 'Technology', pp. 179–204 in the same volume.
60 RS *Year Book* (1991), 367. See also Røstvik and Fyfe, 'Ladies, gentlemen, and scientific publication'.
61 Our interview with Debbie Vaughan, 24 February 2016. Vaughan has worked in publications sales and marketing since the late 1980s.
62 Kathleen Lonsdale had earlier served as vice-president, but that role was in the gift of the president.
63 John Taylor is another somewhat shadowy figure in this story; and another person we have unfortunately not been able to interview.
64 Our (second) interview with Phil Hurst, 19 July 2016.
65 Our (second) interview with Phil Hurst, 19 July 2016.
66 Our (second) interview with Charles Lusty, 20 July 2016.
67 Our interview with Charles Lusty, 25 February 2016.
68 Our (second) interview with Charles Lusty, 20 July 2016. Lusty was congratulated for renegotiating production costs, see RS PEC/14(97), 5 November 1997.
69 Our (second) interview with Phil Hurst, 19 July 2016. See also RS Annual Report (1995), 21.
70 Our interview with Peter Cooper, 3 June 2016, via telephone.
71 Publishing Board minutes [hereafter PUB], 10 July 1997, PUB/14(97) in RS CMB/417.
72 Dee Wood, 'ESPERE', *Ariadne: Web Magazine for Information Professionals* 7 (c. 1997), www.ariadne.ac.uk/issue/7/espere/; also, our interview with Phil Hurst, 26 October 2015.
73 PUB/14(97), 10 July 1997, in RS CMB/417.
74 Our (second) interview with Charles Lusty, 20 July 2016.
75 'Publication Review ...', 29 March 1995, pp. 1 and 7, RS C/31(95).
76 Royal Society advertisement, *The Guardian*, 14 October 1995 [Purdon collection].
77 Our interview with Charles Lusty, 25 February 2016.
78 A. Klug, 'Address of the president, 1997', *N&R* 52.1 (1998): 181–90, at 189.
79 'Publications Review ...', 29 March 1995, p. 16, RS C/31(95).
80 Our interview with Jamie Joseph, 11 July 2016, via email. Joseph worked as an editorial coordinator for *Proceedings B* from 2002 to 2008.
81 'Publication Review ...', 29 March 1995, pp. 1 and 7, RS C/31(95).
82 Our (second) interview with Phil Hurst, 19 July 2016.
83 *Transactions B* editorial board, 10 April 1995, RS CMB/365.
84 Robbins, *History of Oxford University Press IV*, 294 (re the lack of profits by 1994).
85 Our (second) interview with Phil Hurst, 19 July 2016.
86 RS PMC/24(93), 21 July 1993.
87 RS PEC/7(97), 23 April 1997.
88 *Transactions A* editorial board, 11 May 1994, in RS CMB/365.
89 Barbara Gastel, 'From the Los Alamos preprint archive to the arXiv: An interview with Paul Ginsparg', *Science Editor* 25, no. 2 (2002): 42–3.
90 Matthew Cobb, 'The prehistory of biology preprints: A forgotten experiment from the 1960s', *PLOS Biology* 15, no. 11 (2017): e2003995; Daniel Garisto, 'Preprints make inroads outside of physics', *APS News* 28, no. 9 (2019).
91 *Transactions A* editorial board, 11 May 1994, in RS CMB/365.
92 RS Annual Report (1993), 18–19, re the report on the Scientific Information System.
93 Peter Swinnerton-Dyer, 'A system of electronic journals for the United Kingdom', *Serials* 5, no. 3 (1992): 33–5.

94 RS PMC/29(93), 11 November 1993.
95 See 'Publications Review: Implementation', 3 May 1995, RS C/52(95); and RS PUB/17(96), 14 October 1996.
96 Historic webpages are archived by the Internet Archive 'Wayback Machine', https://web.archive.org/web/19961119081823/http://britac3.britac.ac.uk:80/rs/rshome.html.
97 RS PEC/12 (96), 10 June 1996.
98 'Publication Review ...', 29 March 1995, p. 1, RS C/31(95); RS PUB/10(98), 14 January 1998.
99 Royal Society, 'The future of scholarly scientific communication: Conference 2015' (London: Royal Society 2015), 28–30.
100 Jane Henley and Sarah Thomson, 'JournalsOnline: The online journal solution', *Ariadne: Web Magazine for Information Professionals* 12 (c. 1997) http://www.ariadne.ac.uk/issue/12/cover/.
101 The directors were Robert Campbell and (as a non-executive) Roger Elliott, the Society's former physical secretary; both absented themselves from the discussion and vote. See RS PUB/6(97), 9 January 1997 [in RS CMB/417]. The treasurer was asked to release £3,000 for the implementation. On Navigator, see John Cox, 'The shifting sands of technological change: The middle man as navigator', *Serials* 6, no. 4 (1993); Ben Jeapes, 'Blackwell's electronic journal navigator', *The Electronic Library* 15, no. 3 (1997): 189–91.
102 RS PUB/8(99), 29 January 1999 [in RS CMB/417].
103 David Hardy Whiffen and Donal H. Hey, *The Royal Society of Chemistry: The first 150 years* (London: The Royal Society of Chemistry, 1991), 152–4.
104 Email from Charles Lusty to Aileen Fyfe, 11 February 2019.
105 John Taylor, 'JSTOR: An electronic archive from 1665', *N&R* 55.1 (2001): 179–81. In 2015, the Society decided to re-digitise its historic journal holdings to better standards than had been possible in 1999; in this case, it funded and managed the project itself (with external grants), and the result is hosted on its own server. For the *Transactions*, see https://royalsocietypublishing.org/loi/rstl
106 Email from Phil Hurst to Aileen Fyfe, 4 June 2019.
107 Ian Russell, quoted in John Murphy, 'Championing the case of smaller publishers', *Research Information*, 12 October 2006. This interview also has details of Russell's academic and professional background. He left the Society because of the crippling commute from Bristol.
108 Email from Phil Hurst to Aileen Fyfe, 4 June 2019.
109 Vincent Larivière, Stefanie Haustein, and Philippe Mongeon, 'The oligopoly of academic publishers in the digital era', *PLOS ONE* 10, no. 6 (2015): e0127502, 4–5, and 10–11.
110 Our interview with Stuart Taylor, 27 February 2017.
111 Our interview with Phil Hurst, 19 July 2016.
112 Royal Society, 'Science and public affairs' [announcement c. 1986], from journal backmatter, https://royalsocietypublishing.org/cms/asset/37da60a1-6981-4f56-acf3-1087cd3a14a1/back.pdf. See also C. Røstvik, 'Crossword puzzles and gee-whizz factor' (2 November 2017), https://arts.st-andrews.ac.uk/philosophicaltransactions/crossword-puzzles-and-gee-whizz-factor-the-royal-society-and-ba-journal-science-and-public-affairs-in-the-1980s-and-90s.
113 RS PMC/26(89), 28 November 1989.
114 Royal Society, 'Science and public affairs'.
115 RS PMC/26(89), 28 November 1989.
116 The consistent losses were noted in RS PMC/29(90), 28 November 1990. On the new arrangements with the BAAS, see RS Annual Report (1991), 17.
117 Our interview with Charles Lusty, 25 February 2016. The Society was acting on advice from a market research consultant, c. 2003.
118 RS *Review of the Year* (2003), 9.
119 RS *Review of the Year* (2003), 9.
120 John Durant and Nicola Lindsey, 'GM Foods and the media' (November 1999), appendix 5 in House of Lords Select Committee on Science and Technology, 'Science and Society (Third Report, 1999–2000)' (London: HMSO, 2000).
121 'Guidance for editors: The Royal Society's recommendations', in 'Science and Society (Third Report, 1999–2000)', ch. 7, box 3.
122 A. Klug, 'Address of the president, 1999', *N&R* 54.1 (2000), 101.
123 Royal Society, 'Position statement on open access' (24 November 2005), https://web.archive.org/web/20060207171805/http://www.royalsoc.ac.uk/page.asp?id=3882. By

2018, a more nuanced understanding of its own history modified the way the Society phrases its claims about its 'prestigious scientific legacy'; for example it claims that the *Transactions* had 'established the fundamental principles of scientific priority and peer review' but without the implication that peer review had been in place since 1665. See Royal Society, 'Benefits for authors', Royal Society website, https://royalsociety.org/journals/authors. Cf. Noah Moxham and Aileen Fyfe, 'The Royal Society and the prehistory of peer review1665–1965', *Historical Journal* 61, no. 4 (2018): 863–89.
124 Klug, 'Address of the president, 2000', 101.
125 All three were used on the webpage for the Society's Publications in December 1998, https://web.archive.org/web/19981207050316/http://www.pubs.royalsoc.ac.uk.
126 Claire Robinson, 'The Royal Society: Best of British science or corporate rent boy?' *GMWATCH Number* 6 (8 February 2003), https://gmwatch.org/en/news/archive/2003/9419-gmwatch-number-6-822003; see also Claire Robinson, 'Royal Society under fire', Institute of Science in Society, http://www.i-sis.org.uk/RSUF.php.
127 The controversy started with the press release for a paper on the possible environmental benefits of genetically modified, herbicide-tolerant crops. For instance, Tim Radford, 'Scientists grow 'bird-friendly' GM sugar beet', *The Guardian* (15 January 2003), https://www.theguardian.com/science/2003/jan/15/gm.food. The paper was Alan M. Dewar et al., 'A novel approach to the use of genetically modified herbicide tolerant crops for environmental benefit', *ProcB* 270.1513 (2003): 335–40. It extended to the evaluation of the field trial results, which were ultimately published in *Transactions B*, as a theme issue on 'The farm scale evaluations of spring-sown genetically modified crops', *PTB* 358 (2003), issue 1439 [16 October 2003].
128 Bob May, 'Moment of truth for GM Crops', *The Guardian* (10 April 2003), https://www.theguardian.com/science/2003/apr/10/sciencenews.science.
129 Report of the RS Trustees (2005–06), 6.
130 Our interview with Charles Lusty, 25 February 2016.
131 Our interview with Stuart Taylor, 27 February 2017.
132 Web page for launch of *Interface*, April 2004, https://web.archive.org/web/20040416192741/http://www.pubs.royalsoc.ac.uk/interface_homepage.shtml.
133 Submission data since 1994 provided by Stuart Taylor, 5 June 2019.
134 Our interview with Stuart Taylor, 27 February 2017.
135 Memorandum from the Royal Society, in House of Commons Select Committee on Science and Technology, 'Scientific Publications: free for all? (Tenth Report, 2003–04)', vol. 2, appendix 64.
136 Our interview with Stuart Taylor, 27 February 2017.
137 'RS publishing – key performance indicators 2014', February 2015, p. 2 [shared with us by Stuart Taylor].
138 'RS publishing – key performance indicators 2013', February 2014, p. 5 [shared with us by Stuart Taylor].
139 'RS publishing – key performance indicators 2014', February 2015, pp. 1–2 [shared with us by Stuart Taylor].
140 Our interview with Stuart Taylor, 27 February 2017.
141 Our interview with Stuart Taylor, 27 February 2017.
142 'RS publishing – key performance indicators 2014', February 2015, pp. 1–2 [shared with us by Stuart Taylor].
143 'Geographical spread of end users 2008 as Nov 08.xls' [shared with us by Stuart Taylor].
144 'RS publishing – key performance indicators 2014', February 2015, p. 1 [shared with us by Stuart Taylor].
145 'RS publishing – key performance indicators 2014', February 2015, p. 3 [shared with us by Stuart Taylor].
146 'RS publishing – key performance indicators 2014', February 2015, p. 5 [shared with us by Stuart Taylor].
147 RS *Review of the Year* (2003), 9.
148 'Developing world access', https://royalsociety.org/journals/librarians/purchasing/packages/developing-world-access/. Also email from international sales manager, Maria Romano, to Peter Collins, 23 July 2014 [shared with us by Collins].
149 Budapest Open Access Initiative, 14 February 2002, available at https://www.budapestopenaccessinitiative.org/read.
150 For a history of the open access movement, see Robert-Jan Smits and Rachael Pells, *Plan S for Shock: Science. Shock. Solution. Speed* (London: Ubiquity Press, 2022), sections 1.5–1.9.

151 Sally Morris, former CEO of ALPSP, quoted in Murphy, 'Championing the case of smaller publishers'.
152 Royal Society, 'Position statement on open access' (24 November 2005), https://web.archive.org/web/20060207171805/http://www.royalsoc.ac.uk/page.asp?id=3882.
153 Memorandum from the Royal Society, House of Commons Select Committee on Science and Technology, 'Scientific publications: Free for all? (Tenth Report, 2003–04)', vol. 2, 231–2.
154 Memorandum from the Royal Society, 'Scientific publications: Free for all? (Tenth Report, 2003–04)', vol. 2, 229–30.
155 Research Councils UK, 'RCUK position statement on access to research outputs' (June 2005), https://web.archive.org/web/20160427083932/http://www.rcuk.ac.uk/documents/documents/2005statement-pdf.
156 Royal Society, 'Position statement on open access' (24 November 2005).
157 R. Wray, 'Keep science off web, says Royal Society', *The Guardian* (25 November 2005).
158 Open letter to Lord Martin Rees [president of the Royal Society], 7 December 2005, available at https://web.archive.org/web/20051210023301/http://www.frsopenletter.org.
159 Rees open response to fellows, 7 December 2005, available at https://web.archive.org/web/20051229231045/http://www.royalsoc.ac.uk:80/downloaddoc.asp?id=2510.
160 S. Pincock, 'Royal Soc. attacked on open access', *The Scientist* (9 December 2005), http://www.the-scientist.com/?articles.view/articleNo/23528/title/Royal-Soc--attacked-on-open-access. Also, email from Phil Hurst to Aileen Fyfe, 4 June 2019.
161 Neil W Roach, James Heron and Paul V McGraw, 'Resolving multisensory conflict: A strategy for balancing the costs and benefits of audio-visual integration', *ProcB* 273 (2006): 2159–68.
162 Email from Phil Hurst to Aileen Fyfe, 4 June 2019.
163 J. Boone, 'Royal Society tests new system of free access to papers', *Financial Times* (20 June 2006).
164 Royal Society, 'The Royal Society launches trial of new "open access" journal service', press release (Royal Society, 21 June 2006), https://web.archive.org/web/20060823180457/http://www.royalsoc.ac.uk/news.asp?id=4838.
165 S. Pincock, 'Royal Society tries open access', *The Scientist* (22 June 2006), quoting Bob Ward.
166 P. Suber, comments on Royal Society press release, 21 June 2006, available at https://web.archive.org/web/20190206210925/http://legacy.earlham.edu/~peters/fos/2006/06/royal-society-adopts-oa-hybrid-model.html. See also S. Pincock, 'Royal Society tries open access', *The Scientist* (22 June 2006).
167 Our interview with Michael Brady FRS, 23 January 2017, by Skype.
168 David Glover, Christine Holt, Louise Johnson and Peter Parham, 'Introducing Open Biology', *Open Biology* 1, no. 1 (2011): 110001.
169 RS *Year Book* (2016), 445; email from Stuart Taylor to Aileen Fyfe, 4 June 2019.
170 Royal Society 'Article processing charges – FAQs', Royal Society website, https://royalsocietypublishing.org/rsos/charges. For list of countries, see Research4life, 'Eligibility for access to Research4Life', https://www.research4life.org/access/eligibility.
171 Glover et al., 'Introducing Open Biology'.
172 RS *Year Book* (2017), 445.
173 Kayla Graham to EveryONE: PLOS ONE community blog, 2014, https://blogs.plos.org/everyone/2014/01/06/thanking-peer-reviewers/.
174 Data supplied by Stuart Taylor.
175 Our interview with Stuart Taylor, 27 February 2017.
176 RS *Year Book* (2017), 445.
177 'RS publishing – key performance indicators 2014', February 2015, p. 1 [shared with us by Stuart Taylor]. The actual quantity of Royal Society published material that was openly available somewhere was much higher, since the Society permitted authors to deposit their papers in repositories (i.e. green OA).
178 An APC of £900 per accepted article was introduced in January 2018.
179 Our interview with Charles Lusty, 25 February 2016.
180 Our interview with Stuart Taylor, 27 February 2017.
181 Our interview with Michael Brady, 23 January 2017.
182 'RS publishing – key performance indicators 2014', February 2015 [shared with us by Stuart Taylor].

183 Email from Spencer Barrett to Camilla Røstvik, 30 May 2017.
184 Publishing staff, from Stuart Taylor, personal communication 4 June 2019; RS *Trustees' Report* (2011), 5.
185 Our interview with Georgina Mace, 22 January 2016.
186 Royal Society Diversity Report 2015 (issued August 2016), available at https://royalsociety.org/diversity. On earlier criticisms of the gender balance at the Society, see comments in Atiyah, 'Address of the president, 1995', 107.
187 Email from Spencer Barrett to Camilla Røstvik, 30 May 2017; see also Spencer C.H. Barrett, 'Proceedings B 2016: The year in review', *ProcB* 284 (2017), https://doi.org/10.1098/rspb.2016.2633.
188 RS *Trustees' Report* (2006), 6–7.
189 Collins, *The Royal Society and the Promotion of Science*, ch. 11
190 RS *Trustees' Report* (2011), 6.
191 See, for instance, the discussions from 2014 to 2016, reported in 'Report by Planning and Resources Committee', 21 September 2016, RS C/57(16).
192 Our interview with Stuart Taylor, 27 February 2017.
193 Our interview with Stuart Taylor, 27 February 2017.

Reflections
Learning from 350 years

The idea of publishing scientific knowledge in periodical form has been with us for over three and a half centuries. Scientific journals became a vital, even dominant component of the mechanisms by which researchers share their findings with the world, check each other's work and build their careers. In the absence of universal scientific governance, they serve as an important organising principle for an entire system of free enquiry. They have helped define and police disciplinary communities, enabled the professionalisation of science, and helped to provide a structure for scholarly careers and the allocation of scientific prestige. Journals are emblematic of the open-endedness and revisability of scientific research and claims to knowledge, and of science's communal basis and the ideal of rational progress it claims to embody. They are invested with the full weight of science's cultural and social authority, while remaining essentially impenetrable (and, in many practical respects, inaccessible) to outsiders.

But that is what scientific journals have become. It is not what they did in 1665, nor what they did for most of the period in between. By investigating the full 350-year history, we have discovered how journals came to perform their modern functions – but we have also learned that they used to have different functions and purposes. And some of the practices and policies developed to serve those older purposes lived on long after the original need had disappeared.

It would be hard to claim that the *Philosophical Transactions* is a 'typical' or 'archetypal' scientific journal. It is not. From the late eighteenth century onwards, there has been an increasingly complex landscape of other scientific and scholarly periodicals, about whose roles, purposes and practicalities we still know too little. There is still so much work to do, to understand properly the varieties of scholarly periodicals over time, and the relations between periodicals co-existing within a given time. What the longevity of the *Transactions* can provide is the broad canvas on which the transformations in the communication of knowledge become more starkly visible. In the process of telling a 350-year history, we have constantly been making implicit comparisons between earlier and later parts of the story. In this way, big questions somehow seem more urgent than they might have done had we been focusing only on the 1680s, 1840s or 1950s: we have found ourselves investigating the financial models for circulating knowledge; the use of peer review; and the diversity (or not) within the evaluation of scientific research.

Scholars before us have studied the forms of printed product used to construct and communicate scientific knowledge in various periods and places. What we have done is to study the development of the scientific journal from the perspective of those who ran it. Throughout this book, we have aimed to show that, no more than science itself, the scientific journal is not a natural kind existing universally and outside time. It is a human construct that reflects the interests, needs and capabilities of the people who created it, and of those who subsequently adapted and developed it. What, then, are the key lessons from our 350-year history?

1. Scientific journals have been, and can be, many different things

Despite the continuity of its title, the *Philosophical Transactions* has been many different things over its history. Most visibly, its typography, layout and page design have changed over the years; and, since 1997, it has been electronic as well as printed. Its periodicity has also varied. Oldenburg aimed for monthly issues, as did its editors in the 1990s, but for most of the eighteenth and nineteenth centuries, it was much slower, usually appearing in six-monthly parts. For much of the twentieth century, when the *Transactions* was primarily issued as separate papers, one could argue it was not a periodical at all, but a series of monographs. And in 2015, issues of *Transactions A* and *Transactions B* were appearing fortnightly, but they could equally be purchased as stand-alone thematic volumes, and individual papers could be purchased electronically.

More than this, however, the nature of the *Transactions* has changed. In the late seventeenth century, there was nothing obvious nor inevitable about the practical purpose, philosophical implications or social role of the scientific periodical. Successive editors of the early *Transactions* used it to embody at least four different visions of a natural-philosophical periodical between the 1660s and the 1690s. Was it the appropriate venue in which to publish recent ephemeral snippets? To rescue material from unpublished oblivion in the archive? To forge a collaborative research enterprise between geographically distant scholars? For early modern scholars, the pages of a periodical lacked the epistemic rigour to which the new experimental philosophy aspired, and books seemed better able to showcase the extended, systematic treatises in which natural philosophers reported their findings.

By the early eighteenth century, the miscellaneity of content that had characterised Oldenburg's *Transactions* – with its journalistic reporting, its book reviews, and its extracts – had disappeared. The *Transactions* settled into a stable existence as a venue for substantial accounts of self-authored, original research, and from the early nineteenth century, the *Proceedings* published summaries of the Society's meetings. But by the twentieth century, it was the *Proceedings* that catered to the new preference for shorter papers, published more rapidly, leaving the *Transactions* to await its new role. This would eventually prove to be themed issues of commissioned reviews of research in particular fields of science.

The periodical format can perform many different functions for researchers. Oldenburg's *Transactions* combined several of these functions, but the subsequent development of journals – at the Society and beyond – has seen the Society's journals occupying more specific niches, while other niches – notably news, letters and the rapid publication of unrefereed short announcements – have come to be filled by other journals. Proposals that the Society should make its journals more like the *Philosophical Magazine*, or *Nature*, or *PNAS* – or even *New Scientist* – remind us of the flourishing of sub-genres that is hidden behind the singular 'scientific journal'.

The comprehensive coverage offered by the (early) *Proceedings* enabled the *Transactions* to become more selective. Publication in the *Transactions* carried more prestige for authors, and this fed into the professionalisation of science (and of academia) in the later nineteenth century. By the twentieth century, the *Proceedings* had also become more selective, as the Royal Society tried to ensure that its journals were seen as publishing work that was 'significant' and not merely 'worthwhile'. The need to make such distinctions was a consequence of the substantial growth of research output, but also of the growing profile of research, and the training of researchers, in an expanded university system.

By that point, the intended authors and readers of scientific journals had changed dramatically. The *Transactions* had been created in a world of independently wealthy gentlemen scholars, but the *Proceedings* became a tool for professional academics. The geographic focus shifted, too. The very early *Transactions* had reported material from across Europe for mostly British readers, but for most of the eighteenth and nineteenth centuries it was predominantly a periodical showcasing material from British scholars, for British scholars. Even the scope of that 'British' community shifted: in the late seventeenth and eighteenth centuries, the weekly meetings provided a focus for Royal Society activity, and so the members had to be located in or near London to be able to participate; but improvements in transport and communications in the nineteenth and twentieth centuries, and the declining importance of the weekly meeting, enabled more participation from those in the English provinces, Scotland or the wider British academic world. By 2015, it was even possible for the Society to appoint an editor based in Canada.

The use of the English language, and the challenges of international trade, had initially limited the overseas readership of the *Transactions*, but strategic gifting in the eighteenth and nineteenth centuries had placed it – and the *Proceedings* – in the libraries of learned institutions across Europe and the British dominions. The transition to the commercial model of publishing in the mid-twentieth century depended on reaching an international market, especially in North America. This increased international circulation drove a slow increase in submissions from international authors, which became a flood after the 1990 relaunch. The material published in the *Transactions* and the *Proceedings* was, by the twentieth century, most likely to come from authors outside the Society, rather than fellows themselves. This meant that the journals no longer showcased Royal Society activity, as they had done in the late eighteenth or early nineteenth centuries, but (hopefully) demonstrated the Society's ability to identify and attract the best research.

These shifts in the authors, readers and content of the journals demonstrate why it is essential to take the historical context seriously when seeking to understand the meaning or purpose of scientific journals at any particular point in time.

2. There are many different ways of editing a scientific journal; and peer review was not originally about quality control

Throughout this book, we have drawn attention to the practice and purpose of editorial processes. Editors and peer reviewers are familiar elements of modern academic publishing, but they are no more natural nor inevitable than scientific journals themselves. The *Transactions* was run by an individual editor for its first eight decades; and in 2015, all four *Transactions* and *Proceedings* journals had individual editors. Yet for almost 240 years in between, the Royal Society preferred not to have named editors in charge.

The creation of an editorial committee – the Committee of Papers – in 1752 was a conscious effort to create the impression of collective responsibility for the production of the *Transactions*. At that time, and for decades after, entrusting editorial matters to a single 'editor' was seen as risky, both because of possible bias in decision-making, but also, more pragmatically, because it put the *Transactions* at the mercy of ill-health, negligence or incompetence. Collective editorial processes spread the workload as well as the responsibility. However, as we have seen in the cases of Joseph Banks and George Stokes, this did not entirely prevent a single person from having substantial, even dominant, influence within those collective processes.

By the twentieth century, the use of editorial committees for actual decision-making had come to be seen as cumbersome, but editorial workload and responsibility remained distributed. The unpaid labour of referees, committee chairmen and associate editors remained central to the decision-making process, even as editorial staff were taking on increasing amounts of the administration and correspondence. When individual editors were reintroduced in the 1980s, their role was very different from that of Henry Oldenburg or Edmund Halley (or, indeed, of Norman Lockyer at *Nature*): they were not all-powerful autocrats, but figure-heads and facilitators of a distributed editorial process, in which certain work was done by paid staff, but intellectual, evaluative work (and responsibility) was the preserve of referees and editorial board members.

We have argued that it is meaningless to claim that Oldenburg 'invented' peer review in 1665, and we have shown that refereeing was only one element of a set of editorial practices that became increasingly complex as the nineteenth century wore on. These practices were not originally intended to evaluate the quality of papers submitted to the Society, but to establish an eligibility bar that would protect the Society from the reputational risk of being seen to publish material that was trivial, mistaken or already in print. In the nineteenth century, these practices (including refereeing) became a mechanism not merely to determine eligibility but to discriminate between papers suitable for the *Transactions* or the *Proceedings*. That two-tier system of publishing built the prestige and reputation of the *Transactions*, but it also attempted (with little success) to protect the Society's finances from the strain of publishing too many, too lengthy and too lavishly illustrated papers.

The practical aspects of refereeing seem to have changed very little, despite the moves from handwritten letters to printed report forms, but the significance and purpose of refereeing in the twenty-first century is very different to what it had been in the 1830s. In the nineteenth and early twentieth centuries, refereeing was a process associated with the collective decision-making of learned societies. Critics might claim it was anachronistic or inefficient, but in the twentieth century, the Royal Society used it even more, scrutinising submissions for the *Proceedings* as well as the *Transactions*, and trying to sift through the increasing bulk of research papers to identify those that might be 'significant' rather than merely 'worthwhile'. In the 1950s, David Martin could present refereeing as a guarantee of quality, and a unique and valuable service provided by learned society publishers. He had assumed that this function could not be replicated by commercial journal publishers, but he was wrong. By the 1990s, in its new guise of 'peer review', it had become an accepted and standard practice for evaluating research in many areas of academic life, from journals to grant proposals and research excellence frameworks.

The Royal Society's long-standing insistence on papers being communicated via a fellow, who was supposed to know what a 'significant' paper would look like, had ensured that it had maintained a very low formal rejection rate even in the twentieth century. But this was increasingly out of step with the practices of journals that publicised their high rejection rates as a proxy for 'quality'. The dramatic increase in rejections at the Society's journals after 1990 was, therefore, hailed as cause for celebration despite the workload it represented. But the advent of electronic publishing, with a capacity greater than the printed issue, raised questions about the necessity for such selectivity. Online megajournals could publish all 'sound' research, rather than needing to select 'significant' research. The use of 'objective peer review' for *Royal Society Open Science* appears to go against the grain of a century or more of Royal Society rhetoric about refereeing and peer review, but the ambition to publish everything that seems technically sound and of some interest can be seen as reviving the role occupied by the *Proceedings* in its early years or, indeed, by the *Transactions* in its first century. It reminds us that different purposes for a journal may require different editorial regimes.

3. The history of subsidised, or sponsored, circulation of knowledge is much longer than that of profit-driven journal publishing

One part of the Oldenburg story that is not mythical is that he hoped to make money from the publication and sale of the *Transactions*. When this is coupled with the well-known profitability of certain twenty-first-century scientific journal publishers, it becomes easy to assume that journal publishing has always been profitable. The 350-year history of the *Transactions* clearly shows that this is not true. Oldenburg did earn a modest amount of money in payments from the printer, but he was probably the last person to have done so until the 1950s. Public sales of the *Transactions* may possibly have been higher in the early eighteenth century than they were in the nineteenth century (when so many fellows and institutions received free copies), but even so, Hans Sloane claimed to have spent £1,500 supporting the *Transactions* during his 19-year editorship. In 2015 terms, that is equivalent to about £10,000 a year. By the 1890s, Royal Society treasurers expected to support the journals by about £1,700 a year, equivalent to about £177,000 in 2015 money. The success of the Society's request for government support in 1895 made clear that nobody expected the costs of producing and distributing scholarly research journals to be met by public sales.

The inability of sales income to cover production and distribution costs should not, however, be seen as an indication of commercial failure or incompetence. The very fact that the *Transactions* continued to be published for so long despite the fact that it cost its sponsors significant amounts of money clearly demonstrates a very different underlying vision from that with which we are now familiar. It grew out of other forms of scholarly communication, notably personal correspondence networks, that were equally expensive and laborious to maintain, and equally vital to the work of pre-modern science. Sloane, for example, said that he thought of his personal expenditure on the journal as a service to the Royal Society and to a wider natural-philosophical community. And in the 1930s, William Bragg used similar language when he described publishing as a service performed by the Society for the wider scientific community. The cost of providing that service now included the time devoted by fellows to evaluating and selecting the papers to be published, and the salaries and overheads of the Society's nascent publishing team, as well as the costs of print, paper and illustrations.

The *Transactions* became part of a complex economy of information exchange, prestige, personal relationships, gifts and services rendered that underwrote the dynamics of much early modern science, and that would continue to influence scientific publication practices surprisingly far into the nineteenth and early twentieth centuries. Long before the 1950s, there were opportunities when the Society could potentially have profited from its ownership of the *Transactions*, but chose not to do so: it permitted entrepreneurial editors to create commercially successful abridgements of the periodical; it allowed authors and third-party publications to reprint its text and images on generous terms; its public retail prices were set at a level that precluded any attempt to recoup the costs of producing the free copies; and it offered subscription rates for the *Proceedings* that were discounted below cost-price. These were conscious choices that were intended to improve the circulation of the knowledge published under the Society's aegis.

Copies of the *Transactions* had always been strategically disseminated gratis to important correspondents, authors, potential contributors and patrons, even during the era of private ownership. After taking formal control in 1752, the Royal Society extended and institutionalised this approach; giving copies to the king, the English universities, and the royal academies elsewhere in Europe ensured that the Society's name and activities were made known to scholars and patrons. Given the delays and difficulties that European scholars reported in obtaining the *Transactions* throughout the eighteenth century, strategic gifting – with shipping organised and paid for by the Royal Society – could be more effective than the commercial book trade.

This strategic gifting grew over the nineteenth century to include more and more institutions in Britain, Europe and further afield, and some of the gifts developed into regular exchanges that enhanced the Society's library holdings. It cost the Society money to produce the copies and to ship them, and it meant that a substantial proportion of the likely purchasers of the Society's journals did not, in fact, need to purchase them. As we have seen, by the 1890s, this model for the circulation of knowledge was straining the resources of learned societies, and the Royal Society led the successful search for government support. The ever-increasing output of scientific research, and the increasingly global ambitions for circulation, meant that the limits of the sponsored model of knowledge circulation seemed to be approaching – at least, in a world of printed paper.

The post-war emergence of the rhetoric of 'self-help' for learned journals was a response both to these pre-existing strains, and to the emergence of a new and apparently successful model of international commercial journal publishing. The Royal Society's own transition to a sales-focused publishing system was remarkably rapid. In 1949, it had been watching with interest as Cambridge University Press increased sales to North American institutions and, by 1955, it had created its own sales and marketing department. In retrospect, the boom in institutional sales reflected the buoyant state of research institution budgets during the Cold War, and its decline from the 1970s onwards created new challenges as the Royal Society's publishing team sought to sustain their new approach to circulation and sales.

From the 1950s onwards, the existence of significant income from sales meant that the Society no longer needed to subsidise its publishing programme. But it was not until the 1970s, and, even more, the 1990s, that the Society actively sought to generate additional income from its publishing activities. This marks a significant shift from seeing the journals as a valuable service whose costs needed to be covered somehow (whether by Society funds, external grants, or sales), to seeing them as a form of income generation. The 1995 'twin objectives' required the journals to disseminate science and to make money without – apparently – recognising the potential for tension between those aims. The rise of the movement for open access from 2005 onwards argued that making money could be seen as impeding dissemination, by locking knowledge behind subscriptions and paywalls. The Society's willingness to grant free digital access to institutions in the Global South can be seen as a revival, in modern form, of its old philanthropic approach to circulation – but it was predicated upon the success of selling access to affluent institutions in the Global North.

By 2015, the Society had launched two online-only journals that offered free access to all readers with internet access, but the publishing team was still developing a fair and sustainable way of covering the costs of those journals. Whether the Society could return to its pre-1955 model of seeking a coalition of funding sources to support journals that are free for authors and for readers, remained to be seen.

4. Learned societies matter

Our focus upon the Royal Society as a publishing organisation has emphasised the importance of learned societies in the creation of norms of scholarly publishing. Historians of science are, of course, familiar with the role of learned societies and similar organisations, particularly in the period before the expansion of university research. But the contemporary discourse on scholarly communications tends to presume that commercial publishing firms have always been the key players they now are. As we have shown, by the time that commercial publishers became significantly involved in publishing (and not merely printing) scientific research journals, most of the editorial practices associated with academic publishing were firmly established. These range from the use of referees, and the emphasis on originality and significance, to the tradition of not paying authors or reviewers, all of which were consequences of the particular path taken in the historical development of scientific journals within the context of learned societies.

Within the Royal Society itself, we have seen that publishing has varied in prominence over the centuries. Initially, the *Transactions* was not formally part of Society activity, but by the mid-eighteenth century, the reputation of the Society was felt to be so bound up in the *Transactions* that Council took on its management. Through the late eighteenth and nineteenth centuries, weekly meetings and the publication of the *Transactions* (and later the *Proceedings*) were the key activities of the Society. Spending time, labour and money on publishing made sense.

By the late nineteenth century, however, the range of Society activities was expanding: it was giving advice to government, organising expeditions, and making grants to researchers and students. As these other activities expanded during the twentieth century, and especially after 1960, publishing moved into the background, coming to attention only sometimes for its finances, or when the need for reform seemed particularly urgent. One of the consequences of this benign neglect is that the publishing staff – as the people whose work focused day-to-day on the publications – could become increasingly influential in the strategic development of the journals. Lacking formal authority, they had to work by influencing committees and persuading the fellows who served on them. David Martin was clearly very good at this, and by the early twenty-first century, most of the ideas for journal development, from new journals to online journals and from open access to experiments with peer review, seem to have come from the staff. The autonomy of the staff still has its limits, however; the director of publishing still ultimately reports to the Society's Council.

In the 1950s, and again in the 1990s, Royal Society leaders argued that the Society should be 'a guardian of the interests of the scientific process'.[1] In contrast to commercial publishers, the Society as a community of well-respected researchers should be able to take the wider perspective on technological and commercial changes. Its efforts to provide a sustainable model for learned society publishing in the 1950s were significantly successful, encouraging learned societies to streamline their editorial and publishing processes, improve their sales, and thus release them from reliance on external subsidy. In the long term, however, this led to an increasingly commercial approach at the bigger society publishers, to the extent that learned societies could be seen in the 2010s as 'one of the biggest barriers to open access'.[2] The Royal

Society's efforts to guard 'the interests of the scientific process' since the digital revolution have been heavily influenced by its own desire for an income stream that would underpin its political independence.

Societies and mission

The twentieth century was, in many ways, a period of uncertain purpose for Royal Society publishing, despite the success of the *Proceedings* and despite the commercial transformation. What is the appropriate role for learned society journals in an age of international research and global media conglomerates? What specifically is the role of the Royal Society's broad journals in an age of specialised research? And what is the role of the Royal Society, whose claim to leadership of learned society publishers is far less clear now than it was in the 1950s? These questions run deeper than the much-discussed transition from paper-based to digital communication and storage of research.

Such questions are difficult to discuss publicly, particularly for an organisation whose identity as the UK's national academy of science, and as a publisher, is so tightly tied to its history. The Society has been involved, to a greater or lesser extent, in publishing scientific research for over 350 years now. To stop, or even to change dramatically, is nearly unthinkable. As Stuart Taylor, director of publishing, remarked, 'The fact that we not only launched the first science journal in the world, but that we've been doing this for three and a half centuries, does weigh heavily on me.... I'm very conscious of the history, in an old organisation you have to be careful what you change and what you throw away – you have to try to take a long view.'[3] But publishing simply 'because we've always done it (this way)' is not a good reason for continuing. Why, and for what purpose?

When William Bragg asked 'Why do we publish?' in 1938, his answer focused on the motivations of individual scientists. For the Society itself, he seems to have taken it for granted that a desire to be 'of service' was answer enough. Since then, there have been hints of uncertainty in committee reports and administrative reviews, but these are questions more comfortably aired behind closed doors. The Society's celebrations of the *Transactions* in 2015 were not the time or place for serious, critical questioning of the future of the Society's own role in publishing.

What, then, might now be the distinctive purpose of learned society publishing? The 1950s' claim that learned societies were uniquely qualified to guard the high standard of British research, due to their members' role as referees, is surely dead and gone. In the intervening years, academics have shown themselves entirely willing to act as referees for journals that are not run by societies, and societies themselves no longer depend only on their members for refereeing. The Royal Society could conceivably attempt to reclaim a unique role as a guardian of high prestige in research publications by once more utilising its eminent fellowship as pre-publication evaluators. But fellows might not welcome the workload, and in an age of greater consciousness about the lack of diversity in science in general, and the Royal Society in particular, a radical overhaul of the demographic of the fellowship would be needed before such a move could become politically possible.

And yet it may be the case that the Society is in the process of rediscovering a mission and a purpose for its publishing. We were revising and editing this book while moves to more open forms of scholarly communication were becoming articles of public policy (in the UK and Europe especially).[4] It is difficult to imagine that commercial publishers will be the driving forces behind the creation of a system for accrediting and circulating research that is free to both authors and readers; but learned societies used to do this as a service to the scholarly community. For half a century, the Royal Society managed funds from external sponsors, as well as its own investments, to support the non-commercial circulation of scientific research findings by learned societies. The new technological possibilities for production and circulation might make it possible for equivalent collaborations between sponsors and societies to be sustainable once more.

But, like many learned societies, the Royal Society has enjoyed the freedom of action that has been provided by publication surpluses since the closing decades of the twentieth century. To return to viewing publishing as a cost-neutral activity (let alone a subsidised one), would require a significant reorientation. Whether that would be more or less difficult than what the Royal Society did in 1752 is an interesting question that we shall have to leave to future historians. The transition is, however, under way. By 2015, the Society already had two fully open access journals, but *Proceedings A*, *Proceedings B*, *Interface* and *Biology Letters* were operating on a hybrid publishing model (selling subscriptions, and also allowing authors to pay for specific articles to be open). In 2019, the Society launched its first review of publishing strategy since 1995, and the recommendations included transitioning these hybrid journals 'to fully open access over the next five years'. This was agreed by Council in July 2020.[5] By 2021, over half the articles published by the Society were open access.[6]

For the Royal Society specifically, one of the challenges of the twentieth century was the increasing tendency to specialised research.

While other publishers and other societies created increasingly specialised journals, the Royal Society resolutely resisted suggestions to split the *Transactions* or the *Proceedings* further. As generalist journals, they could seem out of place or out of date. But with the rising rhetorical emphasis on interdisciplinary research, the Royal Society's breadth of scope may become an asset. The launch of the new journals *Interface* and *Interface Focus* suggests that the Society has, belatedly, recognised the value in Henry Armstrong's 1902 argument that no other society has 'so favourable a platform for the discussion of borderland problems'.[7] And the flood of submissions to these journals – which outstripped those to *Proceedings A* in 2015 – suggests a real demand for such a platform among contemporary scientists.

It is, of course, far too soon to say how these developments at the Royal Society will play out in the long term. They may prove unsuccessful experiments, or they may be the beginning of a new age of confident and influential Royal Society publishing. As historians, we are much better at analysing the past than predicting the future. But if there is a lesson that we can take from the history of Royal Society publishing, it is that things keep changing. They will surely continue to do so, and we should not expect the modes of communicating and storing scientific research in 2065 to look like those of 1965. But we have also seen that institutions that survive down the centuries – and these include universities, the Royal Society and the *Transactions* – only do so by adapting to the changing times. They change slowly, and perhaps unwillingly, but their longevity is due to their ability to incorporate occasional innovation into the structures and processes that are the accumulated legacy of their historical development. The history of the *Transactions* also shows how rarely this was consciously planned.

If the current challenges facing academic publishing have been precipitated by new technological possibilities, the difficulty of coming up with widely applicable, fair and sustainable solutions has made evident just how many underlying differences there are in the ways that scientific research is made public, in different times, places and disciplines. That diversity has tended to be masked by the apparent universality of 'the scientific journal' and its articles, even in fields that also publish books. The supposed coherence of scientific journals has been underpinned by the Oldenburg myth, promoted both by the academic publishing industry and by scientific organisations themselves. The myth that modern scientific journals perform the same functions as their forerunners from the period of the scientific revolution creates the assumption that the particular ways in which research journals now function are necessary and essential elements of scientific research.

Our 350-year history has shown the historical variety of forms, functions and practices of scientific publishing, and will encourage future historians of science and scholarship to question the meaning and significance of journal publication at specific points in time. We have also uncovered the long-drawn-out processes whereby journals acquired their ways of working and their cultural and epistemic freight. Some of our current practices have a long historical ancestry (for example, peer review), but others are much more recent (for example, commercial circulation). A better understanding of the history, rather than the myth, of scientific journals is crucial to help us decide which practices and functions should be included in any new vision for academic publishing.

Notes

1 'Publication Review Group: Final Report', 29 March 1995, pp. 1 and 7, RS C/31(95).
2 Martin Paul Eve, 'Learned societies in the humanities, open access, and paying for disciplinary goods' (2 January 2017), https://eve.gd/2017/01/02/learned-societies-open-access-and-paying-for-disciplinary-goods.
3 Our interview with Stuart Taylor, 27 February 2017.
4 Robert-Jan Smits and Rachael Pells, *Plan S for Shock: Science. Shock. Solution. Speed* (London: Ubiquity Press, 2022), Part 2.
5 Email from Stuart Taylor to Aileen Fyfe, 25 September 2020.
6 S. Taylor, 'Halfway to open' (17 January 2022), blogpost, https://royalsociety.org/blog/2022/01/halfway-to-open.
7 'Memorandum from Dr Armstrong', RS CMP/08, 6 November 1902.

Bibliography

Archival sources

Royal Society archives, held at the Royal Society, London
Council Minutes: original (CMO) from 1663 to 1832; printed versions (CMP) from 1832 onwards.
Committee Minute Books (CMB), including minutes for the Committee of Papers (from 1752), Library Committee (from 1832), Finance Committee (from 1834), Publications Committee (1935–68), Publications Management Committee (from 1989), Publishing Board (from 1996) and a variety of ad hoc committees.
Accounts Books, including early accounts of the Society 1660–1768 (AB/1); ledgers and cashbooks from 1867 to 1992 (AB/2); and copies of the annual accounts (AB/3).
Correspondence, including Early Letters and Letter Book Original from the seventeenth and eighteenth centuries; Miscellaneous Correspondence (MC) from nineteenth and twentieth centuries; New Letter Book (NLB) from 1880s.
Officers' Minutes: from the mid-twentieth century, the officers of the Society met separately from Council, as an executive group. The papers and minutes of their meetings are in OM2 and OM3, respectively.
Papers submitted to the Society, including Classified Papers, Archived Papers.
Referee Reports (RR), from 1832.
Personal Papers of various fellows and officers of the Society, including Robert Boyle, Charles Blagden, James Jeans, AGC Egerton, Henry Dale, Howard Florey and David Martin.
Society administration: Domestic Manuscripts (DM), 1662–1884; Modern Domestic Archives (MDA), 1870–1981; Records Management Audit Offsite (RMA) from 1960s on.
Journal Book Original (JBO) from 1720 to 1824.
Miscellaneous Manuscripts (MM)
Manuscripts (General) (MS)

For the most recent material (1994 onwards), we thank Stuart Taylor and his team for providing us with access to spreadsheets and reports currently held by the Publishing division.

We have made extensive use of the following serials and occasional reports:
Philosophical Transactions (from 1665); *Proceedings of the Royal Society* (from 1831); *Year Book of the Royal Society* (from 1898); *Biographical Memoirs* [formerly *Obituary Notices*] *of Fellows of the Royal Society* (from 1932); *Notes and Records of the Royal Society* (from 1938); *Royal Society News* (from 1980); *Annual Report of the Royal Society* (from 1979, latterly *Report of the Trustees of the Royal Society*); *Diversity Data Report* (from 2015).

British Library
Personal papers of Hans Sloane and Thomas Birch.
Ledgers for William Bowyer, printer to the Society 1762–8.

Cambridge University Library
Cambridge University Press archive (CUP), including correspondence (Pr_578) and sales and order records.
Personal papers of George Stokes (MS.Add.7656) and Richard Owen (MS.Add.5354).

City of Westminster Archives Centre, London
Harrison & Sons Ltd archive (collection 1272), including ledgers and correspondence (but very little relating to the Royal Society).

National Archives, Kew
Register Books of the Stationers' Company.

Royal Society of Chemistry (formerly, the Chemical Society)
Annual reports.

St Bride Printing Archive, London
Taylor & Francis archive, including correspondence, ledgers (known as 'journals'), cash books and income/expense accounts, 1830–77.

Tyndall Correspondence Project
We have benefited enormously by being granted early access to the files of the John Tyndall Correspondence project, run by Bernard Lightman (York University, Canada). All citations to Tyndall letters are derived from the files of this project; they are cited with their original archival reference and, where available, with their Tyndall reference number (up to the year 1865).

Oral sources

During the course of this research, Røstvik interviewed current and former staff of the Society, and fellows who have been closely involved in the running of the journals. Ethical approval for this research was granted by the University of St Andrews UTREC (University Teaching and Research Ethical Committee). The following people were interviewed:

Michael Brady FRS, chairman of the Publishing Board until 2010.
Peter Cooper, management, various managerial posts, 1975–94; Deputy Executive Secretary, 1994–8.
Helen Eaton, Commissioning Editor for *Philosophical Transactions B*, 2012–13; Senior Commissioning Editor for *Transactions* journals and *Biographical Memoirs*, 2013–present.
Bruce Goatly, editorial staff from 1977; later head of publishing until 1994.
David Griffin, various roles in International Relations, Specialized Research Grants and 'Occasional Publications', 1961–91.
Nick Haines, copy editor, 1991–5.
Phil Hurst, editorial manager, later publisher, 1996–present.
Jamie Joseph, editorial coordinator, then publishing editor for B journals, 2002–8.
Charles Lusty, production manager, later head of publishing operations, 1996–2021.
Georgina Mace FRS, editor of *Transactions B*, 2008–10.
Chris Purdon, editorial assistant, later acting head of publishing, 1986–95.
Claire Rawlinson, publishing editor, later commissioning editor, 2005–12.
Stuart Taylor, publishing director, 2006 to present.
Deborah Vaughan, sales and marketing staff, later marketing manager, 1989–present.

We were also able to draw upon oral history interviews carried out by Peter Collins now held in RS archives, for example, with Peter Warren, former executive secretary.

Printed sources

Adams, Robyn, and Lisa Jardine. 'The return of the Hooke folio', *Notes and Records of the Royal Society* 60, no. 3 (2006): 235–9.
Agar, Jon. *Science in the Twentieth Century and Beyond*. Cambridge: Polity Press, 2013.
Agar, Jon. 'What is science for? The Lighthill report on artificial intelligence reinterpreted', *British Journal for the History of Science* 53, no. 3 (2020): 289–310.
Alter, Peter. *The Reluctant Patron: Science and the state in Britain, 1850–1920*. New York: Berg, 1987.
Anderson, Ronald. 'The referees' assessment of Faraday's electromagnetic induction paper of 1831', *Notes and Records of the Royal Society* 47, no. 2 (1993): 243–56.
Andrade, E.N. da Costa. 'The birth and early days of the Philosophical Transactions', *Notes and Records of the Royal Society* 20, no. 1 (1965): 9–27.
Andrade, E.N. da Costa, and Kathleen Londsdale. 'William Henry Bragg, 1862–1942', *Biographical Memoirs of Fellows of the Royal Society* 4, no. 12 (1943): 276–300.
Archambault, Éric, and Vincent Larivière. 'History of the journal impact factor: Contingencies and consequences', *Scientometrics* 79, no. 3 (2009): 635–49.
Ashworth, William J. 'The calculating eye: Baily, Herschel, Babbage and the business of astronomy', *British Journal for the History of Science* 27, no. 4 (1994): 409–41.
Asser, Paul Nijhoff. 'A new society of publishers', *Scholarly Publishing* 2, no. 3 (April 1971): 301–5.
Astbury, R. 'The renewal of the Licensing Act in 1693 and its lapse in 1695', *The Library* 5th series, 33 (1978): 296–322.
Atiyah, Michael Francis. 'Address of the president, Sir Michael Atiyah, given at the anniversary meeting on 29 November 1991', *Notes and Records of the Royal Society* 46, no. 1 (1992): 155–69.
Atiyah, Michael Francis. 'Address of the president, Sir Michael Atiyah, O.M., given at the anniversary meeting on 30 November 1995', *Notes and Records of the Royal Society* 50, no. 1 (1996): 101–13.
Atkinson, Dwight. *Scientific Discourse in Sociohistorical Context: The Philosophical Transactions of the Royal Society of London, 1675–1975*. London: Routledge, 1998.
Avramov, Iordan. 'An apprenticeship in scientific communication: The early correspondence of Henry Oldenburg (1656–63)', *Notes and Records of the Royal Society* 53, no. 2 (1999): 187–201.
Ayres, Peter. *Women and the Natural Sciences in Edwardian Britain*. London: Palgrave, 2020.
Babbage, Charles. *Reflections on the Decline of Science in England*. London: Fellowes, 1830.
Baddam, Benjamin. *Memoirs of the Royal Society: Being a New Abridgment of the Philosophical Transactions ... 1665 to ... 1735 ... the Whole Carefully Abridg'd from the Originals, and the Order of Time Regularly Observ'd*, 10 vols. London: 1738–41.
Baldwin, Melinda. '"Keeping in the race": Physics, publication speed and national publishing strategies in Nature, 1895–1939', *British Journal for the History of Science* 47, no. 2 (2014): 257–79.
Baldwin, Melinda. 'Tyndall and Stokes: Correspondence, referee reports and the physical sciences in Victorian Britain'. In *The Age of Scientific Naturalism: Tyndall and his contemporaries*, edited by Bernard Lightman and Michael S. Reidy, 171–86. London: Pickering & Chatto, 2014.
Baldwin, Melinda. 'Credibility, peer review, and Nature, 1945–1990', *Notes and Records of the Royal Society* 69, no. 3 (2015): 337–52.
Baldwin, Melinda. *Making 'Nature': The history of a scientific journal*. Chicago, IL: University of Chicago Press, 2015.
Baldwin, Melinda. 'Scientific autonomy, public accountability, and the rise of "peer review" in the Cold War United States', *Isis* 109, no. 3 (2018): 538–58.
Ball, Philip. 'Forging patterns and making waves from biology to geology: A commentary on Turing (1952) "The chemical basis of morphogenesis"', *Philosophical Transactions of the Royal Society B* 370, no. 1666 (2015).
Banks, David. *The Birth of the Academic Article: Le Journal Des Sçavans and the Philosophical Transactions, 1665–1700*. Sheffield: Equinox, 2017.
Barrett, Spencer C.H. 'Proceedings B 2016: The year in review', *Proceedings of the Royal Society B: Biological Sciences* 284, no. 1846 (2017): 2016–633.
Barton, Ruth. *The X-Club: Power and authority in Victorian science*. Chicago, IL: University of Chicago Press, 2018.

Bazerman, Charles. *Shaping Written Knowledge: The genre and activity of the experimental article in science*. Madison: University of Wisconsin Press, 1988.

Beale, Norman, and Elaine Beale. *Echoes of Ingen Housz: The long lost story of the genius who rescued the Habsburgs from smallpox and became the father of photosynthesis*. Salisbury: The Hobnob Press, 2011.

Beckman, Jenny. 'The publication strategies of Jöns Jacob Berzelius (1779–1848): Negotiating national and linguistic boundaries in chemistry', *Annals of Science* 73, no. 2 (2016): 195–207.

Beckman, Jenny. 'Editors, librarians, and publication exchange: The Royal Swedish Academy of Sciences, 1813–1903', *Centaurus* 62, no. 1 (2020): 98–110.

Beeton, Barbara. 'Developments in technical typesetting: TeX at the end of the 20th century'. In *A Century of Science Publishing: A collection of essays*, edited by Einar H. Fredriksson, 191–202. Amsterdam: IOS Press, 2001.

Belknap, Geoffrey. *From a Photograph: Authenticity, science and the periodical press, 1870–1890*. London: Bloomsbury, 2016.

Belknap, Geoffrey. 'Natural history periodicals and changing conceptions of the naturalist community, 1828–65'. In *Science Periodicals in Nineteenth-Century Britain: Constructing scientific communities*, edited by Gowan Dawson, Bernard Lightman, Sally Shuttleworth and Jonathan R. Topham, 172–204. Chicago, IL: University of Chicago Press, 2020.

Bhagavantam, S. 'Chandrasekhara Venkata Raman, 1888–1970', *Biographical Memoirs of Fellows of the Royal Society* 17 (1971): 564–92.

Biagioli, Mario. 'From book censorship to academic peer review', *Emergences: Journal for the Study of Media & Composite Cultures* 12, no. 1 (2002): 11–45.

Bijker, Wiebe E., Thomas P. Hughes, and Trevor J. Pinch, eds. *Social Construction of Technological Systems: New directions in the sociology and history of technology*. Cambridge, MA: MIT Press, 1987.

Bijker, Wiebe E. *Of Bicycles, Bakelites and Bulbs: Toward a theory of sociotechnical change*. Cambridge, MA: MIT Press, 1995.

Birch, Thomas. *The History of the Royal Society of London for Improving of Natural Knowledge*. 4 vols. London: A. Millar, 1756–7.

Black, Alistair, Dave Muddiman, and Helen Plant. *The Early Information Society: Information management in Britain before the computer*. Aldershot: Ashgate, 2007.

Blanning, T.C.W. *The Pursuit of Glory: Europe, 1648–1815*. London: Viking, 2007.

Bluhm, R.K. 'Remarks on the Royal Society's finances, 1660–1768', *Notes and Records of the Royal Society* 13, no. 2 (1958): 82–103.

Bond, T. Christopher. 'Keeping up with the latest transactions: The literary critique of scientific writing in the Hans Sloane years', *Eighteenth-Century Life* 22 (1998): 1–17.

Bornmann, Lutz, and Rüdiger Mutz. 'Growth rates of modern science: A bibliometric analysis based on the number of publications and cited references', *Journal of the Association for Information Science and Technology* 66, no. 11 (2015): 2215–22.

Boyle, Robert. *New Experiments Touching Cold*. London: John Crooke, 1665.

Boylston, Arthur. 'Thomas Nettleton and the dawn of quantitative assessments of the effects of medical interventions', *Journal of the Royal Society of Medicine* 103 (2010): 335–9.

Brayshay, Mark, Philip Harrison, and Brian Chalkley. 'Knowledge, nationhood and governance: The speed of the Royal Post in early-modern England', *Journal of Historical Geography* 24, no. 3 (1998): 265–88.

Brock, Claire. *The Comet Sweeper: Caroline Herschel's astronomical ambition*. London: Icon Books, 2017.

Brock, William H., ed. *H. E. Armstrong and the Teaching of Science 1880–1930*. Cambridge: Cambridge University Press, 1973.

Brock, William H. 'The development of commercial science journals in Victorian Britain'. In *The Development of Science Publishing in Europe*, edited by A.J. Meadows, 95–122. Amsterdam: Elsevier, 1980.

Brock, William H., and A.J. Meadows. *The Lamp of Learning: Taylor & Francis and the development of science publishing*. London: Taylor & Francis, revised edn 1998.

Bromage, Sarah, and Helen Williams. 'Materials, technologies and the printing industry'. In *The Cambridge History of the Book in Britain. Volume 7: The twentieth century and beyond*, edited by Andrew Nash, Claire Squires and I.R. Willison, 41–60. Cambridge: Cambridge University Press, 2019.

Broman, Thomas. 'Criticism and the circulation of news: The scholarly press in the late seventeenth century', *History of Science* 51 (2013): 125–50.

Broman, Thomas. 'The profits and perils of publicity: *Allgemeine Literatur-Zeitung*, the Thurn und Taxis Post, and the periodical trade at the end of the eighteenth century', *Notes and Records of the Royal Society* 69, no. 3 (2015): 261–76.

Brown, D., and Hans Sloane. *Catalogus Plantarum Quae in Insula Jamaica: Sponte Proveniunt, Vel Vulgò Coluntur Cum Earundem Synonymis & Locis Natalibus, Adjectis Aliis Quibusdam Quae in Insulis Maderae, Barbados, Nieves, & Sancti Christophori Nascuntur, Seu Prodromi Historiae Naturalis Jamaicae Pars Prima*. London: D. Brown, 1696.

Burckhardt, Frederick, and James A. Secord, eds. *The Correspondence of Charles Darwin*. 30 vols. Cambridge: Cambridge University Press, 1985–2022.

Burnham, John C. 'The evolution of editorial peer review', *Journal of American Medical Association* 10, no. 263 (9 March 1990): 1323–9.

Campbell, Robert, Ed Pentz, and Ian Borthwick, eds. *Academic and Professional Publishing*. Oxford: Elsevier Science, 2012.

Capshew, James H., and Karen A. Rader. 'Big science: Price to the present', *Osiris* 7 (1992): 2–25.

Carey, Daniel. 'Compiling nature's history: Travellers and travel narratives in the early Royal Society', *Annals of Science* 54 (1997): 269–92.

Cartwright, Mary. 'Note on A. G. Cock's paper 'Chauvinism in Science': The International Research Council, 1919–1926', *Notes and Records of the Royal Society* 39, no. 1 (1984): 125–8.

A Catalogue of Books in Quires, and Copies, Being Part of the Stock of the Late Mr William Innys. London, 1757.

Chadwyck-Healey, Charles. *Publishing for Libraries: At the dawn of the digital age*. London: Bloomsbury, 2020.

Chambers, Neil, ed. *Joseph Banks and the British Museum: The world of collecting, 1770–1830*. Abingdon: Pickering & Chatto, 2007.

Chambers, Neil. *The Scientific Correspondence of Sir Joseph Banks, 1765–1820*. 6 vols. London: Pickering & Chatto, 2007.

Chang, Ku-ming Kevin, and Alan Rocke, eds. *A Global History of Research Education: Disciplines, institutions, and nations, 1840–1950*, History of Universities Series, vol. 34. Oxford: Oxford University Press, 2021.

Christie, Samuel Hunter, and John Bostock. '[Report on Faraday's] experimental researches in electricity', *Abstracts of the Papers Printed in the Philosophical Transactions of the Royal Society of London* 3 (1837): 113–19.

Clark, Ronald W. *A Biography of the Nuffield Foundation*. London: Longman, 1972.

Clarke, Imogen. 'Negotiating progress: Promoting "modern" physics in Britain, 1900–1940'. PhD, University of Manchester, 2012.

Clarke, Imogen. 'The gatekeepers of modern physics: Periodicals and peer review in 1920s Britain', *Isis* 106, no. 1 (2015): 70–93.

Clarke, Imogen, and James Mussell. 'Conservative attitudes to old-established organs: Oliver Lodge and Philosophical Magazine', *Notes and Records of the Royal Society* 69, no. 3 (2015): 321–36.

Clarke, Sabine. 'Pure science with a practical aim: The meanings of fundamental research in Britain, circa 1916–1950', *Isis* 101, no. 2 (2010): 285–311.

Clifford, F.W. 'World list of scientific periodicals'. *Nature*, Supplement to 13 Sept. 1924 (1924): 401–2.

Cobb, Matthew. 'The prehistory of biology preprints: A forgotten experiment from the 1960s', *PLOS Biology* 15, no. 11 (2017): e2003995.

Cock, A.G. 'Chauvinism and internationalism in science: The International Research Council, 1919–1926', *Notes and Records of the Royal Society* 37, no. 2 (1983): 249–88.

Coles, Bryan R. 'The scientific, technical and medical information system in the UK: A study on behalf of the Royal Society, the British Library and the Association of Learned and Professional Society Publishers'. London: British Library, 1993.

Coley, N.G. 'The Animal Chemistry Club; Assistant society to the Royal Society', *Notes and Records of the Royal Society* 22, no. 1 (1967): 173–85.

Collins, Peter. *The Royal Society and the Promotion of Science since 1960*. Cambridge: Cambridge University Press, 2015.

'Comrades of old: Royal Society staff pensioners, biographical sketches'. Privately circulated for The Royal Society Former Staff Association, 2015.

'Conversazione to mark the 300th anniversary of the publication of the *Philosophical Transactions of the Royal Society*, Hooke's *Micrographia*, and Evelyn's *Sylva*', *Notes and Records of the Royal Society* 21, no. 1 (1966): 12–19.

Cook, Alan. *Edmond Halley: Charting the heavens and the seas*. Oxford: Clarendon Press, 1997.

Cox, Brian. 'The Pergamon phenomenon 1951–1991: Robert Maxwell and scientific publishing', *Learned Publishing* 15 (2002): 273–8.

Cox, John. 'The shifting sands of technological change: The middle man as navigator', *Serials* 6, no. 4 (November 1993).

Craik, Alex D.D. 'The hydrostatical works of George Sinclair (c. 1630–1696): Their neglect and criticism', *Notes and Records: The Royal Society Journal of the History of Science* 72, no. 3 (2018): 239–73.

Crilly, Tony. 'The Cambridge Mathematical Journal and its descendants: The linchpin of a research community in the early and mid-Victorian age', *Historia Mathematica* 31, no. 4 (2004): 455–97.

Cromartie, Alan. *Sir Matthew Hale, 1609–1676: Law, religion and natural philosophy*. Cambridge: Cambridge University Press, 1995.

Cronin, Blaise, and Helen Barsky Atkins. *The Web of Knowledge: A Festschrift in honor of Eugene Garfield*. Medford, NJ: Information Today Inc, 2000.

Csiszar, Alex. 'Objectivities in print'. In *Objectivity in Science: New perspectives in science and technology studies* edited by Flavia Padovani, Alan Richardson and Jonathan Y. Tsou, 145–69. Dordrecht: Springer, 2015.

Csiszar, Alex. 'How lives became lists and scientific papers became data: Cataloguing authorship during the nineteenth century', *British Journal for the History of Science* 50 (2017): 23–60.

Csiszar, Alex. 'Peer review: Troubled from the start', *Nature* 532, no. 7599 (2016): 306–8.

Csiszar, Alex. *The Scientific Journal: Authorship and the politics of knowledge in the nineteenth century*. Chicago, IL: University of Chicago Press, 2018.

Csiszar, Alex. 'Proceedings and the public: How a commercial genre transformed science'. In *Science Periodicals in Nineteenth-Century Britain: Constructing scientific communities*, edited by Gowan Dawson, Bernard Lightman, Sally Shuttleworth and Jonathan R. Topham, 103–34. Chicago, IL: University of Chicago Press, 2020.

Cuvier, Georges. 'Eloge historique de Sir Joseph Banks, lu le 2 Avril 1821', *Memoires de l'académie des sciences de l'institut Francais* 5 (1827): 204–30.

Dale, Henry Hallett. 'Francis Alexander Towle, 1874–1932', *Obituary Notices of Fellows of the Royal Society* 1 (1932): 85–7.

Dale, Henry Hallett. 'Henry George Lyons, 1864–1944', *Obituary Notices of Fellows of the Royal Society* 4, no. 13 (1944): 795–809.

Darnton, Robert. *Mesmerism and the End of the Enlightenment in France*. Cambridge, MA: Harvard University Press, 1995 (first published in 1968).

Darwin, Charles. *Life of Erasmus Darwin*. Edited by Desmond King-Hele. Cambridge: Cambridge University Press, 2003.

Darwin, Charles Galton. 'The "reading" of papers at meetings of Royal Society', *Notes and Records of the Royal Society* 2, no. 1 (1939): 25–7.

Davies, John D. Griffith. 'Ronald Winckworth 1884–1950', *Notes and Records of the Royal Society* 8, no. 2 (1951): 293–6.

Dawson, Gowan. '"An independent publication for geologists": The Geological Society, commercial journals, and the remaking of nineteenth-century geology'. In *Science Periodicals in Nineteenth-Century Britain: Constructing scientific communities*, edited by Gowan Dawson, Bernard Lightman, Sally Shuttleworth and Jonathan R. Topham, 137–71. Chicago, IL: University of Chicago Press, 2020.

Dawson, Gowan, and Jonathan R. Topham. 'Scientific, medical, and technical periodicals in nineteenth-century Britain: New formats for new readers'. In *Science Periodicals in Nineteenth-Century Britain: Constructing scientific communities*, edited by Gowan Dawson, Bernard Lightman, Sally Shuttleworth and Jonathan R. Topham, 35–64. Chicago, IL: University of Chicago Press, 2020.

Dawson, Gowan, Bernard Lightman, Sally Shuttleworth and Jonathan R. Topham, eds. *Science Periodicals in Nineteenth-Century Britain: Constructing scientific communities*. Chicago, IL: University of Chicago Press, 2020.

De Beer, Gavin. *The Sciences Were Never at War*. London: Nelson, 1960.
Deazley, Ronan. *On the Origin of the Right to Copy: Charting the movement of copyright law in eighteenth century Britain (1695-1775)*. Oxford: Bloomsbury Publishing, 2004.
Delbourgo, James. *Collecting the World: Hans Sloane and the origins of the British Museum*. London: Allen Lane, 2017.
Despeaux, Sloan Evans. 'Fit to Print? Referee reports on mathematics for the nineteenth-century journals of the Royal Society of London', *Notes and Records of the Royal Society* 65, no. 3 (2011): 233–52.
Donovan, Arthur. *Antoine Lavoisier: Science, administration and revolution*. Cambridge: Cambridge University Press, 1993.
Douglas, Kimberly. 'The serials crisis', *The Serials Librarian* 18, no. 1–2 (1990): 111–21.
Edgerton, David. *Science, Technology and the British Industrial 'Decline', 1870–1970*. Cambridge: Cambridge University Press, 1996.
Edgerton, David. *Warfare State: Britain, 1920–1970*. Cambridge: Cambridge University Press, 2005.
Edgerton, David, and John V. Pickstone. 'United Kingdom'. In *The Cambridge History of Science. Volume 8: Modern science in national, transnational, and global context*, edited by David N. Livingstone, Hugh Richard Slotten and Ronald L. Numbers, 151–91. Cambridge: Cambridge University Press, 2020.
Eliot, Simon, and Jonathan Rose, eds. *Companion to the History of the Book*. Oxford: Blackwell, 2007.
Eve, Martin Paul. 'Learned societies in the humanities, open access, and paying for disciplinary goods'. In *Open Access*, edited by Martin Paul Eve, 2017. https://eve.gd/2017/01/02/learned-societies-open-access-and-paying-for-disciplinary-goods.
Feather, John. 'From rights in copies to copyright: The recognition of authors' rights in English law and practice in the sixteenth and seventeenth centuries'. In *The Construction of Authorship: Textual appropriation in law and literature*, edited by Martha Woodmansee and Peter Jaszi, 191–209. Durham, NC: Duke University Press, 1994.
Feingold, Mordechai. 'Mathematicians and naturalists: Sir Isaac Newton and the Royal Society'. In *Isaac Newton's Natural Philosophy*, edited by Jed Z. Buchwald and I. Bernard Cohen, 77–104. Cambridge, MA: MIT Press, 2001.
Feingold, Mordechai, and Giulia Giannini. *The Institutionalization of Science in Early Modern Europe*. Leiden: Brill, 2019.
Ferry, Georgina. 'The exception and the rule: Women and the Royal Society 1945–2010', *Notes and Records of the Royal Society* 64, suppl. 1 (2010): S163–S172.
Fontes da Costa, Palmira. *The Singular and the Making of Knowledge at the Royal Society of London in the Eighteenth Century*. Newcastle-upon-Tyne: Cambridge Scholars Publishing, 2009.
Forbes, Eric Gray, Lesley Murdin, and Frances Willmoth, eds. *The Correspondence of John Flamsteed, the First Astronomer Royal*. 3 vols. Boca Raton, FL: CRC Press, 1995–2001.
Ford, Katherine. 'The role of the Royal Society in Victorian literary culture'. PhD, Reading University, 2015.
Fox, Robert. *Science Without Frontiers: Cosmopolitanism and national interests in the world of learning, 1870–1940*. Corvallis, OR: Oregon State University Press, 2016.
Fransen, Sietske. 'Antoni Van Leeuwenhoek, his images and draughtsmen', *Perspectives on Science* 27, no. 3 (2019): 485–544.
Fransen, Sietske. 'Latin in a time of change: The choice of language as signifier of a new science?', *Isis* 108, no. 3 (2017): 629–35.
Fransen, Sietske, Niall Hodson, and Karl A.E. Enenkel, eds. *Translating Early Modern Science*. Leiden: Brill, 2017.
Fraser, Kevin J. 'John Hill and the Royal Society in the eighteenth century', *Notes and Records of the Royal Society* 48, no. 1 (1994): 43–67.
Fredriksson, Einar H. *A Century of Science Publishing: A collection of essays*. Amsterdam: IOS Press, 2001.
Fredriksson, Einar H. 'The Dutch publishing scene: Elsevier and North-Holland'. In *A Century of Science Publishing: A collection of essays*, edited by E.H. Fredriksson, 61–76. Amsterdam: IOS Press, 2001.
Fulford, Tim, and Sharon Ruston, eds. *The Collected Letters of Sir Humphry Davy*. Oxford: Oxford University Press, 2020.

Fyfe, Aileen. *Steam-Powered Knowledge: William Chambers and the business of publishing, 1820–1860*. Chicago, IL: University of Chicago Press, 2012.

Fyfe, Aileen. 'Journals, learned societies and money: *Philosophical Transactions*, ca. 1750–1900', *Notes and Records of the Royal Society* 69, no. 3 (2015): 277–99.

Fyfe, Aileen. 'Editors, referees and committees: Distributing editorial work at the Royal Society journals in the late nineteenth and twentieth centuries', *Centaurus* 62, no. 1 (2020): 125–40.

Fyfe, Aileen. 'The production, circulation, consumption and ownership of scientific knowledge: Historical perspectives'. In *CREATe Working Paper*. Glasgow: CREATe: UK Copyright and Creative Economy Centre, 2020.

Fyfe, Aileen. 'The Royal Society and the noncommercial circulation of knowledge'. In *Reassembling Scholarly Communications: Histories, infrastructures, and global politics of open access*, edited by Martin Paul Eve and Jonathan Gray, 147–60. Cambridge, MA: MIT Press, 2020.

Fyfe, Aileen. 'Self-help for learned journals: Scientific societies and the commerce of publishing in the 1950s', *History of Science* 60, no. 2 (2022) 255–79.

Fyfe, Aileen. 'From philanthropy to business: The economics of Royal Society journal publishing in the twentieth century', *Notes and Records of the Royal Society*, published online 3 August 2022, https://doi.org/10.1098/rsnr.2022.0021.

Fyfe, Aileen, and Anna Gielas. 'Introduction: Editorship and the editing of scientific journals, 1750–1950', *Centaurus* 62, no. 1 (2020): 5–20.

Fyfe, Aileen, and Noah Moxham. 'Making public ahead of print: Meetings and publications at the Royal Society, 1752–1892', *Notes and Records of the Royal Society* 70, no. 4 (2016): 361–79.

Fyfe, Aileen, Kelly Coate, Stephen Curry, Stuart Lawson, Noah Moxham, and Camilla Mørk Røstvik. *Untangling Academic Publishing: A history of the relationship between commercial interests, academic prestige and the circulation of research*. St Andrews: University of St Andrews, 2017. http://doi.org/10.5281/zenodo.546100.

Fyfe, Aileen, Julie McDougall-Waters, and Noah Moxham. 'Credit, copyright, and the circulation of scientific knowledge: The Royal Society in the long nineteenth century', *Victorian Periodicals Review* 51, no. 4 (2018): 597–615.

Fyfe, Aileen, Julie McDougall-Waters, and Noah Moxham. '*Philosophical Transactions*: 350 years of publishing at the Royal Society (1665–2015)'. [Exhibition catalogue] London: The Royal Society and the University of St Andrews, 2014. http://hdl.handle.net/10023/6058.

Fyfe, Aileen, Flaminio Squazzoni, Didier Torny, and Pierpaolo Dondio. 'Managing the growth of peer review at the Royal Society journals, 1865–1965', *Science, Technology and Human Values* 45, no. 3 (2020): 405–29.

Galison, Peter, and B.W. Hevly. *Big Science: The growth of large-scale research*. Stanford, CA: Stanford University Press, 1992.

Gardiner, J.C. 'Review of self-help for learned journals', *Proceedings of the Botanical Society of the British Isles* 5 (1963/4).

Garisto, Daniel. 'Preprints make inroads outside of physics', *APS News* 28, no. 9 (2019).

Gascoigne, John. *Science in the Service of Empire: Joseph Banks, the British state and the uses of science in the age of revolution*. Cambridge: Cambridge University Press, 1998.

Gascoigne, John. *Joseph Banks and the English Enlightenment*. Cambridge: Cambridge University Press, 2003.

Gastel, Barbara. 'From the Los Alamos preprint archive to the ArXiv: An interview with Paul Ginsparg', *Science Editor* 25, no. 2 (2002): 42–3.

Gay, Hannah. 'A questionable project: Herbert Mcleod and the making of the fourth series of the Royal Society Catalogue of Scientific Papers, 1901–25', *Annals of Science* 70 (2013): 149–74.

Gay, Hannah, and John W. Gay. 'Brothers in science: Science and fraternal culture in nineteenth-century Britain', *History of Science* 35 (1997): 425–53.

Geiger, Roger L. 'Science, universities, and national defense, 1945–1970', *Osiris* 7 (1992): 26–48.

Gibelin, Jacques. *Abrégé des Transactions Philosophiques de la Société royale de Londres*. Paris: Buisson, 1787–91.

Gielas, Anna M. 'Early sole editorship in the Holy Roman Empire and Britain, 1770s–1830s'. PhD, University of St Andrews, 2019.

Gielas, Anna. 'Turning tradition into an instrument of research: The editorship of William Nicholson (1753–1815)', *Centaurus* 62, no. 1 (2020): 38–53.

Gitelman, Lisa. *Paper Knowledge: Toward a media history of documents*. Durham, NC: Duke University Press, 2012.
Glover, David, Christine Holt, Louise Johnson, and Peter Parham. 'Introducing Open Biology', *Open Biology* 1, no. 1 (2011): 110001.
Goddard, Richard. *'Drawing on Copper': The Basire family of copper-plate engravers and their works*. Maastricht: Maastricht University Press, 2016.
Goetze, Heinz. *Springer Verlag: History of a scientific publishing house*, Part II *1945–1992*. Berlin; Heidelberg: Springer, 1996.
Goldgar, Anne. *Impolite Learning: Conduct and community in the Republic of Letters, 1680–1750*. New Haven, CT: Yale University Press, 1995.
Golinski, Jan. 'Humphry Davy: The experimental self', *Eighteenth-Century Studies* 45, no. 1 (2011): 15–28.
Goodare, Jennifer Rose. 'Representing science in a divided world: The Royal Society and Cold War Britain'. PhD, University of Manchester, 2013.
Gooday, Graeme. *Domesticating Electricity: Expertise, uncertainty and gender, 1880–1914*. London: Pickering & Chatto, 2008.
Gooday, Graeme. 'Periodical physics in Britain: Institutional and industrial contexts, 1870–1900'. In *Science Periodicals in Nineteenth-Century Britain: Constructing scientific communities*, edited by Gowan Dawson, Bernard Lightman, Sally Shuttleworth and Jonathan R. Topham, 238–73. Chicago, IL: University of Chicago Press, 2020.
Gordin, Michael D. *Scientific Babel: How science was done before and after global English*. Chicago, IL: University of Chicago Press, 2015.
Graham, Kayla. 'Thanking our peer reviewers'. *EveryONE: PLOS ONE community blog*, 2014. https://everyone.plos.org/2014/01/06/thanking-peer-reviewers/.
Graham, Loren R. 'Big Science in the last years of the Big Soviet Union', *Osiris* 7 (1992): 49–71.
[Granville, Augustus Bozzi]. *Science Without a Head; or, the Royal Society Dissected. By One of the 687 F.R.S.* London: T. Ridgway, 1830.
Granville, Augustus Bozzi. *The Royal Society in the XIXth Century; Being a Statistical Summary of Its Labours During the Last Thirty-Five Years*. London: Printed for the author; sold by John Churchill, 1836.
Granville, Augustus Bozzi. *Autobiography of A.B. Granville; Being Eighty-Eight Years of the Life of a Physician. Edited with a brief account of the last years of his life*. Edited by Paulina Granville, 2 vols. London: H.S. King, 1874.
Greenaway, Frank. *Science International: A history of the International Council of Scientific Unions*. Cambridge: Cambridge University Press, 2006.
Griffin, Robert J. 'Anonymity and authorship', *New Literary History* 30 (1999): 877–95.
Gross, Alan G., Joseph E. Harmon, and Michael S. Reidy. *Communicating Science: The scientific article from the seventeenth century to the present*. New York: Oxford University Press, 2002.
Guénon, Anne-Sylvie. 'Les publications de l'académie des sciences'. In *Histoire et mémoire de l'Académie des sciences: Guide de recherches*, edited by Académie des Sciences, Éric Brian and Christiane Demeulenaere-Douyère, 113–28. Paris: Tec & DOC Lavoisier, 1996.
Guerrini, Anita. *The Courtiers' Anatomists: Animals and humans in Louis XIV's Paris*. Chicago, IL: University of Chicago Press, 2015.
Gunther, Robert T. *Early Science in Oxford*. 14 vols. Oxford: Clarendon Press, 1923–45.
Hahn, Roger. *The Anatomy of a Scientific Institution: The Paris Academy of Sciences, 1666–1803*. Berkeley: University of California Press, 1971.
Haines, Joe. *Maxwell*. London: McDonald & Co., 1988.
Hall, A. Rupert, and Marie Boas Hall, eds. *The Correspondence of Henry Oldenburg*. 13 vols. Madison: University of Wisconsin Press, 1965–86.
Hall, A. Rupert, and Laura Tilling, eds. *The Correspondence of Isaac Newton* (paperback edn). 7 vols. Cambridge: Cambridge University Press, for the Royal Society, 2008.
Hall, Marie Boas. *Promoting Experimental Learning: Experiment and the Royal Society, 1660–1727*. Cambridge: Cambridge University Press, 1993.
Hall, Marie Boas. *All Scientists Now: The Royal Society in the nineteenth century*. Cambridge: Cambridge University Press, 2002.
Hall, Marie Boas, and Patricia H. Clarke. *The Library and Archives of the Royal Society: 1600–1990*. London: The Royal Society, 1992.
Hardy, G.H. 'Letter to the editor', *Notes and Records of the Royal Society* 3, no. 1 (1940): 96.

Harris, Michael. *London Newspapers in the Age of Walpole: A study of the origins of the modern English press*. Madison, NJ: Fairleigh Dickinson University Press, 1987.

Harrison, Andrew John. 'Scientific naturalists and the government of the Royal Society 1850–1900'. PhD, Open University, 1989.

Harrison, Cecil Reeves, and H.G. Harrison. *House of Harrison: Being an account of the family and firm of Harrison and Sons, printers to the King*. London: Harrison & Sons, 1914.

Heilbron, John. *Physics at the Royal Society During Newton's Presidency*. Los Angeles: University of California, 1983.

Heilbron, John. 'A mathematicians' mutiny with morals'. In *World Changes: Thomas Kuhn and the nature of science*, edited by Paul Horwich, 81–129. Cambridge, MA: MIT Press, 1993.

Henley, Jane, and Sarah Thomson. 'JournalsOnline: The online journal solution', *Ariadne: Web Magazine for Information Professionals* 12 (c. 1997). http://www.ariadne.ac.uk/issue/12/cover.

Hewitt, Martin. *The Dawn of the Cheap Press in Victorian Britain: The end of the 'taxes on knowledge', 1849–1869*. London: Bloomsbury, 2013.

Hewitt, Rachel. *Map of a Nation: A biography of the Ordnance Survey*. London: Granta, 2010.

Higgitt, Rebekah. 'Instruments and relics: The history and use of the Royal Society's object collections c.1850–1950', *Journal of the History of Collections* 31, no. 3 (2018): 469–85.

Hill, Archibald Vivian. 'J. D. Griffith Davies 1899–1953 (Assistant Secretary of the Royal Society, 1937–1946)', *Notes and Records of the Royal Society* 11, no. 2 (1955): 129–33.

Hill, Archibald Vivian. 'Memories and reflections'. Unpublished PDF, Churchill College, Cambridge, 1974.

Hill, John. *A Review of the Works of the Royal Society*. London: R. Griffiths, 1751.

[Hill, John]. *A Dissertation Upon Royal Societies: In three letters from a nobleman on his travels, to a person of distinction in Sclavonia*. London: printed for John Doughty, 1750.

Hinds, Peter. *'The Horrid Popish Plot': Roger L'Estrange and the circulation of political discourse in late seventeenth-century London*. Oxford: Oxford University Press, 2010.

Hobbs, Andrew. *A Fleet Street in Every Town: The provincial press in England, 1855–1900*. Cambridge: Open Book Publishers, 2018.

Hodgkin, Dorothy M.C. 'Kathleen Lonsdale, 28 January 1903–1 April 1971', *Biographical Memoirs of Fellows of the Royal Society* 21 (1975): 447–84.

Holman, Valerie. *Print for Victory: Book publishing in Britain 1939–1945*. London: British Library, 2008.

Hooke, Robert. *Animadversions on the first part of the Machina Coelestis*. London: John Martyn, 1674.

Hooke, Robert. *Lampas*. London: John Martyn, 1677 [1676].

Hooke, Robert. *Lectures and Collections*. London: John Martyn, 1678.

Hooke, Robert. *Lectures: De Potentia Restitutiva*. London: John Martyn, 1678.

Hooper, Mark. 'Scholarly review, old and new', *Journal of Scholarly Publishing* 51, no. 1 (2019): 63–75.

Hoppit, Julian. 'Sir Joseph Banks's provincial turn', *The Historical Journal* 61, no. 2 (2017): 403–29.

Hoskin, Michael. *Discoverers of the Universe: William and Caroline Herschel*. Princeton, NJ: Princeton University Press, 2011.

House of Commons Select Committee on Science and Technology. 'Scientific Publications: Free for all? (Tenth Report, 2003–04)'. London: HMSO, 2004.

House of Lords Select Committee on Science and Technology. 'Science and Society (Third Report, 1999–2000)'. London: HMSO, 2000.

Hughes, Jeff. '"Divine right" or democracy? The Royal Society "revolt" of 1935', *Notes and Records of the Royal Society* 64, suppl. 1 (2010): S101–S117.

Hughes, Jeff. 'Doing diaries: David Martin, the Royal Society and scientific London, 1947–1950', *Notes and Records of the Royal Society* 66, no. 3 (2012): 273–94.

Hughes, Jeff. 'Mugwumps? The Royal Society and the governance of post-war British science'. In *Scientific Governance in Britain, 1914–79*, edited by Don Leggett and Charlotte Sleigh, 81–99. Manchester: Manchester University Press, 2016.

Hughes, Thomas P. *Networks of Power: Electrification in Western Society*. Baltimore, MD: Johns Hopkins, 1983.

Hughes, Thomas P. *Human-Built World: How to think about technology and culture*. Chicago, IL: University of Chicago Press, 2004.

Hume, Robert D. 'The value of money in eighteenth-century England: Incomes, prices, buying power – and some problems in cultural economics', *Huntington Library Quarterly* 77, no. 4 (2014): 373–416.
Hunter, Michael. *Establishing the New Science: The experience of the early Royal Society*. Woodbridge: Boydell & Brewer, 1989.
Hunter, Michael, and Edward B. Davis, eds. *The Works of Robert Boyle*. 14 vols. London: Pickering & Chatto, 1999–2000.
Iliffe, Rob. '"In the warehouse": Privacy, property and priority in the early Royal Society', *History of Science* 30, no. 1 (1992): 29–68.
Iliffe, Rob. 'Author-mongering: The editor between producer and consumer'. In *The Consumption of Culture 1600–1800: Image, Object, Text*, edited by Ann Bermingham and John Brewer, 166–92. London: Routledge, 1995.
James, Frank A.J.L., ed. *The Correspondence of Michael Faraday*. 6 vols. London: Institution of Electrical Engineers, 1991–2012.
James, Frank A.J.L., and Sharon Ruston. 'New studies on Humphry Davy: Introduction', *Ambix* 66, nos. 2–3 (2019): 95–102.
Jardine, Lisa. *The Curious Life of Robert Hooke: The man who measured London*. London: HarperCollins, 2003.
Jardine, Lisa. *Going Dutch: How England plundered Holland's glory*. London: Harper Perennial, 2008.
Jardine, Lisa. *Temptation in the Archives: Essays in Golden Age Dutch culture*. London: UCL Press, 2015.
Jeapes, Ben. 'Blackwell's electronic journal navigator', *The Electronic Library* 15, no. 3 (1997): 189–91.
Jeffery, G.B. 'Louis Napoleon George Filon, 1875–1937', *Obituary Notices of Fellows of the Royal Society* 2, no. 7 (1939): 501–9.
Jenkins, Bill. 'Commercial scientific journals and their editors in Edinburgh, 1819–1832', *Centaurus* 62, no. 1 (2020): 69–81.
Johns, Adrian. 'Flamsteed's optics and the identity of the astronomical observer'. In *Flamsteed's Stars: New perspectives on the life and work of the first Astronomer Royal*, edited by Frances Willmoth, 77–106. Woodbridge: Boydell & Brewer, 1997.
Johns, Adrian. *The Nature of the Book: Print and knowledge in the making*. Chicago, IL: University of Chicago Press, 1998.
Johns, Adrian. 'Miscellaneous methods: Authors, societies and journals in early modern England', *British Journal for the History of Science* 33, no. 2 (2000): 159–86.
Jones, Allan. 'J.G. Crowther's war: Institutional strife at the BBC and British Council', *British Journal for the History of Science* 49, no. 2 (2016): 259–78.
Jones, H. Kay. *Butterworths: History of a publishing house*. London: Butterworth & Co., 1980.
Jones, Helen. *Duty and Citizenship: The correspondence and papers of Violet Markham, 1896–1953*. London: Historian's Press, 1994.
The Jubilee of the Chemical Society of London. Record of the proceedings together with an account of the history and development of the Society, 1841–1891. London: Harrison & Sons, 1896.
Katz, Bernard. 'Archibald Vivian Hill, 26 September, 1886–3 June 1977', *Biographical Memoirs of Fellows of the Royal Society* 24 (1978): 71–149.
Katzen, May F. 'The changing appearance of research journals in science and technology'. In *The Development of Science Publishing in Europe*, edited by A.J. Meadows, 177–214. Amsterdam: Elsevier, 1980.
Kermes, Hannah, Stefania Degaetano, Ashraf Khamis, Jörg Knappen, and Elke Teich. 'The Royal Society Corpus: From uncharted data to corpus'. In *Proceedings of the LREC 2016*. Portoroz, Slovenia, 2016.
[King, William]. *The Transactioneer with Some of His Philosophical Fancies: In two dialogues*. London: no imprint, 1700.
Klug, Sir Aaron. 'Address of the president, Sir Aaron Klug, O.M., P.R.S., given at the anniversary meeting on 1 December 1997'. *Notes and Records of the Royal Society* 52, no. 1 (1998): 181–90.
Klug, Sir Aaron. 'Address of the president, Sir Aaron Klug, O.M., P.R.S., given at the anniversary meeting on 30 November 1999', *Notes and Records of the Royal Society* 54, no. 1 (2000): 99–108.
Knoche, Michael. 'Scientific journals under National Socialism', *Libraries & Culture* 26 (1991): 415–26.

Kranzler, David. *Holocaust Hero: Solomon Schonfeld*. Jersey City, NJ: Ktav Publishing House, 2004.

Kronick, David A. *A History of Scientific and Technical Periodicals: The origins and development of the scientific and technical press, 1665–1790*. Metuchen, NJ: Scarecrow Press, 1976.

Kronick, David A. 'Authorship and authority in the scientific periodicals of the seventeenth and eighteenth centuries', *Library Quarterly* 48, no. 3 (1978): 225–75.

Kronick, David A. 'Notes on the printing history of the early "Philosophical Transactions"', *Libraries & Culture* 25, no. 2 (1990): 243–68.

Kronick, David A. *Scientific and Technical Periodicals of the Seventeenth and Eighteenth Centuries: A guide*. Metuchen, NJ: Scarecrow Press, 1991.

Kronick, David A. 'Medical "publishing societies" in eighteenth-century Britain'. *Bulletin of the Medical Library Association* 82, no. 3 (1994): 277–82.

Kroupa, Sebestian. 'Ex epistulis Philippinensibus: Georg Joseph Kamel SJ (1661–1706) and his correspondence network', *Centaurus* 57, no. 4 (2015): 229–59.

Kusukawa, Sachiko. 'The *Historia Piscium (1686)*', *Notes and Records of the Royal Society* 54, no. 2 (2000): 179–97.

Kusukawa, Sachiko. 'Picturing knowledge in the early Royal Society: The examples of Richard Waller and Henry Hunt', *Notes and Records of the Royal Society* 65, no. 3 (2011): 273–94.

Kusukawa, Sachiko, Felicity Henderson, Alexander Marr, Sietske Fransen, Katherine Reinhart, and Judith Weik. *Science Made Visible: Drawings, prints, objects*. London: The Royal Society, 2018. https://issuu.com/crassh/docs/des5543_1_science_made_visible_exhi.

Larivière, Vincent, and Cassidy R. Sugimoto. 'The journal impact factor: A brief history, critique, and discussion of adverse effects'. In *Springer Handbook of Science and Technology Indicators*, edited by Wolfgang Glänzel, Henk F. Moed, Ulrich Schmoch and Mike Thelwall, 3–24. Cham: Springer International Publishing, 2019.

Larivière, Vincent, Stefanie Haustein, and Philippe Mongeon. 'The oligopoly of academic publishers in the digital era', *PLOS ONE* 10, no. 6 (2015): e0127502.

Larmor, Joseph, ed. *Memoir and Scientific Correspondence of the Late Sir George Gabriel Stokes*. 2 vols. Cambridge: Cambridge University Press, 1907.

Lee, Carole J., Cassidy R. Sugimoto, Guo Zhang, and Blaise Cronin. 'Bias in peer review', *Journal of the American Society for Information Science and Technology* 64, no. 1 (2013): 2–17.

Lefebvre, Lex. 'The story of STM', *Serials* 7, no. 1 (March 1994): 53–5.

Levine, Joseph. *Between the Ancients and the Moderns: Baroque culture in Restoration England*. New Haven, CT: Yale University Press, 1999.

Levine, Joseph M. *Dr. Woodward's Shield: History, science, and satire in Augustan England*. Ithaca, NY: Cornell University Press, 1991 (first published in 1977).

Lewis, Cherry. 'Doctoring geology: The medical origins of the Geological Society'. In *The Making of the Geological Society of London*, edited by Cherry Lewis and Simon Knell, 49–92. London: Geological Society of London, 2009.

Lewis, Cherry, and Simon Knell. *The Making of the Geological Society of London*. London: Geological Society of London, 2009.

Lewis, John L. *125 Years: The Physical Society and the Institute of Physics*. London: Taylor & Francis, 1999.

Lightman, Bernard. '"Knowledge" confronts "nature": Richard Proctor and popular science periodicals'. In *Culture and Science in the Nineteenth-Century Media*, edited by Louise Henson, Geoffrey Cantor, Gowan Dawson, Richard Noakes, Sally Shuttleworth and Jonathan R. Topham, 199–210. Aldershot: Ashgate, 2004.

Lightman, Bernard, Michael S. Reidy, and Roland Jackson, eds. *The Correspondence of John Tyndall*. Pittsburgh, PA: University of Pittsburgh Press, 2016–.

Lipkowitz, Elise S. 'Seized natural-history collections and the redefinition of scientific cosmopolitanism in the era of the French Revolution', *British Journal for the History of Science* 47, no. 1 (2014): 15–41.

Low, Sampson. *The English Catalogue of Books*. 9 vols. London: Sampson Low, 1863–1915.

Lowthorp, John. *The Philosophical Transactions Abridged, and Disposed under General Heads*. 3 vols. For Thomas Bennet, Robert Knaplock, Richard Wilkin, 1703–8.

Luna, Paul. 'Technology'. In *History of Oxford University Press: Volume IV*, edited by Keith Robbins, 179–204. Oxford: Oxford University Press, 2017.

Lynall, Gregory. *Swift and Science: The satire, politics, and theology of natural knowledge, 1690–1730*. Basingstoke: Palgrave Macmillan, 2012.

Lyons, Henry. 'The Society's finances Part I: 1662–1830', *Notes and Records of the Royal Society* 1, no. 2 (1938): 73–87.

Lyons, Henry. 'The Society's finances Part II: 1831–1938', *Notes and Records of the Royal Society* 2, no. 1 (1939): 47–67.

Lyons, Henry. 'On the finances of the Royal Society for the ten years 30th November 1929 to 30th November 1939'. In *Year Book of the Royal Society*, 222–7. London: Royal Society, 1940.

Lyons, Henry, ed. *The Record of the Royal Society of London*. 4th edn. London: The Royal Society, 1940. Reprint, 1992.

Mabe, Michael, and Anthony Watkinson. 'Journals (STM and Humanities)'. In *The Cambridge History of the Book in Britain. Volume 7: The twentieth century and beyond*, edited by Andrew Nash, Claire Squires and Ian Willison, 484–98. Cambridge: Cambridge University Press, 2019.

Macfarlane, Bruce. 'The spirit of research', *Oxford Review of Education* 47, no. 6 (2021): 737–51.

Mackenzie, Eneas. *A descriptive and historical account of the town and county of Newcastle-Upon-Tyne*. 2 vols. Newcastle-upon-Tyne: Mackenzie and Dent, 1827.

MacLeod, Christine. *Inventing the Industrial Revolution: The English patent system 1660–1800*. Cambridge: Cambridge University Press, 1988.

MacLeod, Christine. 'Strategies for innovation: The diffusion of new technology in nineteenth-century British industry', *Economic History Review* 45 (1992): 285–307.

MacLeod, Roy M. 'The Royal Society and the government grant: Notes on the administration of scientific research, 1849–1914', *The Historical Journal* 14, no. 2 (1971): 323–58.

MacLeod, Roy M. 'Whigs and savants: Reflections on the reform movement in the Royal Society, 1830–48'. In *Metropolis and Province: Science in British culture, 1780–1850*, edited by Ian Inkster and J.B. Morrell, 55–90. London: Hutchinson, 1983.

MacLeod, Roy M. 'The mobilisation of minds and the crisis in international science: The Krieg der Geister and the Manifesto of the 93', *Journal of War & Culture Studies* 11, no. 1 (2018): 58–78.

MacLeod, Roy M., and E. Kay Andrews. 'The Origins of the D.S.I.R.: Reflections on ideas and men, 1915–1916', *Public Administration* 48 (1970): 32–.

MacPike, Eugene Fairfield, ed. *Correspondence and Papers of Edmond Halley*. Oxford: Oxford University Press, 1932.

Marks, Robert H. 'Learned societies adapt to new publishing realities: A review of the role played by U.S. societies'. In *A Century of Science Publishing: A collection of essays*, edited by E.H. Fredriksson, 91–5. Amsterdam: IOS Press, 2001.

Martin, David C. 'The Royal Society's interest in scientific publications and the dissemination of information', *Aslib Proceedings* 9, no. 5 (1957): 127–41.

Martin, David C. 'Former homes of the Royal Society', *Notes and Records of the Royal Society* 22, no. 1/2 (1967): 12–19.

Martin, David C. 'The retirement of Mr J. C. Graddon, assistant editor to the Society', *Notes and Records of the Royal Society* 27, no. 2 (1973): 325–6.

Maslen, Keith, and John Lancaster, eds. *The Bowyer Ledgers: The printing accounts of William Bowyer father and son*. London: Bibliographical Society, 2017 (first published in 1991).

Mason, Joan. 'Hertha Ayrton (1854–1923) and the admission of women to the Royal Society of London', *Notes and Records of the Royal Society* 45, no. 2 (1991): 201–20.

Mason, Joan. 'The admission of the first women to the Royal Society of London', *Notes and Records of the Royal Society* 46, no. 2 (1992): 279–300.

Mason, Joan. 'The women fellows' jubilee', *Notes and Records of the Royal Society* 49, no. 1 (1995): 125–40.

Massey, Harrie, and Harold Thompson. 'David Christie Martin, 1914–1976', *Biographical Memoirs of Fellows of the Royal Society* 24 (1978): 390–407.

Mayer, Uwe. '"Kein Tummelplatz, Darauff Gelehrte Leut Kugeln Wechseln". Principles and practice of Mencke's editorship of the *Acta Eruditorum* in the light of mathematical controversies', *Archives internationals d'histoires des sciences* 63, nos. 170–1 (2013): 49–59.

McClellan III, James. *Science Reorganized: Scientific societies in the eighteenth century*. New York: Columbia University Press, 1985.

McClellan III, James. 'Specialist control: The publications committee of the Académie Royale des Sciences (Paris), 1700–1793', *Transactions of the American Philosophical Society* 93, no. 3 (2003): 1–134.

McConnell, Anita. *Jesse Ramsden (1735–1800): London's leading scientific instrument maker*. Aldershot: Ashgate, 2007.

McGucken, William. 'The Royal Society and the genesis of the Scientific Advisory Committee to Britain's War Cabinet, 1939–1940', *Notes and Records of the Royal Society* 33, no. 1 (1978): 87–115.

McKitterick, David. *A History of Cambridge University Press. Volume 3: New worlds for learning, 1873–1972*. Cambridge: Cambridge University Press, 2004.

Meadows, A.J. *Science and Controversy: A biography of Sir Norman Lockyer*. London: Macmillan, 1972.

Meadows, A.J. *The Scientific Journal*. London: Aslib, 1979.

Meadows, A.J., ed. *The Development of Science Publishing in Europe*. Amsterdam: Elsevier, 1980.

Meadows, A.J. *Victorian Scientist: The growth of a profession*. London: British Library, 2004.

Miller, David Philip. 'The Royal Society of London, 1800–1835: A study in the cultural politics of scientific organization'. PhD, University of Pennsylvania, 1981.

Miller, David Philip. 'Between hostile camps: Davy's presidency of the Royal Society, 1820–1827', *British Journal for the History of Science* 16 (1983): 1–48.

Miller, David Philip. 'Joseph Banks, empire and "centres of calculation" in late Hanoverian London'. In *Visions of Empire: Voyages, botany, and representations of nature*, edited by David Philip Miller and Peter Hans Reill, 21–37. Cambridge: Cambridge University Press, 1996.

Miller, David Philip. 'The "Hardwicke Circle": The Whig supremacy and its demise in the 18th-century Royal Society', *Notes and Records of the Royal Society* 52, no. 1 (1998): 73–91.

Miller, David Philip. 'The usefulness of natural philosophy: The Royal Society and the culture of practical utility in the later eighteenth century', *British Journal for the History of Science* 32 (1999): 185–201.

Milne, E.A. *Sir James Jeans: A biography*. Cambridge: Cambridge University Press, 1952.

Miranda, Robert N. 'Robert Maxwell: Forty-four years as publisher'. In *A Century of Science Publishing: A collection of essays*, edited by E.H. Fredriksson, 77–89. Amsterdam: IOS Press, 2001.

Morley, Frank. *Self-Help for Learned Journals: Notes compiled for the Nuffield Foundation*. London: The Nuffield Foundation, 1963.

Morrell, Jack B. 'Thomas Thomson: Professor of chemistry and university reformer', *British Journal for the History of Science* 4 (1969): 245–65.

Morrell, Jack, and Arnold Thackray. *Gentlemen of Science: Early years of the British Association for the Advancement of Science*. Oxford: Oxford University Press, 1981.

Motte, Benjamin. *A Reply to the Preface Publish'd by Mr Henry Jones*. London: J. Jones, 1722.

Moxham, Noah. 'Making it official: Experiments in institutionality and the employees and publications of the Royal Society, 1675–1705'. PhD, Queen Mary University of London, 2011.

Moxham, Noah. 'Fit for print: Developing an institutional model of scientific periodical publishing in England, 1665–ca. 1714', *Notes and Records of the Royal Society* 69, no. 3 (2015): 241–60.

Moxham, Noah. 'Authors, editors and newsmongers: Form and genre in the *Philosophical Transactions* under Henry Oldenburg'. In *News Networks in Early Modern Europe*, edited by Joad Raymond and Noah Moxham, 463–92. Leiden: Brill, 2016.

Moxham, Noah. 'Natural Knowledge, Inc.: The Royal Society as a metropolitan corporation', *British Journal for the History of Science* 52, no. 2 (2019): 249–71.

Moxham, Noah. 'The uses of licensing: Publishing strategy and the imprimatur at the early Royal Society'. In *Institutionalisation of Sciences in Early Modern Europe*, edited by Mordechai Feingold and Giulia Giannini, 266–91. Leiden: Brill, 2019.

Moxham, Noah. '"Accoucheur of literature": Joseph Banks and the *Philosophical Transactions*, 1778–1820', *Centaurus* 62, no. 1 (2020): 21–37.

Moxham, Noah, and Aileen Fyfe. 'The Royal Society and the prehistory of peer review, 1665–1965', *The Historical Journal* 61, no. 4 (2018): 863–89.

Muddiman, Dave. 'Red information scientist: The information career of J.D. Bernal', *Journal of Documentation* 59, no. 4 (2003): 387–409.

Murphy, John. 'Championing the case of smaller publishers', *Research Information*, 12 October 2006.

Mussell, James. *Science, Time and Space in the Late Nineteenth-Century Periodical Press*. Aldershot: Ashgate, 2007.

Nash, Andrew, Claire Squires, and Ian Willison, eds. *The Cambridge History of the Book in Britain. Volume 7: The twentieth century and beyond*. Cambridge: Cambridge University Press, 2019.

Neeley, Kathryn A. *Mary Somerville: Science, illumination, and the female mind*. Cambridge: Cambridge University Press, 2001.

Newman, Benjamin. 'Authorising geographical knowledge: The development of peer review in the Journal of the Royal Geographical Society, 1830–c.1880', *Journal of Historical Geography* 64 (2019): 85–97.

Nicholson, William. *The Life of William Nicholson, 1753–1815: A memoir of enlightenment, commerce, politics, arts and science*. Edited by Sue Durrell. London: Peter Owen, 2018.

Noel, Marianne. 'Building the economic value of a journal in chemistry: The case of the Journal of the American Chemical Society (1879–2010)' [in French], *Revue française des sciences de l'information et de la communication*, no. 11 (2017).

Oeuvres Completes de Christiaan Huygens. Edited by Société Hollandaise des sciences. 22 vols. The Hague: Martinus Nijhoff, 1888–1950.

Oudshoorn, Nelly, and Trevor Pinch, eds. *How Users Matter: The co-construction of users and technology*. Cambridge, MA: MIT Press, 2003.

Parkinson, Richard, ed. *The Private Journal and Literary Remains of John Byrom*. 2 vols. Manchester: Chetham Society, 1854.

Paulus, Kurt. 'ALPSP: Bring on the half century!' *Learned Publishing* 25, no. 3 (2012): 167–74.

'P.D.R'. 'Robert William Frederick Harrison (1858–1945)', *Notes and Records of the Royal Society* 4, no. 1 (1946): 113–20.

Peiffer, Jeanne, Maria Conforti, and Patrizia Delpiano, eds. *Les journaux savants dans l'europe des XVIIe et XVIIIe siècles: Communication et construction des savoirs [Scholarly Journals in Early Modern Europe: Communication and the construction of knowledge]* Vol. 63. Archives Internationales d'Histoire des Sciences. Turnhout: Brepols, 2013.

Piazzi Smyth, Charles. 'Solar science at the pleasure of *secret* referees', *Nature* (13 April 1871): 468–9.

Pietsch, Tamson. *Empire of Scholars: Universities, networks and the British academic world, 1850–1939*. Manchester: Manchester University Press, 2013.

Pietsch, Tamson. 'Geographies of selection: Academic appointments in the British academic world, 1850–1939'. In *Mobilities of Knowledge*, edited by Heike Jöns, Peter Meusburger and Michael Heffernan, 157–83. Cham: Springer International Publishing, 2017.

Pontille, David, and Didier Torny. 'From manuscript evaluation to article valuation: The changing technologies of journal peer review', *Human Studies* 38, no. 1 (2015): 57–79.

Pooley, Julian. '"A laborious and truly useful gentleman": Mapping the networks of John Nichols (1745–1826), printer, antiquary and biographer', *Journal for Eighteenth-Century Studies* 38, no. 4 (2015): 497–509.

Potter, Jane. 'The book in wartime'. In *The Cambridge History of the Book in Britain. Volume 7: The twentieth century and beyond*, edited by Andrew Nash, Claire Squires and I.R. Willison, 567–79. Cambridge: Cambridge University Press, 2019.

Potts, Jason, John Hartley, Lucy Montgomery, Cameron Neylon, and Ellie Rennie. 'A Journal is a club: A new economic model for scholarly publishing', *SSRN* (2016). http://dx.doi.org/10.2139/ssrn2763975.

Price, Derek J. de Solla. *Little Science, Big Science*. New York: Columbia University Press, 1963.

Price, Derek J. de Solla. *Science since Babylon*. revised edn. New Haven, CT: Yale University Press, 1975 (first published in 1961).

Priestley, Joseph. *Experiments and Observations Relating to Various Branches of Natural Philosophy, with a Continuation of the Experiments Upon Air. The Second Volume*. London: for J. Johnson, 1781.

Rall, Jack A. 'Nobel Laureate A. V. Hill and the refugee scholars, 1933–1945'. *Advances in Physiology Education* 41, no. 2 (2017): 248–59.

Ranford, Paul. 'Sir George Gabriel Stokes, Bart (1819–1903): His impact on science and scientists'. *Philosophical Transactions of the Royal Society A* 378, no. 2174 (2020).

Raven, James. 'Location, size and succession: The book shops of Paternoster Row before 1800'. In *The London Book Trade*, edited by Robin Myers, Michael Harris and Giles Mandelbrote, 89–126. London: Oak Knoll Press & the British Library, 2003.

Raven, James. *Publishing Business in Eighteenth-Century England*. London: Boydell & Brewer, 2014.
Raymond, Joad. *Pamphlets and Pamphleteering in Early Modern Britain*. Cambridge: Cambridge University Press, 2003.
Raymond, Joad, ed. *The Oxford History of Popular Print Culture*. Oxford: Oxford University Press, 2011.
Raymond, Joad, and Noah Moxham, eds. *News Networks in Early Modern Europe*. Leiden: Brill, 2016.
The Record of the Royal Society of London. 3rd edn. London: The Royal Society, 1912.
Richards, Pamela Spence. '"Aryan librarianship": Academic and research libraries under Hitler', *Journal of Library History* 19 (1984): 231–58.
Richards, Pamela Spence. 'The movement of scientific knowledge from and to Germany under National Socialism', *Minerva* 28, no. 4 (1990): 401–25.
Richardson, Martin. 'Journals'. In *The History of Oxford University Press. Volume IV: 1970 to 2004*, edited by Keith Robbins. Oxford: Oxford University Press, 2017.
Richmond, Marsha L. 'Women in the early history of genetics: William Bateson and the Newnham College Mendelians, 1900–1910', *Isis* 92, no. 1 (2001): 55–90.
Richmond, Marsha L. 'Women as Mendelians and geneticists', *Science & Education* 24, no. 1 (2015): 125–50.
Richter, Derek. 'How many more new journals?' *Nature* 186 (2 April 1960).
Rivington, Charles A. 'Early printers to the Royal Society 1663–1708', *Notes and Records of the Royal Society* 39, no. 1 (1984): 1–27.
Rix, Herbert. *Sermons, Addresses and Essays, with an Appreciation by Philip H. Wicksteed*. London: Williams & Norgate, 1907.
Robbins, Keith, ed. *The History of Oxford University Press. Volume IV: 1970 to 2004*. Series editor Simon Eliot. Oxford: Oxford University Press, 2017.
Robinson, Henry William. 'The administrative staff of the Royal Society 1665–1861', *Notes and Records of the Royal Society* 4, no. 2 (1946): 193–205.
Robinson, H.W., and W. Adams. *The Diary of Robert Hooke 1672–1680*. London: Taylor & Francis, 1935.
Rogers, Pat. *Hacks and Dunces: Pope, Swift and Grub Street*. Abridged edn. London: Methuen, 1980.
Roos, Anna Marie. *Web of Nature: Martin Lister (1639–1712), the first arachnologist*. Leiden: Brill, 2011.
Rose, Edwin D. 'Specimens, slips and systems: Daniel Solander and the classification of nature at the world's first public museum, 1753–1768', *British Journal for the History of Science* 51, no. 2 (2018): 205–37.
Rose, Edwin D. 'From the South Seas to Soho Square: Joseph Banks's library, collection and kingdom of natural history', *Notes and Records: The Royal Society Journal of the History of Science* 73, no. 4 (2019): 499–526.
Rose, Edwin D. 'Publishing nature in the age of revolutions: Joseph Banks, Georg Forster, and the plants of the Pacific', *Historical Journal* 63, no. 5 (2020): 1132–59.
Røstvik, Camilla M., and Aileen Fyfe. 'Ladies, gentlemen, and scientific publication at the Royal Society, 1945–1990', *Open Library of Humanities* 4, no. 1 (2018): 1–40. http://doi.org/10.16995/olh.265.
Rousseau, George S., ed. *The Letters and Papers of Sir John Hill 1714–1775*. New York: AMS Press, 1982.
Rousseau, George S. *The Notorious Sir John Hill: The man destroyed by ambition in the era of celebrity*. Bethlehem, PA: Lehigh University Press, 2012.
Rowland, J.F.B. 'Economic position of some British primary scientific journals', *Journal of Documentation* 38, no. 2 (1982): 94–106.
Rowland, J.F.B. 'The scientist's view of his information system', *Journal of Documentation* 38, no. 1 (1982): 38–42.
Rowlinson, J.S., and Norman H. Robinson. *The Record of the Royal Society: Supplement to the fourth edition, for the years 1940–1989*. London: The Royal Society, 1992.
Royal Society. 'The future of scholarly scientific communication: Conference 2015'. London: Royal Society, 2015.
Rusnock, Andrea. *The Correspondence of James Jurin*. Amsterdam: Rodopi, 1996.

Rusnock, Andrea. 'Correspondence networks and the Royal Society, 1700–1750', *British Journal for the History of Science* 32, no. 2 (1999): 155–69.
Scheiding, Tom. 'Paying for knowledge one page at a time: The author fee in physics in twentieth-century America', *Historical Studies in the Natural Sciences* 39, no. 2 (2009): 219–47.
Schobesberger, Nikolaus, Paul Arblaster, Mario Infelise, André Belo, Noah Moxham, Carmen Espejo, and Joad Raymond. 'European postal networks'. In *News Networks in Early Modern Europe*, edited by Raymond Joad and Noah Moxham, 19–63. Leiden: Brill, 2016.
Schofield, Robert E. *The Enlightened Joseph Priestley: A study of his life and work from 1773 to 1804*. University Park: Pennsylvania State University Press, 2004.
Secord, James A. *Victorian Sensation: The extraordinary publication, reception and secret authorship of vestiges of the natural history of creation*. Chicago, IL: University of Chicago Press, 2000.
Secord, James A. 'How scientific conversation became shop talk'. In *Science in the Marketplace: Nineteenth-century sites and experiences*, edited by Aileen Fyfe and Bernard Lightman, 23–59. Chicago, IL: University of Chicago Press, 2007.
Secord, James A. *Visions of Science: Books and readers at the dawn of the Victorian age*. Oxford: Oxford University Press, 2015.
Shapin, Steven. 'Pump and circumstance: Robert Boyle's literary technology', *Social Studies of Science* 14 (1984): 481–520.
Shapin, Steven. '"O Henry". Review of the correspondence of Henry Oldenburg', *Isis* 78, no. 3 (1987): 417–24.
Shapin, Steven. *A Social History of Truth*. Chicago, IL: University of Chicago Press, 1994.
Shapin, Steven, and Simon Schaffer. *Leviathan and the Air-Pump: Hobbes, Boyle and the experimental life*. Princeton, NJ: Princeton University Press, 1985.
Sharp, Evelyn. *Hertha Ayrton: A memoir*. London: Edward Arnold and Co., 1926.
Sherman, Brad, and Leanne Wiseman. 'Fair copy: Protecting access to scientific information in post-war Britain', *The Modern Law Review* 73, no. 2 (2010): 240–61.
Singh, Rajinder. 'Sir C.V. Raman, Dame Kathleen Lonsdale and their scientific controversy due to the diffuse spots in X-rays photographs, *Indian Journal of History of Science* 37, no. 3 (2002): 267–90.
Singleton, Alan. 'Learned societies and journal publishing', *Journal of Information Science* 3, no. 5 (1981): 211–26.
Sioussat, George L. 'The "Philosophical Transactions" of the Royal Society in the libraries of William Byrd of Westover, Benjamin Franklin, and the American Philosophical Society', *Proceedings of the American Philosophical Society* 93, no. 2 (1949): 99–113.
Slauter, Will. 'The paragraph as information technology. How news traveled in the eighteenth-century Atlantic world', *Annales. Histoire, Sciences Sociales* 67, no. 2 (2012): 253–78.
Slauter, Will. 'Upright piracy: Understanding the lack of copyright for journalism in eighteenth-century Britain', *Book History* 16, no. 1 (2013): 34–61.
Slauter, Will. 'Introduction: Copying and copyright, publishing practice and the law', *Victorian Periodicals Review* 51, no. 4 (2018): 583–96.
Slauter, Will. 'Copyright and the political economy of news in Britain, 1836–1911', *Victorian Periodicals Review* 51, no. 4 (2018): 640–60.
Smits, Robert-Jan, and Rachael Pells. *Plan S for Shock: Science. Shock. Solution. Speed*. London: Ubiquity Press, 2022.
Soll, Jacob. *The Information Master: Jean-Baptiste Colbert's secret state intelligence system*. Ann Arbor: University of Michigan Press, 2009.
Sorrenson, Richard. *Perfect Mechanics: Instrument makers at the Royal Society of London in the eighteenth century*. Boston, MA: Docent Press, 2013.
Stationers' Company of London, and George Edward Briscoe Eyre. *A Transcript of the Registers of the Worshipful Company of Stationers: From 1640–1708 A.D*. 3 vols. London: Privately printed, 1913.
Stroup, Alice. *A Company of Scientists*. Berkeley: University of California Press, 1990.
Suarez, Michael, and Michael Turner, eds. *The Cambridge History of the Book in Britain. Vol. V: 1695–1830*. Cambridge: Cambridge University Press, 2009.
Sussex, Frederick Augustus Duke of. '[Presidential Address 1832]'. *Abstracts of the papers printed in Philosophical Transactions of the Royal Society of London* 3 (1830–1837): 140–55.

Swinnerton-Dyer, Peter. 'A system of electronic journals for the United Kingdom', *Serials* 5, no. 3 (1992): 33–5.
Taylor, John. 'JSTOR: An electronic archive from 1665', *Notes and Records of the Royal Society* 55, no. 1 (2001): 179–81.
Tierney, James. 'Periodicals and the trade, 1695–1780'. In *The Cambridge History of the Book in Britain. Vol. V: 1695–1830*, edited by Michael Suarez and Michael Turner, 479–97. Cambridge: Cambridge University Press, 2009.
Todd, Lord. 'Address at the memorial service for Sir David Christie Martin (1914–1976) at St Columba's Church of Scotland, London on 9 February 1977', *Notes and Records of the Royal Society* 32, no. 1 (1977): 1–3.
Topham, Jonathan R. 'Scientific publishing and the reading of science in early nineteenth-century Britain: An historiographical survey and guide to sources', *Studies in History and Philosophy of Science* 31A (2000): 559–612.
Topham, Jonathan R. 'A textbook revolution'. In *Books and the Sciences in History*, edited by Marina Frasca-Spada and Nicholas Jardine, 317–37. Cambridge: Cambridge University Press, 2000.
Topham, Jonathan R. 'Anthologizing the book of nature: The circulation of knowledge and the origins of the scientific journal in late Georgian Britain'. In *The Circulation of Knowledge between Britain, India, and China*, edited by Bernard Lightman and Gordon McOuat, 119–52. Boston, MA: Brill, 2013.
Topham, Jonathan R. 'The scientific, the literary and the popular: Commerce and the reimagining of the scientific journal in Britain, 1813–1825', *Notes and Records: The Royal Society Journal of the History of Science* 70 (2016): 305–24.
Topham, Jonathan R. 'Redrawing the image of science: Technologies of illustration and the audiences for scientific periodicals in Britain, 1790–1840'. In *Science Periodicals in Nineteenth-Century Britain: Constructing scientific communities*, edited by Gowan Dawson, Bernard Lightman, Sally Shuttleworth and Jonathan R. Topham, 65–102. Chicago, IL: University of Chicago Press, 2020.
Toribio, Pablo. 'The Latin translation of *Philosophical Transactions* (1671–1681)'. In *Translation in Knowledge, Knowledge in Translation*, edited by Rocío G. Sumillera, Jan Surman and Katharina Kühn, 123–44. Amsterdam: John Benjamins, 2020.
Ultee, Maarten. 'Sir Hans Sloane, scientist', *British Library Journal* 14 (1988): 1–21.
Valle, Ellen. ''Reporting the doings of the curious': Authors and editors in the Philosophical Transactions of the Royal Society of London'. In *News Discourse in Early Modern Britain: Selected papers of CHINED 2004*, edited by Nicholas Brownlees, 71–90. Bern: Peter Lang Publishing, 2006.
Vickery, Brian. 'The Royal Society scientific information conference of 1948', *Journal of Documentation* 54, no. 3 (1998): 281–3.
Vittu, Jean-Pierre. 'La formation d'une institution scientifique: le journal des savants de 1665 à 1714 [Premier Article]', *Journal des Savants* (2002): 179–203.
Wardhaugh, Benjamin, ed. 'Charles Hutton and the "dissensions" of 1783–84: Scientific networking and its failures', *Notes and Records of the Royal Society* 71, no. 1 (2017): 41–59.
Wardhaugh, Benjamin. *The Correspondence of Charles Hutton: Mathematical networks in Georgian Britain*. Oxford: Oxford University Press, 2017.
Watts, Iain. '"We want no authors": William Nicholson and the contested role of the scientific journal in Britain, 1797–1813', *British Journal for the History of Science* 47, no. 3 (2014): 397–419.
Watts, Iain. 'Philosophical intelligence: Letters, print and experiment during Napoleon's continental blockade', *Isis* 106, no. 4 (2015): 749–70.
Weedon, Alexis. *Victorian Publishing: The economics of book production for a mass market, 1836–1916*. Aldershot: Ashgate, 2003.
Weinberg, Alvin M. 'Impact of large-scale science on the United States', *Science* 134, no. 3473 (21 July 1961): 161–4.
Werrett, Simon. 'Introduction: Rethinking Joseph Banks', *Notes and Records: The Royal Society Journal of the History of Science* 73, no. 4 (2019): 425–9.
Werskey, Gary. *The Visible College: A collective biography of British scientists and socialists of the 1930s*. London: Allen Lane, 1978.
Westfall, Richard S. *Never at Rest: A biography of Isaac Newton*. Cambridge: Cambridge University Press, 1980.

Whewell, William, and John William Lubbock. '[Report on Airy's] on an inequality of long period in the motions of the Earth and Venus', *Abstracts of the Papers Printed in the Philosophical Transactions of the Royal Society of London* 3 (1830–7): 108–13.

Whiffen, David Hardy, and Donald H. Hey. *The Royal Society of Chemistry: The first 150 years*. London: The Royal Society of Chemistry, 1991.

White, Walter. *The Journals of Walter White, Assistant Secretary to the Royal Society*. London: Chapman and Hall, 1898.

Williams, Harold, ed. *The Correspondence of Jonathan Swift*. 5 vols. Oxford: Clarendon Press, 1963.

Williams, Peter. 'Twenty-five years – and still centre stage: A brief review of the history of ALPSP with a clarification of the aims and objectives of the association', *Learned Publishing* 10, no. 1 (1997): 3–6.

Wills, Hannah. 'Charles Blagden's diary: Information management and British science in the eighteenth century', *Notes and Records: The Royal Society Journal of the History of Science* 73, no. 1 (2019): 61–81.

Wills, Hannah. 'Joseph Banks and Charles Blagden: Cultures of advancement in the scientific worlds of late eighteenth-century London and Paris', *Notes and Records: The Royal Society Journal of the History of Science* 73, no. 4 (2019): 477–97.

Wilson, David B., ed. *The Correspondence between George Gabriel Stokes and Sir William Thomson, Baron Kelvin of Largs*. 2 vols. Cambridge: Cambridge University Press, 1990.

Wilson, Jennifer. 'Dame Kathleen Lonsdale (1903–1971): Her early career in X-ray crystallography', *Interdisciplinary Science Reviews* 40, no. 3 (2015): 265–78.

Winterburn, Emily. 'Learned modesty and the first lady's comet: A commentary on Caroline Herschel (1787): An account of a new comet', *Philosophical Transactions of the Royal Society A:* 373, no. 2039 (2015): 2014-0210.

Wiseman, Leanne, and Brad Sherman. 'Facilitating access to information: Understanding the role of technology in copyright law'. In *The Evolution and Equilibrium of Copyright in the Digital Age*, edited by Susy Frankel and Daniel Gervais, 221–38. Cambridge: Cambridge University Press, 2014.

Wood, Dee. 'ESPERE', *Ariadne: Web Magazine for Information Professionals* 7 (c. 1997), www.ariadne.ac.uk/issue/7/espere/.

Yale, Elizabeth. *Sociable Knowledge: Natural history and the nation in early modern Britain*. Philadelphia: University of Pennsylvania Press, 2016.

Yeo, Richard. *Notebooks, English Virtuosi, and Early Modern Science*. Chicago, IL: University of Chicago Press, 2014.

Zuckerman, Harriet, and Robert K. Merton. 'Patterns of evaluation in science: institutionalisation, structure and functions of the referee system', *Minerva* 9, no. 1 (1971): 66–100.

Index

A

Abridgements of the *Transactions* 133–7, 138–9, 141, 237, 243–4, 263, 267, 331, 568, 599
abstract journals 4, 388, 488
abstracts and summaries of scientific research
 provided by authors 315, 390
 as published items 44, 88, 138, 188, 217, 237, 241–2, 264–5, 267, 269, 278, 288, 304, 309, 311, 314–15, 320, 343, 350, 363–4, 389–90, 392, 442–3, 487, 572, 594
 in RS minute books 106, 108, 165, 187, 263–4, 278–9, 287, 303
Abstracts of Papers Communicated to the Royal Society (1936) 442–3, 458
Abstracts of the Papers Printed in the Philosophical Transactions (1830–32) 265, 267
academic careers 10, 210, 247, 370, 386–7, 430, 499, 594
Academic Press 476
acceptance rate *see* rejection rate
Academia Naturae Curiosorum 115, 132, 221
Acta Eruditorum (Leipzig) 72, 88, 115, 122, 124, 132, 138
Adam, Neil 499–500
advertising
 as source of income for journals 415, 490
 of the *Transactions* in newspapers 172, 229, 263, 335
 see also marketing
Airy, George Biddell 278, 282–3, 284–5, 289, 302, 314, 325
Alden Press 557
Allestry, James, printer 35–9, 54, 65
American Chemical Society 490
American Philosophical Society 175, 183, 220, 222, 234, 457
anatomy 56, 65–6, 77, 106, 285, 289, 296, 340, 526

Andrade, Edward Neville da Costa 508
Annalen der Chemie (Leibig's) 299, 418
Annales de chimie (Paris) 14, 220, 223
Annals of Natural History 332, 334, 374, 392
Annals of Philosophy (Thomson) 217–18, 262
anonymity
 of authors 30, 31, 34, 68, 89–90, 91, 100, 117, 124, 125, 151, 192–3, 195, 211, 258, 443
 of referees 284, 317, 379, 528
anthropology 356, 444
antiquities and antiquarianism 76, 100, 118, 123, 141, 149, 154, 245, 356
appearance of the RS journals *see* format and appearance
Arber, Agnes 395, 460
archive, uses of 206, 309, 441, 461, 564
Arderon, William 154–6
Armstrong, Henry 363, 365–6, 368, 374–80, 387, 391, 393, 396, 411, 438, 506, 526, 534, 604
articles, scientific
 type of material published in the *Proceedings* 311–12, 314–15, 363–4, 388–90, 526, 533, 539
 type of material published in the *Transactions* 21, 24, 28, 33, 45, 55–7, 61, 87, 93, 218, 526, 541, 553–4
 see also length of scientific articles
assistant editor *see* Royal Society staff
assistant secretary *see* Royal Society staff
associate editors 369, 370, 504–5, 516, 525–8, 530, 532, 533, 535, 596
 see also editorial boards
Association of Learned and Professional Society Publishers 514, 569
Aston, Francis 52, 61–71, 77
Astronomical Society (later, Royal) 246, 258, 261, 270, 276, 374, 406, 488

INDEX 627

astronomy 26–8, 77, 93–4, 103, 104, 111, 139, 193, 197, 209, 243, 279, 282, 289, 356
Athenaeum 263, 335
Athenaeum Club 265, 437
Athenian Mercury 120
Atiyah, Michael 558
attribution 28, 31, 89–90, 135, 239, 241, 341, 344–5
Australia 184, 340–1, 347, 408, 422, 438, 464, 525
Austria 372, 418
author charges 490, 580
authors (and authorship) 34, 134–5, 169, 194–5, 209, 303, 365, 438, 464–5, 494, 527
   fellows of the RS as 88, 94, 102, 192, 377, 258, 559
   guidelines for 381, 441, 460, 464, 488
   non-fellows as 88, 97, 194, 303, 365, 378
   overseas 377, 385, 494, 552
   payments to 10, 66, 415
   women as 90, 378, 385, 439, 500–2, 553
   see also anonymity; co-authorship; copyright; manuscripts; prestige; submission process
Auzout, Adrien 27, 30, 46n
Ayrton, Hertha 371–2, 391

B

Babbage, Charles 258–60, 264, 267, 276, 288, 297
back-run of journals 117, 133–4, 177, 243, 267, 338, 482, 486, 490, 567–8, 577
Baddam, Benjamin, editor 136
Baker, Henry 154–5
Baldwin, Charles & Robert, publishers 244
Banks, Joseph 170, 184
   and the circulation of knowledge 217–29, 232, 234–5, 237–47, 335, 336, 340, 342–3, 418, 565
   and the editorial process 195–209, 286, 288, 303, 305, 314, 322, 386
   as president of RS 101, 125, 162, 184–93, 209–11, 257–8, 261, 274, 289, 299, 357, 365, 596
Barrett, Spencer 583–4
Barton, Benjamin Smith 222
Basire, James, and family, engravers 228–9, 272, 418
Bateson, William 378
Bayle, Pierre 115, 138
Beaufort, Francis 265
Benzel, Eric 138
Berlin (Royal) Academy 175, 183, 188, 191, 221, 223
Bernal, John D. 443
Bernoulli, Johann 124
Berthollet, Claude-Louis 239
Berzelius, Jöns Jakob 220, 276
Bevis, John 111
Bianchini, Francesco 103

bias in editorial process 72, 78, 170, 285, 299, 318, 379–80, 386, 441, 505, 596
Bieler, Etienne 385
Big Science 475–6
biochemistry 498, 526
Biographical Memoirs of Fellows of the Royal Society 445, 455, 479, 484–6, 520, 559, 571
biological sciences 296, 371, 395, 498–9, 532, 580
   publishing of 324, 375, 378, 495, 502, 524, 525–6, 539, 551, 553, 572, 574
Biology Letters 569, 572–3, 574, 579, 603
Biot, Jean-Baptiste 223
Birch, Thomas 160, 172
Blackwell 528, 567, 569
Blagden, Charles 184, 445
   and the circulation of knowledge 221–7, 232, 236–9, 243, 418
   and the editorial process 185–6, 192–5, 197–8, 199–204, 206–9, 287, 322
   as secretary to RS 101, 189, 198–9, 209–10
Bone, Quentin 538
book reviews or notices
   in the *Transactions* 26, 33, 68, 88, 90, 93, 106, 110, 135, 138
   in other periodicals 4, 218, 269, 324, 445
book trade 35–7, 39, 41, 133, 138, 234, 263, 458
   RS engagement with 177, 219, 325, 334, 414, 429; *see also* distribution methods
booksellers to the RS 35–9, 72, 73–4, 86–7, 134, 171–4, 176–7, 223, 224, 226, 231, 234–5, 262–3, 335, 341–2, 411–12, 453
Boreham, John 482–4
Bostock, John 278, 283
botany and plants 77, 85–6, 89, 184, 199, 201, 225, 245, 279, 282, 296, 346, 356, 408, 525
Boulton, Matthew 205
Bowles & Gardiner, paper merchant 228, 272
Bowyer, William, printer 135, 226
Boxer, Edward 313
Boycott, Brian 533–4, 538
Boyle, Robert 19–20, 27, 28, 31, 33, 39, 42–4, 53, 57, 72
Bradford, John Rose 371
Bragg, William Henry
   and the editorial process 385, 441, 458, 459–60
   as president 436–7, 443, 455, 457, 460–1, 463, 464, 564, 598, 602
Brande, William 264
Brazil 408, 519, 576
brevity, as a desirable feature of scientific papers 311, 315, 321, 393
British Association for the Advancement of Science 257, 297, 300, 313, 572
British Museum 85, 175, 412, 420, 422
Brooke, Charles 308
Brougham, Henry 194
Brouncker, William (Viscount) 40

Brown, G. Lindor 477, 489, 491
Brown, John 377, 378
Browne, Samuel 89
Brussels, Royal Academy of 189
Bulmer, William, printer 227–9, 235, 260
Burdon-Sanderson, John 315, 318
Burghers, Michael, engraver 67
Busk, George 307, 318
Butterworth Scientific, publisher 476, 482, 486, 489

C

Cable & Wireless 446
Calman, William T. 448–9, 451
Cambrian Printers 557
Cambridge 296–7, 456
 laboratories at 378, 385, 430, 440, 460
 natural philosophers in 62, 63, 94
 University of 175, 283, 302, 365, 370
Cambridge University Press 476, 478, 484, 487, 599
 printing for the RS (after 1954) 495–6, 522–4, 531, 551, 556–7, 562, 599
 publishing for the RS (1937–54) 455–6, 458, 464, 466–7, 476, 478–81, 483–4, 488, 521–4, 531
 quality of typesetting 451–2, 479, 481, 523
 winning the RS contract 410, 411, 437–8, 450–4
Campani, Guiseppe 27, 28
Camper, Peter 193, 238
Canada 308, 338, 408, 422, 438, 464, 583, 595
Cardiff 408, 420, 429
careers in science/academia 5, 8, 12, 210, 247, 297, 426, 436, 438, 592
Caribbean 77, 84, 85–6, 119, 464
Carnegie, Andrew 425
Carpenter, William Benjamin 314, 321, 344, 348
Cartwright, Mary 477, 500
Cassini, Giandomenico 32, 57
Catalogue of Scientific Papers 331–2, 338, 353, 355, 372, 375, 410, 414, 416, 425, 450, 463, 486
Cavendish, Charles 159
Cavendish, Henry 197, 202, 222, 231, 236, 239
Cayley, Arthur 307, 308, 313
Cayley, Dorothy 378
CD-ROM 549, 561, 564, 567
censorship 35, 40, 456, 460–1
Ceylon/Sri Lanka 154, 464
Chadwick, James 461
Chandos, James Brydges, Duke of 135
Chapman, Sydney 387, 459
Chapman, William 205
Charretié, Jean 225
Chemical News 303, 323, 324
Chemical Society (later Royal Society of Chemistry) 246, 299, 334, 336, 365, 379, 406, 417, 514, 515, 517, 518, 567, 569, 570
Chemische Annalen (Crell) 190, 217, 238
Chemische Journal (Crell) 190

Chemist and Druggist 323
chemistry 20, 31, 90, 91, 197, 200, 236, 245–6, 279, 282, 296, 356, 363, 369, 417, 426, 446, 499, 521, 561
China 175, 199, 422, 576
Christie, Samuel Hunter 260, 278
circulation figures for RS journals 271, 334–5, 449–50, 519–20, 556
circulation of knowledge 28–9, 122, 217, 220–6, 242, 487, 577–82, 583
 see also distribution methods for RS journals
Cirillo, Nicola 103
citations 135, 320–1, 346, 454, 464, 487, 488, 516
Clarke, Bryan 538, 539
Clarke, Vivien 535
Clay & Sons, printers 410
Clowes, William, printer 261–2, 353
club, Royal Society as a 10, 155, 160, 186, 288, 553, 561, 601
co-authorship 278, 308–9, 377, 378, 527
co-editorship 52, 61–8, 75–7, 93, 161, 324
Cobbett, William 225
Colbert, Jean-Baptiste 19
Cold War 437, 463, 475, 599
Cole, William 56
collaborative research 61–8, 77
collective editorial decision-making 72, 150, 161–70, 178, 210, 284, 596–7
Collins, John 122
Colwall, Daniel 56
commercial publishers, role in research publishing 13, 133–7, 141, 475–6, 484, 491–3, 514–15
 see also specific firms
commercial viability of scientific journals 13, 137, 415, 475–6, 514, 570
 see also finances of RS journals
Commonwealth 463, 475, 482, 489, 494
communicators and the process of communication (RS editorial process) 109, 162–3, 198, 200, 260
 duties of communicators 196–8, 386, 439–41, 528, 529–30, 385–6
 gatekeeping function 192–8, 303, 314, 439, 444, 494, 499, 527, 531
 involvement of fellows in 306, 308, 378, 385–6, 440, 458, 495, 527
 replaced by direct submission 531–2, 535–7, 542, 552–4
communities of scholars/academics 45, 219–20, 318, 323–4, 343, 364–5, 463, 476, 491–3, 498, 565, 578, 592, 595
 see also scientific community
composition of type *see* typesetting
Comptes rendues (Paris) 263
computers
 in RS offices 523–4, 549, 556, 562–65
 software 524, 550, 551, 554, 562, 567
Cook, James 184
copy-editing 446, 454, 464, 497, 519, 529, 541, 551, 563

copyright 21, 54, 115, 134, 176, 206, 239–42, 332, 345, 486, 487, 493, 578, 581, 585, 599
costs of publishing *see* production costs
Cotes, Roger 94
cover designs 172, 242, 340, 534, 539–40
Cox, Daniel and son, printers 228, 229
Coxe, Daniel 31
Crell, Lorenz 190, 191, 217, 221, 222, 238, 243
criteria for publication 247, 259, 274, 285–7, 289, 311–17, 438–40, 441, 498, 554, 581–2
    see also literary style; novelty; originality; rigour of experiments; significance
Crookes, William 303, 377, 418
Crowfoot, Dorothy *see* Hodgkin, Dorothy Crowfoot
Crowther, James Gerald 443
Cuvier, Georges 276

D

da Costa, Emanual Mendes 152
Dale, Henry 371, 374, 394, 395, 442, 460
Dale, Samuel 86
Dance, Gladys (née Glover) 482–4
Darwin, Charles Galton 394, 444, 459, 498
Darwin, Charles Robert 196, 302, 317, 318, 320, 341
Darwin, Erasmus 194
Darwin, George 405
Darwin, Robert 194, 196
data 97, 127, 140, 238, 283, 396, 438, 441, 506, 555, 564, 565, 567
Davies, Ann (later Horton) 385
Davies, John Griffith 446, 458, 462
Davis & Reymers, booksellers 174
Davis, Charles and Lockyer, booksellers 171–3, 226, 234–5
Davis, Richard, bookseller of Oxford 37–9
Davy, Humphry 257–8
Debraw, John 206
de Bremond, François, translator 138
De Graaf, B. (Nieuwkoop), publishers 486
Deidier, Antoine 96, 99
Denmark 19, 30, 457
Dereham, Thomas 103–4, 137–8, 139
Derham, William 111, 136
Desaguliers, John Theophilus 83, 99, 104–5, 110
de Sallo, Dennis 19, 46n
Diamond, William E. de B. 446, 449–51
digital publishing *see* electronic publishing
diplomacy, scientific 189, 352, 437, 463, 475, 507, 584
distribution methods for RS journals
    diplomatic channels 103–4, 137, 139, 224–5, 238
    gifts and exchanges 103, 174–6, 220, 232–4, 236, 266, 271–2, 333–9, 404, 406–9, 419–23, 422, 449, 482, 519–20, 577
    membership perquisite 171, 173–4, 229–31, 271, 334, 335, 339, 408, 449, 482, 520

sold by the RS 65–6, 262, 335, 411, 481–7, 521, 562, 576
sold via the book trade 172–4, 221, 223–4, 234–6, 262, 271, 333, 334, 341–2, 387, 411, 415, 428–9, 449, 455, 507
    to the Continent 103–4, 116–17, 137–8, 220–36, 336–7
    see also circulation; pricing; sales figures; sales income
Dobbie Typesetting 563
Drach, S.M. 283
Dryander, Jonas 221, 222, 239, 240
Dublin 68, 151, 222, 223
Dulau & Co., publishers 411–12
Dunton, John 120
Durham, Florence 378
Dutch language 19, 126

E

early career researchers 308, 319, 379, 431, 438–9, 442, 462, 482, 528
East Indies 33, 102
École des Mines de France 232
Eddington, Arthur Stanley 394, 447
Edinburgh 116, 189, 194
    see also Royal Society of Edinburgh
Edinburgh Journal of Science (Brewster) 262
Edinburgh Medical Journal 223
editorial boards of journals 495, 503–4, 534, 537–9, 542, 534
editorial community of the RS 77, 196, 284, 300, 303–4, 318, 324, 493, 534
editorial process of the RS 162–7, 185–92, 281–9, 304, 382, 396, 404–6, 445–7, 465, 477–8, 538, 555–6
    criticism of 274, 379–80, 502–5,
    labour and management of 374, 413, 429, 455, 556, 564, 574–5
    see also time taken to publication
editors of RS journals
    appointment of non-secretaries as editors 533–4, 537–9, 583
    secretaries as editors 303, 379, 447, 465, 533
    see also Royal Society officers
editorship
    editorship by committee 72, 76, 274, 303, 503–4
    joint editorship 52–3, 61–8, 72, 75–7, 169–70
    payment for 36, 65, 71, 73–4, 301
    solo editorship 92, 119, 159, 177, 296, 303, 339, 504
Egerton, A.C. 'Jack' 437, 442, 447, 457, 458, 462–4, 503
Elam, Constance 426
electricity 88, 104, 278, 319, 323, 324
electronic (digital) publishing 7, 551, 559, 560–1, 563–8, 568–75, 577–83, 593, 597
electronic mail (e-mail) 530, 549–50
electrotype plates 339, 344, 418
Elliott, Roger 535
Elmsley, Peter, bookseller 234–5
Elsevier 475, 514, 528, 570, 580

Enderby, John 538, 539
engineering 521, 581
English language
  comprehensibility on the Continent 29, 89, 138, 241
  language skills of scientific authors 381, 574
engraving of illustrations *see* Basire, James, and family; illustrations
Entomological Society 246, 336, 406, 415
ephemerality of periodicals 93, 111, 133, 594
epistemic status
  of editorial decisions 105, 110, 166–7, 169, 259, 276–7
  of material published in journals 92, 120, 158, 166–9, 297, 444, 572–3
  of periodicals 43–45, 60, 109, 120, 169, 594, 605
espionage, accusations of 22, 225
etiquette
  among fellows of the Royal Society 41, 43, 68–9, 94–5, 104, 118, 122, 153, 165, 190, 202, 207–8, 278, 441
  among scholars and authors 92–3, 94–5, 103, 122, 152, 188, 192, 319, 320–1
evaluation criteria *see* criteria for publication
evaluation of research 165, 276, 288, 438, 558, 593
  see also criteria for publication; editorial process
Evans, John 390, 403–4, 405–6
Evans, William Geraint 462, 464, 516, 519, 523, 560
Evelyn, John 43
Everett, Joseph D. 308
exchanges of publications *see* distribution methods
executive secretary/director *see* Royal Society staff
experiments 1, 165–6, 169
  fraudulent or 'crank' science 439–40, 532, 553, 555
  referee comments on 285, 312, 318, 381, 393, 441
  reported in *Transactions* 20, 24, 33, 40, 62, 67–8, 89, 92–3, 96, 118, 202–3
  reporting of negative results 312; *see also* significance
  see also replication; Royal Society meetings, experiments at
expertise
  in editorial decision-making 109, 166, 188, 197, 201, 209, 273–6, 277–8, 368, 396
  of publishing staff 262, 341–2, 352, 353, 410, 516, 559–60, 567
  of referees 203, 209, 284, 386, 437, 537
  in science 12, 51, 279, 297, 559
Eyre & Spottiswoode, printers 353, 410

F

Faraday, Michael 274, 277, 278, 285, 286, 312, 314, 319, 336, 461
Farquharson, Marian 371

Fell, Honor 426, 460, 500
Filon, Louis 438–9, 440–1, 466, 499
finances of RS journals 378, 397, 403–9, 413, 455, 465, 467, 478–87
  of electronic publishing and open access 564–5, 577–78, 582
  generation of profit/surplus 35–9, 65–7, 430, 465, 479–80, 486–7, 515, 517, 519, 521, 524, 535, 542–3, 556, 559, 570–1, 576–8, 582–3
  government support 354–5, 357, 414–17, 424–5, 458, 463, 465, 480, 487
  industrial support 426–8, 458, 465, 480, 487
  philanthropic support 354–5, 357, 425–6, 458
  subsidised by editor 75, 86
  subsidised by RS 171, 174, 235, 270–1, 353, 403, 405, 414, 428, 429, 483, 521
  see also Royal Society, finances; sales income from RS publications
first publication, rights to 238–9, 242, 244, 343–5
Fitzgerald, G.F. 380–1
Flamsteed, John 94
Florey, Howard 491
Flower, William Henry 318
Folkes, Martin 84, 99, 100, 102, 106, 125, 149, 153–5, 159–60, 170
Forbes, James David 283
format and appearance of the RS journals 448, 566
  Proceedings 409, 478, 539
  Transactions 35, 117, 132, 141, 219, 227–8, 247, 250n, 260, 271, 342, 345–8, 409, 419, 478–9, 481, 520–1, 554, 539
fossils 103, 118, 157, 193, 340–1, 347, 415, 460
Foster, Michael 297, 301, 302–3, 319, 370, 371, 388, 390, 405
Foulerton, John and Lucy 425, 460
Fowler, Ralph 385, 454
France 28, 57, 109, 130, 189, 197, 208, 227, 232, 258, 516
  see also Académie royale des sciences; *Journal des Sçavans*; Paris
Francis, William 352, 410
  see also Taylor & Francis
Frankland, Edward 308
Franklin, Benjamin 189
French language
  journals/books in 88, 106, 123, 151, 221, 223, 225, 239; *see also Journal des Sçavans*
  published (or not) in the *Transactions*, 29, 105, 183
  scholarly use of 19, 88, 183
  translation of the *Transactions* 138–9, 239
Friedländer and Son, European publishing agents 411–12
funding of science 414, 431, 475, 488–9, 542
  see also research councils

G

Gale, Thomas 73
Galton, Francis 389, 394

Garfield, Eugene 516
Geikie, Archibald 370, 372, 411–14, 416, 424, 430, 431, 466
generalist disciplinary remit
   of mega-journals 581
   of the RS journals 374–6, 441, 506, 515–16, 526–7, 541, 581, 602, 604
   of the *Transactions* 247, 290, 313, 342, 356
   see also specialisation
genetics 389, 526–7, 532, 553, 572–3, 580
Gentleman's Magazine 157, 158, 226
Geographical Society (later Royal) 269, 406
Geological Magazine 323, 374
Geological Society of London 217, 218, 232, 246, 261, 262, 264, 269, 274–5, 318, 323, 339, 379, 406, 415
Geological Survey 302, 370
geology 67, 104, 279, 282, 324, 356
German language
   journals/books 14, 191, 220, 221–2, 418, 528
   material published in the *Transactions* 29, 314, 321
   translation of the *Transactions* 29, 139, 238
Germany and the German lands
   authors from 190–1, 201, 314, 321
   learned institutions in 109, 132, 191, 338
   RS relations with 338, 418–19, 443–4
   see also Berlin (Royal) Academy
Gibelin, Jacques, translator 139, 243, 244
gifts, journals as *see* distribution methods
Gilbert & Rivington, printers 353
Gilbert, Davies 258, 261
Glasgow 42–3, 132, 306, 308, 312, 408, 555
Glover, David 581
Glynn, Ruth 560–1
Goatly, Bruce 516, 519, 522–7, 529, 532–3, 535, 557–8, 560, 562
Gough, Richard 200–1, 205
Gould, William 67
government grant in aid of scientific publishing 414–17, 424–5, 462–3, 480, 487
Graddon, John Courtney 446, 456–7, 462, 477, 479, 493, 504, 560
Graham, George 111
Graham, Thomas 282, 283
Granville, Augustus Bozzi 258–9, 260, 274, 276, 279, 286
Gray, John Edward 320
Gresham College 22, 36, 65, 73
Grew, Nehemiah 43, 52, 55–7, 59–61, 66, 77
Grey, Elizabeth 236–7
Griffiths, Ralph 152, 157–8
growth of science 13, 45, 159, 162, 289–90, 306, 331–2, 350–1, 355, 364, 436, 466, 475–6
growth in numbers of journals 375–6, 475–6, 491, 514
Guardian (Manchester) 443, 573, 579
Günther, Albert 347, 349

H

Halban, Hans von 461
Haldane, J.B.S. 498
Hale, Matthew 33
Halley, Edmond 69, 99, 111, 122, 126, 133, 177, 202, 596
   first editorship 52–3, 70–7, 85, 87, 89, 533
   second editorship 83, 84, 88, 91–5, 100, 103, 110, 124, 127
Hamilton, William 230–1
Hardwicke, Philip Yorke Earl of 159, 164
Hardy, G.H. 385
Hardy, William Bate 371, 385, 395
Harrington, Robert 196–7
Harris, John 118, 121, 160
Harrison & Sons, printers 353–4, 409–12, 417, 429, 448, 450–5, 466, 481
Harrison, Robert 370, 372, 373, 374, 413
Hastings, Warren 230
Hatchett, Charles 240–1
Hauksbee, Francis, the elder 88–9, 91, 110
Hauksbee, Francis, the younger 173
Hazells, printers 410
Heaviside, Oliver 377, 409
Henry, Joseph 332
Herschel, Caroline 194, 308
Herschel, John 258, 261, 276, 277, 283, 284, 317
Herschel, William 193–4, 204–5, 207, 208
Hevelius, Johannes 32, 57, 62, 65, 68–9, 70
hierarchy of journals 260, 287, 311, 313–14, 438, 564, 596
HighWire Press, online hosting service 567
Hill, Archibald Vivian 387, 395, 437, 439–40, 443–4, 451–5, 456, 457, 458, 460, 462, 463, 498
Hill, John 151–64, 170, 171, 274
Hirst, Thomas Archer 315
Hodgkin, Dorothy Crowfoot 500
Hofmann, Wilhelm 297, 308, 322
Hogben, Laurence 498
Hoggan, Frances Elizabeth (née Morgan) 308–9
Holmes, Robert 27, 28
Home, Everard 205, 223
Hooke, Robert 28, 30, 32, 36, 40, 42, 43, 52, 53, 55, 56–61, 65, 68–9, 72, 77
Hooker, Joseph 341, 351, 355, 425
Hopkins, Frederick Gowland 445
Horner, Leonard 317, 320
Hornsby, Thomas 204
Horticultural Society 269
Horton, Ann *see* Davies, Ann
Howard, Henry, Duke of Norfolk 41
Hulme, Nathaniel 203, 207
Hungary 27, 372, 418
Hunt, Henry 65, 73
Hunter, John 193, 201
Hurst, Phil 561–2, 563, 564, 568–9, 570, 579
Hutt, Charles 489
Hutton, Charles 244, 263, 267
Huxley, Andrew 542
Huxley, Julian 443, 457, 458,
Huxley, Thomas Henry 289, 297, 302–3, 306, 308, 318, 324, 331, 347, 348

Huygens, Christiaan 20, 27, 28, 32, 41–2
hypotheses 90, 92, 95–6, 97, 104, 203

I

ICI (Imperial Chemical Industries; formerly Brunner Mond & Co.) 355, 426–8, 430, 465, 479–80, 487
illustrations 66–7, 106, 340, 344, 346–8, 448, 523
    cost of 36, 67, 272, 347, 378, 405
    engraving 67, 177, 228–9, 231, 272, 344, 347, 418
    lithography 260, 272, 293n, 303, 340, 344, 347–8, 403, 479, 486, 523
    wood-engraving 8, 344, 347, 348, 523
impact factor (journal impact factor) 516, 533
imprimatur *see* licensing privilege
index, to the *Transactions* 134, 139, 345
    see also searching the scientific literature
India 89, 102, 199, 201, 230, 232, 408, 438, 441–2, 464, 563, 576
industry and science 12, 426–8, 430, 437, 520
inflation 417, 419, 483, 515, 517, 522, 558
Ingen-Housz, Jan 201–2, 207, 208
ink 9, 227, 448
Innys, John 176–7
Innys, William, bookseller 134, 176–7
Institute of Physics 514, 559, 569, 570
    see also Physical Society
Institute of Scientific Information (Philadelphia) 516
interdisciplinarity 13, 366, 369, 375, 396, 462, 498, 506, 525, 541, 551, 554, 573–4, 604
Interface 572, 574, 579, 580, 603–4
Interface Focus 574, 604
International Association of Scientific, Technical, and Medical Publishers *see* STM Group
International Research Council 419
internationalism, scientific 418, 463, 464, 494, 516
internet and web 3, 563, 564–6, 568, 573, 580, 583, 600
inventions 33, 104–5, 167, 258, 297
Iraq 519
Ireland 62, 63, 68, 151, 222, 223, 308, 411, 422, 464
    see also Royal Irish Academy
Italian language 19, 29, 139–40, 183
Italy 27–8, 29, 103–4, 105, 111, 115, 137–8, 139–40, 223, 231, 516

J

Japan 385, 408, 478, 519, 525, 528, 576
Jeans, James 371, 374, 394, 395, 504
Jenner, Edward 205
Johnson Reprints Co. Ltd 486
Jones, Henry 136
Journal of the Chemical Society 299
journalists and editors, RS attitude to 152, 242, 343, 345, 573

Journal of Natural Philosophy (Nicholson) 217, 236, 240, 251n, 262
Journal of Physiology 390, 527
Journal des Sçavans (Paris, later *Journal des Savants*) 4, 14, 19–21, 26, 28–30, 41, 72, 88, 115, 138, 221
JournalsOnline 567
journals in general *see* print and periodical culture (beyond the RS)
JSTOR (Journal Storage) 567–8, 571, 576
Jurin, James 83, 88, 89, 91, 94, 95, 103, 124, 132, 134, 138, 140
    editorial practice 84, 96–100, 111, 126–32, 139, 161

K

Kamel, Georg Joseph 89
Kegan Paul, publisher 342, 411
Keill, John 122, 124
Kelvin, Lord *see* Thomson, William
Kempe, Alfred 368, 405, 411, 415
Kenya 519
Kikuchi, Taiji 385
King, William 119, 120
Kirwan, Richard 223, 224
knowledge claims 30, 60, 92, 104, 121, 134–5, 193–4, 276
    see also criteria for publication; epistemic status
Kowarski, Lew 461
Kraus Reprint Corporation 486
Krogh, Schack A.S. 457

L

Lagrange, Joseph-Louis 278
Lalande, Jérôme 197
Lamb, Jean 462
Lankester, E. Ray 340, 347, 390, 406
Laplace, Pierre-Simon 278
Larmor, Joseph 370–1, 377, 379, 380–1, 393, 395, 408, 412–14, 430
Latham, John 195
Latimer Trend Ltd, printers 557
Latin language
    published in the *Transactions* 29, 32, 89, 94, 99, 105, 183
    scholarly use of 19, 88, 89, 105, 183
    translation of the *Transactions* 20, 29–30, 32, 45, 138
Lavoisier, Antoine 236
learned societies
    as publishers 10, 379, 415, 476, 488, 507, 514
    cooperation among 487–8, 514, 565
    finances of 355, 415, 424–5, 463, 488–91, 577
    libraries of 220, 232–4
    meetings of, reported in press 218–19, 226, 237, 263–5, 324–5, 343
    mission to circulate knowledge 217, 476, 493, 514–15, 603
    specialist remit 244–7, 283, 300, 313, 375–6
    see also communities of scholars; voluntary labour

Leeuwenhoek, Antoni van 56, 57, 83, 87, 89, 94, 126, 133, 138, 175, 445
Leibniz, Gottfried Wilhelm 95–6, 121–4
length of scientific articles 345–6, 393–4
  in *Proceedings* 269, 314–15, 325, 345, 348, 364, 388–90, 392
  in *Transactions* 31, 100, 314, 328n, 346, 350–1, 392
  see also brevity
Leopoldina 175
Leske, Nathaniel, translator 139
Letters to the editor 310, 324, 497
  see also preliminary notes
Lever, William 425
Lewis, Walter 452
libraries
  public libraries, as gift recipients 137, 338, 340, 345, 408, 420, 429
  of universities, as customers 429, 466–7, 482, 484, 517, 520–1, 567
  of universities, as gift recipients 175, 336, 406–8, 415, 420, 422, 482, 599
  as ways to access journals 234, 333, 336, 339, 420, 463, 483, 506, 568
licensing
  Licensing Act and its lapse 34, 115
  RS licensing privilege and imprimatur 34–5, 39, 65, 68–9, 134, 135, 140, 153, 167
  University of Oxford privilege 37, 65, 68
Lighthill, James 477, 502–6, 526, 533
Lindemann, Frederick, 1st Viscount Cherwell 387
Linnean Society 217, 218, 232, 245, 261, 318, 335, 371, 406, 412, 415, 446, 488
Lisbon Royal Academy of Sciences 232
Lister, Joseph 368
Lister, Martin 62, 67, 76, 133
Lister, Susanna 67
literary style, as criterion for publication 286–7, 289, 318–19, 381, 437
lithography *see* illustrations
Lloyd, Humphrey 314
Lockyer, Norman 303, 320, 324, 344, 377, 390, 596
London Gazette 82, 119, 353
London Mathematical Society 406
Longman, publisher 269, 412
Lonsdale, Kathleen (née Yardley) 385, 459–60, 461, 477, 500–1
Los Alamos laboratory 565
Louis XIV, King of France 19, 33, 41
Lowthorp, John 135–9, 243–4, 263, 267
Lubbock, John William 270–1, 276, 278, 283
Lusty, Charles 561, 563, 568, 574, 582
Lusty, Robert 489–90, 514
Lyons, Henry 444, 447–8

M

Macclesfield, George Parker, Earl of 159–60, 163–4, 175, 177
Mace, Georgina 583–4
Machin, John 99, 100, 102, 105, 110, 124
Macmillan, publisher 324, 332, 415
magnetism 88, 278, 285, 325
Manchester 167–9, 183, 408, 439
Mann, Theodore 189
Manuscript Central (Thomson Reuters) 562
manuscripts (before publication)
  ownership of the original copy 206, 322, 382
  preparation of 381, 441, 446, 454, 460, 464, 488
  typed 381–2, 464, 524, 530
  see also authors, guidelines for; priority claims; revisions; submissions
Mariotte, Edmé 63
marketing (advertising and publicity) 28, 130, 172, 229, 262–3, 269, 335, 341–2, 411–12, 429, 444, 449, 453, 455–6, 466–7, 484, 489–90, 521, 527, 529, 537, 543, 559, 572
  international 13, 466, 475, 478, 486, 519, 528, 595
mark-up languages 541, 566, 575
Marryat, Dorothea 378
Marsham, Robert 238
Martin, David Christie 462, 464, 476–8, 479–83, 488–93, 495, 497, 499, 505, 507, 515, 516, 517, 523, 542, 553, 582, 584, 597, 601
Martyn, John, printer 35–9, 54, 60, 65, 68, 136
Maskelyne, Nevil 197, 204, 209
Massey, Harrie 495
mathematical journals 82, 296, 297, 300, 308, 313, 526
mathematical papers
  evaluation and correction of 99, 104, 110–11, 205, 209, 278, 285, 289, 306, 313, 346, 498–9
  typesetting of 272, 409–11, 451, 452, 454, 495, 523, 524, 557, 563, 566
mathematics 70, 121–5, 258, 356, 541, 581
  in the *Transactions* 72, 76, 83, 87, 93–4, 97, 133, 135, 139, 243, 426, 526
  at the RS 198–9, 237, 279, 282, 394
Matthews, Paul 533–4
Maty, Matthew 152
Maxton, Julie 584
Maxwell, James Clerk 309, 312, 318, 340, 379, 409
Maxwell, Robert 475, 504
May, Robert 573
McCance, Robert 494–5, 502
McLaren, Anne 561
mechanics 28, 32, 33, 41–2, 57, 71, 93, 135, 139
medical journals 115–16, 136, 140, 221, 223, 297, 313, 569, 570
medicine 84, 94
  in the *Transactions* 72, 76, 87, 89, 90, 91–2, 93, 96–7, 313, 356, 444, 555
  see also smallpox
meetings *see* learned societies; Royal Society meetings
mega-journals 581, 597
Menter, James 516–17

Messel, Rudolph 425, 426, 460
MetaPress, online hosting services 567
meteorology 84, 97, 99, 138, 140, 279, 521
metrics 260
    see also citations; performance indicators
Micheli, Pier Antonio 105
microfilm, microfiche, etc. 486, 549, 567
Microscopical Society, Royal 371
microscopy 57, 356
Miles, Ashley 502–3, 506
Miller, William A. 350
Miller, William Hallowes 302, 308, 312, 321
Mintern Brothers, lithographic printers 348
Miscellanea Curiosa 4, 115
mission for scholarship 217, 220, 290, 332–3, 429–30, 491
molluscs 86, 289, 320, 340, 347, 445
Molyneux family 83
Mond, Alfred 426–7
Mond, Ludwig 355, 425, 426
    see also ICI
money *see* finances of RS journals; Royal Society, finances
Monro, Alexander 116
Moray, Robert 20, 36, 37, 50n, 56
Morley, Frank Vigor 489–90, 493, 495, 497
Mortimer, Cromwell 83, 84, 101, 171
    death of (and its consequences) 150, 157, 159–61
    as editor of the *Transactions* 103, 104–7, 110, 139, 149, 153, 170, 173, 209
Mott, Nevill 495
Motte, Benjamin, bookseller 136
multi-disciplinary *see* generalist
Musgrave, William 52, 67, 68
Mynde, James, engraver 229

N

National Academy of Sciences (USA) 497, 516, 526, 529, 552
natural history 62, 101, 110, 123, 158, 210, 356, 412
    in the *Transactions* 26, 72, 76, 77, 84, 87, 88, 89, 90, 91, 93, 94, 97, 154–5, 346
Natural History Review 324, 356, 374
natural philosophy 3, 12, 19, 44–5, 53, 63, 91, 95–6, 101, 103–4, 120, 124, 202, 593, 598
    in the *Transactions* 24, 29, 38, 65, 71, 77, 88, 93, 110
natural philosophical journals 14, 115, 190, 217, 236, 240, 262
Nature 4, 6, 14, 310, 324–5, 335, 344, 374, 375, 442–3
    business model 332, 345, 415, 476
    editorial process 303, 530
    as model for RS journals 324, 497, 506, 539, 572, 575, 594
    rapid publication 324, 389, 393, 395, 442, 497
navigating the scientific literature *see* searching the scientific literature

navy, admiralty and ships 225, 265, 332, 336, 344, 348
Needham, Dorothy Moyle 395
Needs of Research in Fundamental Science after the War (1945) 462
Netherlands, the 29, 30, 32, 138, 222
Nettleton, John 127–30
New Scientist 506, 572, 594
New Zealand 184, 408, 438, 464
news, scientific 226, 236, 324–5
    conveyed in correspondence 63, 198, 222, 238
    the *Proceedings* as 265, 343
    published in other journals 19–21, 218, 269, 324–5, 344, 374, 445, 476
    published in the *Transactions* 28, 34, 44, 53, 57, 87, 111, 194, 594
    transnational circulation of 21, 28, 29, 190, 220–1, 226
newspapers 35, 77, 116, 140, 263, 323, 372
    editorial practices 90, 190, 222, 225, 345
    production and circulation 7, 35–6, 125–6, 130–1, 260, 523
    reporting of science 237, 573, 579
    see also advertising of the *Transactions*
news-sheet, the *Transactions* as 4, 11, 21, 24, 65, 218
Newton, Isaac 3, 34, 40, 62, 63, 74, 106, 445
    as author 71, 94, 103, 105, 122
    calculus dispute 95, 121–5
    as president of RS 83, 88, 91, 97, 99, 100, 101, 105, 111, 116, 124–5, 140, 153
Nichols, John, bookseller 226–7, 228
Nicholson, John W. 387
Nicholson, William 190, 217, 218, 236, 239–43, 244, 262
Nicol, George, bookseller 235–6, 261
Nicol, William, printer 235, 261
non-commercial circulation *see* distribution methods
Northampton, Spencer Compton, Marquess of 258, 288
Notes and Records of the Royal Society 444–5, 455, 458, 479, 480, 484, 520, 559, 571
Notes on the Preparation of Papers 460, 464, 488
novelty, as criterion for publication 133, 286, 313, 314, 322
Nuffield Foundation 488–90, 507

O

Obituary Notices of Fellows of the Royal Society see *Biographical Memoirs of the Royal Society*
Observations sur la physique see Rozier, François and his *Observations sur la physique*
observatories (and their libraries) 336, 339, 408
Occasional Notices see *Notes and Records of the Royal Society*
offprints *see* separate copies
Oldenburg, Dora Katherina 39, 53–4

Oldenburg, Henry
  as editor of the Transactions 19–51, 56, 57, 70, 138, 154, 157, 183, 184
  editorial practice 26, 28–34, 40–5, 61, 62, 87–8, 89, 99, 133, 190, 218, 343, 445, 593, 594
  financial arrangements 21, 36–40, 59, 65, 67, 71, 75, 117, 125
  'invention' of the *Transactions* 2–4, 9, 10, 11, 596, 597–8, 604
  personal life 19–20, 22, 46, 47n, 51–2, 54
  as secretary to the RS 22, 35, 53–5, 68, 85, 122, 171
open access
  in general 416, 552, 570, 600
  at the RS 551, 568, 570, 577–83, 584, 601, 603
  see also circulation of scientific knowledge; distribution methods
Open Access Scholarly Publishers Association 580
Open Biology 580–2, 584
optics 26, 40, 103, 135, 189, 194, 315, 317
originality, as a criterion for publication 105, 132–3, 158, 189–92, 211, 285–6, 314
Owen, Richard 283, 285, 287, 289, 312, 317, 318, 340–1, 347
ownership of scientific journals 491–3
  see also Transactions, ownership of
Oxford 464, 557
  natural philosophers in 36, 37, 52, 61–68, 76, 77, 87
  printing of the *Transactions* in 37–8, 65–7
  University of 37, 65, 68, 175, 365
Oxford University Press 557, 560–1, 564

P

page charges *see* author charges
page limits (or not) for articles 348, 388, 390, 391, 392, 401n, 539
Paget, James 297
Pakistan 464
palaeontology 289, 526
  see also fossils
paper for RS journals
  cost 227, 346, 417, 478, 522, 539
  paper merchants 172, 228, 272, 448, 522
  quality and material 227, 346, 448, 522–3, 524
  size 26, 35, 129, 219, 228, 267, 270, 390, 409, 539
  supply of (including rationing) 7, 228, 272, 323, 346, 403, 417, 443
Paris 137, 139, 165, 221–5, 239, 241
  Académie (Royale) des Sciences 41, 63, 82, 87, 96, 132, 151, 165–7, 175, 183, 188, 191, 210, 223, 263, 276, 277, 285
  Institut National 223–4, 225, 232
  see also France; *Journal des Sçavans*
Parsons, William *see* Rosse, Earl of
Partridge, Linda 583
pathology 375, 439

patronage 9, 119, 123, 135, 149, 153, 174, 199, 220, 258, 599
  royal patrons 82, 109, 132, 259, 422, the RS as patron 176, 235, 332, 408
Payne & MacKinlay (booksellers) 224
Peacock, George 283, 289
Pears, Charles 238
Pearson, Karl 377, 389–90, 405, 439
peer review 1, 4, 7, 10, 34–5, 40, 260, 593, 595–7, 601, 605
  management of 562, 569, 574, 575
  objective (or technical) peer review 315, 581–2, 583, 597
  in public debates about science 569, 572–3, 579, 582
  use in late twentieth century 530, 552, 553, 573
  see also refereeing
Peierls, Rudolf 495
Pellew, Caroline 378
Pepys, William Haseldine 283
performance indicators (for journal publishing) 310, 333, 345, 464, 493–4, 495–6, 516, 542–3, 545n, 553, 574–5, 577
Pergamon Press 475, 476, 484, 489, 504, 514, 515, 520, 528, 570
  see also Robert Maxwell
periodicals, in general *see* print culture
periodicity and pattern of issue
  of other periodicals 72, 120, 125, 217–18, 236, 240, 262–3, 324, 343, 374, 442, 497, 574, 581
  of the *Proceedings* 265–6, 311, 325, 334, 374, 387–8, 390, 518, 531, 553, 568
  of the *Transactions* 20, 35, 53, 56, 73, 87, 101, 108, 125–6, 130, 139, 141, 149, 161, 218, 237, 239–40, 342, 529, 539–41, 554, 568, 574, 593
Perry, John 371–2
Petiver, James 76, 88, 89
Petty, William 40
pharmacy and pharmacology 323, 375, 527, 444, 561, 569
philanthropic circulation *see* distribution methods
philanthropic support for scientific publishing *see* finances of RS journals
Philippines, the 89, 408
Phillips, John 283
Philosophical Collections 52, 59–60, 61
Philosophical Magazine 14, 243, 374, 380, 393, 394, 594
  editorial 263, 264–5, 387
  printing, pricing, etc. 236, 261–2, 332, 334, 345–6, 410, 415, 455
  see also Taylor & Francis
Philosophical Transactions see *Transactions*
photocopying 408, 487, 549, 565, 581
photography 285, 441, 479, 486, 523, 567
Physical Society 374, 377, 410, 488, 514
  see also Institute of Physics

physics
  in general 200, 307, 356, 394–5, 409, 565
  at the RS 78, 279, 282, 289, 369, 377, 417
  in the RS journals 20, 87, 88–9, 90–1,
      296, 318, 363, 429, 441, 462, 521, 525,
      527, 541
  see also natural philosophy; mechanics
physiology 31, 62, 245–6, 279, 282, 375, 388,
      390, 391, 395, 498, 526, 527
Pickering, Roger 105
Pitcairn, David 231
Pitt-Rivers, Rosalind 500
plagiarism 41, 42
  see also originality
Planta, Joseph 188, 200, 206, 208
PLOS ONE 581
Plot, Robert 52, 61–7, 71, 73, 77
PNAS see *Proceedings of the National Academy
      of Sciences* (Washington)
Poleni, Giovanni 105, 111
politeness *see* etiquette among scholars
Porter, George 535
Porter, Helen 500
Portugal 208, 232
postal service
  for correspondence 128–30, 206–7, 222,
      279, 530, 550
  for manuscripts, books and journals 222,
      283, 321, 339, 408, 454
Powell, Baden 283
Prain, David 419
preliminary notes
  in *Proceedings of the RS* 314, 389–90,
      401n, 497
  in *Nature* 389, 442
  see also letters to the editor
preprints 550, 565
  see also separate copies
presidents of the RS *see* Royal Society officers
prestige
  arising from publication 12, 119, 170, 195–6,
      210, 246–7, 259, 260, 287–8, 297–9,
      325, 378–9, 436, 438, 442
prices and subscription rates
  of the *Proceedings* 334, 350, 412, 429,
      483, 517
  of the *Transactions* 37, 66, 131–2, 133,
      143n, 235, 271, 335, 412, 483, 517
pricing of RS journals 413, 429, 449, 464, 483,
      484, 517
Priestley, Joseph 195, 201–2, 231, 236
prime ministers 225, 424, 542
Pringle, John 188, 198
print and periodical culture (beyond the RS)
  nineteenth and twentieth centuries 247,
      263, 269, 297–9, 303, 323–4, 332–3,
      343–4, 375–6, 393, 410, 415, 445, 494,
      497, 528
  seventeenth and eighteenth centuries 35–6,
      115, 120–1, 126, 129–31, 140–1, 190,
      217–20
printers (and publishers) to the RS

agreements/contracts 172–4, 234–5, 262,
      480–1
  change of 220, 226–8, 234, 260–3, 272,
      352–4, 409, 410–11, 437, 445, 449–55,
      481, 557
  quality of service provided 226–7, 261,
      409–11, 448, 450–1, 454, 481, 523
  see also names of individuals and firms
printing for the RS journals
  costs of 36, 262, 346, 353–4, 433n, 448,
      464, 478, 561–2
  delays 458, 464, 495–6, 531
  schedule 128, 205–6, 238, 455, 531
printing technology
  digital printing 575
  offset lithographic printing 8, 479, 486,
      523–4, 556
  steam printing 260–2, 323, 346, 354, 448
print runs
  of the *Proceedings* 265, 272, 334, 406,
      413–14, 432n, 478, 575
  of the *Transactions* 38, 131, 227, 228, 235,
      250n, 262, 272, 335, 358, 406, 413–14,
      432n, 575
priority claims 41–2, 45, 121–5, 201–2, 320,
      322–3, 373, 382, 389, 442, 456, 461
  see also manuscripts; revisions; speed of
      publication
Proceedings of the National Academy of Sciences
      (Washington) 497, 516, 526, 529, 552
Proceedings of the Royal Society
  creation of, and later reforms to 264–9,
      311, 364, 387–95, 409
  criticisms of 267, 324, 325, 364, 366, 494–5
  editorial processes 287, 296, 301, 304,
      310–11, 325, 380, 382, 385–6, 395,
      396
  length of volumes 457, 568
  Proceedings A vs *B* 364, 391, 394–5, 532
  relationship to the *Transactions* 311–12,
      363–4, 387–9, 391, 396, 537
  *see also Abstracts of the Papers Printed in
      the Philosophical Transactions*; articles,
      length of; back-run; hierarchy of
      journals; periodicity; prices; print run;
      sales figures; speed
Proceedings-type journals, as a genre 263–4,
      269
production costs
  of the *Proceedings* 265, 350, 428
  of the RS journals 270–3, 352–3, 404,
      419–22, 451, 463, 522–4, 541, 561–2,
      575, 580
  of the *Transactions* 171–4, 272–3, 289–90,
      428
professionalisation
  of academia 8, 210, 430, 438, 594
  of publishing 262, 446, 466, 552, 558–60,
      566
  of science 2, 12, 257, 355, 357, 592, 594
profit/loss of RS journals *see* finances of RS
      journals

INDEX   637

proofs and proof-reading 97, 106, 108, 303, 339
   checking of proofs 206, 208–9, 348, 352, 409, 529
   in mathematics 99, 105, 209, 410
   practical logistics 200, 209, 227, 374, 455, 497, 529, 551, 575
propriety *see* etiquette
provincial societies 232, 245, 338, 406
publication, definition of the moment of 241, 342–3, 565
Purdon, Chris 535, 539, 558

Q

quality, intellectual
   associated with refereeing/peer review 286–7, 319–20, 425, 442, 491–2, 497–500, 505, 553, 555–6, 565, 572–3, 574, 595–7
   of the early *Transactions* 118, 152–3, 194, 286–7, 297, 319–20
   of RS journals 438, 442, 444, 493, 499, 529, 553–4, 581, 584
   of scientific journals 424–5, 438, 476, 491, 497, 529
quality of print production
   of the early *Transactions* 66, 171, 219, 220, 226–7, 261, 332–3
   of RS journals 272, 346, 356, 430, 448, 452, 464, 466, 478–9, 481, 522, 529

R

Raman, Chandrasekhara Venkata 441–2
Ramsay, William 344
Ramsden, Jesse 208
ranking *see* hierarchy of journals
Ray, John 67, 76, 85, 86
Rayleigh, 3rd Baron (John William Strutt) 309, 317, 371, 377, 379, 380, 381, 388, 390, 410, 414, 415
Rayleigh, 4th Baron (Robert Strutt) 377
readers and readership
   communities of readers 8, 45, 60, 65, 116, 183, 193, 219–20, 323–4, 345, 420, 438, 504, 592
   examples of individual readers 128, 236
   of the *Proceedings* 265, 345, 483
   of RS journals 415, 437, 438, 441, 442, 483–4
   of the *Transactions* 29, 59, 65, 66, 89, 94, 105, 132, 135, 217, 333, 341, 345, 372, 429
   see also libraries
Rees, Martin 579
refereeing at the RS
   confidentiality of 104, 205, 279, 284, 300, 317, 528
   criticisms/objections to 284, 289, 310, 379–80, 386
   evaluative practices of 285, 305–23, 497–502, 554–6, 573
   financial implications of 287–8, 289–90
   practical logistics of 279–80, 283, 317–18, 374, 380, 497, 530–1, 554–5
   time taken 310, 380–2, 530

   use after 1830 273–81, 287, 306, 369, 382–4, 386, 396
   use (occasional) before 1830 99, 104–5, 110, 166, 188, 196–7, 202–5, 274
   workload of 278–9, 377, 380, 382, 387, 499–500
   see also criteria for publication; peer review; quality, intellectual (associated with refereeing); referee reports; referees; rejection rate; revisions
refereeing beyond the RS 274–5, 276, 318, 530
referee reports
   format of 203, 276–9, 281, 284, 287, 318
   standardisation of 382–4, 391, 404–5, 530–1
referees
   busy 283, 387, 499–500
   gender 308, 372, 459–60, 505
   geographical origins 283, 306, 504, 530
   most active 282, 307, 380, 387, 458–9, 500
   non-fellows as 460, 504, 534, 542, 553
   number of 304, 380, 386, 396, 441, 458
   pool of 277, 279, 282–3, 304–5, 386, 504, 542, 553, 584
rejection of papers
   process for 310, 369, 442, 503
   reasons for 189–90, 440, 441, 581
   see also withdrawal of papers by their authors
rejection rate (and acceptance rate) 188, 214n, 288, 309, 314, 439, 440, 468n, 553
   see also selectivity
replication of experiments/observations 24, 44, 67, 165, 193, 285, 312, 380, 573
repository of knowledge, the *Transactions* as 8, 92, 108, 110, 133, 140
reprinting and excerpting
   of illustrations 340–1, 344–5
   role in the circulation of knowledge 190, 217, 218, 221–2, 226, 237–44, 332, 343–5
   of RS material in other publications 117, 138, 219, 226, 237–9, 240–3, 343–4, 345, 585, 599
   of RS publications 265, 454, 462, 486, 567
   in the *Transactions* 28, 30, 41, 45, 87–8
reputation, academic or scholarly (of individuals) *see* prestige
reputation of journals
   the *Transactions* 365, 378–9, 462
   RS journals 438, 515, 533, 553
   see also hierarchy of journals; *Proceedings*, in relationship to *Transactions*
research council 419, 424, 571, 578, 579
retractions 122, 440
reuse of material *see* reprinting
revisions 104, 165–6, 206, 322–3, 381, 382
   by authors 128–9, 206, 322–3, 381
   by editorial intervention 106, 155, 202, 206–7, 210, 322
   suggested by referees 203–4, 284–5, 300, 317–23, 381, 531
Reymers, Charles 172

rigour of experiments, as criterion for publication 312, 318
Rix, Herbert 370
Roberts, Sydney 452–5
Robinson, Robert 462
Robinson, Tancred 68, 69, 90
Rockefeller Foundation 425, 457, 458
Roget, Peter Mark 260, 286
Roscoe, Henry 308
Rosse, Earl of 258
Rowlinson, John 559
Roy, William 208, 231
Royal Institution 232, 257, 264, 274, 385
Royal Irish Academy 183, 232, 334, 336
Royal Mail *see* postal service
Royal Navy *see* navy
Royal Society
    biological sciences vs physical sciences 356, 364–5, 391, 458, 494, 519, 532
    criticisms of 119–20, 150, 151–8, 184, 199, 258–9, 363, 365–6, 443, 573, 579
    finances 75, 102, 109, 149, 174, 184, 259, 269-71, 351–3, 355, 403, 416, 517, 558
    see also finances of RS journals
    library and archive 5–6, 55–6, 77, 107, 178, 189, 232, 271, 309–10, 314, 336–7, 380, 461, 564
    medals and prizes 191, 204–5, 259, 279, 321, 369
    reform of 157, 158, 159–70, 210–11, 257–60, 264, 272–80, 288, 364, 366–74, 448, 534–5, 558
    relationship to government 303, 353, 354, 356, 371, 414–17, 424–5, 427–8, 430–1, 436, 458, 460–2, 462–3, 480, 495, 515, 542, 558, 569, 571–2, 573, 574, 599, 601
    relationship to other societies 303, 376, 416, 457, 479, 481, 514, 569
    reputation 150, 152, 159, 160–1, 170, 178, 193–4, 220, 258, 363, 374, 420–1, 573
    scholarly mission 352, 357, 436–7, 580, 584
Royal Society committees
    Committee of Papers 106–8, 161–70, 185–92, 198, 202, 205, 209, 269, 274, 276–7, 279–80, 281, 284, 287, 289, 300, 301, 303–5, 310, 322, 346, 350, 366, 368–9, 403, 405, 447, 465, 504, 534, 596
    Finance Committee 270–3, 352, 353, 403, 417, 419, 429–30, 517
    Library Committee 333, 338, 341–2, 352, 353, 403, 419–20, 422, 429
    Publications Committee 448–9, 450, 451, 453, 455, 504
    Publications Management Committee 535, 538, 542, 560, 565
    Publications Policy Reviews 516, 520, 527–9, 533
    Publishing Board 559, 560, 566, 580, 583
    Publishing Review (1995) 551, 562, 582
    scientific committees 276, 279–80, 287, 290, 299, 303
    sectional committees 279, 366, 368–70, 374, 379, 382, 386–7, 388, 390–1, 396, 405, 417, 441, 447, 458, 465–6, 495, 500, 502, 503–4
Royal Society Council
    membership of 198, 288, 299, 306, 436, 462, 477
    power of 101–2, 164, 231, 299, 368, 369, 457
    role in publication decisions 35, 37, 39, 68, 70, 74, 277–9, 303, 310, 369, 442, 447, 503, 534, 601
Royal Society fellowship
    disciplinary rivalries among 100–1, 123–4, 198, 259, 364–5, 377, 394
    election criteria and procedures 101–2, 152–3, 192, 195–6, 210, 259, 262, 288, 297, 371–2, 379, 390, 427, 442
    participation in editorial process 163, 184, 192–8, 302, 303–4, 306, 386–7
    privileged access (or not) to RS journals 193, 209, 385, 390, 552
    size of 131, 184, 212n, 227, 258, 265, 272, 306, 355, 365, 387, 449, 495, 504
    social status of 246, 257, 288, 297, 355, 357, 365
    women fellows 365, 371–2, 457, 477
Royal Society meetings 45, 95, 149, 152, 386–7, 437, 498
    discussion at 53, 95, 104, 162, 185–7, 259, 364, 366–7, 385
    discussion meetings 520, 535, 541, 554
    experiments at 19, 30, 43–4, 45, 51, 53, 55, 61, 63, 72, 77, 89, 102, 111, 149, 367
    organisation of (format/agenda/time) 101, 162, 185–6, 195, 198–202, 260, 301, 303, 364
    reading papers aloud 105, 149, 185, 187, 195–6, 199, 240, 287, 301, 366–7
    relationship to the editorial process 109, 111, 118, 160, 161–3, 185, 246, 260, 366–8, 385
    role in forming shared scholarly norms 285, 299–300, 306, 317, 373
    as source of content for the early *Transactions* 11, 28, 31, 40, 77, 85, 109
Royal Society News 571
Royal Society officers
    presidents 83, 101, 122–5, 153, 162, 164, 169–70, 184, 197, 200–1, 299, 301, 306
    presidential cliques 163–4, 178, 185, 198–210
    secretarial duties 55, 102, 160, 172, 200, 205–6, 226, 264, 278, 280, 290, 299, 300–3, 371, 447, 477, 533
    treasurers 7, 102, 174, 267, 269, 270–1, 354, 395, 405, 426, 456, 558, 563
    treasurers' influence over the RS journals 270–1, 273, 301, 311, 315, 333, 350–1, 368, 391, 403–4, 405–6, 415, 419, 428, 431, 465, 483, 516–18, 519, 521, 598

Royal Society Open Science 581–2, 597
Royal Society publishing (for the *Transactions* and *Proceedings*, see separate entries)
   aims/mission 502, 506–8, 533, 535, 542–3, 551, 554, 562, 570, 582–3, 584–5
   editorial team 445–7, 451, 516, 519, 525, 535
   growth of published output 354, 364–5, 394
   growth of staff 522, 559, 574
   leadership 535, 558, 560–2, 569
   policy and strategy 369–70, 516–18, 527, 532–42, 559
   print unit 477, 479
   sales team 482–3, 515, 518–21, 525, 535, 564
   training and expertise 446, 483, 551, 559–62, 569
Royal Society staff 353, 372, 425–6, 456, 462, 507, 516, 534–5, 561, 584
   assistant editor 437, 445–6, 457, 462, 477, 519, 522
   assistant secretary 301, 302, 305, 333, 335, 370, 373–4, 403, 413, 425, 446, 462
   clerk 53, 69–71, 74, 75, 83, 172–4, 234
   curator of experiments 55, 63, 72
   director of publishing 560–1, 569, 601, 602
   executive secretary/director 516, 534–5, 584
   operator 65–6, 73
   porter 223
Royal Society of Chemistry *see* Chemical Society
Royal Society of Edinburgh 165, 183, 232, 526
Rozier, François and his *Observations sur la physique* 217, 221, 239, 243
Rumford *see* Thompson, Benjamin
Russell, Ian 569, 570, 578
Russia/USSR 97, 109, 232, 373, 418, 475, 519
   Imperial Academy of Sciences, St Petersburg 175, 183, 224, 231
Rutherford, Ernest 394, 443, 552
   involvement in editorial process 377, 385, 395, 440
   as president of the RS 363, 374, 375–6, 386, 387, 404, 429–31, 436
Rutty, William 83, 84, 100–4, 106, 139

S

Sabine, Edward 283, 285, 299, 301, 306, 311, 350, 354
sales team *see* Royal Society publishing
sales figures
   for the *Proceedings* 334, 335, 394, 428
   for the *Transactions* 174, 235–6, 263, 271, 333, 335
sales income from RS publications 271, 334–5, 404, 428–30, 478, 481–4, 515
Salisbury, Edward 477, 489, 490
Sand, Christoph 31–2
Saunders, Edith 378
Saussure, Horace-Bénédict de 222
Savile, Ann 90
Scheele, Carl Wilhelm 191, 236
Schmeisser, Johan Gottfried 201

scholarly community *see* communities of scholars
scholarly etiquette *see* etiquette
scholarly mission *see* mission for scholarship; learned societies
Schonfeld, Solomon 434–4
Schuster, Arthur 371, 372–3, 395, 462
Science and Public Affairs 506, 571
Science Citation Index 516
scientific community, the 535, 542, 570, 575, 577, 598
   see also communities of scholars
scientific information 42, 463–4, 477, 487, 491, 514, 529, 549–50
Scientific Information Conference (1948) 463–4, 475
Scientific Information, Institute of (ISI) 516
Scott, D.H. 415
searching the scientific literature 134–5, 331–2, 375, 463–4
   see also Catalogue of Scientific Papers; index; scientific information
secretaries of the RS *see* Royal Society officers
Sedgwick, Adam 286
selectivity (as editorial policy) 288, 350, 354, 404–5, 555, 581
   see also rejection rate
Self-Help for Learned Journals (1963) 477, 487–93, 495, 517, 599
separate copies (offprints) *see also* preprints
   of abstracts 264
   available for the press 238, 240–2, 343, 573
   circulated via correspondence 321, 339–40, 487, 550, 565
   of illustrations 238, 340–1, 347
   numbers allowed 237–8, 352, 354, 408, 429, 457, 480, 482
   provided to authors 237–9, 243, 339–41, 345, 408
   on sale 341–3, 411, 412, 484
Seppings, Robert 238
serials crisis 515, 517–18, 564
Sharpey, William 302, 304, 306, 312
Sherrington, Charles 375, 385, 386, 395, 420
Shimizu, Takeo 385
significance, as criterion for publication 167–9, 287, 312–13, 315, 317, 318, 363, 388–9, 438, 441, 498, 506, 582, 600
Simson, Robert 132
Sinclair, George 42–3
Slare, Frederick 63, 72
Sloane, Elizabeth Langley Rose 84, 86, 99
Sloane, Hans
   personal circumstances 75, 84, 86, 121
   as president of the RS 100–3, 106, 125, 138, 149, 153, 209
   as secretary-editor of the *Transactions* 53, 72, 75–7, 82–91, 91–4, 99–100, 105, 110, 116–21, 122–4, 134, 137, 161, 177, 598

smallpox 84, 97, 127–32, 140, 205
Smith, Archibald 313
Smith, Frank 536, 538, 541
Smith, Samuel, printer-bookseller 86
Smyth, Charles Piazzi 309–10
sociability 185–6, 192, 300, 317, 373, 443, 498
    see also club, RS as
Society of Antiquaries 175, 186, 226, 229, 243, 245, 261
society publishers see learned societies
Soddy, Frederick 436
software see computers
Somerville, Mary 195, 308
South Africa 408, 464, 580
Southwell, Robert 62
Spain 175, 224
specialisation
    of science 13, 219, 356–7, 364, 437, 475, 487, 491, 498, 506–7, 603
    specialised journals, 13, 21, 313, 324, 332, 356, 374–6, 441, 506, 515–16, 527, 604
    within the RS 273, 377, 498, 529, 541
    see also generalist; learned societies; specialist remit; Royal Society committees, sectional committees
speed of journal publication 127, 166, 247, 265, 310, 311, 314, 325, 342, 348, 368, 374, 388, 393, 395, 449, 495, 497, 507, 528–31, 539, 572, 575
speed of scholarly communication 28–9, 132, 137, 191–2, 206–7, 223
Spottiswoode & Co. see Eyre & Spottiswoode, printers
Spottiswoode, William 351–2
Sprat, Thomas 43
Springer, publishers 528, 580
Stationers Company 35, 54
Stephenson, Marjory 461
stereotype plates 339, 418
Stewart, Gordon 555
STM Group 515
Stockholm Royal Academy of Sciences 175, 276, 338
Stokes, George Gabriel
    as editor of RS journals 306–17, 317–25, 346–7, 348
    personal circumstances 296–8, 300, 302
    as president of the RS 331–2, 345, 355–6, 366
    as secretary of the RS 300–3, 354, 371, 504, 552, 596
Strutt see Rayleigh
submission numbers (for RS journals) 306, 311, 364, 373, 376–8, 382, 394–5, 417, 439, 458, 465, 493–4, 507, 525–32, 543, 549, 552–4, 558, 583, 595, 604
submission process (to RS journals) 524, 531–2, 535–7, 539, 541, 552–3, 562, 583
    see also authors, guidelines for; communication
subscribers to RS journals
    geographical distribution 519, 525, 575–6

    number of 334, 394, 428–9, 455, 466, 484, 516–18, 519–20, 549, 556, 575, 576
subscription rates see prices and subscription rates
subscription, as a mechanism for purchasing periodicals 412, 413, 455, 483, 490, 518, 535, 564, 565–7
surplus stock of RS journals 137, 174, 231–2, 272, 342, 413–14, 418, 449
Sussex, Augustus Frederick Duke of 258, 270, 277–8, 283, 287
Swartz, Olof 206, 207
Sweden 104, 109, 175, 183, 220, 276, 338, 421
Swedish language 183, 191, 220
Swift, Jonathan 119–20
Swinnerton-Dyer, Peter 565
Switzerland 222, 338

T

Taylor & Francis, publisher 7, 332, 344, 352, 410, 415, 476, 484
    as printer to the RS 334, 335, 341, 348, 353, 354, 455
    see also Richard Taylor, as printer to the RS
Taylor, Brook 124
Taylor, Geoffrey I. 387
Taylor, John 561–2, 567, 569
Taylor, Richard
    as printer to the RS 262–3, 263–5, 267, 270–1, 272, 288
    as scholarly printer 261–2, 264, 346, 352
    see also Taylor & Francis, as printer to the RS
Taylor, Stuart 561, 569–70, 574–5, 577, 580, 582, 584, 602
technological change 7, 13, 260, 323, 347, 455, 487, 522, 524, 541, 549–51, 556–7, 559, 562–3, 577, 601, 604
    see also electronic (digital) publishing
technology see computers; illustrations; printing technology; typesetting
TechSet, typesetters 563
Theobald, Richard 518–20, 525, 532, 535
Thompson, Benjamin, Count Rumford 203, 224
Thomson Reuters 562
Thomson, J.J. 381
Thomson, Thomas 217–18, 219, 222, 237, 262
Thomson, William, Lord Kelvin 296–7, 306, 308, 312, 317, 318, 344, 377
time taken to publication (in RS journals) 126, 129, 310, 368, 380, 387, 395, 464, 495, 497, 507, 529, 531, 539, 549, 575–6
Todd, Robert Bentley 283
Topping, Michael 201
Torricelli, Evangelisto 32–3
Towle, Francis, assistant secretary 373–4, 394, 425, 446, 448
Transactioneer (1700) 91, 117–21, 134, 161

Transactions, Philosophical (of the Royal Society)
  availability 133, 137–41, 175, 234, 341, 566–8, 577
  buying 37, 133–4, 234, 335, 342, 412, 450, 483
  correspondence, published in 31, 33, 74, 85–6, 88, 93
  criticisms of 91, 116, 117–21, 150, 151–8, 324
  early oversight by RS fellows 39–43, 69, 71, 93, 118, 154
  early support from RS 60, 61, 71, 74, 77, 83, 99–100
  foreign languages printed in 29, 32, 183
  length of issues/volumes 66, 87, 108, 131, 202, 289–90, 457, 518–19, 568
  overseas content 61, 87–9, 94, 97–9, 183, 218, 308, 321, 373, 377
  ownership of 53–5, 123, 134, 141, 176–7, 242, 486
  as public face of the RS 109, 117, 118, 125, 159, 220
  RS corporate responsibility or approbation for 124, 150, 159, 166–9, 277, 443–4
  vision and purpose 19–21, 24, 53, 55, 60, 72, 91–3, 109, 110, 111, 296, 396–7, 466
  see also abridgements; articles, length of; availability; book reviews; distribution; format and appearance; periodicity; prices; production costs; translations; *Proceedings*, relationship to the *Transactions*; reputation of journals; sales figures; sales income
translations of the *Transactions* 29–30, 32, 45, 117, 133, 137–40, 238–9, 241, 243
  see separate languages
treasurers of the RS *see* Royal Society officers
Triewald, Marten 104
Trübner & Co., publisher 341–2, 343, 411
truth claims 90, 167, 169, 192, 277, 285, 372
Turing, Alan 498–9
Turpin Distribution (and precursors) 521, 567
Tyndall, John 318–19, 321, 322–3, 339, 340, 344
typescripts *see* manuscripts
typesetting
  by computer 523–4, 550, 551, 556–7, 561, 562–3, 566
  cost of 39, 261, 345, 452, 522, 575
  by hand 129, 205, 208, 238, 239, 451, 454
  by machine 323, 346, 479, 523
  quality of 409, 450, 523
  skill of compositors 262, 322, 346, 352, 353, 410, 415, 478, 523, 568
  see also mathematical papers, typesetting of
typewriters 302, 370, 479, 523, 524
  see also manuscripts, typed
Tyson, Edward 56, 63, 66, 72

U

universities 37, 356, 542, 558, 578
  academic staff of 345, 355, 378, 498, 524
  expansion of 12, 356, 364–5, 429, 439, 594, 600
  see also Cambridge; Oxford
university libraries *see* libraries
University Grants Committee 424, 565
university presses 333, 466, 562
  see also Cambridge University Press; Oxford University Press
USA (including colonial precursors) 27, 218, 224, 231, 232, 336, 338, 346, 422, 438, 457, 464, 465, 475, 478, 481, 486, 487, 516, 519, 521, 529
  authors from 189, 191, 442
  distribution/circulation of RS journals in 466, 484, 486, 520, 526, 599
  learned institutions in 175, 183, 220, 222, 234, 466–7, 490
USSR *see* Russia

V

van Helmont, F.M. 31–2
Venezuela 408, 422
Vince, Samuel 209
vision
  for RS publishing 332, 366, 390, 437, 463, 502, 506, 598, 605
  for the *Transactions* 24, 26, 59, 77, 84, 91, 110, 116, 296
voluntary labour, in the editorial process 3, 7, 10, 69–70, 302, 331, 366, 370, 373, 430, 489, 573–4, 583, 596

W

Wall, Patrick D. 564
Waller, Richard 53, 72–3, 75–7, 85, 91, 93, 121, 124, 133
Wallis, John 41, 68–9, 70, 76
war, effects on RS publishing and activities
  Anglo-Dutch Wars 22, 40
  American Civil War 346
  American independence 189, 232
  First World War 371–3, 404, 406, 417–19
  Napoleonic Wars 221, 224–6, 232
  Second World War 443, 445, 456–8, 461–2, 466, 478, 480
Ward, John 158, 160, 172–4
warehouse arrangements for RS journals 38, 231, 272, 335, 344, 413, 418, 450
Waring, Edward 209
Warren, Peter 516, 534, 558
Waterston, John 379
Watson, William 194
Wedgwood, Josiah and Thomas 195
Weld, Charles 302
Wellcome Trust 425, 579, 580
Wheatstone, Charles 313
Wheldale, Muriel 378

Whewell, William 276, 277, 278
Whitaker, William 127–8, 130
White, Walter 302–3, 305
Whitworth, Joseph 355
Wiley, publisher 528
Wilkins, John 46
Williams, Samuel 191
Williams & Norgate, publishing agent 411
Willoughby, Hugh 159
Willughby, Francis 67, 68
Wilson, C.T.R. 415
Wilson, Patrick 197
Winckworth, Ronald 437, 445–6, 447, 451, 456–7, 458, 462
Winthrop, John 56
withdrawal of papers by their authors 188, 309, 314, 369
  see also rejection of papers
women
  as authors 90, 194–5, 308–9, 372, 378, 385, 395, 459, 500–2, 553
  as fellows of RS 371–2, 426, 461–2, 477
  involved in RS editorial process 459–60, 500, 583
Woodward, John 95, 103, 118–19, 121, 123
Wren, Christopher 41, 57, 445

X
X-club 299

Y
Yardley, Kathleen *see* Lonsdale, Kathleen
Yarrow, Alfred 425
Year Book of the Royal Society 381, 445, 455, 458, 479, 484, 495, 518, 520
Young, James 355
Young, Thomas 240–1

Z
Zoological Society 246, 261, 412, 415, 488
zoology 279, 282, 348, 445, 525, 526
Zuckerman, Solly 443

Lightning Source UK Ltd.
Milton Keynes UK
UKHW020901281022
411205UK00004B/19